MODERN DIESEL TECHNOLOGY: HEAVY EQUIPMENT SYSTEMS
2ND EDITION

Robert Huzij, Angelo Spano, Sean Bennett

DELMAR
CENGAGE Learning·

Australia • Brazil • Japan • Korea • Mexico • Singapore • Spain • United Kingdom • United States

Modern Diesel Technology: Heavy Equipment Systems, 2nd Edition
Robert Huzij, Angelo Spano, Sean Bennett

VP, General Manager, Skills and Planning: Dawn Gerrain

Senior Content Developer: Sharon Chambliss

Product Assistant: Leah Costakis

Vice President, Marketing: Brian Joyner

Marketing Director: Michele McTighe

Market Development Manager: Jon Sheehan

Senior Production Director: Wendy Troeger

Production Manager: Mark Bernard

Content Project Manager: Christopher Chien

Art Director: Jackie Bates/GEX

Cover Image: © iStockphoto.com/Dazman

Cover Inset Image: © Cengage Learning 2014

For product information and technology assistance, contact us at
Cengage Learning Customer & Sales Support, 1-800-354-9706

For permission to use material from this text or product, submit all requests online at **www.cengage.com/permissions**. Further permissions questions can be e-mailed to **permissionrequest@cengage.com**

Library of Congress Control Number: 2013938819

ISBN-13: 978-1-133-69336-9

ISBN-10: 1-133-69336-9

Delmar
5 Maxwell Drive
Clifton Park, NY 12065-2919
USA

Cengage Learning is a leading provider of customized learning solutions with office locations around the globe, including Singapore, the United Kingdom, Australia, Mexico, Brazil, and Japan. Locate your local office at: **international.cengage.com/global**

Cengage Learning products are represented in Canada by Nelson Education, Ltd.

To learn more about Delmar, visit **www.cengage.com/delmar**

Purchase any of our products at your local college store or at our preferred online store **www.cengagebrain.com**

Notice to the Reader
Publisher does not warrant or guarantee any of the products described herein or perform any independent analysis in connection with any of the product information contained herein. Publisher does not assume, and expressly disclaims, any obligation to obtain and include information other than that provided to it by the manufacturer. The reader is expressly warned to consider and adopt all safety precautions that might be indicated by the activities described herein and to avoid all potential hazards. By following the instructions contained herein, the reader willingly assumes all risks in connection with such instructions. The publisher makes no representations or warranties of any kind, including but not limited to, the warranties of fitness for particular purpose or merchantability, nor are any such representations implied with respect to the material set forth herein, and the publisher takes no responsibility with respect to such material. The publisher shall not be liable for any special, consequential, or exemplary damages resulting, in whole or part, from the readers' use of, or reliance upon, this material.

Printed in the United States of America
3 4 5 6 7 20 19 18 17 16

Brief Table of Contents

Table of Contents

Preface for Series

The Modern Diesel Technology (MDT) series of textbooks debuted in 2007 as a means of addressing the learning requirements of schools and colleges whose syllabi used a modular approach to curricula. The initial intent was to provide comprehensive coverage of the subject matter of each title using ASE/NATEF learning outcomes and thus provide educators in programs that directly target a single certification field with a little more flexibility. In some cases, an MDT textbook exceeds the certification competency standards. An example of this is Joseph Bell's *MDT: Electricity & Electronics*, in which the approach challenges the student to attain the level of understanding needed by a technician specializing in the key areas of chassis electrical and electronics systems—in other words, higher than that required by the general service technician.

The MDT series now boasts nine textbooks, some of which are going into their second edition. As the series has evolved, it has expanded in scope with the introduction of books addressing a much broader spectrum of commercial vehicles. Titles now include *Heavy Equipment Systems, Mobile Equipment Hydraulics*, and *Heating, Ventilation, Air Conditioning & Refrigeration*. The latter includes a detailed examination of trailer reefer technology, subject matter that falls outside the learning objectives of a general textbook. While technicians specializing in all three areas are in demand in most areas of the country, there are as yet no national certification standards in place.

In addition, the series now includes two books that are ideal for students beginning their study of commercial vehicle technology. These two titles (*Preventive Maintenance and Inspection* and *Diesel Engines*) are written so that they can be used in high school programs. Each uses simple language and a no-nonsense approach suited for either classroom or self-directed study. That some high schools now option programs specializing in commercial vehicle technology is an enormous progression from the more general secondary school "shop class" which tended to lack focus. It is also a testament to the job potential of careers in the commercial vehicle technology field in a general employment climate that has stagnated for several years. Some forward-thinking high schools have developed transitional programs partnering with both colleges and industry to introduce motive power technology as early as Grade 10, an age at which many students make crucial career decisions. When a high school student graduates with credits in Diesel Technology or Preventive Maintenance Practice, it can accelerate progression through college programs as well as make those responsible for hiring future technicians for commercial fleets and dealerships take notice.

As the MDT series has evolved, textbooks have been added that target specific ASE certifications, providing an invaluable study guide for certified technicians who are adding to their qualifications along with college programs that use a modular learning approach. *Electronic Diesel Engine Diagnosis* (ASE L2), *Truck Brakes, Suspension, and Steering Systems* (ASE T4 and T5), and *Light Duty Diesel Engines* (ASE A9) detail the learning outcomes required for each ASE certification test.

Because each textbook in the MDT series focuses exclusively on the competencies identified by its title, the books can be used as a review and study guide for technicians prepping for specific certification examinations. Common to all of the titles in the MDT series, the objective is to develop hands-on competency without omitting any of the conceptual building blocks that enable an expert understanding of the subject matter from the technician's perspective. The second editions of these titles not only integrate the changes in technology that have taken place over the past five years, but also blend in a wide range of instructor feedback based on actual classroom proofing. Both should combine to make these second editions more pedagogically effective.

Sean Bennett 2012

Preface

ABOUT THIS TEXT

In approaching this 2nd Edition revision of Modern Diesel Technology (MDT) *Heavy Equipment Systems*, the author team accepted the reality that the first edition was not used as a stand-alone textbook by Heavy Equipment specialty programs. For this reason, we made the decision to drop a couple of chapters that more or less repeated content already covered in *Heavy Duty Truck Systems*, which tends to be used as a companion to this book. The trade-off was a real plus because it opened up space to add three brand-new chapters that are totally Heavy Equipment specific. One of these, the hoisting and rigging chapter, is essential to anyone working with heavy off-road machinery, and it came about as a result of feedback requests from our reviewers, a reinforcement of just how important it is to listen to our users. Because of this, we have located this chapter immediately following the chapter on safety.

In the modern post-secondary environment, it is necessary to use multiple textbooks and online resources along with OEM service literature in order to meet increasingly more exacting exit competencies...and employer expectations of entry-level employees. MDT *Heavy Equipment Systems* has no current competitors because it competes in a relatively small market in which the alternatives are extremely out of date. The author team opted to keep this book as up to date as possible, but there is only just so much that can be packed in, given the page count limitations we are required to work within, so we have only addressed the technology that is unique to heavy equipment. The bottom line is that there are current textbooks that adequately cover subject matter such as diesel engines, electronic system management, and multiplexing, along with those chassis systems that are common to both commercial vehicles, agricultural, and heavy off-road equipment.

TECHNOLOGICAL CHANGES

Technology continues to change at an astounding rate. The spectrum of technology used in the heavy equipment field is much broader than in the automotive or heavy commercial vehicle fields. Because a price tag of $5 million is not unheard of for a single piece of equipment, there is a major incentive to maximize machine longevity: for this reason, we see equipment repaired rather than replaced with more frequency than in other motive power areas. This emphasizes the importance of repair techniques for heavy equipment technicians, rather than the diagnose/replace techs that are more common in the truck service and repair industry. In a current open-face mining operation, it is not unusual to observe 30-year-old machinery operating alongside brand-new equipment.

The hydromechanical controls common a generation ago have generally given way to computer processed, data bus-driven, pilot-actuated circuits. To work on the hydraulic circuits in modern equipment, it is a requirement that the repair technician understand the electronics that manage the hydraulics. The challenge we face as educators in this field in keeping up with technological changes is a daunting one, and a good education program should be based on more than reliance on a single textbook: it should be complemented by online references and OEM support literature. In addition, both instructors and their students can benefit from observation and participation in on-line discussion threads such as in iatn (www.iatn.net) troubleshooting discussions: the trick is to teach students and technicians into how to manage the excess of information available, and learn how to discard what is not accurate. The approach used in this book is to deliver the fundamentals in each subject area, before progressing to the components and circuits specific to off-road equipment.

WHAT'S NEW TO THIS EDITION?

As indicated earlier in the preface, this book does not pretend to cover everything the novice heavy equipment technician is required to know before pulling a wrench. What has been added or removed

from this second edition has been done with the objective of optimizing its content, and maximizing its usefulness to the heavy equipment technician. The chapter on safety has been enhanced with additional emphasis on hydraulic safety related to high oil temperatures and high pressure hazards. Also in this chapter, a new section has been added that explains the importance of OPS, ROPS, and FOPS systems on off-road equipment. Following the chapter on safety, a new chapter on hoisting and rigging techniques has been added, explaining safety and handling techniques that are essential and distinct to this field. The hydraulics section, which consists of four chapters in this revised edition, has been enlarged to address the importance and complexity of modern hydraulic systems. More emphasis has been placed on interpreting hydraulic symbols and schematics, and the sections on maintenance and diagnostics, troubleshooting basics, and bleeding procedures for hydraulic pumps have all been expanded. The chapter on final drives has been enlarged to include a new section on double reduction drives along with a technical updating of the existing drive systems. The chapter on tracks and undercarriages has been updated to reflect some of the new maintenance practices used on track equipment. Electronic system management has been covered within the core chapters and further enhanced in a new chapter titled *Machine and Equipment Electronics*, which takes a closer look at how off-road industries use multiplexing and devices such as haptic joysticks, which are not used on highway vehicles. The book ends with a brand-new chapter on AC-drive systems: the AC drivetrain provides a good model for increasingly used hybrid drive technology and the system described in this book is adaptable to AC trolley drive, which has significant appeal in modern open-face mining operations because of its ability to significantly reduce fuel and downtime costs.

THE OFF-ROAD EQUIPMENT INDUSTRY

The prolonged economic downturn that began in 2008, and continues until the time of writing of this edition, has demonstrated how the off-road industry is somewhat insulated from economic ups and downs. Through all of the past five years, there has been a shortage of skilled workers in the mining and construction industries; some of this may be attributable to an unwillingness of domestic workers to move to where the work is, but it also suggests that as a society we are producing an excess of liberal arts majors and not nearly enough motive power technicians.

When it comes to the design, construction, and quality of off-road equipment, North American manufacturers continue to lead the way worldwide. Caterpillar has reported increased profits in each of the past five years, while many industries have stagnated. In addition, industry analysts suggest that the shortage of heavy equipment repair technicians will become more acute over the next few years, underlining the importance of educators in attracting workers of the future into their programs and instilling the skills and work ethics in future mechanical technicians. Any technician working in motive power today understands the importance of staying current in their field, and that means embracing education throughout a career. Learning is and always has been a lifelong process that must be passed on to our young trades persons both in school and on the job.

Sean Bennett
Robert Huzij
Angelo Spano

ACKNOWLEDGMENTS

The authors and publisher wish to thank the following individuals who provided feedback during the production process:

David Conant, Nashville Auto-Diesel College, Nashville, TN
Pat Dillard, Waste Management, Nashville, TN
Casey Eglinton M.Ed., Western Technical College, La Crosse, WI
Arlis Elkins, Texas State Technical College, Marshall, TX

INSTRUCTOR RESOURCES

Time-saving instructor resources are available at the Instructor Companion Web site for the text or on CD. Either delivery option offers the following resources: PowerPoint chapter presentations with selected images, an Electronic Test Bank, an Image Gallery containing images from the book, an Instructor's Guide which includes an answer key, and MS Word documents of the end of chapter review questions.

CHAPTER

1 Safety

Learning Objectives

After reading this chapter, you should be able to:

- Evaluate potential danger in the workplace.

- Describe the importance of maintaining a healthy personal lifestyle.

- Outline the personal safety clothing and equipment required when working in a service garage or field location.

- Distinguish between different types of fire and identify the fire extinguishers required to suppress small-scale fires.

- Outline some procedures required to use shop jacking and hoisting equipment safely.

- Recognize the potential danger when using different types of shop equipment and the importance of using exhaust extraction piping.

- Identify what is required to work safely with chassis electrical systems and shop mains electrical systems.

- Outline the safety procedures required to work with oxyacetylene cutting and welding equipment and how to safely use arc welding stations.

Key Terms

Canadian Standards Association (CSA)

chain hoist

cherry picker

come-along

electric shock resistant (ESR) footwear

falling-object protection system (FOPS)

operator protective structures (OPS)

personal protective equipment (PPE)

pinhole injection injury

rollover protection system (ROPS)

scissor jack

single-phase main

Standards Australia

static charge

static discharge

three-phase main

Underwriters Laboratories (UL)

INTRODUCTION

The mechanical repair trades are physical by nature, and those employed as technicians probably have higher than average levels of personal fitness. Technicians are required to work safely when handling heavy equipment, and to safely handle materials that can be hazardous.

Modern living requires that we assess risk on a minute-by-minute basis and then strategize how to handle a given situation to avoid potential danger. After all, the simple act of crossing a city street requires some planning. There may be a law that prohibits jaywalking so the first decision you make is whether or not you observe that law. Most cities (in North America) accord the pedestrian some rights over vehicles when it comes to crossing a street, but to be eligible for those rights, you may have to observe such things as crosswalk paths and stop/go lights. The result is that you get to choose whether you make the act of crossing a busy city street one of indescribable adventure and danger or a relatively safe procedure that millions successfully perform daily.

The same can be said of working safely in a repair shop environment. There are a multitude of rules and regulations, most of them posted for maximum exposure. But ultimately, it is up to the individual whether he or she chooses to observe those rules. The great thing about the free world we live in, is you get to decide. Nevertheless, if you are planning on a career in the heavy service-repair industry, you should attempt to stack your decisions in favor of your ultimate safety.

A major heavy equipment manufacturer monitored accidents over a five-year period in one of its assembly plants and came up with the following conclusion: a line-production employee's risk of serious injury (defined as one that required some time off work) during the first year of employment was equal to that of years two through six combined. In simple terms: if you can survive your first year injury-free, thereafter, your risk diminishes significantly.

Instructors of mechanical technology often complain that it is difficult to teach safe work practices to entry-level technicians. The difficulty arises from the fact that entry-level motive power students may be motivated to learn transportation technology but tend to turn off when asked to learn about the health and safety issues that accompany working life. It is sometimes difficult to teach safe work practices to persons who have never been injured and may harbor an illusion that they are immune from injury. On the other hand, an injured person probably has acquired, with the injury, some powerful motivation to avoid a repeat.

Shake Hands with Danger!

"Shake Hands with Danger!" *www.youtube.com/watch?v=3_gEVILWVUM* is a film about workplace safety made by Caterpillar many years ago. It uses some painfully realistic shop scenarios to drive its message home. The film's title is the chorus in a catchy, drawled country-and-western ballad that repeats itself many times over during the film, usually just before someone is about to sustain a deadly or limb-severing experience. The hard message of the film is that most lapses in safety are caused by stupidity. This is accurate, and the great feature of this cautionary and often humorous presentation is that helps us identify some of the dangerous shortcuts we take on a daily basis without giving a thought to the consequences. Every technician working with heavy on- and off-highway equipment should see this film at least once during his or her career. I know of one dealership that shows this film once a year to all their service personnel and they claim a 50% lower overall accident rate as a result.

A Healthy Lifestyle

Repairing on- and off-highway heavy equipment certainly requires more physical strength than working at a desk all day, but it would be a tough call to say it was in itself a healthy occupation. Lifting a clutch pack or pulling a high load on a torque wrench for sure requires some muscle power but does not parallel lifting weights in a gym where the repetitions, conditions, and movements are carefully coordinated to develop muscle power. Jerking on a torque wrench while attempting to establish final torque on main caps during an in-chassis engine job can tear muscle as easily as it can develop it. It pays to think about how you use your body and to use your surroundings to maximize leverage and minimize wear and tear. Make a practice of using hoists to move heavier components, even if you know you could manually lift the component: you may believe it is macho to manhandle a 150-pound clutch pack into position, but all it needs is a slight twist of the back while doing so and you can sustain an injury that can last a lifetime. There is nothing especially macho about hobbling around, suffering in the throes of chronic back pain for years.

Part of maintaining a healthy lifestyle means eating properly and making physical activity a component of everyday living. What this means for each individual will differ. Team sports are not just for teenagers, and whether your sport is hockey, baseball, basketball, or football, there are plenty of opportunities to compete at all ages and at a range of levels. If team sports are not your thing, there are many individual pursuits that you can explore. Working out in a gym, hiking, and canoeing are good for your mind as well as your body, and even golf gets you outside and walking. Because of the physical nature of repair technology, it makes sense to routinely practice some form of weight conditioning, especially as you get older.

PERSONAL PROTECTIVE EQUIPMENT

Personal protective equipment (PPE) covers a range of safety apparel we use to protect our bodies. Some of this personal safety equipment, such as safety boots, should be worn continually in the workplace, while other equipment, such as hearing protection, may be worn only when required—for instance, when noise levels are high.

Safety Boots

Safety boots or shoes are required footwear in a repair shop. In most jurisdictions, legislation mandates the use of safety shoes with steel shanks, steel toes, and **Underwriters Laboratories (UL)** *www.ul.com* certification in the United States, **Canadian Standards Association (CSA)** *www.csa.ca* certification in Canada, and **Standards Australia** *www.standards.org.au/* in Australia. That said, beyond any legal requirements, common sense dictates the use of protective footwear in a shop environment. Given the choice, especially in a heavy equipment service and repair facility, safety boots (see **Figure 1-1**) are a better choice than safety shoes because of the additional support and protection to the ankle area. To learn more about footwear standards and how footwear is tested to meet those standards, visit the Web site of the American Society for Testing and Materials (ASTM) *www.astm.org*.

There is a range of options when it comes to selecting a pair of safety boots, cost being a major factor. If you are going to work on a car under a tree over a weekend, it may be that a low-cost pair of safety boots will suffice, but for the professional mobile-equipment technician who wears this footwear daily for the lifetime of the boot, it pays to invest a little more. Generally, the higher the cost of the footwear, the longer it will last, and, more importantly, the greater the comfort level. Because safety footwear standards are slightly different in the United States and Canada, they are briefly outlined here. Australian and New Zealand standards conform to the U.S. ASTM standard.

ASTM Class 75. ASTM Class 75 approved footwear must meet an internal clearance height of 0.5 inches (12 mm) when impacted with a force of 75 ft-lbf (102 joules); this is equivalent to having a 40-pound object dropped from a height of 2 feet onto the toe of the boot. The crush standard must be met whether the toe protective cap is of steel or composite construction.

CSA Class 1. CSA Class 1 approved footwear must meet an internal clearance height of 12 millimeters (0.5 in) when impacted with a force of 100 joules (74 ft-lbf); this is equivalent to having a 20-kilogram weight dropped from a height of 0.5 meters onto the toe of the boot. The crush standard must be met whether the toe protective cap is of steel or composite construction.

AS/NZ. The Australia and New Zealand standard pertaining to safety boots is known as AS/NZ 2210.3 and conforms to ASTM Class 75. This requires that they meet an internal clearance height of 0.5 inches (12 mm) when impacted with a force of 75 ft-lbf (102 joules).

Electric Shock Resistant (ESR) Footwear. Any mechanical technician working around high-voltage potential chassis equipment should wear **electric shock resistant (ESR) footwear**. ESR footwear is designated by an orange omega on a white rectangle in both the United States and Canada. The standards are as follows:

- ASTM ESR footwear: must withstand the application of 14,000 volts at 60 hertz for 1 minute with zero current flow or leakage exceeding 3.0 milliamperes under dry conditions.
- CSA ESR footwear: must withstand the application of 18,000 volts at 60 hertz for 1 minute with zero current flow or leakage exceeding 1.0 milliamperes under dry conditions.

Safety Glasses

Many shops today require all employees to wear safety glasses while on the shop floor. This is really just common sense. Eyes are sensitive to dust, metal shavings, grinding and machining particulate, fluids, and fumes. They are also more complex to repair than feet. Perhaps the major problem, when it comes to making a habit of using safety glasses, is the poor

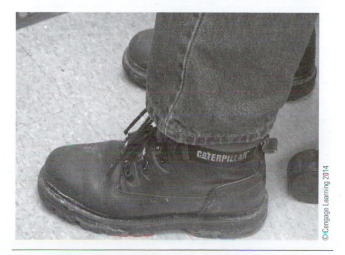

Figure 1-1 UL-approved safety boots.

quality of most shop-supplied eyewear. Shops supply safety glasses because liability issues mandate their availability. Low-cost, mass-produced, and easily scratched plastic safety glasses may meet the employer's obligation to make safety glasses available to employees, but these glasses are unlikely to encourage the daily use of eye protection.

The solution is to not to depend on your employer to provide this essential safety apparel. Get out of the mindset that safety glasses should be free. Free safety glasses are uncomfortable and, although they may protect eyes from flying particulate, they may actually impair vision. Spend a little more and purchase a good quality pair of safety glasses. Even if you do not normally wear eyeglasses, after a couple of days, you will forget you are wearing them. **Figure 1-2** shows three methods of protecting your eyes in the workplace.

Hearing Protection

Two types of hearing protection are used in shops. Hearing muffs are connected by a spring-loaded band and enclose the complete outer ear. This type of hearing protection is available in a range of qualities, determined by the extent to which they muffle sound. Cheaper versions may be almost useless, but good quality hearing muffs can be the most effective when noise levels are high and exposure is prolonged.

A cheaper and generally effective alternative to hearing muffs is a pair of ear sponges (plugs). Each sponge is a malleable cylindrical or conical sponge that can be shaped for insertion into the outer ear cavity; after insertion, the sponge expands to fit the ear cavity. The disadvantage of hearing sponges is that they can be uncomfortable when used for prolonged periods. They are designed for one time use and should be disposed of after removal: reuse can introduce contaminants into the outer ear. **Figure 1-3** shows some typical shop hearing protection devices.

For high-noise-level activities, such as riveting or operating dynamometers, the use of both hearing muffs and sponges should be considered; however, you should be aware that insulating the brain from sound can be disorientating, resulting in loss of balance.

CAUTION *Damage to hearing is seldom incurred by a single exposure to a high level of noise; it is something that more typically results from years of exposure to excessive and repetitive noise levels. Protect your hearing! And note that hearing can be as easily damaged by listening to music at excessive volume as by exposure to buck riveting.*

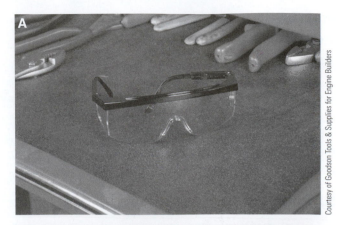
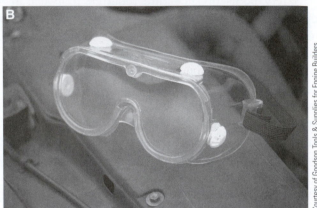
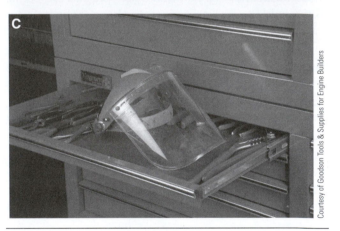

Figure 1-2 (A) Safety glasses, (B) splash goggles, and (C) face shield.

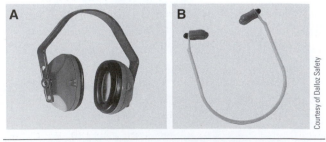

Figure 1-3 Typical (A) ear muffs and (B) ear plugs.

Courtesy of Goodson Tools & Supplies for Engine Builders

Courtesy of Dalloz Safety

Gloves

A wide range of gloves can be used in shop applications to protect the hands from exposure to dangerous or toxic materials and fluids. The following are some examples.

> **CAUTION** *Never wear any type of glove when using a bench-mounted, rotary grinding wheel. There have been cases where a glove has been snagged by the abrasive wheel, dragging the whole hand with it.*

Vinyl Disposable. Most shops today make vinyl disposable gloves available to service personnel. These protect the hands from direct exposure to fuel, oils, and grease. The disadvantage of vinyl gloves is that they do not breathe, and some find the sweating hands that result uncomfortable. Most shop-use vinyl gloves today are made of thin gossamer that minimizes the loss of tactile awareness.

Mechanic's Gloves. These artificial fiber, stretchable gloves have become popular in recent years. They are thin and fit tight to the hand, and so therefore provide the technician with some tactile sense. They are suitable for working on light-duty jobs.

Cloth and Leather Multipurpose Gloves. A typical pair of multipurpose work gloves consists of a rough leather palm and a cotton back. They can be used for a variety of tasks ranging from lifting objects to general protection from cold when working outside. This category of work glove can also provide some insulation for the hands when performing procedures such as buck riveting. You should cease to use this type of glove after it becomes saturated with grease or oil because they can become a fire hazard.

Welding Gloves. Welding gloves are manufactured from rough cured leather. They are designed to protect the hands from exposure to the high temperatures created in welding and flame cutting processes. Such gloves should be task-specific. The rough leather they are made from absorbs grease and oil easily, reducing their ability to insulate. Avoid using welding gloves rather than tools to handle heated steel because the gloves will rapidly harden and require replacement.

Some shops recommend the use of welding gloves when performing heavy suspension work because of their ability to protect fingers. If you perform both welding and suspension work in your job, make sure you use separate gloves for each. The gloves you use for suspension work will absorb oil and grease, after which they should not be used for welding.

Dangerous Materials Gloves. Gloves designed to handle acids or alkalines should be used for the specific task only. Gloves in this category are manufactured from nonreactive, synthetic rubber compounds. Care should be taken when washing up after using this type of glove.

> **CAUTION** *Never wear leather gloves to handle refrigerants. Leather gloves rapidly absorb liquid refrigerant and can adhere to the skin.*

Back Care

Back injuries are said to affect 50% of repair technicians, at some point in their careers, seriously enough for them to have to take some time off work. A bad back does not have to be an occupational hazard. Most of us begin our careers in our twenties when we have sufficient upper-body strength to be able to sustain plenty of abuse. As we age, this upper-body strength gradually diminishes and bad lifting practices can take their toll.

The best strategy for taking care of your back is to observe some simple rules for lifting heavy items. These rules can be summarized as follows:

- Keep your back erect while lifting.
- Keep the weight you are lifting close to your body.
- Bend your legs, and lift using the leg muscles.

Figure 1-4 shows the correct method of using your back while lifting.

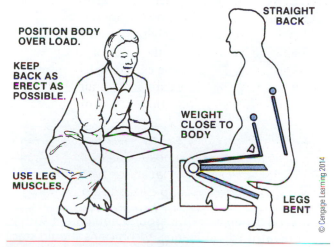

Figure 1-4 Use your leg muscles—never your back—when lifting any heavy load.

Back Braces. A back brace may help you avoid injuring your back. Wearing a back brace makes it more difficult to bend your back, "reminding" you to keep your back straight when lifting. You may have noticed that the sales personnel in one national hardware and home-goods chain are all required to wear back braces. As a mobile equipment technician, you will be required to use your back for lifting, so you should consider the use of a back brace. Body shape plays a role when it comes to back injuries: if you are either taller than average height or overweight, you will be more vulnerable to back injuries.

Coveralls

Many shops today require their service employees to wear a uniform of some kind. These may be work shirts and pants, shop coats, or coveralls. Uniforms have a way of making service personnel look professional. Given the nature of the work, the uniform of choice in truck, bus, and heavy equipment service facilities should be coveralls. The coveralls should preferably be made out of cotton for reasons of comfort and safety. When ordering cotton coveralls for personal use, remember to order at least a size larger than your usual nominal size; unless otherwise treated, cotton shrinks when washed. **Figure 1-5** shows a well-organized workspace around a wheel loader being worked on, free of clutter: this along with wearing the proper PPE is essential to minimize workplace injuries.

CAUTION *Avoid wearing any type of loose-fitting clothing when working with machinery. Shop coats, neckties, and untucked shirts can all be classified as loose-fitting clothing.*

Figure 1-5 When working around equipment such as this Komatsu wheel loader, technicians should stack the odds in the favor of safety by ensuring that all PPE is worn and used properly.

When artificial fibers are used as material for coveralls, they should be treated with fire retardant. Cotton smolders when exposed to fire (that is, when it is not saturated with oil, fuel, or grease, which cause it to be highly flammable). Cleanliness is essential: oily shop clothing not only looks unprofessional, it can be dangerous! Artificial fibers, when not treated with fire retardant, melt when exposed to high temperatures and may fuse to the skin.

CAUTION *Note that, even when treated with fire retardant, some artificial fibers will burn vigorously when exposed to direct flame for a period of time.*

Butane Lighters

There are few more dangerous items routinely observed on the shop floor than the butane cigarette lighter. The explosive potential of the butane lighter is immense, yet it is often stored in a pocket close to where it can do the most amount of damage. A chip of hot welding slag will almost instantly burn through the plastic fuel cell of a butane lighter. Owners of these devices will often compound the danger they represent by lighting torches with them. If you must have a lighter on your person while working, purchase a Zippo!

Hair and Jewelry

Long hair and personal jewelry produce some of the same safety concerns as loose-fitting clothing. If it is your style to wear long hair, it should be secured behind the head and you should consider wearing a cap. Although wearing body jewelry has become a recent trend, you should consider removing as much as possible of this while at work. Bear in mind that body jewelry is often made of conductive metals, so you should consider both the possibility of snagging jewelry and creating some unwanted electrical short circuits.

Safety Helmets

Safety helmets are essential in shops servicing and repairing large off-road equipment, especially when working under the equipment. The helmet should be adjusted to fit snug to the head and certified to meet UL, AS, and CSA certification standards.

Spill Response

Liquids spilled onto the shop floor are a fact of life in a shop environment. However, they should be regarded as an avoidable hazard. Make a point of immediately attending to and cleaning up any liquids spilled onto the shop floor. Ground-up absorbents are available in all shops and should be used. They are manufactured from wood bark, clay, and artificial substances. Until they become saturated with liquid, the absorbents can be regarded as harmless. Once used, absorbents must be disposed of in an environmentally responsible manner.

FIRE SAFETY

Service and repair facilities are usually subject to regular inspections by fire departments. This means that obvious fire hazards are identified and neutralized. Although it should be stressed that fire fighting is a job for trained professionals, any person working in a service shop environment should be able to appropriately respond to a fire in its early stages. This requires some knowledge of the four types of fire extinguisher in current use.

Fire Extinguishers

Fire extinguishers are classified by the types of fire they are designed to suppress. Using the wrong type of fire extinguisher on certain types of fire can be extremely dangerous and actually accelerate the spread of the fire you are attempting to control. Every fire extinguisher will clearly state the types of fire it is designed to suppress using the following class letters. In suppressing a fire, your role as a technician is to assess the risk required and intervene only if there is minimal risk.

Class A. A Class A fire is one involving combustible materials such as wood, paper, natural fibers, biodegradable waste, and dry agricultural waste. A Class A fire can usually be extinguished with water. Fire extinguishers designed to suppress Class A fires use foam or multipurpose dry chemical, usually sodium bicarbonate.

Class B. Class B fires are those involving fuels, oil, grease, paint, and other volatile liquids, flammable gases, and some petrochemical plastics. Water should not be used on Class B fires. Fire extinguishers designed to suppress Class B fires make use of smothering: they use foam, dry chemical, or carbon dioxide. Trained fire personnel may use extinguishers such as Purple K (potassium bicarbonate) or halogenated agents to control fuel and oil fires.

Class C. Class C fires are those involving electrical equipment. First intervention with this type of fire should be to attempt to shut off the power supply; assess the risk before handling any switching devices. When a Class C fire occurs in a vehicle harness, combustible insulation and conduit can produce highly toxic fumes. Great care is required when making any kind of intervention in vehicle chassis or building electrical fires. Fire extinguishers designed to suppress electrical fires use carbon dioxide, dry chemical powder, and Purple K.

Class D. Class D fires are those involving flammable metals. Some metals, when heated to their fire point, begin to vaporize and combust. These metals include magnesium, aluminum, potassium, sodium, and zirconium. Dry powder extinguishers should be used to suppress Class D fires. **Figure 1-6** provides a visual guide to selecting fire suppression equipment, and **Figure 1-7** shows a cabinet used to safely store combustibles.

SHOP EQUIPMENT

Technicians must become familiar with an extensive assortment of shop equipment. Some items can be dangerous if you are not trained how to use them. Make a practice of asking for help if you are not familiar with how to operate any equipment.

Lifting Devices

Many different types of hoists and jacks are used in heavy equipment shops. These can range from simple pulley and chain hoists to a variety of hydraulically actuated hoists. Weight-bearing chains on hoists should be routinely inspected (this is usually required by law): abrasive wear, deformed links, and nicks are reasons to place the equipment out of service. Hydraulic hoists should be inspected for external leaks before using, and any drop-off while in operation should be reason to take the equipment out of service. Never rely on the hydraulic circuit alone when working under equipment on a hoist. After lifting, support the equipment using a mechanical sprag or stands. Remember that this also applies when working on any kind of hydraulically actuated devices on heavy equipment.

CAUTION *Never rely on a hydraulic circuit alone when working underneath raised equipment. Before going under anything raised by hydraulics, make sure it is mechanically supported by stands or a mechanical lock.*

	Class of Fire	Typical Fuel Involved	Type of Extinguisher
Class **A** Fires (green)	**For Ordinary Combustibles** Put out a Class A fire by lowering its temperature or by coating the burning combustibles.	Wood Paper Cloth Rubber Plastics Rubbish Upholstery	Water*[1] Foam* Multipurpose dry chemical[4]
Class **B** Fires (red)	**For Flammable Liquids** Put out a Class B fire by smothering it. Use an extinguisher that gives a blanketing, flame-interrupting effect; cover whole flaming liquid surface.	Gasoline Oil Grease Paint Lighter fluid	Foam* Carbon dioxide[5] Halogenated agent[6] Standard dry chemical[2] Purple K dry chemical[3] Multipurpose dry chemical[4]
Class **C** Fires (blue)	**For Electrical Equipment** Put out a Class C fire by shutting off power as quickly as possible and by always using a nonconducting extinguishing agent to prevent electric shock.	Motors Appliances Wiring Fuse boxes Switchboards	Carbon dioxide[5] Halogenated agent[6] Standard dry chemical[2] Purple K dry chemical[3] Multipurpose dry chemical[4]
Class **D** Fires (yellow)	**For Combustible Metals** Put out a Class D fire of metal chips, turnings, or shavings by smothering or coating with a specially designed extinguishing agent.	Aluminum Magnesium Potassium Sodium Titanium Zirconium	Dry powder extinguishers and agents only

*Cartridge-operated water, foam, and soda-acid types of extinguishers are no longer manufactured. These extinguishers should be removed from service when they become due for their next hydrostatic pressure test.

Notes:

(1) Freezes in low temperatures unless treated with antifreeze solution, usually weighs over 20 pounds (9 kg), and is heavier than any other extinguisher mentioned.

(2) Also called ordinary or regular dry chemical (sodium bicarbonate).

(3) Has the greatest initial fire-stopping power of the extinguishers mentioned for class B fires. Be sure to clean residue immediately after using the extinguisher so sprayed surfaces will not be damaged (potassium bicarbonate).

(4) The only extinguishers that fight A, B, and C classes of fires. However, they should not be used on fires in liquefied fat or oil of appreciable depth. Be sure to clean residue immediately after using the extinguisher so sprayed surfaces will not be damaged (ammonium phosphates).

(5) Use with caution in unventilated, confined spaces.

(6) May cause injury to the operator if the extinguishing agent (a gas) or the gases produced when the agent is applied to a fire is inhaled.

Figure 1-6 Guide to fire extinguisher selection.

Jacks. Many types of jack are used in heavy equipment service facilities. Before using a jack to raise a load, make sure that the weight rating of the jack exceeds the supposed weight of the load. Most jacks used in service-repair shops are hydraulic, and most use air-over-hydraulic actuation because this is faster and requires less effort. Bottle jacks are usually hand-actuated and designed to lift loads of up to 10 tons; they are so named because they have the appearance of a bottle. Air-over-hydraulic jacks capable of lifting up to 30 tons are also available.

Using hydraulic piston jacks should be straightforward: they are designed for a perpendicular lift only. The jack base should be on a level floor and the lift piston should be located on a flat surface on the equipment to be lifted. Never place the lift piston on the arc of a leaf spring or the radius of any suspension device on the mobile equipment. After lifting the equipment, it should be supported mechanically using steel stands. It is acceptable practice to use a hardwood spacer in conjunction with a shop jack: make sure it is exactly level. Whenever using a jack, ensure that the vehicle being jacked can roll either forward or backward: parking brakes should be applied and wheel chocks used on the axles not being raised.

Figure 1-7 Store combustible materials in approved safety cabinets.

Cherry Pickers. **Cherry pickers** come in many shapes and sizes. Light-duty cherry pickers can be used to raise a heavy component, such as a cylinder head, from an engine whereas heavy-duty cherry pickers (see **Figure 1-8**) can lift a large-bore diesel engine out of a chassis. Most cherry pickers have extendable arms.

CAUTION *As the arm of a cherry picker is lengthened, the weight it can lift reduces significantly. Ensure that the weight you are about to lift can be sustained without toppling the cherry picker.*

Scissor Jacks. The popularity of **scissor jacks** has increased recently because of their versatility. Scissor jacks should only be used at one end of a vehicle: the end not being lifted has to be skidded as the lift angle

increases, so it is important that the brakes not be applied during the lifting procedure. When the vehicle or equipment has been hoisted to the required height, engage the mechanical lock on the jack, and then chock the set of wheels at the end of the vehicle still on the floor. Never chock the wheels of a vehicle being lifted on a scissor jack until after the lift is completed.

Chain Hoists. These are often called chain falls. **Chain hoists** can be suspended from a fixed rail or a beam that slides on rails or be mounted on any number of different types of A-frames. Chain hoists in shops in most jurisdictions are required to be inspected periodically. An inspection on a mechanical chain hoist will check for chain link integrity and the ratchet teeth and lock. Electromechanical units will require an inspection of the mechanical and electrical components. Where a chain hoist beam runs on rails, brake operation becomes critical. Some caution is required when braking on the beam because this can cause a pendulum effect on the object being lifted.

Come-alongs. **Come-alongs** describe a number of different types of cable and chain lifting devices that are hand-ratchet actuated. They are used both to lift objects and to apply linear force to them. When used as a lift device, come-alongs should be simple to use, providing the weight being lifted is within rated specification. However, come-alongs are more often used in service shops to apply linear force to an object or component. Great care should be taken, ensuring that the anchor and load are both secure and that the linear force does not exceed the weight rating of the device.

GENERAL SHOP PRECAUTIONS

Because every service facility is different, the potential dangers faced in each shop will differ. In this section, we will outline some general rules and safety strategies to be observed in service and repair shops.

Exhaust Extraction

Diesel engines should be run in a shop environment using an exhaust extraction system. In most cases this will be a flexible pipe or pipes that fit over the exhaust stack(s). Take care when climbing up to fit an exhaust extraction pipe over the vertical exhaust stack(s); use a ladder when you cannot get a secure foothold elsewhere. When parking mobile heavy equipment in and

Figure 1-8 Typical heavy-duty cherry picker.

© Cengage Learning 2014

Figure 1-9 Shop exhaust extraction piping.

out of service bays, park the unit in the bay and shut off the engine. Avoid running an engine without the extraction pipe(s) fitted to the stack(s). **Figure 1-9** shows the exhaust extraction piping used in a diesel engine testing and repair facility.

WARNING *Diesel engine exhaust fumes are known by the State of California to cause respiratory problems, cancer, birth defects, and other reproductive harm in humans. Avoid operating diesel engines unless in a well-ventilated area. When starting an engine outside a shop, warm the engine before driving it into the shop to reduce the contaminants emitted directly into the shop while parking the unit.*

Workplace Housekeeping

Sloppy housekeeping can make your workplace dangerous. Clean up oil spills quickly. You can do this by applying absorbent grit: this not only absorbs oil but makes it less likely that a person could slip and fall on an oil slick. Try to organize parts in bins and on benches when you are disassembling components. Besides adding to the danger potential of your work environment, it indicates a lack of professionalism to see parts and tools strewn all over the shop floor.

Components Under Tension

On mobile heavy equipment, numerous components are under tension, sometimes deadly tensional loads. Never attempt to disassemble a component that you suspect is under high tensional load unless you are exactly sure of how to proceed. Consult service literature and ask more experienced co-workers when you

are unsure of a procedure. Some examples of components that are under tensional load are:

- Air spring brake chambers
- Roll-up van doors
- Elliptical suspension spring packs
- Suspension torsion springs
- Electronic unit pump (EUP) assemblies on some engines
- Roll-up garage doors

Compressed Fluids

Fluids in both liquid and gaseous states can be extremely dangerous when proper safety precautions are not observed. Because of residual pressures in circuits, equipment does not have to be actually running to represent a serious safety hazard. Compressed fluids can be generally divided into pneumatic and hydraulic equipment, but you will see when you read through the next section that many fluid power systems make use of both compressed air and hydraulics.

Pneumatics Safety

Compressed air is used on some heavy equipment chassis both in brake and auxiliary systems. Additionally, compressed air is used to drive both portable and non-portable shop tools and equipment. Most shop powered handtools use pneumatic pressure such as the 0.5-inch drive impact wrench shown in **Figure 1-10**.

WARNING *Always wear safety glasses when coupling and uncoupling or using pneumatic tools.*

Some examples of chassis systems that use compressed air:

- Air brake circuits and actuators
- Pneumatic suspensions
- Pilot control systems on tankers
- Air start systems

Some examples of shop equipment that uses compressed air:

- Pneumatic wrenches
- Pneumatic drills
- Shop air-over-hydraulic presses
- Air-over-hydraulic jacks
- Air-over-hydraulic cylinder hoists

Hydraulic System Safety. Vehicle and shop hydraulic systems use extremely high pressures that can be

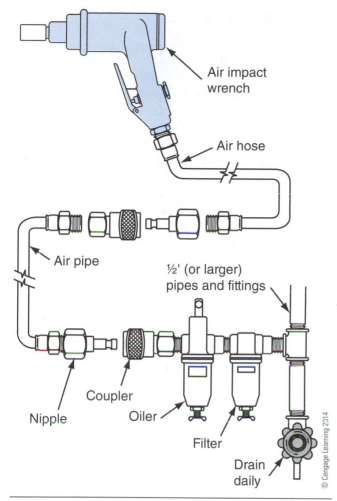

Figure 1-10 Typical setup for a 0.5-inch drive impact gun.

- Automatic transmission control circuits
- Air-over-hydraulic brake circuits
- Clutch control circuits

Some examples of shop equipment using hydraulic circuits are:

- Jacks and hoists
- Presses
- Bearing and liner pullers
- Suspension bushing presses

HYDRAULICS PERSONAL SAFETY

Three general categories of personal injury may occur when working with high-pressure hydraulics:

- Burns. Burn injuries are caused by the spraying of hot, high-pressure hydraulic fluid.
- Abrasions and bruises. This type of injury is caused by the flailing of high-pressure hydraulic lines that have come loose.
- Pinhole leak injection. This category of injury results from barely visible or invisible leakage from high-pressure hydraulic circuits; it is especially dangerous because often the injured person does not become fully aware of the extent of injury until it is too late to do anything about it. This type of injury is not uncommon. Read the next section on pinhole leak injury.

lethal when mishandled. Never forget that idle circuits can hold residual pressures, and many circuits use accumulators. The rule when working with hydraulic circuits is to be absolutely sure about potential dangers before attempting to disassemble a circuit or component. Because a high percentage of off-road equipment makes extensive use of hydraulics, it is essential that every technician properly understand the potential dangers of working around extreme high-pressure circuits. Make a point of memorizing the personal safety elements of working around hydraulics.

WARNING *Always wear safety glasses when working close to shop or vehicle hydraulic circuits and check service literature before attempting a disassembly procedure. Ask someone if you are not sure rather than risk injury.*

Some examples of chassis systems that use hydraulic circuits are:

- Wet-line kits for dump and auxiliary circuits
- High-pressure fuel management circuits

Pinhole Injection Injury. A major leak in a hydraulic circuit is rapidly identified because it makes a mess. A pinhole leak may be microscopic in size and a **pinhole injection injury** occurs when high-pressure droplets are injected into human flesh often without the knowledge of the injured person. The injection takes place in much the same manner as a hypodermic syringe injection. Pinhole droplet injection can occur when the injured person is wearing heavy leather gloves. Often no more than a minor stinging sensation is felt at the time of injury; it is not until several hours later when tissue damage has occurred that the injury becomes painful. This is often too late.

Make sure you are aware of the early symptoms of pinhole leak injection. If you suspect this type of injury, seek immediate medical attention and be aware that some hospital emergency departments may not be familiar with pinhole injection injuries. This will require you to do some explaining, but do it anyway; it is preferable to losing a limb.

WARNING *Pinhole hydraulic leaks can cause maiming and death if not identified. A pinhole injection injury may occur unbeknownst to the injured person: this type of injury must be identified shortly after the injection intrusion into the flesh. Make sure you understand the section on pinhole hydraulic leaks that appears ahead of this warning.*

CAUTION *Never use a bare or gloved hand when investigating hydraulic leaks, especially in bundled clusters of hoses. Wear safety glasses and gloves and use a cardboard strip when attempting to source small leaks in blind clusters of hydraulic hose.*

CHASSIS AND SHOP ELECTRICAL SAFETY

Mobile heavy equipment vehicles today can use numerous computers, all networked to a central data backbone using multiplexing technology. The computers used to control chassis subsystem components, including the engine, transmission, brakes, dash electronics, collision warning systems, and satellite communications systems, function on low-voltage electrical signals and use thousands of solid-state components. While some of these electronic subcircuits are protected against transient voltage spikes, others are not. An unwanted high-voltage spike caused by static discharge or careless placement of electric welding grounds can cause thousands of dollars worth of damage.

Some heavy equipment chassis are equipped with electrical isolation switches. These should be opened any time major service or repair work is performed on a vehicle. When any type of electric welding is performed on a chassis, ensure that the ground clamp is placed close to the work. Placing a welding ground clamp on the front bumper when you are welding at the rear of the chassis is not only capable of causing electronic damage but also of taking out bearings and journals in the major powertrain components of the vehicle. Although electricity can be relied on to take the shortest path to complete a circuit, sometimes it will experiment with determining which is the shortest path. Pulsing electricity through crankshaft journals and transmission bearings causes arcing that results in costly damage.

CAUTION *Whenever performing electric arc welding or cutting on a chassis, make sure you place the ground clamp as close to the work area as possible to avoid creating chassis electronic or arcing damage.*

Static Discharge

When you walk across a plush carpet, your shoes "steal" electrons from the floor. This charge of electrons accumulates in your body, and when you go to grab a door handle, this excess of electrons discharges itself into the door handle, creating an arc as it does so. This is known as **static discharge**. Accumulation of a **static charge** is influenced by factors such as relative humidity and the type of footwear you are wearing. Getting a little zap from the static charge that can accumulate in the human body is seldom going to produce any adverse effects to human health but it can damage sensitive solid-state circuits.

Picture a fuel tanker transport running down an interstate. In the same way your body steals electrons from a carpet, so does the tanker steal electrons from the atmospheric air it is forcing itself through; however, the charge differential that can be accumulated by the tanker is much greater and can exceed 50,000 volts. This type of charge differential can be highly dangerous and produce a spark that can easily ignite fuel vapors. This potential danger accounts for the legal requirement to ground out a tanker chassis before undertaking any load or unload operation.

Static Discharge and Computers. Static charge accumulation in the human body can easily damage computer circuits. Because some pieces of equipment today have a dozen, sometimes more, computer-controlled circuits, it is important for technicians to understand the effects of static discharge. The reason that static discharge has not caused more problems than it has in the service repair industry is due primarily to:

■ Technicians' footwear of choice, usually rubber-soled boots.
■ The tendency of shop floors to be concrete-surfaced rather than carpeted.

Neither of the above factors is conducive to static charge accumulation, but technicians should remember that any carpeted flooring is conductive, and due precautions should be taken.

That said, it is good practice, when troubleshooting requires you to access electronic circuits, to use a ground strap before separating sealed connectors, connecting breakout Tees/boxes, or accessing the data bus beyond just connecting an electronic service tool (EST) to it. A ground strap "electrically" connects you to the device you are working on, so that an unwanted static discharge into a shielded circuit is unlikely. Special care should be taken when working with modules that require you to physically remove and replace solid-state components such as PROM chips from an ECM motherboard.

Chassis Wiring and Connectors

Every year, millions of dollars worth of damage to mobile equipment is created by service technicians who ignore OEM precautions regarding working with chassis wiring systems. Perhaps the most common abuse is puncturing wiring insulation with test lights and digital multimeter (DMM) leads. When you puncture the insulation on copper wiring, in an instant that wiring becomes exposed to both oxygen (in the air) and moisture (relative humidity). The chemical reaction almost immediately produces copper oxides that then react with moisture to form corrosive cupric acid. The acid begins to eat away the wiring, first creating high resistance, and ultimately consuming the wire. The effect is accelerated when copper-stranded wiring is used because the surface area over which the corrosion can act is so much greater.

> **CAUTION** *Never puncture the insulation on chassis wiring. Read the section that immediately precedes this if you want to know why!*

The sad thing about this type of abuse is that it is so easily avoided. There are so many ways that a repair technician can access wiring circuits using the correct tools, it is just stupidity not to use them. Use breakout Tees, breakout boxes, and test lead spoons.

Breakout Tees. A breakout Tee allows you to access an energized circuit. Most breakout Tees allow you to access a single wire circuit: to "Tee" into the circuit, you separate it at a connector; then connect the breakout Tee. This allows you to test the circuit status while it is electrically active.

Breakout Box. A breakout box is a troubleshooting device required by some OEMs. A breakout box does the same thing as a breakout Tee except that, when it is connected into the circuit, you can check the status of a large number of circuit wires and/or terminals. A troubleshooting tree in interactive diagnostic software will often direct a technician to install a breakout box into a circuit, usually at a connector block, and then perform a sequence of DMM driven tests, the results of which may have to be entered into the diagnostic computer.

Test Lead Spoons. Test lead spoons have the potential to save so much unnecessary expenditure, it is surprising that they are not distributed free of charge to anyone working on mobile equipment. Test lead spoons are designed for insertion into weatherproof connectors and terminal blocks without creating damage to either the terminal or the wiring. Get a set from your tool supplier.

Circuit Test Lights. Test lights have few uses on modern equipment and have the potential to inflict damage on electrical wiring when misused. A circuit test light consists of a sharp spike intended to abuse wiring insulation, a lightbulb, a wire, and a ground clamp. If you connect the ground clamp on the tool to a good ground, then any time the spike contacts a voltage differential, the bulb illuminates. Some are specified for working on electronic circuits, and these will illuminate with much smaller voltage potentials. Although this category of circuit test light will not create a current overload in a sensitive circuit, the spike can create all kinds of damage to connector blocks and wiring insulation.

> **CAUTION** *Be aware of the damage you can create using a circuit test light. Many can cause current overloads in sensitive electronic circuits, and even those rated for working on electronics can damage insulation, insulated connectors, and connector blocks.*

Tech Tip: Maybe you should not dispose of your circuit test light in the garbage; after all you paid good money for it. Instead, place it in some remote recess of your toolbox so that the next time you are confronted with an equipment wiring problem, you can better resist the temptation to set about vandalizing its electrical circuits!

Mains Electrical Equipment

Mains electrical circuits, unlike standard electrical circuits, operate at pressures that can be lethal, so you have to be careful when working around any electrical equipment. Electrical pressures may be **single-phase main** operating at pressure values between 110 and 120 volts *or* **three-phase main**, operating at pressures between 400 and 600 volts. In some jurisdictions, repairs to mains electrical equipment and circuits are required to be undertaken by certified personnel. If you

© Cengage Learning 2014

Figure 1-11 High-voltage three-phase electrical outlet.

attempt to repair electrical equipment, make sure you know what you are doing! **Figure 1-11** shows a typical three-phase electrical outlet.

Take extra care when using electrically powered equipment when the area you are working in is wet. And remember that a transient spike of AC voltage driven through a chassis data bus can knock out electronic equipment networked to it. Electrical equipment can also be dangerous around vehicles because of its potential to arc and initiate a fire or explosion.

> **CAUTION** *Do not undertake to repair mains electrical circuit and equipment problems unless you are qualified to do so.*

Some examples of shop equipment using single-phase mains electricity are:

- Electric hand tools
- Portable electric lights
- Computer stations
- Drill presses
- Burnishing and broaching tools

> **CAUTION** *Take care when using trouble lights with incandescent bulbs around volatile liquids and flammable gases. These are capable of creating sufficient heat to ignite flammables. Many jurisdictions have banned the use of this type of trouble light, and they should never be used in garages in which gasoline, propane, and natural gas-fueled vehicles are present. Best bet: use a fluorescent-type trouble light in rubber-insulated housing!*

Some examples of shop equipment using high-voltage, three-phase electricity are:

- Most welding equipment
- Dynamometers
- Lathes and mills
- Large shop air compressors

OXYACETYLENE EQUIPMENT

Technicians use oxyacetylene for heating and cutting on a daily basis. This equipment is used less commonly for braising and welding. Some basic instruction in the techniques of oxyacetylene equipment safety and handling is required. The following information should be understood by anyone working in close proximity with oxyacetylene equipment.

Acetylene Cylinders

Acetylene regulators and hose couplings use a left-hand thread. The regulator gauge working pressure should *never* be set at a value exceeding 15 psi (100 kPa): acetylene becomes extremely unstable at pressures higher than 15 psi. The acetylene cylinder should always be used in the upright position. Using an acetylene cylinder in a horizontal position will result in the acetone draining into the hoses.

It is not possible to determine with any accuracy the quantity of acetylene in a cylinder by observing the pressure gauge because it is in a dissolved condition. The only really accurate way of determining the quantity of gas in the cylinder is to weigh it and subtract this from the weight of the full cylinder, often stamped on the side of the cylinder.

> **CAUTION** *It is a common malpractice to set acetylene pressure at high values. Check a welder's manual for the correct pressure values to set for the equipment and procedure you are using.*

> **CAUTION** *Never operate an acetylene cylinder in anything but an upright position. Using acetylene when the cylinder is horizontal results in acetone exiting, and the acetylene can destabilize the remaining contents of the cylinder.*

Oxygen Cylinders

It should be noted that oxygen cylinders tend to pose more problems than acetylene when exposed to fire. They should always be stored in the same location in a service shop when not in use (this should be identified to the Fire Department during an inspection) and not left randomly on the shop floor.

Oxygen regulators and hose fittings use a right-hand thread. The cylinder pressure gauge will accurately indicate the oxygen quantity in the cylinder, meaning that the volume of oxygen in the cylinder is approximately proportional to the pressure.

Oxygen is stored in the cylinders at a pressure of 2,200 psi (15 MPa) and the handwheel actuated valve ''forward-seats'' to close the flow from the cylinder and ''back-seats'' when the cylinder is opened: it is important to ensure, therefore, that the valve is fully opened when in use. The consequence of not fully opening the valve is leakage past the valve threads. **Figure 1-12** shows a high-pressure oxygen cylinder.

> **CAUTION** *Never use oxygen as a substitute for compressed air when cleaning components in a shop environment. Oxygen can combine with solvents, oils, and grease, resulting in an explosion.*

Regulators and Gauges

A regulator is a device used to reduce the pressure at which gas is delivered: it sets the working

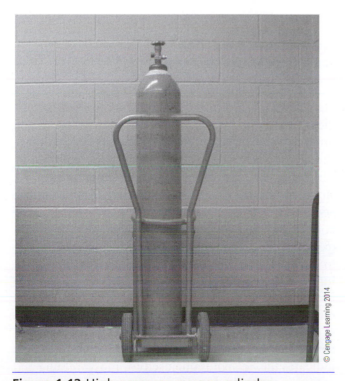

Figure 1-12 High-pressure oxygen cylinder.

© Cengage Learning 2014

pressure of the oxygen or fuel. Both oxygen and fuel regulators function similarly in that they increase the working pressure when turned clockwise (CW). They close off the pressure when backed out counterclockwise (CCW).

Pressure regulator assemblies are usually equipped with two gauges. The cylinder pressure gauge indicates the actual pressure in the cylinder. The working pressure gauge indicates the working pressure, and this should be trimmed using the regulator valve to the required value while under flow.

Hoses and Fittings

The hoses used with oxyacetylene equipment are usually color-coded. Green is used to identify the oxygen hose, and red identifies the fuel hose. Each hose connects the cylinder regulator assembly with the torch. Hoses may be single or paired (Siamese). Hoses should be routinely inspected and replaced when defective. A leaking hose should never be repaired by wrapping with tape. In fact, it is generally bad practice to consider repairing welding gas hoses by any method: simply replace when they fail.

Fittings couple the hoses to the regulators and the torch. Each fitting consists of a nut and gland. Oxygen fittings use a right-hand thread, and fuel fittings use a left-hand thread. The fittings are machined out of brass that has a self-lubricating characteristic. Never lubricate the threads on oxyacetylene fittings.

Torches and Tips

Torches should be ignited by first setting the working pressure under flow for both gases, and then opening the fuel valve *only* and igniting the torch using a flint spark lighter. Set the acetylene flame to a clean burn (no soot); then open the oxygen valve to set the appropriate flame. When setting a cutting torch, set the cutting oxygen last. When extinguishing the torch, close the fuel valve first, and then the oxygen: the cylinders should be shut down finally and the hoses purged. **Figure 1-13** shows an oxyacetylene station with a cutting torch.

Welding, cutting, and heating tips may be used with oxyacetylene equipment. Consult a welder's manual to determine the appropriate working pressures for the tip/process to be used. There is a tendency to set gas working pressure high. Even when using a large heating tip often described as a rosebud, the working pressure of both the acetylene and the oxygen should be set at 7 psi (50 kPa). **Figure 1-14** shows a typical cutting torch.

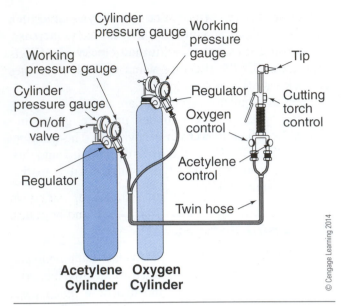

Figure 1-13 Oxyacetylene station set up with a cutting torch.

Figure 1-14 Oxyacetylene cutting torch.

Backfire

Backfire is a condition where the fuel ignites within the nozzle of the torch, producing a popping or squealing noise: it often occurs when the torch nozzle overheats. Extinguish the torch and clean the nozzle with tip cleaners. Torches may be cooled by immersing in water briefly with the oxygen valve open.

Flashback

Flashback is a much more severe condition than backfire: it takes place when the flame travels backward into the torch to the gas-mixing chamber and beyond. Causes of flashback are inappropriate pressure settings (especially low-pressure settings) and leaking hoses/fittings. When a backfire or flashback condition is suspected, close the cylinder valves immediately beginning with the fuel valve. Flashback arresters are usually fitted to the torch and will limit the extent of damage when a flashback occurs.

Eye Protection

Safety requires that a #4 to #6 grade filter be used whenever using an oxyacetylene torch. The flame radiates ultraviolet light that can damage eyesight. Even UV-rated sunglasses are not sufficient protection.

Oxyacetylene Precautions

Take the following precautions when working with oxyacetylene:

- Store oxygen and acetylene upright in a well-ventilated, fireproof room.
- Protect cylinders from snow, ice, and direct sunshine.
- Remember that oil and grease may ignite spontaneously in the presence of oxygen.
- Never use oxygen in place of compressed air.
- Avoid bumping and dropping cylinders.
- Keep cylinders away from electrical equipment where there is a danger of arcing.
- Never lubricate the regulator, gauge, cylinder, and hose fittings with oil or grease.
- Blow out cylinder fittings before connecting regulators: make sure the gas jet is directed away from equipment and other people.
- Use soapy water to check for leaks: NEVER use a flame to check for leaks.
- Thaw frozen spindle valves with warm water: NEVER use a flame.

Adjustment of the Oxyacetylene Flame

To adjust an oxyacetylene flame, the torch acetylene valve is first turned on and the gas ignited. At the point of ignition, the flame will be yellow and produce black smoke. Next, the acetylene pressure should be increased by using the torch fuel valve. This will increase the brightness and reduce the smoking. At the point the smoking disappears, the acetylene working pressure can be assumed to be correct for the nozzle jet size used. Now, the torch oxygen valve is turned on. This will cause the flame to become generally less luminous, and an inner blue luminous cone surrounded by a white-colored plume is formed at the tip of the nozzle. The white-colored plume indicates excess acetylene. As more oxygen is supplied, this plume reduces until there is a clearly defined blue cone with no white plume visible. This indicates the *neutral* flame used for most welding and cutting operations.

Oxidizing Flame. If after setting a neutral flame, the oxygen supply is increased, the blue cone will become smaller and sharper in definition and the outer envelope will become streaky. This is known as an *oxidizing flame* indicating excess oxygen. For most welding procedures an oxidizing flame should be avoided, but in some special applications, a slightly oxidizing flame is required.

Carburizing Flame. A *carburizing flame* is indicated by the presence of a white plume surrounding an inner blue cone. A carburizing flame should also be avoided in most welding operations, although a very slight carburizing flame is used in certain special applications.

Electric Arc Welding

Electric arc welding and cutting processes are used extensively in truck and heavy equipment service garages. Arc welding stations either receive or generate high-voltage charges, and then condition them to lower-voltage, high-current circuits. Before attempting to use any type of arc welding equipment, make sure you receive some basic instruction and training. The following types of welding stations are non-specialized in application and are found in many repair shops:

- Arc welding station: uses a flux-coated, consumable electrode, often known as *stick welding*.
- Metal inert gas (MIG): uses a continuous reel of wire that acts as the electrode, around which inert gas is fed to shield the weld from air and ambient moisture.
- Tungsten inert gas (TIG): uses a non-consumable tungsten electrode surrounded by inert gas, and filler rods are dipped into the welding puddle created.
- Carbon-arc cutting: arc is ignited using carbon electrodes to melt base melt while a jet of compressed air blows through the puddle to make the cut.

Figure 1-15 shows a typical arc welding electrode holder. Because less training and faster weld speeds

are possible with MIG welding processes, stick arc welding is less commonly used in today's shops. However, when specialty alloy and tempered materials are required to be welded, arc welding electrodes (sticks) are commonly selected.

OPERATOR PROTECTIVE STRUCTURES

Operator protective structures (OPSs) are required on a wide range of vehicles operating in warehouses, on highways, off highways, and in mines. Generally they can be categorized as:

- Rollover protection systems (ROPSs)
- Falling-object protection systems (FOPSs)

ROPS and FOPS are mandated by most jurisdictions operating off-highway equipment ranging from forklift trucks to open-face mining machines. Even in jurisdictions in which ROPS and FOPS are not mandated, operators wary of litigation usually have these critical safety systems installed. Applications in which ROPS and FOPS are installed include:

- Rock trucks
- Dozers
- Excavators
- Backhoes
- Loaders
- Graders
- Tractors
- Lawnmowers
- Dump trucks
- Skid steers

Testing of OPS

It is recommended that OPS be tested to physical destruction. For this reason, it is important that ROPSs and FOPSs be engineered for the specific application they are to be installed in.

Importance of Seatbelts. It should be stressed that ROPSs and FOPSs are engineered with the assumption that the operator is wearing a seatbelt. In the case of ROPSs especially, a rollover event could potentially seriously injure an operator not wearing a seatbelt even if the ROPS cage is not compromised. Wearing seatbelts in off-road equipment is not usually legally mandated (as it sometimes is in the case of on-highway vehicles) so seatbelt usage becomes an employer responsibility and one that requires operator training.

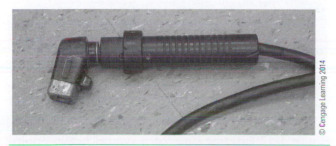

© Cengage Learning 2014

Figure 1-15 Arc welding electrode holder.

Rollover Protection System

Rollover protection systems (ROPSs) are engineered reinforcements installed onto or into a vehicle to prevent crushing or collapse of a vehicle cab when it is involved in a rollover event. ROPSs are designed to reduce injuries to operators and they are required to be maintained in fully functional condition.

A ROPS functions in two ways:

- By increasing structural strength of a cab roof to protect the operator in the event the vehicle rolls
- By absorbing and dissipating the energy loaded onto a roof when a vehicle rolls over

In vehicles such as forklift trucks with an open operator cabin, ROPSs consist of an open tubular cage that encloses the operator. In vehicles with a cab enclosure, a ROPS reinforces the cab with a cage structure to protect the operator.

Falling-Object Protection System

A **falling-object protection system (FOPS)** is an engineered reinforcement installed onto or into a vehicle cabin to reduce injury to an operator in the event of a falling object. The object could be skids in a warehouse environment or a rock in a mining environment.

Who Uses OPS?

Mining companies are leaders in the implementation of workplace safety strategies. Their workplace, health, and safety systems are subject to high levels of scrutiny because they often operate in a high-risk working environment.

ROPSs and FOPSs are covered by SAE J1164, but it should be noted that this J-standard is not currently up to date. There is some motivation on OEMs and employers to ensure that OPS is maintained at a high performance standard due to the consequences of litigation when accidents do occur.

> **CAUTION** *Modifications to OPS, ROPS, and FOPS should not be attempted without first consulting the manufacturer. Modifying engineered safety equipment in even small ways can have lethal consequences. In addition, when undertaking repairs on OPS, ROPS, and FOPS equipment, the technician should consult the manufacturer.*

Shop Organization

Nothing invites accidents more than a dirty, disorganized shop environment. Because of the size and weight of the equipment and components worked on in off-road service facilities, safety has to be foremost in the mind from day one on the job. Some key factors:

- Technicians must use PPE at all times.
- Technicians should make a habit of cleaning up after themselves even in larger operations in which janitorial staff work.
- Ensure you do not operate equipment on which you have not been trained.
- Sub-assembly operations should be allocated a separate area of the service facility where possible. This allows technicians to concentrate on the task at hand without worrying about potential safety hazards created by their fellow workers. **Figure 1-16** shows a dedicated track assembly station.
- The area in front of shop doors should be kept clear of equipment and parts at all times. Visibility can be limited in some mobile heavy equipment and technicians should be provided with as much space as possible to navigate from the yard into the service bay. **Figure 1-17** shows an outside view of a typical heavy equipment shop.

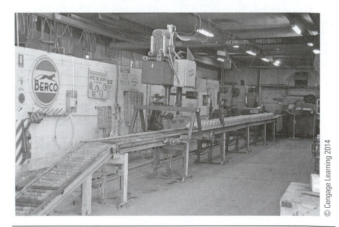

© Cengage Learning 2014

Figure 1-16 Dedicated track assembly station in a Komatsu service facility.

© Cengage Learning 2014

Figure 1-17 Exterior view of a typical heavy equipment shop.

ONLINE TASKS

1. Use an Internet search engine to identify some personal safety equipment suppliers. Check the features offered in safety equipment such as:
   - Safety glasses
   - Safety boots
   - Hearing protection
   - Coveralls
   - Gloves

2. Paste this URL into your favorites/bookmarks folder: *www.osha.gov*. This is one of the best maintained, government-operated Web sites: it is up to date with an abundance of useful links.

Shop Tasks

1. Take a tour of the shop garage making a note of every type of fire extinguisher. List the types of fire each will suppress. Note the location of fire hazards and the assigned place to store combustibles such as oxyacetylene equipment, fuels, window-washer concentrate, and grease and waste barrels.

2. With the guidance of an instructor, run through the process required to change an oxyacetylene torch body to the fuel and oxygen hoses. Do the same with the attachments such as the cutting head, welding head, and rosebud (heating) head.

3. Under the guidance of an instructor, set the pressures on an oxyacetylene torch for cutting, ignite the fuel, and set the oxygen to produce a neutral flame.

4. Identify the welding equipment in your shop/ garage. Identify the types of electrode used with arc welding equipment and interpret the AWS codes (such as E-7018). Do the same with MIG welding wire, and identify the shielding gas used.

Summary

- To work safely in a service garage environment, it is important to evaluate potential dangers and respond appropriately.

- Working as a service technician is more physical than many occupations, so it is important to maintain healthy personal lifestyle.

- Personal safety clothing and equipment such as safety boots, eye protection, coveralls, hearing protection, and different types of gloves are required when working in a service garage.

- Technicians should learn to distinguish between the four different types of fires and identify the fire extinguishers required to suppress them.

- Jacks and hoists are used extensively in service facilities and should be used properly and inspected routinely.

- The potential danger of exhaust emissions should be recognized so that equipment, when inside, is run with exhaust extraction piping.

- It is important to identify what is required to work safely with chassis electrical systems because of the costly damage caused by simple errors.

- Shop mains electrical systems are used in portable power and stationary equipment and can be lethal if not handled properly.

- Oxyacetylene cutting and welding equipment is used extensively in service facilities and technicians should understand how the equipment is set up.

- Care should be taken when using any of the arc welding processes because of the potential for personal injury (electric shock) and equipment damage (high-voltage spikes).

Review Questions

1. Which of the following describes the phrase, "Shake Hands with Danger"?

   A. a way of life in any modern repair shop

   B. a safety video presentation made by Caterpillar

   C. a sure way of avoiding accidents

   D. something that only beginners do

2. When will a worker be more likely to be injured?

    A. during the first day on the job   C. during the second to fourth year of employment

                                    D. during the year before retirement

    B. during the first year of employment

3. Which of the following should be used when lifting a heavy object?

    A. Keep your back erect while lifting.   C. Bend your legs and lift using the leg muscles.

                                    D. All of the above.

    B. Keep the weight you are lifting close to your body.

4. What is Purple K?

    A. a new type of stimulant   C. a toxic gas

    B. a dry powder fire suppressant   D. a type of paint

5. Which of the following is usually a requirement of a safety shoe or boot?

    A. UL certification   C. steel toe

    B. steel sole shank   D. all of the above

6. What type of gloves are recommended by some shops when working with heavy suspension components?

    A. synthetic rubber   D. Gloves should never be worn when working with suspension components.

    B. vinyl disposable

    C. leather welding gloves

7. Which of the following is a false statement?

    A. Never wear gloves when using a bench grinder.   C. Disposable vinyl gloves help protect hands from being exposed to high temperatures.

    B. Wear leather gloves during exposure to oil and fuel.   D. Welding gloves harden when repeatedly working with refrigerant.

8. Which type of fire can be safely extinguished with water?

    A. Class A   C. Class C

    B. Class B   D. Class D

9. When attempting to suppress a Class C fire in a chassis, which of the following is good practice?

    A. Disconnect the batteries.   C. Avoid inhaling the fumes produced by burning conduit.

    B. Use a carbon dioxide fire extinguisher.   D. All of the above.

10. When buck riveting steel panels, which of the following personal protection items are required?

    A. safety glasses   C. work gloves

    B. hearing protection   D. all of the above

11. What sealing compound should be used on brass oxyacetylene fittings?

   A. nothing   C. multipurpose thread sealant

   B. never-seize   D. Teflon® tape

12. When oxyacetylene hoses are color-coded, which color is used to indicate the oxygen hose?

   A. blue   C. red

   B. green   D. black

13. To tighten an acetylene regulator fitting, which of the following should be true?

   A. Turn clockwise.   C. Torque oxygen fitting first.

   B. Turn counterclockwise.   D. Ensure the regulator valve is open.

14. Which type of oxyacetylene flame should be set for most heating and welding procedures?

   A. neutral   C. carburizing

   B. oxidizing   D. white

15. What is the maximum safe acetylene flow pressure that should be set at the regulator?

   A. 5 psi   C. 105 psi

   B. 15 psi   D. 150 psi

16. Technician A says that many jurisdictions have banned the use of trouble lights with incandescent bulbs because they can ignite flammable substances. Technician B says that fluorescent trouble lights should be used in service garages. Who is correct?

   A. Technician A only   C. Both A and B

   B. Technician B only   D. Neither A nor B

17. Which of the following compressed gases in a cylinder tends to be the most dangerous in the event of a fire?

   A. oxygen   C. propane

   B. acetylene   D. argon

18. When performing arc welding on the rear of a piece of mobile heavy equipment, where should the ground clamp be fixed?

   A. on the front bumper   C. on the battery positive post

   B. on the battery ground post   D. as close to the weld area as possible

19. What is the minimum level of eye protection required when using an oxyacetylene cutting torch?

   A. UL-rated safety goggles   C. lens with grade 4-rated filter

   B. any shades with 100% UV   D. lens with grade 12-rated filter
      protection

20. Technician A says that it is okay to work on a defective mains electrical system, providing you have learned to identify the three color codes used in the wiring. Technician B says that only qualified electricians should work on three-phase, high-voltage circuits and components. Who is correct?

   A. Technician A only   C. Both A and B

   B. Technician B only   D. Neither A nor B

21. Technician A says that it is recommended that the manufacturer be consulted before repairing FOP equipment. Technician B says that FOP equipment can be modified providing the job is tested to destruction after the modification is complete. Who is correct?

A. Technician A only   C. Both A and B

B. Technician B only   D. Neither A nor B

22. Technician A says that when ESR safety boots are worn, it is safe to handle exposed high-voltage wires. Technician B says that ESR safety boots must be certified to sustain 14,000 volts for up to one minute. Who is correct?

A. Technician A only   C. Both A and B

B. Technician B only   D. Neither A nor B

23. What is used to identify ESR footwear?

A. green triangle   C. skull and crossbones

B. orange omega on white   D. red image of a flame
    rectangle

24. What should you do if you suspect a hydraulic pinhole injury in your hand?

A. Immediately go to a hospital.   C. Clean with alcohol and cover with a Band-Aid.

B. Book an appointment with   D. Accept the fact that you will lose your hand.
    your family physician.

25. Technician A says that you should always use thick leather gloves when attempting to trace a blind leak in hydraulic hose. Technician B says that hydraulic pinhole injection can take place in flesh without the knowledge of the injured person. Who is correct?

A. Technician A only   C. Both A and B

B. Technician B only   D. Neither A nor B

2 Hoisting and Rigging Systems

Learning Objectives

After reading this chapter, you should be able to:

- Identify hoisting and rigging hazards.
- Outline proper hoisting and rigging procedures.
- Determine load weights.
- Identify wire rope, sling, and chain characteristics.
- Inspect wire rope, slings, and chains.
- Outline industry standards for using rigging hardware.
- Understand the importance of rigging workload limits.
- Outline industry crane operational practices and the basic of signaling.

Key Terms

center of gravity	reeving	winches
chain hoists	safe working load (SWL)	wire rope
eye bolt	shackle	working load limit (WLL)
fiber rope	sheave	
hook	sling configuration	

INTRODUCTION

For technicians working with heavy equipment, disassembly of large components is a daily responsibility. Technicians must be trained in the use of overhead cranes and hoisting equipment to a level of competency and understand the importance of safety. Most accidents occur from improper use and lack of understanding of hoisting and rigging procedures rather than equipment malfunctions. This chapter provides some of the fundamentals of safe rigging and hoisting techniques along with some tips on the proper use of this equipment.

Heavy equipment such as mobile cranes is maintained by technicians. In order to properly maintain any type of hoisting apparatus, the technician must understand how the equipment functions. Inspections on cables, blocks, hooks, and sheaves, as well as structural components must be carried out on a scheduled and logged time frame. This chapter will review the guidelines and criteria to properly inspect and identify defective equipment.

HOISTING AND RIGGING HAZARDS

It is important that technicians involved with hoisting and rigging activities are trained in both safety and operating procedures. Hoisting equipment should be operated only by trained personnel. The cause of most rigging accidents can often be traced back to a lack of knowledge on the part of a rigger.

INTRODUCTION TO HOISTING AND RIGGING

A safe hoisting and rigging operation requires the rigger to understand:

- The weight of the load and rigging hardware.
- The lift capacity of the hoisting device.
- The **working load limit (WLL)** of the hoisting rope, slings, and hardware.

When the weights and capacities of components are known, the rigger must then determine how to lift the load so that it is stable and safe to move. Training and experience enables the technician to recognize hazards that can have an impact on a hoisting operation. The technician must also be aware of elements that can affect hoisting safety, factors that reduce capacities of lifting equipment, and safe practices in rigging, lifting, and landing loads. He must also be familiar with the proper inspection and use of slings and other rigging hardware.

Most crane and rigging accidents can be prevented by technicians when following basic safe hoisting and rigging practices. When a technician is working with a rigger or a rigging crew, it is vital that the operator is aware of all aspects of the lift and that a means of communication has been agreed upon, including what signals will be used.

ELEMENTS AFFECTING HOISTING SAFETY

Key variables that can impact a safe lift are:

Working Load Limit Not Known. Assume nothing. Ensure that you know the working load limits of the equipment being used. Never exceed these limits.

Questionable Equipment. Do not use equipment that is suspected to be unsafe or unsuitable for a lift. When in doubt, wait until its suitability has been verified by a competent person.

Weather Conditions. When the visibility of riggers or hoist crew is impaired by snow, fog, rain, darkness, or dust, extra caution must be exercised. For example, operate in "all slow," and if necessary, postpone the lift. At sub-freezing temperatures, be aware that loads may be frozen to the ground or structure they are resting on. In extreme cold conditions avoid shock-loading or impacting the hoist equipment and hardware, which may have become brittle.

Hazardous Wind Conditions. Technicians working outside in the elements should never carry out a hoisting or rigging operation when winds create hazards for workers, the general public, or property. Assess load size and shape to determine whether wind conditions may cause problems. For example, even though the weight of the load may be within the capacity of the equipment, loads with large wind-catching surfaces may swing or rotate out of control during the lift in high or gusting winds. Swinging and rotating loads not only present a danger to riggers, there is also the potential for the forces to overload the hoisting equipment.

Defective Components. Examine all hardware, tackle, and slings before use. It is industry practice to destroy defective components because when it is merely tagged or discarded, it may be picked up and used by someone unaware of its defects.

Electrical Contact. One of the most frequent causes of death of riggers is electrocution. An electrical path can be created when a part of the hoist, load line, or load comes into close proximity to an energized overhead power line. When a crane is operating near a live power line and the load, hoist lines, or any other part of the hoisting operation could encroach on the minimum permitted distance (see **Table 2-1**), specific measures must be taken to avoid negligence and injury. On mobile cranes, consider the boom length when it is lowered, and never allow the boom to reach the minimum distance to the electrical power lines.

Hoist Line Not Plumb. The working load limits of hoisting equipment apply only to freely suspended loads on plumb (true vertical/perpendicular to the ground) hoist lines. If the hoist line is not plumb during load handling, side loads are created, which can destabilize the equipment and cause structural failure or cause equipment to tip with very little warning.

Factors that Reduce Capacity

The working load limits of hoisting and rigging equipment are based on ideal conditions. These ideal

TABLE 2-1: MINIMUM ALLOWABLE WORKING DISTANCE TO HIGH-VOLTAGE WIRES	
Normal Phase-to-Phase Voltage Rating	**Minimum Distance**
750 volts or more, but no greater than 150,000 volts	10 feet (3 meters)
Over 150,000 volts, but no greater than 250,000 volts	15 feet (4.5 meters)
More than 250,000 volts	20 feet (6 meters)

Note: Wind can blow hoist lines, power lines, or your load to cross the minimum allowable distance.

(Data courtesy of Construction Safety Association of Ontario.)

circumstances are seldom achieved in the field. Riggers must therefore recognize the factors that can reduce the capacity of the hoisting equipment.

Swing. Swinging of suspended loads creates additional dynamic forces on the hoist in addition to the weight of the load. On mobile cranes these additional dynamic forces are difficult to calculate and account for, and may cause a crane to tip over or hoisting hardware failure. The force of the swinging action makes the load drift away from the machine centerline, increasing the radius and side-loading on the equipment (see **Figure 2-1**). The load should be kept directly below the boom point or upper load block. This is best accomplished by controlling the load's movement with slow operation. Although pendant cranes cannot tip over, the load momentum from an abrupt stop-and-go motion can cause the load to impact surrounding equipment. Operating a crane with care and precision takes training and experience.

Condition of Equipment. The rated working load limits apply only to equipment and hardware in good condition. Any equipment damaged in service should be taken out of service and destroyed or repaired if applicable.

Weight of Hardware. The rated load of hoisting equipment does not account for the weight of hook blocks, hooks, slings, equalizer beams, and other parts of the lifting equipment. The combined weight of these items must be added to the total weight of the load, and the capacity of the hoisting equipment, including design safety factors, must be large enough to account for the extra load to be lifted.

Dynamic Forces. The working load limits of rigging and hoisting equipment are determined for static loads. The design safety factor is applied to account, in part, for the dynamic motions of the load and equipment. To ensure that the working load limit is not exceeded during operation, allow for wind loading and other dynamic forces created by the movements of the machine and

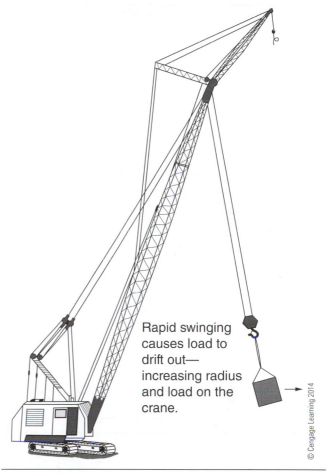

Rapid swinging causes load to drift out—increasing radius and load on the crane.

© Cengage Learning 2014

Figure 2-1 Rapid swinging will cause side-loading on equipment.

its load. Avoid sudden snatching, swinging, and stopping of suspended loads. Rapid acceleration and deceleration also increases these dynamic forces.

Slings

After the hoist rope, slings are the most commonly used piece of rigging equipment. The following precautions should be observed when working with slings.

- Never use damaged slings. Inspect slings regularly to ensure their safety. Check **wire rope** slings for kinking, wear, abrasion, broken wires, worn or cracked fittings, loose seizings and splices, crushing, flattening, and rust or corrosion. Special attention should be given to the areas around thimbles and other fittings.

- Slings should be marked with an identification number and their maximum capacity on a flat ferrule or permanently attached ring with the capacity of the sling when used for a vertical load, or at a 45-degree angle. Ensure that everyone is aware of how the rating system works (see **Figure 2-2**).

- Avoid sharp bends, pinching, and crushing. Use loops and thimbles at all times. Corner pads that prevent the sling from being sharply bent or cut can be made from split sections of large-diameter pipe, corner saddles, padding, or blocking (see **Figure 2-3**).

- Never allow wire rope slings to lie on the ground for long periods of time. Damp wet surfaces, rusty steel, or corrosive substances will damage the sling and should be put out of service.

In many cases the wire rope rusts in the core of the cable and may go undetected on a regular inspection.

- Avoid dragging slings out from underneath loads. The weight of the load can damage the sling and cable.

- Keep wire rope slings away from flame cutting and electric welding. Never make slings from discarded hoist rope. Slings should be made from certified manufacturers following hoisting and rigging standards. Avoid using single-leg wire rope slings with hand-spliced eyes. The load can spin, causing the rope to un-lay and the splice to pull out. Use slings with Flemish spliced eyes and never wrap a wire sling completely around a hook (see **Figure 2-4**). The sharp radius will damage the sling. Wire rope slings have a predetermined bending radius. The larger the rope diameter, the larger the bend radius required: a ⅜-inch sling requires a minimum radius of 4 inches.

- Avoid bending the eye section of wire rope slings around corners. The swage point causes a sharp radius to occur and the bend will weaken the splice or swaging (see **Figure 2-5**).

- When using multi-leg slings always ensure that the sling angle is greater than 45 degrees and preferably greater than 60 degrees. When the horizontal distance between the attachment points on the load is less than the length of the shortest sling leg, then the angle is greater than 60 degrees and generally safe (see **Figure 2-6**).

Multi-Leg Slings. With slings having more than two legs lifting an uneven load, it is possible for some of the legs to be subjected to the full load while the others do no more than balance it. There is no way of

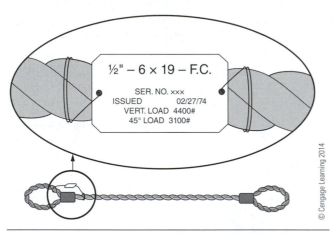

Figure 2-2 Sling identification tags.

½" – 6 × 19 – F.C.

SER. NO. xxx
ISSUED   02/27/74
VERT. LOAD 4400#
45° LOAD 3100#

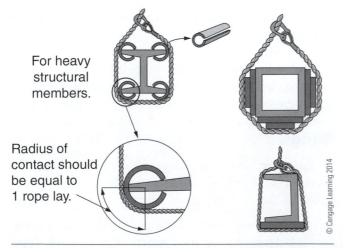

For heavy structural members.

Radius of contact should be equal to 1 rope lay.

Figure 2-3 Avoid making sharp corners with slings.

NO!

Figure 2-4 Never double wrap a wire rope sling on a hook.

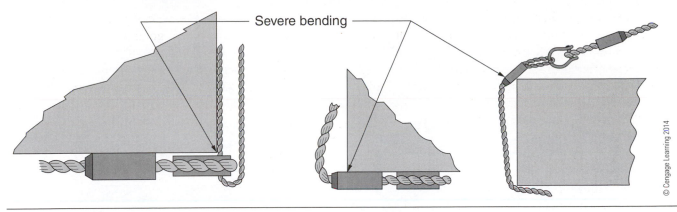

Figure 2-5 Severe bending of spliced area can cause sling to fail.

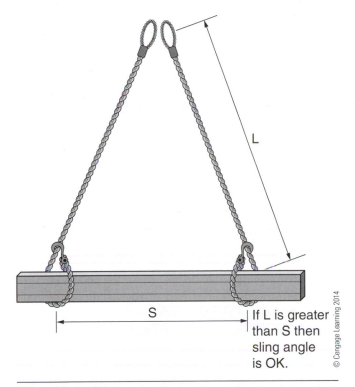

Figure 2-6 Distance between the attachment points should be less than the length of the sling leg.

If L is greater than S then sling angle is OK.

knowing that each leg is carrying its fair share of the load. As a result, when lifting rigid objects with three- or four-leg bridle slings, make sure that at least two of the legs can support the total load. In other words, consider multi-leg slings used on uneven loads as having only two legs.

When using multi-leg slings to lift loads in which one end is much heavier than the other (some legs simply provide balance), the tension on the most heavily loaded leg(s) is more important than the tension on the lightly loaded legs. In these situations, slings should be selected to support the most heavily loaded leg(s). In these instances, do not treat each leg

as equally loaded (in other words, do not divide the total weight by the number of legs). Keep in mind that the motion of the load during hoisting and travel can cause the weight to shift onto different legs. This will result in increases and decreases on the load of any leg.

When using choker hitches, forcing the eye down toward the load increases tension in the sling, which can result in rope damage. Use thimbles and **shackles** to reduce friction on the running line such as those in **Figure 2-7**.

Whenever two or more rope eyes must be placed over a hook, install a shackle on the hook with the shackle pin resting in the hook and attach the rope eyes to the shackle. This will prevent the spread of the sling legs from opening up the hook and prevent the eyes from damaging each other under load (see **Figure 2-8**).

RIGGING, LIFTING, AND LANDING LOADS

Loads should be rigged to prevent any parts from shifting or dislodging during the lift. Suspended loads should be securely slung and properly balanced before they are set in motion.

It is critical to keep the load under control at all times. When removing or lifting large components such as a dump box from a large rock truck, the use of one or more taglines can help prevent uncontrolled motion and help land the load safely. Caution must be taken so loads do not drop as they are safely landed on blocks or safety stands. This must be done before the load is being unhooked.

Lifting beams should be plainly marked with their weight and designed working loads, and be used only for their intended purpose. Never wrap the hoist rope around the load. Attach the load to only the hook, with

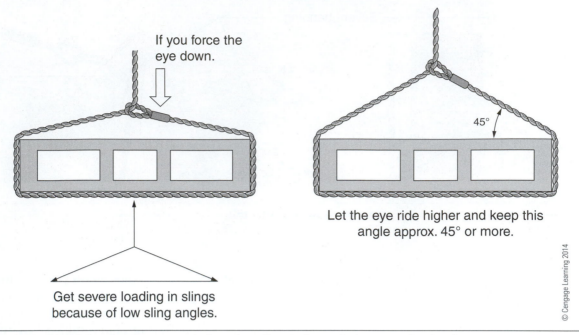

If you force the eye down.

Get severe loading in slings because of low sling angles.

45°

Let the eye ride higher and keep this angle approx. 45° or more.

© Cengage Learning 2014

Figure 2-7 Increased tension is generated as the sling is forced closer to the load.

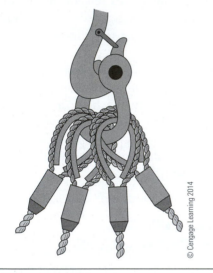

© Cengage Learning 2014

Figure 2-8 Use a shackle when connecting multiple slings to a hook.

slings or other rigging devices. The load line should be brought over the load's **center of gravity** before the lift is started. Hands must be kept away from pinch points as slack is being taken up and proper hand protection should be worn. When handling wire rope, be careful of broken strand wires that can puncture the skin.

Ensure everyone stands clear of loads being lifted, lowered, and freed of slings. As slings are being withdrawn, they may catch under the load and suddenly fly loose. Before making a lift, check to see that the sling is properly attached to the load and never

work under a suspended load. Never make temporary repairs to a sling that is unsafe to use. Secure or remove unused sling legs of a multi-leg sling before the load is lifted (see **Figure 2-9**).

Never point-load a hook unless it is designed and rated for that use. Begin a lift by raising the load slightly to make sure that the load is free and that all sling legs are supporting the load. Avoid impact loading caused by sudden jerking during lifting and lowering. This can stress the sling beyond its rated capacity. Take up slack on the sling gradually and avoid lifting or swinging a load over other workers.

When using two or more slings on a load, ensure that they are all made from the same material and have the same load rating. Prepare adequate blocking or safety stands in a predetermined landing area before loads are lowered.

Determining Weight of the Load

A key step in rigging and hoisting is determining the weight of the load. You can obtain the load's weight from manufacturer's specifications, service information systems, and other dependable sources. If weight information is not provided, you can calculate its approximate weight by using the following formula.

Calculating Weight. The standard weight for steel is: 1 square foot 1 inch thick weighs approximately

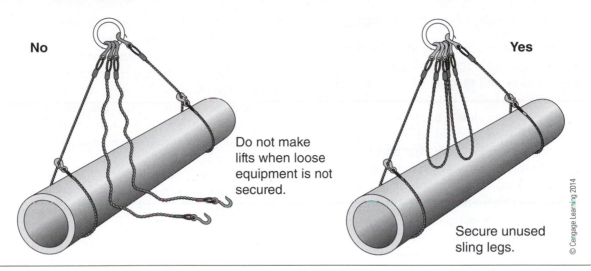

No

Do not make lifts when loose equipment is not secured.

Yes

Secure unused sling legs.

© Cengage Learning 2014

Figure 2-9 When using multiple leg slings, secure unused legs.

40 pounds. For example, applying the standard weight for steel to calculate the weight of a 14-liter engine, we must first get approximate dimensions. We can approximate its size as a rectangle and make the assumption it is solid steel.

engine length: 4.5 feet

engine height: 3 feet

engine width: 12 inches

Now the formula can be calculated as follows:

$$12 \times 3 \times 4.5 \times 40 = 6,480 \text{ pounds}$$

Because the assumption was made that the engine was solid steel, the formula produces a result that is higher than the actual weight, providing a good margin of safety. The lighter weight is due to the fact that many engine components use materials such as aluminum and brass that are lighter than steel. It is a good practice to calculate the weight heavier than actual weight so that the rigging equipment performs well within its lifting capacity.

Calculating the Weight of Various Shapes of Steel. To estimate the weight of various shapes of steel, it helps to envision the steel object as a flat plate. Visually separate the parts, or imagine flattening them into rectangles. For instance, angle iron has a structural shape that can be considered a bent plate. If you flatten angle iron, the result is a plate. For example, 5 × 3 × ¼-inch angle iron would flatten out to approximately a ¼-inch plate that is 8 inches wide. Once you have figured out the flattened size, the volume and weight can be calculated using the previous formula. Because the calculation for the

standard weight of steel is expressed in square feet per inch of thickness, the 8-inch width must be divided by 12 to get the fraction of a foot that it represents. The ¼-inch thickness is already expressed as a fraction of 1 inch. This produces the following formula:

$$40 \text{ pounds} \times 8" \div 12" \times \frac{1}{4}" = 6.67 \text{ pounds}$$

To calculate the weight of a circular or spherical piece of steel (the tube of a hydraulic cylinder) the square foot area must be calculated. To calculate the square foot area of a cylinder, we must first calculate the circumference (the distance around the edge of the circle) and then the area. The formula to calculate the circumference of a circle is diameter × 3.14(π). If we were to use an example of a hydraulic cylinder tube measuring 1 foot in diameter and 6 feet long and a wall thickness of 0.75 inches, the formula would be as follows:

circumference: 1 × 3.14 = 3.14 feet

area: 3.14 × 6 = 18.84

Using the original formula, the approximate weight of the cylinder is calculated as follows.

$$18.84 \times 0.75 \times 40 = 565.2 \text{ pounds}$$

Hydraulic cylinders will have mounting hardware welded to each end of the tube. This extra weight should also be approximated and added to the total weight of the cylinder. The time taken to calculate the approximate weight of any object is time well spent and may save a serious accident through failure of lifting hardware.

HARDWARE, WIRE ROPE, SLINGS

A technician must be able to rig a load to ensure its stability when lifted. This requires knowledge of safe sling configurations and the use of related hardware such as shackles, eyebolts, and wire rope clips. Determining the working load limits of the rigging equipment as well as the weight of the load is also a fundamental requirement of safe rigging practice. Do not use any equipment that is suspected to be unsafe or unsuitable for the load. The working load limits of all hoisting and rigging equipment hardware are based on ideal conditions seldom achieved in the field. It is therefore important to recognize the factors such as wear, improper sling angles, point loading, and center of gravity that can affect the rated working load limits of equipment and hardware.

Slings

Slings are often abused in service facilities and used in severely worn condition. Wear damage is typically caused by failure to provide blocking or softeners between slings and load, thereby allowing sharp edges or corners of the load to cut or fray the slings. Pulling slings out from under loads causes abrasion and/or kinking. Shock loading increases the stress on a sling that may already be overloaded. Traffic running over slings, specially tracked equipment or lift trucks, contributes to wear. Because of these and other conditions, as well as errors in calculating loads and estimating sling working angles, it is recommended that workload limits be based on a design factor of at least 5:1. For the same reasons, slings must be carefully inspected before each use.

Sling Angles. The rated capacity of any sling depends on its size, its configuration, and the angles formed by the sling to the horizontal position.

For example, using a two-leg sling to lift 1,000 pounds, each leg at a sling angle of 90 degrees will be subject to a 500-pound stress load. However, the load on each leg will increase as the working angle decreases. By the time the working angle has increased to 30 degrees, the working load will be 1,000 pounds on each leg! Sling angles should always be greater than 45 degrees whenever possible (see **Figure 2-10** and **Figure 2-11**). The use of any sling at angles lower than 30 degrees is extremely dangerous. This is especially true when an error of only 5 degrees in sling angle can reduce your lifting capacity greater than that of your slings.

Low sling angles also create large, compressive forces on the load that may cause buckling on softer, longer materials. Some load tables list sling angles as low as 15 degrees but the use of any sling at an angle less than 30 degrees is extremely dangerous. Not only are the loads in each leg high at these low angles but an error in measurement as little as 5 degrees can affect the load in the sling drastically. For example, the data in **Table 2-2** illustrates the effect of a 5-degree error in angle measurement on the sling load. Notice that there is almost a 50% error in the assumed load at the 15-degree sling angle.

Sling Configurations

Slings are manufactured from various materials such as wire rope and nylon web, and are available in various configurations suitable for different purposes. This section will describe the most common configurations used and the applications associated with them. Note that the term "sling" covers a wide variety of fiber ropes, wire ropes, chains, and webs.

Single Vertical Hitch. A single vertical hitch supports a load by a single vertical leg. The total weight of the load is carried by the leg at a sling angle of 90 degrees.

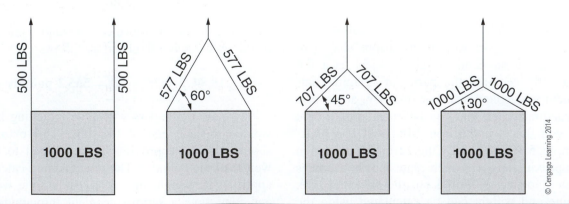

Figure 2-10 Load effect on sling in ratio to sling angle.

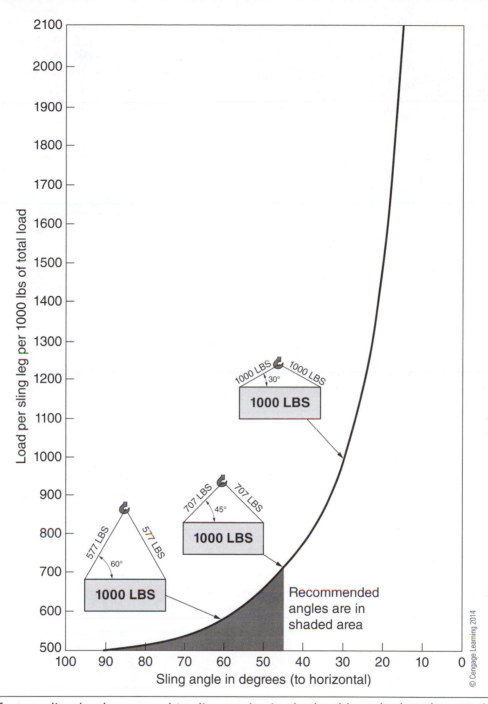

Figure 2-11 Effect on sling load compared to sling angle. Angle should not be less than 45 degrees.

TABLE 2-2: EFFECT OF SLING ANGLES MEASUREMENT ERROR ON LOAD				
Assumed Sling Angle	**Assumed Load Pounds per Leg**	**Actual Angle Is 5° Less Than Assumed Angle**	**Actual Load Pounds per Leg**	**Error %**
90°	500	85°	502	0.4
75°	518	70°	532	2.8
60°	577	55°	610	5.7
45°	707	40°	778	9.1
30°	1,000	25°	1,183	18.3
15°	1,932	10°	2,880	49

© Cengage Learning 2014

As explained earlier, sling angle is measured from a horizontal plane and the weight of the load can equal the working load limit of the sling. End fittings or eyelets can vary; thimbles should be used within eyelets to protect the cable. Mechanical-Flemish splices provide the best securing method with wire rope eye splices. The single vertical hitch must not be used for lifting loose material, or anything difficult to balance because it provides no control over the load and may allow it to rotate. Single vertical hitches should be used on items equipped with lifting eyes or shackles (see **Figure 2-12**).

Bridle Hitch.   Two, three, or four single hitches can be used together to form a bridle hitch for hoisting when the component to be lifted is equipped with proper lifting hardware or attachments. Also used with a wide assortment of end fittings, bridle hitches provide excellent load stability when the load is distributed equally among the legs and the hook is directly over the load's center of gravity. To distribute the load equally it may be necessary to adjust the leg lengths with a turnbuckle. Proper use of a bridle hitch requires that sling angles be carefully measured to ensure that individual legs are not overloaded. In many cases the load may not be distributed evenly when a four-leg sling lifts a particular load. In cases like this, assume that the load is carried by two legs only and ''rate'' the four-leg sling as a two-leg sling. This will ensure that

Figure 2-12 Single vertical hitch sling.

the capacity of the sling is within its **safe working load (SWL)** limit (see **Figure 2-13**).

Single Basket Hitch.   This type of sling is used to support a load by attaching one end of the sling to the hook, then passing the other end under the load and attaching it to the hook. Ensure that the load does not turn or slide along the rope during a lift and that the center of gravity is below the sling (see **Figure 2-14**).

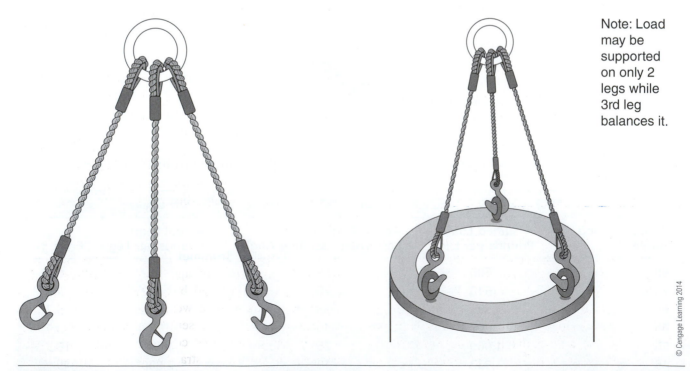

Note: Load may be supported on only 2 legs while 3rd leg balances it.

Figure 2-13 Three-leg bridle hitch on a single ring.

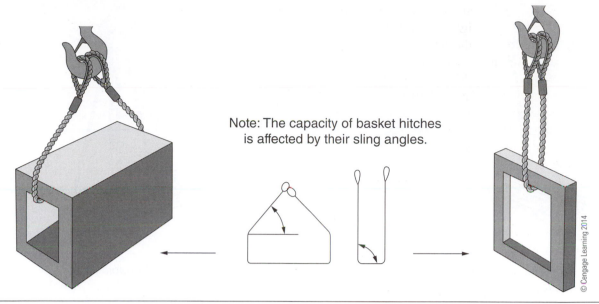

Note: The capacity of basket hitches is affected by their sling angles.

Figure 2-14 Single basket hitch.

CAUTION *The capacity of basket hitches is affected by their sling angles.*

Note: Load may be carried by only two legs while the other legs merely balance it.

Double Basket Hitch. This sling consists of two single basket hitches placed under the load. On smooth surfaces, the legs will tend to draw together as the load is lifted. To counter this, brace the hitch against a change in contour, or other reliable means to prevent the slings from slipping closer to the vertical. You must keep the legs far enough apart to provide balance, but not so far apart that they create angles below 60 degrees from the horizontal. On smooth surfaces, a double wrap basket hitch may be a better choice.

Double Wrap Basket Hitch. The double wrap basket hitch is a basket hitch wrapped completely around the load, compressing it rather than merely supporting it. The double wrap basket hitch can be used in pairs like the double basket hitch. This method is excellent for handling loose material, pipe, rods, or smooth cylindrical loads because the sling is in full 360-degree contact with the load and draws it together as the weight of the load is placed on the sling (see **Figure 2-15**).

Single Choker Hitch. This hitch forms a noose in the rope. It does not provide full 360-degree contact with the load, and therefore should not be used to lift loads difficult to balance or lift loosely bundled items. Choker hitches are useful for turning loads and for resisting a load that wants to turn. Using a choker hitch with two legs can provide stability for longer loads. Like the single choker, this configuration does not completely grip the load. Lift the load horizontally with slings of even length to prevent the load from sliding out (see **Figure 2-16**). Loosely bundled loads should be lifted using a double wrap choker hitch.

Double Wrap Choker Hitch. This hitch is formed by wrapping the sling completely around the load and hooking it into the vertical part of the sling. The hitch is in full 360-degree contact with the load and also draws the load tightly together as the double wrap basket hitch. The difference between the two hitches is that the basket hitch provides greater lifting capacity based on same sling size and material (see **Figure 2-17**).

Endless Slings or Grommet Slings. These slings are useful for a variety of applications. Endless chain slings are manufactured by attaching the ends of a length of chain with a welded or mechanical link. Endless web slings are sewn. An endless wire rope sling is made from one continuous strand wrapped into itself to form a six-strand rope with a strand core. The end is tucked into the body at the point where the

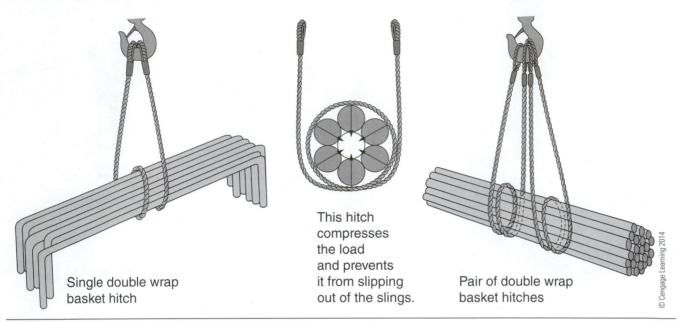

Single double wrap
basket hitch

This hitch
compresses
the load
and prevents
it from slipping
out of the slings.

Pair of double wrap
basket hitches

© Cengage Learning 2014

Figure 2-15 Double wrap basket hitch.

strand was first laid into itself. These slings can be used in a number of configurations, as vertical hitches, basket hitches, choker hitches, and combinations of these basic arrangements. They are flexible but tend to wear more rapidly than other slings because they are not normally equipped with fittings and may deform when bent over hooks or choked (see **Figure 2-18**).

Braided Slings.   Braided slings are usually fabricated from six to eight small-diameter ropes braided to form a single rope that provides a large bearing surface, excellent strength, and flexibility in every direction. They are easy to handle and almost impossible to kink. These slings can be used in all the standard configurations and combinations mentioned but are especially useful for basket hitches where low bearing pressure is desirable or where the bend radius is extremely sharp.

Center of Gravity

Center of gravity is the point around which an object's weight is evenly balanced. The entire weight may be considered concentrated at this point. A suspended object will always move until its center of gravity is directly below its suspension point. To make a level or stable lift, the crane or hook block must be directly above this point before the load is lifted. Thus a load that is slung above and through the center of gravity will not topple or slide out of the slings. It is always important to rig the load so that it is stable. The load's center of gravity must be visualized and placed

directly under the main hook and below the lowest sling attachment point before the load is lifted.

If an object is symmetrical in shape and uniform in composition, its center of gravity will lie at its geometric center. The center of gravity of an oddly shaped object, however, can be difficult to locate. One way to estimate its location is to guess where the center of gravity lies. Rig the load accordingly; make a trial lift and then watch the movement of the suspended load. The center of gravity will always move within the constraints of the rigging and toward the vertical between the hook and the ground. Adjust the sling suspension for the best balance and stability. The center of gravity always moves toward the lowest point on the ground. For this reason, the sling attachment points on your load should be located above the center of gravity whenever practical. If the sling attachment points lie below the center of gravity, your load could flip over or topple. When the center of gravity is closer to one sling leg than to the other, the closer sling leg bears a greater share of the weight. When a load tilts after it is lifted, the tension increases on one sling leg and decreases on the other sling leg. If the load tilts, land the load and rig it again to equalize the load on each leg (see **Figure 2-19**).

Working Load Limits

Knowledge of working load limits (WLL) is essential to the use of ropes, slings, and rigging hardware. The working load limit should be stamped, pressed, printed, tagged, or otherwise indicated on all rigging equipment.

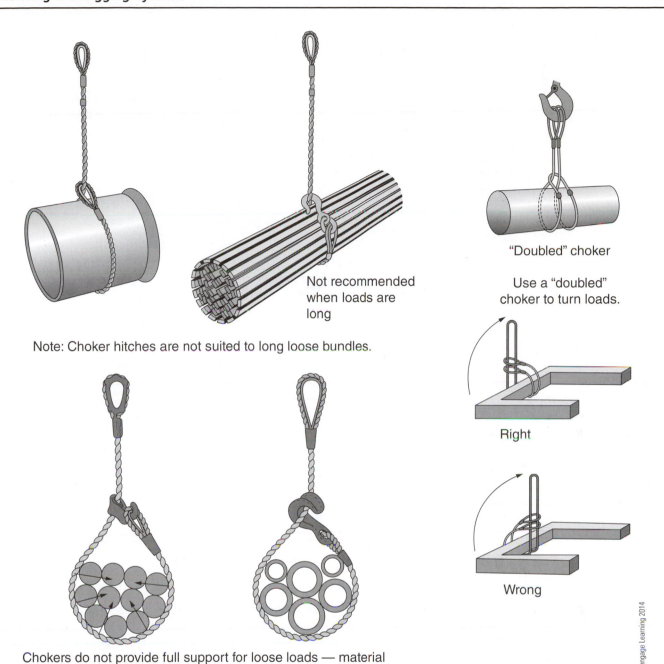

Note: Choker hitches are not suited to long loose bundles.

Not recommended when loads are long

"Doubled" choker

Use a "doubled" choker to turn loads.

Right

Wrong

Chokers do not provide full support for loose loads — material can fall out.

© Cengage Learning 2014

Figure 2-16 Choker hitches.

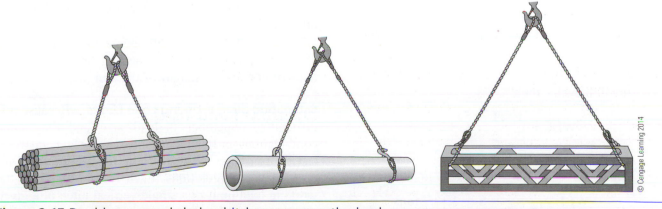

© Cengage Learning 2014

Figure 2-17 Double wrapped choker hitches compress the load.

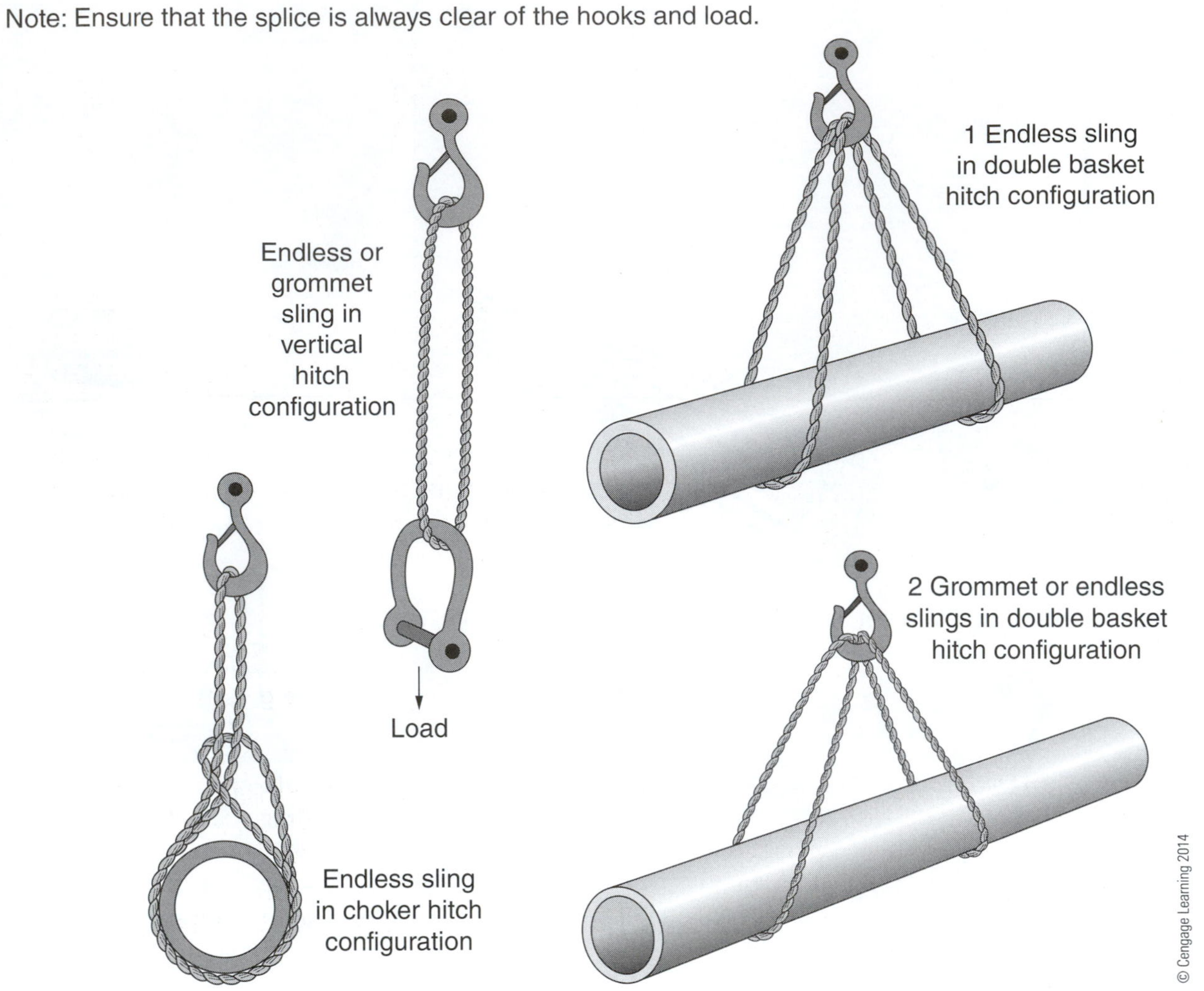

Note: Ensure that the splice is always clear of the hooks and load.

Figure 2-18 Endless slings or grommet slings.

Field Calculation Formula. The **field calculation formula** can be used to compute the working load limit of a wire rope in tons (2,000 pounds). The formula applies to new wire rope of improved plow steel and a design factor of 5 to 1.

$$\textbf{WLL} = \textbf{DIAMETER} \times \textbf{DIAMETER} \times \textbf{8}$$
or
$$\textbf{D}^2 \times \textbf{8}$$

(DIAMETER = nominal rope diameter in inches)

Example

(a) ½-inch diameter rope

$$\textbf{WLL} = \frac{1}{2} \times \frac{1}{2} \times 8 = 2 \textbf{ tons}$$

(b) ⅝-inch diameter rope

$$\textbf{WLL} = \frac{5}{8} \times \frac{5}{8} \times 8 = 3.125 \textbf{ tons}$$

Sling Angle and Working Load Limit. In rigging tables, sling capacities are related to set angles of 90 degrees, 60 degrees, 45 degrees, and 30 degrees. Measuring and determining angles can be difficult; however, you can determine three of them with readily available tools. When a 90-degree angle is formed at the crane hook, you can measure this with a square. It forms two 45-degree angles at the load. A 60-degree angle can also be easily identified. With a two-leg bridle hitch, a 60-degree angle is formed when the distance between the attachment points on the load equals the length of the sling leg.

Estimating Sling Working Load Limits. Because it is difficult to remember all load, size, and sling angle combinations provided in tables, some general rules can be used to estimate working load limits for common sling configurations. Each rule is based on the working load limit of a single vertical hitch of a given size and material and on the ratio **H/L. H** is the vertical

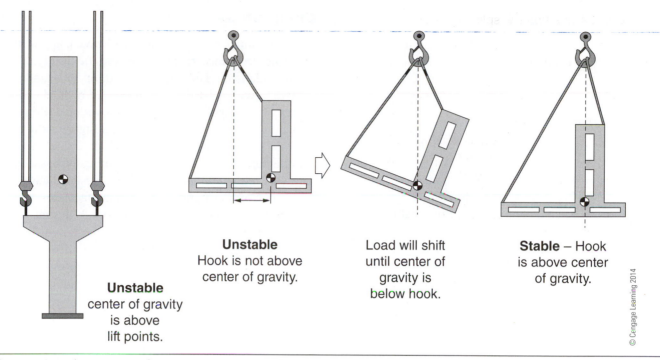

Unstable
center of gravity
is above
lift points.

Unstable
Hook is not above
center of gravity.

Load will shift
until center of
gravity is
below hook.

Stable – Hook
is above center
of gravity.

© Cengage Learning 2014

Figure 2-19 The center of gravity always moves toward the lowest point on the ground.

distance from the saddle of the hook to the top of the load. **L** is the distance, measured along the sling, from the saddle of the hook to the top of the load attachment point. If you cannot measure the entire length of the sling, measure along the sling from the top of the load to a convenient point and call this distance **l**. From this point measure down to the load and call this distance **h**. The ratio **h/l** will be the same as the ratio **H/L** (see **Figure 2-20**).

The ratio of height (H) to length of the sling (L) provides the reduction in capacity due to the sling angle.

This gives the value of reduced capacity. If there are more than two slings and the load is shared equally by all legs, we can increase the reduced capacity.

Capacity is increased (multiplied) by the number of legs.

Note: The smaller the sling angle, the lower the working load limit.

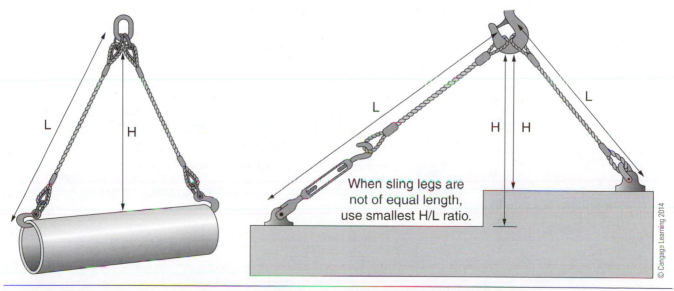

When sling legs are
not of equal length,
use smallest H/L ratio.

© Cengage Learning 2014

Figure 2-20 Estimating working load limits for common sling configurations.

TYPES OF SLINGS

A rigger is required to be able to identify the main types of sling used in the heavy equipment service industry.

Wire Rope Slings

The use of wire rope slings for lifting materials provides several advantages over other types of slings. Although they are not as strong as chain, they are light in weight and have very good flexibility. Broken or frayed wires indicate fatigued or failing cables. Properly fabricated wire rope slings are safe for general lifting use. All wire rope slings should be made of alloyed plow steel with independent wire rope core to reduce the risk of crushing. All OEMs will have a list of criteria in selecting the proper rope construction for a given application.

It is recommended that eyes in wire rope slings be equipped with thimbles. Thimbles are preformed steel inserts within the eye to protect the wire rope when connected to a hook. The eyelets should be formed with a Flemish splice and secured by swaged or pressed mechanical sleeves or fittings. With the exception of socket type connections, this is the only method that produces an eye as strong as the rope itself.

The capacity of a wire rope sling can be greatly affected by being bent sharply around pins, hooks, or parts of a load. The wire rope industry uses the term "D/d ratio" to express the severity of a bend. "D" is the diameter of curvature that the rope is subjected to and "d" is the diameter of the rope.

Wire rope slings should be inspected frequently for broken wires, kinks, abrasion, and corrosion. Inspection procedures and replacement criteria should be followed according to the manufacturer's guidelines. These criteria must be followed regardless of sling type or application.

Chain Slings

Chain slings are best suited to heavier applications requiring resistance to abrasion and flexibility. Alloy chain grade is marked with an 8, 80, or 800; grade 100 is marked with a 10, 100, or 1000. Alloy steel chain is the only type which can be used for overhead lifting and is marked with the letter "A." This designates the chain to be of heat-treated alloy steel, which is the strongest manufactured chain. As with all slings and associated hardware, chain slings must have a design factor of 5 to 1. In North America, chain manufacturers usually give working load limits based on a design factor margin of 3.5 or 4 to 1. Always check with manufacturers to determine the design factor in which their working load limits are calculated. If the design factor is less than 5 to 1, calculate the working load limit of the chain by multiplying the catalogue working load limit by the manufacturer's design factor and dividing by 5.

Catalogue WLL × Manufacturer's Design Factor Ú 5 = WLL (based on design factor of 5 to 1)

Example

½" Alloy Steel Chain

Catalogue WLL = 13,000 pounds

Design Factor = 3.5

13,000 pounds × 3.5 Ú 5 = 9,100 pounds

Therefore, the chain sling must be de-rated to 9,100 pounds.

Never tie a knot in a chain sling to shorten the reach. Slings can be supplied with grab hooks to reduce the length of the reach. Inspect chain slings for inner link wear and wear on the outside of the link barrels (see **Figure 2-21**).

Manufacturers publish tables of allowable wear for various link sizes as well as supplying wear

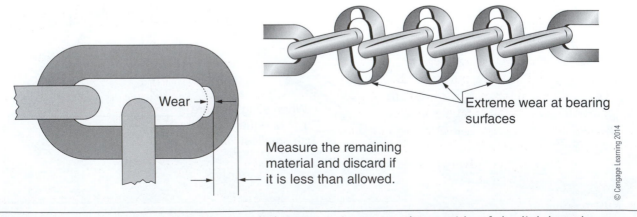

Figure 2-21 Inspecting chain slings for inner link wear and wear on the outside of the link barrels.

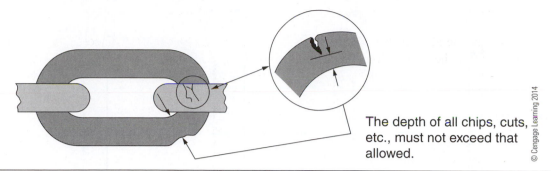

The depth of all chips, cuts, etc., must not exceed that allowed.

© Cengage Learning 2014

Figure 2-22 Inspect chains for nicks and gouges that may cause stress concentrations and weaken links.

gauges to inspect the chain for wear. This will indicate when a sling must be replaced or, if possible, links replaced. Gauges and specification tables from a particular manufacturer should be used only with that brand of chain because exact dimensions of a given nominal size can vary from one manufacturer to another.

CAUTION *Inspect all links for wear at contact surfaces.*

Check chain slings for nicks and gouges that may cause stress concentrations and weaken links (see **Figure 2-22**). Remove the chain from service if any large nicks or gouges are evident within any link. Any repairs must be done according to manufacturers' specifications and procedures. Never use repair links or mechanical couplings to splice broken lengths of alloy steel chain. Never use a chain if the links are stretched or do not move freely.

Tech Tip: A chain is only as strong as its weakest link.

Synthetic Web Slings. Synthetic slings are typically made of two materials—nylon and Dacron-polyester. Synthetic web slings offer a number of advantages over metal slings for rigging purposes. Because of their softer material and greater width, they have fewer tendencies to scratch finely machined or polished surfaces than wire rope or chain slings and due to their flexibility, they tend to mold themselves to the shape of a load. Other advantages of synthetic slings are:

- Rust-proof.
- Non-sparking.

- Minimal twisting and spinning during lifting.
- Ease of rigging due to light weight.
- Elasticity and stretch under load.
- Ability to absorb heavy shock loads.

Synthetic sling termination eyelets are made with standard eyes or twisted eyes on each end. Standard eyes are assembled and sewn to form a flat body sling on the same plane as the sling body. These eyes can be made to the full web width or tapered by being folded and sewn narrower than the web width. Twisted eye terminations have the eye openings 90 degrees from the web width. This configuration is used when the sling application is predominately used as a choker. This prevents the sling from twisting and grabs the load more uniformly. Twisted eye terminations are also manufactured in a full web or tapered end design (see **Figure 2-23** and **Figure 2-24**).

In place of sewn eyes, slings are also available with metal end fittings. Both aluminum and steel

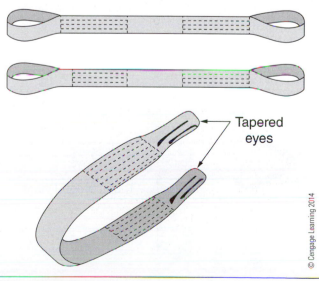

Tapered eyes

© Cengage Learning 2014

Figure 2-23 Standard eye and tapered eye slings.

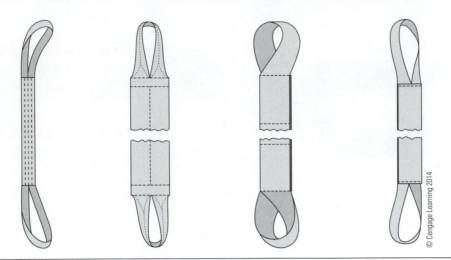

Figure 2-24 Twisted eye slings.

triangular ends are available. Combination hardware for either end of the sling is available to allow the sling to be used in all three combinations: vertical, basket, and choker. These attachments help reduce wear in the sling eyes and thus lengthen sling life (see **Figure 2-25**).

Synthetic web slings are susceptible to cuts from repeated use around sharp-cornered objects or fray from rough-surfaced loads. Protective devices are offered by most sling manufacturers to minimize these effects. Strips of leather, nylon, or other materials sewn on the body of a sling can protect against this wear (see **Figure 2-26**). Leather pads are the most resistant to wear and cutting but are subject to weathering and deterioration.

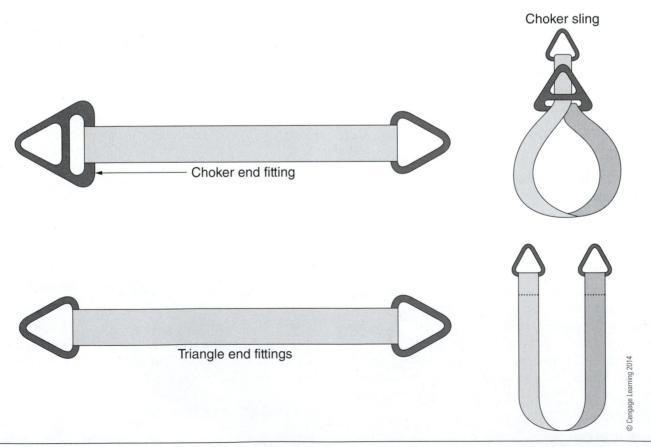

Figure 2-25 Metal end eyes.

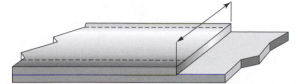

Regular. This is the type that is sewn on to give fixed protection at expected wear points. They can be sewn anywhere on the sling, at any length on one side or on both sides.

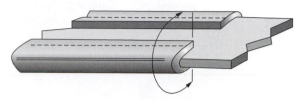

Edgeguard. A strip of webbing or leather is sewn around each edge of the sling. This is necessary for certain applications where the sling edges are subject to damage.

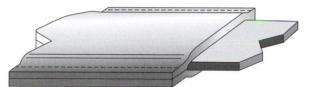

Sleeve. Sometimes called sliding sleeve or tube type wear pads, these pads are ideal for handling material with sharp edges because the sleeve doesn't move when the sling stretches and adjusts to the load. Sleeves cover both sides of the sling and can be shifted to points of expected maximum wear.

© Cengage Learning 2014

Figure 2-26 Protective covers prevent wear on slings.

Endless or Grommet Type Slings

Endless grommet slings are made from one piece of webbing lapped and sewn to form a continuous ring. They can be used as vertical, bridle, baskets, hitches, or choke configurations. Their light weight and flexibility make them a popular source in many shops. Because their load contact points can be shifted with every use, wear is evenly distributed and sling life extended (see **Figure 2-27**).

RIGGING HARDWARE

It is imperative that a technician knows what hardware to use when lifting. Knowing working load limits (WLL) of hardware and matching it with the rope or chain used with it makes for safe hoisting practices. Hardware must be of adequate strength for the application used. Forged alloy steel load-rated hardware is the only recommended hardware for overhead lifting. Load-rated hardware is always stamped with working load limits. Do not use any hardware that does not have a WLL stamped on it. Hardware should be inspected regularly and before each lift. Much like any inspection, look for

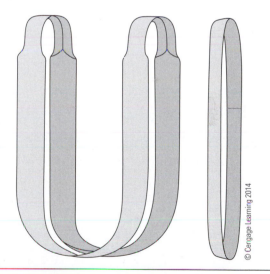

© Cengage Learning 2014

Figure 2-27 Endless grommet type slings.

abnormal wear, cracks, corrosion, deformation, or any obvious damage. Replace and dispose of any damaged hardware. **Figure 2-28** shows the critical hook inspection areas.

Hoisting Hooks

All hoisting hooks are made to accommodate safety latches and therefore should be equipped

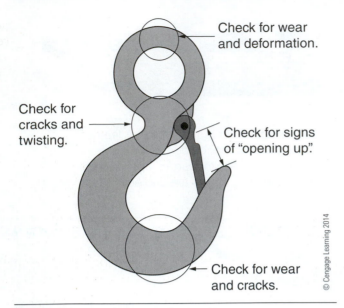

Check for wear and deformation.

Check for cracks and twisting.

Check for signs of "opening up".

Check for wear and cracks.

© Cengage Learning 2014

Figure 2-28 Hook inspection points.

with one. Sliding hooks and grab hooks, as you would find on a length of chain used for securing loads on a truck, do not use safety latches. These chains are not to be used for overhead lifting. As mentioned earlier, only forged alloy steel with WLL stamped or marked on the saddle should be used. When a load is placed on a hook, the load should be focused in the middle of the hook. A load close to the tip will load the hook to the weakest end and reduce the safe working load limit considerably.

Shackles

Shackles are available in various types and sizes and like hooks only forged alloy steel should be used.

Never replace shackle pins with standard grade bolts; shackle pins are designed and manufactured to match the individual shackle capacity. Inspect for wear, distortion, and mouth opening (see **Figure 2-29**). Inspect crown area for irregular wear and do not use a shackle where it will be pulled or loaded at an angle. This severely reduces its capacity and tends to spread the legs.

Eye Bolts

The two types of **eye bolts** used in industry today are the shoulder type and shoulderless type eye bolt. Shoulderless eye bolts should be used only for vertical lifting because angle loading can bend or break the eye bolt (see **Figure 2-30**). Even with shoulder type eye and ring bolts will lose some capacity when loaded on an angle. Shoulder type eye bolts should make contact with the working surface and be properly torqued (see **Figure 2-31**). In many cases it is necessary to stack hardened washers under the shoulder of the eye bolt to ensure a firm, uniform contact with the working surface. The force or load on the eye should always follow the plane of the eye, never in the other direction (see **Figure 2-32**). This will cause the eye to bend and break and is particularly important with bridle slings, which always develop an angular pull on eye bolts unless a spreader bar is used. Eye bolts are usually too small to insert a hook and should be used with a shackle. Never point-load a hook into an eye bolt. Damage to both the eye and hook will occur. Never **reeve** a thread through a pair of eye bolts. This causes extreme side loading on the eyes. Instead, attach a separate sling to each bolt using shackles.

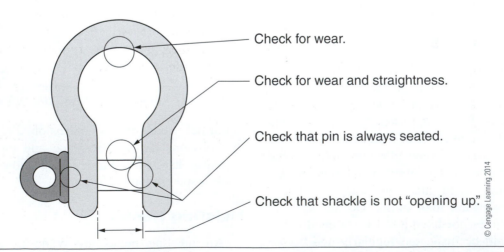

Check for wear.

Check for wear and straightness.

Check that pin is always seated.

Check that shackle is not "opening up".

© Cengage Learning 2014

Figure 2-29 Shackle inspection points.

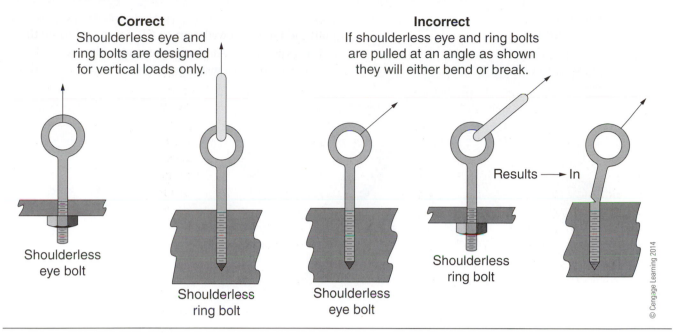

Correct
Shoulderless eye and ring bolts are designed for vertical loads only.

Incorrect
If shoulderless eye and ring bolts are pulled at an angle as shown they will either bend or break.

Shoulderless eye bolt

Shoulderless ring bolt

Shoulderless eye bolt

Results → In

Shoulderless ring bolt

© Cengage Learning 2014

Figure 2-30 Shoulderless type eye bolt.

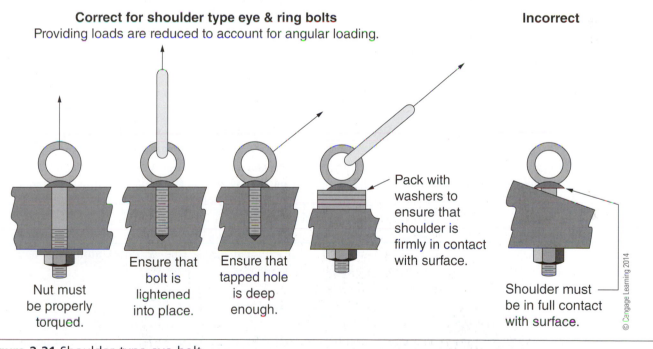

Correct for shoulder type eye & ring bolts
Providing loads are reduced to account for angular loading.

Incorrect

Nut must be properly torqued.

Ensure that bolt is lightened into place.

Ensure that tapped hole is deep enough.

Pack with washers to ensure that shoulder is firmly in contact with surface.

Shoulder must be in full contact with surface.

© Cengage Learning 2014

Figure 2-31 Shoulder type eye bolt.

Wire Rope Clips

Wire rope clips are used for fabricating termination ends or securing the "dead end" of a rope used with a wedge and socket. U-bolt clips must be attached with the U-section in contact with the dead end side of the rope. Tightening and retightening the nuts is required to make sure the cable has enough clamping force not to slip. Following manufacturer's procedures is very critical. **Table 2-3** shows the number of clips required and the torque specifications for the diameter of rope used. **Figure 2-33** shows step-by-step instructions on attaching wire rope clips.

Spreader and Equalizing Beams

Spreader beams are usually used to support long loads during lifts where the use of slings would develop

Correct orientation —
Load is in the plane of the eye.

Incorrect orientation —
When the load is applied to the eye in this direction it will bend.

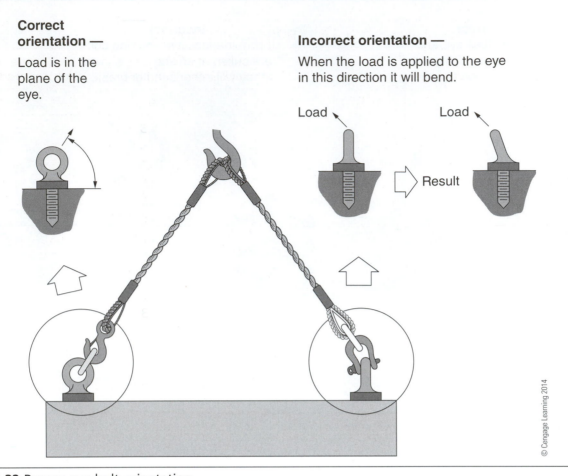

Load   Load

Result

Figure 2-32 Proper eye bolt orientation.

too low an angle and exceed the lifting capacity of the sling. Special applications for lifting components from machinery are also a very common instance where they are used. Spreader beams eliminate the hazard of load tipping, sliding, or bending. Equalizer beams, as the name implies, are used to equalize loads on sling legs and keep equal load on dual hoist cranes when making tandem lifts.

Spreader and equalizer beams are manufactured for a specific application. When they are used for something other than their specific application, make sure the beams have the capacity for the load. The capacity of beams with multiple attachment points changes based on the distance between the attachment points. For example, if the distance between attachment points is doubled, the capacity of the beam is reduced to half (see **Figure 2-34**).

Lever-Operated Hoists or Come-Alongs

Come-alongs are a portable means of lifting or pulling loads short distances. They can be used vertically, horizontally, or on an angle. They are very

useful when used as a second leg to level an uneven lift. When using come-alongs in this or any application, make sure the capacity of the unit falls within the lifting range. The section on chain hoists "inspections and designated chain specifications" equally applies to come-alongs. Never use an attachment to assist cranking to lift a load. This will place the come-along over its capacity. Use a come-along of a larger capacity.

Chain Hoists. Chain hoists (also known as *chain falls* [slang] and *block and tackle* [Australia]) are also a useful hoisting device because a load can be stopped and held stationary at any point. With their slow rate of travel, chain hoists also allow precise vertical placement. When properly rigged, a vertical line between the upper and lower hooks should be gauged. Never install the hook of a chain hoist into an eyelet that is too small. They are intended for use in a vertical or near-vertical position only. If the gear housing is resting against an object while under load, it can be damaged or broken as well. Ensure that the hoist is hanging freely and the chain falls are not obstructed. Chain hoists should be inspected on regular intervals

TABLE 2-3: INSTALLATION OF WIRE ROPE CLIPS						
Rope Diameter		Minimum Number of Clips	Amount of Rope Turn Back From Thimble		Torque for Unlubricated Bolts	
In.	mm		In.	cm	lb-ft	N-m
⅛	3.18	2	3¼	8.3	—	—
³⁄₁₆	4.76	2	3¾	9.5	—	—
¼	6.35	2	4¾	12.1	15	20
⁵⁄₁₆	7.94	2	5½	14.0	30	41
⅜	9.53	2	6½	16.5	45	61
⁷⁄₁₆	11.11	2	7	17.8	65	88
½	12.7	3	11½	29.2	65	88
⁹⁄₁₆	14.3	3	12	30.5	95	129
⅝	15.9	3	12	30.5	95	129
¾	19.0	4	18	45.7	130	131
⅞	22.2	4	19	48.3	225	305
1	25.4	5	26	66.0	225	305
1⅛	28.6	6	34	86.4	225	305
1¼	31.8	6	37	94.0	360	488
1⅜	34.9	7	44	111.8	360	488
1½	38.1	7	48	121.9	360	488
1⅝	41.3	7	51	129.5	430	583
1¾	44.5	7	53	134.6	590	800
2	50.8	8	71	180.3	750	1017
2¼	57.2	8	73	185.4	750	1017
2½	63.5	9	84	213.4	750	1017
2¾	69.9	10	100	254	750	1017
3	76.3	10	106	269.2	1200	1627

© Cengage Learning 2014

and before each lift. The chain component of chain hoists also falls under the guidelines of chain inspections and specifications.

Before using the hoist, inspect the chain for nicks, gouges, twists, and wear. Check the chain guide for wear. Hooks should be measured for signs of opening up. Ensure that the hooks swivel freely and are equipped with safety latches. If the hoist has been subjected to shock loads or dropped, it should be inspected thoroughly before being put back into service. The load brake can be checked by raising the load a few inches off the ground and watching for creeping. If the hoist chain requires replacement, follow the manufacturer's recommendations. Different manufacturers use different pitches for their load chain. Chain intended for one brand of hoist will not necessary mesh properly with the lift wheel of another brand and the hoist will not operate properly. The load chain on chain hoists is case-hardened to reduce surface wear

and is unsuitable for any other use. Load chains are brittle and will stretch only 3% before failing, whereas Grade 8 alloy chain will stretch at least 15%. A load chain removed from a hoist should be destroyed by cutting it into short pieces.

CRANE OPERATIONS

Electric Chain Hoists, Electric Wire Rope Hoists, and Pendant Cranes

Electric chain and wire rope hoists are typically suspended on a fixed beam or on a rolling trolley. On larger cranes the trolleys are motorized, but on smaller cranes they are moved by tugging gently on the pendant and riding along the rail. Most heavy equipment service facilities use pendant cranes. These electric cranes come in varying sizes and capacities (2 to 200 tons). The trolley rides along a

STEP 1

APPLY FIRST CLIP — One base width from dead end of wire rope — U-bolt over dead end — live end rests in clip saddle. Tighten nuts evenly to recommended torque.

STEP 2

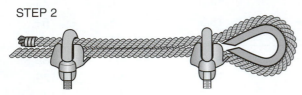

APPLY SECOND CLIP — Nearest loop as possible — U-bolt over dead end — turn on nuts firm but DO NOT TIGHTEN.

STEP 3

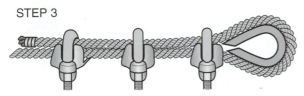

ALL OTHER CLIPS — Space equally between first two.

STEP 4

Apply

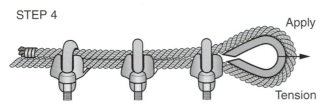

Tension

Apply tension and tighten all nuts to recommended torque.

STEP 5

Apply

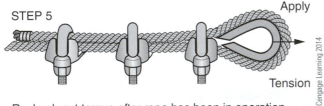

Tension

Recheck nut torque after rope has been in operation.

© Cengage Learning 2014

Figure 2-33 Procedure for installing cable clips.

bridge, which is supported and travels on a set of stationary rails. The crane movements are usually dictated on how the crane is situated in respect to the poles (N.S.E.W.). Apart from these differences, all cranes are quite similar in operating procedures and precautions.

Here are some useful tips when operating a crane:

- Position the hook and block directly over the load.
  - Raise the load gradually to ensure that the lifting ring is seated in the bottom of the hook and the load is safely secured and aligned.
  - Never exceed the working load limit of the hoisting equipment and hardware.
- Inspect pendant control cables for cuts, kinks, or signs of wear.
- Inspect hoist cables for fraying, kinking, crushing, and twisting between the cable and the drum.
- Check hoist drum for proper alignment and stacking of the cable.
- Inspect the block and hook for cracks, bending, or distortion, and the safety latch for proper operation.
- Know and follow manufacturer's instructions and warnings listed on the crane.
- Ensure the intended path of travel is clear of obstructions, including people.
- Ensure proper crane operation and report any deficiencies to the supervisor.
- To prevent crushing hazards, never position yourself between the load and a permanent object.
- Avoid sudden stops and starts. This can increase load momentum.
- Raise the load only high enough to avoid obstructions.
- Never hoist loads over workers. Let people know you are moving with the crane.
- Do not leave loads suspended in the air. If necessary, lock the controls.
- Avoid impacting the trolley and bridge against their "stops."
- Avoid swinging the load or hook when travelling.

Most cranes are equipped with limit switches to prevent maximum allowable position in all crane functions. These devices will automatically stop the crane at its travel limits or maximum allowable up position. Limit switches should be checked daily for correct operation. When the operator of a crane does not have a clear view of the load and its intended path of travel, a signaler must direct operations. Signals for pendant cranes differ from those for mobile and tower cranes since machine movements are different. **Figure 2-35** shows hand signals used for mobile cranes. It is the

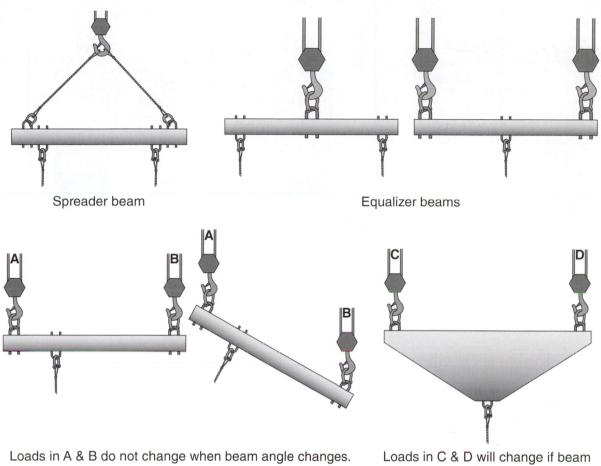

Spreader beam   Equalizer beams

Loads in A & B do not change when beam angle changes.   Loads in C & D will change if beam angle changes.

Combination of spreader and equalizer beams.

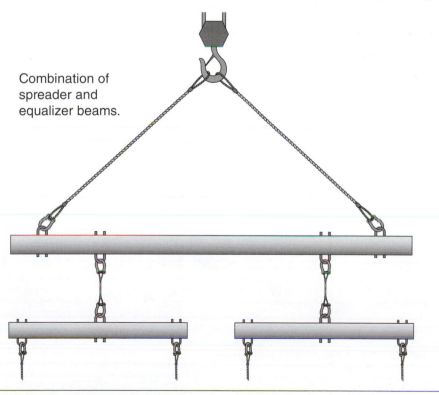

Figure 2-34 Equalizer and spreader type beams.

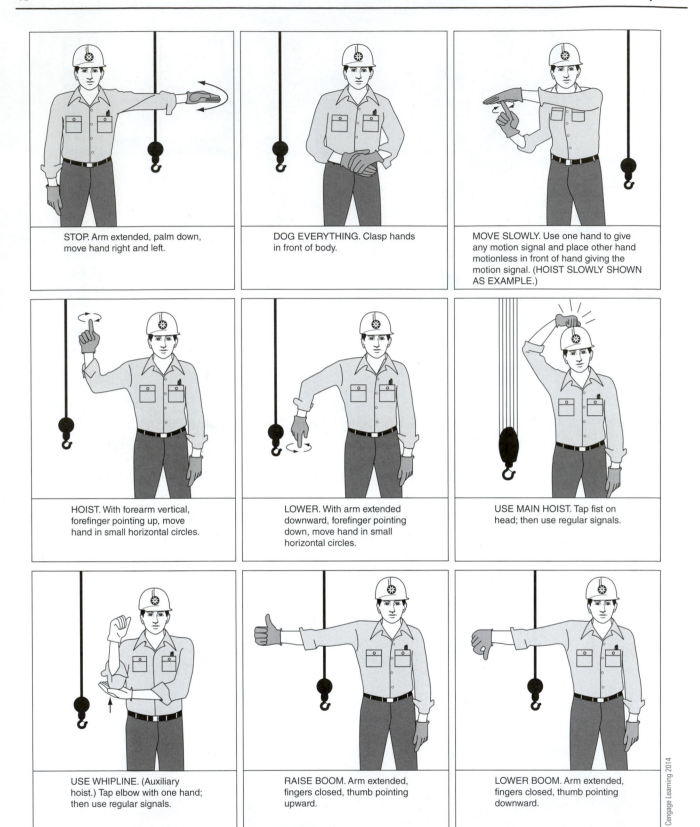

Figure 2-35 Hand signals for crane operation (*continued*).

SWING. Arm extended, point with finger in direction of swing of boom.

RAISE THE BOOM AND LOWER THE LOAD. With arm extended thumb pointing up, flex fingers in and out as long as load movement is desired.

LOWER THE BOOM AND RAISE THE LOAD. With arm extended, thumb pointing down, flex fingers in and out as long as load movement is desired.

TRAVEL. (Rail Mount or trolley) Arm extended forward, hand open and slightly raised, making pushing motion in direction of travel.

EXTENDED BOOM. (Telescoping booms.) Both fists in front of body with thumbs pointing outward.

RETRACT BOOM. (Telescoping booms.) Both fists in front of body with thumbs pointing toward each other.

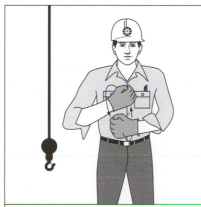

TRAVEL. (Both tracks.) Use both fists in front of body, making a circular motion about each other, indicating direction of travel; forward or backward. (For crawler cranes only.)

TRAVEL. (One track.) Lock the track on side indicated by raised first. Travel opposite track in direction indicated by circular motion of other fist, rotated vertically in front of body. (For crawler cranes only.)

Figure 2-35 Continued.

responsibility of everyone working with cranes to know and understand the signals required for safe pendant crane operation.

ONLINE TASK

Locate each of the following URLs and navigate to pages that provide supporting information to this chapter:

http://web.princeton.edu/sites/ehs/healthsafetyguide/B15.htm

www.oetio.com/Hoisting_and_Rigging_Safety_Manual.aspx

www.sscrane.com/crane-hand-signals-directing-crane-movements/

www.pnl.gov/contracts/hoist_rigging/mobile_cranes.asp

Shop Tasks

1. In a shop class, locate all hoisting chains and slings. Make a thorough inspection of each sling referencing the guidelines covered in this chapter. Record and explain why you have condemned any slings and report them to your immediate supervisor.
2. In a shop class, determine the weight of a common component that is hoisted using the ''Determining Load Weights'' formula covered in this chapter. Then physically weigh the component to determine your accuracy.

Summary

- Safety is the most important consideration when hoisting and rigging.

- A competent technician with the proper training is the key to working safely around heavy equipment: knowing and understanding the proper hoisting and rigging techniques outlined in this chapter is the start to becoming a competent technician.

- Inspect hoisting equipment regularly and before making critical lifts.

- Be aware of potential dangers when hoisting loads around or near other workers.

- Know the working load limits of equipment and hardware used.

- Calculate load weights prior to making a lift.

- Determine the center of gravity of loads being lifted and the proper rigging placements.

Review Questions

1. When a crane is working near high voltage wires exceeding 250,000 volts, what is the minimum distance the boom should be kept clear of the wires?

   A. 10 feet   C. 20 feet

   B. 15 feet   D. 25 feet

2. When lifting with a chain sling, the sling angle should not exceed:

   A. 30 degrees   C. 50 degrees

   B. 40 degrees   D. 60 degrees

3. What do the letters ''WLL'' represent when stamped on a hook or block?

   A. will lift load   C. working limit load

   B. working lift load   D. working load limit

4. When using eye bolts for lifting, the angle the chain or sling connected to it should not be less than:

   A. 45 degrees

   B. 60 degrees

   C. 75 degrees

   D. 80 degrees

5. When inspecting chain slings, what do you look for?

   A. inner link wear

   B. wear on the outside of the link barrels

   C. stretched or kinked links

   D. all of the above

6. When lifting with a chain sling, the sling angle should not be less than:

   A. 35 degrees

   B. 45 degrees

   C. 55 degrees

   D. 65 degrees

7. Field calculation formula for working load limits of wire rope slings is:

   A. WLL = Diameter × Diameter × 8

   B. WLL = Diameter × Diameter × 0.7854

   C. WLL = Radius × Radius × 8

   D. WLL = Radius × Radius × 0.7854

8. Wire rope that is used under normal hoisting conditions is designed with a safety factor of:

   A. 2 to 1

   B. 3 to 1

   C. 4 to 1

   D. 5 to 1

9. When can a crane be used to lift more than its rated capacity?

   A. The load is not more than 10% greater than its rated capacity.

   B. The load is not more than 5% greater than its rated capacity.

   C. The load is not more than 1% greater than its rated capacity.

   D. None of the above. A crane should never lift more than its rated capacity.

10. Synthetic fiber slings have their highest rated capacity when used as a:

    A. single choker hitch

    B. single vertical hitch

    C. single basket hitch

    D. None of the above. Fiber slings should not be used for lifting.

11. When comparing chain slings to wire rope slings, Ω inch chain slings are equivalent to:

    A. ½-inch wire rope slings

    B. ⅝-inch wire rope slings

    C. inch wire rope slings

    D. 1-inch wire rope slings

12. When connecting slings to raise a load, the load center of gravity should be located:

    A. below the sling attachment point

    B. above the sling attachment point

    C. parallel with the sling attachment point

    D. symmetrical to the sling attachment point

13. Come-along chains must be replaced with:

    A. equivalent size cable   C. manufacturer's replacement chain

    B. equivalent size chain   D. a larger chain size

14. When can point-loading a hook into a lifting eye be used?

    A. when the load is lower than the capacity of the hook.   C. only when the hook is not equipped with a safety latch.

    B. only when the load is lifted a few inches.   D. none of the above. Never point load a hook into a lifting eye.

15. Shackles, hooks, and chains used in hoisting applications:

    A. must meet manufacturer's specifications   C. must have a safety factor of 3 to 1

          D. must be manufactured from forged alloy steel

    B. must exceed manufacturer's specifications

3 Introduction to Hydraulics

Learning Objectives

After reading this chapter, you should be able to:

- Explain fundamental hydraulic principles.
- Define *fluids* as they apply to pneumatics and hydraulics.
- Explain the terms hydrostatics and hydrodynamics.
- Interpret Pascal's law and Bernoulli's principle.
- Apply the laws of hydraulics.
- Define laminar flow.
- Explain how pressure and flow are used to manage outcomes in a hydraulic circuit.
- Describe an open-circuit hydraulic system.
- Describe a closed-circuit hydraulic system.
- Calculate force, pressure, and area.
- Identify some of the hydraulic circuits used in heavy equipment.

Key Terms

Bernoulli's principle	hydrodynamics	prime mover
closed-center	hydrostatics	ram
flow	laminar flow	reservoir
fluid	manometer	spool valve
force	open-center	Torricelli's tube
head	Pascal's law	work
head pressure	pressure	

INTRODUCTION

Fluid power is the practice of using gases or liquids in confined circuits to transmit power or multiply human or machine effort. A **fluid** is any substance that has fluidity; that is, it is able to flow and yield to alter shape. For our purposes, any gas or liquid can be classified as a fluid. In mobile hydraulics, we use both pneumatics and hydraulics in different ways. Here are some examples:

Pneumatic circuits:

- Air brakes
- Air suspensions
- Wiper motors
- HVAC controls
- Air starters

Hydraulic circuits:

- Hydraulic brakes
- Hydraulic implements
- Power steering systems
- Powershift and automatic transmissions
- Fuel systems
- Wet-line kits
- Torque converters
- Lift gates

The term *hydraulics* is used to describe fluid power circuits that use liquids, usually specially formulated oils in confined circuits to transmit force or motion. The practice of hydraulics predates recorded history and is thought to have played a role in the construction of the Egyptian pyramids. In this chapter, we will focus on gaining an understanding of basic hydraulic circuits. A heavy equipment technician is required to have a good knowledge of hydraulics to understand steering and brake systems, hydraulic auxiliary equipment, engine operation, fuel systems, and many types of transmissions.

Although hydraulics may have been used by the ancient Egyptians and Romans, it did not really become a science until the time of Blaise Pascal (France 1623–1662). Pascal is so important to the understanding of modern hydraulics that most programs of study begin with what we know today as **Pascal's law**, which states:

Pressure applied to a confined liquid is transmitted undiminished in all directions and acts with equal force on all equal areas, at right angles to those areas.

FUNDAMENTALS

Although we do not often differentiate between them, hydraulics can be divided into two branches, which we know as hydrostatics and hydrodynamics. **Hydrostatics** is the science of transmitting force by pushing on a confined liquid. In a hydrostatic system, transfer of energy takes place because a confined liquid is subject to pressure. Hydrostatic principal is defined as "high-pressure fluid moving at a slow rate of speed." A good example of a hydrostatic circuit is that used to actuate a boom cylinder. **Hydrodynamics**

is the science of moving liquids to transmit energy. Hydrodynamic principal is defined as "low-pressure fluid moving at a high rate of speed." A water wheel or turbine can both be classified as hydrodynamic devices, but the best example on mobile equipment chassis is the torque converter used to impart drive to a powershift transmission.

- Hydrostatics: low fluid movement with high system pressures.
- Hydrodynamics: high fluid velocity with lower system pressures.

Pressure

Pressure is force applied to a specific area. When a confined fluid is subject to pressure, the force applied to the area of confinement will be uniform throughout (Pascal's law). When a liquid confined in a vessel is pushed on, the pressure that results acts evenly on all of the walls of the vessel. This characteristic makes it possible to transmit force or "push" through pipes. In hydraulics, we use liquids rather than gases as hydraulic fluids, because liquids are not easily compressed. Because of this incompressibility, we can relay action almost instantaneously, so long as the circuit is full of liquid and contains no air. Pressure is usually expressed in pounds per square inch or kilopascals.

Creating Pressure. Pressure is created when there is resistance to flow in a circuit. If we get a bicycle pump that has a plunger with a section area of 1 sq. in. and is equipped with a pressure gauge and we draw into that pump a column of water, then trap the water into the pump with a finger, we can say the following:

- If no force is applied to the pump plunger, the gauge will read zero.
- If 10 pounds of linear "push" is applied to the pump, the gauge will read 10 psi.
- If 15 pounds of linear "push" is applied to the pump, the gauge will read 15 psi.
- If the finger closing the pump outlet is removed, the water will be displaced as force is applied to the pump.

We can conclude from this that the pressure of a confined liquid is always in proportion to the applied force. And if the restriction is removed (in our example here, the finger), gauge pressure drops to zero.

Atmospheric Pressure. The Earth's atmosphere extends upward for around 50 miles (80 km) and we know that air exerts a force because of its weight. A column of air measuring 1 sq. in. (6.45 cm) extending

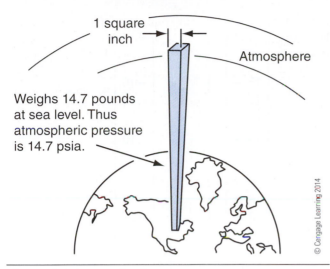

1 square inch

Atmosphere

Weighs 14.7 pounds at sea level. Thus atmospheric pressure is 14.7 psia.

© Cengage Learning 2014

Figure 3-1 Atmospheric pressure.

TABLE 3-1: MATHEMATICAL RELATIONSHIP BETWEEN FORCE, PRESSURE, AND SECTIONAL AREA

To Calculate	Use Formula
Pressure	$P = \dfrac{\text{force}}{\text{area}}$
Force	$F = \text{pressure} \times \text{area}$
Area	$A = \dfrac{\text{force}}{\text{pressure}}$

© Cengage Learning 2014

50 miles (80 km) into the sky would weigh 14.7 pounds (6.67 kg) at sea level as shown in **Figure 3-1**. We express atmospheric pressure at sea level as 14.7 psia (pounds per square inch absolute) or 101.3 kPa, its metric equivalent. If we stood on a high mountain, the column of air would measure less than 50 miles (80 km) and the result would be a lower weight of air in the column. Similarly, if we were below sea level (in a mine for instance), the weight of air would be greater in the column. In North America, we sometimes use the term *atm* (short for *atmosphere*) to describe a unit of measurement of atmospheric pressure; the Europeans use the unit *bar* (short for *barometric* pressure). The relative equivalents of these values appear a little later in this chapter.

Force. **Force** is push or pull effort. The weight of one object placed upon another exerts force on it proportional to its weight. If the objects were glued to each other and we lifted the upper, a pull force would be exerted by the lower object proportional to its weight. Force does not necessarily result in any work done. If you were to push on the rear of a parked railway locomotive, you could apply a lot of force but that effort would be unlikely to result in any movement of the locomotive. The formula for force (*F*) is calculated by multiplying pressure (*P*) by the area (*A*) it acts on.

$$\mathbf{F = P \times A}$$

Calculating Force. In hydraulics, force is the product of pressure multiplied by area. For instance if a fluid pressure of 100 psi acts on a piston sectional area of 50 sq. in. it means that 100 pounds of pressure acts on each square inch of the total sectional area of

the piston. The linear force that results can be calculated as follows:

$$\textbf{Force} = \textbf{100 psi} \times \textbf{50 sq. in.} = \textbf{5000 pounds}$$

Table 3-1 shows the mathematical relationship between force, pressure, and sectional area.

Head Pressure. If you have ever drained fuel out of a tank through its drain valve, you will have observed that the fuel runs out of the tank faster when the tank is full and progressively slower as the tank empties. This occurs because the pressure that forces the fuel from the tank is defined by the weight of the fuel in the tank, and as the tank empties, less weight results in lower pressure. We express this as **head pressure**. By definition, **head** is the vertical distance between two levels in a fluid and it can be expressed using linear or pressure values. For instance, a 10-foot (3.05 m) head of water is equivalent to 4.4 psi (30 kPa). The weight of a fuel or hydraulic oil is less than that of water so we would have to first reckon its specific gravity and then calculate the head required to move it.

Vacuum. A perfect vacuum is a volume that has been evacuated of all matter. That means that no air (mixture of nitrogen and oxygen molecules) can be present in a true vacuum. A true vacuum is difficult to achieve, but we tend to use the word to describe any pressure condition lower than atmospheric pressure. Partial vacuums are used in hydraulics so that when a pump creates a low-pressure area on its inlet side, the static head pressure of our atmosphere acts on the **reservoir** (fluid storage tank) oil to push it toward the pump. In this way, pumps are used to move fluid through circuits whether they are on a wet-line kit, steering gear circuit, or fuel subsystem.

Pressure Scales. There are a number of different pressure scales used today, but all are based on atmospheric pressure. One unit of atmosphere is the equivalent of atmospheric pressure and it can be expressed in all these ways:

$$1 \text{ atm} = 1 \text{ bar (European)}$$
$$= 14.7 \text{ psia}$$
$$= 29.92 \text{ inches Hg (inches of mercury)}$$
$$= 101.3 \text{ kPa (metric)}$$

However, the above values are not precisely equivalent to each other:

$$1 \text{ atm} = 1.0192 \text{ bar}$$
$$1 \text{ bar} = 29.53 \text{ inches Hg}$$
$$= 14.503 \text{ psia}$$
$$1''Hg = 13.6 \text{ inches } H_2O @ 60°F$$

An Italian named Evangelista Torricelli discovered the concept of atmospheric pressure. He inverted a tube filled with mercury into a bowl of the liquid and then observed that the column of mercury in the tube fell until atmospheric pressure acting on the surface balanced against the vacuum created in the tube, as shown in **Figure 3-2**.

At sea level, vacuum in the column in **Torricelli's tube** would support 29.92 inches of mercury. Mercury is usually abbreviated to its elemental symbol Hg. To this day, we use manometers to measure pressure values both below and above atmospheric. To convert 1 psi to its equivalent in inches of mercury, 1 psi is equal to 2 inches Hg (or more precisely, 2.036 inches Hg).

Manometer. A **manometer** is a single tube arranged in a U-shape mounted to a linear measuring scale as shown in **Figure 3-3**. It may be filled to the zero on the calibration scale with either water (H_2O) or mercury (Hg), depending on the pressure range to be measured.

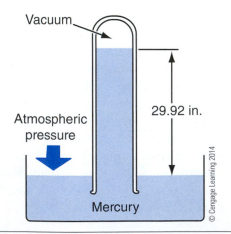

Figure 3-2 Torricelli: Atmospheric pressure expressed in inches of mercury (Hg).

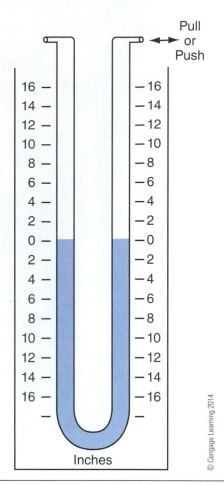

Figure 3-3 Manometer shown at atmospheric pressure: Medium in column can be either H_2O or Hg.

A manometer can measure either push or pull on the fluid within the column. Actual examples of how we use manometer readings are as follows:

- Vacuum restriction on a diesel engine fuel subsystem primary circuit is specified in inches of mercury ("Hg).
- Inlet restriction of a diesel engine air filter system is specified in inches of water ("H_2O).

Note that an inch of mercury is equivalent to 13.6 inches of water, so it is important that you do not mix the two scales. In practice, we tend to use pressure gauges that read values below atmospheric pressure rather than manometers: these pressure gauges display values in inches of mercury on a rotary scale.

Absolute Pressure. Absolute pressure uses a scale in which the zero point is a complete absence of pressure; therefore the indicator needle would rest at 14.7 psi. Gauge pressure has atmospheric pressure as its zero point; therefore the indicator needle would rest at 0 psi. A gauge therefore reads zero when exposed to the atmosphere. To avoid confusing absolute pressure with gauge pressure, absolute pressure is expressed as psia. Gauge pressure is usually expressed as psi or psig.

How Pressure Affects Fluid.   All fluids have a point at which they boil as well as a point at which they will freeze. These calculated points are usually based at atmospheric pressure (14.7 psi.). The most well-known fluid (water) will boil at 212°F (100°C) and freeze at 32°F (0°C) based at atmospheric pressure. It is important to know the freezing and boiling points of a hydraulic fluid due to the fact that it can cause serious effects to a hydraulic system. When a fluid is pressurized as in a pressurized reservoir, the window between the freezing point and the boiling point increases proportionately to the increase in pressure. Inversely, the window between freezing and boiling points decreases if the fluid is exposed to a pressure below atmospheric. This is why water boils at a lower temperature when you are located at a higher altitude. If a hydraulic fluid is subjected to a low pressure (below atmospheric) as the fluid temperature increases, the oil can start to vaporize and cause cavitation within the system. A plugged reservoir vent can easily subject a hydraulic system to this condition.

HYDRAULIC LEVERS

Hydraulic levers can be used to demonstrate Pascal's law: Pressure equals force divided by the sectional area it acts on. Similarly, force equals pressure multiplied by area.

$$F = P \times A$$

To calculate the force exerted by a cylindrical ram, you would use the following formula:

Force exerted = Hydraulic pressure (psi)
× cylindrical area
(diameter² × 0.7854)
Force exerted × cylindrical area (diameter² × 0.7854)

If we build a simple hydraulic circuit such as that shown in **Figure 3-4**, we have the building blocks of a hydraulic jack.

A variation on the hydraulic jack principle can be seen in the type of hydraulic circuit shown in **Figure 3-5**, the basis of a hydraulic ram. A hydraulic **ram** is a linear actuator in which hydraulic force is converted into mechanical energy. The arrangement shown in **Figure 3-5** consists of a pair of unequally sized cylinders connected by a pipe. In the figure, one of the cylinders has a sectional area of 1 sq. in. and the other, 2 sq. in., so the following would be true:

- Applying a force of 100 pounds on the piston in the smaller cylinder would lift a weight of 200 pounds supported on the piston in the larger cylinder.
- Applying a force of 100 pounds on the piston in the smaller cylinder produces a circuit pressure of 100 psi, because the force is being applied to a sectional area of 1 sq. in. P = F/A
- The circuit potential (P) is 100 psi and because this acts on a sectional area of 2 sq. in. it can raise 200 pounds.

$$F = P \times A$$

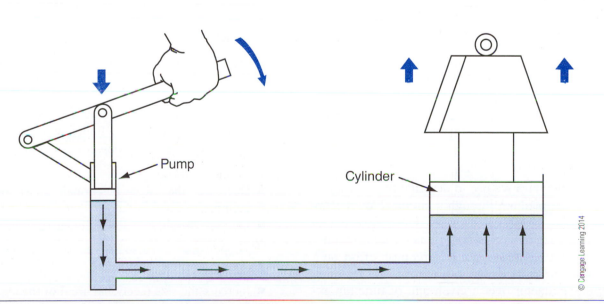

Figure 3-4 Simple hydraulic lever: Pressure multiplies force.

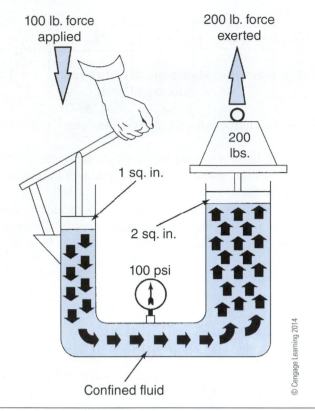

Figure 3-5 Pressure multiplies force.

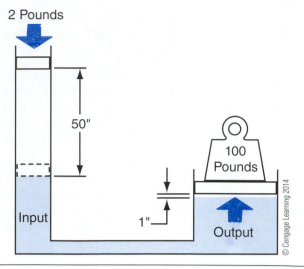

Figure 3-6 Applied force and distance in a hydraulic circuit.

■ Accordingly, if a force of 10 pounds was to be applied to the smaller piston, the resulting circuit pressure would be 10 psi, and the circuit would have the potential to raise a weight of only 20 pounds.

This principle is applied every time we use a hydraulic jack. A hydraulic jack is a simple hydraulic lever as is a hoist lift or a loader arm.

There has to be a trade-off in this type of circuit just as there is with any mechanical lever. **Figure 3-6** shows 2 pounds of force applied to the smaller cylinder. To raise a 100-pound weight 1 inch in the larger cylinder, the piston in the smaller cylinder must move through 50 inches. This is due to the greater area the oil must occupy. This movement is accomplished using multiple strokes and a check valve in typical hydraulic circuits.

Flow

Flow is the term we use to describe the movement of a hydraulic medium through a circuit. Flow occurs when there is a difference in pressure between two points. In a hydraulic circuit, flow is created by a device such as a pump. A pump exerts push effort on a fluid. Flow rate is the volume or mass of fluid passing through a conductor over a given unit of time: an example would be gallons per minute (gpm).

Measuring Flow. Flow can be measured in two ways: velocity and flow rate. The velocity of a fluid in a confined circuit is the speed at which it moves through it. It is measured in feet per second (fps). Flow rate is the volume of fluid that passes a point in a hydraulic circuit in a given time. It is measured in gallons per minute (gpm). Flow rate determines the speed at which a load moves. Fluid velocity and flow rate have to be considered when sizing the hydraulic hoses and lines that connect hydraulic circuit components.

If a small-diameter pipe opens into a larger diameter, then we can conclude:

■ A constant flow rate will result in lower velocity when the diameter increases.
■ A constant flow rate will result in higher velocity when the diameter decreases.
■ The velocity of oil in a hydraulic line is inversely proportional to its cross-sectional area.

We should also add that lower fluid velocities are generally desirable to reduce friction and turbulence in the fluid.

The same will work for hydraulic cylinders. Given an equal flow rate, a small cylinder will move faster than a larger cylinder. If the objective is to increase the speed at which a load moves, then

■ Decrease the size (sectional area) of the cylinder.
■ Increase the flow to the cylinder (gpm).

The opposite would also be true, so if the objective was to slow the speed at which a load moves, then

■ Increase the size (sectional area) of the cylinder.
■ Decrease the flow to the cylinder (gpm).

Therefore, the speed of a cylinder is proportional to the flow it is subject to, and inversely proportional to the piston area. Note, there would also be a decrease or force increase to the cylinder capacity.

Pressure Drop.   In a confined hydraulic circuit, whenever there is flow, a pressure drop results. Again, the opposite applies. Whenever there is a difference in pressure, there must be flow. Should the pressure difference be too great to establish equilibrium, there would be continuous flow. In a flowing hydraulic circuit, pressure is always highest upstream and lowest downstream. This is why we use the term *pressure drop*. A pressure drop always occurs downstream from a restriction in a circuit.

Flow Restrictions.   Pressure drop will occur whenever there is a restriction to flow, and it will occur downstream from the restriction. In hydraulics, a restriction in a circuit may be unintended, such as a collapsed line, or intended, such as a restrictive orifice. The smaller the line or passage that hydraulic fluid is forced through, the greater the pressure drop. The energy lost because of a pressure drop is converted to heat energy.

Work.   **Work** occurs when effort or force produces an observable result. In a hydraulic circuit, this means moving a load. To produce work in a hydraulic circuit, we must have flow. Work is measured in units of force multiplied by distance, for example, pound-feet.

Work = Force × Distance

Energy.   There are many forms of energy, and energy is simply the capacity to perform work. In a hydraulic circuit, the objective is to transfer energy. We transfer energy from one form to another and from one point to another. The idea is to accomplish this as efficiently as possible and not waste too much by transforming it into heat.

In a typical hydraulic circuit, mechanical energy is required to drive a hydraulic pump to create flow and kinetic energy potential in the fluid. Fluid under pressure is the potential energy of a hydraulic circuit. This energy can then be reconverted to mechanical energy to move a load. This mechanical energy is kinetic energy, or the energy of motion. The prime source of the energy used in a hydraulic system may be the heat value of the fuel used in an engine or the electrical energy in a battery. The term **prime mover** is used to describe the machine that creates the mechanical energy required to power a hydraulic pump, although it is sometimes used to describe the pump itself.

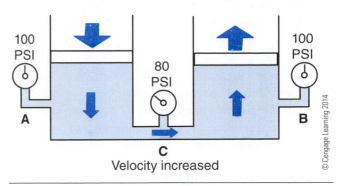

Figure 3-7 Bernoulli's principle.

Bernoulli's principle.   Bernoulli's principle states that if flow in a circuit is constant, then the sum of the pressure and kinetic energy must also be constant. As we said before, when fluid is forced through areas of different diameters, the result is changes in fluid velocity. Fluid flow through a large pipe will be slow, but when the large pipe reduces to a smaller pipe, the fluid velocity will increase. **Figure 3-7** shows a hydraulic circuit consisting of two cylinders connected by a pipe, with force being applied to one of the cylinders. Bernoulli proved that in such a circuit, when force is applied to the cylinder on the left, the pressure measured at the connecting pipe will be less than that measured at either cylinder. This happens because the increase in fluid velocity that occurs in the connecting pipe means that some of the pressure energy has been converted to kinetic energy. In the cylinder on the right, assuming that there have been no frictional losses, the pressure should return to the same value as the cylinder on the left because the kinetic energy has been converted back into pressure.

Laminar Flow.   Flow of a hydraulic medium through a circuit should be as streamlined as possible. Streamlined flow is known as **laminar flow**. Laminar flow is required to minimize friction. Changes in section, sharp turns, and high flow speeds can cause turbulence and crosscurrents in a hydraulic circuit, resulting in friction losses and pressure drops.

TYPES OF HYDRAULIC SYSTEM

Hydraulic systems can be grouped into two main categories:

- Open-center systems
- Closed-center systems

The primary difference between open-center and closed-center systems has to do with what you do with the hydraulic oil in the circuit after it leaves the pump, which is primarily dictated by the type of control valve used.

Open-Center Systems. In an **open-center** system, the pump runs constantly, which means that oil flows through the circuit continuously. The pump typically used is a fixed-displacement pump so it produces continuous flow when it is driven. A valve is used to manage the circuit, and when this valve is in its "open" or neutral position, fluid is allowed to return to the reservoir. An example of an open-center hydraulic system is the power-assisted steering used on highway vehicles. **Figure 3-8** shows an example of an open-center **spool valve**. A spool valve consists of a stationary cylindrical bore within which a spool or piston moves with the objective of directing oil flow.

Closed-Center Systems. In a **closed-center** system, the pump can be "rested" during operation whenever flow is not required to operate an actuator. This means that the control valve blocks flow from the pump when it is in its "closed" or neutral position. A closed-center system requires the use of either a variable displacement pump or proportioning unloading control valves. When a variable displacement pump is used, the pump continues to rotate but it ceases to move fluid when rested. Closed-center systems are becoming more popular due to energy conservation and have many uses on agricultural and mining equipment in applications such as loaders and excavators.

Summary of Basics. This concludes the introduction to some of the basic principles of hydraulics. Having some knowledge of the language of hydraulics is important to understanding not only the hydraulic circuits used on heavy equipment systems, but also transmissions, steering, brakes, suspensions, engines, and fuel systems.

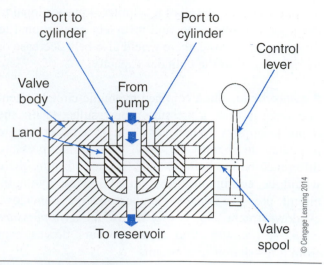

Figure 3-8 Open-center spool valve.

ONLINE TASKS

Unlike many areas of motive power technology, the Internet can be the source of plenty of useful information on the basics of hydraulics. Here are some tips:

1. Visit *www.howstuffworks.com*. Enter some of the key words in this chapter, and see what you come up with.
2. Use a search engine and type in Blaise Pascal. See what you can discover about this genius that has nothing to do with hydraulics.
3. Check out this URL: *http://hydraulics.eaton.com*. See if you can navigate to any information that explains some of the basic principles discussed in this chapter.

Shop Task

Select a piece of mobile equipment in your shop and identify how fluid power principles are used in its various circuits. In this exercise, do not limit yourself to identifying only hydrostatic implement circuits.

Summary

- Fluid power is the practice of using gases or liquids in confined circuits to transmit power or multiply human or machine effort.

- A fluid is any substance that has fluidity: any gas or liquid can be classified as a fluid.

- Fundamental hydraulic principles include Pascal's law, Bernoulli's principle, and how force, pressure, and sectional area are used in hydraulic circuits to produce outcomes.

- Pascal's law states that pressure applied to a confined liquid is transmitted undiminished in all directions and acts with equal force on all equal areas, at right angles to those areas.

- Hydrostatics is the science of transmitting force by pushing on a confined liquid. In a hydrostatic system, transfer of energy takes place because a confined liquid is subject to pressure.

- Hydrodynamics is the science of moving liquids to transmit energy.

- The formula for force (F) is calculated by multiplying pressure (P) by the area (A) it acts on.

- Flow occurs when there is a difference in pressure between two points. In a hydraulic circuit, flow is created by a device such as a pump.

- In a hydraulic circuit, whenever there is flow, a pressure drop results. The opposite also applies: whenever there is a difference in pressure, there must be flow. When the pressure difference is too great to establish equilibrium, continuous flow results.

- Bernoulli's principle states that if flow in a circuit is constant, then the sum of the pressure and kinetic energy must also be constant.

- An open-center system is one in which the pump runs constantly, resulting in continuous oil flow through the circuit: a valve is used to manage the circuit and when this valve is in its "open" or neutral position, fluid is allowed to return to the reservoir.

- A closed-center system is one in which the pump can be "rested" whenever flow is not required to operate an actuator: this means that the control valve blocks flow from the pump when it is in its "closed" or neutral position.

- A closed-center system requires the use of either a variable displacement pump or of proportioning control valves.

- A typical simple hydraulic circuit consists of a reservoir, pump, valves, actuators, conductors, and connectors.

Review Questions

1. Which of the following best describes the operating principles used to raise a loader bucket?

   A. mast ram

   B. hydrostatics

   C. hydrodynamics

   D. hydrology

2. Which of the following is a correct specification of atmospheric pressure at sea level?

   A. 14.7 inches of Hg

   B. 29.8 psi

   C. 101.3 kPa

   D. 29.8 inches of H_2O

3. How many psi is a manometer reading of 12 inches Hg?

   A. 6 psi

   B. 14.7 psi

   C. 24 psi

   D. 101.3 psi

4. Which statement is correct when two unequally sized, hydraulic cylinders in the same circuit are subjected to equal flow?

   A. The larger cylinder will raise less weight.

   B. The smaller cylinder will move faster.

   C. The larger cylinder will require more pressure.

   D. The smaller cylinder will raise more weight.

5. What will occur to the pressure downstream when there is a fixed restriction to the flow in a hydraulic circuit?

   A. remains unchanged

   B. decreases

   C. increases

   D. surges

6. Which statement describes pump operation in an open-center hydraulic circuit?

    A. runs constantly   C. manages to have rest cycles

    B. produces higher operating   D. produces lower operating pressures
       pressures

7. What is the result known as when the application of force produces movement?

    A. horsepower   C. work

    B. power   D. friction

8. When a diesel engine is used to drive a hydraulic pump that powers a lift cylinder, which of the following is correctly known as the *prime mover*?

    A. the diesel engine   C. the lift cylinder

    B. the hydraulic pump   D. the load being moved

9. What force moves the fluid through the circuit upstream to the hydraulic pump when a vented reservoir is used in a hydraulic system?

    A. vacuum   C. hydraulic pressure

    B. atmospheric pressure   D. return pressure

10. Which of the following describes what we call *laminar flow* in a hydraulic circuit?

    A. restricted flow   C. aerated fluid

    B. streamlined flow   D. contaminated fluid

11. Which of the following is a measure of fluid *flow*?

    A. inches of Hg   C. gallons per minute (gpm)

    B. inches of H_2O D. head

12. How much linear force is generated when a hydraulic force of 1000 psi acts on a sectional area of 50 sq. in?

    A. 50 pounds   C. 50,000 pounds

    B. 5,000 pounds   D. 500,000 pounds

13. Who is responsible for the law of hydraulics that states, "pressure applied to a confined liquid is transmitted undiminished in all directions and acts with equal force on all equal areas, at right angles to those areas"?

    A. Newton   C. Bernoulli

    B. Pascal   D. Watt

14. Technician A says that substances in a liquid state are classified as fluids. Technician B says that the air we breathe is by definition a fluid. Who is correct?

    A. Technician A only   C. Both A and B

    B. Technician B only   D. Neither A nor B

15. Which of the following is a measure of *flow rate*?

    A. psi   C. BHP

    B. gpm   D. lb/ft

CHAPTER

4 Hydraulic System Components

Learning Objectives

After reading this chapter, you should be able to:

- Identify the key components required in simple hydraulic circuits.
- Identify the types of reservoirs and accumulators used in hydraulic circuits.
- Describe the function of valves used in hydraulic circuits.
- Explain the operating principles of the different types of valves used in hydraulic circuits.
- Define the term *positive displacement*.
- Identify fixed and variable displacement pumps.
- Outline the operating principles of external and internal gear pumps.
- Outline the operating principles of balanced and unbalanced vane pumps.
- Outline the operating principles of axial and radial piston pumps.
- Outline the operating principles of hydraulic motors.
- Describe the construction of hydraulic conductors and couplers.
- Install hydraulic fittings, couplers, and hoses using the correct procedure.
- Outline the properties of hydraulic fluids.
- Interpret both SAE and ISO viscosity grades.

Key Terms

accumulator	pilot-operated relief valve	ram
cam ring	port plate	reservoir
coupler	positive displacement	rotary actuators
directional control valve	pressure compensator	spool valve
dryseal coupler	pressure-control valve	swashplate
flow compensator	quick-connect coupler	torque limiting compensator
flow-control valve	quick-coupler	viscosity
linear actuator	quick-release coupler	viscosity index (VI)

HYDRAULIC COMPONENTS

Most of the components in a hydraulic circuit are interrelated in much the same way as components in an electrical circuit. The components in the circuits described in this chapter will be consistent with those found in mobile hydraulic systems.

RESERVOIRS

A **reservoir** in a hydraulic system has the following roles:

- Stores hydraulic oil.
- Helps keep oil clean and free of air.
- Acts as a heat exchanger to help cool the oil.

A reservoir is typically equipped with:

- **Filler Cap.** This is the means of replenishing oil in the hydraulic circuit. Depending on the system, the filler cap may be equipped with a vent or be held under pressure.
- **Oil Level Gauge** or **Dipstick**. This provides a means of verifying that there is sufficient oil in the reservoir.
- **Outlet and Return Lines.** These provide a means of conducting the oil out of and back into the reservoir. Although the lines can be positioned on top of or at the sides of a reservoir, the end of each line should be located near the bottom within the reservoir. If an oil return line discharges above the fluid level in a reservoir, foaming can result.
- **Baffle(s).** The function of a baffle or baffles in a reservoir is to separate the return oil from that being drawn out by the pump. However, baffle(s) may be unnecessary on some systems, depending on how the lines and filters are arranged in the circuit. In some applications, baffles also prevent the oil from agitation, which can cause the oil to aerate.
- **Inlet Filter/Screen.** The inlet filter is often just a screen or strainer located in series with the system hydraulic pump.
- **Inspection Cover.** The inspection cover is used to gain access to the internal side of the reservoir for scheduled inspections and clean-out maintenance.
- **Hydraulic Filter.** Many equipment manufacturers locate the main hydraulic filter within the hydraulic tank. The filter housing and lines are built directly into the reservoir. Return oil from the hydraulic system will enter the filters

and be drained directly into the reservoir. The filters are accessed by a large cover bolted directly on the reservoir.

- **Drain Plug.** Drain plugs provide a means of dumping the hydraulic system oil during system service. Some drain plugs are magnetic to attract and hold small metal particles that pass through the system.
- **Reservoir Capacity.** Reservoir capacity is based on the size of the hydraulic system, number of pumps and size, number and size of actuators as well as the pressure that the system is subjected to. These and other factors are considered when reservoir capacity is determined. The general "rule of thumb" is two to three times the pump's capacity for these factors. This would mean that a system using a 40 gpm (gallons per minute) pump would require a reservoir capacity of 80 to 120 gallons depending on particulars. With the sleeker design of modern-day machines, it is no longer possible to use this rule of thumb. As there are factors that determine how big the reservoir needs to be, there are also factors that can allow the reservoir size to be reduced. The main consideration for using a larger reservoir is to limit heat build up; therefore modifications to a hydraulic system that would generate less heat can typically use a smaller reservoir.

- The use of oil coolers, whether air-to-oil or water-to-oil, can reduce oil temperature quickly and return the oil back to the system. This is the most common way of keeping oil temperature constant in machines today.
- High-efficiency hydraulic systems such as flow compensated systems supply only as much oil as the system demands. Pressure compensated systems will cut out flow completely when a preset pressure is reached.
- Torque limiting systems will not allow a system to work at high flow and high pressures. These systems do not generate excessive heat and require smaller reservoirs. Keep in mind that these systems and reservoirs must also have sufficient oil capacity to accommodate all cylinders in their retracted positions. **Figure 4-1** shows the components that make up a typical pressurized reservoir.

HYDRAULIC JACKS

A hydraulic jack consists of a simple hand-actuated pump plunger, a reservoir, two check valves, and an

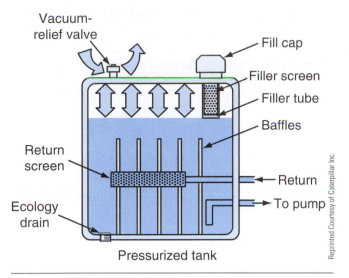

Pressurized tank

Vacuum-relief valve

Fill cap

Filler screen

Filler tube

Baffles

Return screen

Return

To pump

Ecology drain

Reprinted Courtesy of Caterpillar Inc.

Figure 4-1 Components of a typical pressurized reservoir.

output piston. As shown in **Figure 4-2**, when the pump plunger is pulled, reduced pressure is created in the plunger chamber. This causes the atmospheric pressure acting on the oil in the reservoir to push fluid into the plunger chamber through the inlet check valve. While the pump plunger is being pulled through its stroke, load-induced pressure acting on the output piston holds the outlet check valve closed.

When the plunger chamber has been charged with oil (plunger is at the end of its stroke), the pressure in the plunger chamber should be close to the atmospheric pressure in the reservoir. When the pump plunger is forced to retract, the pressure in the plunger chamber exceeds that of the output piston cylinder, which forces the outlet check valve open,

permitting flow and pressure to charge the output cylinder.

Output flow of a simple hydraulic jack is determined by the volume displaced in a single pump cycle. The force developed is defined by the force potential applied to the pump plunger. This simple hydro-mechanical device forms the basis of hydraulic floor jacks and single-acting hydraulic cylinders used on equipment.

ACCUMULATORS

A mechanical coiled spring is a simple **accumulator**. When the spring is compressed, it "stores" energy and can also be used to dampen shock and buffer pressure rise. Accumulators used in hydraulic circuits perform the following roles:

- Store potential energy.
- Dampen shocks and pressure surges.
- Maintain a consistent pressure.

In an actual hydraulic circuit, an accumulator is usually used for a specific function in one of the foregoing roles. An accumulator that stores potential energy is often used to back up a hydraulic pump in the event of pump failure or at system start-up. For instance, in large off-highway dump trucks using hydraulically assisted brakes, in the event of oil supply failure, an accumulator is used to feed several charges of oil into the circuit to effect an emergency stop. Accumulators can also be used to dampen shock loads, control pressure surges, and ensure smooth operation of a hydraulic circuit. Although it uses a self-contained circuit, any vehicle shock absorber is a type of accumulator.

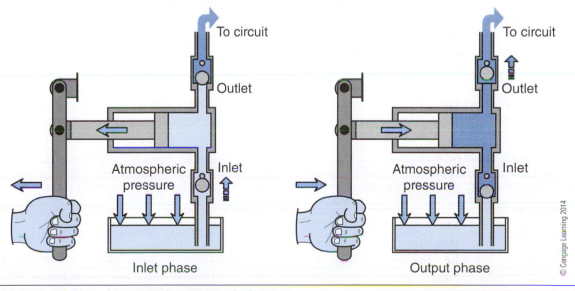

Inlet phase

Output phase

To circuit

Outlet

Atmospheric pressure

Inlet

© Cengage Learning 2014

Figure 4-2 A simple hand-actuated pump and plunger.

In terms of operating principles, accumulators generally fall into one of three categories:

- Gas-loaded
- Weight-loaded
- Spring-loaded

Gas-Loaded Accumulators

Although liquids tend *not* to compress, gases do compress; therefore, gases can be used to load a volume of oil under pressure. In this type of accumulator, the gas (usually an inert gas such as nitrogen) and hydraulic oil occupy the same chamber but are separated by a piston, diaphragm, or bladder. When circuit pressure rises, incoming oil to the chamber compresses the gas; when circuit pressure drops off, the gas in the chamber expands, forcing oil out into the circuit. Most gas-loaded accumulators are pre-charged with the compressed gas that enables their operation. **Figure 4-3** shows a sectional view of a typical gas-loaded accumulator.

Weight-Loaded Accumulators

A weight-loaded accumulator uses a gravity piston and cylinder. Weight is loaded to the piston so that it acts on hydraulic oil in the accumulator chamber. When circuit pressure rises, oil is unloaded into the accumulator chamber thus raising the piston and the weight it supports. When circuit pressure drops off, gravity forces the weight down, thereby charging oil to the circuit. Because the weight is constant, these types of accumulators are used to keep a specific charge pressure of oil throughout their loaded and unloaded cycles. Due to

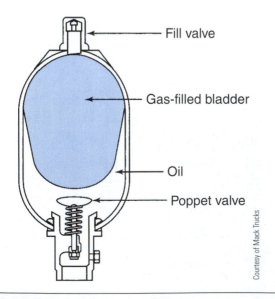

Figure 4-3 Hydraulic accumulator.

their large size, these types of accumulators are not very popular with heavy equipment.

Spring-Loaded Accumulators

A spring-loaded accumulator functions much like the weight-loaded version except that the weight is replaced by a mechanical spring. Springs have a progressive force characteristic—the further the spring is compressed, the greater amount of force is required to fully compress it. Due to this characteristic, spring-loaded accumulators also are not suitable for heavy mobile equipment.

For further details on accumulators, see Chapter 8 section "Hydraulic Accumulators."

PUMPS

Pumps create flow. They aspirate fluid at an inlet and displace it to an outlet. Displacement can take place in two ways: non-positive and positive. A non-positive displacement pump picks up fluid and moves it. Examples of this are water wheels or the coolant pump used on a typical diesel engine. Non-positive displacement pumps can be classified as hydrodynamic: if the outlet of this type of pump is blocked, flow will be reduced or completely stopped.

A **positive displacement** pump not only creates flow but it also backs it up. Fluid picked up by a positive displacement pump is sealed and held while the pump turns. Positive displacement pumps are classified as hydrostatic. They have a positive seal between the inlet and outlet so if the outlet ever became blocked, something would have to give. Most hydraulic circuits use only positive displacement pumps, and we can divide these into two general categories:

- Fixed-displacement pumps
- Variable-displacement pumps

Figure 4-4 shows a comparison between non-positive and positive displacement pumping principles.

When studying pumps, you should remember that hydraulic pumps only create flow: they do not create pressure. Pressure results by creating resistance to flow.

Fixed-Displacement Pumps

A fixed-displacement pump will move the same amount of oil each cycle with the result that the slug volume picked up by the pump at its inlet equals the

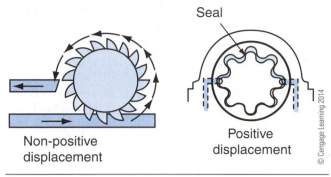

Figure 4-4 Non-positive and positive displacement pumping principles.

slug volume discharged to its outlet per cycle. This means that the speed of the pump determines how much hydraulic oil is moved. Fixed-displacement pumps are often used in low-flow, low-pressure applications, such as the pilot circuit of a larger hydraulic system.

Variable-Displacement Pumps

Variable-displacement pumps are positive displacement pumps designed to vary the volume of oil they move each cycle, even when they are run at the same speed. They use an internal control mechanism to vary the output of oil, usually with the objective of conserving hydraulic energy and reducing flow when demand for oil is minimal. This chapter will cover in greater detail the operating principles of these types of pumps.

Gear Pumps

Gear pumps are widely used in mobile hydraulics because of their simplicity. They are also used to move fuel through diesel fuel subsystems and as engine lube oil pumps. Gear pumps are a type of fixed-displacement pump. Two types of gear pumps are used:

- External gear
- Internal gear; two types are used:
  - Crescent gear
  - Gerotor gear

External Gear Pumps. Two close-fit, intermeshing gears are driven within a housing. One of the gears is keyed to a drive shaft, which drives the second because it is in mesh with it. As the gears rotate, they create a low-pressure area as the teeth come out of mesh. The oil is then forced in from the reservoir due to the weight of the oil and atmospheric pressure acting on it. Oil is then trapped between the teeth and the housing

as the gears rotate. This oil is then moved around the outside of the gear teeth to the outlet of the pump housing. The close mesh of the gear teeth forms a seal that prevents oil from backing up to the inlet. Oil is discharged through the outlet side of the pump housing from which the circuit is charged. This principle is demonstrated in **Figure 4-5**. **Figure 4-6** shows a cutaway of a typical gear-type pump.

Internal Gear Pumps. The internal gear pump also uses a pair of intermeshing gears, but now a spur gear rotates within an annular internal gear, meshing on one side of it. Both gears are divided on the other side by a crescent-shaped separator. These are known as crescent-type gear pumps. The driveshaft turns the spur gear that meshes with the annular internal gear. Both gears will rotate in the same direction. As the gear teeth come out of mesh, oil from the inlet is

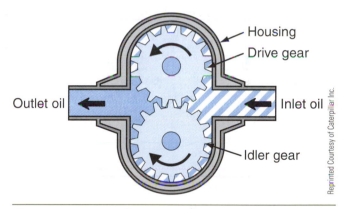

Figure 4-5 External gear pump.

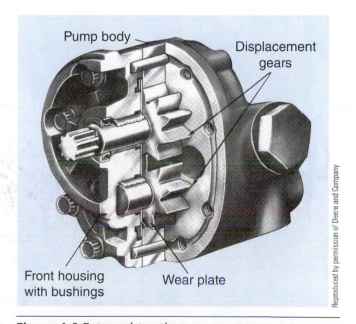

Figure 4-6 External tooth-gear pump cutaway.

trapped between their teeth and the separator, and this oil is unloaded to the outlet. As with the external gear pump, when the gear teeth mesh, a seal is formed to prevent backup. **Figure 4-7** shows the operating principle of an internal gear pump, and **Figure 4-8** is a cutaway showing the internal components.

Gerotor Gear Pumps. A gerotor gear pump is a variation of the internal gear pump. An internal rotor

with external lobes rotates within an outer rotor ring with internal lobes. No separator is used. The internal rotor is driven within the outer rotor ring. The internal rotor has one fewer lobe than the outer rotor ring, with the result that only one lobe is fully engaged to the rotor ring at any given moment of operation. As the lobes on the internal rotor ride on the lobes on the outer ring, oil becomes entrapped: as the assembly rotates, oil is squeezed out to the discharge port. **Figure 4-9** shows the operating principle of a gerotor gear-type pump.

Vane Pumps

Vane pumps are also used extensively in hydraulic circuits. Mobile equipment steering systems commonly use vane pumps. Vane pumps function using a slotted rotor that is fitted with sliding vanes: the rotor turns within a stationary liner known as a **cam ring**. The two types of vane pumps are:

- Balanced
- Unbalanced

Balanced Vane Pumps. In a balanced vane pump, the rotor turns within a stationary liner known as a cam ring. The rotor is fitted with slots within which sliding vanes are inserted. When the rotor is driven, centrifugal force throws the vanes outward so they contact the cam ring. As the vanes follow the contour of the cam ring, fluid from the pump inlet is trapped between the crescent-shaped "chambers" formed between vanes. These chambers are continually expanding and

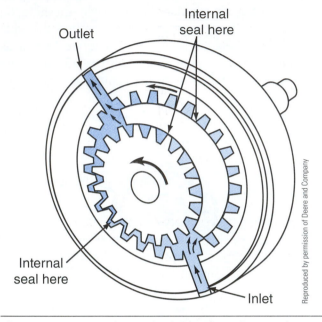

Figure 4-7 Operating principle of an internal gear pump.

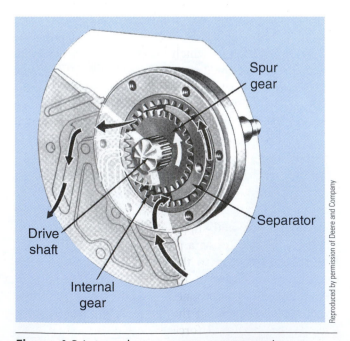

Figure 4-8 Internal gear pump components.

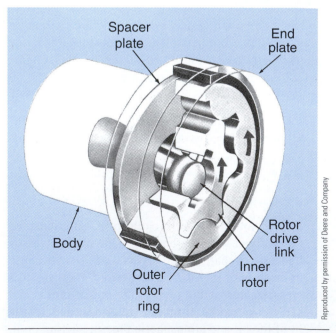

Figure 4-9 Gerotor-type pump.

contracting as the rotor turns: oil from the inlet is trapped in the chamber formed by two vanes. Turning the rotor causes the chamber to contract into the register with the outlet. Once pressure starts to rise, centrifugal force is not sufficient to hold the vanes against the cam ring. To force the vanes to maintain contact against the cam ring, pump outlet pressure is routed to the underside of the vanes through a drilled passage in the rotor. As the pressure rises in the system, the pressure behind the vanes also rises proportionately. This action repeats itself twice per revolution because there is a pair of both inlet ports and outlet ports 180 degrees apart from each other. With two outlet ports 180 degrees apart from each other, there is no side thrust on the rotor and rotating shaft due to the pressure being equal on both sides and cancelling each other out. Balanced vane pumps are always fixed displacement. **Figure 4-10** shows the operating principle of a balanced vane pump.

Unbalanced Vane Pumps. The unbalanced vane pump uses the same principle as the balanced version with the exception that the operating cycle occurs only once per revolution. The location of the axis of the rotor is offset and has only one inlet and one outlet port. In operation, inlet oil becomes entrapped between a pair of vanes when they are expanded and in register with the inlet port. As the rotor turns, the chamber is contracted before the rotation of the pump brings the outlet port into register with the chamber. The disadvantage of the unbalanced vane pump is the radial load placed on one side of the rotor as each chamber formed between the vanes is contracted between fill

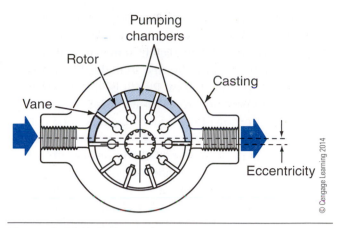

Figure 4-11 Unbalanced vane pump.

and discharge: this can cause the bearing to fail. No such force acted on the opposite side of the pump because the inlet oil was under little or no pressure. **Figure 4-11** shows the operating principle of an unbalanced vane pump.

Piston Pumps

There is a wide variety of piston pumps, beginning with the simplest pumps and progressing to some of the more complex pumps used in hydraulic circuits. There are three general types of piston pumps used:

- Axial piston
- Bent axis piston
- Radial piston

Axial Piston Pumps. In an axial piston pump, the pumping elements are arranged in parallel with the pump driveshaft axis. A cylinder block houses the pumping elements, which consist of a piston and barrel bores machined into the cylinder head. The base of each piston is fitted with a shoe or slipper by means of a ball socket, which rides against a tilted plate known as a **swashplate** or wobble plate. The swashplate does not rotate, but in some instances, the tilt angle can be controlled. Each piston is hollow and the piston shoe has a small drilled passage through its center to allow a small controlled amount of oil to pass and lubricate the contact area between the swashplate face and slipper. Fluid is charged to each pump element through a port plate (or valve plate) as the piston is drawn to the bottom of its travel. The **port plate** is fixed to the pump housing and does not rotate. This component separates the inlet side from the outlet side of the pump as the pistons now change direction and force the oil to the outlet side. The cylinder block and port plate are lapped in

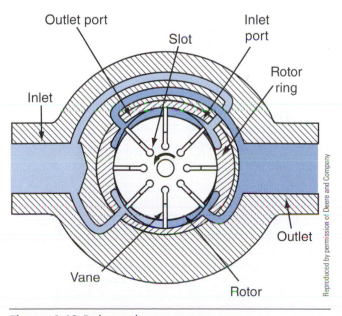

Figure 4-10 Balanced vane pump.

manufacture to form a viscous seal between the two components. A heavy spring located in the center of the barrel keeps pressure against the port plate as it rotates against its face. As the cylinder block rotates (because the pistons follow the tilt of the swashplate), the pistons repeat inward and outward strokes through each revolution of the cylinder block. A shoe plate is fitted between the cylinder block assembly and the pistons. Its purpose is to retract the pistons from their respective bores as they follow the swashplate angle. This creates a low pressure at the inlet side of the pump so that the reservoir can fill each chamber as it extends out from the bore. **Figure 4-12** shows a sectional view of a variable-displacement, swashplate pump: in a fixed-displacement version, the angle (pitch) of the swashplate cannot be altered.

If the angle of the swashplate were fixed (it is in some), the pump would be of the fixed-displacement type. When the swashplate angle is movable and controlled by a servo, the stroke through which the pistons are able to move can be either increased or decreased, and the pump would be classified as variable displacement. The farther the pistons travel, the greater the amount of oil that can be displaced per pump cycle. **Figure 4-13** shows how varying the swash angle can be used to alter the displacement of the pump.

Variable-Displacement, Axial Piston Pump. Much of the current equipment today is designed with variable-displacement-type pumps. Primarily, the axial piston pump design is used. This is due to the fact that swashplate control can easily be established. The transition to variable-displacement pumps is to achieve more energy-efficient equipment and conserve energy waste.

If we consider a hydraulic system in an excavator operating at maximum rpm (this is a typical setting for excavators during operation), the flow rate from a fixed-displacement pump would be at maximum whether the excavator was operating at high or low speeds when lowering pipes into the ground. Both these operation modes require different implement speeds, which are controlled by the operator's input to the control valves; hydraulic flow rate dictates the speed of the actuators. When fixed-displacement pumps are used, a large volume of oil goes unused during slow operations. This amounts to waste energy in the form of heat generated, and pumping losses, resulting in the burning of more fuel. In the fast mode of operation, energy is also lost when the system is forced to work in high-flow and high-pressure conditions. In this operating phase, most of the oil is forced to return to the reservoir through the main relief valve generating excessive heat. When the flow rate of oil can be controlled based on the demand of the system, this conserves energy.

Variable-displacement pumps with some form of control mechanism can be used to control the rate of oil in a hydraulic system. Axial piston pumps are controlled by moving the swashplate. There are two basic swashplate designs currently used: saddle type and pintle type, as shown in **Figure 4-14**. These swashplates implement a control piston and either a bias piston or bias spring. The larger of the two pistons, the control piston is used to "de-stroke" the pump, meaning that it will move the swashplate closer to a perpendicular position with the rotating shaft, and therefore reduce output flow. The bias piston or spring is used to "up-stroke" the pump to maximum displacement (see **Figure 4-15**).

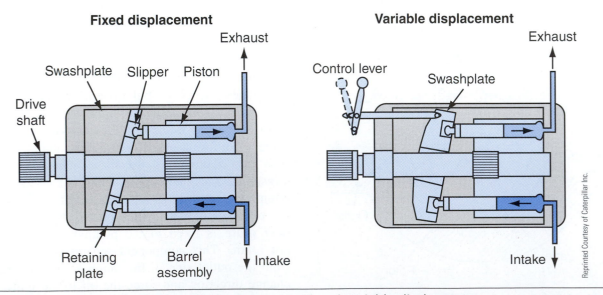

Figure 4-12 Sectional view of a swashplate pump. Fixed and variable displacement.

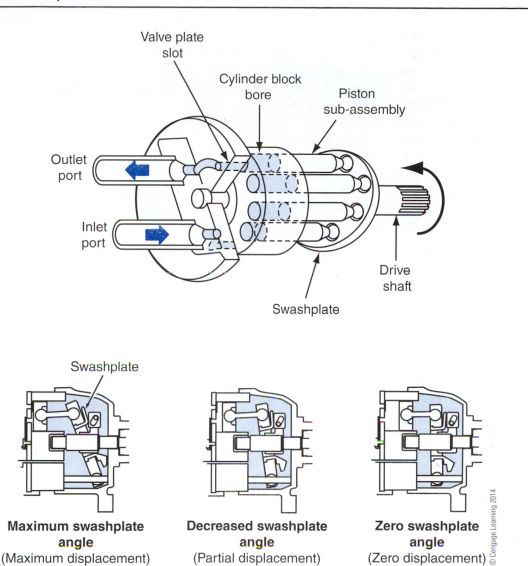

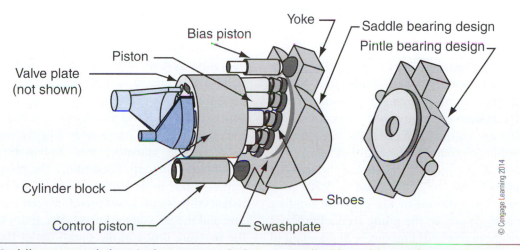

Figure 4-13 Effect of swash angle on displacement.

Figure 4-14 Saddle-type and the pintle-type swashplate controlled by a bias and control piston.

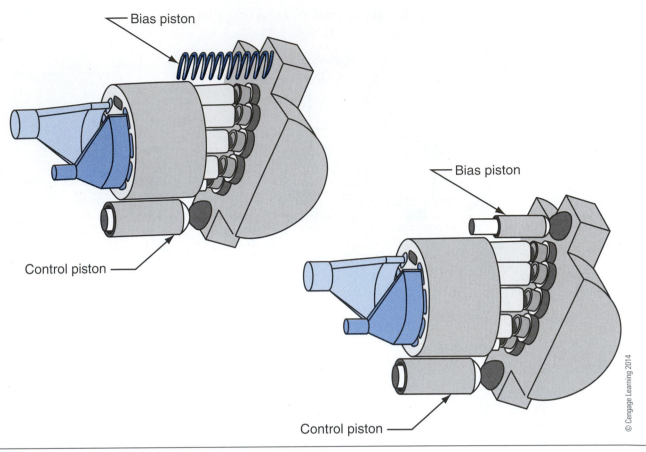

Bias piston

Control piston

Bias piston

Control piston

© Cengage Learning 2014

Figure 4-15 Saddle-type swashplate using a bias piston and bias spring.

To control the position of the swashplate through the control piston and bias piston, compensators are used to direct or drain oil from the pistons controlling the swashplate.

There are three basic types of compensators used:

1. Pressure limiting compensators (PC)
2. Flow limiting compensators (FC)
3. Torque limiting compensators (TLC)

We will review each one and how they are implemented into a hydraulic system.

Pressure Limiting Compensators. High flow combined with high pressure is not desirable. In a system using fixed-displacement pumps, this condition can occur when the system or implement is forced against a resistance that causes the main relief valve to open. If the operator continues to hold the control valve open to that implement, all the pump oil is forced through the relief valve (which may be set as high as 3,200 psi), wasting energy. The purpose of the pressure limiting compensator is to de-stroke the pump (reducing flow) once this pressure is reached. Pressure will maintain in the system but with very little flow.

Figure 4-16 shows a variable-displacement pump with both a pressure and a **flow compensator**. Both compensators are able to sense pump pressure directly from the pump outlet line. Focusing strictly on the **pressure compensator**, pump outlet pressure is sensed on the bottom side of the spool. The compensator spring exerts a mechanical force equivalent to 3,000 psi oil pressure (actual pressure is OEM-specific) and will hold the spool closed until the pressure on the bottom side of the spool can overcome the spring pressure on the top.

Given the same scenario of the pressure in the hydraulic system reaching relief valve pressure on a fixed-displacement pump, this system will force the pressure compensator spool upward overcoming the spring force of 3,000 psi. As the compensator spool is forced upward, a passage is opened to the control piston (large actuator) and forces the piston and swashplate outward, de-stroking the pump. The pump will continue to de-stroke (reduce flow output) until a balanced pressure is reached between the pump outlet line and the pressure compensator spool (3,000 psi). If the pressure drops in the pump outlet line due to less resistance at the actuator and implement, the pressure

**Pump and compensator operation
Engine off**

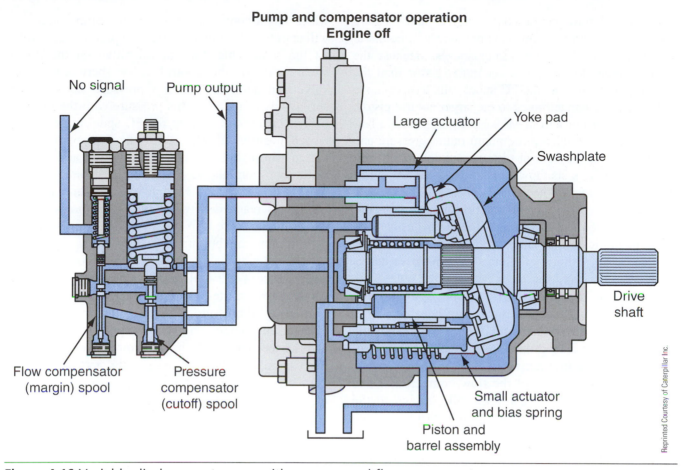

No signal   Pump output

Large actuator   Yoke pad

Swashplate

Drive shaft

Flow compensator (margin) spool   Pressure compensator (cutoff) spool

Small actuator and bias spring

Piston and barrel assembly

Figure 4-16 Variable-displacement pump with pressure and flow compensators.

also drops at the bottom side of the compensator spool and the spring will force the spool downward closing the passage to the control piston. The control piston passage is now able to drain into the case; the bias piston and spring will up-stroke the pump and flow increases. Many PC systems do not use a main relief valve because the PC limits maximum pressure within the system. If a main relief valve is used, it is acting only as a safety relief to the PC.

Flow Limiting Compensators. Flow limiting compensators in many systems are also referred to as "load-sensing" compensators. In either case, they both perform the same function: they control the amount of oil displacement from the pump to the control valve. Either term may be used through this chapter.

Although the pressure compensator de-strokes the pump when high pressure is reached, the pump is still displacing its maximum amount of oil flow whether the machine is working in fast or slow mode, as noted with the previous example. Considering the oil flowing to the control valve in slow mode, some oil is directed to the actuator doing work and the remaining oil is routed back to the reservoir. The oil routed back to the

reservoir is considered waste energy because it is not accomplishing work and consumes energy from the pump and its drive system. If the system were able to deliver the amount of oil based on the operator's demand, this would increase its efficiency further. The flow compensator is able to control output flow anywhere from "0" to "maximum" displacement.

Most FC/PC systems use closed-center control valves. This means pump oil flow is *not* allowed to travel through the control valves in the neutral or hold position. With **Figure 4-16** showing both pressure and flow compensators, we will now focus on the FC spool. Pump outlet oil is directed to the bottom of both the FC and PC compensator spools. Control springs keep the spools biased in one direction. Spring settings are determined by the OEM; we will use approximated figures for these examples. The pressure compensator spring is set to deliver a force of 3,000 psi (larger spring) and the flow compensator spring is set for 300 psi (smaller spring).

Note that the variable-displacement pump is biased to the maximum displacement position. The instant the pump is driven, oil will be displaced through the pump outlet toward the closed-center control valves. The oil cannot travel through the control valves and pressure will instantly start to build. This pressure will be felt at the bottom of both compensator spools. Because the flow compensator spring is set much lower than the pressure compensator, the FC spool will start to move and open the passage routed to the larger control piston. The control piston will de-stroke the pump until a balance is reached between the pump outlet pressure and the pressure at the control piston to maintain 300 psi. This is known as standby pressure (see **Figure 4-17**). Pump output flow to the control valves will be reduced to zero. A small amount of oil flow to maintain this pressure is used to flow through the pump pistons and shoes to lubricate the pump itself. This oil, known as case drain oil, simply falls into the pump case and is returned to the reservoir.

When the operator starts to move a function (control valve), the pump must now be able to up-stroke and displace the amount of oil demanded by the operator. A pressure drop occurs through the control valve when it is first activated. This pressure drop is felt on the bottom side of the flow compensator and the FC spring will close off the oil passage to the control piston and cause it to drain. The bias piston,

which has a continuous port to the pump outlet line, will start to fill again and force the pump to up-stroke. At the same time, the control valve on the downstream side of the control valve (between control valve and actuator) sends a pressure signal to the spring side of the FC. This pressure with the pressure of the spring tries to keep the FC spool biased to the closed position (down). With the FC spool closed, the control piston will continue to drain and the bias piston will continue to fill. The swashplate will continue to up-stroke (see **Figure 4-18**). When a pressure balance is reached between the signal pressure plus the spring pressure (top side of the FC spool), and the pressure on the bottom side of the FC spool, the swashplate will maintain this position (constant flow) (see **Figure 4-19**) until another pressure differential occurs between the FC spool.

The signal pressure is dictated by the resistance the actuator encounters while traveling through its stroke and the speed at which the operator wishes to move it. For example, if the resistance at the actuator was reduced, so would the oil pressure be reduced. Since the FC receives its signal from the actuator side of the control valve, the signal pressure would also be reduced.

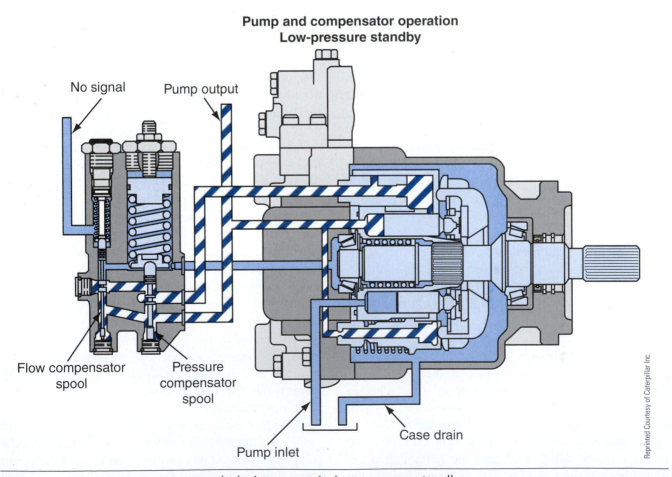

**Pump and compensator operation
Low-pressure standby**

No signal

Pump output

Flow compensator spool

Pressure compensator spool

Pump inlet

Case drain

Reprinted Courtesy of Caterpillar Inc.

Figure 4-17 Flow compensator spool placing pump in low-pressure standby.

**Pump and compensator operation
Up-stroking**

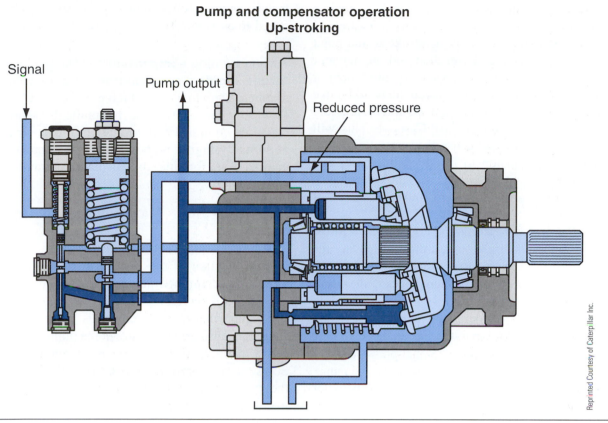

Figure 4-18 Flow compensator spool up-stroking the pump.

**Pump and compensator operation
Constant flow**

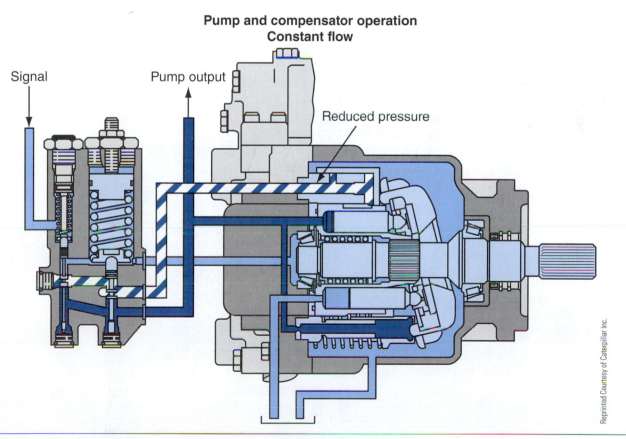

Figure 4-19 Flow compensator spool pump in constant flow.

This reduced pressure would cause the FC spool to move upward and fill the control piston, de-stroking the pump. The pump will only de-stroke to the point where a pressure balance is reached and the actuator will move at that desired speed. If the operator chooses to increase the actuator speed, he would move the control valve further causing an initial pressure drop on the bottom side of the FC spool. This will move the spool down and cause the control piston to drain and the bias piston to fill. Flow increases and therefore actuator speed increases. The control process repeats itself and the system finds its balance or metering point.

When the operator returns the control valve to the neutral position, signal pressure is drained leaving the FC with only the spring force equivalent to 300 psi. The higher pressure on the bottom side of the spool easily shifts the spool upward and oil is routed to the control piston de-stroking the pump to maintain standby pressure. If the operator forces the hydraulic system to its maximum pressure, signal pressure to the FC would be at maximum trying to up-stroke the pump. If the pressure reaches 3,000 psi (pressure compensator setting), while the FC is draining the control piston, the PC spool will be activated, and charge the control piston, forcing the swashplate to de-stroke. This is referred to as high-pressure stall (see **Figure 4-20**).

Torque Limiting Compensators. The purpose of the **torque limiting compensator** is to prevent a hydraulic system to work at high pressure and maximum flow simultaneously. The hydraulic horsepower on machines equipped with torque limiting compensators is normally higher than the rated horsepower of the engine. The formula to calculate hydraulic horsepower is:

$$\frac{\textbf{Pump gpm} \times \textbf{Maximum pressure (psi)}}{\textbf{1714}}$$

The function of the TLC is to reduce pump gpm after a predetermined pressure is reached. Unlike the pressure compensator that will completely de-stroke the pump when a set pressure is reached, the TLC progressively starts to de-stroke the pump at a pressure below the PC setting (approximately 500 to 1,000 psi). So as the pressure rises above the setting of the TLC, the pump gpm is progressively reduced. There are many different forms of TLC made by different pump manufacturers. We will look at a system that utilizes a

Pump and compensator operation
High-pressure stall

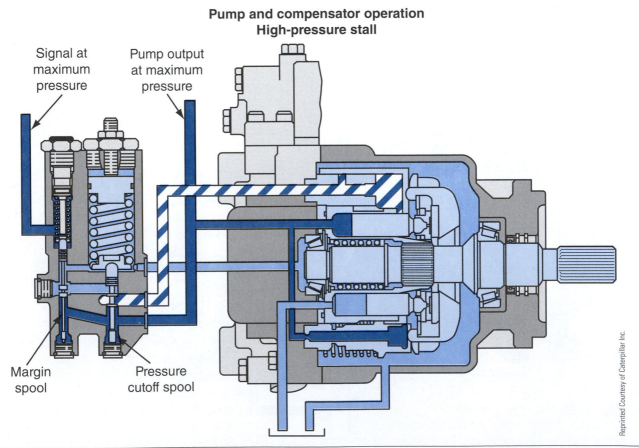

Figure 4-20 Pump de-stroking during high-pressure stall.

Work tool piston pump
Engine off

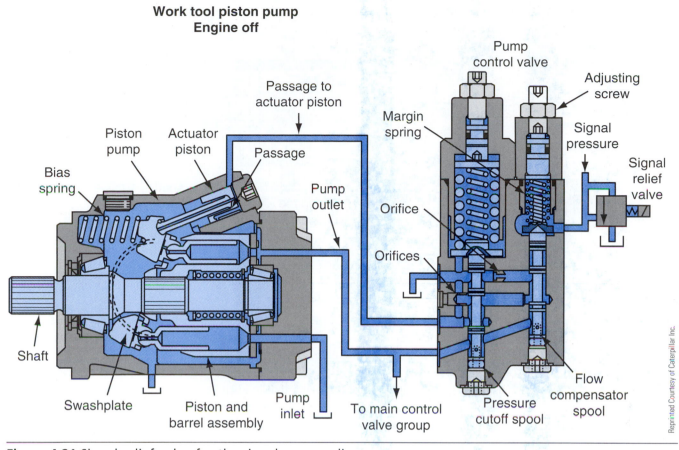

Figure 4-21 Signal relief valve for the signal pressure line.

signal relief valve (see **Figure 4-21**). This relief valve works together with the PC and FC, but primarily with the FC. Reviewing that the FC needs a pressure signal sent to it from the control valve to allow the hydraulic pump to up-stroke, a relief valve is implemented in parallel to the signal line of the FC.

This signal relief valve will not allow signal pressure from the control valve to the FC to rise above the setting of the TLC relief valve. For instance, let us determine the pressure of a TLC relief valve set for 2,000 psi. For purposes of this description, the previous settings of the FC and PC (300 psi and 3,000 psi, respectively) will be used. When an operator is working in conditions of hard rock or material, the resistance at the actuators is high, causing the hydraulic system to work at higher pressures. With this scenario, let us consider what is happening with the hydraulic system equipped with a TLC pump.

The system is working at higher pressures, so the signal sent back to the FC is also high but will not exceed 2,500 psi because of the signal relief valve. We must remember the flow compensator setting (which controls pump gpm) is based on the signal pressure plus the spring pressure on one side of the valve and the pump outlet pressure on the opposite side. Assuming that

this working pressure is equivalent to pump flow of 60% of maximum and the operator continues to demand more flow and pressure, the system will react as follows: The system pressure rises because there is still flow from the pump but the resistance at the actuator also continues to rise. If the TLC (signal relief valve) will not allow a signal pressure higher than 2,500 psi, the pressure on the bottom side of the FC will be higher. This higher pressure will overcome the signal pressure force and the FC spool will move upward and route pump outlet pressure to the control piston, which in turn will de-stroke the pump. The pump will progressively de-stroke based on the pressure differential generated from the pump outlet side.

Bent Axis Piston Pumps. A bent axis piston is much like an axial piston pump except that a bent axis pump does not use a swashplate. The housing is offset at an angle from the drive so the piston and barrel assemblies are displaced from the centerline of the drive (see **Figure 4-22**). The piston assemblies do not implement slippers or shoes. The ball socket of the piston assemblies is secured to the drive shaft flange. The drive shaft utilizes a universal joint to connect with the barrel assembly to rotate it along with the pistons.

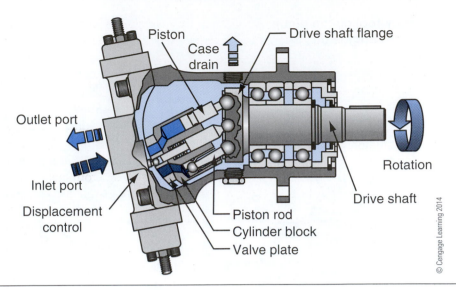

Figure 4-22 Bent axis piston pump.

Because the barrel is fixed to this position, the pistons are forced in and out of the barrel as it rotates. A port plate located on the barrel face designates inlet and outlet of the pump.

Variable-Displacement Bent Axis Piston Pumps.
The variable-displacement version of a bent axis pump is controlled by moving the piston and barrel assembly centerline closer to or farther from the driveshaft centerline. The greater the displacement angle, the greater the output flow. A displacement control servo is used to raise or lower the barrel and piston assembly. Much like the axial piston pump, a bias spring can be used to keep the displacement servo at maximum angle. A pressure and flow compensator must be used to control the oil pressure to the displacement control servo to determine output of the pump.

Radial Piston Pumps.
Radial piston pumps are capable of high pressures, high volumes, and variable displacement. Radial piston pumps operate in two ways:

- Rotating cam
- Rotating piston

Rotating Cam.
In a rotating cam pump, radially opposed pistons are located in a fixed pump body: a central driveshaft is machined with an eccentric cam. As the cam profile is rotated, it forces the pistons outward to effect a pumping stroke. Each piston is loaded by spring force to ride the cam profile. This means that the outward stroke of the piston is effected by cam profile and the inboard stroke by spring force. Oil inlet and outlet passages are located at either end of

each pump element. Spring-loaded valves in the inlet and outlet ports permit oil to flow into and out of the piston bores.

During the inlet stroke, the piston is returned by the spring that loads it to ride the cam profile, creating low pressure in the piston bore. This low pressure in combination with oil pressure unseats the inlet valve, allowing the pump chamber to be charged with oil. When the pump chamber is filled with oil, the inlet valve is closed by its spring.

After the pump chamber has been charged, the cam rotates and acts on the piston, driving it outward through its pumping stroke. Oil in the pump chamber is pressurized and unseats the discharge valve, unloading it into the outlet circuit. The pumping cycles of the pistons work sequentially as the cam rotates. This produces a continuous flow of oil. Four-, six-, and eight-piston versions of the pump are common. Oil output depends on the rotational speed of the pump (assuming it is of the fixed-displacement type).

A common mobile application of cam-actuated radial piston pumps is the high-pressure pump used on recently introduced common rail, diesel fuel injection. **Figure 4-23** shows the components used in a radial piston pump.

When a variable-displacement rotating cam pump is required, a stroke control mechanism has to be used. This functions to hold the pistons away from the actuating cam profile, either shortening or eliminating the piston effective stroke. A stroke control valve directs oil into the center of the pump so that it acts on and hold off the piston. The result is that while the camshaft continues to turn, pump chamber output can be held at nearly zero. When the hydraulic circuit

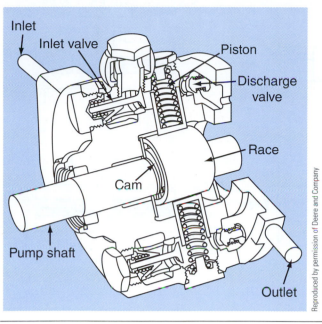

Figure 4-23 Rotating cam type of radial piston pump.

In a fixed-displacement pump, flow is related to pump speed. The faster the rotational speed, the more oil pumped. This means that pump ratings should be specified with flow, pressure, and rpm values.

Pump Drives. On both on- and off-highway mobile heavy equipment, hydraulic pumps are usually driven either directly or indirectly by the vehicle engine. Power-take-off (PTO) devices may be driven directly by the engine or from gearing in the transmission. Alternatively, electrical power produced by the chassis electrical system can be used to drive electric motors located anywhere in the equipment or on devices towed by the equipment. For instance, several remotely located electric motors (this is a common alternative use for diesel engine electric cranking motors) may be used to actuate a number of independent hydraulic circuits.

Due to limits on the supply of air and emissions regulations, most underground tunneling and mining equipment utilizes large electric motors to drive the hydraulic pumps.

requires pump outlet flow once again, hydraulic pressure at the center of the pump is dropped. This allows the piston springs to return the pistons to ride the cam profiles, and pumping action resumes.

Rotating Piston. A rotating piston pump functions similarly to an unbalanced vane pump. Radially reciprocating pistons are arranged within an eccentric cylinder. As the cylinder assembly rotates, the pistons are loaded by centrifugal force against the pumping housing wall. This permits the piston bores to be filled with oil from inlet ports, and as the cylinder is rotated, the pistons are driven inward displacing the oil to outlet ports.

Variable displacement is achieved by adjusting the axis of the outer housing to that of the cylinder. This alters piston travel and, therefore, the volume of oil displaced per cycle.

Pump Performance

Hydraulic pumps are usually rated by delivery rate or, in other words, the volume they can displace over a given time period. This is usually expressed in gallon per minute (gpm) at a specific rpm, such as 1,500 or 1,200 rpm. This specification should be accompanied by a rating of the back-pressure a pump can sustain while producing the gpm rating. As pump delivery pressure increases, internal leakage will also increase with the result that usable volume decreases.

VALVES

Valves are used to manage flow and pressure in hydraulic circuits. There are three basic types of valves used in hydraulic circuits. The general operating principles of these three categories of valves are shown in **Figure 4-24**:

- Pressure control
- Directional control
- Flow control

Pressure-Control Valves

Hydraulic systems are designed to operate at specific working pressures, so **pressure-control valves** are used to regulate pressure in the system or in sections of a system. Many different types can be used, classified as follows:

- Counterbalance valves
- Relief valves
- Pressure reducing valves
- Sequence valves
- Unloading valves

Counterbalance Valves. Counterbalance valves are used to maintain back-pressure in a circuit or component. They are designed to block a fluid flow path until pilot pressure (direct or remote) overcomes the spring

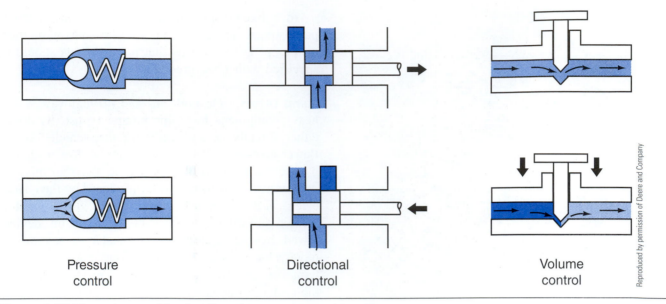

Pressure
control

Directional
control

Volume
control

Reproduced by permission of Deere and Company

Figure 4-24 Three types of hydraulic valves.

tension setting of the valve and opens the spill circuit. These are often used in a double-acting lift ram to prevent sudden or uncontrolled dropping.

Relief Valves. Two general types of pressure-relief valve are used in hydraulic circuits:

Directly operated. A directly-operated relief valve consists of two operating ports, a ball poppet, and a tension spring. The tension spring loads the ball poppet against the valve seat, giving the valve a normally closed operating status. When hydraulic pressure acting on the ball poppet is sufficient to overcome the spring pressure, the ball unseats and permits fluid to spill back to the reservoir, bypassing the circuit, and dropping circuit pressure. When system pressure returns to normal, spring pressure once again loads the poppet onto the valve seat. Directly-operated relief valves may be adjustable. A screw located behind the tension spring permits adjustment of the valve opening pressure. **Figure 4-25** shows an illustration of a directly-operated relief valve.

Pilot operated. A **pilot-operated relief valve** is used to handle large volumes of fluid with little pressure differential. It consists of a main relief valve, through which the main pressure line passes, and a smaller relief valve that acts as a trigger to control the main relief valve. The valve body has a secondary port that directs flow back to the reservoir when the main relief valve opens. A passage in the valve body directs fluid flow to the pilot valve: as inlet pressure increases, so does pressure in the pilot passage. When inlet pressure equals the pilot valve setting, it opens, releasing oil trapped behind the main relief valve and opening it.

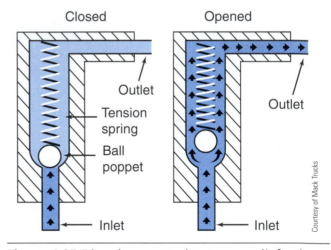

Closed   Opened

Outlet

Tension
spring

Ball
poppet

Outlet

Inlet   Inlet

Courtesy of Mack Trucks

Figure 4-25 Directly operated pressure-relief valve.

Excess oil then spills through the discharge port, preventing further pressure rise. At the moment inlet pressure drops below the valve setting pressure, the valves close. **Figure 4-26** shows an illustration of a pilot-operated relief valve.

Pressure Reducing Valve. Pressure reducing valves are commonly used to supply a secondary circuit with oil at a reduced pressure from the primary circuit. Pump oil is directed through a valve spool that controls a restricted passage. The spool is spring biased to the open position. A sensing passage on the downstream side of the valve spool is routed into a piston chamber with a small slug piston. This will monitor the pressure on the controlled side of the circuit. As long as the pressure does not exceed the spring force, oil is

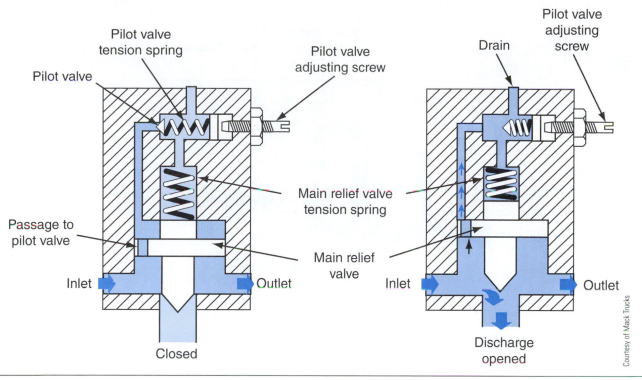

Figure 4-26 Pilot-operated pressure-relief valve.

allowed to flow through to the secondary circuit. If the pressure in the secondary circuit rises above the bias spring setting, the slug piston will shift the valve spool restricting oil flow through the valve. The greater the restriction, the greater the pressure drop through the valve. **Figure 4-27** shows an illustration of a pressure reducing valve.

Sequence Valve. The purpose of a sequence valve is to supply a second actuator only after the first actuator has been satisfied at a preset pressure. Outlet oil from a directional control valve will split the flow of oil between two actuators that will work in series to each other. The second actuator is supplied oil only through the sequence valve. An unloading spool is spring and oil-pressure biased to keep the passage to the second actuator closed. Oil pressure is also routed to the bottom side of the spool as well as the top side of the spool. A small pilot relief valve that sets the opening pressure of the valve is located on the spring chamber side of the spool. The spool remains closed until the preset pressure of the pilot valve is reached and allows oil to drain from the spring chamber side of the spool. A pressure differential is created between the top and bottom side of the spool and the spool is forced to open, allowing oil to flow to the second actuator.

Figure 4-28 shows an illustration of a sequence valve.

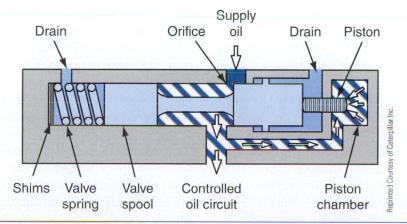

Figure 4-27 Pressure reducing valve.

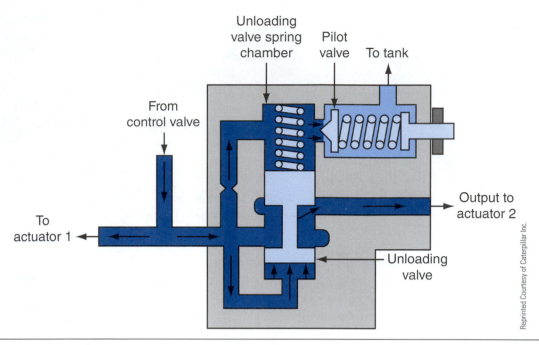

Figure 4-28 Sequence valve.

Unloading Valve. Unloading valves are normally used with fixed-displacement pumps and closed-center control valves. Machine systems that use unloading valves include braking systems and pilot control systems. Some OEMs still use this system for their main implements. It is considerably cheaper to manufacture than using an LS/PC system and still provides some energy efficiency.

An unloading valve can be used to charge a hydraulic circuit to maintain a threshold pressure. Using a brake system as an example, air brake systems fill a reservoir with air pressure to a set value. Once this pressure is reached, the air compressor is unloaded (cut-out) and stops pumping air into the reservoir. A check valve is used so that air cannot leak back through the fill line. When the air pressure drops below a set value, the compressor refills the reservoir (cut-in). An air governor is used to control the compressor's cut-in and cut-out pressures.

Similar to air brake systems, hydraulic braking systems that use a hydraulic pump and unloading valve work in the same manner. The pump will charge a hydraulic accumulator with oil, and once a preset pressure is reached, the unloading valve will redirect pump oil back to the hydraulic reservoir. A check valve is used to keep hydraulic pressure from leaking back. This pressure is stored within the accumulator to be used as needed for the system. If the pressure drops below the setting of the unloading valve, the valve will redirect pump oil to refill the system to its pressure setting.

Figure 4-29 illustrates an unloading valve and accumulator system. The valve has three passages, one to the system separated by the isolation check valve, a pump passage from the hydraulic pump, and a reservoir passage for return oil when the valve is unloaded. The valve works in two modes: charge and unloaded. In charge mode, hydraulic pump oil is directed past the isolation check valve and continues to fill the system and accumulator. As the system fills, pressure starts to rise as the gas in the nitrogen-filled accumulator is compressed. This pressure acts on the unloading piston on the accumulator side. On the pump side, the same pressure is channeled through a restriction on the main poppet as well as on the pilot poppet. The adjustable pilot poppet sets the cut-out pressure of the system. For example, if this poppet was set for 1,000 psi, the system would fill until this pressure setting lifts the poppet off its seat and creates pressure drop behind the main poppet and unloading piston. This pressure drop causes the main poppet to lift and open the port passage to the reservoir and pump oil be directed through it. Simultaneously, the isolation check valve closes and traps pressurized oil on the accumulator side. Due to the larger surface area on the unloading piston, it holds the pilot poppet off its seat until the pressure on the accumulator side drops to a range below the pilot poppet. When this happens, pressure starts to rise behind the main poppet and together with the light spring closes off the tank passage. Oil is then forced past the isolation check valve and refills the system. The advantage to this system is that if an engine

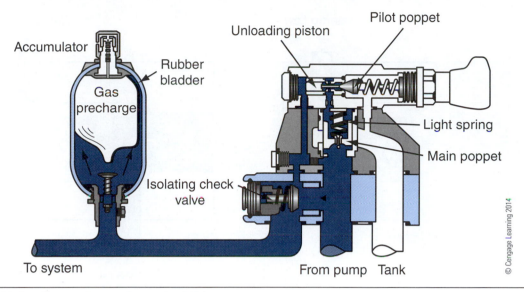

Figure 4-29 Unloading valve and accumulator.

failure was to occur, there would be sufficient stored hydraulic energy to allow the machine to safely stop.

Flow-Control Valves

Flow-control valves are used to either split flow between multiple circuits or control the flow and speed of an actuator. These types are classified as:

- Non-compensated or fixed-orifice flow control
- Pressure compensated or bleed-off flow control
- Flow dividers

Flow-control valves that are used to regulate the speed of an actuator function by defining a flow area and can be classified as adjustable and non-adjustable. They are rated by operating pressure and capacity, and can be used in meter-in or meter-out applications.

Non-Compensated or Fixed-Orifice Flow Control. When a meter-in fixed-orifice valve is used to control flow (see **Figure 4-30**), the valve is placed in

series between the control valve and the actuator. This method of flow control is used in hydraulic circuits that continually resist pump flow, such as that used to actuate a dump box ram. In many cases flow control is required in only a single direction so as not to resist pump flow. The dump box of a haul truck for example would extend and retract slowly with a fixed-orifice flow-control valve. This may not be desirable due to the restricted flow trying to raise the box. A single-direction flow-control valve will allow flow to bypass the restricting orifice in the cylinder extend mode and force the oil through the restriction in the retract mode, controlling the actuator speed (see **Figure 4-31**).

A meter-out flow-control valve is required when an actuator has the potential to run away or over-speed. This run-away condition can cause cavitation to occur in the actuator. The meter-out flow-control valve is placed between the actuator and the control valve

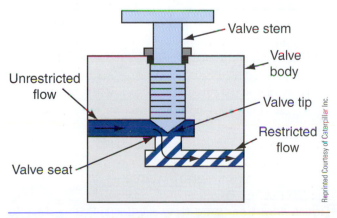

Figure 4-30 Fixed-orifice flow-control valve.

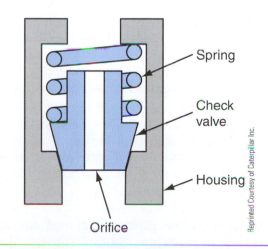

Figure 4-31 Flow-control valve single direction.

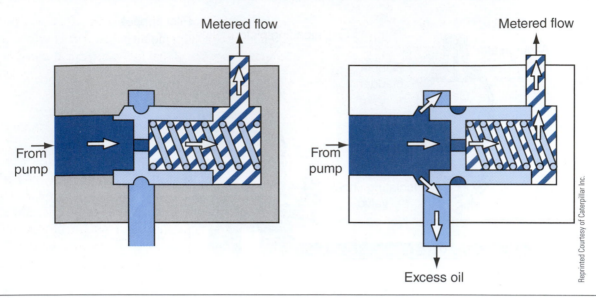

Metered flow

From
pump

Metered flow

From
pump

Excess oil

Reprinted Courtesy of Caterpillar Inc.

Figure 4-32 Pressure compensated bleed-off flow control.

return port. This will control flow away from the actuator allowing the pump to displace sufficient oil, preventing cavitation to occur.

Pressure Compensated or Bleed-Off Flow Control.
One method of bleed-off flow control is to place the valve between the pump outlet and the control valve with a drain line to the reservoir. It meters diverted flow as opposed to working flow. The result is that metered flow is directed back to the reservoir at the load pressure rather than the relief valve pressure (see **Figure 4-32**). A fixed orifice is located through the center of a control piston that dictates the regulated or metered flow rate to the control valve. With pump gpm low, force against the control piston is also low. The control piston spring will force the control piston to restrict the bleed port and keep a constant supply of oil to the metered port. As pump flow increases, force against the control piston increases, compressing the spring and opening the bleed-off port. This maintains a constant flow rate through the valve. The pressure compensated feature of the valve becomes a factor as system pressure increases. When this happens, the control spool would have the tendency to open and bleed-off more oil. However, simultaneously the pressure on the metered flow side also increases and with the help of the spring, forces the control spool to close the bleed port maintaining the flow rate on the metered side constant. Bleed-off control valves are the least precise means of controlling flow.

Flow Dividers.
Flow dividers are used to supply multiple circuits from one source. The two design types used are spool type and gear type.

A gear-type flow divider has a common inlet to two or more gear sections connected by a common shaft controlling rotational speed (see **Figure 4-33**). Much like a gear motor, the size of the gear sections dictates the displacement area, therefore dictating the amount of oil passing through each section. It is possible to allow different displacement gear-sets to split the flow according to the need of each section.

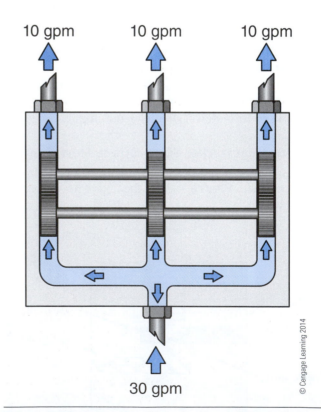

10 gpm   10 gpm   10 gpm

30 gpm

© Cengage Learning 2014

Figure 4-33 Gear-type flow divider splitting flow to three circuits.

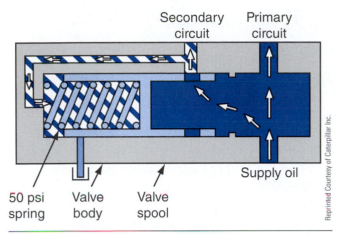

50 psi spring   Valve body   Valve spool   Supply oil

Reprinted Courtesy of Caterpillar Inc.

Figure 4-34 Pressure compensated spool type flow divider.

A spool type flow divider can provide proportional or priority flow control. When a proportional flow is required, a pressure compensated type is used to accurately split the flow between the circuits (see **Figure 4-34**). When a priority flow is required, a set restriction is provided on the priority side that will set the flow rate to the circuit. The remaining oil is directed to the secondary circuit. As flow rate changes, the secondary circuit will receive more or less flow while the priority side stays satisfied.

Directional Control Valves

Directional control valves direct flow of oil through a hydraulic circuit. They include:

- Check valves
- Rotary valves
- Spool valves

The center section of **Figure 4-24** shows the operating principles of directional control valves.

Check Valves. Check valves use a spring-loaded poppet that seats and unseats. They are designed to open to permit flow in one direction but close to prevent flow in the other.

Rotary Valves. A rotary valve uses a rotary spool that turns to open and close oil passages and thereby allows the option of routing oil to different subcircuits. Rotary valves are commonly used in steering circuits on center articulated type machines. This valve is also commonly known as a hand metering unit. As the valve is rotated, a passageway is made between the pump supply port and the actuator, but the oil must first pass a metering unit (gerotor) that rotates with

the valve. Because the metering unit is of a fixed displacement, the amount of oil that is directed to the actuator is proportional to the rotational speed at which the operator rotates the valve and the displacement area of the metering unit.

Spool Valves. Spool valves use a sliding spool within a valve body to open and close hydraulic sub-circuits. They are the primary type of directional control valves used in hydraulic circuits and function to start, actuate, and stop most activity in a typical circuit. A spool valve consists of a body and a cylindrical spool machined with annular recesses and lands. The land areas of the spools in many cases are machined with metering slots. These slots are used to slowly expose the oil passage to the actuator port for finite speed control. Figure 3-8 in Chapter 3 shows a typical spool valve.

Most spool valves are used to control one function or actuator. When multiple spool valves are required, they are manufactured into one valve body called a monoblock design or they are manufactured individually (spool and body) and bolted together. This is called a stack or sectional-type valve. These valves require multiple seals between each section and are prone to leaks and distortion. The advantage to this type of valve is that a single section may be replaced at a lower cost compared to a monoblock valve. Monoblock valves are least likely to leak and typically offer more compact designs. In some cases, a combination of sectional and monoblock designs has been incorporated.

The pump oil path through multiple spool control valves can be classified as being a series connection or series/parallel connection. The most widely used is the series/parallel connection. This allows all spools to have a pump supply of oil when multiple spool functions are operated. Although oil flow will take the path of least resistance when more than one spool is used, it is up to the operator to control the metering through each spool to control the oil flow to the actuators.

When spool valves are used for directional control, they can be classified by characteristic or function as follows:

- Number of spool positions
- Number of flow paths in extreme positions
- Flow routing in the center position
- Method of shifting the spool
- Method of retracting the spool

Table 4-1 shows the directional control options possible in spool valves by type.

TABLE 4-1: SPOOL VALVE DIRECTIONAL CONTROL OPTIONS		
Type	**Classification**	**Operation**
Path of flow	Two-way	Two flow paths in two extreme spool positions
Path of flow	Four-way	Four flow paths in two extreme spool positions
Control	Manually operated	Hand-actuated spool shift
Control	Pilot actuated	Hydraulic pressure shifts spool
Control	Solenoid actuated	Solenoid action shifts spool
Control	Solenoid controlled, pilot actuated	Solenoid shifts pilot spool which directs pilot flow to the main spool
Position	Two-position	Spool has two extreme positions of dwell
Position	Three-position	Spool has two extreme positions plus center position
Position	Four-position	Spool has two extreme positions, center position, plus a float position
Spring	Offset spring	Spring returns spool to normal offset position when shifter force is released
Spring	Center spring	Spring returns spool to center position when shifter force is released
Spool	Open center Closed center Tandem center	The common types of spool valve categorized by flow pattern when the spool is in its center position (three-position valves) or crossover position (two-position valves)
	Partially closed center	Three-position valves
	Semi-open center	Three-position valves
	Float center or motor center	Three-position valves

ACTUATORS

Hydraulic actuators convert the fluid power from the pump into mechanical work. In mobile hydraulic systems, actuators can be grouped as hydraulic cylinders and hydraulic motors. A hydraulic cylinder is a **linear actuator**. A hydraulic motor is a **rotary actuator**.

Hydraulic Cylinders

Three types of hydraulic cylinders are used:

- Piston
- Vane
- Telescopic

Piston-Type Cylinders. In hydraulics, both single-acting and double-acting cylinders are used. In both types, a piston is moved linearly within the cylinder bore when it is subjected to pressurized oil. A single-acting cylinder is hydraulically actuated in one direction only. A double-acting cylinder is hydraulically actuated in both directions. In application, a dump box ram requires only a single-acting cylinder while the boom on an excavator requires a double-acting cylinder.

Single-Acting Cylinders. In a single-acting cylinder, hydraulic pressure is applied to only one side of the piston. Single-acting cylinders may be either outward or inward actuated. When an outward-actuated cylinder has hydraulic pressure applied to it, the piston and rod are forced outward to lift the load. When the oil pressure is relieved, the weight of the load forces the piston and rod back into the cylinder. One side of a single-acting cylinder is dry. The dry side must be vented so that when oil pressure on the pressure side is relieved, air is allowed to enter, thus preventing a vacuum. In an outward-actuated single-acting cylinder, the dry side or vent chamber is on the rod side of the piston.

When an inward-actuated cylinder has hydraulic pressure applied to it, the rod is pulled inward into the cylinder. This means that oil pressure acts on the rod side of the piston, and the vent chamber is on the opposite side as shown in **Figure 4-35**.

A seal on the piston prevents oil leakage from the wet side to the dry side of the cylinder, and a wiper seal is used to clean the rod as it extends and retracts. To prevent dirt from entering the dry side of the cylinder, a porous breather is used in the air vent. In some applications the dry side of the cylinder is vented to the hydraulic reservoir so that the cylinder can siphon oil

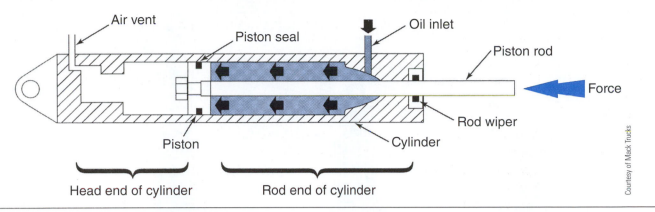

Figure 4-35 Single-acting cylinder.

in and out as it extends and retracts. This keeps the vented side lubricated and prevents condensation and oxidation from occurring within the cylinder.

A **ram** is a single-acting cylinder in which the rod serves as the piston. The rod is slightly smaller than the cylinder bore. A shoulder on the end of the rod prevents the rod from being expelled from the cylinder. A ram-type cylinder eliminates the need for a dry side because oil fills the entire inner chamber when actuated.

Double-Acting Cylinders. Double-acting cylinders can provide force in both directions, which means that oil pressure can be applied to either side of the piston within the cylinder. When oil pressure is applied to one end of the cylinder to extend it, oil from the opposite side returns to the reservoir. This will be reversed when the cylinder is retracted. Good examples of double-acting cylinders are those required to actuate the swing arm and bucket on a backhoe application.

Double-acting cylinders may be balanced or unbalanced. In the balanced double-acting cylinder, the piston rod extends through the piston head on both sides, giving an equal surface area on which hydraulic pressure can act, on both sides of the piston. In an unbalanced double-acting cylinder, a piston rod is located on only one side of the piston. The result is that there is more surface area on which hydraulic pressure can act on the side without the rod. Therefore, the cylinder would have a greater force capacity extending than retracting. As well, the cylinder would be unbalanced, retracting faster than extending, providing the flow rates delivered were equal. Many cylinders are equipped with a cushioning feature. Cushioning prevents the cylinder piston from impacting when it reaches the end of its stroke. Cushioning can be provided in both directions. This is accomplished by using a snubber, which restricts the oil flow on the outlet side of the cylinder as the piston reaches its furthest

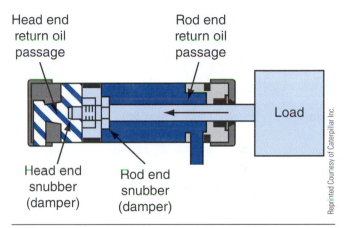

Figure 4-36 Double-acting cylinder with cushioning.

retraction or extension point. **Figure 4-36** shows a sectional view of a typical double-acting cylinder.

Vane-Type Cylinders. Vane-type cylinders may be found in underground drilling rigs to position drilling rods into the rock face. A vane-type cylinder provides restricted rotary motion no greater than 360 degrees clockwise and counterclockwise. It consists of a cylindrical barrel, within which a shaft and vane rotate when pressurized oil enters. As the shaft vane rotates, it closes off an inlet passage in the top plate, leaving only a small orifice to discharge the oil: this slows the rotating vane as it comes to the end of its stroke. Most vane-type cylinders are double-acting. Pressurized oil is directed through one chamber to swing left, the other to swing right. Double-acting vane-type cylinders can be used in applications such as backhoes because they enable a boom and bucket to be swung rapidly from trench to pile; an alternative to a double-acting vane-type cylinder would be to use a pair of cylinders.

Telescopic Cylinders. Telescopic cylinders are used when a long stroke length is required compared to its retracted length. This is accomplished by multiple

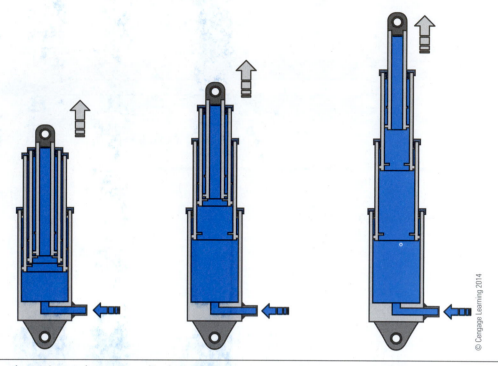

Figure 4-37 Single-acting telescopic cylinder.

sleeve sections, each one smaller than the other inserted within each section. Each section extends in series from largest to smallest (see **Figure 4-37**). The disadvantage to these cylinders is that as each section extends, the effective surface area progressively is reduced, meaning that the force capacity is also reduced with each extending section. In addition, each section will extend faster providing the flow rate stays constant.

HYDRAULIC MOTORS

The function of hydraulic motors is opposite that of hydraulic pumps:

- Pump: draws in oil and displaces it, converting mechanical energy into fluid force.
- Motor: oil under pressure is forced in and spilled out, converting hydraulic energy into mechanical energy.

A pump drives its fluid, while a motor is driven by fluid. **Figure 4-38** shows this principle.

The work output of a motor is in the form of twisting force that we describe as torque. Like pumps, motors can be either fixed displacement or variable displacement. In a fixed-displacement motor, the speed is varied by managing the input flow of oil: this type of motor usually has a fixed torque rating. Variable-displacement motors can have both variable speeds and output torque. Motor force is rated by torque because unlike a cylinder, which exerts linear force,

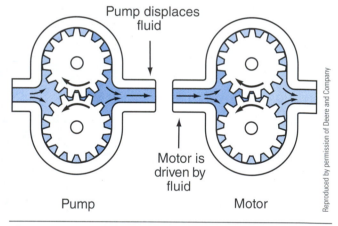

Figure 4-38 Hydraulic pump versus motor principles.

motors rotate. Torque is twisting force. The torque calculation for hydraulic motors is:

Torque/in-lbs

$$= \frac{\textbf{Pressure (psi) in} \times \textbf{Motor Displacement (cir)}}{2\pi}$$

There are three categories of hydraulic motors:

- Gear motors
- Vane motors
- Piston motors

All hydraulic motors rotate; they are driven by incoming hydraulic oil under pressure.

Gear Motors

Gear motors are similar to gear pumps except that they function in reverse. Gear motors can be external or internal.

External Gear. An external-gear motor is driven by pressurized hydraulic oil forced into the motor inlet. This oil acts on a pair of intermeshing gears, turning them away from the inlet with the oil passing between the external gear teeth and the motor housing. Hydraulic oil is then discharged to the motor outlet at a lower pressure. An output or driveshaft is integral with the axis of one of the two gears, allowing the motor to convert hydraulic potential to mechanical energy. **Figure 4-39** shows an external-gear hydraulic motor.

Internal Gear. An internal-gear motor is similar to an internal-gear pump in that an external lobed rotor is driven within an internal lobed rotor ring. The axis of the internal rotor is offset from that of the rotor ring. Pressurized hydraulic oil is delivered to the motor inlet, charging the cavity formed between the inner and outer rotor lobes and forcing both to rotate. This action discharges the oil at lower pressure through the outlet. The motor driveshaft is connected to the inner rotor. **Figure 4-40** shows an internal-gear hydraulic motor.

Vane Motors

Vane motors use both balanced and unbalanced operating principles and function similarly to vane pumps operating in reverse. However, vane-type hydraulic motors are equipped with springs to load the vanes into the liner wall: this is necessary because

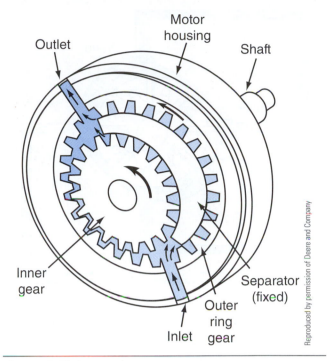

Figure 4-40 Internal-gear motor.

incoming oil at the inlet is at a high enough pressure (especially at start-up) to overcome any amount of centrifugal force developed by motor rotation and prevent the vanes from contacting the pump liner or rotor ring. Once in motion, the combination of spring force and oil pressure routed behind the vane is used to load the vanes into the liner wall, preventing any oil to bypass.

A balanced version of a vane-type motor (see **Figure 4-41**) will locate a pair of diametrically opposed inlet and outlet ports to prevent offset loading of the output shaft connected to the rotor. Balanced vane motors are only capable of fixed displacement, but they operate with higher efficiencies than gear motors and rotational direction can be reversed simply by switching the direction of fluid flow.

Piston Motors

Because piston motors are more complex and require more maintenance than gear and vane motors, they tend to be used in applications where higher speeds and pressures are required, such as for hydrostatic drive transmissions. The three types used are:

- Axial piston
- Bent axis piston
- Radial piston

Axial Piston. Axial piston motors are often used as hydrostatic drive units on off-highway equipment. The pistons in an axial motor are placed parallel to the

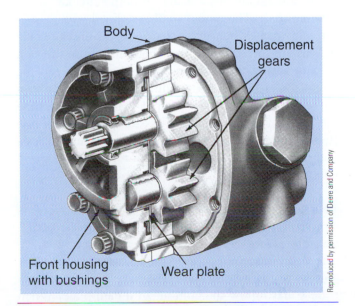

Figure 4-39 External-gear motor.

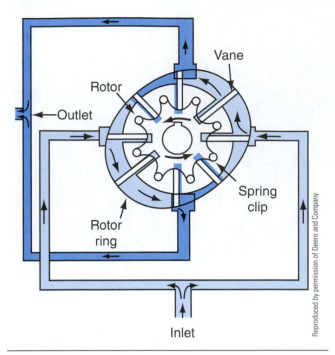

Figure 4-41 Balanced vane motor in operation.

rotating shaft. The operation of this motor is shown in **Figure 4-42**.

The end cap contains ports A and B, each of which can act as either inlet or outlet ports, depending on the desired direction of rotation. The valve plate or port plate separates the A and B ports within the cylinder block and barrel. This designates oil passage to the pistons to rotate in a clockwise or counterclockwise direction. The pistons reciprocate in bores machined into the cylinder block rotor. As fluid enters port A, the piston bore is charged with high-pressure oil that

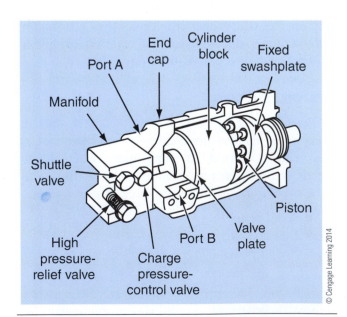

Figure 4-42 Axial piston motor.

forces the piston against a fixed-angle swashplate. This allows the piston to slide down the swashplate face, turning the cylinder block that in turn rotates the output shaft. As the cylinder block continues to turn, other piston bores align with inlet port A, allowing each to be similarly actuated in sequence. Oil is discharged at lower pressure through outlet port B as it comes into register at the end of each piston stroke.

In variable-displacement applications of axial piston motors, the swashplate angle can be adjusted by using a stroke control compensator or displacement control valve. In either case the motor would be equipped with a control piston and bias piston or servos connected to the swashplate. The greater the swashplate angle to the shaft, the larger the amount of oil that has to be displaced, resulting in a decrease in motor output speed. In other words, swashplate angle determines motor displacement per cycle.

Bent Axis Piston Motor. A bent axis piston motor does not use a swashplate to determine the piston stroke. As the name implies, the piston and block assembly is placed on an axis to the main shaft. When the pistons are forced out of the block assembly, the block and piston assembly is forced to turn. The stroke is determined by the angle between the centerline of the output shaft to the centerline of the block assembly. A universal joint is used at the axis point to deliver the force generated to the output shaft. A valve plate is still used to determine oil flow to the inlet and outlet sides of the motor. **Figure 4-43** shows a typical variable-displacement bent axis motor.

Radial Piston Motor. Unlike the axial piston motor, the piston assemblies are placed perpendicular to the rotating shaft. The output shaft is splined with an offset cam lobe. When energized with oil pressure, the pistons are forced out of their bore and cause the cam lobe to rotate; this happens in sequence to keep the motor rotating in the desired direction. A rotary control valve that rotates with the output shaft is used to time the oil delivery in sequence to each piston assembly. Radial piston motors have a high displacement area and are typically used in high-torque, low-speed applications.

Cam Lobe Piston Motor. A cam lobe piston motor is similar to the radial piston motor in that the piston assemblies are perpendicular to the rotating shaft. The piston assemblies are fitted into a rotor that is splined to the output shaft. They are also fitted with rollers that follow the contour of a cam plate. When oil is fed into the pistons through a valve plate, the

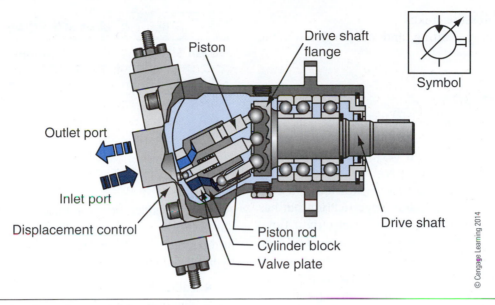

Figure 4-43 Bent axis motor variable displacement.

pistons are forced out of the bore and ramp down the cam plate forcing the piston and rotor to rotate. The inlet and outlet ports for each piston are timed such that they will exhaust when the pistons are forced into the bore (cam plate ramp up) and are pressurized when forced out (cam plate ramp down). These types of motors are currently very popular in skid steer applications (see **Figure 4-44**).

CONDUCTORS AND CONNECTORS

In any hydraulic circuit, the hydraulic medium has to conduct from the various components plumbed into the circuit. In mobile hydraulic equipment, hoses tend to function best as hydraulic conductors because they:

- Allow for movement and flexing
- Absorb vibrations

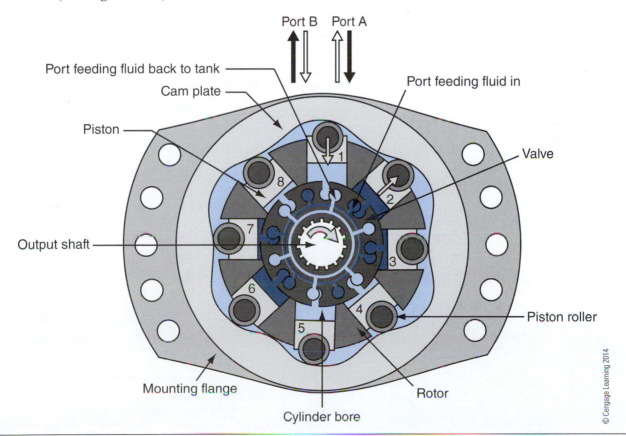

Figure 4-44 Cam lobe motor.

- Sustain pressure spikes
- Enable easy routing and connection on the chassis

Hydraulic Hoses

The selection of hoses in a hydraulic circuit is important because the hoses play a major role in determining how the system performs. The flow requirements of a conductor determine what the internal diameter of a hose should be. For instance, a hose that can withstand the pressure specification but has too small internal diameter will restrict circuit flow, causing overheating and pressure losses. The size of any hydraulic hose is determined by its inside diameter. This is sometimes indicated as *dash size* in ¹⁄₁₆-inch increments: each dash number indicates ¹⁄₁₆ inch. A #4 dash hose is equivalent to ⁴⁄₁₆ inch or ¼ inch. **Table 4-2** shows some dash sizes and their nominal equivalents.

A large internal diameter hose has to be stronger to sustain the working pressures of a hydraulic circuit.

TABLE 4-2: DASH SIZE EQUIVALENTS

Dash Size	Nominal Diameter
# 4	¼ inch
# 6	⅜ inch
# 8	½ inch
# 10	⅝ inch
# 12	¾ inch

Another consideration for hose selection is that the hose must be compatible with the hydraulic fluid used in the system. Four general types of hoses are used in hydraulic circuits:

- Fabric braid hose
- Single-wire braid hose
- Multiple-wire braid hose (up to 6 wire braid)
- Multiple-spiral wire hose (up to 6 wire spirals)

These types of hydraulic hoses are shown in **Figure 4-45**.

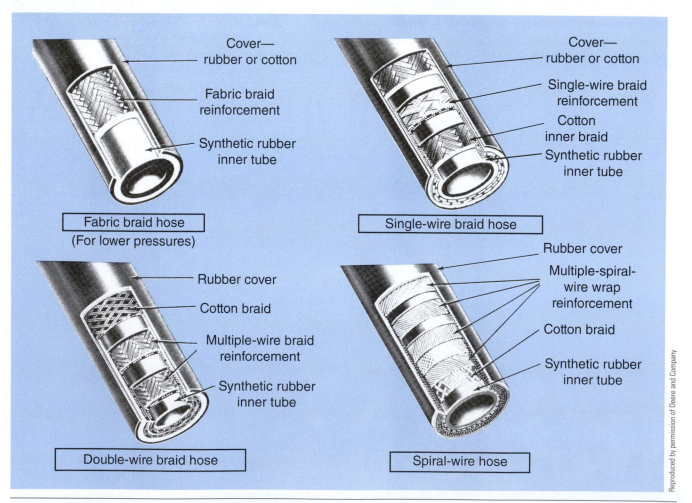

Figure 4-45 Four types of hydraulic hoses.

Fabric Braid Hose. Fabric braid hoses are used for petroleum base hydraulic and fuel oils at low pressures. There are various categories of fabric braid hose, depending on application. They are often used on the low-pressure side of a fuel subsystem or the line that connects a hydraulic reservoir to a pump. They are also commonly used in return circuits. Typically, fabric braid hoses are constructed using a synthetic inner tube with fiber reinforcing, sometimes with a spiral wire to prevent collapse. The outer wrapping can be either synthetic rubber or fiber braid. If fabric braid hose is to be used at lower than atmospheric pressures, the maximum low-pressure specification must be observed. Ratings up to 40 inches Hg inlet restriction are available, but the specified rating drops as the inside diameter increases. Fabric braid line is not recommended for use as pressure lines in mobile hydraulic circuits.

Single-Wire Braid Hose. Single-wire braid hose is commonly used as the conductor for the pressure side of diesel fuel subsystems and for some hydraulic circuits. The inner tube is usually manufactured with synthetic rubber reinforced with a single braid of high-tensile steel wire. The outer wrap consists of a synthetic rubber or braided cotton, designed to be oil and abrasion resistant. Single-wire braid is often used in truck, farm, and heavy equipment hydraulic circuits and fuel subsystem applications.

Multiple-Wire Braid Hose. Double-wire braid hose is commonly used as the conductor for high-pressure hydraulic circuits. The inner tube is usually manufactured with synthetic rubber reinforced with at least two braids of high-tensile steel wire. The outer wrap again consists of a synthetic rubber or braided cotton, designed to be oil and abrasion resistant. Double-wire braid is often used in hydraulic circuits using higher pressure. Up to six wire braids may be used, but spiral-wire hose is more common in high-pressure circuits.

Spiral-Wire Hose. Spiral-wire braid hose is used in high-pressure hydraulic circuits and is the preferred option in systems that have to sustain high pressure surges. The inner tube is usually manufactured with synthetic rubber reinforced with multiple spirals of high-tensile steel wire. The outer wrap consists of a synthetic rubber or braided cotton, designed to be oil and abrasion resistant.

Installing Hoses

Some care should be taken when installing hydraulic hoses because improper installation can result in rapid failure. Ensure that the OEM-specified working pressures are observed when replacing hoses. **Table 4-3** identifies the maximum working pressures of some common hydraulic hose types. **Figure 4-46** provides some tips on routing hydraulic hoses. The following precautions should be observed:

- **Taut hose.** There should always be some slack when a hose is installed to prevent straining even when the coupled ends do not move in relation to each other. Taut hoses tend to bulge and weaken in service, and hoses under pressure can fractionally shrink.
- **Loops.** Angled fittings should be used to avoid creating loops: this should reduce the total length of hose required.
- **Twisting.** Hoses should not be subjected to twisting either during installation or in operation. When installing hoses, tighten the fitting on the hose, not the hose on the fitting.
- **Rubbing.** Use hose clamps and brackets to ensure that hoses do not contact moving parts on the chassis.
- **Heat.** Hoses should be prevented from contacting hot surfaces, such as engine components. Shield hoses when necessary.

TABLE 4-3: HOSE WORKING PRESSURES			
Hose Size	Working Pressure: Single-Wire Braid	Working Pressure: Multiple-Wire Braid	Working Pressure: Multiple-Spiral Wire
¼ inch #4	3,000 psi	5,000 psi	N/A
⅜ inch #6	2,250 psi	4,000 psi	5,000 psi
½ inch #8	2,000 psi	3,500 psi	4,000 psi
⅝ inch #10	1,750 psi	2,750 psi	3,500 psi
¾ inch #12	1,500 psi	2,250 psi	3,000 psi
1 inch #16	800 psi	1,875 psi	3,000 psi
1½ inch #24	500 psi	1,250 psi	3,000 psi
2 inch #32	350 psi	1,125 psi	2,500 psi

Hose installation

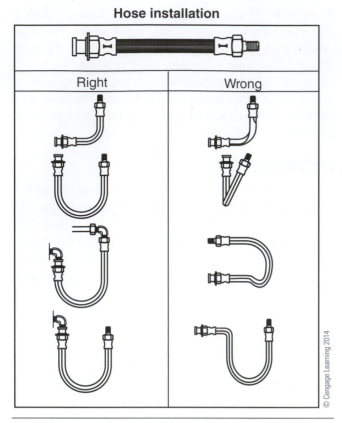

Figure 4-46 Hydraulic hose routing options.

- **Sharp bends.** Bends choke down flow and can lead to premature failure. The maximum bend radius permitted depends on the type of hose used, but generally, tighter bends are permissible in low-pressure systems.

Couplers (Connectors)

Hydraulic hose **couplers** (also known as connectors and fittings) are made of steel, stainless steel, brass, or fiber composites. Hose couplers or fittings can either be reusable or permanent. Hose fittings are installed at the hose ends and the mating end consists of either a nipple (male fit) or a socket (female fit). Some examples of couplers are shown in **Figure 4-47**. Adapters are separate from the hose assembly and are used to couple hoses to other components such as valves, actuators, or pumps.

Permanent Hose Fittings

When a hydraulic hose is fitted with permanent hose fittings, the fittings are discarded with the hose in the event of a hose failure. These fittings are crimped or swagged onto the hose. When a hose fitted with permanent hose fittings fails, the hose must be replaced. If it cannot be replaced as an assembly, one must be made up using stock hose cut to length and fitted with either

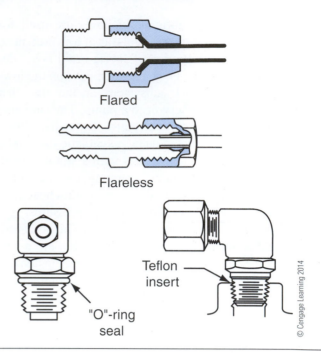

Figure 4-47 Assorted couplers.

crimp-type or reusable fittings. **Figure 4-48** shows some standard permanent hose couplers.

Reusable Hose Fittings

Reusable fittings are common in repair shops: a hose assembly station, some stock hose, and an assortment of reusable fittings can replace many of the hundreds of different types of hoses used on various pieces of equipment. When a hose with reusable fittings wears out, the fittings can be removed and assembled onto new stock hose. Reusable fittings are usually screwed onto hose, although some types of low-pressure hose may use press fits.

Assembling Hose Fittings

Figure 4-49 shows some hose-to-fitting assembly methods. Because of the high pressure used in many hydraulic circuits, the OEM assembly procedure must be precisely observed.

CAUTION *Separation of a fitting and hose at high pressure can be dangerous! Never reuse a suspect fitting. Observe the manufacturer's assembly procedure to the letter.*

Sealing Fittings. Fittings can be sealed to couplers in the following ways:

- Tapered threads
- O-rings
- Nipple and seat (flare)

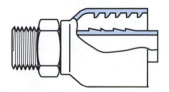

1. SAE straight Thrd. "O" ring

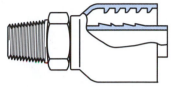

2. Male pipe

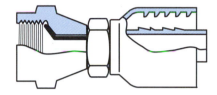

3. 37° JIC

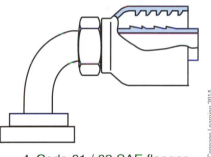

4. Code 61 / 62 SAE flanges

© Cengage Learning 2014

Figure 4-48 Permanent hose couplers.

When making hydraulic connections, ensure that the coupler fittings are compatible with each other.

Adapters. An adapter is separate from the hose assembly. Adapters have the following functions:

- Couple a hose fitting to a component
- Connect hydraulic lines in a circuit
- Act as a reducer in a circuit
- Connect a pair of hoses on either side of a bulkhead

Coupling Guidelines

When making hydraulic connections, the following guidelines should be observed:

- Torque the fitting on the hose, not the hose on the fittings.
- Couple male ends before female ends.
- Ensure the sealing method of each fitting to be coupled is the same.

- Use 45- and 90-degree elbows to improve hose routing.
- Use hydraulic pipe seal compound on the male threads only (on thread-seal unions).
- Use two wrenches when tightening unions to avoid twisting hose.
- Never over-torque hydraulic fittings.

Tech Tip: When tightening the fittings on a pair of hydraulic couplers, always use two wrenches to avoid twisting hoses or damaging adapters.

Pipes and Tubes

Pipes used in hydraulic circuits are generally made from cold drawn, seamless, mild steel. The pipe should never be galvanized because the zinc can flake off and plug up hydraulic circuits. Tubing can also be used, and it has the advantage of being able to sustain some flex, hence its use in vehicle brake systems. Tubing should be manufactured from cold drawn steel if used in moderate- to high-pressure circuits. For low-pressure circuits, copper or aluminum tubing may be used.

Pipe Couplers. Pipe joints are made by threading the OD and then turning it into a tap-threaded bore. The standard threads used in vehicle applications are:

- National (standard) pipe tapered (thread) (NPT)
- National (standard) pipe (dryseal) taper (thread) fuel (NPTF)

An NPT coupling is sealed by the interference contact between the flanks of the mating threads, and consequently pipe sealing compound is required to ensure a seal. In an NPTF dryseal coupling, the roots and crests of the opposing threads contact before the flanks contact, and a seal is created by crushing the thread crests. **Dryseal couplers** are intended for one-off usage and usually do not require thread seal compound.

Tech Tip: When coupling pipe connectors, remove burrs from both the male and female threads; then apply sealing compound to the male threads only.

Tube Couplers. Tube couplers are designed so that the fitting can be tightened while the tube

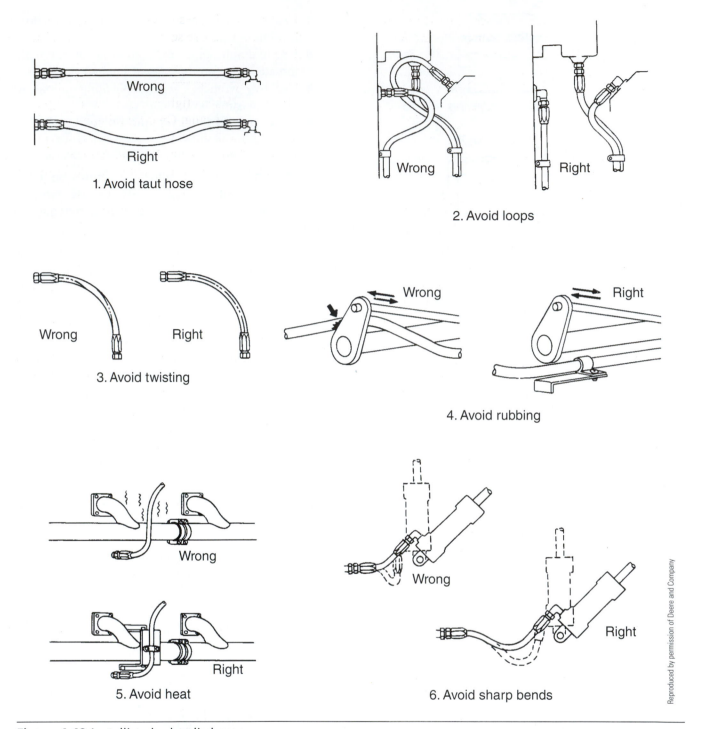

Figure 4-49 Installing hydraulic hoses.

remains stationary. Two types of tube coupling seal are used: flared and flareless. Flared fittings seal by the metal-to-metal crush created when the flared mating surfaces contact between a pair of fittings. Flareless fittings seal by compression, either with or without a ferrule.

Flared Fittings. Flared fittings are used in thin wall tubing and the flare angle is either 45 degrees (SAE standard), or 37 degrees (joint industry committee

or JIC standard). The following types of flare fittings are used:

- **Inverted flare.** A 45-degree flare is rolled to the inside of the fitting body. Commonly used in hydraulic brake circuits.
- **Two-piece flare.** A tapered nut aligns and seals the flared end of the tube.
- **Three-piece flare.** A three-piece flare fitting consists of a body, a sleeve, and a nut that fits

over the tube. The sleeve free floats, permitting clearance between the nut and tube and aligning the fitting. When tightened, the sleeve is locked without imparting twist to the flared tube.

- **Self-flaring.** These fittings use a wedge-type sleeve: as the sleeve is tightened, the fitting is forced into the tube end, spreading it into a flare. This type of fitting combines the advantages of high strength and fast assembly.

> **CAUTION** *Never attempt to cross-couple SAE and JIC fittings: the result will be to damage both.*

Flareless Fittings. Flareless fittings are commonly used in vehicle fluid power circuits, especially pneumatic circuits. They require no special tools for assembly and are reusable. There are three basic types:

- **Ferrule fittings.** These consist of a body, a compression nut, and ferrule. A wedge-shaped ferrule is compressed into the fitting body by the compression nut, creating a seal between the tube and the body.
- **Compression fittings.** These are used with thin-walled tubing and seal by crimping the end of the tube to form a seal.
- **O-ring fittings.** The principle is similar to ferrule type fittings except that a compressible rubber-compound O-ring replaces the ferrule. As the fitting nut is torqued, the O-ring is compressed, forming a seal between the tube and the fitting body. One of the most popular fittings found on current machines is the O-ring face seal (ORFS). The fitting swivel end (female end) rotates on a recessed flat flange. The mating surface accommodates a small O-ring that will form the seal when the two surfaces (flange and seal face) are compressed together. **Figure 4-50** shows a typical ORFS fitting.

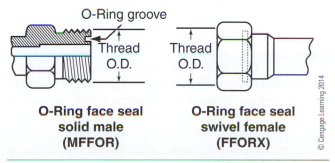

O-Ring face seal solid male (MFFOR)   **O-Ring face seal swivel female (FFORX)**

Figure 4-50 O-ring face seal type fitting.

Several different types of O-ring are used, including round section, square section, D-section, and steel-backed.

> ***Tech Tip:*** When tightening tube fittings, torque only until snug. Over-torquing any kind of line fitting is a direct cause of leaks. Use two wrenches when torquing fittings, one to hold the fitting body and the other to turn the compression nut.

Quick-Couplers. When hydraulic lines have to be frequently connected and disconnected, **quick-release couplers** are used. Quick-release couplers are simple connect/disconnect devices with an integral positive seal that allows coupling and uncoupling without admitting air to the circuit. A typical application is a wet-line kit on a dump trailer that has to be connected to multiple tractor units. A quick coupler is a self-sealing device that shuts off flow when disconnected. This eliminates the need to bleed a system each time a disconnect/connect takes place. **Quick-connect couplers** consist of a male and female coupler. There are four types:

- **Double poppet.** A double-poppet quick connector has a self-sealing poppet in both the male and female halves. When coupled, each poppet is forced off its seat, allowing oil flow through the coupling. When disconnected, the poppets in both the male and female halves are closed by spring force, sealing both sides of the disconnected circuit. A connected double-poppet coupler is locked into position by a ring of balls that are held into position by a spring-loaded sleeve. When the sleeve is retracted, spring pressure on the ring of balls is relieved, permitting the disconnect.
- **Sleeve and poppet.** Sleeve-and-poppet quick-connect couplers have a self-sealing poppet in one half and a tubular valve and seal in the other half of the coupling. This type of quick coupler provides faster sealing on disconnect, minimizing the entry of air or oil leakage from the coupling.
- **Sliding seal.** Sliding seal couplers use a sliding gate that covers a port on each coupling half at disconnect. It tends to be prone to some leakage during connect and disconnect.
- **Double-rotating ball.** Double-rotating ball quick connectors couple by inserting the line

plug into the coupler body while turning a locking valve lever. The action of rotating the locking lever forces back a pair of sealing balls, permitting circuit flow. When disconnected, the locking lever is turned and spring pressure loads the sealing balls into recesses in the locking valve lever, sealing the circuit. Most double-rotating ball quick couplers have an automatic lock release that permits both immediate sealing and hose release in the event of an unintended hydraulic circuit separation. For instance, if such a union is made between a tow unit and a trailer, separation will occur without damage to the hydraulics of either circuit.

HYDRAULIC FLUIDS

Hydraulic fluid is the medium used to transmit force through a hydraulic circuit. Most hydraulic fluids are refined from petroleum-base stocks and are seldom compatible with vehicle brake fluids. Hydraulic fluids used in hydraulic systems are usually specialty hydraulic oils, but engine and transmission oils can also be used. Always check when adding or replacing hydraulic oil. Synthetic hydraulic oils are commonly used in today's hydraulic circuits because they have wider temperature operating ranges and offer greater longevity.

Hydraulic oils must perform the following:

- Act as hydraulic media to transmit force.
- Lubricate the moving components in a hydraulic circuit.
- Resist breakdown over long periods of time.
- Protect circuit components against rust and corrosion.
- Resist foaming.
- Maintain a relatively constant viscosity over a wide temperature range.
- Resist combining with contaminants such as air, water, and particulates.
- Conduct heat.

When minimum operating temperatures are specified for a hydraulic system, fluid operating temperatures, rather than ambient temperatures, are referenced. However, because fluid viscosity increases as temperature drops (the oil thickens), minimum operating temperatures must at least respect the lowest ambient temperatures. If hydraulic oil thickens to the point that the pump inlet becomes restricted, the pump starves for oil, resulting in cavitation and pump damage.

Maximum operating temperatures influence the types of seals used in the circuit components because some seals can harden at high temperatures. Hydraulic fluids thin as temperature increases, resulting in lower lubricity: this results in a loss of lubrication at high load points in the circuit.

Viscosity

Viscosity is a rating of a liquid's resistance to fluid shear, but it is also a measure of its resistance to flow. Petroleum oils tend to thin as temperatures increase and begin to gel as temperatures decrease. In a hydraulic circuit, if the viscosity of hydraulic fluid is too low (fluid is thin), the chances of leakage past seals and poor surface lubrication increase. If viscosity is too high (fluid is thick), more pump force is required to push fluid through the circuit and the operation of actuators may be sluggish.

There are two measures of viscosity: absolute (also known as dynamic) and kinematic (which factors in the density of the oil). In hydraulics, kinematic viscosity tends to be used to rate oils.

SAE Viscosity Grades. The viscosity rating of hydraulic fluids is graded using SAE codes similar to engine or transmission lubricants. Lower SAE numbers are used to denote low-viscosity oils and high SAE numbers are used to denote high-viscosity oils. This means that SAE 10W hydraulic oil will flow more readily than SAE 30 graded oil at a given temperature.

ISO Viscosity Grades. International Standards Organization (ISO) is commonly used to grade hydraulic oil viscosity. ISO 32 is the most commonly used viscosity grade, and this is approximately equivalent to SAE 10 grade. Lighter ISO grades such as ISO 15 and 22 tend to be used in extreme cold-weather applications, while the heavier grades, 46, 68, and 100, are used in industrial high-pressure applications. The correct viscosity for a specific hydraulic system depends on the following variables:

- Starting viscosity at the minimum predicted ambient temperature.
- Desired viscosity range when the system is subjected to predicted extreme duty and load cycles.

Viscosity Index

Viscosity index (VI) is used to rate the oil's change in thickness as temperature changes. If oil thickens at low temperatures and becomes very thin at high temperatures, it has low VI. When oil can retain a relatively consistent viscosity through a wide temperature range, it can be described as having high VI. In a hydraulic circuit, the hydraulic medium is required to be thick enough to prevent leakage while also offering low flow resistance through the circuit, so high VI is essential.

ONLINE TASKS

1. Use any Internet search engine to research hydraulic oil specifications. By using a specific piece of equipment, your geographic location, and the operating environment, outline the reasons you have selected a particular oil.
2. Check out this URL: *www.hydraulic-supply.com*. You may have already come across this when undertaking the online exercise outlined in Chapter 3. Bookmark it.

Shop Tasks

1. Select any hydraulic hose on a piece of equipment in your shop. Without removing it, identify the hose and coupler specifications to the extent that you can either order (from parts) or make up an equivalent hose using in-house assembly equipment.
2. Select a piece of mobile equipment in your shop. Attempt to identify all the pumps in the system by type. Note the location of each pump and its prime mover. Identify any hydraulic motors in the system.

Summary

- The potential energy of a hydraulic circuit is pressurized hydraulic fluid.
- Hydraulic pumps convert mechanical energy into hydraulic potential.
- The prime mover of a hydraulic circuit in mobile heavy equipment is the vehicle engine either directly, by driving the pump, or indirectly, by means of a PTO or chassis electricity.
- Common pumps used in mobile equipment are internal and external gear, unbalanced and balanced vane, and axial and radial piston pumps.
- Valves are used to manage flow and direction through a hydraulic circuit.

- Actuators, such as hydraulic cylinders and motors, convert hydraulic potential into mechanical movement.
- Hydraulic oil is used to store and transmit hydraulic energy through a hydraulic system.
- Hydraulic oil viscosity is rated by SAE and ISO standards.
- Viscosity index rates oil's ability to change in thickness as temperature changes. If oil thickens at low temperatures and becomes very thin at high temperatures, it has low VI.

Review Questions

1. When a hydraulic pump discharges the same slug volume it picks up per cycle, it should be described as:

   A. variable-displacement

   B. constant cycle

   C. positive displacement

   D. non-positive cycle

2. How is hydraulic oil routed through a typical external gear pump?

   A. through the center between the intermeshing gears

   B. around the outside between gear teeth and pump body

   C. through the inlet to the gear output shaft

   D. around the idler gear shaft to the outlet port

3. Which of the following defines a vane-type hydraulic pump that has one inlet port and one outlet port?

   A. balanced

   B. unbalanced

   C. variable-displacement

   D. twin-cycle

4. Technician A says that a hydraulic cylinder is a linear actuator. Technician B says that a hydraulic motor is a rotary actuator. Who is correct?

   A. Technician A only   C. Both A and B

   B. Technician B only   D. Neither A nor B

5. Which of the following is not a function of a hydraulic accumulator?

   A. dampens pressure surges   C. maintains consistent circuit pressure

   B. stores potential energy   D. charges hydraulic actuators

6. What is the definition of *laminar flow*?

   A. high turbulence   C. streamlined flow

   B. vortex flow   D. flow through elbows

7. What amount of linear force will be generated if a hydraulic pressure of 1,000 psi acts on a piston measuring 25 square inches?

   A. 25 pounds   C. 2,500 pounds

   B. 250 pounds   D. 25,000 pounds

8. Which type of gear pump uses a crescent-shaped separator located between an external spur gear and an annular internal gear?

   A. internal gear   C. rotor gear

   B. external gear   D. triple gear

9. Which of the following is used in an axial piston-type hydraulic pump?

   A. rotating cam   C. radial piston

   B. swashplate   D. compressor type

10. How is Hydraulic pump performance rated?

    A. flow in gpm   C. revolutions per minute (rpm)

    B. pressure in psi   D. All of the above.

11. What is usually used to drive mobile equipment hydraulic pumps?

    A. electricity   C. transmission

    B. engine   D. All of the above.

12. What is the primary function of a flow-control valve?

    A. control the speed of an actuator   C. manage circuit resistance

    B. control circuit pressure   D. switch circuit outcomes

13. What type of actuator would be required to actuate an excavator boom?

    A. single-acting   C. separate lift and return rams

    B. double-acting   D. vane-type motor

14. What force is used to turn a hydraulic motor?

    A. electrical

    B. diesel engine

    C. transmission

    D. hydraulic

15. What type of hydraulic hose would be able to sustain the highest working pressures?

    A. spiral wire

    B. single-wire braid

    C. double-wire braid

    D. fabric braid

16. When comparing a section of ⅜-inch single-wire braid hydraulic hose to a ½-inch single-wire braid, which of the following should be true?

    A. The ⅜-inch hose will have a higher specified flow rate.

    B. The ⅜-inch hose will have a higher specified working pressure.

    C. The ½-inch hose will have a lower specified flow rate.

    D. The ½-inch hose will have a higher specified burst pressure.

17. Which of the following threads used in hydraulic fittings is described as dryseal?

    A. UNF

    B. UNC

    C. NPT

    D. NPTF

18. Which of the following would be true of a hydraulic fluid described as being of high viscosity?

    A. thickens as temperature increases

    B. more pump force is required to push it through the circuit

    C. has better resistance to cold-weather gelling

    D. performs better through a wide temperature range

19. What seat angle is used on SAE standard fittings?

    A. 35 degrees

    B. 37 degrees

    C. 45 degrees

    D. 60 degrees

20. Which valve is classified as a pressure-control valve?

    A. check valve

    B. relief valve

    C. priority valve

    D. directional control valve

5 Hydraulic Symbols and Schematics

Learning Objectives

After reading this chapter, you should be able to:

- Identify typical hydraulic circuit symbols.
- Interpret simple hydraulic schematics.

Key Terms

American National Standards Institute (ANSI)

International Standards Organization (ISO)

conductors

SYMBOLS AND SCHEMATICS

When working with vehicle hydraulic schematics, it is important to be able to interpret standard symbols. Symbols are two-dimensional figures that roughly approximate the components they represent in a hydraulic schematic. The symbols used in hydraulic schematics are standardized by the **American National Standards Institute (ANSI)** and by the **International Standards Organization (ISO)**. This helps make life easier because most manufacturers today use either ANSI or ISO symbols in their schematics. The graphics used in hydraulic schematics consist of shapes and marks.

Shapes and Marks

Symbols for the critical components on a hydraulic schematic will be explained in some detail in this section. However, the basic shapes used in hydraulic schematics are:

- Circle/semicircle—indicates a pump or motor symbol.
- Square (or envelope)—represents one position or path through a valve. Two squares together indicate a two-position valve.

- Diamond—indicates a component that conditions fluid in the circuit, such as a filter or heat exchanger.
- Rectangle—indicates a hydraulic cylinder or reservoir.

A symbol is drawn by using one of the four basic shapes above and adding the appropriate marks. Typical marks include:

- Solid lines—indicate the flow route for hydraulic fluid.
- Dashed lines—indicate a pilot line that connects a control circuit to a slave circuit.
- Dotted lines—indicate a return or exhaust circuit.
- Center lines—enclose assemblies.
- Arrows—show the direction of fluid flow or rotational direction of pumps and motors. Arrows at a 45-degree angle indicate an adjustment point. A vertical arrow in a component indicates pressure.
- Arcs—show points of adjustment in the circuit such as the flow-control valve.

Conductor Symbols. Hydraulic pipes, hoses, and tubes that carry fluid in a hydraulic circuit are known as **conductors**. They are represented by lines on a schematic. Solid lines indicate inlet, pressure, or return circuits. Broken lines with long dashes usually indicate pilot or control lines, while lines of short dashes indicate leakage oil. An arced line (that is not semicircular) between two dots indicates a flexible hose. **Figure 5-1** shows how conductors are typically represented on a hydraulic schematic.

A semicircular loop on one line crossing another indicates that the two lines are not hydraulically connected. However, sometimes crossing (but not hydraulically connected) lines simply cross each other. In the event that a pair of intersecting lines are hydraulically connected, a dot is shown at the intersection point. These three arrangements are shown in **Figure 5-2**.

Reservoir Symbols. A horizontal rectangle is used in schematics to represent a reservoir. A pressurized reservoir is indicated when the rectangle is drawn fully enclosed as indicated in **Figure 5-3**. A vented-to-atmosphere reservoir is indicated when the rectangle is open at the top as indicated in **Figure 5-3**.

The way in which lines are shown entering a rectangle symbol is also of significance. A line terminating either above the rectangle or before its base indicates that return flow is above the fluid, whereas one that is drawn contacting the base of the rectangle indicates a line that terminates below the fluid level. All these symbols are shown in **Figure 5-3**.

Lines connected to a reservoir are almost always drawn from the top, regardless of where the connection is made on the actual component. Every reservoir has at least two hydraulic lines connected to it and can have many more. Because numerous components in a hydraulic circuit may depend on a single reservoir, the reservoir symbol is often repeated rather than creating a scramble of return lines all over the schematic. In most cases, the reservoir symbol is the only one to be

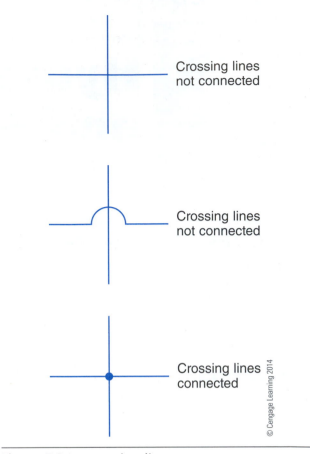

Figure 5-2 Intersecting lines.

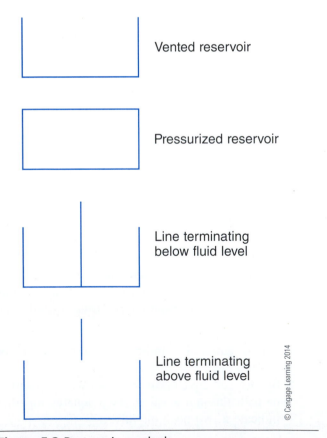

Figure 5-3 Reservoir symbols.

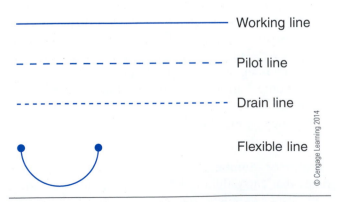

Figure 5-1 Conductor symbols.

repeated on a schematic; you can regard it in the same way a ground symbol is regarded on an electrical schematic.

Pump and Motor Symbols. Pump and motor symbols are similar in appearance. Both use a circle. A hydraulic pump is indicated when the small triangle at the output line (within the circle) is shown pointing outward (see **Figure 5-4**). A hydraulic motor is indicated when the small triangle at the inlet line (within the circle) is shown pointing inward (see **Figure 5-5**). When hydraulic pumps and motors are fixed-displacement, the circle that represents them is shown with just the inlet and outlet lines contacting the circle. When either is variable-displacement, an arrow is drawn through the circle as shown in **Figure 5-4**.

Hydraulic motors appear almost identical to the symbols shown in **Figure 5-4**, except the direction of the triangles is reversed as shown in **Figure 5-5**.

Valve Symbols

When valves are represented in hydraulic schematics, they tend to be a little more tricky. We will take a look at some simple valves used in schematics, beginning with a pressure-relief valve.

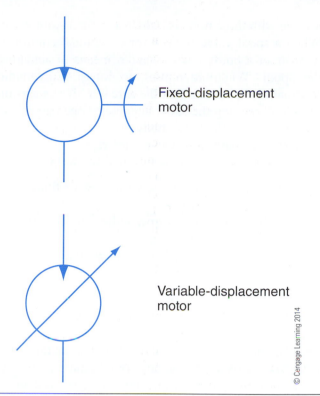

Fixed-displacement motor

Variable-displacement motor

Figure 5-5 Fixed- and variable-displacement hydraulic motor symbols.

Pressure-Relief Valves. Pressure-relief valves are required in even the simplest hydraulic circuits: they usually drain to the reservoir when tripped. A dashed line as shown in **Figure 5-6** represents a pilot line; pressure in this line opposes spring force on the opposite side of the valve. When pilot pressure exceeds the mechanical force of the spring, the spool shifts, opening the valve and returning fluid to the reservoir. This relieves the pressure on the pump side, reducing pilot pressure and allowing spring force to return the relief valve to its normal operating status. Relief valves are used to define maximum circuit pressure.

Directional Control Valve Symbols. Directional control valves are used to manage fluid flow in a hydraulic circuit. There are many types, but generally these valves are classified by the number of valve positions and of ports. The number of ports in a valve is usually referred to as way(s). For example,

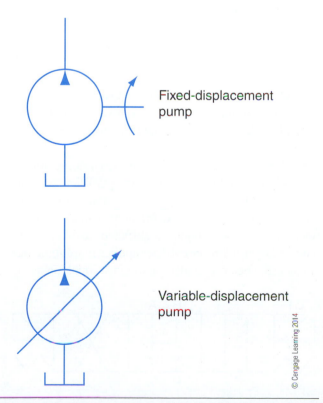

Fixed-displacement pump

Variable-displacement pump

Figure 5-4 Fixed- and variable-displacement pump symbols.

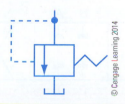

Figure 5-6 Pressure-relief valve symbol.

a valve with three ports is known as a three-way valve. When a spool valve is used for directional control, it consists of a body, a spool, and a means of actuating the spool. When attempting to interpret hydraulic schematics, remember that the actuator always acts on the spool: moving the spool is going to open and close the passages in the valve body.

To understand how valves are represented in hydraulic schematics, you should note the following:

- Boxes with arrows show the flow of fluid when the valve is shifted.
- A box without an arrow indicates flow (if any) in the neutral position.
- A box without an arrow shows the number of ports (ways) in the valve.

Figure 5-7 shows a two-position, two-way valve. It is shown with a mechanical actuator indicated on the right side of the figure and a spring return to neutral on the left side of the figure. We can also tell that this valve is normally closed (NC) in operation because both ports are shown to be blocked in neutral: arrows on the port symbols would indicate that they were open. This schematic shows a valve that would typically be used as a safety gate. When the gate is not closed, then no flow is possible in the circuit that connects to it.

Figure 5-8 shows a more complex type of direction control valve that is electrically actuated. This valve has three positions indicated by its three larger, center boxes, and four ports (ways), indicated in the center box. Some OEMs use the term *envelope* in place of box. It is a closed-center valve, meaning that when the spool is neutral, all the ports are blocked. The small rectangles with diagonal lines on each end identified as C1 and C2 are electrically actuated coils. The spool is shifted by energizing the coils. Springs on either side of the valve, depicted under the

coils, return the valve to the center position when the coils are not energized.

In **Figure 5-8**, the following alpha symbols are used:

- C1—left side coil
- C2—right side coil
- A—connects to an external device
- B—connects to an external device
- P—pump feed
- T—tank (reservoir)

In operation, when C1 is energized, the valve shifts and directs pressure to the B port, and drains the A port to the reservoir. The reverse occurs when C2 is energized: this feeds pressure to the A port and drains the B port to the reservoir.

In **Figure 5-9**, another four-way, three-position directional control valve is shown. It is also electrically actuated (note the coils on either side) with springs to return the spool to the center position when the coils are not energized. The center position is a motor center, sometimes referred to as a *float center*. This allows the actuator to "float" or free-wheel based on the force acting on it and the type of actuator used. This type of directional control valve can be used to control a hydraulic cylinder. It is integrated into the schematic shown a little later in **Figure 5-13**.

Bulldozers and wheel loaders use a four-position directional control valve on the blade and loader circuit (raise, neutral, and lower). The fourth position is a detented position in which it will hold the spool stationary in this position. The ports to the actuator (A and B) will all be connected with the pump port and return port (P and T). This is a true float position allowing the dozer blade or loader bucket to follow the contour of the ground as the machine moves (typically in reverse).

Cylinder Symbol. Cylinders are commonly used actuators in mobile hydraulics technology. When pressure is charged to one side of a cylinder, output movement occurs, providing the applied pressure acting on the sectional area of the cylinder is sufficient to overcome the load. A larger diameter cylinder applies more force than a smaller diameter cylinder, assuming equal pressure is

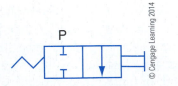

Figure 5-7 Two-position, two-way valve symbol.

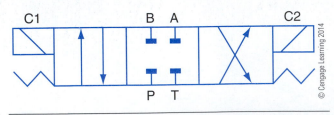

Figure 5-8 Three-position, four-way valve symbol.

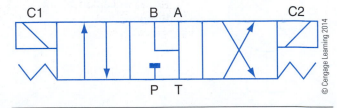

Figure 5-9 Alternate three-position, four-way valve symbol.

applied to both. The symbol for a cylinder is easy enough to identify and is represented in **Figure 5-10**.

Filter, Cooler, and Accumulator Symbols. Filters and strainers are represented in hydraulic schematics by a diamond shape with a vertical broken line through it. The symbol for a heat exchanger or cooler is similar, with a solid colored triangle at the top and bottom that are connected by a vertical line (see **Figure 5-11**). Accumulators are represented by a capsule shape

divided across the center. The accumulator potential is also indicated in the schematic. Both spring-loaded and gas-charged accumulators are shown in **Figure 5-12**.

Simple Hydraulic Circuit Schematic

Figure 5-13 incorporates most of the symbols we have introduced so far into a simple hydraulic circuit built to actuate a cylinder. Note how the directional control valve that manages the circuit is integrated into it.

Figure 5-10 Symbol for a hydraulic cylinder.

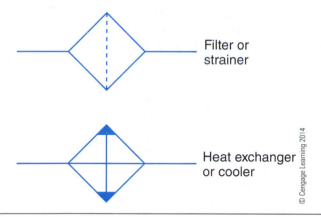

Filter or strainer

Heat exchanger or cooler

Figure 5-11 Symbols for filters and strainers.

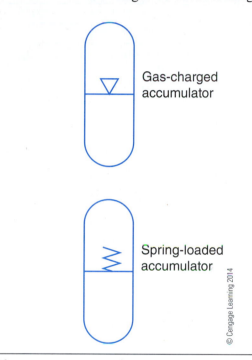

Gas-charged accumulator

Spring-loaded accumulator

Figure 5-12 Accumulator symbols.

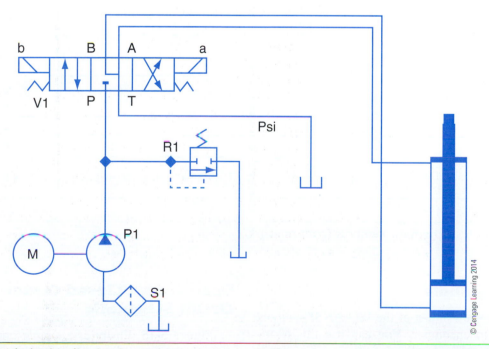

Figure 5-13 Simple hydraulic circuit used to actuate a cylinder.

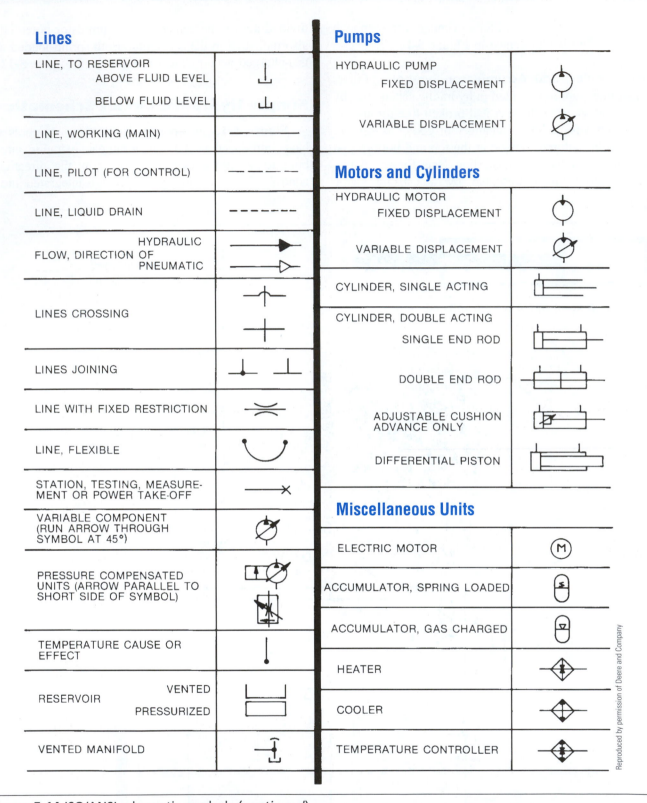

Figure 5-14 ISO/ANSI schematic symbols (*continued*).

ISO/ANSI Symbols

Figure 5-14 shows most of the ISO/ANSI symbols you are likely to encounter. You can use this to interpret some typical hydraulic circuits.

Open Center/Closed Center Circuit Schematic

Figure 5-15 illustrates a simple open-center schematic with a fixed-displacement pump. A control valve

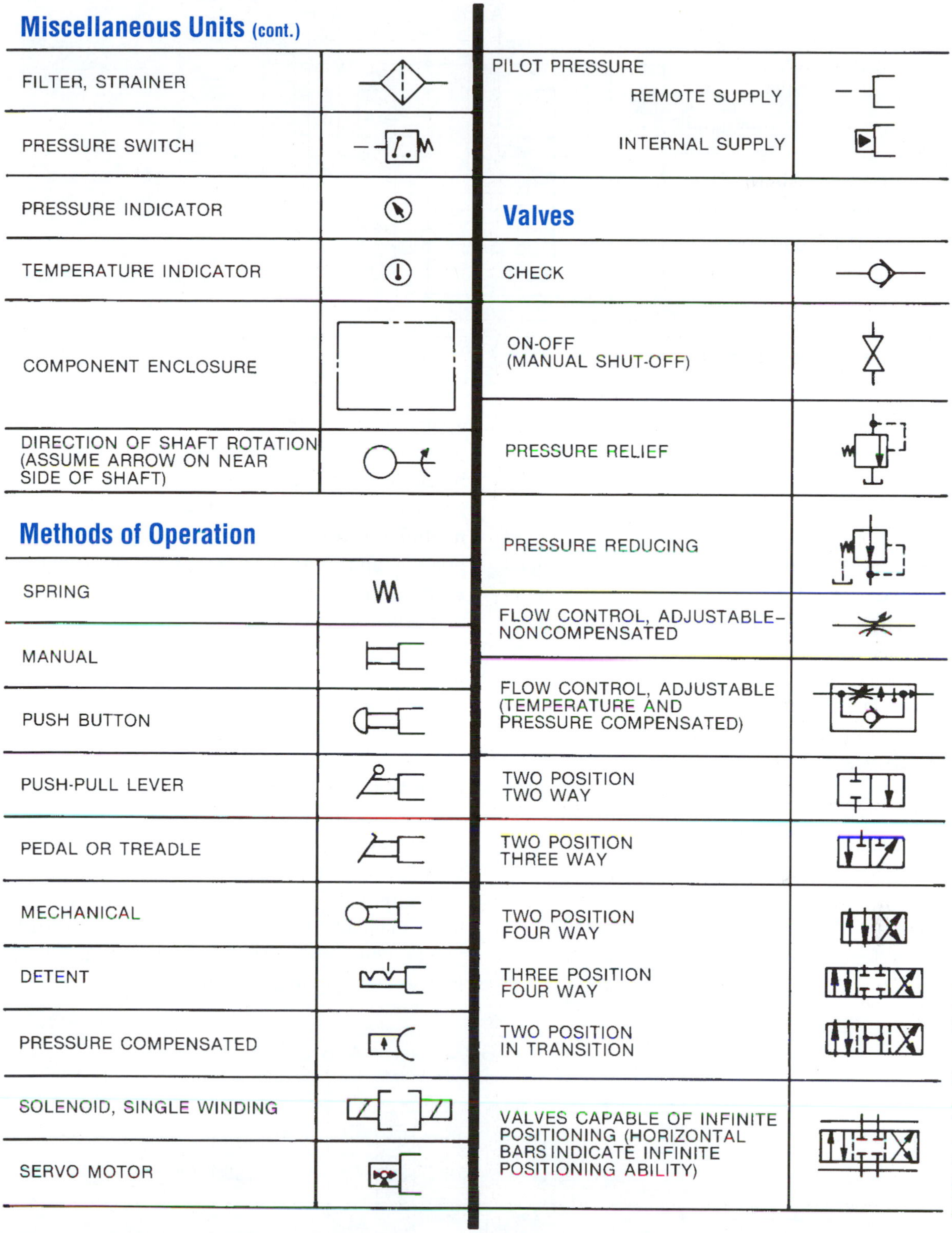

Miscellaneous Units (cont.)

FILTER, STRAINER		PILOT PRESSURE REMOTE SUPPLY	
PRESSURE SWITCH		INTERNAL SUPPLY	
PRESSURE INDICATOR		**Valves**	
TEMPERATURE INDICATOR		CHECK	
COMPONENT ENCLOSURE		ON-OFF (MANUAL SHUT-OFF)	
DIRECTION OF SHAFT ROTATION (ASSUME ARROW ON NEAR SIDE OF SHAFT)		PRESSURE RELIEF	

Methods of Operation

		PRESSURE REDUCING	
SPRING		FLOW CONTROL, ADJUSTABLE—NONCOMPENSATED	
MANUAL		FLOW CONTROL, ADJUSTABLE (TEMPERATURE AND PRESSURE COMPENSATED)	
PUSH BUTTON			
PUSH-PULL LEVER		TWO POSITION TWO WAY	
PEDAL OR TREADLE		TWO POSITION THREE WAY	
MECHANICAL		TWO POSITION FOUR WAY	
DETENT		THREE POSITION FOUR WAY	
PRESSURE COMPENSATED		TWO POSITION IN TRANSITION	
SOLENOID, SINGLE WINDING		VALVES CAPABLE OF INFINITE POSITIONING (HORIZONTAL BARS INDICATE INFINITE POSITIONING ABILITY)	
SERVO MOTOR			

Figure 5-14 *Continued.*

incorporates two manually operated three-position six-way spool valves, spring centered. The main relief valve is also incorporated into the control valve, which is connected to the reservoir line on the downstream side. As mentioned in Chapter 3, most heavy-equipment control valve circuits use a "series/parallel" flow connection through the valve. This is shown on the pump line cross-connection entering the control valve. The cross connects the main relief valve, the open-center passage, and the parallel connection to each spool valve.

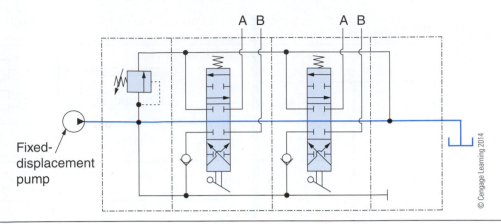

Figure 5-15 Open-center schematic.

Each parallel connection has a one-way check valve called a *load drop* check valve. (The purpose of the load drop check valve is to prevent oil from draining backward when the spool valve is activated. This can occur when there is a pressure differential in the actuator port and hydraulic pump supply port, which causes the load to drop. Pressure must become higher in the pump supply port before the check valve can open and allow oil to flow through.) Lines A and B reflect the connections to each actuator.

When one spool valve is actuated either to extend or retract a cylinder, the spool valve is placed in the top or bottom position. Let us use the top land position and review how the connections align with the ports. In this case, A port would drain through the spool valve back to the reservoir. Pump supply port would not be able to drain to the reservoir and take the parallel connection through the load check valve and then through the spool to B port. This will allow a cylinder to extend or retract depending on the line connections to the actuator. If the resistance at the actuator forces the pressure to exceed that of the pressure-relief valve, the relief valve will open and bleed off enough oil to maintain maximum system pressure. When the operator returns the spool to the neutral position, oil will be trapped at the actuator ports A and B holding the cylinder in that position. Pump supply oil will pass through the center opening and return to reservoir.

The speed of the actuator can be controlled by placing the spool finitely between land areas, allowing some oil to be directed through the parallel port to the actuator and the remainder to drain through the open-center connection back to tank. The metering slots machined into the land areas provide infinite positioning and finite control.

Figure 5-16 illustrates a closed-center control valve with a variable-displacement pump equipped with a flow limiting compensator. A control valve incorporates three manually operated three-position five-way spool valves, which are spring centered. The main relief valve is also incorporated into the control valve, which is

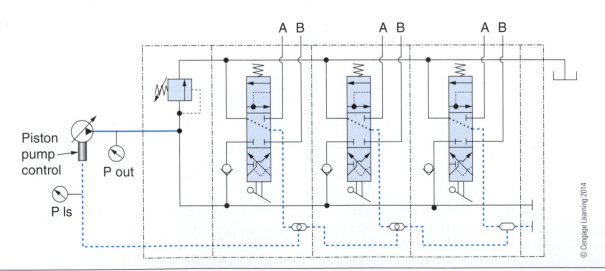

Figure 5-16 Closed-center schematic load sensing.

connected to the reservoir line on the downstream side. This valve also uses a series/parallel flow connection through the valve along with load check valves; an added feature is a shuttle valve used to relay a signal to the flow compensator. Review the section on "Flow Compensators" in Chapter 4 to better understand the operating principles of this system.

In the neutral position (closed center) the pump flow through the valve would be zero due to the flow compensator de-stroking the pump. When the spool has been shifted to the top land section, the pump supply port is aligned with actuator B port, and A port is aligned with the reservoir port. A signal line is connected to the shuttle valve through internal porting of the spool. This signal line lets the flow compensator know the system is demanding flow from the pump, so the pump will up-stroke to meet the flow demand. The actuator will move at whatever speed pump flow determines. The purpose of shuttle valves is to relay the strongest signal to the flow compensator when two or more functions are operated simultaneously.

Load sensing is another term used for this type of system because the pump supplies only what is demanded. This is different from a fixed-displacement pump, where the oil supply is constant (based on pump rpm). The operator must meter the oil between the actuator and return remaining oil to the tank by means of controlling the spool valve.

Equipment manufacturers make available comprehensive and detailed schematics for technicians. These schematics help the technician to troubleshoot and understand the complete hydraulic system equipped on each machine. **Figure 5-17** shows a Caterpillar D8 bulldozer schematic equipped with a load-sensing pressure compensated variable-displacement pump, a pressurized reservoir, and a three-spool closed-center control valve equipped with shuttle valves. These will control the blade lift, blade tilt or angle, and ripper control cylinders. Note the special features on the blade lift control, implementing a four-position spool with float position and a quick-drop valve. This valve is used to regenerate oil from the cylinder rod end to the cylinder head end for rapid lowering of the blade.

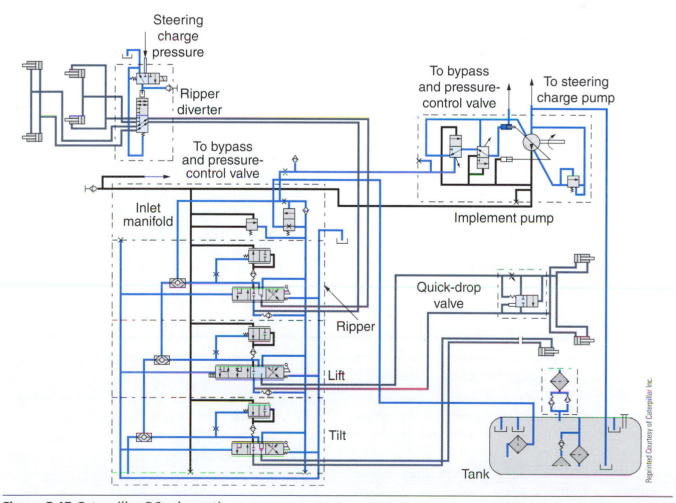

Figure 5-17 Caterpillar D8 schematic.

The ripper control valve utilizes a diverter valve that, when activated, will control one group of cylinders or the other. The diverter valve is bias to the ripper lift cylinders and when activated controls the ripper shank angle. Because one control valve operates both functions, simultaneous ripper lift and angle is not possible.

The hydraulic filter is mounted within the reservoir.

HYDRAULIC FORMULAE

The following are some formulae used to spec out hydraulic circuit performance:

Torque (T) in pound-feet
$$T = HP \times 5252 \div rpm$$

Horsepower (HP)
$$HP = T \times rpm \div 5252$$

Revolutions per minute (rpm)
$$rpm = HP \times 5252 \div T$$

Oil flow velocity (V)
$$V = GPM \times 0.3208 \div A$$
GPM = gallons per minute
A = (pipe id sectional area)

Hydraulic horsepower
$$HP = psig \times GPM \div 1714$$
psig = pounds per square inch gauge

Cylinder thrust force
$$T = A \times psi$$

Cylinder piston speed (S)
$$S = CIM \div A$$
CIM = oil flow into cylinder
A = piston sectional area

Interesting Facts

- Compressibility of hydraulic oils: 0.4% per 1,000 and up to 1.0% @ 3,000 psi applied pressure.
- Compressibility of water: 0.3% per 1,000 psi applied pressure.
- Hydraulic power and temperature: 1 watt consumption raises temperature of 1 gallon oil by 1 degree Fahrenheit per hour.
- Hydraulic power: for every 1 HP of drive power, 1 gpm at 1,500 psi can be produced.
- Fluid velocity: pressure line at 3,000 psi equals equivalent of 25 feet per second.

ONLINE TASKS

1. Use an Internet search engine to locate sites that offer hydraulic schematics and the means to interpret them. **Caution:** Avoid sites that require payment for a download.
2. Check out this URL: *www.sae.org*. Navigate to identify pages that set standards for mobile hydraulic equipment. Explain what a J-standard is.

Shop Tasks

1. Select a piece of mobile equipment equipped with hydraulic implements. Select one implement, for instance, a bucket control cylinder, and create a schematic that routes the circuit from prime mover, to pump, through control valves, and ending up at the implement. Use the circuit symbols identified in this chapter.
2. Select a piece of mobile equipment in your shop equipped with hydraulic implements and identify each type of control valve in the circuit. In each case, identify the means used to shift the spool in each control valve.

Summary

- ANSI and ISO graphic symbols are used to represent hydraulic components and connectors in hydraulic schematics.
- A circle (sometimes a semicircle) is the symbol used to represent pumps or motors in hydraulic schematics.
- A circle symbol with the triangle pointing outward indicates a hydraulic pump.
- A circle symbol with the triangle pointing inward indicates a motor.
- A variable-displacement pump or motor is indicated by an angled arrow drawn through the circle.
- A square (or envelope) is used to represent one position or path through a valve, while two squares together indicate a two-position valve.

- Diamonds indicate a component that conditions fluid in the circuit such as a filter or heat exchanger.

- Rectangular symbols indicate a hydraulic cylinder or reservoir.

- Hydraulic symbols are drawn on a schematic using one of the four basic shapes and adding the appropriate marks.

- Marks are used on hydraulic schematics to show the flow routing of the circuit being represented.

- Solid lines indicate the flow path for hydraulic fluid.

- Dashed lines indicate a pilot line that connects a control circuit to a slave circuit.

- Dotted lines indicate a return or exhaust circuit.

- Center lines are used to enclose assemblies or subcircuits.

- Arrows show the direction of fluid flow or the rotational direction of pumps and motors.

- Arcs are used to show points in a circuit where flow can be modulated such as a flow-control valve.

Review Questions

1. Technician A says that a circle symbol on a hydraulic schematic can represent a hydraulic pump. Technician B says that a circle symbol on a hydraulic schematic can represent a hydraulic motor. Who is right?

   A. Technician A only
   B. Technician B only
   C. Both A and B
   D. Neither A nor B

2. What does a circle symbol with the triangle pointing outward indicate?

   A. gas-loaded accumulator
   B. heat exchanger
   C. hydraulic pump
   D. hydraulic motor

3. Technician A says that dashed lines on a hydraulic schematic indicate a pilot line that connects a control circuit to a slave circuit. Technician B says that dotted lines are used in schematics to indicate a line under constant pressure. Who is right?

   A. Technician A only
   B. Technician B only
   C. Both A and B
   D. Neither A nor B

4. What symbol shape is used to indicate a heat exchanger?

   A. circle
   B. rectangle
   C. square
   D. diamond

5. Technician A says that arrows are used to show the direction of fluid flow in hydraulic schematics. Technician B says that arrows are used to show the rotational direction of pumps and motors in hydraulic schematics. Who is right?

   A. Technician A only
   B. Technician B only
   C. Both A and B
   D. Neither A nor B

6. Technician A says that hydraulic oil is not compressible. Technician B says that hydraulic oil is less compressible than water. Who is right?

   A. Technician A only
   B. Technician B only
   C. Both A and B
   D. Neither A nor B

7. Technician A says that an accumulator is represented on a hydraulic schematic by a rectangle, rounded at the corners. Technician B says that a control valve is represented on a hydraulic schematic by an oval shape. Who is correct?

   A. Technician A only   C. Both A and B

   B. Technician B only   D. Neither A nor B

8. If equal pressure is applied to a large-diameter cylinder and a small-diameter cylinder, which of the two will produce more force?

   A. Both cylinders must produce equal force.
                                 C. The small-diameter cylinder produces more force.
                                 D. None of the above; "pressure" only affects actuator speed.
   B. The large-diameter cylinder produces more force.

9. Which component is represented by a diamond-shaped symbol with solid colored triangles on the top and bottom connected by a vertical line?

   A. filter   C. heat exchanger

   B. strainer   D. safety-relief valve

10. What is represented on a schematic by a pair of unbroken, intersecting lines with a dot at the intersection point?

    A. two hydraulic lines cross but do not connect
                                 C. a regulator is represented by the dot
                                 D. an accumulator is represented by the dot
    B. two hydraulic lines connected

11. Where does the pump oil flow on an open-center multi-spool control valve directs in the neutral position through?

    A. the implements   C. only the last spool

    B. only the first spool   D. all the spools

12. Where is the system main relief valve located on the D8 schematic?

    A. in the implement pump   C. in the inlet manifold

    B. in the reservoir   D. in the ripper diverter valve

6 Hydraulic System Maintenance and Diagnostics

Learning Objectives

After reading this chapter, you should be able to:

- Work safely around hydraulic equipment.
- Outline the benefits of a preventative maintenance program.
- Describe the features of a sound maintenance program.
- Outline a drain, flush, and refill procedure.
- Identify the effects of unwanted restrictions in hydraulic circuits.
- Connect a hydraulic analyzer into a hydraulic circuit.
- Perform some typical hydraulic analyzer tests.
- Use troubleshooting charts and tables to help identify circuit malfunctions.
- Identify the hydraulic circuits found on a typical excavator.
- Outline some diagnostic tests for the track drive, swing, boom, stick, and bucket hydraulic circuits on an excavator.

Key Terms

aeration	kidney-loop machine	residual pressure
boom circuit	margin pressure	standby pressure
bucket circuit	particle counter	stick circuit
case drain	pilot pressure	swing circuit
drift	preventative maintenance (PM)	track circuit
hydraulic analyzer	preventative maintenance inspection (PMI)	

MAINTENANCE

Proper maintenance is key to having a hydraulic system perform properly and to minimize repair downtime. Good maintenance requires some upfront investment in labor, data tracking, and scheduling, plus some low-cost items such as fluid and filters; but the payoff is avoiding costly downtime and major component failure. Properly planned and scheduled maintenance is known as **preventative maintenance (PM)**. A key factor of preventative maintenance is the scheduling of **preventative maintenance inspections (PMIs)**. Adhering to PMIs in a planned maintenance schedule has been proven to minimize costly unscheduled equipment failures.

The type of preventative maintenance schedule that suits a particular operation varies depending on the equipment and the kind of environment in which it operates. Any maintenance program should be flexible enough to provide for monitoring the program itself: this type of evaluation allows the program to be adapted as equipment and operating conditions change. Typical maintenance problems result from:

- Low reservoir oil level
- Clogged oil filters
- Contaminated or improper oil in the system
- Dirt in the hydraulic circuit

Preventative maintenance schedules should be observed. Most routine maintenance on hydraulic circuits takes little time and can eliminate expensive component repair and replacement. This section will take you through some typical maintenance procedures.

Safe Practice

Hydraulic circuits are designed to run at high pressures and to support high loads. It is essential that you work safely around chassis hydraulic equipment. Make sure you understand the safety precautions outlined in Chapter 1 in this book. Here is a quick reminder of some of the basic rules specific to working around hydraulic circuits:

- Never work under any device that is supported only by hydraulics. A raised bucket or dozer blade must be mechanically supported before you work under it. Just as when using a floor jack, you must use some mechanical device for supporting any raised equipment or components.
- Hydraulic circuit components can retain high **residual pressures**: the system does not have to be active for this to be a potential hazard. Try to ensure that pressures are relieved throughout the circuit before opening it up. Crack hydraulic line nuts slowly and be sure to wear both safety glasses and gloves.

Drain, Flush, and Refill

A draining, flushing, and refilling procedure is the only way to remove contaminants and degenerated

hydraulic fluid from a hydraulic circuit. The frequency with which this procedure is performed will depend on the severity of the working conditions and, in many cases, simply draining and replacing the hydraulic oil will be sufficient. Sludging oil, solid deposits, and the appearance of varnishing on components are indicators that draining the system should be followed by system flushing before refilling with new hydraulic oil.

Drain System. Gravity drain the system, clean any sediment from the reservoir, and replace the filter elements. Inspect the old oil. If it shows signs of severe degeneration, flushing the system should be considered. Indicators of severe degeneration of hydraulic oil are a burned odor, excessive stickiness, discoloration, and lacquer flakes in oil.

Flush System. Most hydraulic system flushing should be performed with the specified service hydraulic oil. After draining, fill the system reservoir with the specified oil, bleed the hydraulic pump, and then operate the equipment to circulate the oil through the entire system. All the valves should be actuated to ensure that the flushing oil reaches every part of the circuit. Finally, drain and dispose of the flushing oil. Many large machine systems use hundreds of gallons of oil; this may be an expensive method of flushing the system, but it is often recommended by OEMs. Some OEMs will approve a filter flushing process where the oil within the system is flushed through low-micron filters until the fluid reaches an acceptable particle level count. **Figure 6-1** shows a **kidney-loop machine** equipped with a **particle counter**. This machine may be connected to "loop" the oil from the machine through the filters and back into the machine without draining it. In many cases, this process can be done with the machine running and allowing the oil to circulate through the entire system. The low-micron filters will remove the contaminants and the particle counter will monitor the level of contaminants within the system. When the contaminants reach an acceptable level, the oil and machine can be returned to service. A very strict oil sampling program must be in place to have OEM acceptance.

Fill the System. Remembering that the intrusion of dirt into hydraulic systems probably causes more downtime than any other single problem, ensure that everything used in the refill procedure is clean. A new filter should be used. Fill the reservoir to the prescribed level with new hydraulic oil. Then run the equipment through its entire working cycle several times to ensure that any air is purged from the circuit.

Figure 6-1 Kidney-loop machine and particle counter.

The equipment should run smoothly. Finally, recheck the oil level in the reservoir.

Check System Performance. Before and after repairs, and as part of preventative maintenance, hydraulic systems should be checked. Some OEMs have system-specific checklists and these should be used when available. Maintenance checks should be performed in the following fields.

Check Reservoir. This check should be undertaken regularly because it can be performed in seconds and can identify small problems before they become major problems. Before opening the system filler cap, clean it and the surrounding area with a clean rag.

When dipsticks are used, clean the dipstick with a lint-free cloth before checking the level. Look for:

- Oil **aeration**. Foaming oil or bubbles can indicate an air leak in the system, usually on the suction side of the circuit.
- Change in oil level. Reservoir oil level that requires constant topping up usually indicates an external leak in the system.
- Milky oil. This almost always indicates the presence of water in the system. Often system flushing will be required because hydraulic oil and water emulsify under pressure.

Many reservoirs incorporate the filter and filter lines within the reservoir. Proper inspection of these reservoirs and filter housing requires removal of the reservoir from the machine and disassembling it. Large clean-out and inspection covers are typically implemented with these types of reservoirs and must be removed to gain access to the filter connections and all internal components.

Check Oil Lines and Connections. Both low- and high-pressure sides should be checked for leaks and physical damage. Low-pressure side leaks can result in air being pulled into the system, while high-pressure side leaks will produce external leakage. Pinched lines can create restrictions that result in foaming, overheating, and reduction in hydraulic efficiency. A pinched inlet line can also cause cavitation. In equipment with high working hours (20,000 to 30,000), a common practice during a major overhaul is to replace all flexible oil lines due to increased risk of deterioration of the inner tube. The rubber liner can become brittle and fragments can break away and contaminate the entire system. It is almost impossible to find the specific defective hose, and when age and service hours are high, it is a good PM practice to replace all lines.

Check Circuit Components. A thorough visual inspection should be made of all of the hydraulic circuit components including the oil cooler, valves, cylinders, pumps, and motors. The circuit should be run through its working cycle to check for leaks and whether the flow is sufficient to run the system. Check the oil cooler fins for plugging and test system valve operation, remembering that wear can cause internal leakage. Motors should not be allowed to run hot. If this happens, check that the oil supply is adequate and that the oil cooler is functioning properly. Also check for leaks around the motor hose fittings, shaft, seals, and body mating surfaces.

Flow and Pressure Testing. Various types of **hydraulic analyzers** are available for testing hydraulic circuits. A hydraulic analyzer typically consists of a pressure gauge(s), flow meter, manual gate (choke) valve, and thermometer. **Figure 6-2** shows one type of hydraulic analyzer. **Figure 6-3** shows how the analyzer is plumbed into a circuit to perform a simple pump output test.

Typically, a hydraulic analyzer checks:

- **Flow.** Flow testing determines whether the pump is producing its specified output.
- **Pressure.** Pressure testing verifies the operation of the relief valve and, in closed-center systems, pump performance.
- **Cycle Times.** Cycle times are a measurement of independent implement speeds. Because cycle

Figure 6-2 Hydraulic system analyzer.

© Cengage Learning 2014

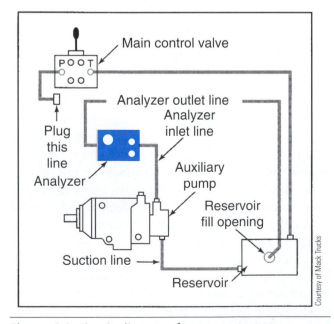

Courtesy of Mack Trucks

Figure 6-3 Circuit diagram for pump test.

times are calculated from specific flow and pressure rates, most OEMs will list a specific time chart for each implement of every machine they manufacture. Cycle times can allow the technician to perform a quick evaluation of the machine performance without installing pressure gauges and flow meters. They can also lead the technician to a specific fault within that system. An example of this is if all implements are operating at a lower speed than normal, the technician would perform a pump efficiency test because the pump is the common component that would affect the speed of the implements. In the case of multiple pumps it may be that only the implements to one pump are slower than normal, therefore isolating the problem to that specific circuit.

> **Note:** It is common practice to test engine performance prior to testing cycle times. Poor engine performance in many cases causes poor hydraulic performance.

- **Temperature.** Hydraulic oil temperature should be measured because system pressure and flow specifications are calculated with the system at operating temperature. An infrared temperature gun is a non-invasive method of quickly checking internal leakage of components as well as testing temperature differentials within a circuit.
- **Leakage.** Leakage testing identifies leakage in a specific circuit component.

When you are required to test a vehicle hydraulic circuit, observe the OEM test cycle. It is recommended that you record the results of each test step using a procedure such as that shown in **Table 6-1** for a pump performance test and **Table 6-2** for a hydraulic circuit performance test. In our example here, we have used a typical excavator system: most manufacturers use system-specific tables for recording performance test data in their service literature. The swing, boom, stick, and bucket circuits are described in a little more detail later in this chapter.

A typical test cycle consists of the following steps:

1. Install the hydraulic tester into the circuit to be tested, between the pump and the control valve. The pressure line should be connected to the inlet port of the tester.
2. Connect the hydraulic tester outlet (return) to the system reservoir.
3. After installation, check the system oil level.

TABLE 6-1: HYDRAULIC PUMP PERFORMANCE TESTS							
Record Pump Flow in Gallons Per Minute (gpm)							
Pressure	0 psi	500 psi	1,000 psi	2,000 psi	2,500 psi	3,000 psi	3,500 psi
Flow in gpm							

TABLE 6-2: HYDRAULIC CIRCUIT TESTS							
Record Pump Flow in Gallons Per Minute (gpm)							
Circuit Test	0 psi	500	1,000	2,000	2,500	3,000	3,500
Boom up							
Boom down							
Stick in							
Stick out							
Bucket in							
Bucket out							
Swing CW (right)							
Swing CCW (left)							

4. Start the engine and gradually close the hydraulic tester gate valve to load the circuit and bring the temperature up to normal operating value.

5. Open the gate valve and record system flow at zero pressure.

6. Move the system control valve into one of its power positions and gradually close the tester gate valve, recording flow in 250 psi increments from zero up to full system pressure.

7. Repeat the above test in each of the system control valve power positions.

....

Tech Tip: When performing steps 6 and 7, ensure the system oil temperature is the same. This may require that you allow the oil to circulate and cool between each test.

....

OEMs identify the test points along with specifications in their service literature. The test points used to diagnose the hydraulic circuits in an excavator are shown in **Figure 6-4** and **Figure 6-5**. In this test profile, a tachometer is connected to monitor engine speed. Note the test locations for both the pumps and control valve.

VIEW OF PUMPS ON ENGINE

1. Location of plugs for checking system pressure and flow from the rear pump. 2. Location of plugs for checking system pressure and flow from the swing pump. 3. Location of plugs for checking system pressure and flow from the front pump. 4. Location of plug for checking the signal pressure to the front and rear pumps.

Figure 6-4 Test points for system checks.

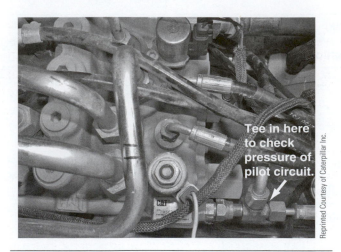

Figure 6-5 Control valve pressure test location.

Diagnosing Test Results. You should make sure that you consult the OEM literature when diagnosing hydraulic circuit test results, but here are some typical guidelines used to test a simple hydraulic cylinder type circuit:

- Pump flow at maximum pressure should be at least 75% of the pump flow measured at zero pressure; OEM specifications can require higher percentage values. A pump that fails to meet 75% of this value is probably worn.
- When flow at each pressure value throughout testing is within 75% of pump flow at zero pressure, the hydraulic circuit components are probably functioning properly.
- When pressure drops off before full load, a leaking circuit is indicated. To determine whether the leakage is occurring in the control valve or in the cylinder, disconnect the cylinder return hose and move the control valve to a power position. If oil leaks at the cylinder return, it is probably leaking internally and is defective. If no oil exits the cylinder return line, the control valve may be at fault.
- A relief valve should open within a window of its specified pressure, so check the OEM specifications. A defective relief valve can be identified when the valve opens before system maximum pressure is achieved. If circuit flow suddenly drops off before system pressure is achieved, a defective relief valve is indicated.

LEAKAGE

Theoretically, a hydraulic pump should be able to turn a hydraulic motor at the same speed, providing both are of equal displacement. To achieve this would require 100% volumetric efficiency. However, this is not achievable due to circuit leakage factors. Circuit leakage in hydraulic circuits can be classified as:

- Internal leakage
- External leakage

Intended Internal Leakage

Any leakage that is designed into a hydraulic circuit can be classified as internal leakage. Intended internal leakage is used to lubricate plungers and pistons, valve spools, pumps, motor components, and pretty much anything that moves in the circuit. Intended internal leakage can also help "guide" a close-fitting component, such as a plunger, in the bore it reciprocates within. Leak paths are built into certain hydraulic components, such as motors, to provide balanced operation. These leak paths are referred to as **case drain** and typically are routed back to the reservoir via the hydraulic filter.

Piston pumps have a specific internal leakage rate that is directed through the case drain circuit. Many OEMs have a case drain leakage test to determine if internal leakage is greater than the acceptable level. This is measured by using a flow meter on the pump case drain outlet line and following the manufacturer's test procedure of load testing the pump and measuring the outcome from the flow meter. If the flow is greater than the acceptable level, the pump may be deemed defective. On many hydraulic systems a pressure tap is connected to the case drain circuit, and instead of a flow measurement, a pressure test of the case drain is measured. Always refer to proper OEM test procedures and guidelines prior to condemning components.

Unintended Internal Leakage

Unintended internal leakage can occur when normal wear opens up flow paths in intended internal leakage circuits, or it can result from internal seal failure. Excessive internal leakage slows down actuators and, in the case of failed seals, can create a leakage path that diverts all of the pump flow around the actuator.

Like piston-type pumps, piston-type hydraulic motors have a specified internal leakage rate. The motor leakage rate can be tested in the same manner as pumps, but as always refer to proper OEM test procedures and guidelines prior to condemning components.

Pressure and Leakage. Whether internal leakage is intended or not, it increases with pressure. In addition, heat is generated at the leakage flow path. Therefore, operating a circuit at higher than specified pressures produces both increased leakage and adds heat to the circuit fluid. While excess internal leakage can result in loss of control or function in a hydraulic circuit, the good news is that the oil is not lost. One way or another, it will be routed back to the reservoir, either by means of the return circuit or by drain passages.

Viscosity and Leakage. The viscosity of hydraulic oil influences internal leakage flow rate. A low-viscosity oil flows more readily and therefore increases internal leakage. The viscosity and viscosity index of the hydraulic oil used are key considerations, both in providing the appropriate intended internal leakage and in preventing excessive internal leakage.

Internal Seal Failure. When an internal seal fails, the result can be a complete failure of the hydraulic circuit. Oil continues to flow through the circuit and heat is generated at the location of the seal failure. A sudden rupture of an internal seal can be dangerous in many types of hydraulic circuit.

External Leakage

All hydraulic circuit external leakage is unintended. External leakage presents less of a challenge in troubleshooting because of the mess it makes. Apparently minor external hydraulic leaks should always be promptly repaired because what begins as seal leakage can progress to a rupture. External leakage can usually be sourced to one of the following:

- Improper assembly
- Poor maintenance
- Shock load on an actuator
- Lines subjected to excessive vibration

CAUTION *Pin-holes in hydraulic lines result in high-pressure oil escaping from the system in minute droplets. These oil droplets can penetrate the skin, often with little external evidence, and the result can be blood poisoning. Use a piece of cardboard to search for leaks in hydraulic lines; never use a bare hand.*

LS/PC AND TORQUE LIMITING SYSTEM TEST

In a load sensing/pressure compensated (LS/PC), torque limiting system, there are four basic tests to check compensator performance:

- Low-pressure standby test
- High-pressure stall test
- Margin pressure test
- Torque limiter test

These tests are critical because they affect the variable pump performance between zero and maximum flow. The following tests are performed with oil at operating temperature, engine at high idle, and all implements in their rest position on the ground. Before performing each test, check for leaks on all connections made to the system.

Low-Pressure Standby Test. A typical **standby pressure** can be from 200 to 500 psi. Therefore, a pressure gauge from 0 to 1,000 psi would be most accurate to install on the pump outlet line. It is recommended to use an auto-ranging gauge that is made up of multiple gauges that will automatically switch to the appropriate gauge as the pressures rise beyond the limits of each gauge.

Start and run the engine at high idle. Be sure not to touch any of the controls including the steering circuit. By doing so, pressure may spike and damage the pressure gauge. Allow the engine speed to stabilize and record your reading. Pressure should be within the limits of the OEM specification. If the pressure is not within specification, do not adjust until all other pressure tests are performed.

High-Pressure Stall Test. This test checks the setting of the pressure compensator that will de-stroke the pump when pressure in the system reaches the specification setting. High-pressure compensator settings may range from 2,500 to 4,000 psi. A 0 to 6,000 psi gauge is recommended for this test installed at the pump outlet line. Always refer to OEM specifications and recommendations.

Start and run the engine at high idle. Slowly move each control valve lever independently to fully retract each cylinder. Do not hold the control valve open with the cylinder fully retracted for more than 10 seconds. This should be substantial time to record your readings for each implement. Wait 30 seconds between each implement to allow oil to cool. If pressure settings are too high or too low, the pressure compensator spring must be adjusted. A counterclockwise turn will lower the spring setting thus lowering the pressure. A clockwise turn will increase spring setting, raising the compensator pressure setting.

Margin Pressure Test. **Margin pressure** is the differential pressure between pump outlet working pressure and signal pressure to the flow compensator. Therefore, two pressure gauges are required for this test. Two 0 to 4,000 psi gauges would be sufficient, one connected at the pump outlet line and one connected to the flow compensator signal line.

Start and run the engine at high idle. Move one implement at a steady rate of speed and record the pressures. The pressure differential should always be as close to standby pressure as possible. This is because the pressure required to move the actuator is sent to the flow compensator via the check-valve network as signal pressure. This pressure plus the spring pressure pushing on the flow compensator spool is additive. Pump outlet pressure routed to the opposite end of the spool must equal this pressure to maintain a balanced position for the swashplate.

Torque Limiter Test. As described in Chapter 4, the torque limiter is responsible for not allowing the hydraulic circuit to work above the horsepower limits of the engine. It is not necessary to be working in high-flow and high-pressure conditions simultaneously. Therefore, the torque limiter is set to inversely control pump displacement to pump outlet pressure.

Prior to this test, an engine performance evaluation must be done. If engine performance is within specification and hydraulic performance is low, then proceed with this test.

Pick one implement with a port relief valve and set to 500 psi below the pressure compensator setting. Example: If the pressure compensator is set for 3,000 psi, adjust port relief valve to 2,500 psi. By doing this, the pressure compensator will not be able to control pump displacement when this circuit is placed under load. With an accurate rpm gauge installed, set engine to 1,000 rpm. Operate the implement against its stop or farthest extension point. Note engine rpm drop. If engine rpm drops more than 150 rpm, torque limiter adjustment is necessary. Adjust torque limiter spring outward to relieve spring tension and retest. The proper setting should allow engine to lug between 100 and 150 rpm. If engine lug is

less than 100 rpm, torque limiter spring tension should be increased. Reset the port relief valve to its original setting.

Note: This is one example and should not be used as standard practice for all machines. Always refer to OEM testing and adjusting procedures for machine-specific testing.

TROUBLESHOOTING CHARTS

Most hydraulic circuit failures follow a pattern, producing either a sudden or a gradual loss of system performance. Most manufacturers produce system troubleshooting trees or charts to help technicians diagnose malfunctions. The use of electronic management of hydraulic circuits has facilitated the sourcing of problems, sometimes by guided interactive tree navigation. The following sequence of tables (**Table 6-3** through **Table 6-14**) identifies some typical hydraulic circuit malfunctions and remedies.

TROUBLESHOOTING PROCEDURE

Any troubleshooting procedure should be specific to the equipment being diagnosed and to a range of other variables such as:

- Hydromechanical controls
- Electromechanical controls
- Electronic (computer) controls
- Equipment OEM

For purposes of study, we will base our troubleshooting profile on an excavator equipped with upper

TABLE 6-3: SYSTEM INOPERATIVE

Cause	Remedy
No oil in system.	Fill system and check for leaks.
Improper oil in system.	Reference OEM specifications. Flush and refill system.
Restricted oil filter.	Change oil and replace filter.
Worn-out pump or circuit components.	Repair or replace defective components. Change oil and flush system.
Leakage in suction line (air in system).	Repair or replace as necessary.
Excessive load.	Check OEM specifications for rated maximum load.
Defective pump drive.	Repair or replace belts and couplings. Check alignment, backlash, and belt tension.
Internal leakage.	Check circuit component operation, including relief valve. Repair or replace as necessary.
External leakage.	Visually inspect circuit lines and components. Repair or replace as necessary.

TABLE 6-4: ERRATIC OPERATION	
Cause	**Remedy**
Aerated system.	Check suction side of circuit for leaks.
Excessively cold oil in system.	Allow sufficient warm-up of oil and consider using a winter-rated oil (lower viscosity).
Restricted or damaged circuit components.	Clean and replace. Flush circuit and change oil.
Restricted lines and or filters.	Clean or replace lines and filters as necessary.

© Cengage Learning 2014

TABLE 6-5: SLUGGISH OPERATION	
Cause	**Remedy**
Oil gelling.	Allow oil some warm-up time. Consider using lower viscosity oil.
Low pump drive speed.	Increase prime mover (drive engine) rpm.
Aerated system.	Check suction side for leaks and repair.
Worn system components.	Check for internal and external leakage. Check for cause of wear. Repair or replace components as necessary.
Restricted filters or lines.	Clean and/or replace filters and lines.
Low oil level.	Check reservoir oil level.
Oil leaks.	Torque fittings and visually check seals and lines.
Improper adjustments.	Check relief valve settings and orifice flow areas per OEM specifications.

© Cengage Learning 2014

TABLE 6-6: SYSTEM OPERATES TOO QUICKLY	
Cause	**Remedy**
Wrong size or improperly adjusted restrictor.	Replace or adjust as necessary.
Prime mover running too fast.	Adjust engine speed to specification.

© Cengage Learning 2014

TABLE 6-7: FOAMING HYDRAULIC OIL	
Cause	**Remedy**
Improper oil, low oil level, contaminated oil.	Check oil specification, correct oil level, flush and refill oil in circuit.
Aerated oil.	Check suction side for air induction.

© Cengage Learning 2014

structure swing, boom, bucket, and stick controls, and we will assume that the controls are hydromechanical. It should be noted that circuit functionality is not changed when electro-hydraulic controls are used. The drive and steer systems of the excavator are part of the **track circuit** and are studied later in this book. When approaching any troubleshooting procedure, remember that the specified oil flow and pressure are required. Oil flow is a function of prime mover (engine) rpm, and pressure readings indicate the resistance to oil flow within the circuit. In general, you should begin a troubleshooting procedure with a visual check and proceed to performance tests before progressing to instrument testing using hydraulic analyzers. The following section briefly describes the circuits integrated into a typical excavator.

Swing Circuit

The upper structure is mounted on the excavator undercarriage and incorporates the cab, engine housing, and boom assemblies. This upper structure is hydraulically rotated by the **swing circuit** and is identified in **Figure 6-6**.

The swing motor is capable of rotating the upper structure of the excavator through a 360-degree turn. The oil supply for the swing circuit is usually separate from that used to supply the implements and track drive circuits. In troubleshooting profiles, it is normal practice to refer to swing direction as CW (right) or CCW (left) rotation. **Figure 6-7** shows a cross-section of the excavator swing drive control.

TABLE 6-8: OVERHEATED OIL

Cause	Remedy
Oil cycling through relief valve for prolonged period.	Return control valve to neutral position when not in use.
Contaminated, incorrect, or low oil level.	Replace oil and flush circuit, ensuring reservoir oil level is correct.
Prime mover running too fast.	Reduce engine speed.
Restricted lines or filters.	Clean and/or replace lines and filters.
Insufficient heat exchanger ratings or restricted heat exchanger grills.	Clean contamination from heat exchanger air flow passages. Check heat exchanger specifications.
Mechanical malfunction/binding/lockup of components.	Locate problem and repair or replace.
Excessive internal leakage.	Locate defective component and repair or replace.

© Cengage Learning 2014

TABLE 6-9: LEAKAGE AT PUMP OR MOTOR

Cause	Remedy
Failed seal(s).	Replace, check shaft alignment.
Loose or failed components.	Tighten and/or replace components.

© Cengage Learning 2014

TABLE 6-10: NOISY PUMP

Cause	Remedy
Incorrect or contaminated oil, low oil level.	Check OEM oil spec, clean and flush oil, fill to correct level.
High inlet restriction in suction line.	Clean and/or replace suction line.
Worn or damaged pump.	Repair or replace pump.

© Cengage Learning 2014

TABLE 6-11: HYDRAULIC LOAD DROPS, CONTROL VALVE IN NEUTRAL

Cause	Remedy
Internal or external cylinder leaks.	Repair or replace worn components.
Control valve not centering on release.	Check linkage and for spool binding.
Load check valve defective.	Repair as necessary.

© Cengage Learning 2014

TABLE 6-12: LEAKING CONTROL VALVE

Cause	Remedy
Loose tie-bolts. Worn or damaged seals.	Torque to specification. Replace seals.

© Cengage Learning 2014

TABLE 6-13: BINDING CONTROL VALVE

Cause	Remedy
Tie-bolts too tight.	Back off and torque to specification.
Valve linkage misaligned.	Align linkage.
Internally damaged control valve.	Replace control valve.

© Cengage Learning 2014

TABLE 6-14: HYDRAULIC CYLINDER LEAKAGE

Cause	Remedy
Damaged or worn seals.	Replace seals.
Damaged rod.	Replace cylinder assembly.

© Cengage Learning 2014

Boom Circuit

The **boom circuit** is actuated by a pair of synchronized hydraulic cylinders. The cylinders are double-acting and are used to raise or draw back the boom assembly. An excavator boom assembly is shown in **Figure 6-8**.

Figure 6-6 Excavator upper structure controlled by swing circuit.

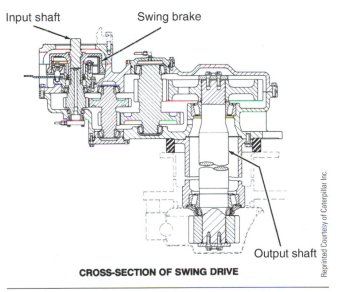

CROSS-SECTION OF SWING DRIVE

Figure 6-7 Cross-section of swing drive: 1. Input shaft. 2. Swing brake. 3. Output shaft.

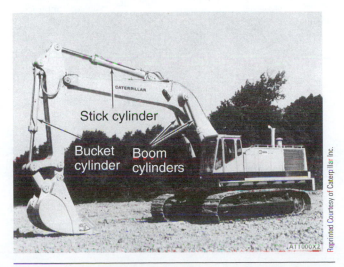

Figure 6-8 Excavator boom, stick, and bucket controls.

Boom operation is mastered by a control valve assembly. **Figure 6-9** shows a control valve body that is integrated into a valve assembly consisting of boom control valve, bucket control valve, and right track motor control valves.

Stick Circuit

The **stick circuit** is actuated off the boom and uses a single, double-acting cylinder to vary the bucket pivot angle. You can see this by referencing **Figure 6-8**. The cylinders used as actuators in the bucket, stick, and boom circuits are similar in both design and operation. A sectional view of a typical cylinder is shown in **Figure 6-10**.

Bucket Circuit

The excavator bucket is mounted on a pivot on the stick implement and is moved by a double-acting cylinder actuated by the **bucket circuit** as shown in **Figure 6-8**. The bucket is scooped inward as the bucket cylinder is extended and moved outward to dump as the bucket cylinder is retracted.

Excavator Circuit Features

When troubleshooting a suspected hydraulic circuit in the typical excavator machine we are using as our example, you should keep in mind the following points:

- The swing circuit is usually a closed circuit with a dedicated pump, control valves, relief valve, and motor. The relief valve is located in the pressure supply between the main control valve and the swing motor. The swing circuit will not usually cause the engine rpm to drift below rated speed when it is actuated by itself.
- The system piston pumps are used to charge the track motors. A typical arrangement uses the front axial piston pump to supply the right track motor and the rear axial piston pump to supply the left track motor. The common components to both track circuits are the speed and direction control valve, the main relief valve, and the swivel. The speed and direction control valve manages the main control valves (one for each track) by using **pilot pressure**. Pilot pressure moves the stem in the main control valve that then routes oil pressure from one of the piston pumps to a track motor. Hydraulic oil to and from the track motors must pass through the swivel, the only component in each track circuit that is common to both circuits other than the brake valve. A track control valve is shown in **Figure 6-11**.

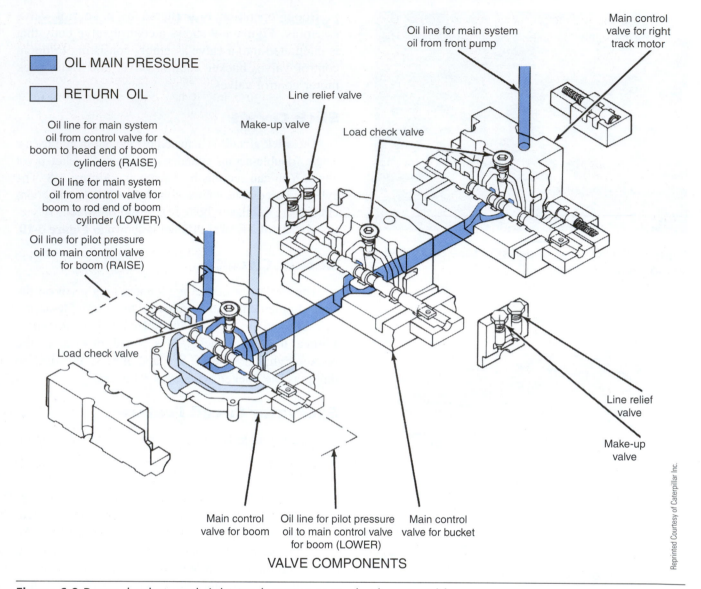

OIL MAIN PRESSURE

RETURN OIL

Line relief valve

Make-up valve

Load check valve

Oil line for main system oil from front pump

Main control valve for right track motor

Oil line for main system oil from control valve for boom to head end of boom cylinders (RAISE)

Oil line for main system oil from control valve for boom to rod end of boom cylinder (LOWER)

Oil line for pilot pressure oil to main control valve for boom (RAISE)

Load check valve

Line relief valve

Make-up valve

Main control valve for boom

Oil line for pilot pressure oil to main control valve for boom (LOWER)

Main control valve for bucket

VALVE COMPONENTS

Reprinted Courtesy of Caterpillar Inc.

Figure 6-9 Boom, bucket, and right track motor control valve assembly.

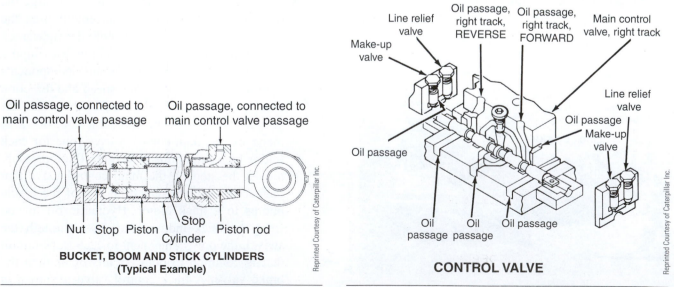

Oil passage, connected to main control valve passage

Oil passage, connected to main control valve passage

Nut   Stop   Piston   Stop Cylinder   Piston rod

BUCKET, BOOM AND STICK CYLINDERS
(Typical Example)

Reprinted Courtesy of Caterpillar Inc.

Figure 6-10 Typical bucket, boom, and stick cylinders.

Line relief valve

Oil passage, right track, REVERSE

Oil passage, right track, FORWARD

Main control valve, right track

Make-up valve

Line relief valve

Oil passage

Oil passage
Make-up valve

Oil passage

Oil passage   Oil passage   Oil passage

CONTROL VALVE

Reprinted Courtesy of Caterpillar Inc.

Figure 6-11 Main (right) track control valve.

Because the brakes are activated by spring force, hydraulic pressure from the brake valve must relieve the spring force before the excavator can be moved.

- A drop in engine rpm or an increase in implement load can cause the piston pumps to decrease output. This is achieved by a signal to both pumps that alters pump swash angle and puts them in stroke-back mode. The signal is sourced from the underspeed valve or the summing valve, which senses a sudden increase in load from the implement. **Figure 6-12** shows a sectional view of a pump selector valve.
- To ensure that the specified signal pressure is sent to the axial piston pumps, the pilot pressure must be correct. Pilot pressure can be adjusted and set by correctly shimming the pilot relief valve. Check this value to specifications and adjust if required.
- When there is a problem that is common to all the hydraulic circuits on the machine, check the engine stall speeds first, using the manufacturer service literature.

Typical Excavator Circuit Problems

Now we will take a look at some typical hydraulic circuit problems, using our typical excavator model. The information in this section references Caterpillar service literature, but it has been adapted to share commonality with other OEM hydromechanically controlled systems.

Visual Checks

Begin a visual check of the equipment with the engine turned off and any hydraulic implements neutralized: lower booms and buckets to the ground. Then:

1. Check the reservoir oil level. Slowly loosen the reservoir cap to relieve any residual pressure.
2. Remove and inspect strainers and filter elements for obvious restriction.
3. Closely inspect all lines and hoses checking for kinks, crushing, and leakage.
4. Inspect external control linkages for bends, breakage, and excessive play.

Performance Tests

When performance testing, you are required to start the engine and work the implements. During this phase of troubleshooting, you should be looking for evidence of leakage and obvious malfunctions, such as a valve or actuator that is not working properly. The speed of rod movement or motor output torque can be used to determine the performance of actuators in the system. Use **Table 6-1** and **Table 6-2** (located earlier in this chapter) to guide you through the tests.

Tech Tip: A key to accurate troubleshooting is developing the ability to navigate through OEM hydraulic schematics. Familiarize yourself with the schematic(s) before attempting to source a problem, and you will find you spend less time locating problems.

Using the excavator as our example, you should raise and lower the boom, open and close the bucket, and swing the boom both ways. Do this several times using the following general guidelines:

1. Observe the cylinders that actuate the boom, stick, and bucket. The movement of each should be smooth as they are extended and retracted.
2. While cycling the cylinders, listen for pump noise. Note any irregularities.

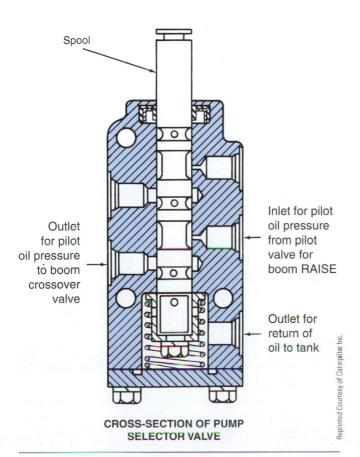

Spool

Outlet for pilot oil pressure to boom crossover valve

Inlet for pilot oil pressure from pilot valve for boom RAISE

Outlet for return of oil to tank

Reprinted Courtesy of Caterpillar Inc.

CROSS-SECTION OF PUMP SELECTOR VALVE

Figure 6-12 Cross-section of a pump selector valve.

3. Listen for the sound of relief valves tripping. Check the relief valve opening pressures to specification. Relief valves that trip below specification cause system malfunctions: relief valve opening pressures may differ for each implement and the swing circuits.

Typical Problems and Causes

Once again, we will use our typical excavator to look at some possible hydraulic problems and their sources.

Problem: Cylinder extend and retract cycle times exceed those shown in the specifications.

Possible Causes:

1. High idle engine rpm set too low. This results in a reduction of pump output flow. *Note*: The rated engine rpm (full load) is the speed at which maximum horsepower transfer is available— in hydromechanically managed engines, this is often lower than engine high idle speed by 5% to 20%. Check service literature on what is required to adjust the rated speed of a specific engine; this may be programmable (electronically controlled engines) or require governor trim adjustment.
2. Relief valve trip pressure in the pilot system incorrectly (low) set.
3. Damaged piston pumps.
4. Swashplates in the implement pumps seized or set at a minimum angle.

Problem: Cycle time required to extend or retract a single cylinder exceeds that shown in the specifications.

Possible Causes: Note that this type of problem is usually sourced in the control circuit of the affected cylinder.

1. Check the linkage between the control lever and the pilot valve, making sure that it is properly connected and not bent or damaged in any way.
2. Turned spool in the pilot valve. A spool can sometimes turn in its bore, blocking pilot fluid delivery to the main control valve.
3. Spool or housing damage in the main control valve. This type of problem can be caused by the loss of too much oil.
4. Cylinder damage or leaking piston seals.
5. Front piston pump malfunction, if the problem is in the boom or bucket circuits.
6. Rear piston pump, if the problem is in the stick circuit.

Problem: Drop in engine rpm (often close to stall) when the load on the hydraulic system is increased.

This condition can usually happen only when the piston pumps do not stroke back and there is no reduction in their output.

Possible Causes:

1. Low relief valve pressure setting in the pilot system.
2. Severe loss of engine power output.
3. Damaged actuator or servo valve in one or both piston pumps.

Problem: With the engine running at rated rpm, there is insufficient force (pressure) at the implements.

Possible Causes:

1. Low relief valve pressure in the pilot system. This results in modulation at the main control valves, resulting in insufficient pilot pressure to shift the spools in the main control valves to full travel.
2. Main pressure relief valve set too low.
3. Internal pump damage limiting pressure. Check the flow from the case drain on the pump.

Problem: Aerated hydraulic oil.

Possible Causes:

1. Leak in the oil line between the reservoir and the pump.
2. Failure to bleed the hydraulic system after assembly, inspection, or testing (see **Photo Sequence 1**).
3. Relief valve malfunction causing it to repeatedly open and close.
4. Leakage at the cylinder seals.

Problem: Noisy pump operation.

Note that noise coming from the area in which the pumps are located may not necessarily be caused by a pump. Check the pump drives. Noise from a pump drive can reverberate through the connecting shaft and housings, making it seem as though the noise is in the pump.

Possible Causes:

1. Oil aeration. This could be the cause if all the pumps are noisy.
2. Pump damage. If the noise is sourced from only one pump, it is probably caused by wear or damage to that pump.

Problem: High hydraulic oil temperature.

Bleeding a Hydraulic Pump

After any hydraulic pump replacement, it is important that the pump be bled of any air that may be trapped in the inlet hose between the hydraulic reservoir and the pump. This procedure outlines a general set of steps used to bleed off the trapped air that may cause any aeration or cavitation within the pump. It is by no means intended to replace specific manufacturer procedures or guidelines. It must also be noted that the equipment using hydraulic pumps is infinitely variable; this procedure may not be applicable to all situations.

© Cengage Learning 2014

P1-1 This procedure should be implemented after a hydraulic pump, inlet hose, or hydraulic reservoir is drained or replaced or any time air is suspected to have entered the hydraulic system. The example we are using is on a Caterpillar 248B skid steer loader.

© Cengage Learning 2014

P1-2 Before any work is started on a machine, proper safety procedures and lockout systems must be implemented. First, raise the boom high enough to engage the safety arm to lock the boom in the raised position. Remove the holding pin and slowly lower the safety arm in place.
Note: Do not let the safety arm fall into place as it may damage the cylinder rod.

© Cengage Learning 2014

P1-3 Because the pump is located below the operator's cab, it must be raised to access the pumps and hydraulic components. To raise the cab, first remove the cab bolts located at the front corners of the cab.

© Cengage Learning 2014

P1-4 This machine is equipped with gas-filled cylinders and will raise with very little effort. For machines not equipped with gas-filled cylinders, an overhead crane is necessary to raise the cab.

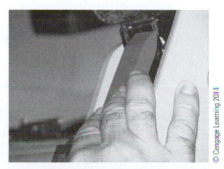

© Cengage Learning 2014

P1-5 Once the cab is in the raised position, make sure the safety latch is in place to prevent a sudden drop of the cab.

© Cengage Learning 2014

P1-6 After any work is completed and it is necessary to bleed the pump, locate the inlet line to the pump and remove the hose clamp.

(Continued)

P1-7 Insert a dull-tipped screw-driver between the flex-hose and inlet flange fitting of the pump.

P1-8 & P1-9 Fill the hydraulic reservoir to the specified fluid level as indicated by the gauge. Be sure to use the recommended viscosity grade as indicated by the OEM.

With the reservoir at the full level, oil should start to flow through the pump inlet line, displacing air through the opening created by the inserted tool. When the oil flows at a steady rate and there is no evidence of air in the line, remove the tool and reinstall the hose clamp. On certain machines it may be necessary to slightly pressurize the reservoir to allow oil flow to the hydraulic pump. A hydraulic pump may be mounted higher than the reservoir.

The machine is now safe to start at low idle to allow oil to circulate through the system. At this point the technician should be looking for any possible oil leaks and any other abnormal sounds and functions.

It is important to double-check the hydraulic fluid level after the cab and boom have been lowered to the ground and are in proper checking position.

High oil temperature can be caused by a number of factors. For instance, removing baffles in the control compartment can allow heated air to act on the pumps and inhibit heat exchanger operation. Check the air-flow paths on the equipment.

Possible Causes:

1. Oil aeration.
2. Restriction of air flow through the heat exchanger (oil cooler).
3. High ambient air temperatures.
4. Internal restriction in the heat exchanger—restricting the flow of oil through the cooler because of sludging or flow-control valve problem.

Problem: Low hydraulic oil temperature.

Low oil temperature can cause problems. The temperature of the oil during operation must enable it to flow freely through the lines and components.

Possible Causes:

1. Cooler flow-control valve malfunction permitting excessive oil flow to the cooler.
2. Cooler, reservoir, and other hydraulic components insufficiently protected from low ambient temperatures. Note OEM recommendations for operating equipment in northern and subarctic conditions.

Problem: Improper modulation of all the implements.

When troubleshooting this condition, remember that the only part of the hydraulic system that is common to all implements is the pilot circuit. A defect in the pilot system can cause surging in the implement circuits and prevent the correct operation of the piston pumps.

Possible Causes:

1. Leakage in the system that prevents the pressure reaching specified values.
2. Low oil temperature.
3. Low opening pressure of the pilot system relief valve.
4. Aerated hydraulic oil.
5. Damaged or worn pilot pump.

Problem: The modulation of one implement is not normal.

When this occurs in just one circuit, check for a root cause that is specific to the malfunctioning circuit.

Possible Causes:

1. Bent or disconnected linkage to a pilot valve spool.
2. Turned or binding spool in a pilot valve.
3. Air in circuit.
4. Binding main control valve spool.

Problem: Machine will not move when either the FORWARD or the REVERSE pedal is depressed.

Focus on the components in the track control and drive circuits, including the front and rear piston pumps (one pump for each track motor), main relief valve, line relief valves, swivel, motors (one for each track), makeup valves, main control valves, pilot control valves, and overspeed valves. Note that the only components common to both track motors are the main relief valve, the swivel, and the track brake valve (a signal pressure from the track brake valve is required to release the brakes before the machine can be moved).

Possible Causes:

1. Pressure in the pilot system not reaching that specified to release the track brakes.
2. Machine safety lever is located in the SAFE position.
3. Stop has been turned in on the track brake valve, preventing signal pressure from releasing the track brakes.
4. Dirt in the track brake valve.
5. Leaking seal in the brake line section of the swivel.
6. External object preventing the tracks from turning.

Problem: Machine moves but does not steer.

Possible Causes: This problem must be caused by either the steering valve or its linkage. First ensure that the linkage is adjusted properly and not bent, broken, or disconnected. If there is no problem with the linkage, check inside the steering valve to be sure the lever is not disconnected.

Problem: Excessive hydraulic cylinder **drift** (movement off preset position).

Cylinder drift is usually caused by circuit leakage. In stick and bucket circuits you should check the main control valves, line relief valves, makeup valves, and cylinders. In the boom circuit, check these components plus the boom vent valve, boom check and relief valve, and the rate of boom lower valve.

Possible Causes:

1. Leakage in and around the seals on the cylinder pistons.
2. Leakage at the line relief or makeup valves in the main control valve module for the circuit with drift.
3. Spool in the main control valve not correctly centered, usually caused by a broken spring or binding valve spool.
4. Small amount of pilot (signal) pressure at the end of the spool in the main control valve causing an off-center spool and back leakage.
5. In the case of boom cylinder drift, the cause can be any of those above or a defective boom vent valve, boom check valve, or relief valve.

Problem: Boom will not raise or lower, but all other implements operate correctly.

Possible Causes:

1. Stuck spool in the boom vent valve, causing the boom check valve to block and preventing head-end oil to exit the cylinder for boom.
2. Seized spool in the main control valve caused by a defect in the spool or body of the valve.
3. Turned spool in the pilot valve bore.
4. Disconnected, bent, or out-of-adjustment control linkage between the lever and pilot valve.

Problem: High oil level in the final drive housing.

Possible Causes:

1. Excessive oil added to the housing (service problem).
2. Oil leakage from the track motor.
3. Oil leakage from the track brake.

Problem: High oil level in the pump drive housing.

Possible Causes:

1. Excessive oil added to the housing (service problem).
2. Leakage past pump seals.

Problem: Surge results when the boom or stick control valve for the boom or stick is activated.

Possible Causes:

1. Summing valve not working properly.
2. Horsepower output of the engine has been reprogrammed or otherwise altered.
3. Line from the summing valve to the pumps is not open.

Problem: Noisy line relief valves.

Possible Causes:

1. Low relief valve set pressure.
2. High main relief valve set pressure.

Alternate Systems

Most hydraulic circuits used on off-highway equipment share the same operating principles no matter where they are manufactured or by which OEM. Once you understand the operation of one piece of equipment, most others can be likened to it. **Figure 6-13** is a complete hydraulic circuit schematic of a simple backhoe loader: you should be able to recognize the various implement circuits identified on the schematic.

ONLINE TASKS

1. Locate each of the following URLs and navigate to pages that provide troubleshooting information:
   - *www.cat.com*
   - *www.deere.com*
   - *www.komatsuamerica.com*
   - *www.jcb.com*
   - *www.liebherr.com*
   - *http://www.manitowoccranes.com*
   - *www.casece.com*

2. To get relevant service and troubleshooting information, you usually have to have access to an OEM data hub, which is limited access and password protected. If your college or workplace has data hub access, select a piece of equipment, identify a hypothetical problem, and download the troubleshooting data. For instance, the Caterpillar subscription site is *https://www.imap.cat.com*. You will be prompted for passwords at logon, and will not proceed beyond this point unless you have access.

Shop Tasks

1. Locate the hydraulic schematics that outline the hydraulic circuits on a piece of equipment. Make a copy and identify the location of the pump(s), control valve(s), and actuators. Note all the circuit test locations on your copy.
2. Use OEM service literature to connect a hydraulic analyzer into the hydraulic circuits on a piece of equipment. Complete a troubleshooting test profile such as those used in **Table 6-1** and **Table 6-2** in this chapter.

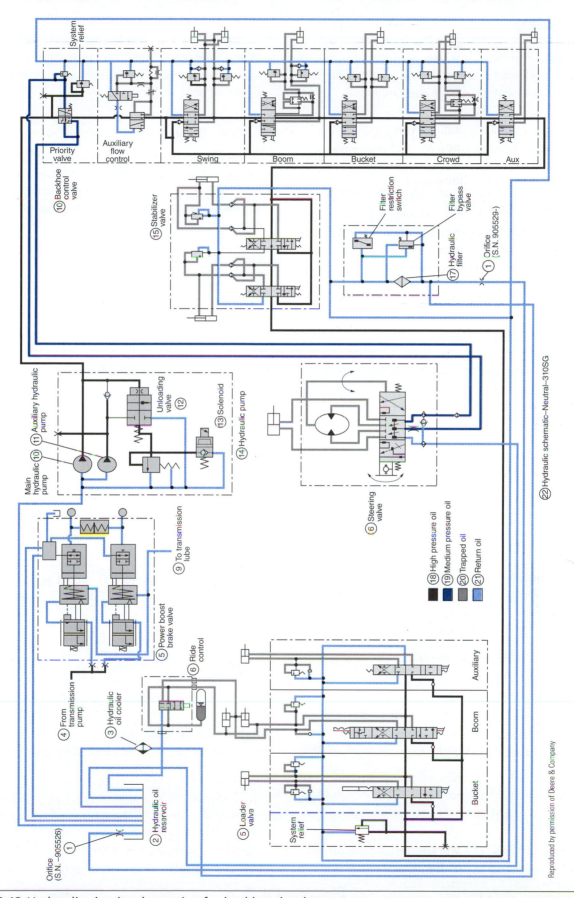

Figure 6-13 Hydraulic circuit schematic of a backhoe loader.

Summary

- Maintenance procedures on mobile equipment hydraulic systems begin with ensuring the system is clean both inside and outside the circuit.

- Routine replacement of hydraulic fluid, sometimes accompanied by system flushing, is recommended to minimize system malfunctions and downtime.

- Hydraulic circuit testers are used to analyze hydraulic circuit performance.

- A typical general-purpose hydraulic circuit tester consists of pressure gauge(s), flow meter, flow-control valve, and thermometer.

- When hydraulic analyzer test data falls outside specification, use OEM troubleshooting tables and charts to source the problem(s).

- A key to troubleshooting is developing the ability to navigate through OEM hydraulic schematics.

- A typical excavator uses hydraulics in the track drive, swing, and implement circuits: the circuits may be complementary so, to ensure that you are following the correct procedure, the OEM service literature should be referenced.

- The implement circuit on a typical excavator consists of boom, stick, and bucket circuits: each circuit uses double-acting hydraulic cylinders.

Review Questions

1. What would cause a hydraulic pump to produce 0 psi and only 60% of the OEM-specified flow?

   A. defective pressure-relief valve

   B. worn pump

   C. defective reservoir vent cap

   D. worn unloading valve

2. What would cause the boom to not raise or lower on a piece of equipment, but all other implements appear to operate correctly?

   A. aerated hydraulic oil

   B. out-of-adjustment control

   C. low pump relief valve setting

   D. oil leakage from track brake linkage between the lever and pilot valve

3. A track-driven excavator will not move when either the FORWARD or REVERSE pedal is depressed, but all the implements functioning normally. Which of the following would be the LEAST likely cause of the problem?

   A. insufficient pilot pressure to release the track brakes

   B. machine safety lever located in the SAFE position

   C. stop fully turned in on the track brake valve

   D. aerated hydraulic oil

4. When operating a backhoe, excessive hydraulic cylinder drift occurs. Technician A says that cylinder drift can be caused by circuit leakage. Technician B says that the main control valves should be checked. Who is correct?

   A. Technician A only

   B. Technician B only

   C. Both A and B

   D. Neither A nor B

5. Technician A says that aerated hydraulic oil can cause a circuit to operate erratically. Technician B says that excessively cold hydraulic oil can cause a system to operate erratically. Who is correct?

   A. Technician A only

   B. Technician B only

   C. Both A and B

   D. Neither A nor B

6. Which two pressure values denote *margin pressure*?

    A. pump outlet and pump inlet   C. pump outlet and system pressure

    B. pump outlet and standby   D. pump outlet and case drain
       pressure

7. A hydraulic bucket load drops when the control valve is moved to neutral. Technician A says that this could be caused by a control valve that binds and does not center. Technician B says this could be caused by a seized mechanical linkage. Who is correct?

    A. Technician A only   C. Both A and B

    B. Technician B only   D. Neither A nor B

8. What would cause pressure to drop off in a hydraulic implement circuit significantly before full load is achieved?

    A. defective relief valve   C. seized control valve spool

    B. internal leak in the circuit   D. cold hydraulic fluid

9. The oil level in a final drive housing of a track-driven excavator is excessively high. Technician A says that oil leakage from the track motor could be the cause. Technician B says that oil leakage from the track brake could be the problem. Who is correct?

    A. Technician A only   C. Both A and B

    B. Technician B only   D. Neither A nor B

10. The boom on an excavator will not raise or lower, but all the other implements appear to work properly. Technician A says that this could be caused by a seized spool in the main control valve. Technician B says that this could be caused by a turned spool in the pilot valve bore. Who is correct?

    A. Technician A only   C. Both A and B

    B. Technician B only   D. Neither A nor B

CHAPTER 7

Hydraulic Brakes

Learning Objectives

After reading this chapter, you should be able to:

- Describe the fundamentals of off-road hydraulic brake systems.
- Identify the major components that make up off-road hydraulic brake systems.
- Describe the principles of operation of off-road hydraulic brake systems.
- Outline the maintenance and repair procedures associated with off-road hydraulic brake systems.
- Identify the basic troubleshooting procedures for off-road hydraulic brake systems.

Key Terms

accumulator	drum	rotor
brake application pressure	foot brake valve	servo brake
brake pressure	hygroscopic	slave cylinder
brake shoes	master cylinder	system pressure
brake valve	metering	variable-pressure reducing valve
caliper	modulated brake pressure	wheel cylinder
charge valve	park brake valve	
check valve	pressure switch	

INTRODUCTION

Over the last 30 years, off-road equipment manufacturers have used a wide variety of brake designs. Modern-day off-road brake systems have evolved to become what might be considered technological marvels in brake design. Brake systems on most off-road equipment are able to operate in conditions that would overload and wear out a conventional on-road brake system in a very short time. In most cases, there are no paved roads for the equipment to operate on; dirt and gravel roadways are the norm in this environment. Given the grades and roadways, the ability to safely slow down and stop equipment that weighs as much as 200 tons is a daunting task.

In the past, both air brakes and air-over-hydraulic brakes were commonly used on off-road equipment, but these have fallen out of favor because of the high maintenance costs associated with these brake designs. Modern equipment design favors the use of hydraulic brake systems that are classified as internal wet disc brakes. These systems use a brake design that has the brakes located inside the axle assembly, which keeps the dirt out. The design has evolved so remarkably that these brakes may last for many thousands of hours before they need to be overhauled. It is not uncommon for the brake friction discs to last up to 10,000 hours of operation before they need to be replaced.

FUNDAMENTALS

Brake design has changed dramatically since the invention of the wheel. As equipment speed and weight increased, the brake system design had to keep pace with these changes. Brake systems are used to control the speed of the equipment as well as to bring it to a safe stop when required. No matter what the design of the brake system, the principle of operation remains similar. All brake systems rely on applying some form of friction to slow or stop the equipment. We know that energy cannot be destroyed, but it can be converted to another form of energy. The kinetic energy of moving equipment has to be converted to heat energy when it is braked, after which the heat energy has to be dissipated to atmosphere. A brake application by the operator forces friction material against a metal surface to slow down or stop the equipment. The friction device is usually located in the drive axle and, depending on the axle design, the brake system can be located either inboard or outboard on the axle assembly.

Early on, it was discovered that it made more sense to wear out the friction lining than to wear out costly drums or rotors. Brake systems initially used complex linkages that took advantage of lever principles to transmit the force of the operator's foot to the wheel ends. As brake systems evolved, the use of hydraulics to transmit the operator's foot pressure to the wheel ends became increasingly popular. Early hydraulic brake designs used a fluid confined in a hydraulic circuit to transmit the mechanical force of an operator's foot pressure to a master cylinder: this is converted to hydraulic pressure that is always proportional to the mechanical pressure. Modern equipment uses some form of hydraulic assist to amplify the operator's foot pressure (force). The wheel-end application force used on such systems is determined by how much **modulated brake pressure** is delivered to the wheel ends.

The nature of off-road equipment operating conditions requires that brake systems have a number of safety features built into them. The use of two independent brake circuits on equipment is desirable. An independent mechanical park brake is also required to safely park the equipment. In some industries, such as mining, the equipment must be equipped with three independent brake circuits. The third circuit (emergency brake) is a combination of the wheel ends and the park brake circuit.

In off-road equipment, brake systems have become more complex over the past 10 years. The use of complex hydraulic brake circuits, with variable-displacement pressure-compensated pumps with ABS circuits, is becoming common in off-road equipment. These modern brake circuits monitor the **brake application pressure** and will apply the brakes automatically if the pressure falls below safe operating pressure. Not only will the brakes apply when the **brake pressures** are too low, but many systems will also automatically apply the brakes if there is a loss of electrical power or if transmission clutch or torque pressure falls below safe operating levels. On many types of equipment, the brakes cannot be released after an emergency application until the operator activates a brake interlock circuit. This interlock brake circuit ensures that the operator is in the cab when the brakes are released. To understand and repair these brake systems, a technician should have a good understanding of basic hydraulics and know how to interpret electric/hydraulic schematics. This chapter will cover the basic principles of hydraulic brakes, including shoe-type brakes and external **caliper** disc brakes, as well as taking an in-depth look at various types of hydraulically applied internal wet disc brake systems.

BASIC HYDRAULIC BRAKE SYSTEM COMPONENTS

To fully understand the complex brake systems that are used on larger off-road equipment, we need to examine the basic brake systems and their components that are still found on some smaller off-road equipment. **Figure 7-1** shows the master cylinder that is typically found on off-road utility vehicles, such as a four-wheel-drive Toyota Land Cruiser, which is commonly used by personnel in the mining industry to get around the job site. Note that the front and rear brakes use separate circuits incorporated into the design of the master cylinder. Each section of a dual master cylinder has its own brake reservoir. This provides a degree of safety; if one of the brake circuits were to fail, the other would still be operational. A warning light located in the operator's compartment alerts the driver to a loss of a brake circuit.

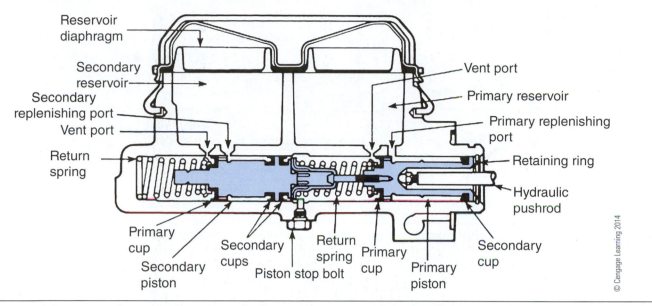

Figure 7-1 Cutaway of a typical master cylinder used on some of the smaller types of off-road equipment.

Dual Circuit Master Cylinder

Master cylinder design has not changed much over the last 40 years. The last major change incorporated the use of two independent circuits in the master cylinder, known as a *tandem master cylinder*. Since the development of the dual circuit master cylinder, very few changes have taken place in the design. However, the material quality used in the internal components, especially in the seals and cups, has improved over the years. **Figure 7-1** shows a cutaway of a typical dual circuit master cylinder. Note that the single bore houses both the primary and secondary pistons for both circuits. They are separated by seals, which prevent one circuit's fluid from entering the other circuit. If a failure were to occur in either circuit, the other circuit would not be affected.

The master cylinder, as its name suggests, controls the flow of hydraulic brake fluid to the wheel end actuators, which are often referred to as **slave cylinders** or **wheel cylinders** in the manufacturer's service literatures. The actual wheel end forces that are produced by the master cylinder are totally dependent on the force that is applied to the **master cylinder** by the operator. In **Figure 7-1**, each hydraulic chamber is connected to its own reservoir. In the "brakes released" position, the primary and secondary pistons are held in the retracted position by a spring. When the operator activates the brakes, the primary piston is forced to move forward, thereby closing the inlet port, which allows brake pressure to build in the primary circuit. The increase in primary

brake pressure forces the secondary piston to move forward, closing off the inlet port and causing the pressure in the secondary circuit to increase. When the operator releases the force on the brake pedal, pressure in the brake circuit drops instantly as the pistons retract. The compensating ports located in the piston chambers help vent the brake fluid back into the reservoirs.

Hydrostatic Drive Brake Master Cylinder

Hydrostatic drive equipment often incorporates a conventional master cylinder brake circuit that is capable of dissipating drive pressure from the drive circuit during a brake application. **Figure 7-2** shows a cutaway of this type of master cylinder. The operator controls the brake pedal application through a mechanical brake pedal in the operator's compartment. A boot on the plunger end of the master cylinder seals the unit from the environment. When the brakes are not engaged, oil from the reservoir keeps the valve spool chamber and the brake plunger chamber filled with oil. In the released position, the valve spool blocks the flow of drive signal pressure from returning to the tank. When the operator depresses the brake pedal, the valve spool is forced forward, bleeding off the drive system signal pressure back to the tank. Note that both fluid chambers in the master cylinder are still open to the tank. As the valve spool continues its forward travel, it pushes against the brake plunger blocking the forward chamber from the tank side.

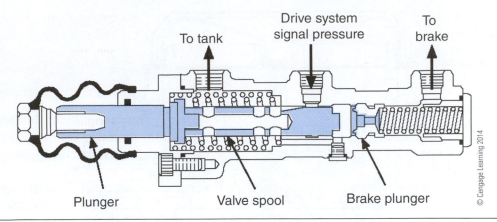

Figure 7-2 Cutaway of a typical master cylinder used on some of the smaller hydrostatic drive equipment.

Continued brake pedal travel builds pressure in this chamber; this pressure applies the brakes.

When the operator releases the brake pedal, the valve spool and brake plunger are forced back by return springs into the released position, dissipating brake pressure from the circuit and blocking the port that allows the drive system signal pressure to be bled off to the tank. This design allows the dissipation of drive system signal pressure to coincide with the application of the brakes on this type of drive system. Many equipment manufacturers that use hydrostatic drives incorporate this type of master cylinder design into their equipment.

Dual Master Cylinder

Conventional dual master cylinders are not always suited to all applications of off-road equipment. Independent control of brakes on each side of the equipment requires the use of a specialized type of master cylinder. In many equipment designs, two separate master cylinders are required so that each brake circuit on the equipment can be controlled independently. The master cylinders used on both sides of the brake circuit are identical in design. A separate brake pedal is used on each master cylinder, allowing the operator to apply the brakes to each side of the equipment independently. This design gives the equipment greater steering control when operating in tight quarters.

Both master cylinders have their own supply port, which is connected to a common brake reservoir, or in some cases, dual reservoirs. A return spring in each master cylinder keeps the valve stem in the neutral position when the brakes are not activated.

When the operator activates the brake pedal, the plunger shown in **Figure 7-3** moves to the left, forcing the valve stem to block the flow of oil to the reservoir.

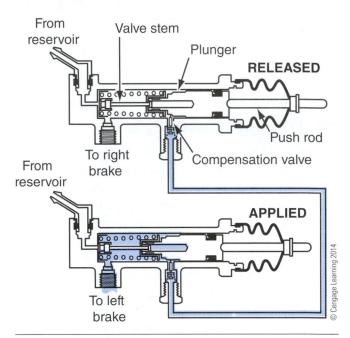

Figure 7-3 Cutaway of a typical dual master cylinder configuration with the brakes applied on the left and right brake circuits independently.

This action allows pressure to increase in the brake circuit. At the same time the increase in pressure in the chamber forces the compensator valve to unseat, allowing oil to flow through the bridge pipe to the other master cylinder where it is blocked by the other compensator valve. When both left and right brakes are activated simultaneously, both compensator valves open to allow the pressure to equalize in both brake circuits. This permits equal brake engagement on both sides of the equipment. The compensator valves act as **check valves** when independent brake application is required on either side, and are forced off their seats to equalize the brake pressure when the operator activates both brakes at the same time.

When the brake pedal(s) is released, the return spring forces the plunger and valve stem to return to the neutral position, opening the supply port back to the tank. The pressure in the system forces the plunger back and the pressure in the brake circuit is equalized.

Boost-Assist Master Cylinder

On large equipment where brake pressures need to be considerably higher than what can be achieved with normal operator foot pressure, additional hydraulic pressure can be used to multiply brake force without exerting extra pedal effort on the part of the operator. This boost-assist piston applies additional pressure to the master cylinder plunger to achieve a higher brake pressure than is possible with operator foot pressure. Depending on the equipment manufacturer, the oil pressure used for the power-assist side can come from a separate pump or from a hydraulic circuit that incorporates a reduced pressure circuit. In **Figure 7-4** the master cylinder is shown in the applied position. The master cylinder piston is held in this position by a return spring. In this piston, the boost piston rear seal keeps the inlet oil from entering the master cylinder. The two spring chambers in the master cylinder are open to the reservoir, which keeps the pressure equalized. Since the brake line openings are located in the spring chamber area, they remain filled with oil. Note that when the boost piston is in the released position, the center valve is also open to the reservoir.

When the operator applies foot pressure to the brake pedal, the boost piston is forced to move to the left, allowing pressurized oil to enter the chamber behind the boost piston that applies hydraulic pressure to add additional force to the input plunger. The input plunger forces the master cylinder plunger to move to the left, causing the center valve to close the opening to the reservoir. When the center valve closes the opening to the reservoir, the pressure starts to build in the spring chamber in front of the master cylinder plunger. During normal braking action, the flapper valve located on the right side of the boost piston closes. If the operator depresses the brake pedal too quickly, the inlet oil is not able to fill the boost-assist side of the chamber quickly enough.

To ensure smooth pedal effort during a quick brake application, the flapper valve allows the oil to flow from the low-pressure side of the boost piston to the inlet side of the boost piston, keeping the inlet side of the piston constantly filled with oil. Once satisfactory pressure is reached on the inlet side of the boost piston, the flapper valve closes. The pressure applied to the boost piston is proportional to the effort exerted by the operator's foot on the brake pedal. Under certain conditions, a small hole in the boost piston allows oil to flow to the reservoir. When the operator applies more force to the brake pedal, the metal-to-metal contact surface between the input piston and the boost piston is increased, slowing down the oil leakage back to the reservoir. During boost piston movement, the oil on the low-pressure side of the boost piston is sent to the reservoir. When the normal oil level is exceeded, the excess oil flows back to the tank through an overflow port located in the reservoir.

Continued brake pedal travel allows the pressure in the spring chamber to build and, because this chamber is open to the brake wheel ends, the brakes are applied. In **Figure 7-4** note that the center valve seal blocks the flow of oil back to the reservoir when the spring chamber is pressurized. If a brake application becomes

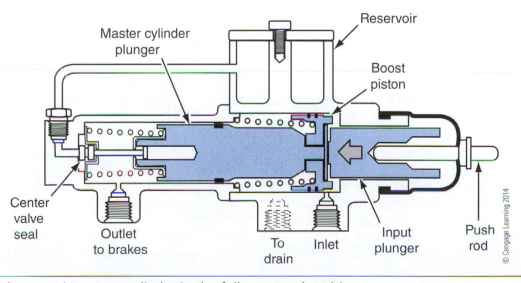

Figure 7-4 A boost-assist master cylinder in the fully engaged position.

necessary when the engine is not running, the operator's foot effort on the pedal still can apply the brakes even though no power assist is available. During initial release of the brake pedal, the boost pressure on the right side of the boost piston moves the input piston back to the right. At this point the pressurized oil from the inlet port flows to the low-pressure side of the boost piston through the hole in the center of the piston. During this step, the oil from the inlet port flows to the reservoir until the boost piston returns to its original position, which blocks the inlet port.

Hydraulic-Assist Two-Stage Master Cylinder

Equipment manufacturers use a wide variety of master cylinder designs to accommodate equipment component layouts. This particular application has the brake pedal connected to a bell crank lever and cable, which controls the movement of the brake booster plunger. When the operator depresses the brake pedal, the booster plunger forces the power piston to move. **Figure 7-5** shows the master cylinder in the applied position; the return spring is keeping the power piston and booster connector against the left side of the master cylinder. The booster connector has a spring inside that keeps the plunger piston in the retracted position. Plunger travel is limited by a snap ring located on the end of the plunger piston. As long as the plunger piston is in the released position, inlet oil can flow into the spring chamber by way of a notch in the plunger piston.

Because the spring chamber is open to the return, oil circulates through the system. Oil can flow to the main chamber of the master cylinder through an orifice located at the left end of the main chamber. During this stage the main chamber and the second stage chambers (as well as the brake lines) are connected hydraulically to the reservoir, keeping the pressure equalized in the system. This design uses two pistons to apply the brakes and has an external reservoir that maintains the fluid necessary to engage the brakes, regardless of brake pad wear.

The complex operation of this brake design necessitates that the brake engagement be shown in two steps to clearly explain its operation. The first step, the start of engagement, is explained in detail so that you can fully understand the brake engagement process. A relief valve located in the master cylinder housing limits the amount of pressure available to act on the power piston. Further travel of the main piston blocks the inlet orifice passage to the reservoir; this causes the pressure to build in the main chamber and second stage chamber. At this point, pressure in the main chamber activates the service brakes until the second stage piston seats against the end seal. Once this occurs, oil from the main chamber is blocked from acting on the service brakes, at which point the oil pressure from the second stage chamber of the master cylinder acts to apply the service brakes. Further brake pedal travel allows the pressure to build up in the main chamber until the pressure is high enough to unseat the second stage piston, allowing the oil to flow to the service brakes. An external line connected on the brake line side of the master cylinder allows the brake pressure to act on a bypass valve located in the master cylinder housing above the second stage piston chamber. The bypass valve limits the amount of brake pressure that can be developed. If the pressure were to rise above a predetermined point, the bypass valve would open,

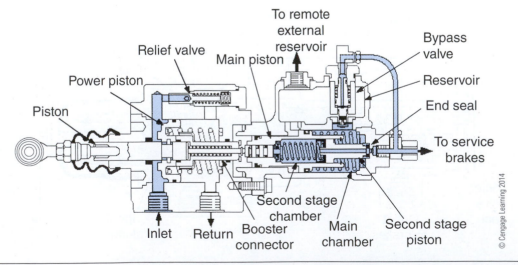

Figure 7-5 A hydraulically assisted master cylinder with the master cylinder in the fully engaged position.

forcing the brake pressure to return to second stage pressure. Brake pressures are generally low at this point of initial engagement, to provide smooth control of braking during partial braking.

In **Figure 7-5** the master cylinder is in the fully applied position, which forces the fluid to act on the bypass valve. When this pressure reaches a pre-determined point, the bypass valve opens, forcing the brake engagement to go to second stage engagement. The fluid flows from the main chamber to an external reservoir causing a pressure drop in the main chamber, which forces the second stage piston to seat on the end seal. The pressurized oil in the second stage chamber cannot escape because of the sealing action of the cupped seal on the piston. Because this area (second stage) is approximately 30% of the area of the main chamber, the pressure flowing to the service brakes will be much higher than it was during the first stage of brake application.

When the operator releases the brake pedal, the plunger piston is forced back by spring force. At this point a passage through the power piston opens, allowing the oil behind the power piston to return to the reservoir. The power piston and booster connector are forced back by a return spring, and the pistons in the master cylinder are forced back by a combination of oil pressure and spring pressure. This brake design allows the brakes to be applied by the operator even if the engine is not running. However, the pressure will be limited to the pedal effort since no external oil pressure is available to apply an assist to the power piston. The power piston will bottom out on the booster connector, forcing the main piston to build a limited amount of brake pressure.

Dual Power-Assist Master Cylinder

Modern equipment design has incorporated the use of independent dual brake circuits built into one master cylinder for the front and rear brakes for obvious safety reasons: Failure of a brake circuit that has separate front and rear brake circuits still allows the operator to stop the equipment safely. **Figure 7-6** shows a typical dual power-assist master cylinder in the released position. This design can be broken down into two sections: the power cylinder and the master cylinder. The operation of each section will be described in detail in this chapter.

In the released position, the brake pedal is fully released and return springs keep the internal components in their respective positions. Oil is present in the chamber between the load feedback piston and the plunger. Oil from the inlet also passes around the servo piston and flows back to the tank by way of the power cylinder chamber. The master cylinder houses two separate brake compartments, primary and secondary, which are open to the reservoir when the brake assembly is in the released position. This allows the pressure in both circuits to be equalized with the brake line full.

To activate the service brakes the operator depresses the brake pedal; this forces the servo piston to shift to the left, blocking the flow of inlet oil to the tank. This action allows the pressure to increase

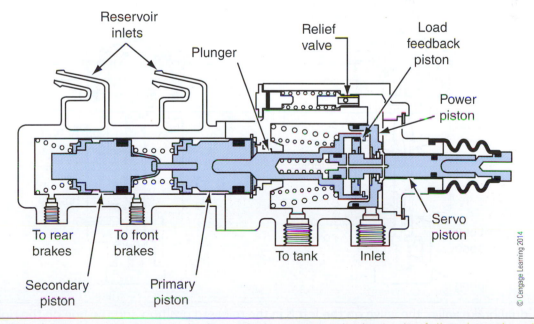

Figure 7-6 A dual power-assist master cylinder with the master cylinder in the fully released position.

quickly behind the power piston to increase the assist force that is applied to the plunger, which then acts on the primary piston. A relief valve located in the power cylinder housing limits the pressure that is available to act on the power piston. The power piston also has an area where the pressure acts on a load feedback piston, giving the operator feedback from the brake application. Both primary and secondary pistons in the master cylinder are connected together mechanically.

When the primary and secondary pistons are forced to the inboard, they block the inlets. This forces the pressure to rise in the two chambers that are applying the brakes. Chamber pressure in the front brake chamber applies pressure to the secondary piston face, forcing the pressure to equalize in both front and rear chambers. When the operator releases the brake pedal, the servo piston will return to its original location because the force of the return springs allows oil to flow around the servo piston and back to the tank.

Hydraulic Brake-Control Valve

Brake-control valves are used to control the flow of fluid to the service brakes on equipment that uses hydraulic brakes to get pressurized oil supply from a hydraulic circuit. This circuit can be part of a main hydraulic system, or it can be a dedicated brake hydraulic circuit designed just for this purpose. There are a wide variety of **brake valves** in use on off-road equipment to suit many different applications. To gain an understanding of how these brake valves work, a commonly used brake valve design will be explained.

Figure 7-7 shows a cutaway view of a typical hydraulic brake-control valve. Note that this valve incorporates what is commonly referred to as a dual circuit pedal. In other words, it has two independent brake circuits, one on top of the other, in the same valve body. Other designs use two separate valves connected to the same pedal, or two separate pedals (often found on loaders) where one pedal applies the brakes only and the other brake pedal applies the brakes as well as disengaging the transmission. This design allows the transmission to disengage automatically whenever this pedal is used to apply the brakes during truck loading operations. Another common application for dual pedals is on equipment that requires the brakes to be applied independently on each side of the equipment. The modern backhoe is a

BRAKE PEDAL OFF   **BRAKE PEDAL APPLIED**

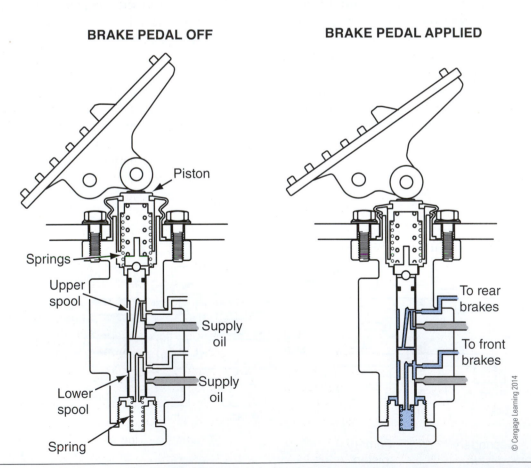

Figure 7-7 A dual brake valve: the valve on the left is in the released position and the one on the right is in the applied position.

good example of equipment that often uses this brake design.

Figure 7-7 shows two supply ports that have pressurized oil present at the inlets. The brake pedal on the left is shown in the released position with no oil flowing through the valve. A spring at the bottom of the valve keeps the internal spools centered in the brake valve body. As long as the two internal valves remain in this position, the supply oil is blocked from flowing to the wheel end brakes. The upper spring returns the pedal to the retracted position when the operator releases the pedal.

When the operator steps on the brake pedal, the upper piston forces the top spring to compress, moving the two spools down. When this takes place, the oil from the inlet passages flows to the brakes. Small passages located between the inlet side and the outlet side of the spools act as restrictions to limit the amount of oils that can flow. The more the pedal is depressed, the more oil will flow. This allows the brake pedal to function as a **variable-pressure reducing valve**, which limits the pressure flowing to the wheel end brakes. The more the pedal is depressed, the higher the pressure will be at the wheel end brakes.

A small internal passage in each spool allows the oil to flow from the modulated pressure side of the circuit to the underside of the spools. This provides a resistance to the pedal movement and gives the operator physical feedback by providing a slight resistance. When the force on the upper springs is lower than the oil pressure underneath the spools, the pressure forces them to move up. By **metering** the flow of oil, the brake valve works like a variable-pressure reducing valve.

When the operator releases the brake pedal, the oil pressure below the spools and the lower spring returns the spools to the neutral position by sending the flow of oil from the wheel end brakes back to the reservoir, while at the same time blocking the flow of oil from the inlet side to the brakes.

Dual Pedal Brake Valve

Figure 7-8 shows a typical brake valve that is used on a variety of backhoe loaders or small skid steer loaders that require independent side-to-side braking. With the pedal in the released position, the return springs keep the modulator spool and both brake selector spools in the released position. The oil flows from the supply side through the modulator valve and back to the reservoir through a small hole drilled through the modulator valve. Both brake spools in the released position allow the brake side to flow back to

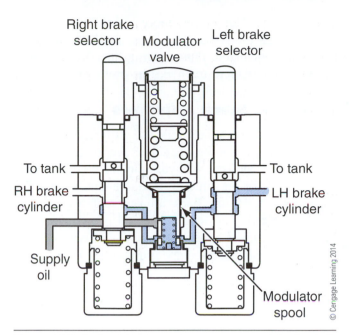

Figure 7-8 A dual brake valve cutaway with the left brake beginning engagement.

the reservoir. Two springs are used in the modulator valve; the lower spring is used to return the modulator valve to the released position when the operator releases the brakes.

When the operator begins to apply the brakes to the left side of the equipment as shown in **Figure 7-8**, the selector spool is forced down, blocking the return to the tank passage. At the same time, it opens a passage from the outlet side of the modulator valve to the service brakes on one side. When this occurs the modulator spool moves down, allowing some of the supply oil to flow through a cross-drilled hole into the lower spring cavity of the outlet port. The position of the modulator valve causes a restriction to the lower spring cavity.

When the springs in the modulator valve are further compressed, the valve is forced to shift further; this increases the oil pressure at the outlet ports. The amount of restriction is determined by the force applied to the brake pedal, which will modulate the pressure, acting like a variable-pressure reducing valve. The modulator valve is capable of supplying oil to both brake valves simultaneously, but only the left side is being utilized in this example.

If the operator were to apply both brakes together, the pressure would equalize in both brake circuits just as it would if one or the other brake circuits were to be engaged separately due to the action of the modulator valve. For example, when brake line pressure increases in a brake circuit, the pressure of the oil in the modulator valve rises above the upper spring pressure, forcing it up. This causes more restriction in the oil flow through the cross-drilled hole until the opposing forces are equalized, thus metering the flow of oil to the service brakes to maintain the pressure.

Hydraulic Slack Adjusters

Equipment manufacturers use different means of compensating for brake wear. This section explains the operation of one type of hydraulic slack adjuster that Caterpillar uses on some of its equipment. **Figure 7-9** shows the hydraulic slack adjuster cutaway in two different views, one in the released position and the other in the brakes-applied position. This type of slack adjuster is designed to compensate for brake wear. The slack adjusters are located in series between the brake valve and the wheel end brakes. Each wheel end has its own hydraulic slack adjuster.

When the operator applies the service brakes, pressurized oil flows into the slack adjuster through the inlet port from the master cylinder or brake valve, depending on the equipment brake circuit design. This slack adjuster will work on either hydraulic brake circuit. Continued wear of brake material will require

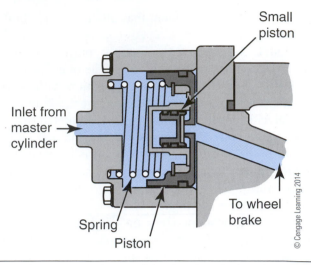

Figure 7-10 Cutaway of hydraulic slack adjuster in the applied position.

that more oil be displaced with each brake application as the wear increases. In **Figure 7-10** the oil pressure from the brake valve forces the large piston to the end of the adjuster. The oil now fills the oil passage that leads to the small piston. The small piston will move away from the seat if the pressure in the oil passages is higher than the pressure in the brakes, forcing oil to flow to the brakes.

When the service brakes are released, the oil flow is vented to the inlet port side of the slack adjuster. The lower brake pressure returns the smaller piston to the seat. This forces the service brakes to exhaust the oil from the wheel end brakes through the slack adjuster. This action forces the large piston back into the re-leased position.

Diverter Valve

For various reasons, many applications on off-road equipment require the use of a diverter valve in the brake circuit. Equipment with spring-applied park brakes requires a diverter valve so that the park brakes can be released in the event the engine can no longer be run safely (see **Figure 7-11**). Off-road haulage trucks often use an electric supplemental drive pump, shown in **Figure 7-12**, to provide emergency hydraulic pressure for towing purposes.

Drum Hydraulic Brakes

Although hydraulic drum brakes are not as popular on off-road equipment today as in the past, they still account for a number of brake systems currently in use. The basic system consists of a drum assembly generally made from cast iron and is bolted or held in place on a rotating assembly that includes the wheel,

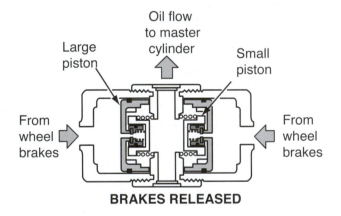

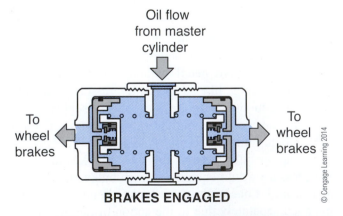

Figure 7-9 Cutaway of hydraulic slack adjusters: one is in the released position and the other is in the applied position.

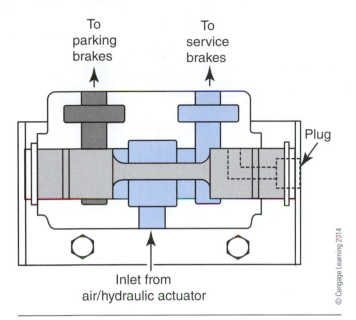

Figure 7-11 Cutaway of a diverter valve used on off-road equipment with spring-applied park brakes.

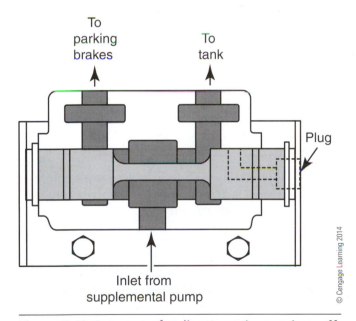

Figure 7-12 Cutaway of a diverter valve used on off-road equipment with a connection for a supplemental electric hydraulic pump.

hub assembly, and drive axle. A backing plate is required to hold the shoes, wheel cylinder, adjusters, and mounting linkage in place. In some of the smaller equipment, the use of a mechanical park brake that is integrated into the rear brakes of the equipment is also possible. The shoes are lined with a friction material that is forced to rub on the inside of the drum when the brakes are actuated.

On off-road equipment, the shoes are generally pinned on the bottom and have a wheel cylinder at the top, which forces the shoes out against the inside of the drum during brake actuation. When the brakes are applied, the friction material converts the kinetic energy of the moving equipment to heat energy. Hydraulic pressure is applied to the wheel cylinders from an actuating device, such as a master cylinder or brake hydraulic system, to force the brake lining against the inside of the drum. **Figure 7-13** shows what takes place when the brake shoe is forced against the drum. Frictional drag created by the rotation of the drum wedges the shoe tighter into the drum using a principle known as self-energization.

In our example of a non-servo drum brake, each shoe is anchored to the backing plate at the bottom to an anchor pin. The wheel cylinder applies actuating force to the toe of the shoe when the equipment is moving forward to energize the leading shoe. In this design, this energizing concept applies to only one shoe at a time and is dependent on the direction of travel. When traveling in reverse, the trailing shoe becomes self-energized. This design does not make the best use of this principle because only one shoe can be energized at a time: which shoe is energized under braking depends on the direction of travel.

Some brake designs have an additional feature that uses the servo principle as well as the self-energizing principle. These designs are often found on smaller types of equipment. The principle of operation is shown in **Figure 7-14** where servo action takes place when hydraulic force is applied to the toe ends of the **brake shoes** during forward motion. The primary shoe reacts to the rotational forces first because it has a weaker return spring. The shoe lifts off the anchor and is forced against the inside of the drum. The shoe can then pivot on the heel of the shoe, causing it to attempt to rotate with the drum, an action that transfers force to the secondary shoe, by means of a connecting link, In this way, both shoes are forced into the drum,

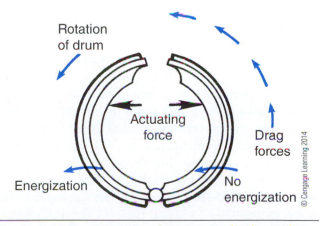

Figure 7-13 An example of non-servo brake action.

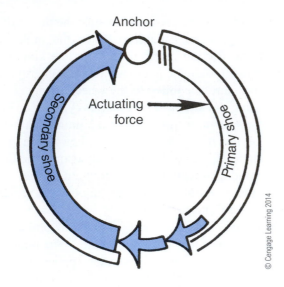

Figure 7-14 The self-energizing concept of a servo action.

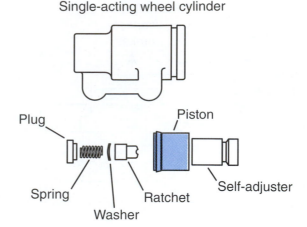

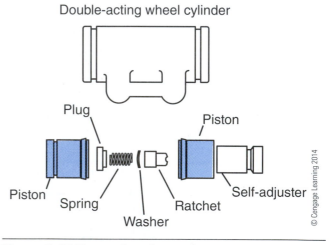

Figure 7-15 Illustration shows the two most common types of wheel actuators in use: single-acting and double-acting wheel cylinders.

meaning that they work together. This combination of rotational and actuating forces causes the complete shoe assembly to become energized as a single unit.

Wheel Cylinders

Hydraulic drum brakes, regardless of whether they are a non-servo or **servo brake** design system, must have a wheel cylinder to actuate them. The wheel cylinders use hydraulic pressure from a brake valve to force the shoes out against the drum. There are two styles of wheel cylinders in use on these systems (see **Figure 7-15**). The single-acting wheel cylinder has only one piston and actuates one shoe of an anchor pin pivoted design. Double-acting wheel cylinders actuate two shoes at one time, one from each end of the wheel cylinder. A specific amount of brake fluid is sent to the wheel cylinder at one time, and as the brake lining wears out, the wheel cylinder travel becomes inadequate to apply sufficient force to stop the equipment. Regardless of design, wheel cylinders rely on cups to seal the pressurized brake fluid inside the housing behind the piston. When the operator releases the brake pedal, the brake shoe retractor springs pull the shoes back toward the anchor pins and into the released position. Note that some of the older off-road drum brake systems used a residual check valve in the master cylinder to maintain a residual pressure of approximately 12 psi in the wheel cylinder. This was required to take up any slack and prevent wheel cup sag, which could lead to leaks or air getting into the wheel cylinder. The residual pressure is insufficient to apply the brakes because it takes a greater amount of

pressure to overcome the return spring tension. All wheel cylinders have a rubber boot on the piston end that protects the piston area from dirt and water.

Self-Adjusting Mechanism

A self-adjusting mechanism is necessary to keep the brake shoes close to the drum and to take up the extra clearance caused by drum and shoe wear. Brake adjustment can only take place when the machine is moved in reverse. The adjusting lever, cable, and overtravel spring are key components, as shown in **Figure 7-16**. When the brakes are activated in the reverse direction and the brake shoes are adjusted correctly, the cable tension and spring force are equal. These equalized forces keep the adjusting lever from moving. During normal operation, the brake lining and drum will wear and the upper cable travel will increase, causing the tension to decrease. This causes

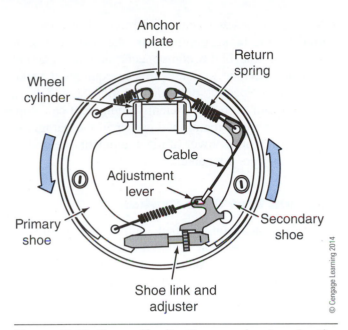

Figure 7-16 The self-adjusting mechanism in the forward direction. No self-adjusting takes place in forward.

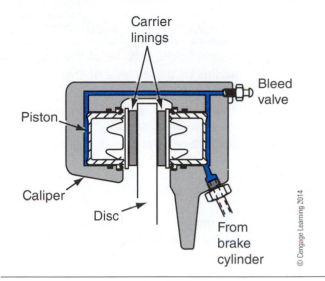

Figure 7-17 Cutaway of a fixed-caliper design.

the cable tension to become lower than the spring force. When this occurs, the spring force pulls the adjusting lever back and down, turning the adjusting screw slots. This causes the brake shoes to move out until the tension on the adjusting lever spring and the cable is in balance.

Hydraulic External Caliper Brakes

External hydraulic disc brake systems come in two basic designs: the fixed caliper and the sliding caliper. They are easily identified by observing whether or not the calipers have pistons on both sides of the rotor. If there are pistons on both sides of the rotor (as shown in **Figure 7-17**), they are the fixed-caliper design. Both brake designs have their own shortcomings and wear characteristics. Regardless of the caliper design, the hydraulic portions of the brake systems are generally identical in operation.

The sliding-caliper design is shown in **Figure 7-18**; it has hydraulic pistons on one side only. In operation, the sliding caliper pulls the opposite pad and mounting hardware toward the rotor each time the piston on one side is activated. This design is superior to the drum brake for a number of reasons. The most obvious is the difference in how the caliper clamps the **drum** or rotor. The disc brake caliper squeezes the rotor, and as it heats up, the **rotor** expands into the caliper. When a drum brake heats up, the drum expands away from the shoes, which can lower braking force creating what is known as brake fade. The clamping forces that the

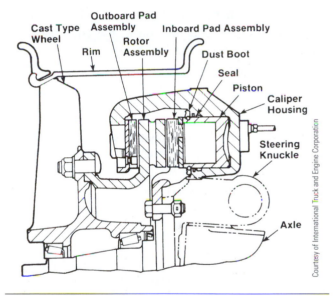

Figure 7-18 Cutaway of a sliding-caliper hydraulic disc brake system.

caliper transfers to the rotor are not affected by servo action as on a drum brake. When disc brakes heat up, their braking capabilities are not compromised by brake fade. In addition, their design produces better cooling because the friction surfaces on the rotors are exposed to air flow and can dissipate heat rapidly unlike drum brake friction surfaces, which are contained inside the drum.

Disc brakes have many obvious advantages over drum brakes when used on off-road equipment. They are more resistant to brake fade, provide better cooling capability, have better tolerance to dirt and water, require less maintenance, do not need to be adjusted, and

TABLE 7-1: SWEPT AREA CALCULATION

R_p	Radius to the outside of the pad
R_r	Radius to the outside of the rotor
W_p	Pad width
Note	If R_p nearly equals R_r, use the following formula to calculate swept area:
Formula	Disc Brake Swept Area = $6.28W_p$ $(2R_r - W_p)$

© Cengage Learning 2014

Note: 6.28 represents $\pi \times 2$ (3.14 × 2)

have a greater swept area than drum brakes of the same size. Look at the formula shown in **Table 7-1**.

$$\textbf{Disc Brake Swept Area} = \textbf{6.28} \times \textbf{4 (2} \times \textbf{15} \times \textbf{4)}$$
$$= \textbf{25.12} \times \textbf{26}$$
$$= \textbf{653.12 sq. in. per rotor}$$

When engineers design equipment, they design the components to have a total swept area that factors in the weight, speed, and traction of the equipment. To see how this impacts the brakes on some equipment, we can calculate the swept area of a similar size drum brake system to see what the difference is in swept area between the two designs. The example below shows a similar diameter dimension and pad width to show what the difference looks like.

In our example, we will use the following values to complete the formula:

R_p	15 inches
R_r	15 inches
W_p	4 inches
Note	If R_p nearly equals R_r, use the following formula to calculate swept area:
Formula	Disc Brake Swept Area = $6.28W_p$ $(2R_r - W_p)$

Drum Brake
$$\textbf{Swept Area} = \textbf{3.14} \times \textbf{D (Drum Diameter)}$$
$$\times \textbf{L (Pad Width)}$$
$$= \textbf{3.14} \times \textbf{30 inches} \times \textbf{4 inches}$$
$$= \textbf{378.8 sq. in. per drum}$$

It is not hard to see which brake system has the higher swept area.

Now that we know a little bit about what makes one brake system inherently superior to another, we will examine a few other factors that affect the equipment's stopping ability. A number of factors affect brake performance. Brake performance is limited to the force applied by the piston, the deflection of the drum or

rotor, and associated brake hardware. The wear incurred by the brake components, the operating temperatures of the brake components, and tire traction should not become major factors in equipment brake performance. Ideally, only tire traction should be a limiting factor on how efficiently equipment can stop. If any other factor mentioned earlier in the paragraph limits the stopping power of the equipment, the brakes are probably not adequate.

Internal Wet Disc Brakes

Brake systems on off-road equipment were always considered an area that required a lot of maintenance until the introduction of internal wet disc brakes. No matter how well external disc brakes worked, they had one major hurdle that was difficult to overcome. The equipment operated in less than ideal conditions, which often led to a short brake component life in many applications. The problem stemmed from the fact that the rotors and calipers were exposed to the road grime that the equipment operated in, leading to short brake component life and often higher maintenance costs.

The introduction of the internal wet disc brake on off-road equipment changed that; brakes that were sealed internally in the axle housing, away from the road grime, proved to be the answer to high maintenance costs associated with external disc brakes.

The internal wet disc brake system is highly suited to applications where road grime contamination is extremely likely to occur. Braking is achieved through the use of a number of friction discs that are alternately sandwiched between the same number of steel discs that are held stationary by external teeth or notches splined to the wheel spindle. The friction discs are splined internally to the rotating wheel. The sealed system operates in an oil bath, which cools the brake components as well as provides a means of lubrication. Some manufacturers mount the internal wet disc brake components right next to the differential, and others mount them at the wheel end of the axle. This depends entirely on equipment design and application. Lubrication and cooling on some designs are often integrated with the lubrication and cooling for the axle assembly and require the use of specialized lubricants. In heavy-duty applications, a separate cooling circuit that has its own pump and cooler can be used to dissipate the heat generated by this type of brake system. This brake system requires very little in the way of maintenance. No adjustments are generally required because the design compensates for wear automatically.

When the brakes are in the released position, as shown in **Figure 7-19**, a number of return springs (see **Figure 7-20**) keep the discs and plates separated, thus allowing the wheels to turn with very little resistance. Depending on equipment application, the brake cavity is either filled with oil or has an independent brake cooling circuit that circulates fluid through the brake

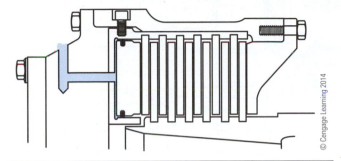

Figure 7-19 The internal wet disc brake in the released position.

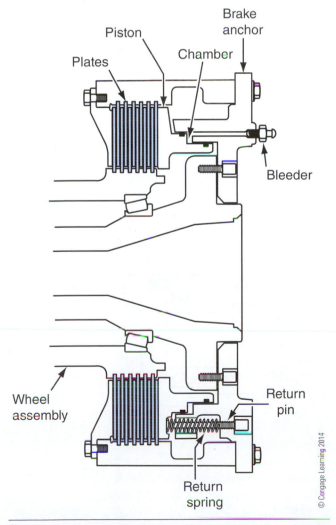

Figure 7-20 Cutaway view of an internal hydraulic wet disc brake assembly.

component cavity. The steel discs are externally splined to the axle, and the friction discs are splined to the rotating wheel. When the equipment is in motion, the friction discs rotate with the wheel; but they are not in contact with the stationary steel discs, which are splined to the spindle. A minimal amount of friction is generated as the constant circulation of oil in the housing keeps things well lubricated.

When the operator applies the brakes, oil is routed to the wheel end through a foot brake valve and enters the wheel end through an inlet port located externally on the inside of the wheel assembly. High-pressure oil enters the piston cavity and forces the piston to move to the right, squeezing together the rotating friction discs and stationary steel discs. This squeezing effect of the friction discs, which are splined to the rotating wheels, slows down the equipment. When sufficient hydraulic force is exerted on the piston, the wheel will lock to a stationary position. The discs are constantly immersed in an oil bath, which absorbs the heat energy created by the discs. Heavy-duty applications of this design have a separate cooling circuit that is constantly circulating oil through a cooler and return filter system to keep the oil clean and cool.

Inboard Hydraulic Wet Disc Brake

The internal wet disc brake design has many variations that are based on equipment application. Some off-road equipment applications do not have room in the wheel end to accommodate a wet disc brake, necessitating a design variation of this brake system. Many wheel loaders and farm tractors fall into this category; they have the wet disc brakes located in the final drive right next to the differential. With this design the clutch and plate piston are held stationary by pins fastened to the outer housing of the axle assembly. A friction disc is located between the two stationary members and is splined to the rotating drive axle (sun gear shaft), as shown in **Figure 7-21**. The amount of friction and the number of steel plates used vary and depends on the size of the equipment and its application. When the brakes are in the released position, the piston is held away from the disc by return springs.

When the operator applies the brakes (see **Figure 7-22**), hydraulic oil is routed behind the clutch piston, actuating it to compress the disc(s) together between it and the plate. Because the sandwiched friction disc turns with the wheels, the equipment is braked. The brake and differential share the same oil for lubrication and cooling and require the use of a specific type of lubricant. If lubricant must be added when servicing this type of equipment, be sure to use the

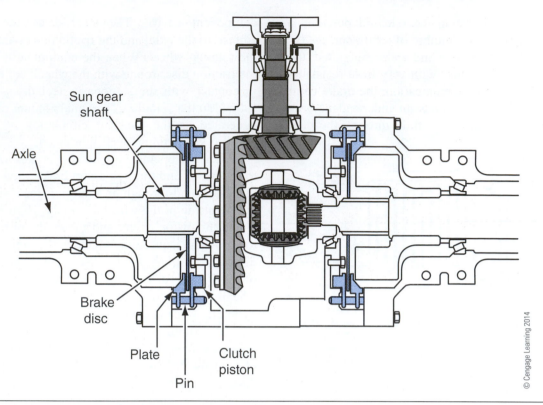

Figure 7-21 Cutaway of a typical inboard internal wet disc brake.

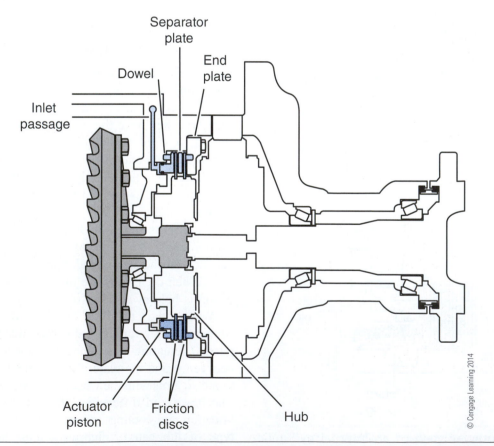

Figure 7-22 Cutaway of a typical inboard internal wet disc brake system in the applied position.

correct type; conventional differential lubricant may not be compatible with the brake friction material and could result in early brake failure and severe differential component damage. This brake design does not require any brake adjustment during its life other than periodic oil changes based on the equipment manufacturer's recommendations.

Mechanical Internal Park Brake

Equipment that is placed in the park position must not be parked using hydraulic pressure only for obvious safety reasons. If hydraulic pressure is lost for any reason, the equipment would move and cause property damage or injury to anyone in the immediate area. Manufacturers have come up with many variations of a mechanical park brake on off-road equipment. **Figure 7-23** shows a typical mechanical multi-disc park brake that uses an external mechanical lever to apply the park brake.

This type of mechanical brake has a caliper that locks the bevel pinion to the housing to prevent any movement when parked. It is actuated by a mechanical level on the outside of the housing. The caliper assembly is mounted to the inside of the housing, and the friction discs are mounted to the pinion shaft with

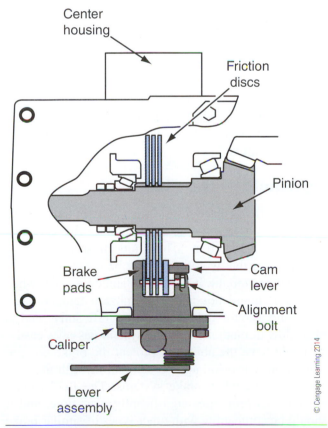

Figure 7-23 Cutaway of a typical internal mechanical park brake system.

© Cengage Learning 2014

internal spines. When the operator sets the park brake, the pads in the caliper squeeze the friction discs to lock them together and prevent the pinion shaft from rotating.

Spring-Applied Park Brake

Modern equipment generally uses a variation of a hydraulic multi-disc wet disc brake that incorporates a series of extremely heavy-duty springs to apply a mechanical force to the disc. This design requires a constant source of hydraulic pressure to hold the park brake in its released position. By the nature of its design, a measure of safety is built in because when the equipment is shut down or loses hydraulic brake pressure, the park brakes automatically apply. The park brake example shown in **Figure 7-24** can be found on large articulating haulage trucks and is located on the output transfer drive.

The park brake consists of two sets of discs: steel and friction. The steel discs are externally splined to the axle housing and do not rotate; the friction discs are splined to the rotating shaft. A hydraulic piston is located in the assembly with a pressure plate, which uses a number of heavy-duty springs on one side of it, and the hydraulic piston on the other side (see **Figure 7-25**). The large heavy springs force the pressure plate against the piston, which locks the friction discs to the stationary steel disks and prevents any movement of the axle shaft whenever the equipment's park brake is engaged. Because this design requires hydraulic pressure to release the brakes, the brakes are automatically applied whenever the equipment is not running. When the operator wants to release the park brake, hydraulic oil is delivered to the hydraulic side of the piston, which must build up sufficient force to overcome the mechanical force of the springs to release the park brake. This ensures that the park brakes do not release when hydraulic pressure is too low, adding a measure of safety. The down side is that the park brakes can drag if insufficient pressure is available to fully release them.

Combination Service/Park Brake

Some manufacturers combine the park brake and service brake in the wheel end brake assembly, shown in **Figure 7-26**. This is accomplished by using a separate circuit and piston to apply the park brake, which uses the same discs and plates that the service brakes use. This lowers cost and complexity, and at the same time increases reliability and safety. This design has a spring-applied wheel end brake in each wheel end of

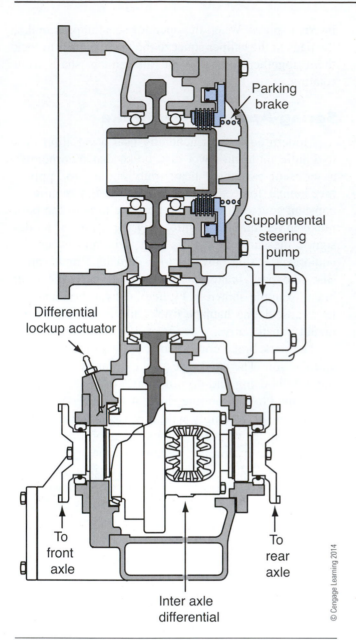

Figure 7-24 Cutaway of a typical internal spring-applied park brake system located on the output of a transfer case drive assembly.

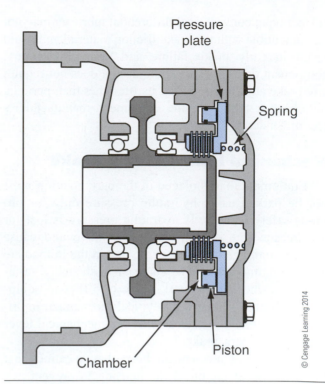

Figure 7-25 Close-up view of a typical internal spring-applied park brake system.

the multiple spring pressure that is acting on the park brake piston, which in turn exerts mechanical force on the service brake piston. When the operator applies the park brake, hydraulic oil is released from the cavity between the two pistons, allowing the heavy park brake springs behind the park brake piston to apply pressure to the service brake piston: this action clamps the stationary steel discs and friction discs together to apply the brakes. A cutaway view of this system is shown in **Figure 7-27**.

Hydraulic Brake Fluids

Brake fluids used in a brake system must have certain characteristics to be suitable for a heavy equipment application. The fluid must not boil at temperatures that would normally be encountered in a typical brake circuit. It must not change viscosity when cold and should not freeze at subzero temperatures. The fluid should not compress significantly, and it must flow through small brake system passages easily. It should have the ability to protect the brake components from corrosion and be compatible with material commonly found in brake systems. The fluid should not alter its properties significantly over time, and it should be compatible with other types of similar brake fluids. A good fluid should not form any gum or deposits on brake components during its life.

the equipment, making it almost failsafe. It is very unlikely that all four wheel ends would fail at the same time.

The park brake circuit uses a series of heavy springs located around the park brake piston, which is positioned behind the service brake piston. When the operator releases the park brakes, he or she activates the park brake release circuit, which sends hydraulic oil pressure to the hydraulic cavity between the park brake piston and the service brake piston. This forces the service brake piston (the only one to directly contact the discs) to release the force it is exerting on the discs. The oil pressure exerts a force to overcome

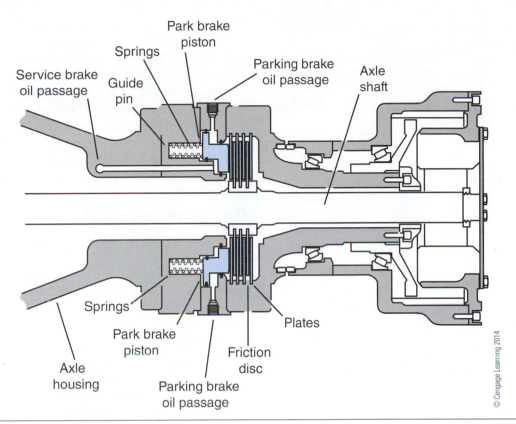

Figure 7-26 Close-up view of a typical internal spring-applied park brake system.

Figure 7-27 Close-up view of typical combination spring-applied park brake/service brake system.

In North America, brake fluid standards are quantified by two groups: the Society of Automotive Engineers (SAE) and the Department of Transportation (DOT). Conventional brake fluid has three categories: DOT 3, DOT 4, and DOT 5. The first two brake fluids, DOT 3 and DOT 4, are formulated from a polyglycol base, which has some unique characteristics. One characteristic, unique to this type of brake fluid, is its ability to absorb water (**hygroscopic**). This is a desirable characteristic that ensures that if any moisture gets into the brake system, it will quickly be absorbed by the brake fluid. This prevents corrosion from forming and also prevents the water from freezing in the brake lines during cold-weather operation. The disadvantage of a hygroscopic brake fluid absorbing water is that it lowers the boiling point over time. Any hygroscopic brake fluids that have been sitting on a shelf must be completely sealed in their original containers, or they can absorb any moisture present in the air and become unusable. One of the most important qualities a brake fluid can have is its resistance to boiling. Let's use DOT 3 to demonstrate how water in the brake fluid lowers the boiling point. DOT 3 brake fluids have a dry boiling point of approximately 400°F (205°C). Once the brake fluid absorbs some water, the boiling point will quickly fall to around 284°F (140°C). It's easy to see that the wet brake fluid would pose a problem when the brake fluid gets hot. Once boiling occurs in brake fluid in a wheel caliper, the gas bubbles formed are compressible and cause a pressure drop, significantly reducing the stopping ability.

DOT 5 brake fluid is formulated from a silicone base and does not absorb water; in other words, it is non-hygroscopic. When using this type of fluid in a brake circuit, it must be kept free from water; any water entering the brake system will cause corrosion. The other negative feature is that it is slightly compressible. This makes it unsuitable for brake systems

TABLE 7-2: DOT BRAKE FLUID BOILING POINT PARAMETERS			
	DOT 3	DOT 4	DOT 5
Dry boiling point (degrees Fahrenheit)	401°F	446°F	500°F
Dry boiling point (degrees Celsius)	205°C	230°C	260°C
Wet boiling point (degrees Fahrenheit)	284°F	311°F	356°F
Wet boiling point (degrees Celsius)	140°C	155°C	180°C

© Cengage Learning 2014

that utilize ABS circuits. Whenever any service work is to be performed on a brake system, identify the type of brake fluid in use so that any fluid that may need to be added to the brake system is compatible with the existing fluid. Refer to **Table 7-2**.

HYDRAULIC BRAKE CIRCUITS

Modern off-road equipment design engineers have used basic hydraulic principles when designing brake circuits to safely control the speed and stopping ability of equipment. Conventional hydraulic brake circuits use fluid that is confined in a brake hydraulic circuit to transmit hydraulic force to the wheel end brakes. The distance from the master cylinder to the wheel ends can be considerable, making hydraulic brakes a desirable option. With minor exceptions, hydraulic

brake circuits have not changed much over the years. Mechanical actuation of a master cylinder by the operator's boot creates hydraulic pressure, which is then transmitted to slave cylinders. Slave cylinders are often boost-assisted on heavy-equipment brake systems to help increase the force at the wheel ends where the foundation brakes are located. The mechanical force applied to the foundation brakes is directly proportional to the hydraulic pressure delivered by the master cylinder to actuate the slave cylinders.

Basic Brake Circuit

The basic hydraulic brake circuit uses a master cylinder to supply control pressure that is used to actuate slave cylinders. With the brake pedal in the released position, the return spring holds the plunger in the master cylinder in the neutral position. The check valve at the end of the master cylinder is open to the reservoir return circuit. Another check valve located at the end of the slave cylinder is also open to the reservoir as well as to the wheel end brakes. This design ensures that the pressure is equalized in the brake circuit and all lines are kept full of fluid. A return spring forces the power piston over to the left side of the brake valve. The primary piston blocks the inlet port preventing boost pressure from entering the rear of the power piston when in the released position.

To apply the service brakes (see **Figure 7-28**), the operator applies pedal pressure to the master cylinder, which immediately closes the port to the reservoir and

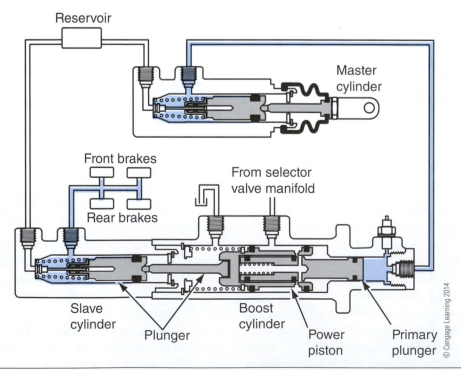

Figure 7-28 Cutaway view of a typical dual hydraulic service brake circuit with the brakes applied.

builds pressure in the master cylinder. The resulting pressure acts on the inlet port of the boost cylinder and forces the primary piston, power piston, and slave piston to move inboard, closing off the check valve at the end of the slave cylinder chamber that is open to the reservoir. This allows the inlet in the boost cylinder to open, routing hydraulic pressure to act on the power piston. Boost pressure behind the piston assists in developing the pressure necessary to apply the brakes.

Hydraulic Wet Disc Brake Circuit

This section explains the operation of a hydraulically applied internal wet disc brake system used on off-road equipment. This system uses the hydraulically controlled internal wet disc brakes that are ideally suited for use on off-road equipment operating in an extremely abrasive environment such as a quarry or mining operation. This section provides an overview of how this brake system functions, along with a more detailed explanation of the operation of each component. The section concludes with a brief discussion of basic diagnostic and repair procedures.

Figure 7-29 shows a pictorial view of a basic hydraulic brake circuit. When the operator activates the tandem **foot brake valve**, pressurized oil is routed to the brake wheel ends to apply the multi-disc brakes. A hydraulic pump provides a constant flow of oil to the brake circuit, sending the oil flow to the **charge valve** that regulates the hydraulic brake pressure in the circuit.

An **accumulator** charge valve is used to maintain adequate brake pressure in the brake circuit by controlling the flow of pressurized oil in the brake circuit. Once maximum pressure is achieved in the brake circuit (kick-out pressure), the charge valve redirects the pump flow back to the tank.

When the oil pressure falls below a threshold value (kick-in pressure), the charge valve redirects the flow of oil back into the brake circuit, charging the circuit backup to the kick-out pressure.

Two accumulators are used in this basic brake circuit: one for the front service brakes and one for the rear service brakes. The pre-charged nitrogen accumulators store hydraulic energy, which is used by the brake circuit to apply the brakes. In the event of an engine failure, the accumulators store sufficient energy to apply the brakes. A separate **park brake valve** is used to activate the spring-applied park brakes. Note that two independent circuits are used for the service brakes: one for the front brake and one for the rear. A third circuit is used to actuate the park brakes. This design gives the equipment three independent brake circuits, making it extremely reliable and safe.

The park brake consists of a spring-applied hydraulically released brake that can be mounted on any component that is mechanically connected to the wheels. Many equipment manufacturers mount it on the output shaft of the transmission, on the transfer case, or sometimes on the differential input shaft. The park brake in this circuit (see **Figure 7-30**) has an electric hydraulic **pressure switch** that monitors the brake pressure; if it were to fall below a predetermined pressure, the pressure switch would send electrical power to the park brake solenoid to actuate the park brake. The park brake solenoid valve cuts off the oil flow to the spring-applied park brake. When the park brake is in the applied position, the park brake solenoid is not energized. The park brake does not receive any

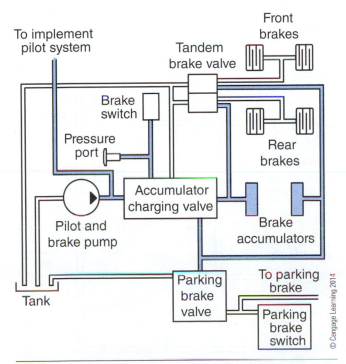

Figure 7-29 Pictorial view of a typical basic hydraulic brake circuit.

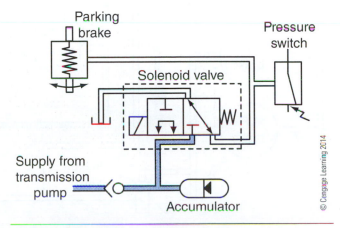

Figure 7-30 Schematic of the park brake circuit.

oil in this position; the oil is allowed to drain back to the tank through the park brake valve.

In some applications, off-road equipment brakes are used in severe-duty applications. In other words, the brakes are used to a point where they require a cooling circuit to keep the brake oil temperature within safe limits. In these circumstances, an extra cooling circuit can be designed into the equipment's brake system to cool the oil. In **Figure 7-31** oil is circulated through the wheel end brake clutch pack during operation, continually removing any contamination formed as a result of disc wear. This continuous flow of oil also cools the brake components; before it enters the wheel ends it is filtered and cooled. In this instance, the brake cooling circuit has a separate pump dedicated to this purpose. Some equipment manufacturers use excess oil from the main hydraulic circuit to cool the brakes. Pressures in the brake cooling circuits are generally low (5 to 25 psi) in comparison to the brake activation pressures.

Brake Components

Brake Pump:. A hydraulic pump is necessary to supply oil for an internal wet disc brake circuit. A gear pump with an output of approximately 5 to 12 gpm (gallons per minute) is generally used for this purpose because of its low cost. The pump is driven directly by the engine, or in some cases it is mounted to the torque converter and driven by gearing in the torque converter (see **Figure 7-32**). The brake pump requires very little in the way of maintenance other than making sure it gets a clean, unrestricted supply of hydraulic oil.

Accumulator. The piston-type accumulator used on brake systems is pre-charged with dry nitrogen. The amount of pre-charge can vary considerably from

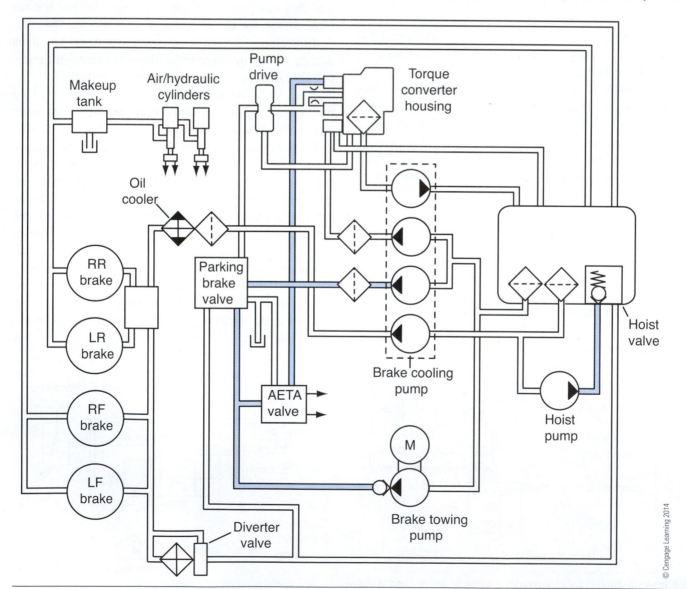

Figure 7-31 Schematic of a typical brake cooling circuit.

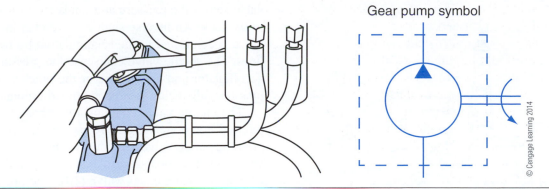

Gear pump symbol

Figure 7-32 Schematic of a typical brake pump and its location. In this case, it is mounted to the torque converter.

application to application. The amount of pre-charge and the charge procedure are covered in the maintenance and troubleshooting section in this chapter. The accumulator contains a free-floating piston that has high-pressure O-rings, which separate the dry nitrogen on one side and hydraulic oil on the other side. Its purpose is to store pressurized oil that can be used to apply the brakes when required. See **Figure 7-33** for the location of the two brake accumulators.

Because hydraulic oil does not compress measurably, the pressurized hydraulic oil loses its pressure quickly when released into the brake system. Nitrogen gas on one side of the accumulator piston helps maintain the hydraulic pressure in the accumulator circuit so that a number of brake applications can be used before the pressure gets low enough to start the accumulator charge process. Without any nitrogen pressure in the accumulator, the brake pressure will fall almost immediately. In the charged mode, the pressure on both sides of the

accumulator equalizes. When the operator applies the brakes, some of the hydraulic pressure is used up. The higher pressure on the nitrogen side of the piston forces the piston down, increasing the pressure on the hydraulic side until both are equal. The accumulators should be checked for nitrogen pre-charge on a regular interval. The piston seals can develop a leak, causing oil to enter the gas side of the accumulators or gas to enter the oil side. If the pre-charge is low, it is an indication that the nitrogen gas is leaking into the oil side.

Brake Charge Valve. The brake charge valve (see **Figure 7-34**) supplies oil flow when required to the brake circuit accumulators. The purpose of the charge valves is to maintain a predetermined amount of pressure in the brake accumulator circuit when the engine is running (see **Figure 7-35**). The flow control valve (8) regulates the flow of oil into the brake circuit. When the oil pressure requirements are met, the flow-control

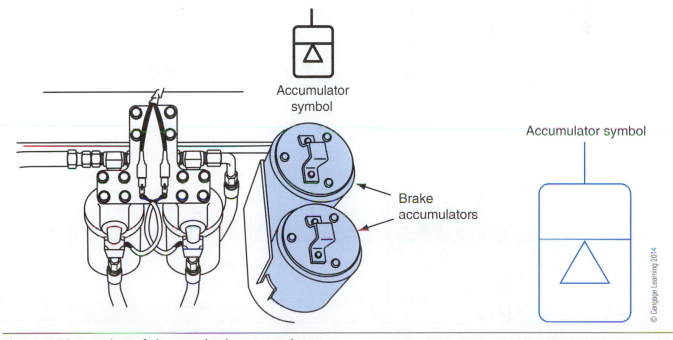

Accumulator symbol

Accumulator symbol

Brake accumulators

Figure 7-33 Location of the two brake accumulators.

valve senses the pressure and sends the flow of oil back to the tank. An inverse shuttle valve (12) in the charge valve is connected to the accumulators by two separate lines. The shuttle valve can sense the pressure in either accumulator and charge one or the other whenever the pressure falls below a predetermined amount.

The charge valve contains a number of specialized internal components, an inverse shuttle valve (12), a pressure-relief valve (6), cut-in and cut-out pressure valves (11), a flow-control valve (8), a flow-control orifice (9), and a check valve (10). The pressure-relief valve limits the maximum pressure in the brake circuit. A flow-control valve regulates the oil flow to the brake system, allowing any excess oil to flow back to the tank. A check valve stops the oil flow from the accumulator side of the circuit from returning to the tank.

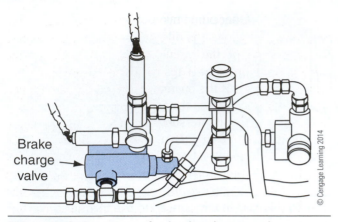

Figure 7-34 Location of a brake charge valve.

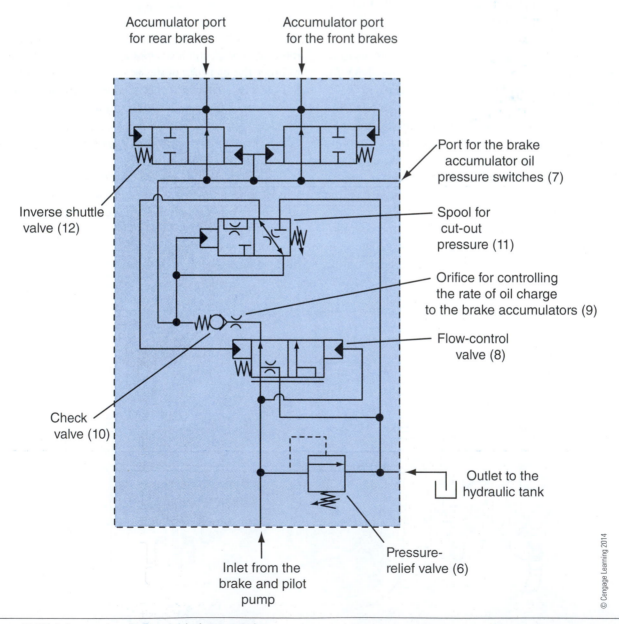

Figure 7-35 Schematic of a typical charge valve.

The cut-in and cut-out pressure circuit has a spool valve that is capable of sensing low charge pressure (cut-in pressure). When this occurs, the spool shifts, allowing the oil to charge the accumulator circuit. A flow-control valve also shifts at the same time to stop the flow of oil to the inverse shuttle valve. This allows the inverse shuttle valves to charge the two accumulators separately. A port (7) connects to two brake accumulator pressure switches: one pressure switch activates a low brake pressure light in the operator's compartment, and the other pressure switch sets the park brake if the brake pressure falls below a predetermined amount.

Brake Control Valve.

When the operator steps on the brake pedal, as shown in **Figure 7-36**, the upper piston forces the top spring to compress, moving the two spools down. When this takes place, the oil from the inlet passages flows to the brakes. Small passages located between the inlet side and the outlet side of the spools act as restrictions, limiting the amount of oils that can flow to the brakes. The more the pedal is depressed, the more oil will flow. This allows the brake pedal to function as a variable-pressure reducing valve by limiting the pressure flowing to the wheel end brakes. The more the pedal is depressed, the higher the pressure

will be at the wheel end brakes. A small internal passage in each spool allows the oil to flow from the modulated pressure side of the circuit to the underside of the spools. This provides a resistance to the pedal movement, which gives the operator feedback by providing a slight resistance. When the force on the upper springs is lower than the oil pressure underneath the spools, it forces them to move up. By metering the flow of oil, the brake valve acts like a variable-pressure reducing valve.

When the operator releases the brake pedal (shown on the left of **Figure 7-36**), the oil pressure below the spools and the lower spring returns the spools to the neutral position by sending the flow of oil from the wheel end brakes back to the reservoir and at the same time blocking the flow of oil from the inlet side to the brakes.

Accumulator Pressure Switch

The accumulator pressure switch is a hydraulic electric valve that monitors the accumulator pressure and triggers a park brake application when the pressure in the accumulators gets below a predetermined value. A second accumulator pressure switch turns on a low accumulator pressure light in the operator's compartment when the pressure gets below a predetermined pressure. The brake

BRAKE PEDAL OFF   **BRAKE PEDAL APPLIED**

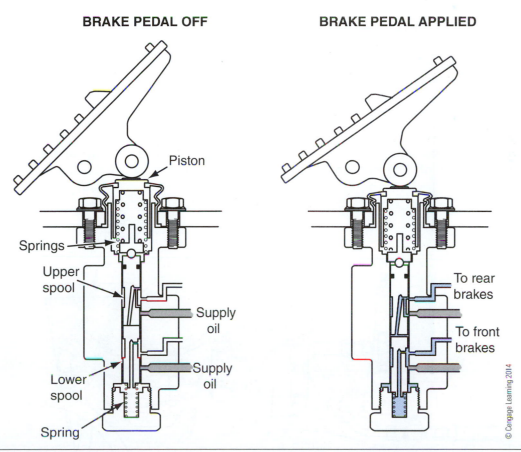

© Cengage Learning 2014

Figure 7-36 Cutaway of two brake valves. The one on the left is in the released position. The one on the right is in the applied position.

accumulator pressure switch will prevent the operator from releasing the brakes before a safe predetermined accumulator pressure is reached. Refer to **Figure 7-37**.

After the operator starts the engine, the charge valve will charge the brake accumulators until the pressure reaches the low limit of the pressure switch. Once the pressure reaches the pressure switch's low limit value, the electrical contacts open in the pressure switch and turn off the low brake pressure indicator light in the operator's compartment. The accumulator pressure switch also triggers a signal to the power train ECM that allows the park brakes to be released. In the event of a low brake pressure problem during operation, the accumulator pressure switch will turn on a brake impending indicator light in the operator's compartment and apply the park brake once the pressure falls below the switch's low limit value. The operator will not be able to release the park brake until the safe accumulator pressure is above the low limit value. When this occurs, the operator will have to cycle the park brake switch on and off before

the brakes can be released. An electrical interlock circuit prevents the accidental release of the park brake should it be applied inadvertently by a loss of brake pressure.

Park Brake Control Valve

The park brake control valve is an electric hydraulic valve that is activated by an electric control button in the operator's compartment. Shown in **Figure 7-38**, the valve controls the flow of oil to the spring-applied park brake. The park brake circuit cannot release the park brake until brake pressure is above a predetermined minimum pressure. In the event that the brake pressure were to fall below the predetermined safe operating pressure, the park brake would be set automatically by the accumulator brake pressure switch.

When the park brake switch in the operator's compartment is set to the off position, the solenoids on the park brake control valve are energized and the valve allows the oil to flow to the park brakes, disengaging them. When the operator activates the park brake, the solenoids on the park

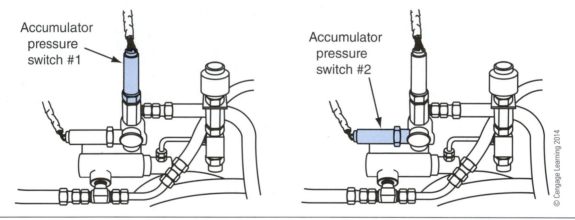

Figure 7-37 Location of two accumulator pressure switches. The one on the left is for the indicator light in the operator's compartment. The one on the right applies the park brake if the pressure drops below a predetermined value.

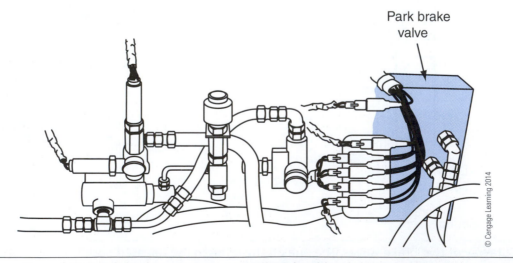

Figure 7-38 Location of the park brake valve on an R1700G load haul dump loader.

brake valve are de-energized. This also takes place when the operator turns the key switch to the off position or if brake pressure falls below a predetermined value. This allows the oil to drain from the park brakes back to the tank, activating the spring-applied park brakes.

Figure 7-39 shows the schematic view of the park brake valve in the on position. Solenoid (2) is de-energized, and the flow of oil from the pilot relief valve at port (5) is blocked by the valve spool (10). In this position the park brakes are applied by spring pressure. Oil from the park brakes is vented back to the tank through ports (9) and (7) through valve spool (10). When the operator releases the park brakes, solenoid (2) is energized and opens the port at valve spool (10), allowing oil to flow from the pilot port relief valve at port (5) to port (9) releasing the spring-applied brakes.

Service Brake Operation

The service brakes on this equipment are a hydraulically applied internal wet disc brake system. The park brake is a spring-applied oil pressure released system. Both systems use the same wheel end brake components to apply the brakes. Since the park brakes

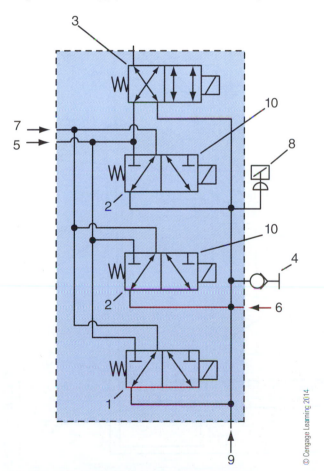

Figure 7-39 Hydraulic schematic for a park brake valve from an R1700G load haul dump loader.

© Cengage Learning 2014

are spring engaged, the service brakes cannot be applied when the park brake is engaged because of the design of the system (see **Figure 7-40**). After the engine is started, the brake/pilot pump (54) sends oil flow to the check valve (42) and on to the brake accumulator charge valve (59). The charge valve will charge the accumulators (17, 18) until the predetermined cut-out pressure is reached. The oil then flows to the service brake valve (39). When the operator actuates the service brake pedal, modulated oil pressure is directed to the wheel end brakes. The farther the pedal is depressed, the more pressure is sent to the wheel end brakes.

Park Brake Operation

When the operator applies the park brake (see **Figure 7-40**), the park brake solenoids are de-energized, which causes the park brake control valve (16) to redirect the flow of oil from the wheel end brake to the hydraulic tank. The park brake springs force the pistons to apply pressure to the friction discs, thus stopping the equipment.

When the operator releases the park brake, the park brake solenoids are energized to redirect the oil flow from the tank to the wheel end park pistons, which force the piston to retract against spring pressure and release the park brakes. As long as there is oil pressure behind the park brake pistons opposing the park brake spring pressure, the park brakes will remain in the released position. If for any reason brake pressure or engine power is lost, the park brakes will apply immediately.

TESTING/ADJUSTING AND TROUBLESHOOTING

This section provides examples of service and diagnostic procedures that are normally carried out by qualified technicians. When performing this type of work, always consult the manufacturer's service literature for the specific procedures to follow. Before any service or testing can be performed on any equipment, a technician must carefully read and understand the service literature that covers the service or testing procedures to be performed. Failure to follow the manufacturer's recommended procedures can lead to serious injury or equipment damage. Unless you have the proper training to carry out any service work or diagnostic testing, the work is best left to qualified technicians.

Performing service work on hydraulic systems requires that technician follow established safety procedures when working with pressurized hydraulic circuits. Disposal of hydraulic fluids must always be carried out in accordance with local laws and established approved, safe practices.

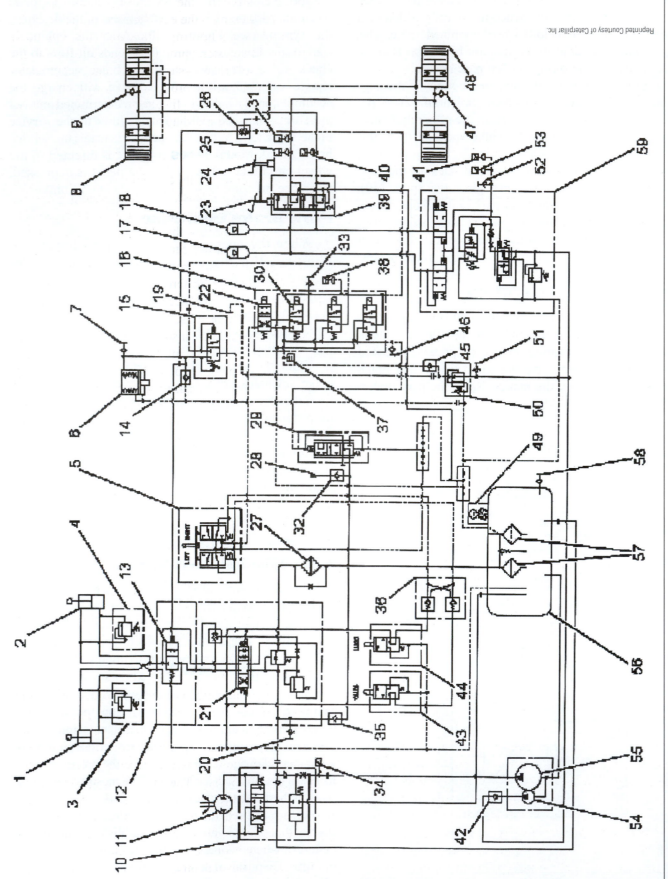

Figure 7-40 The R1700G loader brake schematic.

Tech Tip: High-pressure oil can remain in a hydraulic brake circuit for an indefinite period of time after the engine has been shut down: this is known as residual pressure. Failure to relieve residual pressure can result in serious injury or death. Never use your hands, even if you are wearing gloves, to detect any hydraulic leaks in a hydraulic circuit; use cardboard for this purpose. High-pressure fluid escaping from a pin-sized hole can cause severe injury or death. High-pressure oil can penetrate the skin, causing severe injury. Always seek medical attention immediately if this type of injury is suspected.

Before any work can begin, the equipment must be parked on level ground and have wheel chocks installed on both sides of the tires. Be sure that any implements attached to the equipment are lowered to the ground. The steering frame lock must be installed along with any appropriate lockouts or tags that are required. If any steering lock mechanisms or transmission locks are present on the equipment, ensure that they are engaged. To relieve any residual pressure in the main hydraulic system, turn on the key switch and activate the steering and dump hoist controls. Depress the brake pedal several times to relieve the brake **system pressure**. If the system has a pressurized hydraulic tank, depress the hydraulic tank relief valve, to release the pressure. Be sure you have a suitable container to catch any fluid that may be released. Visually inspect the complete brake system for any signs of oil leakage or damaged components. Before any testing takes place, be sure the oil is at operating temperature. As with all diagnostic procedures, start at the beginning, and this usually means the tank. Check the oil level; look for indications of air in the oil (aerated oil). Place a small amount of oil in a clear container and visually examine it for air bubbles. *Note*: This must be performed immediately after the equipment is shut down because bubbles disperse within minutes. Inspect the filter element for excessive dirt; use a magnet to determine if the particles are metallic. Check the condition of all hydraulic lines and fittings for obvious damage. In diagnostics, there are usually more than one root causes to a performance complaint. Before performing specific tests, read over the troubleshooting section in the service literature to review any common problems or symptoms before beginning diagnostic procedures. Ensure that the accumulator nitrogen charge pressure is correct. Be sure that the oil you are using is the correct viscosity for the operating conditions and temperatures in which the equipment is operating. The following troubleshooting chart outlines the most common brake system problems encountered on internal wet disc brake systems. Use **Table 7-3** as an example of a diagnostic procedure. Always refer to the manufacturer's service literature for the specific equipment that you are diagnosing.

Brake System Pressure Release

This procedure is an example of the steps that should be taken to release the brake system pressure; always consult the equipment's service literature for the specific procedures that are to be used on the equipment that you are servicing. Not all procedures are identical; they vary from equipment to equipment. They may even be different on the same type of equipment that was produced at a different location or time. The equipment must be parked on level ground and have wheel chocks installed on both sides of the tires. Be sure that any implements attached to the equipment are lowered to the ground. The steering frame lock must be installed along with any appropriate lockouts or tags that are required. If any steering lock mechanisms or transmission locks are present on the equipment, ensure that they are engaged.

Procedure

- Set the park brake and stop the engine before beginning.
- Make at least 80 applications to the service brake to relieve the accumulator hydraulic pressure. Observe the brake pressure gauge on the dashboard of the equipment. Note that there is a gauge for each brake circuit, front and rear. The gauges should read 0 after all the pressure is relieved with the brake pedal depressed.

Brake Pump Flow Test

The brake pump test is designed to determine the condition of the pump and its ability to pump oil. Pump flow tests are conducted at two different pump speeds to assess the pump's condition. The difference between pump flow of two different operating pressures is called *flow loss*. The specifications used in this example (see **Table 7-4** and **Table 7-5**) are for reference purposes and do not represent actual flows for any given pump. Always refer to the manufacturer's service literature for the equipment in question.

To perform this test on the equipment, a flow meter must be installed to measure the pump output. An example of actual flow loss is shown in **Table 7-6**.

TABLE 7-3: HYDRAULIC BRAKE TROUBLESHOOTING	
Problem	**Possible Cause**
Service brakes are slow to apply.	• Nitrogen gas charge is low • Hydraulic fitting or lines are leaking • Brake discs are worn beyond limits • Aeration of the brake fluid • Damaged hydraulic brake lines
Accumulator pressure will not reach cut-out pressure at engine idle.	• Low brake pump output, excessive brake pump wear
No front service brakes.	• Brake lines to the front brakes are pinched or twisted • Internal damage to the front brake components • Defective service brake control valve
No rear service brakes.	• Brake lines to the rear brakes are pinched or twisted • Internal damage to the front brake components • Defective service brake control valve
Park brake does not engage.	• Defective park brake control switch • Sticky park brake solenoid control valve
Service brakes do not hold.	• Internal brake discs worn beyond permissible tolerances
No warning system activation.	• Defective pressure sensor or switch • Blown circuit fuse or display
Service brakes do not apply evenly.	• Uneven brake disc wear • Brake control valve is sticking during use • Low brake pressure to either brake circuit
Excessive force required to depress the brake control valve.	• Sticking brake control valve due to oil contamination
Service brakes not releasing completely.	• Wheel end brake component damage • Sticking brake control valve due to oil contamination • Obstruction under the brake pedal; dirt or rocks preventing the pedal from releasing completely
Brake pedal kick-back during brake application.	• Air in the brake system

© Cengage Learning 2014

TABLE 7-4: DETERMINING PUMP FLOW LOSS
Method of Determining Flow Loss
Pump Flow at 690 kPa (100 psi)
Pump Flow at 6,900 kPa (1,000 psi)
Flow Loss
Example of Determining Flow Loss
217.6 L/min (57.5 US gpm)
196.8 L/min (52 US gpm)
Loss = 20.8 L/min (5.5 US gpm)

© Cengage Learning 2014

Brake Accumulator Charge and Test Procedure

This procedure is an example of the steps that should be taken to check and charge a brake accumulator. Always refer to the equipment's service literature for the specific procedures that are to be used on the equipment that you are servicing. Not all procedures are identical; they vary from equipment to equipment. They may even be different on the same type of equipment that was produced at a different location or time. The equipment must be parked on level ground and have approved wheel chokes installed on both sides of the tires. Be sure that any implements attached to the equipment are lowered to the ground. The steering frame lock must be installed along with any appropriate lockouts or tags that are required. If any steering lock mechanisms, such as transmission locks, are used on the equipment, ensure that they are engaged.

Tech Tip: The only gas that can be used to charge accumulators is dry nitrogen gas. No other gases are permitted for this procedure; gases, such as oxygen, may create explosions and should never be used for this purpose. Be sure that the nitrogen cylinders have the appropriate connections to match the hoses

TABLE 7-5: DETERMINING PUMP FLOW LOSS

Method of Determining Flow Loss

$\dfrac{\text{Flow Loss (L/min or US gpm)}}{\text{Pump Flow at 690 kPa (100 psi)}}$	× 100	=	Percent of Flow Loss

Example of Determining Flow Loss

$\dfrac{\text{(20.8 L (5.5 gpm))}}{\text{(217.6 L/min (57.5 US gpm))}}$	× 100	=	9.5%

© Cengage Learning 2014

TABLE 7-6: ACTUAL PUMP FLOW LOSS

Example of Actual Pump Flow Loss

$\dfrac{\text{Pump Flow at 6,900 kPa (1,000 psi)}}{\text{Pump Flow at 690 kPa (100 psi)}}$	× 100	=	Percent of Flow Loss

© Cengage Learning 2014

and gauges. Before beginning this procedure, make sure that any residual hydraulic pressure is relieved from the brake system.

To ensure that the nitrogen charge pressure is correct, the hydraulic pressure must be completely relieved with the accumulator piston bottomed out in the accumulator. To ensure that the correct pressure is present in the accumulator, a temperature correction chart should be consulted before beginning this procedure. In our example (see **Table 7-7**) we will use 800 psi as the desired charge pressure at 70°F (21°C).

To check the nitrogen charge in the accumulator, connect a 0 to 2,000 psi (0 to 13,800 kPa) pressure

TABLE 7-7: PRESSURE/TEMPERATURE RELATIONSHIP

Accumulator Pre-charge Pressure/Temperature Relationship for 5,500 kPa (800 psi)

20°F (27°C)	725 psi (5,000 kPa)
30°F (21°C)	740 psi (5,105 kPa)
40°F (4°C)	760 psi (5,235 kPa)
50°F (10°C)	775 psi (5,340 kPa)
60°F (16°C)	785 psi (5,420 kPa)
70°F (21°C)	800 psi (5,500 kPa)
80°F (27°C)	815 psi (5,605 kPa)
90°F (32°C)	830 psi (5,735 kPa)
100°F (38°C)	845 psi (5,840 kPa)
110°F (43°C)	860 psi (5,920 kPa)
120°F (49°C)	875 psi (6,000 kPa)

© Cengage Learning 2014

gauge to the connection on the top of the accumulator. (Refer to the service literature for the equipment being serviced for the correct procedures.) If the reading on the gauge is high, release some of the excess nitrogen. If the pressure is low, follow the general steps in the service literature for the equipment being serviced to charge the accumulator.

The steps used to charge an accumulator are outlined below.

Procedure

- Connect the appropriate hose to the nitrogen cylinder to valve (3) as shown in **Figure 7-41**.
- Close valve (3) and open the nitrogen cylinder. Adjust screw (1) on the regulator (9) until the pressure on gauge (7) is at the correct charge pressure.

Note: Be sure to compensate for ambient temperature variations.

- Open the valve to the accumulator (3) and proceed to charge the accumulator. Once the pressure is stabilized in the accumulator, close valve (3). Pressure on gauges (4) and (7) should be the same.
- Repeat the above procedure if the pressures do not match.
- Close valve (3) and close the nitrogen cylinder valve.
- Remove the chuck and test equipment and reinstall the protective cap onto the accumulator.

When installing a new or reconditioned accumulator, perform the following steps to purge the accumulator of

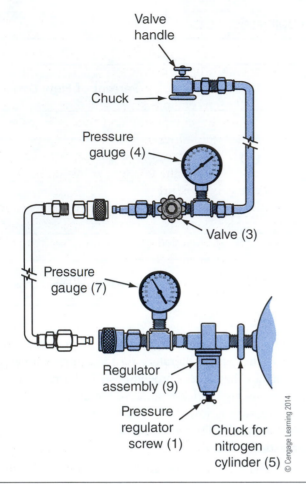

Figure 7-41 The charge procedure hardware required to charge a brake.

any air trapped inside. Then perform the steps to charge an accumulator outlined earlier. All of the air must be purged from the gas side of the accumulator before any nitrogen is introduced into the accumulator. Pour approximately 2 quarts (1.9 L) of SAE 10W oil into the gas side of the accumulator. This is necessary to lubricate the top piston seal and ensure that all of the air is removed from the accumulator.

Procedure

- Install the accumulator on the equipment.
- Refer to **Figure 7-41** to identify the equipment that is needed to connect the nitrogen cylinder with the appropriate gauges and adapters to properly charge the accumulator.
- Install the chuck (5) to the accumulator nitrogen charge valve with the other end of the charge hose pointing into a suitable container capable of holding the oil from the nitrogen side of the accumulator.
- Start and run the engine until it reaches normal operating temperature. As the charge valve

charges the bottom section of the accumulator, the air and oil in the top of the accumulator will be forced out into the container. When the oil ceases to flow from the hose, close valve (3) completely.

- Stop the engine and relieve the brake system pressure as outlined earlier in this chapter under the heading "Brake System Pressure Release" or in OEM service literature.

Now follow the steps to charge the accumulator to complete the procedure.

Accumulator Charge Valve Test Procedure

The following procedure is an example of the steps that must be used to check accumulator charge valve operation. Always consult the equipment service literature for the specific procedures that are to be used on equipment that you are servicing. Not all procedures are identical and they may vary from equipment to equipment; in some cases they may even be significantly different for the same model of equipment that was manufactured at a later or earlier time. The equipment must be parked on level ground and have approved wheel chocks installed on both sides of the tires. Be sure that any implements attached to the equipment are lowered to the ground. The steering frame lock must be installed along with any appropriate lockouts or tags that are required. If any steering lock mechanisms, such as transmission locks, are present on the equipment, ensure that they are engaged.

Procedure

- Before proceeding, ensure that the engine is not running.
- Depress the service brake pedal until all the pressure is removed from the brake accumulators. The number of times that you must depress the brake pedal will vary from equipment to equipment. In our example, we will use 75 applications.
- Connect a pressure gauge (0–25,000 kPa or 0–3,600 psi) to the gauge connection pressure tap shown in **Figure 7-42**.
- Observe the pressure reading before starting the engine. The pressure should be 0 psi.
- Start the engine and build up the charge pressure until it reaches cut-out pressure. Refer to the manufacturer's service literature for the correct specifications. In our example, we will use 14,480 kPa (2,100 psi). Shut off the engine at this point.

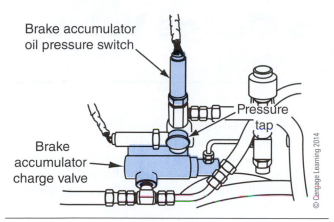

Brake accumulator oil pressure switch

Pressure tap

Brake accumulator charge valve

© Cengage Learning 2014

Figure 7-42 Location at which a gauge must be connected to check accumulator charge valve operation on a CAT R1700G load haul dump loader. Brake accumulator oil pressure switch for the alert indicator, pressure tap (brake accumulator oil), brake accumulator charging valve.

- Turn the key to the on position without starting the engine.
- Apply the service brakes repeatedly and count the number of times you apply the brakes while checking the pressure on the service gauge.
- Once the gauge needle drops to 8,300 kPa (1,200 psi) the accumulator oil pressure gauge on the dashboard should illuminate. If this occurs as described, then the accumulator low oil pressure alert is working properly.
- Apply the service brakes and continue counting the brake applications while observing the gauge pressure. At one point the pressure will drop off rapidly; this is an indication of how much nitrogen pre-charge pressure is actually in the accumulator.
- Add up the number of brake applications required to get to the point where the gauge drops off rapidly. Check the manufacturer's service literature to determine the number of applications that is correct for the equipment you are servicing. In our example, we will use 75 brake applications as the minimum amount.
- If the cut-out pressure is not correct, refer to the manufacturer's service literature for the adjustment procedures. Adjustment procedures will vary from equipment to equipment. Not all charge valves have a cut-out pressure adjustment that can be readily performed in the field. It may be necessary to replace the charge valve in some cases (see **Photo Sequence P2-6**). (Accumulator Charge Valve Pressure Check).
- The cover for spring 4 in the diagram must be removed to gain access to the shim pack (5).

- Add or remove shims according to requirements. One shim added or removed will change the pressure by approximately 33 psi (230 kPa).
- Reinstall the brake charge valve and recheck the pressures as outlined previously.
- After the cut-out pressure is within specification, run the equipment and apply the brake pedal as many times as it takes to get the pressure to the point where it will start to charge backup. Observe the pressure at this point. This is the cut-in pressure.
- If the pressure difference between the cut-in and cut-out pressure is not within specifications, shims must be added or removed in location (6) (see **Photo Sequence P2-6**).
- Remove the charge valve and add or remove the correct amount of shims to achieve the correct spread in pressures.
- Reinstall the charge valve and retest the pressures for correct values. Reinstall any additional hardware that was removed to access the brake charge valve.

ONLINE TASKS

1. Use an Internet search engine to research hydraulic brake oil specifications. Identify a piece of equipment in your particular location, paying attention to the operating temperature conditions that exist in your area. Outline the reasoning behind selecting this particular type of hydraulic oil.
2. Check out this URL: *www.hydraulic-supply.com/html/productline/mfgprod/vickers-hydraulics.htm* to discover important facts and classifications of hydraulic oil used on off-road hydraulic brake circuits.

Shop Tasks

1. Select a piece of equipment in your shop that has a hydraulically applied brake system and identify the components that make up the brake circuit. Identify the type of brake hoses and lines that are used and the manufacturer's specifications for these components.
2. Select a piece of mobile equipment in your shop. Identify the type and model of the equipment and list all the components and the corresponding part numbers of a typical hydraulically applied brake circuit. Use the correct parts book for that particular piece of equipment.

PHOTO SEQUENCE 2 Hydraulic Brakes Charge Valve Pressure Check

The following example is a general procedure used to check brake charge valve pressure and operation on equipment that has hydraulically applied internal wet disc brakes. It is not intended to replace service literature which should always be referenced, and is only shown as an example of the steps required to perform this type of procedure. Before beginning this procedure, be sure to ensure that the oil flow for the brake circuit is within specifications. Restriction to flow in any hydraulic circuit determines the correct pressure values.

CAUTION *Never use your hands to check for leaks in any high-pressure hydraulic circuit. Use a small piece of cardboard. A pin-size leak can allow high pressure oil to unknowingly penetrate skin. If oil is accidentally injected into the skin, immediate medical attention is required. Also note that oil in a hydraulic circuit can become extremely hot during operation. This can lead to severe burns if accidentally splashed onto yourself or other personnel in the work area.*

© Cengage Learning 2014

P2-1 Depending on the employer, some form of tagging or lock-out procedure must be used to protect the technicians while work is being performed on the equipment.

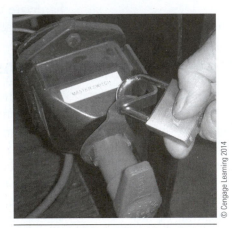

© Cengage Learning 2014

P2-2 Some employers require you to actually lock out the main power disconnect on the equipment. This lock can be installed and removed only by the technician who is performing the work on the equipment. It must be removed on completion of the work.

© Cengage Learning 2014

P2-3 Proceed by lowering the brake system pressure by activating the brake valve as many times as necessary to get the brake pressure to go to zero. The foot brake valve is located in the operator's compartment.

(Continued)

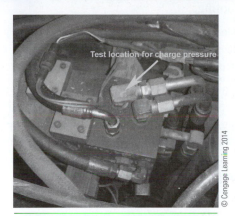

P2-4 Connect a suitable gauge to the fitting. A 0 to 3,000 psi (20,700 kPa) hydraulic pressure gauge must be used. At this point, the gauge should read zero after it is connected and tightened.

P2-5 Have someone start the engine and observe the reading on the pressure gauge as it increases. A short pause in oil pressure rise indicates the accumulator nitrogen precharge pressure. Pressure will continue to rise to a point where it will pause again. This pause point is the cut-in pressure. Continue observing the pressure rise until it reaches its cut-out pressure and compare to specifications.

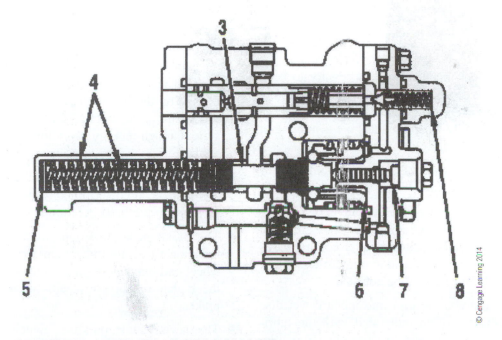

ACCUMULATOR CHARGING VALVE

P2-6 If the cut-out pressure is not within specifications, the following adjustments can be made. Relieve the oil pressure in the brake circuit as outlined previously. The brake charge valve must be removed from the equipment to make the necessary adjustments.

Summary

- All brake systems rely on friction to slow or stop the equipment. Brake systems convert the kinetic energy of moving equipment to heat energy when the brake pedal is depressed.

- On heavy equipment, the use of two independent brake circuits is usually required. In addition, the use of an independent mechanical park brake is also required to safely park the equipment. In some industries such as mining, the equipment is required by law to be equipped with three independent brake circuits.

- On large equipment where brake pressures need to be considerably higher than what can be achieved under what would be considered normal operator foot pressure, additional hydraulic pressure can be used to multiply brake force without exerting extra pedal effort on the part of the operator.

- On most off-road equipment, brake shoes are generally pinned on the bottom and have a wheel cylinder at the top of the foundation assembly. The function of the wheel cylinder is to force the shoes into contact with the inside of the drum during brake actuation.

- Hydraulic drum brakes, regardless of whether they are a non-servo or servo brake design system, require a wheel cylinder to actuate them. The wheel cylinders use hydraulic pressure from a brake valve to force the shoes out against the drum.

- A self-adjustment mechanism is necessary to keep the brake shoes close to the drum and take up any clearance caused by drum and shoe wear. Adjustment can occur only when the machine is moving in the reverse direction.

- External hydraulic disc brake systems come in two basic designs; fixed-caliper and sliding-caliper.

- Disc brakes have many obvious advantages over drum brakes on off-road equipment. They are more resistant to brake fade, provide better cooling, have better tolerance to contamination, require less maintenance and do not need to be adjusted.

- Lubrication and cooling on some inboard wet disc brake systems are often integrated with the lubrication and cooling for the axle assembly and require the use of specialized lubricants. In heavy-duty applications, a separate cooling circuit that has its own oil pump and cooler can be used to dissipate the heat generated by this type of brake system.

- Inboard hydraulic wet disc brakes allow the brake and differential to share the same oil for lubrication and cooling.

- When equipment is parked, it must not rely on hydraulic pressure alone: a mechanical park brake is required.

- Many mechanical park brakes consist of two sets of discs. Steel disks are externally splined to the housing and do not rotate. Friction discs are splined to a rotating shaft. A pressure plate has a number of heavy-duty springs on one side of it and the hydraulic piston on the other side. Either one can lock the friction and steel discs together.

- In North America, brake fluid standards are quantified by two groups: the Society of Automotive Engineers (SAE) and the Department of Transportation (DOT). Conventional brake fluid has three categories: DOT 3, DOT 4, and DOT 5.

- DOT 3 brake fluids have a dry boiling point of approx 400°F (205°C). Once the brake fluid absorbs some moisture, the boiling point will quickly fall to around 284°F (140°C).

- DOT 5 brake fluid is formulated from a silicone base and does not absorb water; in other words it is non-hygroscopic.

- Two accumulators are used in a typical off-road brake circuit; one for the front service brakes and one for the rear service brakes. Nitrogen precharged accumulators store hydraulic energy, which is used by the brake circuit to apply the brakes. In the event of an engine failure, the accumulators store enough energy to allow the brakes to be applied.

- The purpose of the accumulator charge valve is to maintain a predetermined amount of hydraulic pressure in the brake accumulator circuit.

- The brake pedal functions as a variable-pressure reducing valve, limiting the pressure flowing to the wheel end brakes. The more the pedal is depressed, the higher the pressure will be at the wheel end brakes.

- The accumulator pressure switch is a hydraulic electric valve that monitors the accumulator pressure and triggers a park brake application when the pressure in the accumulators gets below a predetermined value.

- When the park brake switch in the operator's compartment is set to the off position, the solenoids on the park brake control valve are energized and the valve allows the oil to flow to the spring-applied park brakes.

- On spring-applied park brake systems, when there is oil pressure acting on the park brake pistons opposing the park brake spring pressure, the park brakes remain in the released position.

- High-pressure oil can remain in a hydraulic brake circuit for an indefinite period of time after engine shut down. Failure to relieve residual pressure can result in serious injury or death.

- The only gas that can be used to charge accumulators is dry nitrogen gas.

- To ensure that the nitrogen charge pressure is correct, the hydraulic pressure must be totally relieved with the accumulator piston bottomed out in the accumulator.

- Not all charge valves have a cut-out pressure adjustment that can be performed in the field.

Review Questions

1. How many independent brake circuits are generally found on heavy equipment?

   A. one

   B. two

   C. three

   D. four

2. Which statement describes the hydraulic brake system principle on a non-servo drum brake?

   A. Actual braking force at the wheels is dependent on mechanical force.

   B. Application force is multiplied hydraulically by the wheel cylinders.

   C. Actual braking force at the wheels is not dependent on mechanical force.

   D. Application force is multiplied hydraulically by the operator foot.

3. Which brake fluid does not absorb moisture?

   A. DOT 2

   B. DOT 3

   C. DOT 4

   D. DOT 5

4. What component provides hydraulic pressure to operate hydraulic brake boosters?

   A. power steering pump

   B. reserve electric motor and pump

   C. dedicated hydraulic pump

   D. all of the above

5. Which brake shoe in a non-servo drum brake assembly is energized during braking in reverse?

   A. leading shoe

   B. both shoes

   C. trailing shoe

   D. neither shoe

6. What type of parking brake is used on off-road equipment with hydraulic brake systems?

   A. hydraulic canister brake driveline mounted

   B. driveline-mounted hydraulic applied drum brake

   C. cable and lever actuated secondary shoe brake

   D. driveline-mounted, mechanically applied brake

7. What occurs when the operator depresses the brake pedal of the loader with hydraulic internal wet brakes?

A. Hydraulic pressure actuates the mechanical booster assemblies.

B. Hydraulic fluid flows through the quick-release valves.

C. Modulated pressure actuates wheel end brake assembly.

D. Modulated hydraulic pressure actuates the diverter valve.

8. What must be done before forcing a caliper piston back into its bore on a wheel cylinder during servicing?

A. Open the bleed nut and drain the displaced fluid.

B. Open the master cylinder lid to allow the fluid to return to the reservoir.

C. Drain the master cylinder to prevent spillage.

D. Remove the brake line to drain the caliper.

9. What combination of materials is used in a multi-disc brake assembly on heavy equipment?

A. steel and friction discs

B. friction discs and springs

C. steel discs and brass discs

D. brass discs and friction discs

10. What is the purpose of the brake pump in a wet disc brake system?

A. supply oil to the accumulator and the pilot circuit

B. supply oil to hydraulic applied spring release brakes

C. supply oil to the spring applied hydraulic release brakes

D. supply oil to the park brake system calipers

11. What occurs when the internal seal on the spindle of a hydraulic applied wet disc wheel end leaks?

A. Hydraulic oil will enter the differential.

B. Hydraulic oil will enter the wheel end.

C. The differential oil will enter the wheel end brakes.

D. Hydraulic oil will leak externally.

12. What advantage does a multi-disc brake system have over conventional brake systems?

A. more swept area

B. less moving components

C. braking surface area is minimized

D. less cooling required

13. What does the term *charge valve cut-out pressure* refer to on an internal wet disc brake system?

A. when the charge valve stops supply to the accumulator

B. when the charge valve begins to supply the pilot circuit

C. when the charge valve begins to supply the accumulator

D. when the charge valve stops charging the reservoir

14. What type of device is the accumulator pressure switch in a wet disc brake system?

A. hydraulic/electric

B. air brake

C. electric/air

D. air/hydraulic

15. What is the purpose of the accumulator(s) in a hydraulically applied wet disc brake system?

A. to store hydraulic energy for brake application

B. to apply the park brakes if engine power were lost

C. to release the driveline park brake when brake pressure is low

D. to apply the service brakes

8 Spring-Applied Hydraulically Released Brakes

Learning Objectives

After reading this chapter, you should be able to:

- Describe the fundamentals of off-road spring-applied hydraulically released brake systems.

- Identify the major components that make up spring-applied hydraulically released brake systems.

- Describe the principles of operation of spring-applied hydraulically released brake systems.

- Outline the maintenance and repair procedures associated with spring-applied hydraulically released brake systems.

- Identify the basic troubleshooting procedures for spring-applied hydraulically released brake systems.

Key Terms

accumulator

application pressure

brake pressure

brake pump

charge valve

check valve

foot valve

friction discs

park brake valve

pilot pressure

pressure switch

reverse modulation

sequence valve

steel disc

system pressure

wheel end pressure

INTRODUCTION

Chapter 7, "Hydraulic Brakes," covers the basic principles of internal hydraulically applied wet brakes and these should be understood before studying this chapter. Spring-applied, hydraulically released brakes are designed to function with a highest level of redundancy: a loss of hydraulic pressure results in the immediate application of the brakes. **Figure 8-1** shows the conditions an excavator equipped with spring-applied, hydraulically released brakes works under.

BASIC BRAKE SYSTEM COMPONENTS

A spring-applied hydraulically released brake system (see **Figure 8-2**) shares similar components that are used in a hydraulically applied brake circuit. The use of a multi-disc wheel end brake design is similar to those used in hydraulically applied brakes. The main difference is that the heavy springs in the assembly are used to actually apply the brakes. Hydraulic pressure is used to hold off the brakes in the released position in this design.

Figure 8-1 Typical excavator that utilizes a spring-applied hydraulically released brake system.

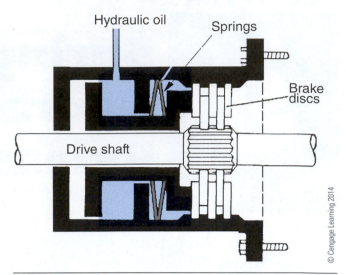

Figure 8-2 Cutaway of a spring-applied hydraulically released brake system used on an excavator final drive.

Brake Pump

The **brake pump** (see **Figure 8-3**) used in this type of brake circuit is usually a heavy-duty gear-type pump capable of supplying a continuous oil flow of up to 19 gpm (gallons per minute) or 72 L/m (liters per minute) at 2,400 rpm. In many applications, this pump also supplies the oil flow for the brake cooling system, as well as the equipment's pilot circuit. Gear pumps are used because of their low cost and ability to operate in less-than-ideal operating conditions. The pumps are generally mounted to the torque converter and are located in a high heat area. In some applications, manufacturers are using variable-flow piston

pumps, which can be de-stroked and thus eliminate the need for a **charge valve** in the brake circuit. The downside to using a variable-displacement pump for the brake circuit is the higher cost when compared to a gear pump. The upside is the cost saving of not having to use a charge valve in the brake circuit to regulate brake pressure.

Hydraulic Accumulators

Although there are different accumulator designs, those used in brake systems are all similar in operation. As shown in **Figure 8-4**, three basic designs can be found in brake systems. The piston-type pneumatic accumulator is probably the most popular design in use, with the bladder-type generally coming in second. A third type, the diaphragm-type is also found on some types of equipment. These accumulators all perform the same function: they store hydraulic energy that can be used by the brake circuit when required. The size of

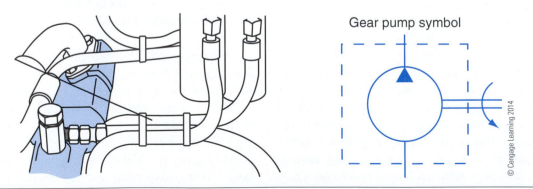

Figure 8-3 Schematic of a typical brake pump. It is mounted to the torque converter.

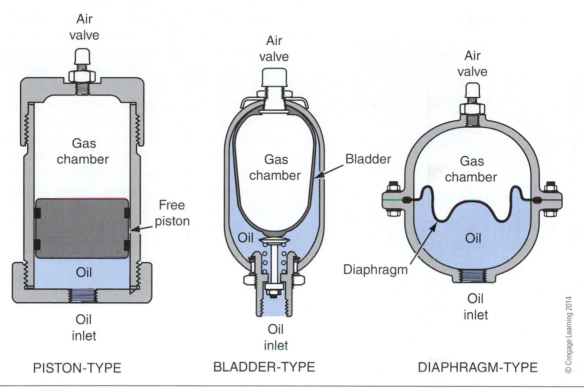

Air
valve

Air
valve

Air
valve

Gas
chamber

Gas
chamber

Bladder

Gas
chamber

Free
piston

Oil

Oil

Diaphragm

Oil

Oil

Oil
inlet

Oil
inlet

Oil
inlet

PISTON-TYPE   BLADDER-TYPE   DIAPHRAGM-TYPE

© Cengage Learning 2014

Figure 8-4 The three most common designs of accumulators used on off-road equipment brake circuits.

the accumulators depends on the design of the brake system; the larger the equipment, the larger the accumulator capacity.

Piston-type accumulators, as shown in **Figure 8-5**, consist of a cylindrical tube that have a free-floating piston inside to keep the hydraulic oil and nitrogen separated. The diaphragm-type of accumulator uses a synthetic rubber diaphragm that separates the hydraulic oil from the nitrogen gas. The bladder-type of accumulator (see **Figure 8-6**) has a metal shell that is fitted with a synthetic flexible bladder. The bladder is pre-charged with nitrogen, and the metal shell contains the hydraulic oil. A poppet valve, located at the bottom hydraulic port, keeps the bladder contained in the accumulator when the hydraulic pressure is discharged. The use of accumulators on a brake circuit enables the use of a much smaller hydraulic pump to keep the system charged, preventing excessive pump operation that requires additional horsepower to turn.

Only approved connectors are to be used to charge accumulators. Always refer to the manufacturer's service literature before attempting to service these components. Only dry nitrogen gas is permitted for use as a charging agent in an accumulator; under no circumstances should air or oxygen be used because they support the potential of an explosion if mixed with oil and high temperatures.

Tech Tip: The accumulator is classified as a high-pressure vessel and must be treated accordingly. Under no circumstance should any welding or brazing be done on the accumulator casing. If physical damage to the casing is discovered, it should be discarded. If damage occurs during operation, the pressure must be relieved before the accumulator is replaced.

- Always read the manufacturer's service literature that comes with the replacement accumulators. Some manufacturers ship the accumulators with a low pre-charge (10 bar/150 psi) to prevent damage during shipping.
- Under no circumstances should tire valves be used to replace the gas valve in an accumulator. Tire valves are not designed to withstand the high pressures used in an accumulator.

Piston-Type Accumulators

Piston-type accumulators have a free-floating piston that keeps the hydraulic oil and high-pressure nitrogen separated. The piston has a high-pressure seal

CHARGING

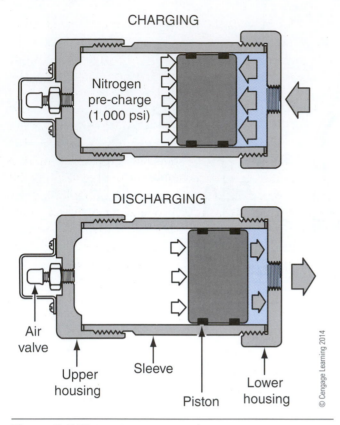

Figure 8-5 Piston-type accumulators.

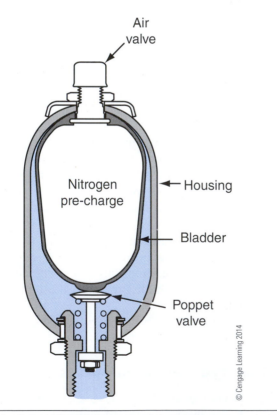

Figure 8-6 A bladder-type accumulator.

that is replaceable. Bladder-type accumulators (shown in **Figure 8-6**) have a flexible bladder that is not serviceable in the field. In most cases, when these accumulators develop a leak, they are bled down and replaced. Accumulators (shown in **Figure 8-4**) must be checked for pre-charge on a regular basis usually during routine maintenance and servicing. Low accumulator pre-charge will cause excessive cycling of the charge valve, leading to high oil temperatures. Always refer to the manufacturer's service procedures before attempting to service a brake system that uses accumulators. The high pressures used in the accumulator can cause severe injury or death if not handled properly. Many manufacturers recommend that the accumulator pre-charge pressure be checked after approximately three months in service, then after six months of service, and then annually.

Figure 8-7 shows the location of the brake charge valve. The charge valve directs pressurized oil to the oil side of the accumulator. As the pressure rises on the oil side of the accumulator, it forces the piston to compress the 1,000 psi (83 bar) pre-charge. As the piston retracts, the pressure in the pre-charge rises with the oil pressure and equalizes to the cut-out pressure. When the brakes are operated the pressure will drop with each brake application. The higher pressure above the piston on the gas side will force the piston down to try to maintain equal pressure on both sides. This continues until the cut-in pressure is reached at 1,600 psi (110 bar). At this point the charge valve will start the charge process over again. The stored energy in the accumulator allows the operator to maintain an adequate oil pressure reserve to ensure that full brake release can be achieved at any time during operation.

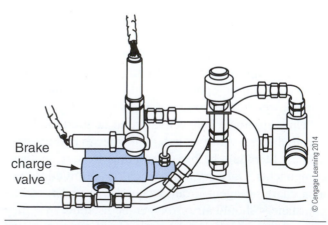

Figure 8-7 Location of the brake charge valve.

Accumulator Charge Valve

The accumulator charge valve, shown in the schematic in **Figure 8-8**, is used to regulate oil flow from the brake pump and to charge the accumulators, which store the energy for use in the brake circuit during equipment operation. The charge valve controls the rate of charge as well as the cycle rate for the brake circuit. In the fully charged position, the oil from the brake pump flows to the dump/hoist and steering pilot circuit. During operation, as shown in **Figure 8-9**, when the pressure falls below a predetermined value (cut-in pressure), the charge valve diverts the oil flow to the accumulators until they are fully charged (cut-out pressure). The charge pressure values are not the same on every type of equipment, but they are generally similar. When the accumulators are fully charged, the oil is diverted to the **pilot pressure** valve and then to the brake cooling circuit.

In **Figure 8-10** the charge valve has completed charging the accumulators to 2,000 psi (138 bar) cut-out pressure. The spring force on spool #2 has been overridden by the oil pressure in line C and shifts the spool so that the oil flow is blocked through line A.

Oil pressure from the pump inlet through line B shifts spool #1 over to send all the oil through a non-restricted port to the dump/hoist/steering pilot circuit.

After a series of brake applications, the accumulator pressure is depleted to where the pressure falls below the cut-in pressure 1,600 psi (110 bar) in our example. This shifts spool #1 over, which restricts the flow of oil to the dump/hoist pilot circuit. The oil from the brake pump now flows through an internal filter and unseats the internal **check valve** sending the oil flow to charge the **accumulators**. Spool #2 shifts to permit the oil to flow through line A, which helps keep spool #1 in the accumulator charge position. The charging continues until the pressure in the accumulators reaches cut-out pressure, 2,000 psi in our example. Once the charge cut-out pressure is achieved, the pressure in line B shifts spool #1 over, sending the pump flow to the dump/hoist/steering pilot circuit.

Hydraulic Return Filter

The hydraulic filter in the brake circuit provides partial flow filtration for the hydraulic oil. The filter,

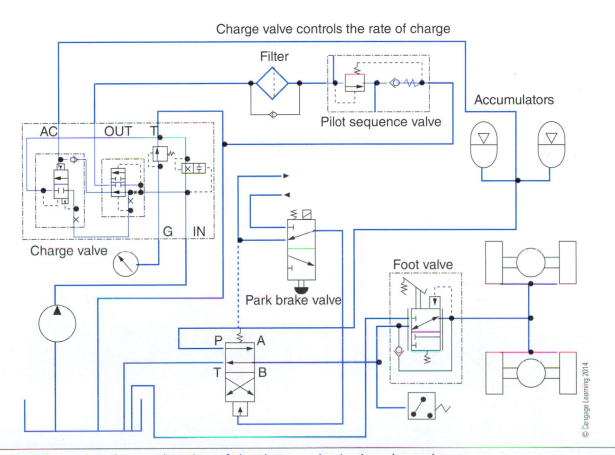

Figure 8-8 Illustration shows a location of the charge valve in the schematic.

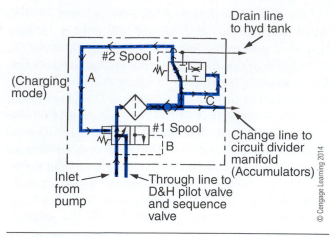

Figure 8-9 Schematic of a charge valve in the charge position.

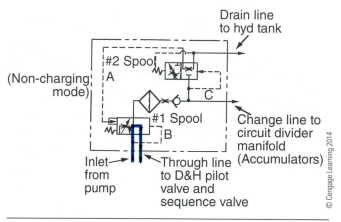

Figure 8-10 Schematic of a charge valve in the fully charged position.

as shown in **Figure 8-11**, has a 25 psi (1.7 bar) bypass relief valve built into the filter. The filter will go into bypass whenever there is a 25 psi difference in pressure across the filter. The filter is also equipped with a visual restrictor gauge to make it easy for the operator to check for dirty filter elements.

Pilot Pressure Sequence Valve

In our example in **Figure 8-12**, the pilot relief valve is a non-adjustable full relief valve that is preset to maintain 200 psi (14 bar) for the dump/hoist/steering pilot circuit. The oil flow for this circuit comes from the accumulator charge valve and in through the inlet port. The inlet and pilot ports are connected together internally. The pilot **sequence valve** is a closed-center valve that regulates the pilot pressure to 200 psi. It is used to operate the dump/hoist/steering valves indirectly through a pilot circuit. There are no provisions made for adjustment to the pilot pressure in this valve. The excess oil left over from the pilot circuit flows to the brake cooling circuit. The brake cooling circuit receives oil only after the pilot circuit flow has been satisfied. An internal check valve ensures that the pressure in the brake cooling circuit never gets above 65 psi (4.5 bar). This is necessary to protect the oil cooler from excessive pressure. Excess oil will open the internal check valve and route surplus oil back to the hydraulic tank. This check valve has no provisions for adjustment or replacement. If the valve fails to deliver the correct pressure, the sequence valve assembly must be replaced.

Accumulator Pressure Switch

The accumulator **pressure switch** in our example is a piston-type pressure switch. It has two sets of contacts that open and close simultaneously when the pressure changes in the brake circuit. In our example, the pressure switch is set to 1,400 psi (196 bar). When the pressure falls below 1,400 psi (196 bar), the normally open (NO) contacts open and the normally closed (NC) contacts close. The NC contacts are connected to an indicator light, placed on the dash in the operator's compartment and alert of an impending

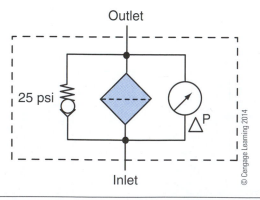

Figure 8-11 Schematic of an in-line hydraulic filter used in the brake circuit.

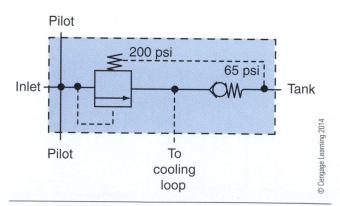

Figure 8-12 Schematic of a pilot pressure sequence valve.

emergency brake application. In **Figure 8-13** the NO contacts control the power to the park brake solenoid. If power is interrupted to the park brake solenoid, the park brake remains in the applied position. When the accumulator pressure falls below 1,400 psi (196 bar), the normally open contact interrupts the flow of power to the park brake solenoid causing the brakes to be applied.

Park Brake Selector Valve

The park brake selector valve is a four-way two-position hydraulic valve that is pilot operated by the park brake control valve. It is also used to apply the park brakes in the event of clutch pressure loss or if the park brake is applied by the operator. In **Figure 8-14**, oil pressure is present at port P during the brakes applied mode, but it cannot flow through port A because it is deadheaded with a plug. When the operator activates the park brake control valve and releases the brakes, oil flows to the pilot-operated port on the selector valve. This pilot port is connected to the clutch pressure in our example, and if the engine is

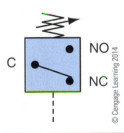

Figure 8-13 Schematic of an accumulator pressure switch.

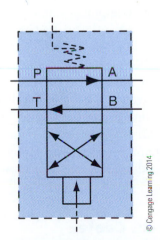

Figure 8-14 Schematic of a park brake selector valve for a spring-applied hydraulically released brake system.

running and there is sufficient clutch pressure at the pilot port, the selector valve will release the brakes. This feature prevents the brakes from being released when there is no clutch pressure. The spring on the other side of the selector valve will not allow the selector valve to release the brakes.

If clutch pressure is present at the pilot port on the selector valve, the park brake control valve will remain in the released position, allowing oil to flow from port P to port B and through the foot-operated brake valve to the wheel ends to release the spring-applied brakes. If clutch pressure were to be lost while the equipment was operating, the spring in the end of the selector valve would shift the spool down and allow the oil to flow back to the tank from the wheel ends, thus applying the spring brakes.

Foot-Operated Brake Control Valve

The brake control valve, as shown in **Figure 8-15**, used on a spring-applied hydraulically released brake system is a closed-center (closed to tank), open-to-pedal **reverse modulation** hydraulic brake valve. Unlike conventional brake systems that use hydraulic pressure to apply the brakes, spring-applied brakes require oil pressure to keep the brakes released, so the brake pedal in this application allows oil to flow to the wheel ends whenever the pedal is in the released position.

The foot-operated brake valve limits the brake **wheel end pressure** to 1,500 psi (103 bar) in our example, while the brake **system pressure** will vary from 1,600 to 2,000 psi (110 to 138 bar). Once 1,500 psi (103 bar) is achieved at the wheel ends, the pressure backs up through a pilot line to the top of the

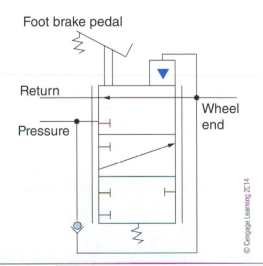

Figure 8-15 Schematic of a spring-applied brake foot control valve.

spool, forcing the spool by overcoming the spring on the bottom of the spool and shutting off oil flow to the wheel ends. If the pressure falls below 1,500 psi (103 bar) the spring at the bottom of the **foot valve** pushes the spool up, allowing oil pressure to increase. By varying the spring tension with the foot pedal, reverse modulation is achieved because oil pressure will always try to balance the varying foot pedal spring tension. When the operator depresses the foot pedal, the spool is forced downward stopping the input flow and gradually allowing the wheel end oil to return to the tank. The more the operator depresses the pedal, the more pressure will be lost at the wheel ends. This applies the brakes further. Completely depressing the foot pedal to the floor releases all the wheel end oil pressure back to the tank and applies the spring brakes. The spring-applied brake pedal works in reverse when compared to the hydraulically applied foot brake valve. The foot-operated brake pedal in **Figure 8-15** has two separate brake circuits: one for each of the front and rear brakes. The previous explanation was given for one side of the brake pedal. The system works exactly the same way for the other side and is controlled by the same foot pedal.

Park Brake Solenoid

An electrically controlled park brake solenoid is used to apply the park brakes in this example. The operator applies the park brakes by de-energizing the solenoid through a park brake button on the dash of the equipment. In **Figure 8-16**, when the solenoid is de-energized, the spool shifts, allowing oil to flow from port P to port A and from port B to port T. When the operator energizes the solenoid, the brakes are released, allowing the oil to flow from port B to port T and from port A to port T.

Driveline Park Brake

Off-road equipment often incorporates a separate mechanical park brake, shown in **Figure 8-17**, that is primarily designed for parking the equipment, although on some equipment it is part of the emergency brake circuit. These units are often found mounted to a

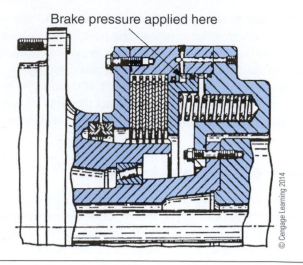

Brake pressure applied here

Figure 8-17 Cutaway of a spring-applied hydraulically released brake system used on an excavator final drive.

driveline on the equipment. Park brakes on off-road equipment are always mechanically applied and hydraulically released. This makes the use of the park brake extremely safe because no hydraulic pressure is required to park the equipment.

The example that will be explained in this section is no exception to this rule; it is mounted to the front driveline of the equipment. Two sets of discs are used in this arrangement. The friction discs are splined to the drive shaft and rotate with it. The stationary discs are splined to a housing that contains the remaining park brake components. Large, heavy internal springs are used to squeeze together the two sets of discs; these springs can develop clamping pressures that can exceed 35,000 psi. When the park brake is engaged, the stationary **steel discs** prevent the friction discs from rotating. With the equipment running, the park brakes are released by hydraulic oil that is directed by the park brake solenoid to the back of an internal piston in the brake assembly located in the housing. The piston compresses the brake application springs, allowing the friction discs to rotate with the driveline. Due to the design of driveline brake assemblies, it does not require full system brake pressure to release the park brakes; a hydraulic pressure-regulating valve is generally used to adjust the release pressure to suit the application.

Hydraulic Oil Cooler

Hydraulic oil coolers are necessary on off-road equipment applications so that the oil temperature can be maintained at a safe temperature. In many severe-duty

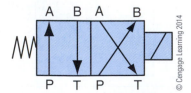

Figure 8-16 Schematic of a park brake solenoid.

applications, the oil temperature can exceed safe operating temperatures. Two designs of oil coolers are generally used for this purpose: the oil-to-air and the liquid-to-liquid designs. Regardless of the type used, cleanliness is a must, and the heat exchanger's cooling fins must be kept clean. On newer equipment the oil temperature is monitored and the operator is alerted when the oil temperature is too high; on older equipment it is often up to the operator to identify oil temperature problems. The most effective way to clean the oil cooler is to use a high-pressure wash or steam. A good cleansing agent should be sprayed onto the cooler fins and allowed to soak before pressure washing. After a cleaning, the coolers can be dried with compressed air. Drying out the coolers is necessary because dirt adheres to the wet surface of the fins more quickly than when the fins have been properly dried. The pressure that flows through the brake cooling circuit, as shown in **Figure 8-18**, is regulated by an additional check valve that limits the pressure in the cooling circuit.

Spring-Applied Hydraulically Released Wet Disc Brake

The spring-applied hydraulically released brake system is used by many manufacturers of off-road equipment. Although the brake system uses

hydraulic components and pressures that are similar to hydraulically applied wet disc brakes, the system differs in one significant way: the brake pressures are reverse modulated. In other words, the pressures are reduced in the brake circuit when the brakes are applied. Conventional hydraulically applied service brakes operate by increasing the pressure in the wheel ends to apply the brakes. The brake components used in the spring-applied service brake system look similar to the components that are used in conventional hydraulically applied service brake systems, but differ in the way they actually function.

Conventional brake valves force pressure into the brake circuit when the brakes are applied; spring-applied service brakes relieve pressure from the brake circuit as the brakes are applied. When checking hydraulic pressures in a spring-applied hydraulically released service brake circuit, you must remember that the hydraulic pressure is progressively reduced as the brakes are applied; and when fully applied, there is no hydraulic pressure at the wheel ends—the friction discs are clamped to the stationary steel discs by mechanical spring pressure. Spring pressures in this type of brake system can exceed 40,000 pounds at each wheel end. Before attempting to service a spring-applied brake system, study the disassembly procedure to ensure that the work will be performed correctly. Failure to follow these manufacturer's recommended procedures can result in serious injury or death.

In **Figure 8-19** the brakes are shown in the applied mode. When sufficient hydraulic pressure is directed into the cavity between the piston and the housing, the pressure pushes back against the brake application springs allowing the friction discs to rotate with the axle. The brake system is designed to monitor the brake pressure so that in the event of low hydraulic pressure, the operator is alerted by an indicator light of an impending emergency brake application.

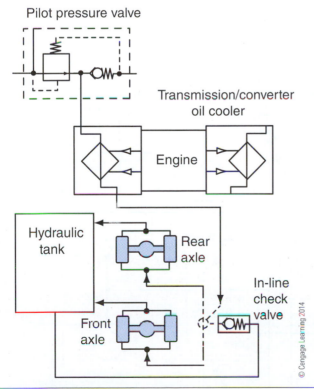

Figure 8-18 Schematic of a typical brake cooling circuit used on a spring-applied hydraulically released brake.

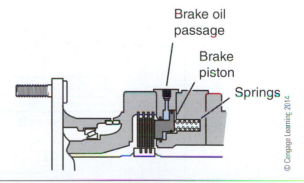

Figure 8-19 Cutaway of a typical spring-applied wet disc brake.

The brakes are designed to go on automatically in the event that the pressure goes below a safe, pre-determined value. This type of brake system is inherently much safer than the conventional hydraulically applied wet disc brake system because each wheel end brake on a spring-applied system can act independently of any other. If a problem occurs in any of the four wheel ends that results in a brake pressure loss, the springs in that particular wheel end can apply the brakes. There are anywhere from 15 to 40 brake application springs in each wheel end brake assembly, so it would be highly unlikely to have a brake application failure on equipment with this type of system. Even if a few brake application springs were to fail in a wheel end, there would be no adverse effect on the stopping ability of the equipment.

SPRING-APPLIED BRAKE SYSTEM OPERATION

A spring-applied brake system, as shown in **Figure 8-20**, operates in reverse when compared to conventional brakes. Each wheel end brake is capable of exerting as much as 80,000 pounds of force from the spring pressure in the wheel ends. The pressure required to release the spring brakes completely is generally in the neighborhood of 1,500 psi (103 bar). **Figure 8-20** shows the general layout of the brake circuit by using a combination of schematic and pictorial views.

With the equipment running, the pump forces oil to flow to the accumulator charge valve where it is directed to the two accumulators until they are fully charged. Once the accumulators are fully charged, the charge valve redirects the oil to a filter before entering the brake pilot sequence valve. The pressure-reducing sequence valve allows 200 psi (14 bar) to flow to the dump/hoist/steering pilot circuit. The remainder of the oil is sent to the brake cooling circuit. When the operator activates the park brake control valve, oil flows to the park brake selector valve, allowing oil from the accumulators to flow to the brake control valve. The oil flows through the brake valve when the pedal is in the released position and then on to the wheel ends where it forces the hydraulic brake piston to retract the brake application springs, thereby releasing the service brakes.

To activate the service brakes, the operator steps on the brake pedal. This progressively reduces the hydraulic pressure in the wheel ends and applies the brakes. The return oil from the foot valve is redirected back to the tank. The more force the operator applies

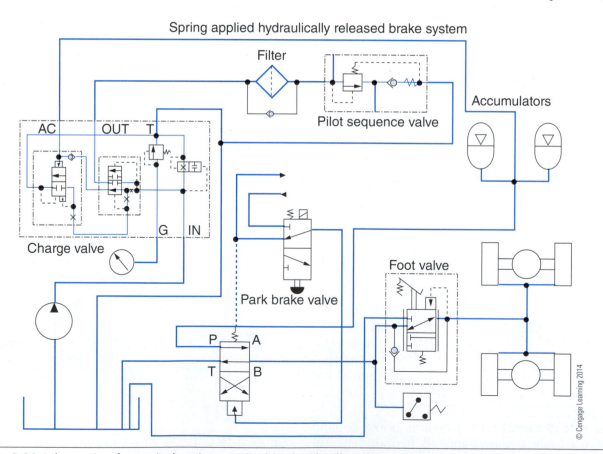

Figure 8-20 Schematic of a typical spring-applied hydraulically released brake circuit.

© Cengage Learning 2014

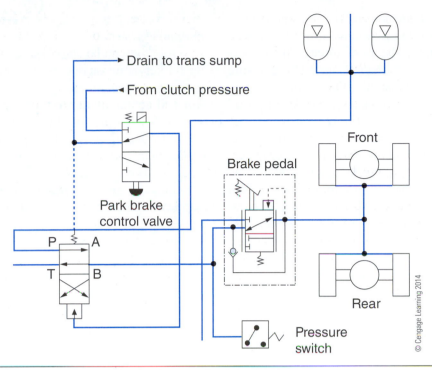

Figure 8-21 The park brake circuit in the brakes applied position.

to the brake pedal, the greater the pressure drop in the service brake circuit. Unlike a conventional brake pedal, the spring-applied brake pedal operates in reverse. It increases the pressure to the wheel end brakes when it releases the brakes and decreases the pressure when it applies the brakes.

To park the equipment, the operator must activate the park brake control valve (see **Figure 8-21**) that is located in the operator's compartment. This action stops the oil flow to the park brake selector valve. The spool in the park brake selector valve shifts and allows the wheel end oil pressure to return to the tank through the foot valve. This blocks oil flow from the accumulators and prevents it from flowing to the wheel end brakes. The brake application springs in the wheel ends can now apply the wheel end brakes.

To release the park brake, the operator activates the **park brake valve**, which redirects oil to the park brake selector valve. This shifts the spool in the selector valve and allows oil from the accumulators to flow through the foot valve to the wheel end, releasing the brakes. In the event that hydraulic pressure is lost for any reason, the worst outcome is that the brakes would mechanically apply. To prevent the operator from operating the equipment with low hydraulic pressure, which could cause brake drag, a brake pressure switch monitors accumulator hydraulic pressure. If hydraulic pressures fall below a predetermined amount, a warning light illuminates on the dashboard in the operator's compartment.

Brake Cooling Circuit

In **Figure 8-22**, the oil flow comes in from the charge valve "out" port and passes through a filter before entering the pilot pressure sequence valve.

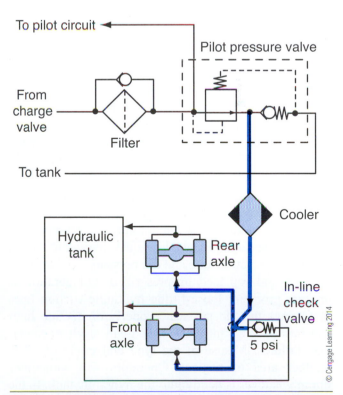

Figure 8-22 A brake cooling circuit from a spring-applied hydraulically released brake circuit.

The pilot pressure sequence valve is considered a full-flow relief valve that maintains a predetermined amount of pressure for the pilot circuit. In our example, this pressure is 200 psi (14 bar). This pressure is generally not field-adjustable. Once pilot pressure is achieved, excess oil flows past a check valve and limits the pressure in the cooling circuit to 65 psi (4.5 bar). The lower pressure is necessary to protect the cooler from excessive pressure. The cooler lowers the temperature of the cooling oil before it enters the wheel ends. The oil then flows to the wheel end cooling circuit where another check valve, rated at 5 psi (0.3 bar) in our example, prevents pressures higher than the check valve rating on the downstream side of the wheel end. The cooling oil flows around the inside of the brake disc compartment absorbing heat from the discs before returning to the tank. Over-pressurization of the wheel end cooling circuit may result in a blown wheel end seal. Some manufacturers route the oil flow through the park brake circuit to dissipate extra heat before returning it back to the tank.

MAINTENANCE

The cleanliness of the hydraulic system cannot be overemphasized. To get a long, trouble-free life out of the brake components, it is very important to maintain the hydraulic system according to the manufacturer's recommendations. Excessive heat is the enemy in all cases when it comes to hydraulic circuits. Regular filter change intervals should be adhered to and followed up with periodic oil analysis. Following the manufacturer's recommended maintenance procedures will reduce the incidence and severity of common problems that develop over time because of poor maintenance practices. By following the prescribed maintenance practices, small problems can be detected and corrected before a major breakdown occurs.

One of the most common problems encountered in equipment hydraulics is maintaining the correct level of oil in the tank. Plugged or dirty hydraulic filters, both on the suction side and the pressure side, must be cleaned and replaced as prescribed. Oil aeration is another problem; loose-fitting hydraulic suction lines or O-rings that do not seal properly on suction lines can introduce air into the oil and can result in cavitation failures. Misuse of the type and grade of oil is another area of neglect. Operators will sometimes dump whatever is handy into the hydraulic tank, in the mistaken belief that all oil is the same. As simple as these problems sound, they are responsible for most hydraulic system failures.

Problems can be caused when the system is opened up for servicing or repairs. It is extremely difficult to maintain a clean environment in the field. Keeping the dirt and contamination out of the hydraulic system is essential; even the small particles that are hard to see with the naked eye can score valves and plug orifices. Using clean containers to transport oil to the job site is of the utmost importance. When performing work on a job site you should always clean up the work area before opening a hydraulic system. This will minimize the chance of contamination. When changing or adding oil, make sure the area around the tank opening is clean and the container that is being used does not have any contamination on the top that could fall into the tank.

Accumulator maintenance often requires that the pre-charge be checked periodically. Most accumulators are serviceable in the field. Always read over the manufacturer's recommended service procedures before attempting to service an accumulator. In general when checking the pre-charge, if you observe that the pressure is higher on the gas side than it should be, then it is likely that the seals on the piston are not sealing effectively and oil will be present in the gas side. If the pressure on the gas side is lower than normal, check for an external leak at the gas charge fitting or cylinder end seals.

Tech Tip: High-pressure oil can remain in a hydraulic brake circuit for a long period of time after the equipment is parked. Failure to properly release residual pressure can result in serious injury or death if the manufacturer's recommended procedures are not followed. Never use your hands, even if you are wearing gloves, to detect any hydraulic leaks in a hydraulic circuit. Use a cardboard for this purpose. High-pressure oil can penetrate the skin, causing severe injury. Always seek medical attention immediately if this type of injury is suspected. When checking nitrogen pre-charge in an accumulator, always follow the manufacturer's recommended procedures for the system that you are servicing. There are significant differences in the way the systems behave when checking accumulator pre-charge.

Tech Tip: The only gas that should be used to charge accumulators is dry nitrogen gas. No other compressed gases are permitted for this procedure; other types of gases, even compressed air, can lead to explosions and should never be used. Be sure that the nitrogen bottles have the correct connections to match the hoses and gauges that are going to be used for this procedure. If there is any visible physical damage to the accumulator, it must be replaced immediately.

CAUTION *Before attempting to replace an accumulator, discharge all pressure in the accumulator following the manufacturer's recommended procedures.*

TESTING AND ADJUSTING

Testing and Recharging

This procedure is only an example of the steps that should be taken to check and charge a bladder-type brake accumulator. Always refer to the equipment service literature for the specific procedures that are to be used on the equipment that you are servicing. Not all procedures are identical; they vary from accumulator to accumulator. They may even be different on the same type of equipment that was produced in a different location or time. The equipment must be parked on level ground and have wheel chocks installed on both sides of the tires. Be sure that any implements attached to the equipment are lowered to the ground. The steering frame lock must be installed along with any appropriate lockouts or tags that are required. If any steering lock mechanisms, such as transmission locks, are used on the equipment, ensure that they are engaged.

To check the nitrogen pre-charge in the accumulator(s), all the oil pressure must be relieved in the brake circuit and the piston must bottom out at the lower end cap. Once you are confident that the fluid side of the accumulator is free from pressure, you are ready to begin.

Procedure

- Before attempting to remove the protective guard from the top of the accumulator, which protects the gas valve (see **Figure 8-23**), clean the surrounding area with compressed air.

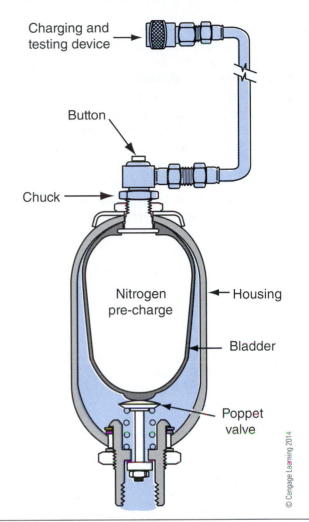

Figure 8-23 A typical bladder-type accumulator along with the connectors and hardware used to charge it.

- Make sure the appropriate connector and test gear are selected before attempting to connect the gas valve on the accumulator. *Not all connectors have the same threads.* Once the connection is tightened and checked for leaks, open the appropriate test valve and record the pressure on the gauge. This reading will be the actual pre-charge in the accumulator.

- As in all cases when performing this type of service work, refer to the manufacturer's specifications for the correct pre-charge. Always refer to the manufacturer's temperature correction chart, it may be necessary to compensate for temperature differences. Checking and pre-charging the accumulator in cold or hot ambient temperatures without compensating for temperature difference will result in either overcharging or undercharging the accumulators. Once the pre-charge charge levels are correct, remove the test gear and replace the protective cap.

- Connect the nitrogen bottle to the adapter hose with the appropriate gauge. Now close the bleeder valve.
- Next, open the shut-off valve on the nitrogen bottle and slowly pre-charge the accumulator. Monitor the charge pressure carefully while performing this step. *When charging bladder or diaphragm accumulators, never allow the pre-charge to exceed specifications as this could lead to bladder or diaphragm damage.*
- If the pre-charge is too high, you can release some of it by carefully opening the bleeder valve. Pre-charge pressure can change significantly because of temperature differences. Always allow the gas temperature to stabilize before rechecking the pre-charge and removing the adapter and nitrogen bottle. See **Table 8-1** for examples of the pressure/temperature relationship.
- Reinstall the protective cap and check for leaks by applying a soap-and-water mixture. Any leaks will form small bubbles around the leak area. Leaks must be repaired before the equipment is returned to service.

Brake Performance Test

Performance brake tests must be done anytime the brakes are serviced. Brake tests must be performed in a suitable area that will not compromise the safety of anyone in the immediate area. Ensure that there is adequate room, both front and rear, to allow for any unintended equipment movement. The equipment must be tested on level ground. This procedure, as with any drivability testing, should be attempted only by qualified personnel.

TABLE 8-1: PRESSURE/TEMPERATURE RELATIONSHIP	
Accumulator Pre-charge Pressure/Temperature Relationship for 800 psi (5,500 kPa)	
20°F (–7°C)	725 psi (5,000 kPa)
30°F (–1°C)	740 psi (5,105 kPa)
40°F (4°C)	760 psi (5,235 kPa)
50°F (10°C)	775 psi (5,340 kPa)
60°F (16°C)	785 psi (5,420 kPa)
70°F (21°C)	800 psi (5,500 kPa)
80°F (27°C)	815 psi (5,605 kPa)
90°F (32°C)	830 psi (5,735 kPa)
100°F (38°C)	845 psi (5,840 kPa)
110°F (43°C)	860 psi (5,920 kPa)
120°F (49°C)	875 psi (6,000 kPa)

© Cengage Learning 2014

The equipment must be brought up to operating temperature with the brake system fully charged.

Procedure

- Release all brakes before beginning any tests. To test the service brakes, fully depress the brake pedal and observe that the brake gauges are showing the correct pressures. Refer to the equipment specifications in the service literature for the correct values.
- The equipment must be placed in the appropriate gear for this test. Refer to the equipment's service literature for the correct gear selection and rpm range. Not all brake systems on equipment are tested the same way.
- With the service brakes applied, depress the accelerator until the test rpm is reached. Observe and record the rpm in gear. The equipment should not move.

CAUTION *Do not keep the equipment in this operating condition for more than 20 seconds at a time.*

- On equipment that incorporates an emergency park brake circuit, locate the test switch on the dashboard in the operator's compartment that will be used for this procedure. This park brake test switch test drains the hydraulic brake fluid from the spring-applied park brake circuit while the emergency brakes are in the released position. This provision allows the operator to isolate the park brake circuit so it can be checked separately.
- Place the equipment in the correct gear for this test and slowly increase the test rpm until the desired rpm is reached. Refer to the manufacturer's service literature for the specifications. Do not allow the converter to stall for more than 20 seconds at a time without a cool-down period.
- The equipment should not move during any of the brake tests. If the equipment does not pass any of the specified tests, it must be repaired before being placed back into service.

Accumulator Charge Valve Test Procedure

This procedure is to be used as an example of a procedure that must be followed to check accumulator charge valve operation. Always refer to the equipment service literature for the specific procedures that are to be used on the equipment that you are testing. Not all procedures are the same and the procedures may vary from equipment to equipment; in some cases they may even be different for the same model of equipment that was manufactured at a later or earlier date.

As with any brake tests, the equipment must be parked on level ground and have wheel chocks installed on both sides of the tires. Any implements attached to the equipment must be lowered to the ground. The steering frame lock must be installed along with any lockouts or tags that are required. If any steering lock mechanisms, such as transmission locks, are present on the equipment, ensure that you engage them.

Procedure

- Ensure that the engine is shut down for this next step. Continually depress the service brake pedal until all hydraulic pressure is depleted from the brake accumulators. The number of times that you depress the brake pedal will vary from equipment to equipment.
- Connect a pressure gauge (0 to 3,600 psi or 0 to 25,000 kPa) to an appropriate location in the charge valve circuit. The connection point for the gauge will vary; always refer to the equipment service literature for the correct location.
- Observe and record the pressure reading before starting the engine. The pressure should be 0 psi. If there is still pressure showing on the gauge, repeat the bleed-down procedure.
- Start the engine and allow the charge pressure to build up until it reaches cut-out pressure. Once maximum charge pressure is showing on the test gauge, shut down the engine. Observe and record the values shown on the test gauge for comparison later. Compare them to the manufacturer's specifications in the service literature.
- The next step requires the key to be in the off position. Apply the service brakes repeatedly and note the number of times you apply the brakes. Each time the brake pedal is depressed, the pressure on the test gauge will drop slightly.
- As soon as the gauge reaches the low accumulator oil pressure level, the accumulator oil pressure indicator light on the dashboard should come on. If this occurs, then the accumulator low-oil-pressure warning system is working correctly. Refer to the equipment service literature for the correct low accumulator pressure value.
- Apply the service brakes continually while counting the brake applications and observing the test gauge pressure. At one point the pressure

will drop off rapidly; this is an indication of how much nitrogen pre-charge pressure is actually in the accumulator.
- Calculate the number of brake applications it took to get to the point where the gauge dropped off rapidly. Check the manufacturer's service literature to determine the correct pre-charge value.
- Start the engine and run the equipment until the brake cut-out pressure is reached. If the cut-out pressure reading is not correct, refer to the manufacturer's service literature for the adjustment procedures. Adjustment procedures will vary from equipment to equipment. Not all charge valves have a cut-out pressure adjustment that can be readily performed in the field. It may be necessary to replace the charge valve in some cases.

Brake Pump Flow Test

To test the flow capabilities of a brake pump, a standard hydraulic pump test is used to verify the pump's condition and its ability to pump oil under pressure. This pump flow test is conducted at two different pump speeds to assess the pump's condition. The difference is calculated between the pump flows at two different operating pressures and is referred to as flow loss. Hydraulic pumps lose efficiency when they wear internally, and a large pumping loss will be present if the pump has excessive wear. The specifications used in this example (**Table 8-2** and **Table 8-3**) are for reference purposes only and do not represent actual flows for any specific pump. Refer to the manufacturer's service literature for the equipment being tested.

To perform this particular test on the equipment, a suitable flow meter must be installed to the pump outlet to measure the pump output. An example of actual flow loss is shown in **Table 8-4**.

TABLE 8-2: DETERMINING PUMP FLOW LOSS
Example of Determining Pump Flow Loss
Pump Flow at 100 psi (690 kPa)
Pump Flow at 1,000 psi (6,900 kPa)
Flow Loss
Example of Determining Flow Loss
217.6 L/min (57.5 US gpm)
196.8 L/min 52 US gpm)
Loss = 20.8 L/min (5.5 US gpm)

© Cengage Learning 2014

TABLE 8-3: DETERMINING PUMP FLOW LOSS
Example of Determining Pump Flow Loss
$\dfrac{\text{Loss (L/min or US gpm)} \times 100}{\text{Pump Flow at 100 psi}}$ = Percent of Flow Loss   (690 kPa)
Example of Determining Pump Flow Loss
$\dfrac{\text{(20.8L (5.5 gpm))}}{\text{(217.6 L/min (57.5 US gpm))}} \times 100 = 9.5\%$

© Cengage Learning 2014

TABLE 8-4: ACTUAL PUMP FLOW LOSS
Example of Calculations Used to Determine Actual Pump Flow Loss
$\dfrac{\text{Pump Flow at 1,000 psi} \times 100}{\text{(6,900 kPa)}}$ = Percent of Flow Loss   Pump Flow at 100 psi   (690 kPa)

© Cengage Learning 2014

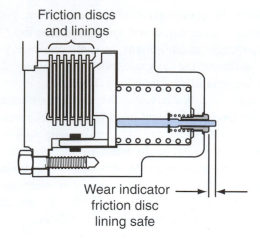

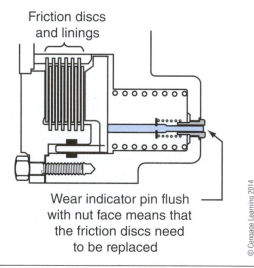

Figure 8-24 Location of the inspection port used to perform a disc brake wear check on a spring-applied hydraulically released brake system.

Checking Internal Friction Disc Wear

This test is generally performed after the equipment fails a brake test. One of the wear conditions that will lead to the equipment's inability to pass a brake test is excessive wear to the friction disc lining. On some spring-applied brake designs, it is almost impossible to accurately determine the friction disc wear without dismantling the wheel ends. On brake systems that have no access ports to the wheel end disc chambers that provide a place to check wear easily, you must eliminate all other possibilities that could lead to a failed brake test first, and then be prepared to dismantle the wheel ends.

Fortunately, most models of equipment have an access port in the wheel end that allows the friction discs to be checked for wear (see **Figure 8-24**). The procedures explained in this chapter are shown as examples only. To perform these checks on equipment, always refer to the equipment service literature test procedure for the equipment in question. The procedures vary somewhat from equipment to equipment.

CAUTION *Before performing any checks or adjustments to the brake system, the equipment must be parked on level ground and have wheel chocks installed on both sides of the tires. Any implements attached to the equipment, such as a*

bucket, must be lowered to the ground. The steering frame lock must be installed along with any lockouts or tags that are required. If any steering lock mechanisms, such as a transmission lock, are present on the equipment, ensure that they are engaged.

The following checks can be performed only while the equipment is shut down with all power turned off and with the brakes in the applied position. Because the brakes are spring applied, they will automatically be in the applied position when the engine is powered down. When performing these tests, be sure to use a container to capture any fluids that might be lost during any adjustments to the brake system. Follow approved fluid disposal procedures according to local regulations and employer policies.

Visual Inspection. Inspect the brake system for oil leaks that can cause brake system performance problems and lead to brake component wear. Inspect the axle housing breather plugs; when plugged, these can create excessive pressure in the axle housing, which can cause oil seal leakage. During normal operation, the friction discs will wear and cause the oil to become contaminated with friction lining material. On some brake systems that do not use an auxiliary cooling circuit, the replacement of the oil at regular service intervals is highly recommended. With this design, the oil that lubricates and cools the brake lining is contained in the wheel end and does not get circulated through a filter and cooled as in some other designs. If excessive brake lining wear is suspected, the oil and friction discs should be replaced. On some equipment the oil in the complete axle assembly must be replaced. In many cases, whenever a friction disc change is performed on a wheel end brake, both wheel end friction discs must be replaced at the same time.

Procedure to Check Friction Disc Wear

- Locate the wear indicator plug on the wheel end and clean the area completely before removing the inspection plug.
- Refer to the equipment service literature to be sure that the correct wear indicator pin is selected to perform the next step. Failure to use the correct wear pin can lead to inaccurate measurements.
- Place the wear indicator pin into the access hole until it bottoms out on the reaction plate. If the pin fits flush or is below the housing face, the friction discs need to be replaced. If the pin projects above the housing face, as shown in **Figure 8-25**, then the friction discs are still good.
- This wear check procedure must be performed on all four wheels to determine the condition of the friction discs. When replacing the inspection plug, make sure to replace the seal or O-ring as this compartment is normally filled with oil.

On most brake designs it is recommended that the air be bled from the spring-applied brake wheel ends any time the brakes are disassembled. Air in the piston cavity will prevent the brakes from releasing completely, which can lead to excessive friction disc wear. Always refer to the manufacturer's recommended procedures for the equipment in question. The example given in this section requires that a minimum of two individuals perform the air purge. As always,

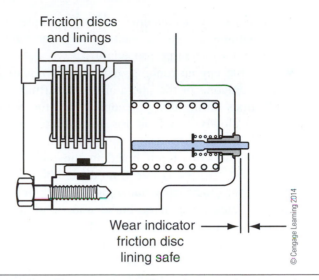

Figure 8-25 Location of and procedure used to determine friction disc wear on a spring-applied hydraulically released brake system.

before any service work is performed, the equipment must be parked on level ground and have wheel chocks installed on both sides of the tires. Any implements attached to the equipment must be lowered to the ground. The equipment steering frame lock must be installed along with any lockouts or tags that are required. If any steering lock mechanisms, or transmission locks, are present on the equipment, ensure that you engage them. Failure to follow the manufacturer's recommended procedures for performing repairs or tests on the equipment can lead to serious injury or death. To perform work on brake systems, the individuals performing the repairs or adjustments must be qualified technicians who are familiar with the manufacturer's recommended procedures before attempting to make any adjustments or checks to the brake system. Under no circumstance should anyone perform any repairs to a brake system without the proper training.

Procedure to Bleed Air from the Brakes

- Ensure that the hydraulic reservoir is full before beginning an air purge.
- Locate the air purge screw on the wheel end in question. Have someone remain in the operator's compartment during the procedure. Start the engine and release the park brake. The engine must remain running during the air purge procedure so that the brake accumulator pressure is maintained at or near maximum pressure.
- To perform this type of procedure, a clear tube that fits onto the purge valve snugly is mandatory. Obvious reasons are that the oil can be

transferred to a container without spillage and air bubbles are easily seen in the clear tubing during the air purge procedure.

- With the engine running, park brakes released, and someone in the operator's compartment at all times during this procedure, open the bleed screw slowly and look for air bubbles in the fluid. Be sure to use a suitable container to capture the oil. Keep bleeding the wheel end until the oil flow is free from air bubbles.
- Close the purge valve and apply the park brakes. Now release the park brake and continue to bleed the air as before until the fluid is free from air. Perform this procedure to each wheel end, one at a time, until all the air is bled from the brake system. Failure to remove all the air from the wheel ends could prevent the brakes from releasing fully, causing premature wear to the friction discs.

Pressure Testing the Brake Cooling Circuit

The brake cooling circuit on spring-applied hydraulically released brakes can develop internal leaks that can result in hydraulic oil flowing into the axle housing. A leak in the brake cooling circuit will result in oil over-filling the axle housing and eventually running out of the axle housing vent. If not found during routine axle servicing, the high oil level will generally cause a seal failure in the axle housing face seals due to excessive oil pressure build-up. To locate the source of the oil leak, each individual brake cooling circuit in each wheel end must be pressure tested individually. A low-air-pressure regulator is required, along with the appropriate fitting to mate to the cooling inlet and outlet ports on each wheel end as shown in **Figure 8-26**. It is advisable to pressure test each wheel end individually after any service work has been performed to the internal brake components. A leak found at this point is easily repaired, compared to discovering it after the equipment has been placed back in service.

CAUTION *Before performing any service or repairs to a brake system, the equipment must be parked on level ground and have wheel chocks installed on both sides of the tires. Any implements attached to the equipment, such as a bucket, must be lowered to the ground. The steering frame lock must be installed along with any lockouts or tags that are required. If any steering lock mechanisms, such as transmission*

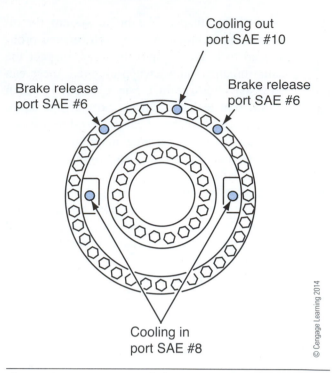

Figure 8-26 Location of the test ports that must be used to pressure test the brake circuit.

locks, are present on the equipment, ensure that you engage them. Failure to follow the manufacturer's recommended procedures can result in personal injury or death.

Procedure to Test for Internal Leaks

- The test equipment for this procedure, as shown in **Figure 8-27**, consists of a 30 psi pressure gauge, air shut-off valve, air regulator, pipe tee, pipe nipple, reducing bushing, and an air line connection.
- This test must be performed with the engine shut down and the park brakes applied. Install a blind plug to seal the cooling outlet port of the wheel end (see **Figure 8-27**). Connect the pressure test equipment to the inlet port of the cooling circuit (see **Figure 8-28**).
- Set the regulator to 12 psi and slowly open the air valve until 12 psi of air is showing on the gauge.
- Close the air valve between the air regulator and the gauge and observe the pressure on the gauge.
- The air pressure should hold for approximately 15 seconds. If it fails to hold, the wheel assembly must be disassembled and repaired. Test both wheel ends on each axle assembly to determine which one is leaking. It is quite possible that both wheel ends are leaking internally.

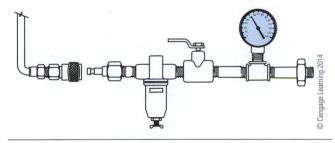

Figure 8-27 Equipment required to perform a pressure test on the wheel end brake cooling circuit.

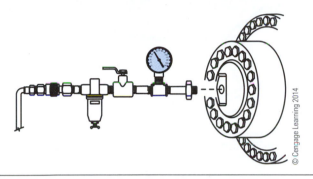

Figure 8-28 Test equipment connected to the brake cooling circuit.

Repairing the Brake Assembly

Table 8-5 indicates the parts and quantities needed to overhaul a wheel end brake.

The following is shown as an example of the general steps that are used to disassemble a spring-applied wheel end brake assembly once it has been removed from the equipment. The example used is not intended to replace the equipment service literature. This actual service work must also include the service literature for the axle assembly that is being serviced. As with any service procedures, following the steps in the exact sequence given is important. The large springs used in this brake system pose a hazard to anyone who does not follow the correct disassembly procedures as outlined in the equipment manufacturer's service literature. Application spring tension in this type of brake system can reach 250,000 pounds of force.

Having all the necessary tools on hand before attempting to perform this type of repair work is essential. The illustrations used in this example are generic and may not be exactly the same as the equipment that you are repairing. Proper disassembly of the front cover and backing plate on these types of brake systems is essential to prevent component damage or personal injury. Proper cleaning of the brake assembly is required before any disassembly work can proceed.

Procedure

- Note that the front cover on the brake assembly, as shown in **Figure 8-29**, is under spring pressure for approximately the first inch of bolt travel. The spring pressure can be neutralized by applying approximately 1,500 psi to the

TABLE 8-5: PARTS LIST			
Description	**Qty.**	**Description**	**Qty.**
No. 1 Multi-disc brake assembly	1	No. 16 Outer piston seal assembly	1
No. 2 Cooling outlet plug	1	No. 17 Outer cover cap screws	12
No. 3 Cooling outlet plug O-ring	1	No. 18 Outer cover washers	12
No. 4 Bleeder screw	2	No. 19 Brake outer cover	1
No. 5 Plug	2	No. 20 Friction disc	3–6
No. 6 O-ring	2	No. 21 Reaction plate	3–6
No. 7 Actuating fluid plug	1	No. 22 Inner piston seal	1
No. 8 Actuating plug O-ring	1	No. 23 Brake piston	1
No. 9 Plug	2	No. 24 Wear indicator pin	1
No. 10 Brake housing	1	No. 25 Indicator pin O-ring	2
No. 11 Inlet O-ring	1	No. 26 Indicator guide O-ring	1
No. 12 Outlet O-ring	1	No. 27 Indicator guide	1
No. 13 Brake apply spring	15	No. 28 Warning sticker	2
No. 14 Outer cover O-ring seal	1	No. 29 Sticker cover	2
No. 15 Piston pressure ring	1	No. 30 Sticker cover screw	2

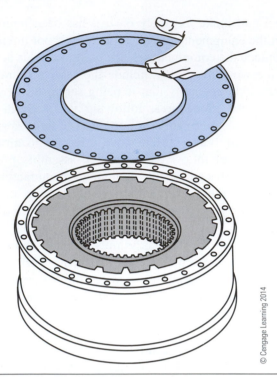

Figure 8-29 Front cover removed from the spring-applied brake assembly.

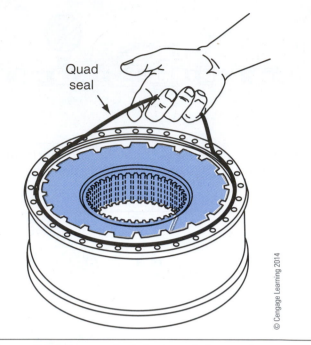

Figure 8-30 Removal of the quad ring seal.

brake inlet piston port with a portable hydraulic power pack. Although it is not necessary to do this, it makes the removal of the brake front cover somewhat easier. If no portable power pack is available, the front cover can still be removed by loosening the front cover bolts in a criss-cross pattern, half a turn at a time, until the spring pressure is dissipated.

- The next step requires the removal of the quad ring seal as shown in **Figure 8-30**. Be sure to carefully inspect the quad ring and corresponding groove in the housing for any signs of damage.
- Carefully remove the steel and friction discs from the housing and make a mental note of the number of discs and the sequence in which they are installed as shown in **Figure 8-31**.
- The next step requires that the brake assembly be turned over to gain access to the backing plate bolts. Caution must be exercised with the following steps. The backing plate bolts are under a great deal of spring force and should never be removed with an impact. The backing plate bolts must be removed in the correct order to prevent any stress to the bolts.
- The next step requires the removal of every second bolt from the backing plate cover, as shown in **Figure 8-32**, slowly releasing the

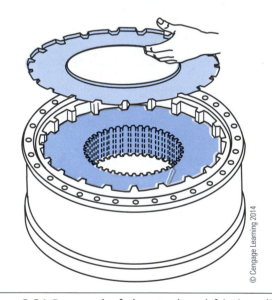

Figure 8-31 Removal of the steel and friction discs.

backing plate by turning each remaining bolt a quarter of a turn at a time in a criss-cross pattern until the backing plate is fully released. Remove all but three bolts from the assembly. The three bolts are left in place to keep the assembly from falling apart when it is flipped over.

- After flipping the brake assembly over, remove the remaining three bolts from the housing and separate the housing from the backing plate. Remove the springs from the brake housing as shown in **Figure 8-33**.

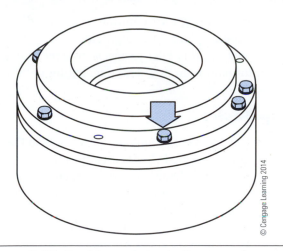

Figure 8-32 Backing plate bolts to release the application spring tension.

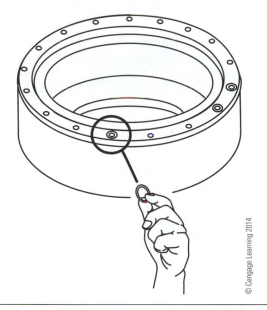

Figure 8-34 Port O-ring being removed from the brake housing.

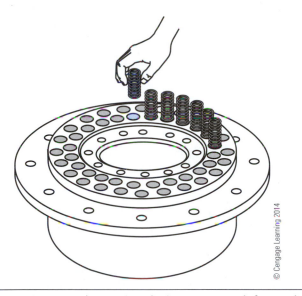

Figure 8-33 Brake spring being removed from the brake housing.

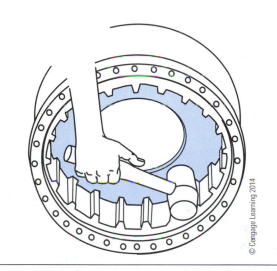

Figure 8-35 Brake piston removal procedure.

- Next, remove the O-rings from the backing plate and the port O-rings from the brake housing as shown in **Figure 8-34**.
- Using a soft blow hammer, gently tap the piston to remove it from the housing as shown in **Figure 8-35**. Be careful that the brake piston does not fall during the final removal stages from the housing; it could easily be damaged. Carefully remove the outer piston seal and expander from the piston and examine the piston and sealing rings for any damage or wear. Next, remove the inner seal and expander ring from the brake housing and inspect the parts for wear or damage.

Clean all the brake parts that will be reused and inspect them for any signs of wear or damage that may affect braking performance. Be sure that all parts are free from nicks or burrs and that no cracks are visible on any of the components. Only the correct OEM replacement parts should be used to reassemble the brake assembly. The seals are delicate and should be handled with extreme care during the installation process. Follow the manufacturer's recommended installation procedure during assembly.

Reassemble the components in the reverse order of disassembly, paying particular attention to the way the O-rings and expander rings are installed on the piston and inner housing. **Figure 8-36** shows the correct orientation of the piston seals once correctly installed.

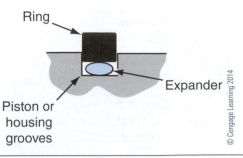

Figure 8-36 Correct orientation of the sealing rings on the brake piston.

ONLINE TASKS

1. Use an Internet search engine to research how heat is dissipated in an off-road spring-applied hydraulically released brake system. Identify a piece of equipment in your particular location, paying attention to the spring-applied brake cooling circuit and note how heat energy from the spring-applied hydraulically released brake system is dissipated. Outline the reasoning behind the type of cooling circuit used on this particular piece of equipment.

2. Check out this URL: *www.mico.com/*. Look for information on new brake technology.

Shop Tasks

1. Select a piece of equipment in your shop that has a spring-applied hydraulically released brake system and identify the components that make up the brake circuit. Identify the type of cooling circuit used; note how oil is directed to this circuit and what cooling pressures are used.

2. Select a piece of mobile equipment in your shop. Identify the type and model of the equipment and list all the components and the corresponding part numbers of a typical spring-applied hydraulically released brake circuit. Identify and use the correct parts book for the particular piece of equipment being worked on.

Summary

- Brake systems on off-road equipment must be able to operate in extremely abrasive conditions that would quickly wear out conventional external brake systems.

- Many off-road equipment manufacturers favor the use of internal spring-applied hydraulically released wet brake systems. These systems incorporate a brake design that has the brakes located inside the axle assembly, which keeps out abrasive road dirt.

- Brake systems convert the kinetic energy of the moving equipment to heat energy when the equipment is braked.

- The wheel end clamping forces used by spring-applied hydraulically-released brakes are dependent on the size and number of springs used in the wheel ends to apply mechanical force to the friction lining.

- Many manufacturers use a brake interlock circuit in preventing brake release until the operator manually resets the interlock.

- Spring-applied hydraulically released brake circuits apply the brakes when brake pressures become too low, and apply the brakes if there is a loss of electrical power or if transmission clutch or torque pressure falls below safe operating levels.

- A spring-applied hydraulically released brake system shares similar components with those used in a hydraulically applied brake circuit.

- Brake pumps that are used on a spring-applied hydraulically released brake circuit are generally heavy-duty gear-type pumps.

- The diaphragm type of accumulator uses a synthetic rubber diaphragm that separates the hydraulic oil from the nitrogen gas.

- An accumulator is a high-pressure vessel and must be treated accordingly. Under no circumstance should welding or brazing of any kind be done on the accumulator casing. If physical damage is discovered to the casing, it must not be used.

- A tire valve should not be used to replace the gas valve in an accumulator because they are not designed to withstand high pressures.

- Piston-type accumulators have a free-floating piston that separates the hydraulic oil and high-pressure nitrogen. The piston has a high-pressure seal that is replaceable.

- The stored energy in the brake accumulator allows the equipment to maintain adequate oil pressure reserve, ensuring a full brake release can be achieved at all times during operation.

- The accumulator charge valve is used to regulate oil flow from the brake pump to charge the accumulators that store the energy for use in the brake circuit during equipment operation.

- The hydraulic filter in the brake circuit provides partial flow filtration of the hydraulic oil.

- The pilot sequence valve used in a spring-applied hydraulically released brake circuit is a closed-center valve.

- The accumulator pressure switch is a piston-type pressure switch that has two sets of contacts that open and close simultaneously when the pressure falls below a predetermined value in a brake circuit.

- The park brake selector valve is a four-way two-position hydraulic valve that is pilot operated by the park brake control valve. It is also used to apply the park brakes in the event of clutch pressure loss or when the operator wants to apply the park brake manually.

- Park brakes on off-road equipment are always mechanically applied and hydraulically released, providing a fail-safe factor.

- Hydraulic oil coolers are necessary on many off-road brake applications so that the oil temperature can be maintained at a safe temperature.

- When checking hydraulic pressures in a spring-applied hydraulically released service brake circuit, note that hydraulic pressure is progressively reduced as the brakes are applied, and when fully applied there is no hydraulic pressure in the wheel ends. In park mode, the friction discs are clamped to the stationary steel discs by mechanical spring pressure.

- To activate the service brakes in a spring-applied hydraulically released brake circuit, the operator depresses the pedal to gradually reduce the hydraulic pressure at the wheel ends applying the brakes. The return oil from the foot valve is redirected back to the tank. The more force the operator applies to the brake pedal, the greater the pressure drop in the service brake circuit.

- To prevent the equipment from operating with low hydraulic brake pressure, which could cause the brakes to drag, a brake pressure switch monitors the accumulator brake pressure.

- The pilot pressure sequence valve used in a brake circuit is a full-flow relief valve that maintains a predetermined amount of pressure in the pilot circuit.

- One of the most common problems encountered in equipment hydraulics is maintaining the correct level of oil in the tank.

- Plugged or dirty hydraulic filters, on either the suction or the pressure side, must be cleaned and replaced per the manufacturer's recommendations.

- Keeping dirt and contamination out of the hydraulic system is essential.

- Never use your hands, even if you are wearing gloves, to locate hydraulic leaks in a hydraulic circuit.

- To check the nitrogen pre-charge in accumulator(s), all the oil pressure must be relieved in the brake circuit and the piston must bottom out at the lower end cap.

- The only gas that should be used to charge brake accumulators is dry nitrogen. No other compressed gases are permitted for this procedure.

- If there is any visible physical damage to the accumulator, it must be replaced immediately.

- Checking and pre-charging the accumulator in cold or hot ambient temperatures without compensating for temperature difference will result in either overcharging or undercharging the accumulators.

- Performance brake tests must be performed anytime the brakes are serviced. Brake tests must be performed in a suitable area that will not compromise the safety of anyone in the immediate area.

- A brake pump flow test is conducted at two different pump speeds to assess the pump's condition. The difference is calculated between the pump flows at two different operating pressures and is referred to as flow loss.

- Most wet disc brake systems have an access port in the wheel end that allows the friction discs to be checked for wear.

- On internal wet disc brake designs, it is recommended that the air be bled from the spring-applied brake wheel ends anytime the brakes are disassembled.

- The brake cooling circuit on spring-applied hydraulically released brakes can develop internal leaks that could result in hydraulic oil flowing into the axle housing.

- To locate the source of the oil leak in a spring-applied brake cooling circuit, each individual brake cooling circuit in each wheel end must be pressure tested individually.

- Application spring tension in spring-applied brake systems can reach 250,000 pounds of force and failure to follow the manufacturer's recommended disassembly procedures can lead to personal injury or death.

- Only the original OEM replacement parts should be used to overhaul any brake assembly.

Review Questions

1. What does the term *cut-in pressure* refer to when referring to a spring-applied hydraulically released brake system?

    A. the pressure that the accumulator charge valve starts to charge the accumulator

    B. the pressure that the accumulator charge valve stops to charge the accumulator

    C. the pressure that the sequence valve starts to supply oil to the cooling circuit

    D. the pressure that the pilot circuit starts to supply oil to the sequence valve

2. What function is performed first by the pilot sequence valve in a spring-applied hydraulically released brake system?

    A. provides oil to the accumulator circuit

    B. controls the cut-in pressure

    C. controls the cut-out pressure

    D. provides oil flow for the steer/hydraulic control circuit

3. What purpose does the 65 psi check valve serve in the pilot sequence valve?

    A. provides a higher pressure oil flow for the brake cooling circuit

    B. provides a lower pressure oil flow for the brake cooling circuit

    C. provides a lower pressure for the pilot circuit control circuit

    D. maintains a safe pressure in the wheel ends

4. What could be responsible for the brake charge valve cycling too often in a spring-applied hydraulically released brake system?

    A. high accumulator pre-charge pressure

    B. low accumulator pre-charge pressure

    C. high wheel end brake pressure

    D. high pilot circuit pressure

5. What is the purpose of the accumulator pressure switch?

    A. monitor the pilot pressure

    B. monitor the cooling pressure

    C. monitor brake pressure

    D. monitor cut-out pressure

6. What component controls the wheel end pressure in the spring-applied hydraulically released brake circuit?

    A. the brake valve

    B. spool 2 in the charge valve

    C. the pilot sequence valve

    D. a pressure-regulating valve in the charge valve

7. What type of oil pump must be used in a spring-applied hydraulically released brake circuit if a charge valve is not used?

    A. gear pump

    B. variable-displacement piston pump

    C. fixed-displacement piston pump

    D. gerotor pump

8. What feature does a low-pressure return filter have when used in a spring-applied hydraulically released brake system?

    A. bypass valve

    B. pressure regulator

    C. full flow with no bypass

    D. two check valves

9. What component receives oil first from the brake pump in a spring-applied hydraulically released brake system?

    A. the accumulator charge valve

    B. the accumulators

    C. the pilot sequence valve

    D. the brake valve

10. What component regulates the pressure that flows through the wheel ends of the brake circuit on a spring-applied hydraulically released brake system?

    A. accumulator pressure switch

    B. the pilot sequence valve

    C. a check valve

    D. brake valve

11. How is brake spring pressure regulated in the spring-applied brake circuit?

    A. by the brake valve

    B. by the charge valve

    C. by the accumulators

    D. by the pilot sequence valve

12. What purpose do the accumulators serve in a spring-applied hydraulically released brake system?

    A. prevent the brake pump from cycling too often

    B. provide emergency brake capabilities

    C. ensure the brakes will always apply

    D. allow the brakes to be applied even if the engine is not running

13. What must be done first when work has to be performed on a spring-applied hydraulically released brake system?

    A. Release the accumulator hydraulic pressure.

    B. Release the emergency brakes.

    C. Drain the hydraulic tank.

    D. Release the accumulator pre-charge.

14. Which pressure is manually adjustable on a spring-applied brake system charge valve?

    A. cut-in pressure

    B. cut-out pressure

    C. both A and B

    D. service brake pressure

15. What is the purpose of the park brake pressure switch in a spring-applied brake system?

    A. turn on a light in the operator's compartment

    B. turn off a light in the operator's compartment

    C. activate the park brake

    D. activate the charge circuit

16. What is the location of the safety relief valve in a spring-applied brake system?

    A. inside the pilot sequence valve

    B. inside the charge valve

    C. inside the brake valve

    D. before the charge valve

Air and Air-Over-Hydraulic Brake Systems

Learning Objectives

After reading this chapter, you should be able to:

- Identify types of off-highway equipment using air and air-over-hydraulic brakes.
- Identify the components used in pneumatic brake systems.
- Explain the operation of single- and dual-circuit air brake systems.
- Identify the major components of an air compressor.
- Outline the operating principles of the valves and controls used in air brake systems.
- Explain the operation of an air brake chamber.
- Outline the functions of the hold-off and service circuits in spring brake systems.
- Describe the operation of S-cam and wedge-actuated drum brakes.
- Describe the operating principles of slack adjusters.
- List the components and describe the operating principles of air disc brake systems.
- Describe the major components and operation of parking and emergency brake systems.
- Identify the major components of air-over-hydraulic brake systems.
- Describe the operation of typical air-over-hydraulic ABS systems.
- Outline some typical maintenance and service procedures for air-over-hydraulic brake systems.
- Interpret air-over-hydraulic system schematics.

Key Terms

aftercooler	caliper	disc brakes
air dryer	check valve	double check valve
air/hydraulic master cylinder	coefficient of friction	drum brakes
air-over-hydraulic intensifier	compressor	dual-circuit application valve
brake chamber	coupler	foot valve
brake shoe	dash control valve	load proportioning valve (LPV)

pop-off valve	supply tank	wet tank
ratio valve	system pressure	wheel cylinder
relay valve	treadle valve	
safety pressure-relief valve	two-way check valve	

INTRODUCTION

Air and air-over-hydraulic brakes are not commonly used in today's off-highway heavy equipment, but a few applications do remain. Nevertheless, technicians working on off-highway heavy equipment should have a good understanding of both air and air-over-hydraulic brakes because off-highway equipment tends to have an operational life much longer than that of highway trucks. Some examples of equipment using air brakes are:

- Scrapers
- Wheel loaders
- Articulated compactors
- Articulated rock/mining/quarry trucks
- Off-highway construction dump trucks
- Off-highway logging trucks and trailer/multi-trailer trains

The first part of this chapter focuses on the principles of brake system pneumatics, and the final part explains air-over-hydraulic brakes. Air-over-hydraulic brake systems have many components in common with air-only and hydraulic-only brake systems; for this reason, it makes sense to study them last.

AIR-ACTUATED BRAKE SYSTEMS

In some ways, air brake operation is similar to hydraulic brake operation, except that compressed air is used in place of hydraulic fluid to actuate the brakes. The potential energy of an air brake or air-over-hydraulic brake system is in the compressed air itself. The retarding effort delivered to the wheels is unrelated to the mechanical force applied to the brake pedal. The source of the potential energy of an air-actuated brake system is the vehicle engine, which drives an air **compressor**. The compressor discharges compressed air to the vehicle air tanks.

When actuated by the operator, the brake pedal meters (precisely measures) some of this compressed air out to the **brake chambers** and actuators. Brake chambers convert air pressure into mechanical or hydraulic force to actuate foundation brakes. Foundation brakes consist of shoes and drums or rotors and

calipers that convert the mechanical force applied to them into friction. In other words, braking effort at the wheel is achieved in the same way as in a hydraulic brake system. The force that actuates the foundation brakes is converted to friction, which retards movement. The friction is converted to heat energy, which then must be dissipated (transferred) to the atmosphere.

Air brakes are used almost exclusively on highway heavy-duty trucks and trailers. They are not common in today's off-highway equipment, except for some off-highway tractors and trailers and multi-trailer trains of the type used in logging. A small number of off-highway rock trucks continue to use air-actuated brakes.

Overview of Air Brake Basics

On-highway air brake systems are governed by legislation introduced in 1975 called Federal Motor Vehicle Safety Standard No. 121 (known as FMVSS 121 in the United States, CMVSS 121 in Canada). Over the years, the law has been modified in small ways to keep up to date with technology, but from the beginning it required all highway vehicles using air brakes to use a dual-circuit application circuit. However, FMVSS 121 does not apply to vehicles that run exclusively off-highway. This means that some air-brake-equipped heavy equipment vehicles use a single-circuit brake system; in reality, they are few in number. When a dual-circuit brake system is used, the system is divided into two subcircuits known as primary and secondary circuits. Off-highway equipment using air brakes is not mandated to use computer-controlled anti-lock brake system (ABS). When ABS is used, it is important to note that it does not significantly change the basic brake system that will be introduced in this chapter. **Figure 9-1** shows the layout of a current air brake system on a Caterpillar 775D quarry truck. If you are familiar with air brake circuits on highway equipment, you will note that the off-highway system shown here is similar. It is a dual-circuit system, but it is simpler in general design: for instance, the system shown here does not include a **supply tank**. This is an OEM preference. In some cases, a supply tank is used on off-highway equipment. As we guide you through air brake

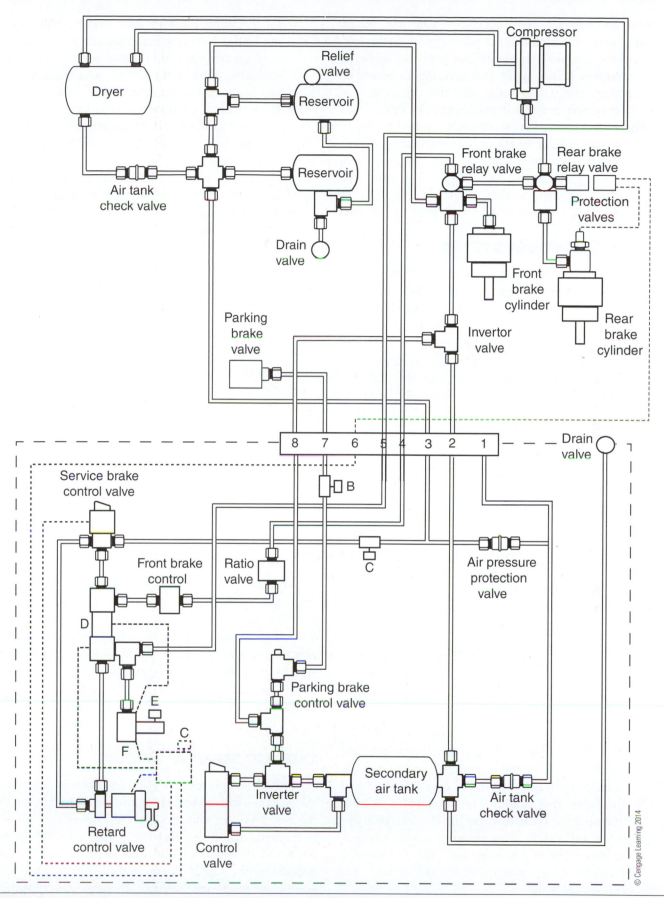

Figure 9-1 Air brake schematic from a Caterpillar 775D quarry truck.

system components, remember that you will see a broader range of components in off-highway equipment because the statutory standards that apply to on-highway equipment do not apply. In describing air brakes in this chapter, we will assume that the equipment is heavily optioned, but this is not always the case.

Figure 9-2 shows the layout of the air brake system used on a Caterpillar 621G on an articulating wheel tractor and scraper unit. Note that the layout here is more similar to that used on a highway tractor and semi-trailer combination.

AIR BRAKE SUBSYSTEMS

Although off-highway equipment is not required to have separate service and parking brake systems, combined with a dual-service application circuit, it makes sense to do this for safety reasons. Just as in the case of dual-circuit hydraulic brakes, dual-circuit pneumatic service brakes provide redundancy in the event of the complete loss of a circuit. In addition, having a means of mechanically applying parking brakes provides a safety factor in the event of a complete loss of system air pressure.

The term *service brakes* is used to describe the running brakes on a vehicle; the running brakes are applied by a **foot valve** or hand valve. The term *parking/emergency brakes* is used to describe the mechanically applied parking brakes on a vehicle, which are controlled by the operator using dash valves.

Assuming that a vehicle is designed to couple to a trailer or other type of articulated towed unit, a complete air brake system is made up of the following subsystems:

- Air supply circuit. The air supply circuit is responsible for charging the air tanks with clean, dry, filtered air. It consists of an air compressor, a governor, an **air dryer**, a supply tank, a low-pressure switch, and a system safety (pop-off) valve.
- Primary circuit. In an articulated combination, the primary circuit is usually responsible for actuating the tractor rear wheel brakes and actuating the trailer service brakes when applied from the foot valve. The foot valve is also known as a **treadle valve** or **dual-circuit application valve**. The primary circuit components consist of a primary tank, the foot valve, a pressure gauge, and quick-release or rear axle **relay valves**.
- Secondary circuit. In an articulated combination, the secondary circuit is usually responsible

for actuating the front axle service brakes and the trailer brakes when actuated by the trailer hand control valve on the tractor. It consists of a secondary tank, a foot valve, a pressure gauge, and quick-release or relay valves.

- **Dash control valves** and the parking/emergency circuit. The dash control circuit consists of two or three dash-mounted valves, which manage the parking and emergency circuits. The parking/emergency circuit consists of spring brake assemblies, and the interconnecting plumbing.
- Trailer circuit. The trailer service and parking/emergency brakes are controlled by the tractor unit. When tractor and trailer air systems are required to be routinely connected and disconnected, they are coupled by a pair of hoses and **couplers**. The trailer brake circuit consists of couplers, air hoses, air tanks, pressure protection valves, and relay valves.
- Foundation brakes. The function of foundation brakes is to convert the air pressure supplied by the application circuits (primary and secondary) into the mechanical force required to stop a vehicle. The parking and emergency brakes may also use the foundation brakes to effect mechanical application of the brakes. Foundation brakes consist of brake actuators (chambers or rotochambers), slack adjusters, S-cams or wedges, shoes, pads, linings, drums, and discs. Additionally, vehicles equipped with ABS will have wheel speed sensors mounted on the wheel assemblies.

Note the location of the dual-circuit air brake system components in the figures used in this book. Our approach to the study of air brake systems begins by introducing and explaining the operation of each of the subcircuits, starting with the air supply circuit and working through to the foundation brakes. Next, a more detailed description of the critical air brake components and valves is provided.

THE AIR SUPPLY CIRCUIT

The air supply circuit is responsible for charging the air tanks with clean, moisture-free air to be used by the brake system and to supply the remaining air requirements of the vehicle. The potential energy of an air brake system is compressed air. The air is compressed by the supply circuit and then distributed and stored in reservoirs (tanks) at pressures typically around 125 psi (860 kPa). Valves controlled by the operator can then deliver compressed air to the brake

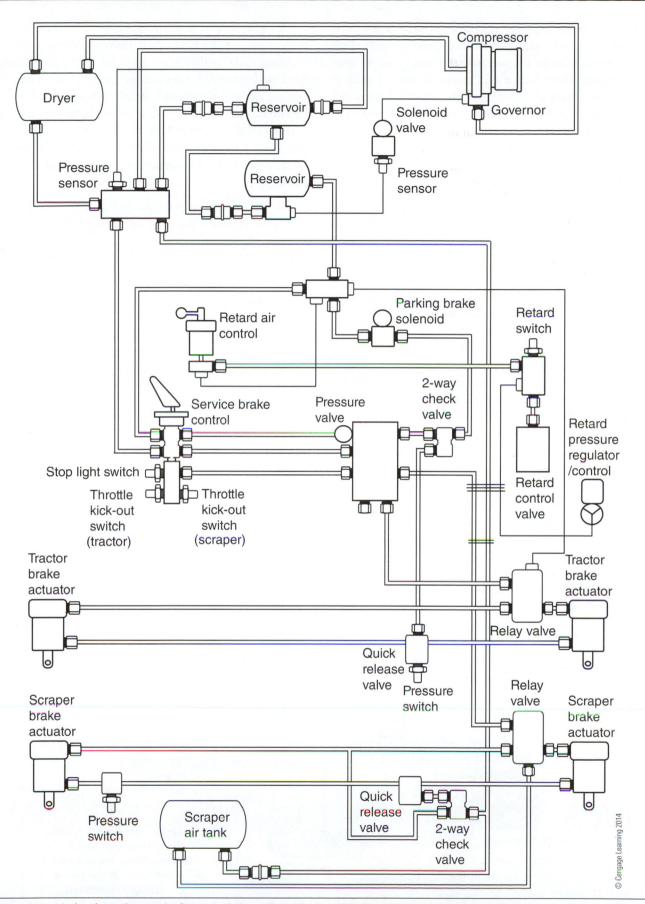

Figure 9-2 Air brake schematic from a Caterpillar 621G articulated tractor and scraper.

system components. Most equipment today is designed to have its parking/emergency brakes fully applied by mechanical force in the event that there is no system air pressure. Therefore, air is required first to release the parking/emergency brake circuit and then to effect service braking when needed. **Figure 9-3** shows a typical air supply circuit.

Air Compressor

The air compressor is driven by the engine. Compressors may be either single- or multiple-cylinder. An air compressor drive usually comes directly from an engine accessory drive gear.

The components of an air compressor are much like those of an engine. The compressor crankshaft is supported by main bearings and drives the connecting rods, which are fitted with piston assemblies. When the crankshaft is rotated by the engine accessory drive, the pistons reciprocate in bores and compress the air required by the supply circuit.

The internal components of the compressor are shown in an exploded view in **Figure 9-4**.

Governors

The functions of the air governor are to monitor system air pressure and manage the loaded and unloaded cycles of the compressor. The governor assembly can be mounted directly to the air compressor (not generally recommended because of higher temperatures close to the compressor) or remotely, often on the firewall of the engine compartment. **Figure 9-5** shows a cutaway view of a typical air governor.

Air pressure in any air brake system must be carefully managed at what is called *system pressure*. System pressure falls within the range of pressures between governed pressure (maximum system pressure) and compressor cut-in pressure (minimum system pressure). Governed pressure is usually set between 120 and 130 psi (827.37 and 896.32 kPa) on most vehicles, but you should note the following, which may or may not be adhered to by manufacturers of heavy off-highway equipment:

- FMVSS 121 specifies that setting governor cut-out pressure at any value between 115 and 135 psi (792.90 and 930.79 kPa) is acceptable.
- Air governor pressure is typically set at around 125 psi (861.84 kPa).

The governor measures system pressure via a signal line that runs from the supply tank and acts on the lower portion of the governor piston. The governor spring acts in opposition to this piston. Governor spring tension is adjusted to define the governor cut-out pressure value. When the system pressure achieves the specified maximum, an air signal is sent from the governor unloader port to the unloader pistons in the compressor cylinder head. When the unloader pistons are actuated, the inlet valves are held open and the compressor is no longer capable of compressing air.

Air Dryers

An air dryer is a common option on an air brake system, and most equipment today uses one, especially when operating in humid or cold conditions. **Figure 9-6** shows the location of an air dryer in the supply circuit.

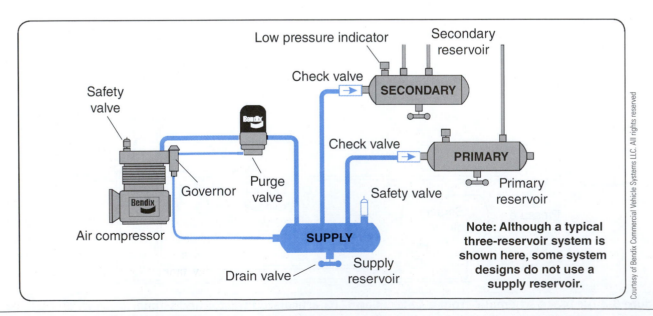

Figure 9-3 Air supply circuit components.

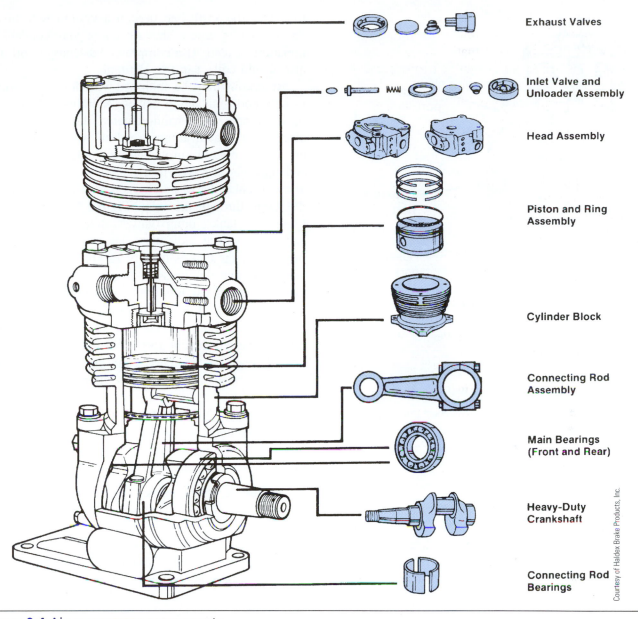

Exhaust Valves

Inlet Valve and Unloader Assembly

Head Assembly

Piston and Ring Assembly

Cylinder Block

Connecting Rod Assembly

Main Bearings (Front and Rear)

Heavy-Duty Crankshaft

Connecting Rod Bearings

Courtesy of Haldex Brake Products, Inc.

Figure 9-4 Air compressor components.

Air from the compressor is either atmospheric or ported boost air from the turbo, both of which contain some percentage of vaporized moisture. This moisture can harm air system valves if it is not removed. The air dryer is located between the compressor discharge and the supply tank. Compressed air is heated to temperatures up to 300°F (150°C) or higher when using a boost air source. Air dryers use two methods to separate moisture from the compressed air.

The air dryer functions primarily as a heat exchanger, to reduce the air temperature, dropping it to a value cool enough to condense any vaporized moisture it receives. Actually, first-stage cooling of the hot compressed air occurs in the discharge line that connects the compressor to the dryer. In some older applications, copper lines were used for this purpose.

Current steel-braided discharge lines still play a role in cooling compressed air. They do not cool as efficiently as copper lines, but they do outlast copper.

Air dryers often use external fins to increase cooling efficiency. When compressed air enters the dryer, it first passes through an oil filter that removes trace quantities of oil that have been discharged by the compressor. Next, the air is passed through a desiccant pack designed to absorb moisture. Some air dryers use only one of these principles to remove moisture from air, whereas others use both. Water and oil separated and condensed in the dryer drain to a sump. Dry air exits the dryer and is routed to the supply tank. **Figure 9-7** shows a cutaway view of a typical air dryer.

The water and oil that drain to the sump must be discharged or purged from the dryer. A signal from the

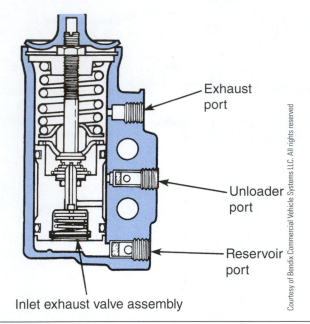

Figure 9-5 Cutaway view of a governor showing the connection ports.

unloader port of the governor (the same signal that "unloads" the compressor at governor cut-out) is used to trigger a purge valve in the air dryer sump. Any liquid in the sump is blown out during the dryer purge cycle.

Supply Tank

The supply tank (or *reservoir*) is the first tank to receive air from the compressor. It is not mandatory to equip off-highway equipment with a supply tank, but when used, it is located downstream from the compressor. The supply tank is sometimes referred

to as the **wet tank** because, in a system with no air dryer, this is where the air cools and condensing moisture collects. If a compressor discharged oil, this also would collect in the wet tank.

Reservoirs must be equipped with draining devices, usually known as dump valves. A draining device can be as simple as a hand-actuated drain cock or an electrically heated automatic drain valve. Many reservoirs are pressure-protected by single **check valves**. A single check valve is a one-way device that will allow a tank inlet to be charged with air but will not discharge through the inlet. **Figure 9-8** shows some typical air tanks, dump valves, single check valves, and low-pressure indicators. The maximum working pressure of the tank is usually 150 psi (1,034.21 kPa). To ensure that this pressure is not exceeded, a **safety pressure-relief valve** is fitted. This is often known as a **pop-off valve**. If the governor malfunctions and does not signal compressor cut-out, the safety pressure-relief valve will trip at 150 psi (1,034.21 kPa) to prevent damage to the chassis air circuits. **Figure 9-9** shows a typical safety pressure-relief valve.

Other Supply Circuit Valves. Other types of valves can be used in the supply circuit. The same valves may be used elsewhere in the circuit. Their functions are briefly described here:

- **Double check valve**: Used when a component or circuit must receive or be controlled by the higher of two possible sources of air pressure.
- **Pressure protection valve**: A normally closed, pressure-sensitive control valve. Some are externally adjustable. Can be used to delay the

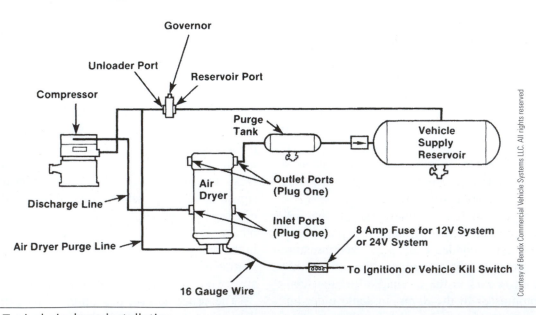

Figure 9-6 Typical air dryer installation.

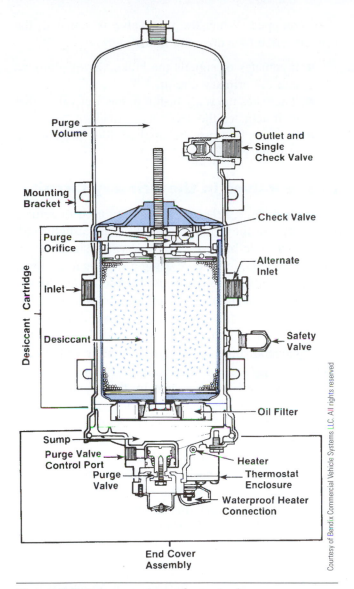

Figure 9-7 Cutaway view of an air dryer.

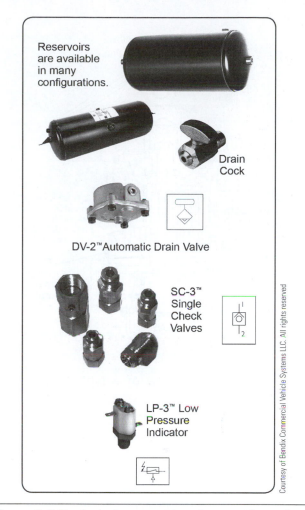

Figure 9-8 Typical air reservoirs, dump valves, check valves, and low-pressure indicators.

Figure 9-9 Safety pressure-relief valve, also known as a pop-off valve.

charging of a circuit or reservoir, or to isolate a circuit or reservoir in the event of a pressure drop-off.

- **Pressure reducing valve**: Used in any part of a circuit that requires a constant set pressure lower than system pressure.

Low-Pressure Indicator

In an air brake system, if the system air pressure drops too low, the brakes will not function properly. A low-pressure indicator is a pressure-sensing electrical switch that closes any time the pressure drops below its preset value. The low-pressure indicator should be specified to close at 50% governor cut-out pressure. In a typical system, this will be at approximately 60 psi (413.69 kPa). When the switch closes, a light and a buzzer are actuated.

ACTUATION CIRCUIT

Air stored in the supplied circuit is made available to the actuation circuit. In an air-only system, the actuation circuit either directly or indirectly, pilots air to actuate the foundation brakes. In an air-over-hydraulic system, the actuation circuit uses air as a signal to manage the foundation brakes.

Foot Valves

A dual-circuit foot valve is shown in **Figure 9-10**. The primary portion of the foot valve is actuated mechanically. This means that it is directly connected to

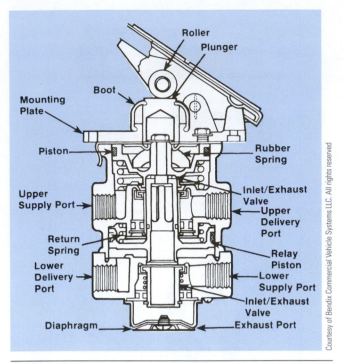

Figure 9-10 Cutaway view of a typical dual-circuit foot valve.

if so equipped. When the foot valve is actuated, the primary circuit air does the following:

- It actuates or signals the brake valves plumbed into the primary circuit.
- It actuates the relay piston in the foot valve; that is, it actuates the secondary circuit.
- It reacts against the applied foot pressure to provide "brake feel."

Brake Valves in the Primary Circuit

In most cases, relay valves will be used to actuate the brakes on the primary circuit. These usually will be the rear axle brakes. The relay valve signal is delivered from the foot valve using a small-gauge signal line. The small-gauge line is used to reduce lag. Signal lag tends to be greatest when the volume of compressed air is greatest. The relay valve is also connected to a closer supply of compressed air, either by a large-gauge air hose or by direct coupling to the tank itself.

The relay valve is actuated by the signal pressure and is designed to deliver a much larger volume from the local supply of air to the service brake chambers it supplies. The signal pressure delivered to the relay valve and the application pressure it delivers to the brake chambers are designed to be identical in most brake systems. The low volume of signal air (in relation to application air) speeds up application and release times. **Figure 9-11** shows a couple of common types of relay valve.

the treadle. A supply of air from the primary reservoir is made available to the primary supply port at system pressure. Travel of the primary piston will determine how much air is metered out to actuate whatever components are in the primary circuit. Also, the air metered out by the foot valve is used as a signal to actuate relay valves on both the chassis and the trailer,

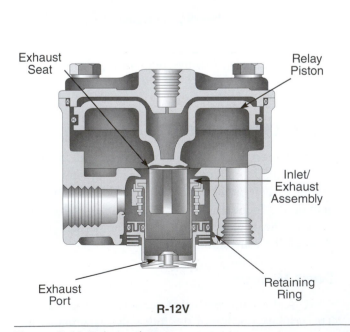

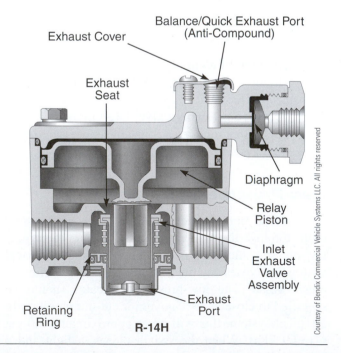

Figure 9-11 Relay valves.

Some systems, mostly much older systems, use quick-release valves in place of the relay valves to route air to the chassis rear service brakes. When a quick-release valve is used to actuate the rear service brakes, the air that acts on the service diaphragms must be routed through the foot valve. This results in extended lag time. To optimize actuation and release times for service brakes, relay valves tend to be used in place of quick-release valves on chassis rear service brakes. **Figure 9-12** shows a typical quick-release valve.

Service Brake Chambers. Service brake chambers are single brake chamber devices and should be differentiated from single hold-off chamber, dual-chamber spring, and rotochamber brake assemblies. When the single chamber on a service brake chamber is charged with air, it actuates the brakes on the unit: the light-duty spring in the chamber functions to retract the brakes only. **Figure 9-13** and **Figure 9-14** show the operation of service brake chambers.

DASH CONTROL AND THE PARKING/EMERGENCY CIRCUIT

The parking and emergency circuits in an air brake system consist of dash-mounted hand control valves, spring brake chambers, and self-applying, fail-safe features. Because all highway vehicles equipped with air brakes must have a means of mechanically applying the parking brakes, so do most off-highway vehicles.

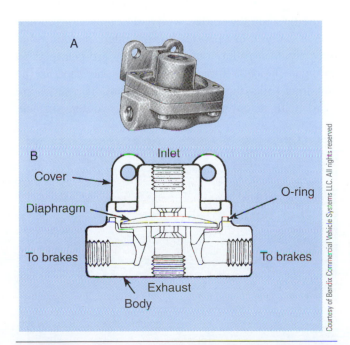

Figure 9-12 Quick-release valve: (A) external view, (B) sectional view.

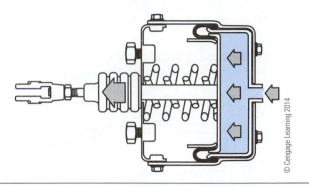

Figure 9-13 Service chamber subject to service brake pressure.

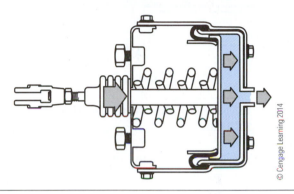

Figure 9-14 Service chamber discharging service brake pressure.

Because air brakes can lose braking capability when air leakage occurs, a fail-safe mechanically actuated emergency brake system is required. Many off-highway, air-brake-equipped vehicles have a single dash control button.

Single Hold-Off Chamber Brakes

Single hold-off chamber brakes are designed to function as simple, single chamber park and service braking units. These units are equipped with a powerful spring that applies mechanical force to the slack adjuster to apply the brakes. To release the unit, hold-off air is applied to the brake chamber. Service braking is accomplished by modulating the hold-off pressure to the unit. **Figure 9-15** shows a single hold-off chamber with air applied to the hold-off diaphragm, releasing the braking force.

Spring and Rotochamber Brakes

Spring brake and rotochamber assemblies are dual-chamber brake actuators designed to perform both service and parking/emergency brake functions; they are used by most air-brake-equipped vehicles today. In these assemblies, service brake functions are

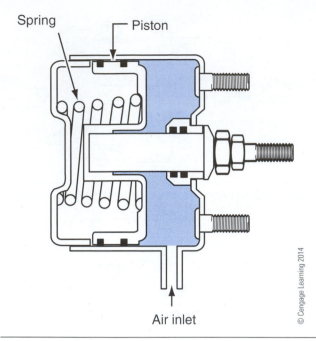

Spring — Piston

© Cengage Learning 2014

Air inlet

Figure 9-15 Single hold-off brake chamber in its released position.

performed by air pressure, and the parking/emergency function happens mechanically by powerful springs. The two chambers are known as service and hold-off (emergency) chambers. Each chamber receives its own air supply that is directed to act on a diaphragm. When the service chamber receives air pressure, it performs service braking. The hold-off diaphragm must have air supplied to it in order to release—that is, cage—the main spring. Without air in the hold-off chambers, a vehicle with a properly functioning brake system will not move.

When the chassis system air pressure is zero, no air is acting on the hold-off diaphragms on the equipment, so the main springs are fully applied to the spring chamber pushrod. This means the system is in park mode. When air pressure builds in the system and the parking brakes are released, air is directed to the hold-off chambers and the force of the main springs is "held-off" the brake chamber pushrod. At least 60 psi (414 kPa) must be present in all the spring brake hold-off chambers before the vehicle parking brakes are considered to be released.

Dash Control Valves. Dash control valves are push-pull air valves that enable the operator to release and apply the vehicle parking brakes. Specifically, these valves manage pressure in the hold-off circuit of the equipment. Most of these valves are pressure sensitive, so they will automatically move from the applied to exhaust position if the supply pressure drops below a certain threshold. In many off-highway applications,

hold-off circuit air is controlled by a single dash valve such as that shown in **Figure 9-16**.

FOUNDATION BRAKES

So far, we have covered the components that manage vehicle braking. The foundation brakes are the components that perform the braking at each wheel end. They consist of a means of actuating the brake, a mounting assembly, friction material, and a drum or rotor onto which braking force can be applied. Foundation brakes can be divided into three types: S-cam type, disc, and wedge brake assemblies (see **Figure 9-17**).

S-Cam Foundation Brakes

In an S-cam brake, the air system is connected to the foundation brake by a slack adjuster. The actuator is a brake chamber. Extending from the brake chamber is a pushrod. A clevis is threaded onto the pushrod, and a clevis pin both connects the pushrod to the slack adjuster and allows it to pivot. The slack adjuster is splined to the S-camshaft, which is mounted in fixed brackets with bushing-type bearings.

When force from the brake chamber pushrod is applied to the slack adjuster, the S-camshaft is rotated. The S-camshaft extends into the foundation assembly in the wheel and its S-shaped cams are located between two arc-shaped **brake shoes**. The foundation assembly is mounted to the axle spider, which is bolted to a flange at the axle end. The brake shoes may either be mounted on the spider, with a closed anchor pin that acts as a pivot, or open-anchor mounted, meaning that they pivot on rollers onto which they are clamped by springs. Opposite the anchor end of the shoe is the actuating roller that rides on the S-cam. The shoes are lined with friction facings on their outer surface.

Reprinted Courtesy of Caterpillar Inc.

Figure 9-16 Dash parking brake control button.

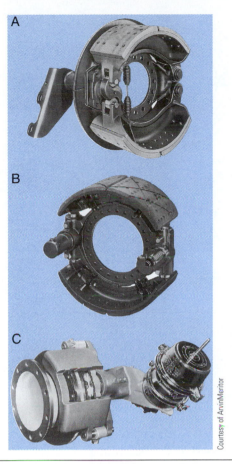

Figure 9-17 Foundation brake types: (A) S-cam brake, (B) wedge brakes, (C) disc brakes.

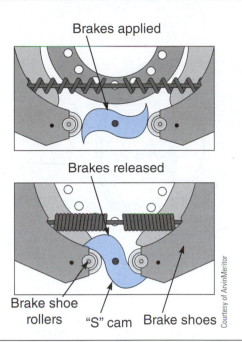

Figure 9-18 S-cam brake operation.

Attached to the wheel assembly and rotating around the fixed brake shoes is a drum. The drum is integral with the wheel hub and rotates with it. When brake torque is applied to the S-camshaft, the S-shaped cams are driven into actuating rollers on the brake shoes, forcing the shoes into the drum. This action is shown in **Figure 9-18**.

When the shoes are forced into the drum, the action creates friction that retards drum movement. In mechanical terms, the energy of motion is converted into friction, which must be dissipated by the drum as heat.

S-Camshafts. The function of the S-camshaft is to convert the brake torque applied to it by the slack adjuster into the linear force that spreads the brake shoes and drives them into the brake drum. The slack adjuster is splined to the S-camshaft and retained by washers and snap rings.

S-camshafts are supported on the axle housing and on the brake spider by nylon bushings that permit the shaft to rotate to transmit brake torque to the shoes. The bushings should be lubricated with chassis grease on a routine basis.

Spiders. The brake spider is bolted or welded to the axle end and provides a mounting for the foundation brake components. It must be tough enough to sustain both the mechanical forces and the heat to which it is subjected. The brake shoes are mounted to the spider by means of either open or closed anchors. The anchor is the pivot end of the shoe.

Brake Shoes. Brake shoes are arc shaped to fit to the inside diameter of the drum that encloses them. They are mounted to the axle spider assembly. Each shoe pivots on its own anchor pin retained in a bore in the brake spider. Today, most shoes are of an open-anchor design. This means that the anchor end of the shoe does not enclose the anchor pin that forms the pivot.

A single retraction spring loads the brake shoe rollers onto the S-cam profile. The function of the retraction spring is to hold the shoes away from the inside surface of the brake drum when the brakes are not being applied. When the brakes are applied, the S-cam rotates, acting on the brake shoe rollers and forcing the brake shoes into the drum. Closed-anchor shoes operate identically to open-anchor shoes. However, on the anchor end of the shoe, a pair of eyes encloses the spider-mounted anchor pin on either side. The disadvantage of the closed-anchor brake shoe design is that more maintenance time is required to perform brake jobs, mainly because the anchor pins tend to seize in their spider bores.

Friction faces are attached to the shoes. These are graded by their **coefficient of friction** rating, which is given a letter code. Coefficient of friction describes the aggressiveness of any friction surface.

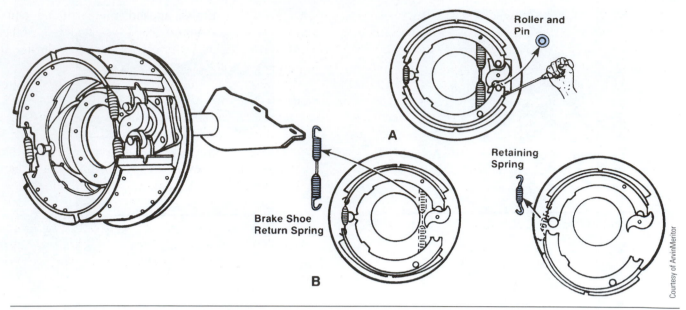

Figure 9-19 Assembling S-cam foundation brakes.

Figure 9-19 shows how a typical S-cam foundation brake is assembled.

Brake Drums. Brake drums are manufactured from cast alloy steels or fabricated steel. The two materials have different coefficients of friction, so they should never be mixed. Brake drums are manufactured in several different sizes. The friction facings used in brake linings today, which do not use asbestos, tend to be harder. This means that the friction facings on the shoes last longer but can also be much tougher on brake drums. It has become routine in some applications for the brake drums (along with the shoes) to be changed at each brake job. Brake drums tend to harden in service. After they show heat checks (small cracks) and heat discoloration (blue shiny areas), they should be replaced rather than machined to oversize.

Wedge Brake Systems

In this brake design, the slack adjuster and S-camshaft are replaced by a wedge-and-roller mechanism that spreads the shoe against the drum. Wedge brake systems may be single- or double-actuated. In a single-actuated wedge brake system, the shoes are mounted on anchors. A single wedge acts to spread the shoes, forcing them into the drum. Dual-actuated wedge brakes tend to be used in heavy-duty applications. The dual-actuated wedge brake assembly has an actuating wedge located at each end of the shoe. The actuating principle of a wedge brake assembly is shown in **Figure 9-20**.

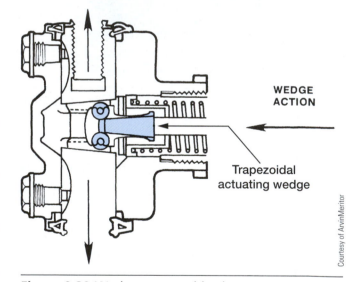

Figure 9-20 Wedge-actuated brake system.

The brake chambers in a wedge foundation brake system are attached directly to the axle spider. The wedge-and-roller actuation mechanism is enclosed within the actuator and chamber tube. When air is delivered to the service chamber, the actuating wedge is driven between rollers that act on plungers that force the shoes apart. When the service application air is exhausted from the service chambers, retraction springs pull the brake shoes away from the drum. The actuating wedges usually are machined at a 14-degree pitch, but they also may be machined at 12 or 16 degrees. They must be carefully checked on replacement.

On all wedge brakes, a self-adjusting mechanism is contained in the wedge brake actuator cylinder.

It consists of a serrated pawl and spring that engages to helical grooves in the actuator plungers. A ratcheting action prevents excessive lining-to-drum clearance occurring as the friction linings wear. Despite the fact that dual-actuated wedge brakes have higher braking efficiencies than S-cam brakes, they are rarely used today due to their much higher overall maintenance cost. The automatic adjusting mechanism and plungers tend to seize when units are not frequently used. This means the foundation brake assembly must be disassembled to repair the problem.

Air Disc Brakes

Air **disc brakes** have higher efficiencies than **drum brakes**; they also have higher initial, operating, and maintenance costs. **Figure 9-21** shows a cutaway view and operating principle of air disc brakes. Although the operating principles remain the same, the actuating mechanisms used differ somewhat between manufacturers.

The air disc brake assembly consists of a caliper assembly, a rotor, and an actuator mechanism. The actuator mechanism is a brake chamber, usually acting on an eccentric lever to amplify its force and convert the linear force produced by the chamber into brake torque to a pair of threaded tubes that act on the pads. The objective of an air disc brake system is to apply clamping force to a caliper assembly within which the pads are mounted. Running between the friction pads of the caliper is a rotor that is bolted to, and therefore rotates with, the wheel assembly. When the power beam is actuated, the threaded tubes force the brake pads into the rotor. Because the caliper is floating, when the inboard-located pad is forced into the rotor, the outboard friction pads are simultaneously forced

into the outboard side of the rotor with an equal amount of force. Some older air-actuated disc brakes use brake chambers that act on a slack adjuster to actuate the caliper assembly.

Air disc brakes provide the advantage of better brake control and higher braking efficiency than S-cam actuated brakes. They also are immune to brake fade and boast a slight overall weight advantage over S-cam brake assemblies. Air disc brakes are an excellent option in severe service applications where braking frequency and loads are at their highest.

> **CAUTION** *Spring brake chambers should be regarded as being potentially lethal. Always observe the OEM precautions when working on or around spring brake chambers.*

Rotochambers

Rotochambers use the same principles as typical spring brake chambers, except that in place of the pancake diaphragm, a rolling-type diaphragm is used. This produces a constant output force throughout the pushrod stroke and provides a much longer stroke. Rotochambers are available in a number of sizes and provide a wide range of output forces. They are more likely to be found in off-highway equipment where their larger physical size and heavier weight are not factors. **Figure 9-22** shows a typical rotochamber assembly.

Slack Adjusters

The linear force produced by the brake chamber must be converted into brake torque (twisting force). Rotation of the S-cam forces the shoes into the drum to effect braking. Slack adjusters connect the brake chamber to the S-cam shaft. **Figure 9-23** demonstrates the role of the slack adjuster in the actuation of the brakes.

The slack adjuster removes excess free play that occurs as the brake friction facings wear. Two general types of automatic slack adjuster are available.

Manual Slack Adjusters. A manual slack adjuster is located on the S-camshaft by means of internal splines on a gear. External teeth on the gear mesh with a worm gear on the adjusting wormshaft. The wormshaft is locked into position by a locking collar that engages to a hex adjusting nut. This locking collar prevents rotation of the wormshaft and sets the adjustment. As friction face wear occurs, slack results; that is, the

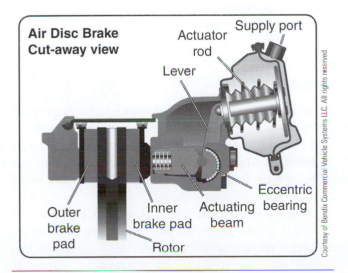

Figure 9-21 Air disc brake cutaway.

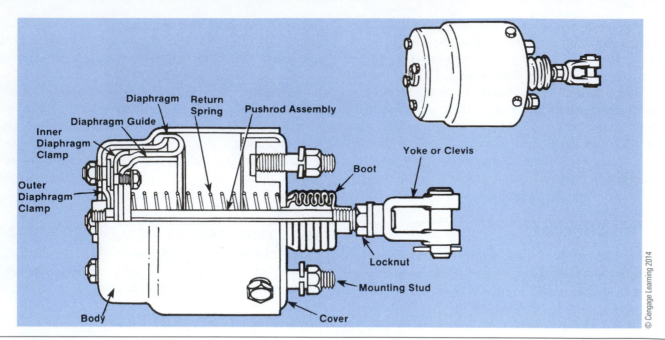

Figure 9-22 A typical rotochamber assembly.

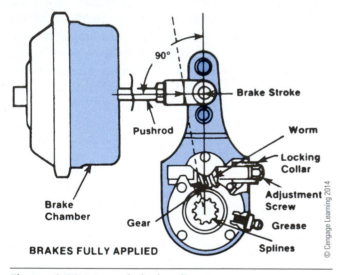

Figure 9-23 Manual slack adjuster.

chamber-actuated pushrod has to travel farther to apply the foundation brake. This slack is removed by adjusting the brakes.

Stroke-Sensing Automatic Slack Adjuster. The stroke-sensing automatic slack adjuster responds to increases in pushrod stroke to actuate an internal cone clutch and rotate an adjusting wormshaft. This wormshaft meshes with the external teeth of the gear engaged to the splines of the S-camshaft. This type of slack adjuster operates on the assumption that if the applied stroke dimension is correctly maintained, the drum-to-lining clearance will be correct. **Figure 9-24** shows a cutaway view of an automatic slack adjuster.

Clearance-Sensing Automatic Slack Adjusters. Clearance-sensing automatic slack adjusters respond only to changes in the return stroke dimension, so they function almost opposite to stroke-sensing models. They "adjust" only when the return stroke dimension increases; if the applied stroke dimension becomes excessive, no adjustment will occur. Clearance-sensing automatic slack adjusters work on the assumption that if the shoe-to-drum clearance is normal, the stroke should be within the required adjustment limits.

Tech Tip: Slack adjusters without a grease fitting should not be assumed to be functional. Slack adjusters should be lubricated with a low-pressure grease gun.

Gauges

In air brake circuits, air pressure gauges display the air pressure in the primary and secondary circuits so the operator can monitor system and application pressures. Either individual gauges or a two-in-one gauge can be used to display the system pressure in the primary and secondary circuits.

When a two-in-one gauge is used, a green needle indicates primary circuit pressure and a red needle indicates secondary circuit pressure. If pressure loss occurs in either circuit, the operator is alerted to the condition after the pressure drops below 60 psi.

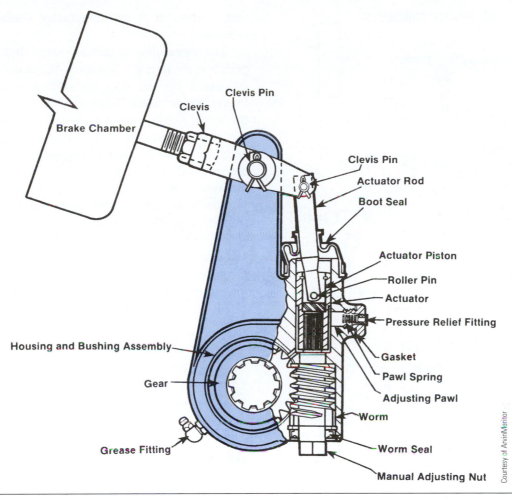

Figure 9-24 Stroke-sensing automatic slack adjuster.

The alert is usually visible (gauges/warning lights) and audible (buzzer). Some systems also have an application pressure gauge. An application pressure gauge tells the operator the exact pressure value of a service application.

AIR-OVER-HYDRAULIC BRAKES

The control circuit used in an air-over-hydraulic brake system works similarly to that in an air-only system. The main focus in this section will be what occurs outside of the control circuit.

Air-Over-Hydraulic Components

Air-over-hydraulic brake systems combine both air and hydraulic principles to braking. The objective is to combine the advantages of both pneumatic and hydraulic brake systems to optimize braking. Many variations of air-over-hydraulic brake systems are used on equipment, but the following statements tend to be true of all systems:

- Pneumatic circuit: This is used as the control circuit. It masters a hydraulic slave circuit.
- Hydraulic circuit: This is the actuation circuit. Hydraulic force is therefore used to actuate the foundation brakes.

The schematic shown in **Figure 9-25** is a simplified Caterpillar air-over-hydraulic brake system identifying the air, hydraulic, and electrical components.

Figure 9-26 is a working schematic of the air-over-hydraulic brake system used on a Caterpillar 950E wheel loader. The components of the system are identified in the text that follows the figure.

Wheel Loader Brake Circuit

Referencing **Figure 9-26**, you can see that the wheel brakes on this wheel loader are caliper-actuated disc brakes. Each of the four wheels on the vehicle is equipped with a disc brake assembly, and in addition, a secondary/parking brake is located on the front driveshaft. On this unit, the secondary/parking brake uses a drum-and-shoe principle: the brake chamber

AIR AND BRAKE CIRCUITS

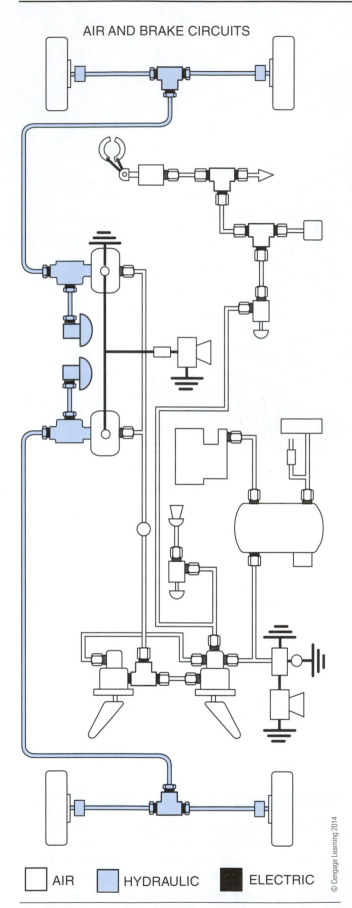

☐ AIR   ■ HYDRAULIC   ■ ELECTRIC

© Cengage Learning 2014

Figure 9-25 Simplified air-over-hydraulic brake system showing all the key components.

that actuates it is spring-actuated (default) and air-released.

The wheel disc brakes are controlled by a pair of pedals in the operator's station. When the right pedal is depressed by the operator's boot, only the wheel brakes are actuated. When the left pedal is depressed, the wheel loader transmission is first disengaged, then the wheel brakes are actuated. On releasing the left pedal, the transmission is engaged before the brakes are released on the unit to prevent rolling when on an incline.

The secondary/parking brake on the unit is controlled manually by a valve located underneath the steering wheel. Because air pressure acts to release the secondary/parking brake, it will begin to apply (spring pressure) automatically if system air pressure gets too low. The brake system on this wheel loader can be subdivided into four separate circuits:

1. Air supply for the air control system
2. Air circuit for the secondary/parking brake circuit
3. Air control circuit for the wheel brakes
4. Hydraulic slave circuit for the wheel brakes

Operating Principles

Several variations of air-over-hydraulic brakes are used on mobile equipment. In general, the control circuit is managed by air pressure and the application circuit by hydraulics. Depressing the brake pedal moves the application valve plunger. This allows modulated system pressure air to flow through quick-release valves and to be directed to a combination **air/hydraulic master cylinder**. Some OEMs refer to the air/hydraulic master cylinder as an **air-over-hydraulic intensifier**. An air/hydraulic master cylinder divides the air and hydraulic sections of the circuit. In addition, separate air/hydraulic master cylinders may be used for the front and rear brakes. Air pressure applied to the booster moves an actuating piston, which creates hydraulic pressure for the foundation brake actuators.

The amount of hydraulic pressure applied to the foundation brakes depends on the modulated (by the foot valve) air pressure applied to the air/hydraulic master cylinder. The resulting hydraulic actuation pressure is routed to the **wheel cylinders**, which effect vehicle braking in the same manner as in a hydraulic brake system.

Hydraulic pressure forces the pistons of each wheel cylinder outward. Outward movement of the pistons applies brake pressure on the brake drum or disc **calipers**. A brake-lining wear switch may be mounted

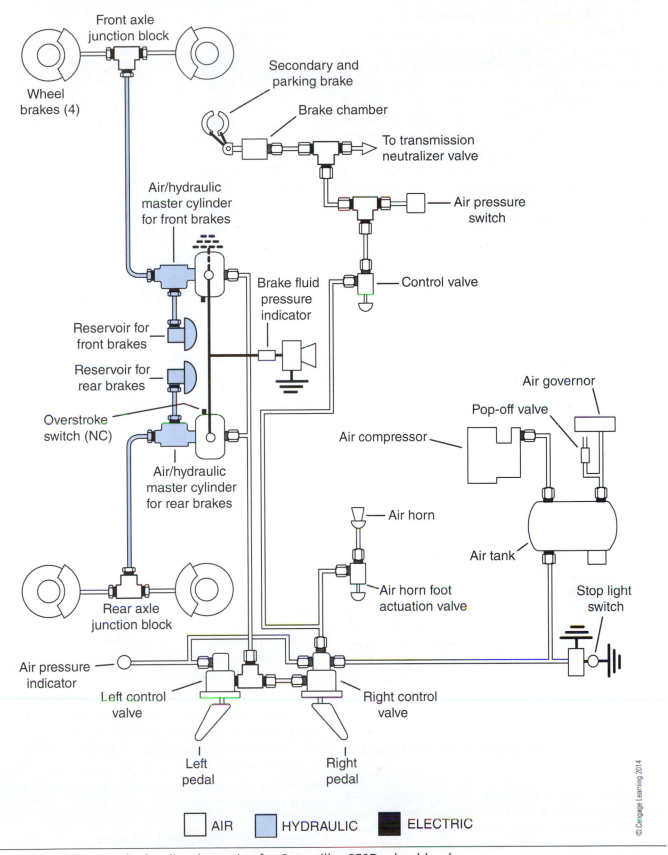

Figure 9-26 Air-over-hydraulic schematic of a Caterpillar 950D wheel loader.

in each air/hydraulic master cylinder to indicate excessive travel of the air piston. Depending on the type of equipment, a front-wheel limiting valve can be used on the front axle brakes. This limiting valve proportions fluid pressure during light braking, both to balance the braking and to reduce brake lining wear. A **load proportioning valve (LPV)** is integrated into some systems and is used to modulate hydraulic (application) pressure to front and rear brakes based on vehicle loads. This can control suspension dive on rough terrain on an unloaded piece of equipment.

Air Control Circuit

The setup of an air control circuit is similar to that of an air-brake-equipped vehicle. Typically it consists of the following components:

- Engine driven compressor
- Air dryer circuit
- Governor
- Air storage reservoirs
- Foot (treadle) valve
- Quick-release valve(s)
- Air/hydraulic master cylinder (air-over-hydraulic intensifier)

Most modern air-over-hydraulic systems are dual-circuit systems and may incorporate an anti-lock brake system (ABS). Dual-circuit systems provide the brake circuit with a safety redundancy factor, meaning that a sudden leak in one portion of a system should not result in a total loss of braking. Other fail-safe features include the usual buzzers and lights that warn of low air pressure in the system and overstroke in the air/hydraulic master cylinder. When the air/hydraulic master cylinder overstroke alert is illuminated on the operator interface, the usual cause is hydraulic circuit leakage or low hydraulic oil level. **Figure 9-27** shows the air control circuit components of a typical air-over-hydraulic brake system; the schematic does not show the location of the air/hydraulic (master) cylinders.

Hydraulic Foundation Brake Circuit

Air-over-hydraulic systems may use either a drum- or disc-based foundation brake. The arrangement may be either on all four wheels or on mixed disc and drum systems. There are other variations, however. Some systems use a mixture of hydraulic disc and full air S-cam drums. When a mixed air-only plus an air-over-hydraulic system is used, a delay valve is required to keep the hydraulic brakes from coming on before the slower-acting air brakes. The hydraulic foundation

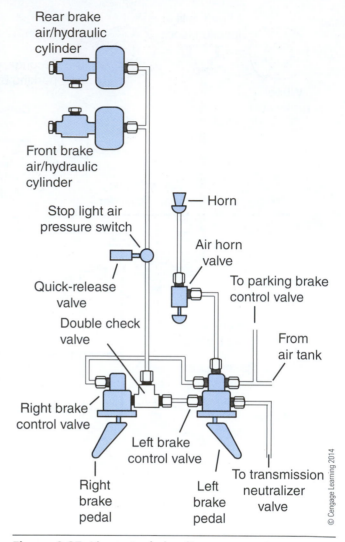

Figure 9-27 Air control circuit components.

brake circuit typically consists of the following components:

- Air/hydraulic master cylinder
- Metering and proportioning valves
- Wheel cylinders and calipers
- Drum or disc foundation assemblies

Figure 9-28 shows a typical hydraulic foundation brake circuit. Some OEMs refer to this as a hydraulic slave circuit. Note the position of the front and rear air/hydraulic master cylinders in the circuit: these units divide the air control (pneumatic) and hydraulic actuation circuits.

Secondary/Parking Brake

A common type of parking brake on air-over-hydraulic systems is the spring brake canister brake. Its parking brake action is obtained by forcing a wedge assembly between the plungers in the wheel cylinder

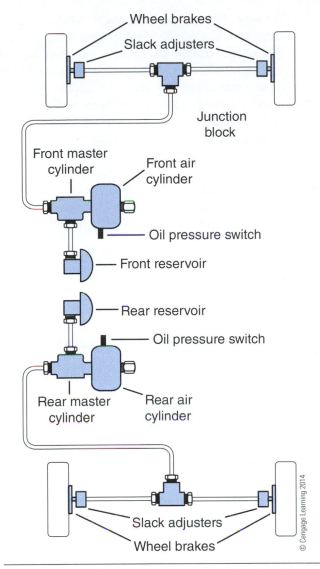

Figure 9-28 Hydraulic foundation brake circuit.

apply (spring pressure) automatically if system air pressure gets too low, providing an operational safety factor. **Figure 9-29** shows a simplified secondary/parking brake circuit that uses a spring-apply/air-release principle.

Air-Over-Hydraulic ABS

When ABS is incorporated into an air-over-hydraulic brake system, it functions within the hydraulic slave circuit of the system. Because of this, it operates similarly to ABS in a hydraulic-only air brake circuit. Because the ABS is integrated into the hydraulic section of the circuit, the cycling frequency is high speed, similar to hydraulic-only systems. ABS management may be data bus driven, enabling directional stability in poor traction conditions.

Bleeding an Air-Over-Hydraulic System

To bleed a typical air-over-hydraulic system used on a two-axle vehicle, follow this procedure:

Procedure

1. Adjust the air pressure in the air reservoir to that specified by the OEM. If the air pressure drops below this specified pressure during the bleeding operation, start the engine.
2. When the air pressure is at the specified system pressure, stop the engine.

on each rear wheel. Engagement of the wedge assembly by the spring brake expands the plunger in the wheel cylinder, forcing the brake shoes into the brake drum. Cable-actuated parking brakes are also found in some air-over-hydraulic systems. This type of unit is mounted at the rear of the transmission so that it acts on the rear wheels through the driveshaft. The parking brake is incorporated in each rear wheel and is cable-controlled from the cab area. For a transmission-located parking brake to function, both rear wheels must be firmly on the ground.

Also commonly used on air-over-hydraulic brake systems is a spring-actuated secondary/parking brake that uses system air pressure on the unit to hold off the spring pressure in the air chamber. This type of secondary/parking brake is controlled manually by an operator dash valve. Because air pressure acts to release the secondary/parking brake, it will begin to

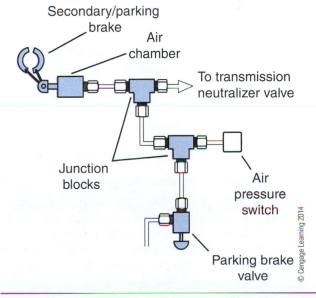

Figure 9-29 Schematic for a spring-apply/air-release secondary/parking brake.

3. Bleed the air from each air bleeder screw in the following order:
   - Front brakes and air/hydraulic master hydraulic cylinder plug
   - Right front wheel air bleeder screw
   - Left front wheel air bleeder screw
   - Rear brakes and air/hydraulic master hydraulic cylinder plug
   - Right rear wheel air bleeder screw
   - Left rear wheel air bleeder screw

4. Fill the brake reservoir with fluid. Refill frequently during bleeding so no air enters the brake lines.

5. Attach a vinyl tube to the bleeder screw. Insert the other end of the tube into a bottle partially filled with brake fluid.

6. Depress the brake pedal and at the same time turn the bleeder screw to the left.

7. Tighten the bleeder screw when the pedal reaches the full stroke. Release the pedal.

8. Repeat until no bubbles are observed.

ONLINE TASKS

1. Access the Bendix Web site at *www.bendix.com/en/products/products_1.jsp* and navigate through it. This site contains useful general and service information on air brake systems. Although much of this information applies to on-highway equipment, you should nevertheless find it beneficial.

2. Because Caterpillar still manufactures equipment with air-over-hydraulic brake systems, you can obtain useful information on their online service information system (SIS). If your college or workplace has SIS data hub access, select a piece of equipment (begin with a Cat 775D quarry truck), and download the brake information and schematics. The Caterpillar SIS sites are *https://www.IMAP@cat.com* (subscription) or *www.cat.com/cda/layout* (open).

3. Use an Internet search engine to locate mobile machinery using air-only actuated brakes: make sure you check out equipment manufactured by off-shore OEMs.

Shop Tasks

1. Locate a piece of equipment equipped with air brakes and manual slack adjusters. Adjust brake-stroke travel to the manufacturer's specifications.

2. Locate a piece of equipment equipped with air brakes and automatic slack adjusters. Check brake-stroke travel to the manufacturer's specifications.

3. Locate a piece of equipment equipped with air-over-hydraulic brakes and, either using the procedure outlined earlier in this chapter or an OEM procedure, bleed the brake circuit using the correct sequence.

Summary

- An air dual-circuit brake system is composed of a supply circuit, primary circuit, secondary circuit, parking/emergency control circuit, trailer circuit, and foundation brake assemblies. Off-highway equipment is not required to meet the redundancy standards of highway vehicles, so their air brake systems may be simplified.

- Air compressors are single-stage, reciprocating-piston air pumps that are either gear- or belt-driven.

- Air dryers help eliminate moisture and contaminants from the air system.

- The potential energy of compressed air is changed into mechanical force in an air brake system by slack adjusters and brake chambers.

- The most common type of foundation brake assembly used on current air brake systems is the S-cam type.

- Slack adjusters multiply the force applied to them by the brake chamber into brake torque.

- Brake torque applied to the S-camshafts results in the shoes being forced against the drum.

- Air disc brakes operate by using an air-actuated caliper to squeeze brake pads against both sides of a rotor.

- Wedge brakes use a drum and a pair of shoes, and air-actuated wedges are used to force the shoes against the drum.

- Service braking applies to the running brakes on equipment. Service braking is managed by the operator using a treadle or foot valve.

- The parking/emergency system manages the hold-off circuit. The hold-off circuit uses system air pressure to "hold off" the mechanical park system of a piece of equipment, usually spring brake or rotochamber units.

- Spring brakes and rotochambers are twin chamber units used for air-actuated service braking and mechanically actuated parking/emergency braking.

- In a typical air-over-hydraulic brake system, the control circuit can be compared with the control circuit in typical air-only brake systems, while the actuation circuit can be compared with that in a typical hydraulic brake circuit.

- The air control circuit in air-over-hydraulic brake systems is charged with compressed air by a compressor with the system-pressure operating window

managed by a governor. The service brakes are operator-actuated by a foot (treadle) valve. The modulated air that exits the foot valve is delivered to an air-actuated, hydraulic master cylinder.

- When the hydraulic master cylinder receives a pneumatic signal from the foot valve, it forces brake fluid out to actuate the foundation brakes.

- Air-over-hydraulic brakes may use disc- or drum-type foundation brakes.

- A typical air-over-hydraulic ABS would add sensors, an electronic control module, and a modulator to the system—all of which function similarly to those in hydraulic-only brakes.

Review Questions

1. Which air brake system component functions as a lever?

   A. brake chamber

   B. rotor

   C. caliper

   D. slack adjuster

2. Which component sets the system cut-in and cut-out pressures?

   A. service reservoir

   B. governor

   C. air dryer

   D. supply tank

3. What is the function of a system safety pop-off valve?

   A. It releases excess pressure from the compressor.

   B. It pressure-protects air in the supply tank.

   C. It alerts the operator when air pressure has dropped below system pressure.

   D. It relieves pressure from the supply tank in the event of governor cut-out failure.

4. The function of the foot valve or dual-circuit application valve is to:

   A. increase the flow of air to the service reservoirs

   B. reduce the flow of air from the service reservoirs

   C. meter air to the service brake circuit

   D. meter air to the brake hold-off chambers

5. The potential energy of an air brake system is:

   A. compressed air

   B. foot pressure

   C. spring pressure

   D. vacuum over atmospheric

6. Which component is a requirement on every air tank?

   A. an air dryer

   B. a safety valve

   C. a drain cock

   D. a pressure protection valve

7. The function of an air brake chamber is to:

    A. convert the energy of compressed air into mechanical force

    B. multiply the mechanical force applied to the slack adjuster

    C. reduce the air pressure for an application

    D. increase the air pressure for service parking brakes

8. What determines the amount of leverage gained by a slack adjuster?

    A. the slack adjuster length

    B. the pushrod stroke length

    C. the S-cam geometry

    D. the number of internal splines

9. S-cam type brake shoes, retraction, and fastening hardware are mounted to:

    A. a spider

    B. a backing plate

    C. the drum

    D. the axle spindle

10. Technician A says that air disc brakes usually produce higher mechanical braking efficiencies than equivalent S-cam brakes. Technician B says that the reason wedge brakes produce higher mechanical braking efficiencies than equivalent S-cam brakes is that they are usually double-actuated. Who is right?

    A. Technician A only

    B. Technician B only

    C. Both A and B

    D. Neither A nor B

11. Which of the following is an advantage of a rotochamber when compared with a spring brake chamber?

    A. lightweight

    B. less costly

    C. smaller in size

    D. longer stroke

12. In a typical air-over-hydraulic brake system, depressing the treadle causes which of the following?

    A. Hydraulic fluid flows through quick-release valves.

    B. Air pressure forces the pistons of each wheel cylinder outward.

    C. An LPV redistributes air pressure to front and rear brakes.

    D. Air pressure is sent to the air/hydraulic master cylinder assemblies.

13. What term is used by some OEMs to describe an air/hydraulic master cylinder?

    A. air-over-hydraulic intensifier

    B. slave valve

    C. hydraulic calipers

    D. treadle valve

14. If the air/hydraulic master cylinder overstroke alert illuminates, what is the most likely cause?

    A. low air pressure

    B. governor improperly set

    C. low hydraulic fluid

    D. excessively high hydraulic pressure

10 Introduction to Powertrains

Key Terms

amboid gear	gears	plain bevel gear
backlash	helical cut	planetary gear
crown gear	herringbone gear	rack-and-pinion gear
direct drive	hypoid gear	spur gears
drive gear	laws of leverage	straight cut spur
driven gear	overdrive	torque
gear pitch	pinion gear	underdrive
gear ratio	pitch diameter	worm gear

INTRODUCTION

The four main purposes of a powertrain are to:

1. Connect and disconnect power from the engine.
2. Modify speed and torque.
3. Provide a means for reverse.
4. Equalize power distribution to the drive wheels.

An engine provides torque to propel a vehicle. A clutch or torque converter provides a way to connect or disconnect the torque produced by the engine to the transmission. Within a transmission, gear ratios allow for different speed and torque ratios in order to propel the vehicle under varying load conditions. The transmission provides a means of reversing engine input rotation to reverse the direction of the vehicle. Many heavy-duty equipment applications have an equal number of forward and reverse speeds. Differentials equalize power distribution to the final drives (see **Figure 10-1**). In four- or six-wheel-drive applications, the power distribution is to both left- and right-hand final drives as well as front to rear final drives as shown in **Figure 10-2**.

GEAR FUNDAMENTALS

The purpose of **gears** or gearing is to transmit rotating (twisting) force known as **torque**. Almost all gears need to be supported by a shaft to allow an axis for them to rotate, either by rotating independently from each other or together as a unit (see **Figure 10-3**).

Figure 10-1 Front-end loader showing the components of a powertrain.

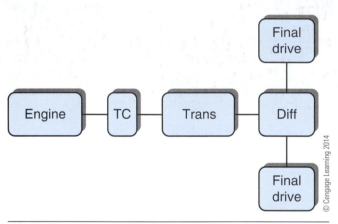

Figure 10-2 The engine produces the torque needed to propel a vehicle and the various components that transfer the torque to the final drives.

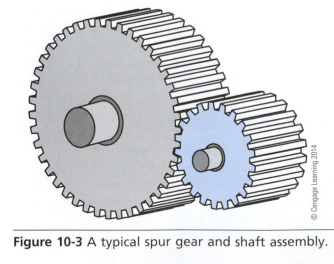

Figure 10-3 A typical spur gear and shaft assembly.

Gears working together can act as a lever to multiply torque or to directly transfer torque to another shaft through mating gears. They can also provide a speed increase or decrease, greater or lower than the original input speed. Because gears act as levers, the same **laws of leverage** principles apply (see **Figure 10-4**).

A longer lever provides a mechanical advantage and a greater amount of torque is achieved. This is similar to a small-diameter gear driving a large-diameter gear. The action is inversely proportional: if a large-diameter gear is driving a small-diameter gear, torque is reduced when speed increases. To define these principles further, we must understand how torque is achieved. As stated earlier, torque is twisting force. It is calculated by multiplying the force applied to a lever by the length of the lever. In **Figure 10-5** and **Figure 10-6**, a pipe wrench is attempting to rotate

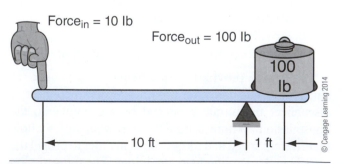

Figure 10-4 Applying a small force at the end of a long lever will produce a larger force on the small lever.

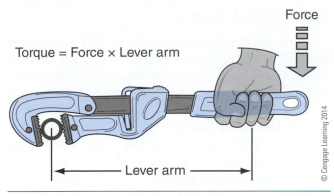

Torque = Force × Lever arm

Figure 10-5 Torque is equal to the amount of force multiplied by the length of the lever.

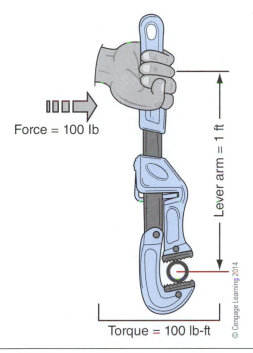

Force = 100 lb

Lever arm = 1 ft

Torque = 100 lb-ft

Figure 10-6 100 pounds of force applied to a 1-foot lever equals 100 pound-feet of torque.

a pipe. By applying a force of 100 pounds to the lever, which in this case is the pipe wrench, at a distance of 1 foot we multiply the force by the distance of the lever. Therefore the torque applied would be calculated as follows:

Torque = Force × Distance
Force = 100 pounds
Distance = 1 foot
Torque = 100 pounds × 1 foot
Total Torque = 100 pound-feet

There are two ways to increase torque: by increasing the force applied, or by increasing the length of the lever. If we increase the force applied to 200 pounds with a 1-foot lever, a total torque of 200 pound-feet

is achieved. If the lever is extended to 2 feet, less force (100 pounds) is required to achieve a torque of 200 pound-feet (see **Figure 10-7**).

If the same amount of force is applied to the extended lever, a torque increase will result: 200 pounds of force × 2-foot lever = 400 pound-feet of torque.

Gears come in a large variety of types and styles to suit the needs of the application. They are usually classified by the type and/or the design of the teeth. The two most common types are the **straight cut spur** and **helical cut**. Straight cut spur teeth are arranged parallel to the supporting shaft. **Spur gears** in mesh will always have at least one pair of teeth in mesh. This tooth design has lower torque capability than helical cut teeth and tends to be noisier than other gear tooth designs. Straight cut spur teeth under load create radial loads on their respective supporting shafts. Radial loading, which is a force perpendicular (90 degrees) to the shaft, occurs when gears push away from each other (**Figure 10-3** and **Figure 10-8**).

Helical cut gear teeth are cut at an angle or helix from the parallel of the supporting shaft. This tooth design has greater torque capability and is quieter because there is always more than one pair of teeth in mesh and the helical design offers greater surface area for tooth contact. Greater torque capability occurs because the paired teeth on the parallel plane are in full mesh; the paired teeth approaching have started to

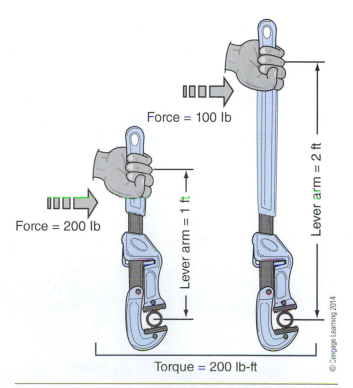

Force = 100 lb

Force = 200 lb

Lever arm = 1 ft

Lever arm = 2 ft

Torque = 200 lb-ft

Figure 10-7 Torque doubles if either the force or the length of the lever is doubled.

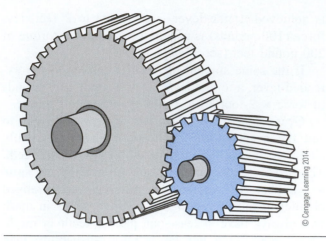

Figure 10-8 Helical spur gear set.

mesh, and the teeth that are coming out of mesh are still in partial mesh. Overall, this produces increased tooth contact and less clattering of gear teeth. Similar to the straight cut spur gear teeth, helical cut gears also produce radial loads on their supporting shafts, but because of their helical design they also produce a side thrust to their respective supporting shafts. This is because helical gears tend to not only push away from each other, but also push away at an angle parallel to their shafts as shown in **Figure 10-8**.

Herringbone Gears

Herringbone gears have double helical-cut gear teeth extending away from each other in opposite directions from the tooth centerline. This design provides the high torque capability of traditional helical cut teeth without the side thrusting. Side thrust is eliminated because, as each double helical comes into mesh, one side thrust cancels out the other one. Although side thrusting is eliminated, the gears still produce a radial load on the shaft (see **Figure 10-9**).

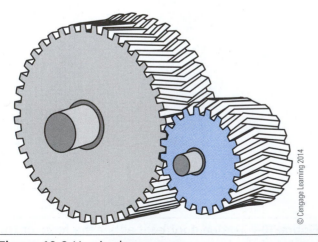

Figure 10-9 Herringbone gears.

Plain Bevel Gears

Plain bevel gears consist of a smaller **pinion gear** matched to a larger **crown gear**. The pinion gear tooth design is a straight cut spur tapered to the edge of the shaft; similarly crown gear teeth are tapered or angled to a smaller thickness at the outer edge as shown in **Figure 10-10**. The pinion gear axis is placed directly on the centerline and 90 degrees to the crown gear axis, allowing the rotating force of the pinion gear to alter its drive by 90 degrees.

Spiral Bevel Gears

Similar to plain bevel gears, spiral bevel gears have a smaller pinion gear matched to a larger crown gear. Spiral bevel gears also provide the means to change the direction of drive input on the pinion gear by 90 degrees; the pinion gear axis is still placed on the centerline axis of the crown gear. The difference between the two configurations is in the gear design. Plain bevel gear teeth are similar to a straight spur gear. However, spiral bevel gears are of a helical-convex design, meaning that instead of traveling from the outer circumference to the center circumference in a straight line, the teeth have a convex shape and are placed angularly to the center of the gear. The advantage of this design over the plain bevel gear is similar to the relationship of the straight spur gear and helical gear designs: more gear tooth contact provides for greater torque capability. The spiral design also allows more than one tooth to be in mesh simultaneously, which results in a quieter running gear arrangement (see **Figure 10-11**).

Hypoid Gears

Hypoid gears resemble spiral gears in gear design. However, hypoid gears have the axis of the pinion gear (the smaller of the two gears) below the centerline of

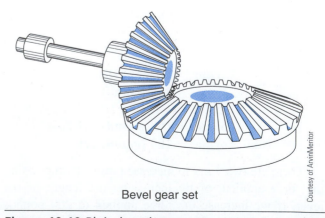

Bevel gear set

Figure 10-10 Plain bevel gear set.

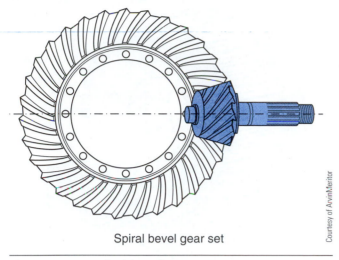

Spiral bevel gear set

Figure 10-11 Spiral bevel gear set.

the crown gear axis. Again, the purpose of this design is to achieve a greater tooth contact area, which results in greater torque capabilities. Hypoid gears also allow more than one gear tooth to be in mesh simultaneously for a smoother, quieter running gear set. Hypoid gear arrangements typically have the convex side of the crown gear as the forward thrust side of tooth contact (see **Figure 10-12**).

Amboid Gears

Amboid gears resemble hypoid gears in every aspect except that the pinion gear axis is mounted above the centerline of the crown gear axis. This also allows greater gear contact and, in many cases, has the concave side of the crown gear tooth as the leading thrust side (see **Figure 10-13**). This configuration allows for high-driveline clearances and is widely used for the rear-drive carrier on tandem-drive axles.

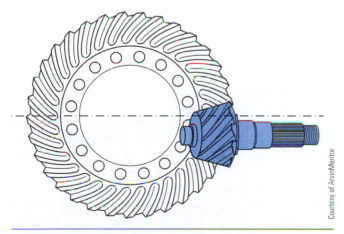

Figure 10-12 Hypoid gearing arrangement.

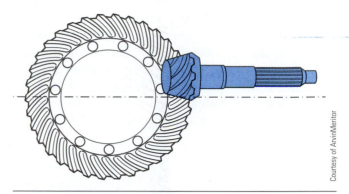

Figure 10-13 Amboid gearing arrangement.

Planetary Gears

Planetary gears are based on the concept of our solar system—the sun in the center with all the planets rotating around it. As shown in **Figure 10-14**, a planetary gear arrangement consists of three different gears: a sun gear, a ring gear, and a set of planetary pinion gears. Like our solar system, the sun gear is located in the center of the gear arrangement. The planetary gears (like the planets) are in mesh with and rotate around the sun gear; they are supported and spaced evenly around the circumference of the sun gear by a planetary carrier. The carrier not only supports the planetary gears but also gives them an axis to rotate on. The ring gear (an internal gear) is meshed with all the planetary gears and supports or holds

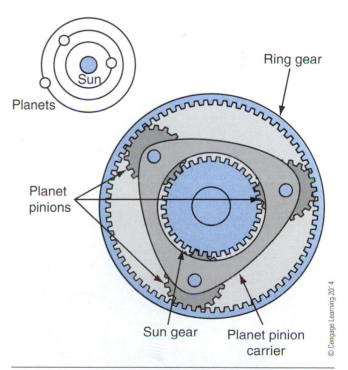

Figure 10-14 Planetary gear sets are based on the design of our solar system.

together the entire gear set. Planetary gears are used for a wide variety of applications, from transmissions to differentials to final drives. This is because one planetary gear set can have as many as seven different gear-ratio configurations and the integrated gear design is capable of high torque and speeds. This will be covered in greater detail in Chapter 11, "Powershift Transmissions."

Worm Gears

Worm gears are used for high-torque/low-speed applications. They consist of a screw gear or worm gear meshed with an external gear (see **Figure 10-15**). They also provide the means to alter the drive input of the worm gear by 90 degrees, similar to the crown and pinion gear designs. This gear application can be seen in many steering gears and gearboxes used with cable winching on cranes.

Rack-and-Pinion Gears

Rack-and-pinion gears are similar to worm gears in their design except that, instead of a worm gear in mesh with an external gear, it uses a rack with a straight gear-tooth design meshed with an external gear as shown in **Figure 10-16**. Rack-and-pinion gears convert rotary motion into linear motion or vice versa, depending on the input or drive component. For example, if the external gear on a rotating shaft is the input or **drive gear,** it will cause the rack (output or **driven gear**) to move in a linear direction. If the rack is the input gear, moving in a linear direction, it will

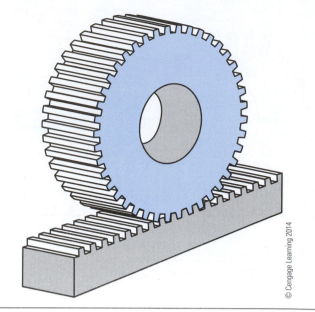

Figure 10-16 Rack-and-pinion gear arrangement.

cause the pinion gear to rotate on its axis or to rotate the shaft to which it is fixed.

GEAR GEOMETRY

Gear geometry is an important factor in gear design; it provides long service life with nominal wear. **Figure 10-17** shows basic gear tooth nomenclature. Outside diameter refers to the gears' outermost diameters, measured directly across the axis from tooth edge to tooth edge. The root diameter is the measurement across the axis within the root or base of the gear teeth. **Pitch diameter** is also the measurement across the axis of the gear, but it is measured approximately between the outside diameter and the root diameter (known as the pitch line). The area within the outside diameter and pitch diameter is referred to as the *addendum*. The area within the root diameter and the pitch diameter is referred to as the *dedendum*. The tooth contact area is the tooth flank. The fillet radius is the machined area between gear teeth at the root of the gear tooth.

Gear Pitch

Gear pitch is a factor of gear design; all gears in mesh are engineered to gear pitch. Gear pitch refers to the number of teeth per given unit of pitch diameter. Here is a simpler example of gear pitch calculation: **Figure 10-18** shows a gear with 36 teeth and a pitch diameter (measured from the pitch line, not the outside diameter) of 6 inches. If you divide the number of teeth by the pitch diameter, the answer equals the gear pitch.

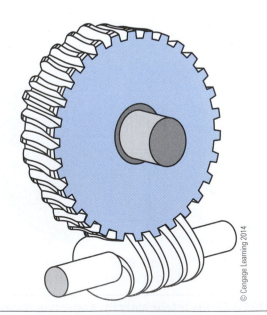

Figure 10-15 Worm gear arrangement.

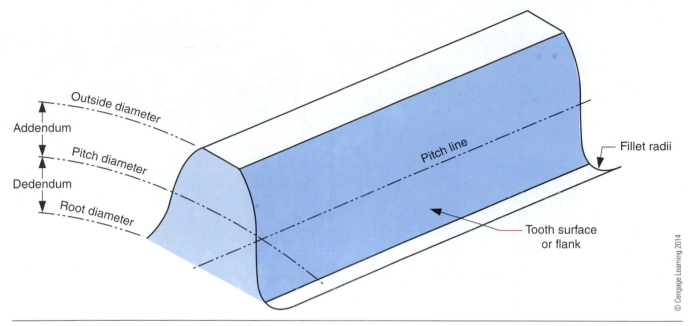

Figure 10-17 Basic gear tooth nomenclature.

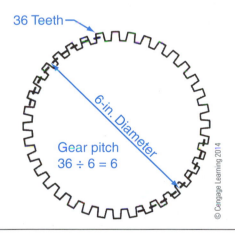

Figure 10-18 Calculating gear pitch.

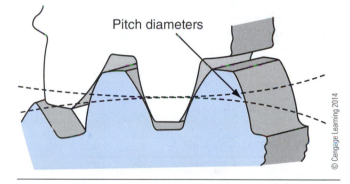

Figure 10-19 Gears rolling at their pitch diameters.

In this example, 36 (number of teeth) divided by 6 (pitch diameter) equals 6; therefore, the gear pitch is equal to 6.

Gears in mesh, regardless of their differences in diameters, must have the same gear pitch. If a larger gear (72 teeth, pitch diameter of 12 inches) is meshed with a gear having a pitch diameter of 6, the larger gear must also have the same gear pitch. In this case, you divide 72 teeth by 12 inches, producing a gear pitch of 6. Congratulations! The gears mesh perfectly.

Gear Backlash

Gear **backlash** is the clearance, or free play, between two gears in mesh. Gears are designed to roll at their pitch diameters (see **Figure 10-19** and **Figure 10-20**).

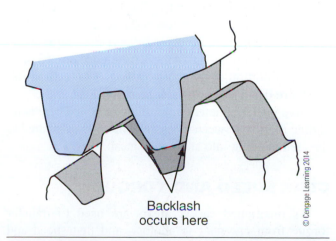

Figure 10-20 Clearance between two gears in mesh is referred to as backlash.

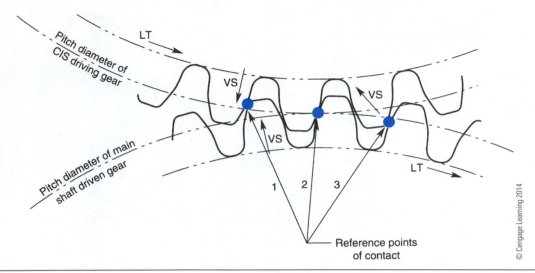

Figure 10-21 Three stages of gear tooth contact: 1. gears coming into mesh, 2. gears in full mesh, and 3. gears coming out of mesh.

Gear teeth are strongest at this point and can transfer the greatest amount of force. If there is too much clearance or backlash between the gears, the pitch diameters are extended from each other and cause the addendum sections of the gear teeth to transfer the force. Because the tooth design is not meant to transfer the total force at this point, excessive tooth loading can cause premature failure. If there is too little backlash, this can cause the teeth to contact the root diameters, which will also result in premature gear failure.

Gear Tooth Contact

The amount of gear tooth contact is dictated by the gear tooth design. **Figure 10-21** shows the three stages of gear tooth contact: 1. gears coming into mesh, 2. gears in full mesh, and 3. gears coming out of mesh. Gears coming into mesh contact the addendum portion of the driven gear and the dedendum portion of the driving gear as shown on reference point 1. The tooth loading or load transfer (LT) is relatively low because the gear teeth in full mesh (reference point 2) support the major force, and the teeth coming out of mesh (reference point 3) also support a portion of the tooth loading. The teeth in full mesh are designed to transfer their full load at their respective pitch diameters in a rolling action. There is no sliding motion between the gear teeth when compared to the teeth coming into and out of mesh.

GEAR SPEED AND TORQUE

As mentioned earlier, gears are used to transfer torque from one gear to another, and that speed and torque are affected inversely and proportionately. This section will show how this effect takes place.

Gear Ratios

Gear ratio describes the speed relationship between two gears in mesh. The speed ratio is the number of rotations of the input gear, compared to the number of rotations of the output gear. This relationship of speed is dictated by the number of teeth on each gear and the speed at which they are rotating. To calculate gear ratios, you must always determine the input member (drive gear) from the output member (driven gear). **Figure 10-22** shows two gears of equal size and equal number of teeth. Because there is no difference between the two gears, the output speed and torque will be the same as the input speed and torque. This is referred to as **direct drive** having a gear ratio of 1:1. If the two gears in mesh have different

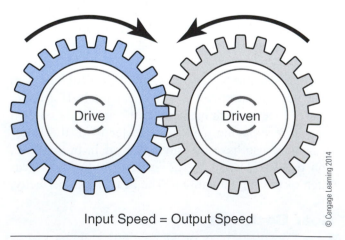

Input Speed = Output Speed

Figure 10-22 When gears have an equal number of teeth in mesh, the input speed will be equal to the output speed.

diameters, as shown in **Figure 10-23**, a small gear (input) is driving a large gear (driven or output). Since it takes the smaller input gear more speed (revolutions) to complete one revolution of the larger output gear, a speed reduction occurs. This type of gear ratio is referred to as an **underdrive**. Inversely, if a large gear (input) is driving a small gear (driven or output), a speed increase occurs because the large gear takes less than one revolution to rotate the driven (small) gear one revolution. This type of gear ratio is referred to as an **overdrive** as shown in **Figure 10-24**.

Gear ratios can be calculated either by the relationship of the number of teeth on the input gear and the number of teeth on the output gear, or by the differential of speeds between the input gear and output gear. To calculate a gear ratio based on the relationship of the number of teeth between the input and output gears, you divide the number of teeth on the driven gear by the number of teeth on the drive gear. This is an example of an underdrive gear ratio when the drive gear has 25 teeth and the driven gear has 125 teeth.

$$\frac{\text{\# of teeth on driven gear}}{\text{\# of teeth on drive gear}} = \frac{125 \text{ teeth}}{25 \text{ teeth}} = 5$$

Therefore, on this combination of gears, for every 5 revolutions of the drive gear, the driven gear will rotate 1 revolution. Expressed as an underdrive gear ratio, this equals **5:1**.

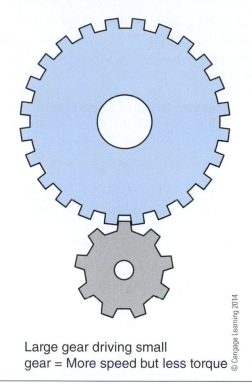

Large gear driving small gear = More speed but less torque

© Cengage Learning 2014

Figure 10-24 A large gear driving a small gear will rotate faster, with less torque.

Note: Whenever the number of teeth on the drive gear is lower than the number of teeth on the driven gear, an underdrive will result.

An example of an overdrive gear ratio is as follows: the drive gear has 80 teeth and the driven gear has 50 teeth. Using the same formula, it is calculated as follows:

$$\frac{\text{\# of teeth on driven gear}}{\text{\# of teeth on drive gear}} = \frac{50 \text{ teeth}}{80 \text{ teeth}} = 0.625$$

Therefore, on this combination of gears, for every 0.625 revolutions of the drive gear, the driven gear will rotate 1 revolution. Expressed as an overdrive gear ratio, this equals **0.625:1**.

Note: Whenever the number of teeth on the drive gear is greater than the number of teeth on the driven gear, an overdrive will result.

A direct drive combination has the same number of teeth on the drive gear as on the driven gear. There is no speed or torque differential.

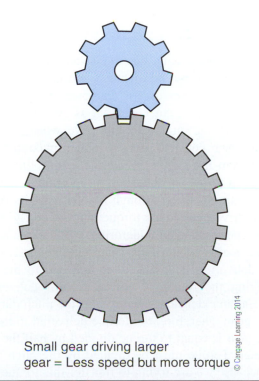

Small gear driving larger gear = Less speed but more torque

© Cengage Learning 2014

Figure 10-23 A small gear driving a large gear will rotate slower, with greater torque.

To calculate gear ratios through speed differentials, the formula is opposite of that using the number of teeth on each of the drive or driven gears. Divide the speed in rpm (rotations per minute) of the drive gear or shaft by the rpm of the driven gear or shaft. In the following example, the drive (input) gear is rotating at a speed of 200 rpm and the driven (output) gear is rotating at a speed 40 rpm. The formula will read as follows:

$$\frac{\textbf{Speed/rpm of drive gear}}{\textbf{Speed/rpm of driven gear}} = \frac{\textbf{200 rpm}}{\textbf{40 rpm}} = \textbf{5.0}$$

This results in an underdrive gear ratio of **5:1**.

....

Note: Whenever the speed of the drive (input) gear is higher than the speed of the driven (output) gear, an underdrive results.

....

The next example shows an overdrive gear ratio using the speed differentials of the following gears and shafts. Drive (input) gear speed is 100 rpm; driven (output) gear speed is 120 rpm.

$$\frac{\textbf{Speed/rpm of drive gear}}{\textbf{Speed/rpm of driven gear}} = \frac{\textbf{100 rpm}}{\textbf{120 rpm}} = \textbf{0.83}$$

This results in an overdrive gear ratio of **0.83:1**.

....

Note: Whenever the speed of the drive (input) gear is lower than the speed of the driven (output) gear, an overdrive results.

....

Because gear speeds are proportionate to their respective gear ratios, calculations of either speed (rpm) and/or ratio can be achieved by combining the two formulas by using at least three of the four known data and a "cross multiplication" calculation.

For example, calculate the output speed of a gear combination with the drive (input) gear having 25 teeth and the driven (output) gear having 125 teeth with an input speed of 200 rpm. The formula is as follows:

$$\frac{\textbf{\# of teeth on driven gear}}{\textbf{\# of teeth on drive gear}} = \frac{\textbf{Speed/rpm of drive gear}}{\textbf{Speed/rpm of driven gear}}$$

Placing the known data in its appropriate slot

$$\frac{125}{25} = \frac{200}{X}$$

Therefore, driven (output) gear X

$$= \frac{25 \times 200}{125} = \frac{5000}{125}$$

Driven (output) gear $X = 40$ rpm

This result indicates that this gear combination, with the input gear rotating at a speed of 200 rpm through the gear ratio of 5:1, has the output gear rotating at a reduced (underdrive) speed of 40 rpm.

Using the same example of gear teeth and speeds, change the known data to drive (input) gear has 25 teeth rotating at a speed of 200 rpm and the output gear is rotating at a reduced speed of 40 rpm. The formula is as follows:

$$\frac{\textbf{\# of teeth on driven gear}}{\textbf{\# of teeth on drive gear}} = \frac{\textbf{Speed/rpm of drive gear}}{\textbf{Speed/rpm of driven gear}}$$

Place the known data in the appropriate slot

$$\frac{X}{25} = \frac{200}{40}$$

Therefore, number of teeth on driven (output) gear X

$$\frac{25 \times 200}{40} = \frac{5000}{40}$$

Number of teeth on gear X

$$= \textbf{125 teeth}$$

Speed and Torque Relationship

We now understand how the speed of two gears in mesh is affected by the diameter and number of teeth on a drive gear versus the diameter and number of teeth on a driven gear. Gears act as levers to transfer torque from the engine to the drive wheels through different gear configurations to obtain maximum efficiency from the engine speed and torque limitations. Speed and torque are inversely proportional, meaning that by gaining speed through a given gear ratio as in an overdrive, we reduce torque, and inversely, by reducing speed (as in an underdrive) we increase torque. Torque is always transferred through the center of the shaft delivering it. If a gear is attached to the shaft delivering the rotational force, it must deliver this torque at an extension equal to half the pitch diameter of the gear. Therefore, the torque delivered to the end of the gear is equal to torque divided by half the pitch diameter. Torque must be divided by the pitch diameter because we are trying to transfer this torque at an extension away from the center of the driving shaft. This is much like trying to lift a pail of water at an extended arm's length compared to lifting it straight up beside you.

Tech Tip: The drive gear of any gear combination will always reduce the original input torque value. The increase in torque is dictated by the diameter of its mating driven gear.

Because torque is transferred to the gear pitch diameter and the driven gear receives this torque at its equivalent pitch diameter, the driven gear will transfer torque to the center of its shaft equal to the torque times half the pitch diameter. The torque must be multiplied because the driven gear is now acting as a lever and fulcrum assembly as shown in **Figure 10-25**. When the fulcrum is farthest from the end of the lever (see **Figure 10-4**), the force is multiplied equal to the length of that lever. A torque increase occurs at the center of its shaft. If the fulcrum is placed closer to the end of the lever, that distance reduces the multiplication factor proportionately to the distance.

Figure 10-26 shows how torque is transferred through two gears of equal size and equal number of teeth. The input torque of the drive gear is equal

to 10 pound-foot at a distance of 1 foot. Therefore, 10 pound-foot is divided by the lever distance of 1 foot, which is equal to 10 pounds of force. This 10 pounds of force is relayed to the drive gear's outer pitch diameter. Because this force is acting at an extension of 1 foot to the center of the driven gear, we multiply the 10 pounds of force by the 1-foot lever, which is equal to 10 lb-ft of torque.

Tech Tip: Idler gears do not change speed or torque ratios between drive and driven gear sets.

Figure 10-27 shows how torque is increased by using a drive gear that is half the diameter of the driven gear. The same input torque of 10 lb-ft divided by the 1-foot distance of the drive gear equals 10 pounds of force. This 10 pounds of force is relayed to the drive gear's outer pitch diameter. Because this force is acting at an extension of 2 feet to the center of the driven gear, we multiply the 10 pounds of force by the 2-foot lever, which is equal to 20 lb-ft of torque. Torque increase is doubled with a speed reduction and decreased to half the input speed due to the driven gear being twice the diameter of the drive gear.

Here is a third example. The drive gear is twice the diameter of the driven gear, such as an overdrive gear ratio. Given: The drive gear has a pitch radius of 2 feet with a torque of 10 lb-ft applied to it. The torque transfer to the driven gear is equal to 10 lb-ft divided by 2 feet, which is equal to 5 pounds of force. This 5 pounds of force is applied to the driven gear at a distance of 1 foot. Therefore, 5 pounds of force is multiplied by 1 foot, which equals 5 lb-ft of torque. The torque is reduced by half and the speed has doubled due to the drive gear being twice the diameter of the driven gear.

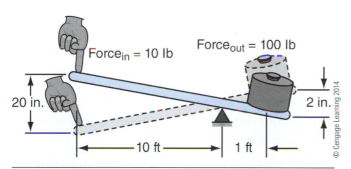

Figure 10-25 The force gained by the length of the lever is proportional to the distance loss.

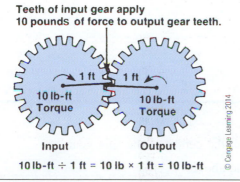

Figure 10-26 When gears are of equal size, input torque equals output torque.

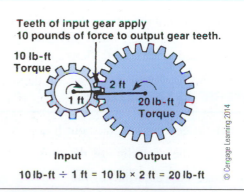

Figure 10-27 When the input gear is smaller than the output gear, a torque increase occurs.

Idler Gears

External gears in mesh rotate in opposite directions. Idler gears are used to transfer rotation in a specific direction without changing the gear ratio. **Figure 10-28** shows an idler gear used to transfer gear rotation of a driven gear in the same direction of the drive gear. Idler gears can also be used to transfer rotational force in place of chain drives or belt drives.

ONLINE TASKS

1. Access the Caterpillar online service information system (SIS) through your college or workplace data hub. Find the final drive gear ratio of a 988G-model front-end wheel loader. The Caterpillar SIS site is *https://login.cat.com.* You will be prompted for a password at logon and you will not get any farther unless you have access.

2. Access the Web site *auto.howstuffworks.com.* Although this Web site is primarily based on automotive technology, it has a wide knowledge base of basic principles that apply to most

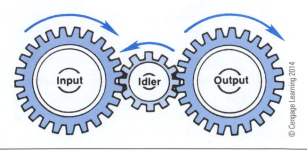

Figure 10-28 Idler gears are used to transfer a specific gear direction without changing speed or torque.

mechanical trades. Use the search function to find "How Helical Gears Work."

Shop Tasks

1. Locate a transmission partially disassembled and calculate the gear ratio of any two gears in mesh. Note which gear is the driving gear and which gear is driven. Identify the type of gearing used within the transmission (e.g., helical, spur, herringbone).

2. Locate a differential partially disassembled and identify the type of gearing used (e.g., hypoid or amboid). With the same gear set, calculate the final drive ratio.

3. Locate a gear within a disassembled transmission and calculate the gear pitch as outlined in this chapter.

Summary

- Gears relay a rotational or twisting force known as torque.

- Combinations of gears can provide a speed increase or torque increase.

- A variety of gear designs are used in heavy equipment dependent on the application of the equipment.

- Gear geometry nomenclature is as follows: outside diameter, root diameter, pitch diameter, pitch line, addendum, dedendum, tooth flank, and fillet radius.

- Gear pitch refers to the number of teeth per unit of pitch diameter.

- Gear backlash refers to the clearance between two gears in mesh.

- The three stages of gear contact are coming into mesh, full mesh, and coming out of mesh.

Review Questions

1. A hypoid gearing arrangement has its pinion gear mounted:

   A. below the centerline of the ring

   B. above the centerline of the ring

   C. in line with the center of the ring gear

   D. none of the above

2. The term *gear backlash* refers to:

   A. the amount of load placed on gears in mesh

   B. the amount of clearance between two gears in mesh

   C. the amount of axial *clearance* between two gears in mesh

   D. the amount of endplay on the transmission output shaft

3. A gear with a gear pitch of 10 will only mesh with a:

   A. 10-inch gear

   B. 5-inch gear

   C. 20 gear pitch diameter

   D. gear having the same gear pitch

4. Spur gears cut at an angle or spiral:

   A. have a greater load capability

   B. run more quietly than helical

   C. both A and B

   D. none of the above

5. What is the gear ratio of two external gears in mesh if the driven gear has 24 teeth and the drive gear has 18 teeth?

   A. 0.75:1

   B. 1.33:1

   C. 4.32:1

   D. 43.2:1

6. In Question 5, the gear ratio is:

   A. an overdrive

   B. an underdrive

   C. a direct drive

   D. an under-overdrive

7. Gear pitch refers to:

   A. the number of teeth per inch around the gear circumference

   B. the number of teeth per inch around the gear circumference divided by the number of teeth

   C. the gear width

   D. the number of teeth around the circumference of the gear divided by the gear diameter

8. Helical gears:

   A. are cut at an angle or spiral

   B. have a greater load capability

   C. run more quietly than spur gears

   D. all of the above

9. The three main components in a planetary gear set are:

   A. the sun gear, ring gear, and pinion gear

   B. the sun gear, ring gear, and planetary gears

   C. the sun gear, pinion gear, and planetary gears

   D. the pinion gear, ring gear, and planetary gears

10. How many gear ratios can be achieved through a planetary gear set?

   A. five

   B. six

   C. seven

   D. eight

CHAPTER

11 Powershift Transmissions

Learning Objectives

After reading this chapter, you should be able to:

- Identify the major components in a typical powershift transmission, including clutch packs, planetary gears, shift manifold, and input/output shafts.

- Trace powerflows through planetary gear sets and calculate their gear ratios.

- Explain the importance of using the correct lubricant and maintaining the correct oil level in a transmission.

- Outline the preventative maintenance inspections (PMI) required by powershift transmissions.

- Identify the differences between planetary shift transmissions and multiple countershaft transmissions.

- Describe procedures for troubleshooting powershift transmissions and torque converters.

- Identify the causes of typical transmission performance problems, such as low power, transmission slipping, and hard shifting.

- Describe hydraulic circuits and oil flows through powershift transmissions.

- Identify the major components of a torque converter and fluid couplings including turbine, impeller/pump, stator, and overrunning clutch.

- Describe the differences between a torque converter and a fluid coupling.

- Describe power and oil flow through a torque converter.

- Identify the major components of a forward/reverse shuttled transmission.

Key Terms

fluid coupling

forward/reverse shuttle

impeller/pump

multiple countershaft transmission

planetary carrier

planetary gear set

planetary pinion gears

planetary shift transmission

powershift

ring gear

rotating clutch pack

shift valves

stationary clutch pack

stator

sun gear

torque converter lockup clutch modulating valve

torque multiplication

transfer case

turbine

vortex oil flow

INTRODUCTION

This chapter introduces powershift transmission concepts that include both planetary and multiple countershaft types as they are used in heavy-duty equipment. Powershift transmissions shift under full power from the engine. In a manual shift transmission, engine torque must be disconnected before a gear selection is completed. A powershift transmission—by means of clutch packs and planetary gearing or countershaft gears in constant mesh—shifts without interrupting drive torque or synchronizing gears. Most manufacturers use a fluid coupling with powershift transmissions. A fluid coupling, such as a torque converter, can absorb variations in a torque load during shifting, but in some instances, a clutch is also used to input engine torque to a powershift transmission. When a driveline clutch is used, it is limited to starting the vehicle from a stationary position. Depending on the application of the vehicle, many manufacturers have eliminated the use of both the torque converter and/or the clutch assembly. Instead, slow and smooth oil modulation to the clutch packs allows engine torque to be transmitted directly to the first gear clutch pack and planetary gear set. Powershift transmissions do not need to break torque from the engine during shifts while the vehicle is in motion.

This chapter will also cover forward/reverse shuttled transmissions and the various types and styles of torque converters, and fluid couplings used with many powershift transmissions. The laws of simple planetary gearing and compound planetary sets will also be discussed in detail.

FORWARD/REVERSE SHUTTLED TRANSMISSIONS

As discussed earlier, some vehicles require an equal number of forward and reverse speeds to accommodate faster cycle times between loads in a loader or bulldozer application. We will look at a wheel loader/backhoe application where this is commonly used. Most manufacturers accomplished this with a **forward/reverse shuttle** incorporated with a four-speed manual or powershift transmission (see **Figure 11-1**). It should also be noted that this application incorporates a torque converter in place of a clutch. This is to allow faster directional changes, and the torque converter acts as a cushioning device when direction changes are made while still moving in the opposite direction or when an operator forces the machine into a dirt pile with the loader bucket.

Note: It is recommended that the vehicle come to a complete stop before a direction change is made.

Forward/Reverse Shuttle

A forward/reverse shuttle consists of two clutch packs made up of multiple discs and plates splined to an input shaft from the torque converter. The clutch packs are hydraulically engaged and spring released by a piston incorporated within the clutch pack assembly. Either one of the clutch discs or clutch plates can be splined to the input shaft to relay input torque from the engine via the torque converter. Two floating gears (forward and reverse gear) on the input shaft are splined to the opposite disc or plate within each clutch pack assembly. For our purposes let's say that the discs are splined to the input shaft and the plates are splined to the forward/reverse floating gears.

Because the clutch packs are spring released, the discs and plates are not in contact with each other. When the engine is started, the torque converter provides rotational torque to the input shaft and the clutch discs that are splined to it. The forward/reverse gears and splined clutch plates are stationary and do not rotate. The forward gear is in constant mesh with the countershaft gears, which are in constant mesh with the output shaft gears. The reverse gear is in constant mesh with the reverse idler shaft, which is in constant mesh with the countershaft gears, which in turn are in constant mesh with the output shaft gears (see **Figure 11-2**).

Forward Direction. In a typical loader/backhoe application, the transmission is a manually operated, four-speed, constant-mesh, synchronized, single-countershaft transmission. When the forward/reverse directional control lever (see **Figure 11-3**) is selected to the forward-drive position, a solenoid-controlled valve directs oil pressure to the forward clutch pack piston. It squeezes together the plates and discs that are splined to the input shaft and forward gear. This causes the forward gear to rotate at the same speed as the input shaft. Because the forward gear is in constant mesh with the transmission countershaft and output shaft gears, they will all rotate within the transmission. The four-speed manual transmission section will drive the vehicle at a selected speed.

Note: No output is possible until a gear/speed selection is made within the manual transmission section. The manual transmission

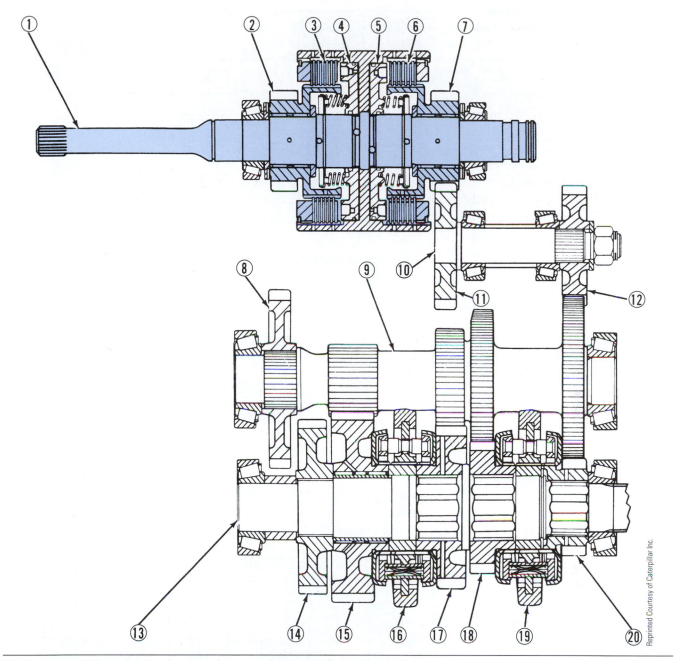

Figure 11-1 Wheel loader/backhoe transmission with four-speed manual transmission and forward/reverse shuttle.

gear speed should be selected before the forward/reverse shuttle. Pre-selecting forward or reverse will cause clashing of the manual transmission synchronizers and gears when attempting to engage the manual section of the transmission. This is equivalent to engaging first gear in a manual transmission without the clutch pedal depressed.

Speed changes can be made on the fly by depressing and holding the clutch disconnect switch located on the shifter handle/knob. The clutch disconnect switch, when depressed, temporarily de-energizes the forward control solenoid valve, causing the forward clutch pack to release. This will place the forward/reverse shuttle in a non-driving neutral condition. Because the transmission is synchronized, the operator can shift to the desired gear or speed without any clashing of gears.

Reverse Direction. When the forward/reverse directional control lever is selected to the reverse drive position, the solenoid-controlled valve directs oil

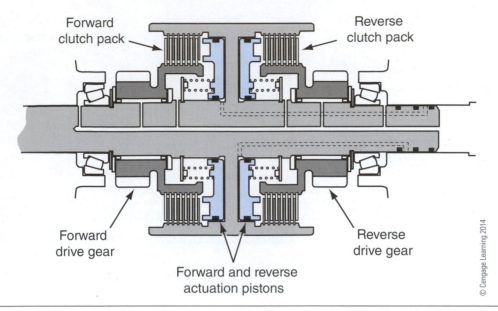

Forward
clutch pack

Reverse
clutch pack

Forward
drive gear

Reverse
drive gear

Forward and reverse
actuation pistons

© Cengage Learning 2014

Figure 11-2 Forward/reverse shuttle consisting of two clutch packs to drive either the forward drive gear or the reverse drive gear.

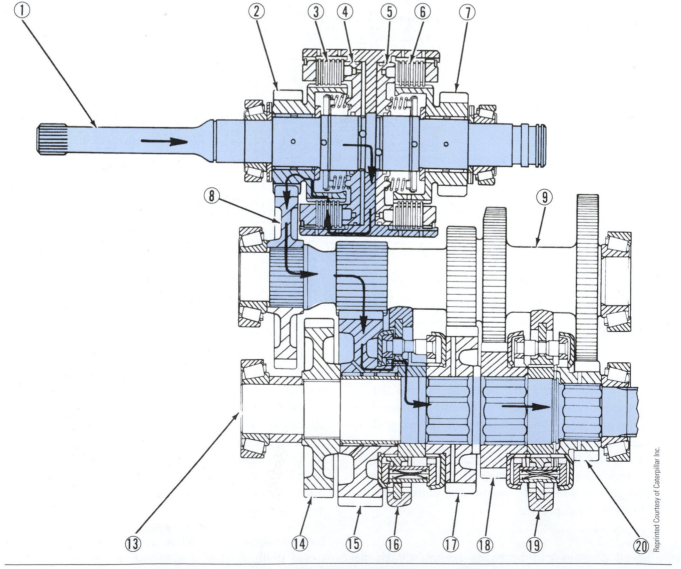

Reprinted Courtesy of Caterpillar Inc.

Figure 11-3 Forward clutch pack will drive countershaft and main output gears.

pressure to the reverse clutch pack piston. It squeezes together the plates and discs that are splined to the input shaft and reverse gear. This causes the reverse gear to rotate at the same speed as the input shaft. Because the reverse gear is in constant mesh with the reverse idler shaft and gear, transmission countershaft, and output shaft gears, they will all rotate in the opposite direction within the transmission. Because the input drive is reversed through a separate reverse idler shaft, all the transmission speeds will drive in the reverse rotation (see **Figure 11-4**). Gear or speed changes can be achieved in reverse direction in the same manner as the forward speeds. The clutch release switch will neutralize the reverse control solenoid valve while depressed.

Note: During normal working condition the manual transmission section is kept at the desired speed/gear to suit the working condition. The operator shuttles between forward and reverse to achieve directional changes of the machine.

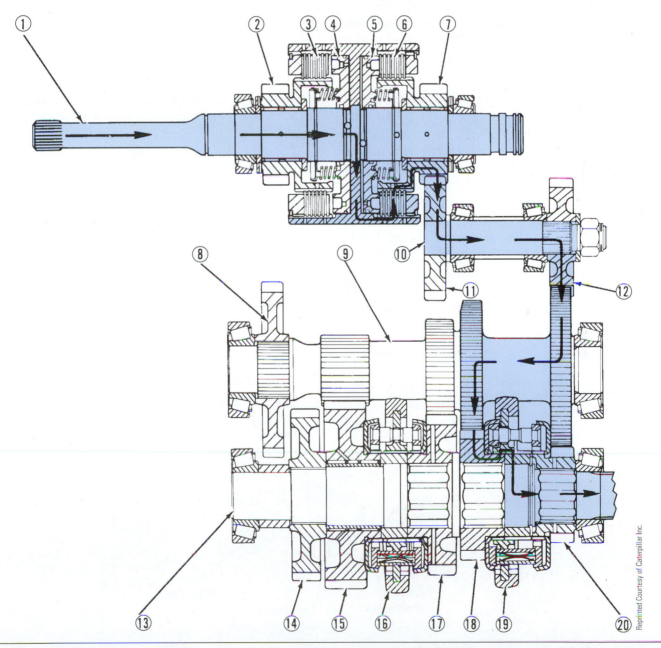

Figure 11-4 Reverse clutch pack engaged with drive reverse idler shaft and gear, countershaft, and main output gears in opposite direction.

Four-Wheel-Drive Option

Many loader/backhoe machines are equipped with four-wheel-drive. This gives the machines more traction for loading or for working in very muddy or slippery conditions. These machines are equipped with a front drive axle as well as the regular rear drive axle. Engine torque must now be simultaneously transferred to the front drive axle and rear drive axle at the same transmission speed or gear ratio. This is accomplished by using a transfer gear to equally transfer the engine torque through the transmission to the front and rear drive shafts (see **Figure 11-5**).

Four-wheel-drive is not always necessary, so the option to engage four-wheel-drive is made available by implementing a clutch pack within the **transfer case** to split the drive between front and rear drive axles.

Transfer Gear and Clutch Pack. The transfer gear is in constant mesh with the transmission output shaft and will rotate in either forward or reverse speed, dictated by the forward/reverse shuttle and manual transmission selected gear. The rear drive axle is the primary drive for the vehicle in this application, so the transfer gear is directly splined to the rear drive yoke and drive shaft.

Note that the transfer gear provides the output from the transmission to the front drive axle and the original transmission output shaft relays the powerflow to the transfer gear. The transfer gear and shaft, together with the main output shaft to the rear drive axle, provide output torque to both drive axles. Refer to **Figure 11-5**.

A multiple-disc clutch pack made up of friction discs and plates similar to the forward/reverse shuttle is implemented to operate the transfer gear and front output shaft. The friction discs are splined to the transfer gear, and the transfer gear is in constant mesh with the output shaft gear of the rear drive axle. The plates are splined to a hub and the output shaft to the front drive axle. A solenoid-controlled valve, activated by a switch in the cab, directs oil to pressurize or drain the clutch pack when two- or all-wheel-drive is selected.

In two-wheel-drive mode, powerflow is transmitted to the transfer gear and the output shaft to the rear drive axle. The friction discs within the clutch pack will rotate with the output shaft but will not transfer any torque to the front axle, because the clutch pack is disengaged. The clutch pack plates will rotate at the same speed as the friction discs and transfer gear and shaft, provided that the front and rear drive shafts are rotating at the same speed (all wheels on the ground).

The only time there is a speed differential between the clutch pack discs and plates is when the vehicle is being propelled forward or reverse with the front wheels off the ground. With the front wheels off the ground, they are not forced to rotate between the friction of the ground and the tire. The clutch plates splined to the forward driveshaft will then remain stationary, and the friction discs splined to the transfer gear and rear output shaft will rotate within the clutch pack. Because there is no synchronization of gears necessary to engage the

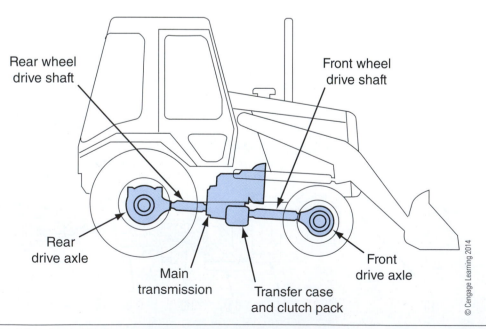

Rear wheel drive shaft

Front wheel drive shaft

Rear drive axle

Main transmission

Transfer case and clutch pack

Front drive axle

© Cengage Learning 2014

Figure 11-5 Wheel loader with four-wheel-drive option: rear drive axle, rear drive shaft, main transmission, transfer gear and clutch pack, front wheel drive shaft, front drive axle.

front-wheel-drive axle, all-wheel-drive mode can be engaged or disengaged while the vehicle is in motion (see **Figure 11-6** and **Figure 11-7**).

There is no speed or gear limitation to switching between two-wheel-drive and all-wheel-drive. The clutch pack allows a smooth transition between the two modes.

Hydraulic Schematic

Figure 11-8 shows the forward/reverse shuttle hydraulic system with the four-wheel-drive option. Oil from the transmission sump is pumped through

the transmission circuit using a crescent-type gear pump (20). A screen-type filter is used at the inlet of the pump to contain any large metal fillings that may be drawn into the pump (19). These pumps are typically driven by either the torque converter or the transmission input shaft. In some applications, the pump may be remotely mounted and driven by another driveline source. An inline filter is used on the outlet side of the pump to contain any smaller particles that may have passed through the suction screen and pump (1). Clean filtered oil is directed to the directional solenoid-controlled selector valve (6) and the AWD (all-wheel-drive) solenoid valve (4). In the

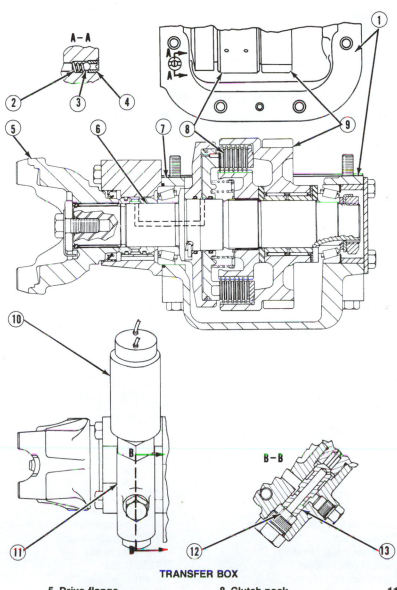

1. Transfer box (drive group).	5. Drive flange.	8. Clutch pack.	11. Solenoid valve.
2. Ball check spring.	6. Transfer drive shaft.	9. Transfer drive gear.	12. Solenoid valve spring.
3. Ball check.	7. Transfer drive group gaskets: two of different thickness.	10. Solenoid.	13. Solenoid valve spool.
4. Ball check seat.			

Figure 11-6 Wheel loader transmission with four-wheel-drive option.

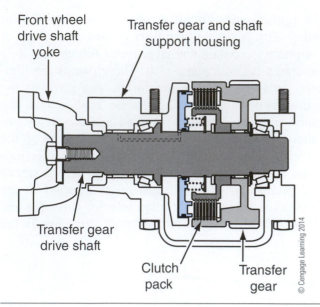

Front wheel drive shaft yoke

Transfer gear and shaft support housing

Transfer gear drive shaft

Clutch pack

Transfer gear

© Cengage Learning 2014

Figure 11-7 Wheel loader with four-wheel-drive option: front wheel driveshaft yoke, transfer gear driveshaft, transfer gear and support housing, clutch pack, transfer gear.

neutral position, oil passage is blocked at the directional selector valve and oil is forced to open the relief valve (5) and flow toward the torque converter. The torque converter supply oil is also drawn from a restricted orifice located within the relief valve; this continuously allows oil availability to the torque converter. The torque converter inlet relief valve (12) will limit the oil pressure within the torque converter to prevent overpressurizing and damaging the converter. Hot oil from the torque converter (13) is sent to the oil cooler (15) and then directed through the transmission lube circuit to cool and lubricate the bearings. This oil will find its way back to the sump (18) where the transmission pump will start the process over again.

When the operator selects either forward or reverse, the direction-selector solenoid valve shifts to either the forward or reverse position. Transmission pump oil then travels through the selector valve to the appropriate forward or reverse clutch pack (7) and (10) and engages the drive to the transmission countershafts. The relief valve will limit maximum clutch pressure

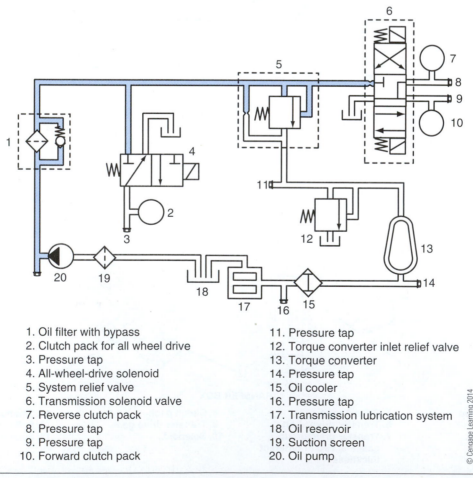

1. Oil filter with bypass
2. Clutch pack for all wheel drive
3. Pressure tap
4. All-wheel-drive solenoid
5. System relief valve
6. Transmission solenoid valve
7. Reverse clutch pack
8. Pressure tap
9. Pressure tap
10. Forward clutch pack
11. Pressure tap
12. Torque converter inlet relief valve
13. Torque converter
14. Pressure tap
15. Oil cooler
16. Pressure tap
17. Transmission lubrication system
18. Oil reservoir
19. Suction screen
20. Oil pump

© Cengage Learning 2014

Figure 11-8 The transmission hydraulic system in neutral.

and relay the oil to the torque converter once the clutch pack has been satisfied. When the operator chooses to travel in the opposite direction by shifting the F-N-R (forward-neutral-reverse) selector valve to the reverse position, the direction-selector solenoid valve will energize and position the valve so that pump oil will be directed to the reverse clutch pack and simultaneously drain the forward clutch pack to the transmission sump. When the F-N-R selector is placed in the neutral position, both the forward and reverse clutch packs are allowed to drain to the sump, disconnecting the drive to the transmission countershafts.

When the operator chooses to engage the AWD function, the selector switch in the cab energizes the AWD solenoid valve (4), shifting it to direct pump oil to the AWD clutch pack and engage the front drive axle. The relief valve also limits maximum clutch pressure to the AWD clutch pack. De-energizing the solenoid valve causes it to shift back to its original position, allowing the clutch pack to drain to the transmission sump and disengaging the drive to the front drive axle.

TORQUE CONVERTERS/FLUID COUPLINGS

Powershift transmissions are usually driven by a fluid coupling, often a torque converter. Torque converters and other types of fluid couplings use a hydrodynamic principle to transfer engine torque to the transmission, without a direct mechanical link between the two components. The term *hydrodynamic* derives from the Greek words "hydro" (meaning water, or in this case oil) and "dynamic" (meaning motion). Engine torque is transferred to the transmission by moving fluid at high velocity onto a driven component that absorbs energy and drives the transmission. This drive median between the engine and transmission is a hydraulic fluid; typically a specialty transmission fluid is used. This is in contrast to a clutch assembly, in which the drive medium between the engine and transmission is a mechanical connection of friction disc(s). **Figure 11-9** demonstrates the hydrodynamic principle by using a garden hose to direct water at a paddle wheel. The water striking the paddle wheel at a high velocity causes it to absorb the energy and rotate. Because the paddle wheel is directly connected to the drive wheel, the drive wheel is forced to rotate. Speed and torque transfer to the drive wheel is directly proportional to the volume and velocity of the water striking the paddle wheel.

A torque converter is a type of **fluid coupling** that allows the engine to rotate independently of

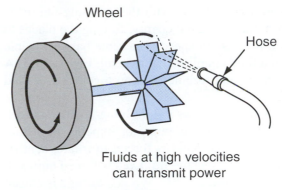

Fluids at high velocities can transmit power

HYDROSTATIC DRIVE
High pressure +
low velocity

TORQUE CONVERTER
Low pressure +
high velocity

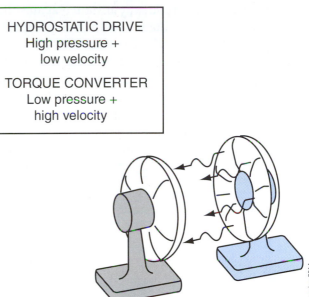

One fan (or turbine) can drive another using fluid force

© Cengage Learning 2014

Figure 11-9 Principles of fluid coupling.

the transmission. (Torque converter/fluid coupling differences will be covered later in this chapter.) If the engine is turning slowly, such as in idling, torque transferred through the torque converter is low, requiring only light brake effort to hold the vehicle stationary. Additional brake pressure would be required to prevent the vehicle from moving if the engine was accelerated. As an engine accelerates and pumps more fluid through the torque converter, greater hydrodynamic force is created, resulting in higher torque transfer.

Fluid Couplings

A fluid coupling consists of a pump or impeller, the component that moves the fluid, and a **turbine**, the component that absorbs the energy of the fluid and transfers it to the transmission.

Figure 11-9 shows how air can be used to transfer energy from one fan to another. As the fan speed increases, so does the velocity of the air. The velocity of the air coming in contact with the opposing fan blade will absorb this increased energy, forcing the blades to rotate and increase in speed. The same basic principle applies to a fluid coupling that uses oil. **Figure 11-10A** shows a pump and turbine rotation together in a fluid coupling.

Figure 11-10B shows a fluid lying level in a bowl. This bowl represents the pump or impeller of a fluid coupling. When the bowl is forced to rotate, centrifugal force acts against the fluid and is forced outward, as shown in **Figure 11-10C**. When a turbine is placed close to the pump, the fluid is transmitted against the turbine and causes it to rotate with the pump/impeller as shown in **Figure 11-10D**. The fluid then returns to the pump/impeller and is redirected to the turbine. To make the fluid coupling more efficient, vanes are placed on both the pump/impeller and the turbine. These vanes increase the efficiency of the fluid coupling by directing the fluid flow against the vanes, which provide a positive surface area to absorb the force. The vanes on the pump/impeller are shaped so

Reprinted Courtesy of Caterpillar Inc.

Figure 11-11 A sectioned fluid coupling drive.

that they direct the fluid more efficiently to the turbine. **Figure 11-11** shows a typical fluid coupling with pump/impeller and turbine.

The engine drives the pump/impeller of a fluid coupling. Once the engine has started, fluid is directed to the turbine within the fluid coupling. The velocity at which the turbine rotates is dictated by engine speed and the resistance of the turbine. When the engine is idling and there is no resistance on the turbine, the two components will rotate at approximately the same speed. If a light load is placed on the turbine, there will be a greater speed differential between the two components. The greatest speed differential is achieved by stopping the turbine. The pump/impeller displaces fluid to the stationary turbine. Simply put, this is what happens:

- The turbine is stationary, so fluid oil becomes very turbulent and loses efficiency.
- Increasing engine speed increases pump/impeller speed, which increases fluid velocity to the turbine.
- The turbine can now overcome the resistance and will start to rotate.
- The load is overcome by the increasing speed, so the turbine increases in speed to a point of almost equal speed.
- The fluid coupling has become more efficient because the fluid is less turbulent.

Note that fluid couplings are never 100% efficient. At peak efficiency there is approximately a 10% speed differential between the pump/impeller and the turbine at what is known as coupling point and efficiency ranges from zero to 90% in the other operating phases.

Torque Converters

Torque converters work under the same principles as a fluid coupling, but with some modifications to

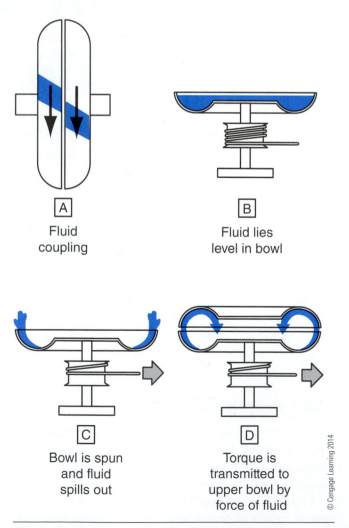

A
Fluid
coupling

B
Fluid lies
level in bowl

C
Bowl is spun
and fluid
spills out

D
Torque is
transmitted to
upper bowl by
force of fluid

© Cengage Learning 2014

Figure 11-10 Operation of a fluid coupling.

make them more efficient. They are equipped with a pump/impeller and turbine much like a fluid coupling, but they are also equipped with a **stator**. Split guide rings, shown in **Figure 11-12**, are designed into the turbine and impeller so that the fluid has an unrestricted fluid path from the **impeller/pump** to the turbine.

A stator is a component used as a fluid lever to assist oil flow within the torque converter. The principles of stators are as follows.

Figure 11-13A shows fluid directed to a flat surface. The energy of fluid in motion is absorbed by the flat surface and is quickly dissipated. There is only a small amount of force felt on the surface. **Figure 11-13B** shows the surface with a slight curvature; the fluid flows more smoothly. There is an increase of energy absorbed over the previous surface, but with less turbulence. The fluid continues to have velocity and acts on the surface area. **Figure 11-13C** shows a surface area with a greater curvature. The fluid energy is completely absorbed by the total surface area, increasing the force acting upon it. The fluid retains its velocity energy that can be used elsewhere.

A stator is located in the center of the torque converter between the pump/impeller and turbine, as shown in **Figure 11-14**. When the pump/impeller is rotating and the turbine is stationary, the fluid within the fluid coupling can be turbulent. A stator redirects the fluid, after it has come in contact with the stationary turbine, back to the pump/impeller at an increased velocity. By redirecting the fluid in the same direction as the pump/impeller rotation and at an increased velocity, the stator increases the force acting on the turbine. This provides an increase in torque output from the turbine/torque converter. This fluid regeneration is known as **torque multiplication**. The fluid regeneration causes the oil to flow in a pattern known as "vortex flow." This vortex flow pattern is a

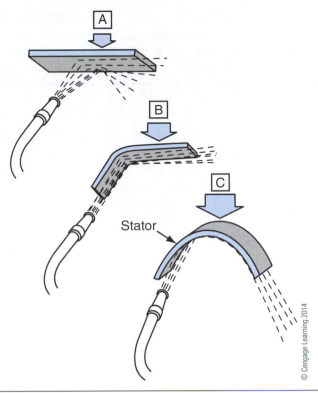

Figure 11-13 (A) Energy of fluid in motion is absorbed by the flat surface and quickly dissipated. (B) Surfaces with a slight curvature allow fluid to flow more smoothly. (C) Fluid energy is completely absorbed by the surface area, increasing the force acting upon it.

continuous spiral from the pump to the turbine, through the stator, and back to the pump/impeller. The split guide rings assist with vortex flow.

Maximum vortex flow occurs at torque converter stall. This is a condition where the turbine has stopped rotating due to an increased load on the vehicle and the pump/impeller is being driven at its maximum rpm. The greatest torque multiplication occurs at this point, making the torque converter more efficient than a fluid coupling. However, this efficiency generates significant heat; it is not recommended that a torque converter remain in a stall condition for any extended period. When the torque converter is in a stall condition with the engine idling, the transmission engaged, and the brakes applied, vortex flow is at a minimum and will not overheat the torque converter. If the operator releases the brakes and starts to accelerate the engine, the velocity of the fluid increases, thus increasing vortex flow. This causes the turbine to rotate and drive the transmission and the vehicle drivetrain. As turbine speed increases, the vortex spiral becomes wider and torque multiplication decreases. Soon the turbine speed will be close to pump/impeller speed, meaning that the vortex flow spiral completely

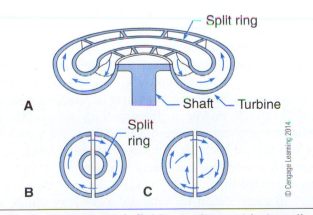

Figure 11-12 Typical fluid coupling with impeller/pump and turbine.

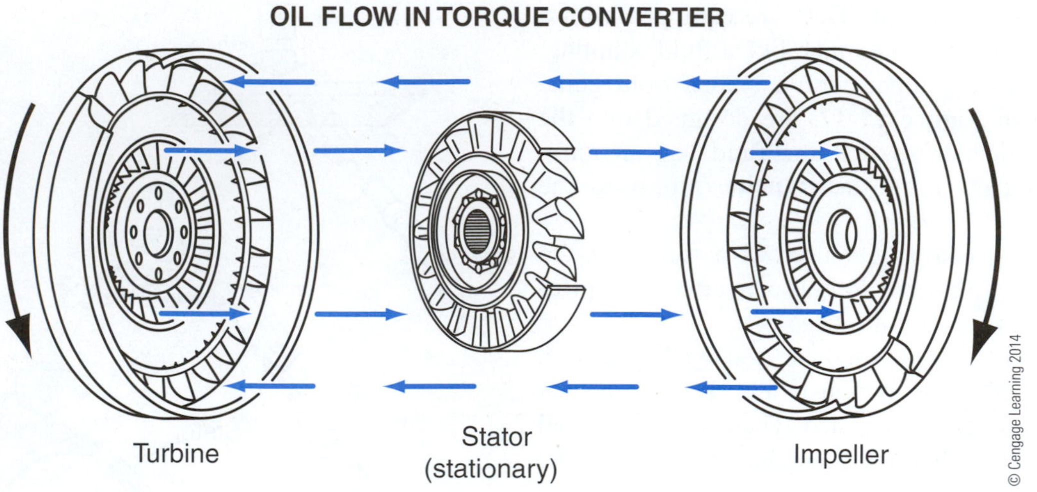

OIL FLOW IN TORQUE CONVERTER

Turbine   Stator   Impeller
                              (stationary)

© Cengage Learning 2014

Figure 11-14 A stator located between two components can redirect fluid back to the impeller.

diminishes, and the fluid flow dynamics change to rotary flow. The transformation between vortex flow and rotary flow occurs because, as turbine speed increases, centrifugal force acting on the fluid increases, meaning that the fluid volume returning to the pump/impeller is reduced. At this moment, vortex flow has completely diminished, and the fluid in the housing travels in a rotary direction together with the pump and turbine (see **Figure 11-15**).

The torque output of a torque converter is infinitely variable between turbine stall and the coupling point at rotary flow. The loads dictate the torque output of the torque converter.

A torque converter is very efficient at maximum vortex flow because it can multiply input torque from the engine. However, efficiency decreases as rotary flow is achieved. The stator, which helps increase vortex flow, can create resistance to rotary flow when the fluid is trying to rotate as a single mass. For this reason many torque converter stators are equipped with an overrunning clutch.

An overrunning clutch on the stator allows the stator to be stationary during vortex flow and to freewheel when rotary flow is achieved (see **Figure 11-16**). Another efficiency factor of the torque converter is the speed differential between the pump and turbine. There is typically a 10% loss between input (pump) and output (turbine) speed at what is known as "coupling phase." Therefore, many torque converters are equipped with a lockup clutch.

A lockup clutch, as shown in **Figure 11-17**, locks the turbine and the torque converter housing together to achieve a 1:1 ratio output between the engine and torque converter. Most torque converters with a lockup clutch are equipped with a stator that incorporates an overrunning clutch. These torque converters are immune to stall speed and lockup speed issues. It should be noted that the application of the vehicle and the manufacturer's decision determines whether or not the torque converter is equipped with a lockup clutch and/or a freewheeling stator. Most bulldozer applications use a stationary stator with no lockup feature because bulldozers typically work under stall conditions and rarely reach rotary flow for any extended time. Rock trucks or scrapers that may have to carry a load for long distances may utilize a lockup feature in the torque converter. Once the vehicle is in top speed and rotary flow is at its

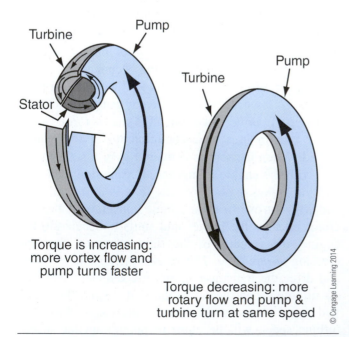

Turbine   Pump

Stator

                          Pump

                Turbine

Torque is increasing: more vortex flow and pump turns faster

Torque decreasing: more rotary flow and pump & turbine turn at same speed

© Cengage Learning 2014

Figure 11-15 Comparison of vortex and rotary flows.

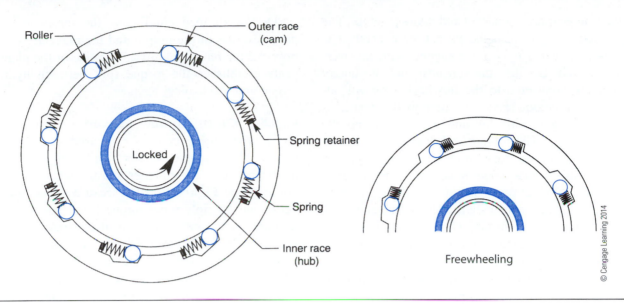

Figure 11-16 Roller-type overrunning clutch.

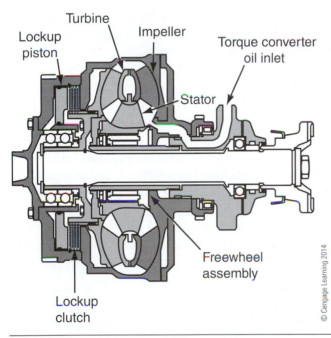

Figure 11-17 Lockup clutch locking together the turbine and torque converter housing.

maximum, locking the torque converter makes the vehicle more efficient to run.

Torque Divider Torque Converter. A torque divider provides both a hydraulic and a mechanical connection by using a planetary gear set from the engine to the transmission. The torque converter provides the hydraulic connection, while the planetary gear set provides the mechanical connection. During operation, the planetary gear set and the torque converter work together to provide torque multiplication as load on the machine increases.

Figure 11-18 shows a typical torque divider as is used in the Caterpillar D8R. The pump/impeller, torque converter housing, and **sun gear** are on a direct mechanical connection to the engine flywheel. The turbine

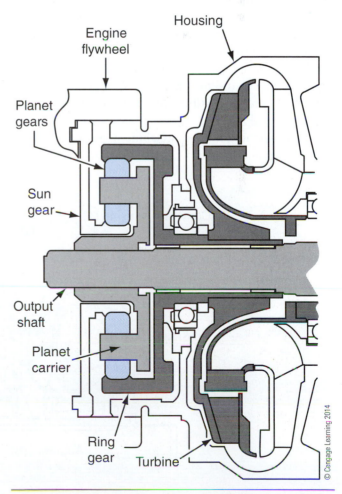

Figure 11-18 Torque divider with planetary gear set.

and the ring gear are connected and rotate together. The output shaft is splined to the **planetary carrier**. The stator is stationary; it is not equipped with an overrunning clutch. Because the sun gear and the pump/ impeller are connected to the flywheel, they will always rotate at engine speed. As the impeller rotates, it directs fluid against the turbine blades, causing the turbine to rotate. The turbine and **ring gear** are connected, so the turbine rotation causes the ring gear to rotate. During no-load conditions, the components of the planetary gear set rotate as a unit, and the planetary gears do not rotate on their shafts. The torque converter is in a rotary flow condition. Therefore, output speed ratio between the engine and torque converter is 1:1.

When the operator places a load on the machine, the output shaft slows down. A decrease in output shaft speed causes the rpm of the planetary carrier to decrease. Decreasing the planetary carrier rotation causes the planetary gears to rotate, which decreases the rpm of the ring gear and the turbine. The torque converter goes into a vortex flow condition and the torque splits with the torque converter (multiplying the torque hydraulically) and the planetary gear set (multiplying the torque mechanically).

An extremely heavy load can stall the machine. If the machine stalls, the output shaft and the planetary carrier will not rotate. This torque converter stall condition causes the ring gear and turbine to rotate slowly opposite to the direction of engine rotation. Maximum torque multiplication is achieved just as the ring gear and turbine begin to turn in the opposite direction. Avoid holding a machine in a stall condition for any extended period.

During all load conditions, the torque converter provides 70% of the output and the planetary gear set provides the remaining 30%. The size of the planetary gears establishes the torque split between hydraulic torque and mechanical torque.

Impeller Clutch Torque Converter. An impeller clutch-type torque converter is used to change the output of the torque converter by allowing the impeller to slip between the impeller and the torque converter housing (see **Figure 11-19**). Because the torque converter can multiply input torque from the engine, undesirable conditions can occur at the final drives of a wheel loader application. In first gear, wheel loaders produce enough torque to cause wheel slippage or wheel spin. If a machine is working in rock conditions, wheel spin can cause tires to be cut and torn apart. The impeller clutch allows the operator to minimize wheel spin by implementing a control feature when the left brake pedal is depressed. Reducing the output torque of the torque converter also provides more engine power for the hydraulic system.

OPERATION OF THE IMPELLER CLUTCH

An impeller clutch solenoid valve controls the fluid flow to the impeller clutch pack. The impeller clutch solenoid is activated by the powertrain Electronic Control Module (ECM). Refer to **Figure 11-19**. When the impeller clutch solenoid valve is not energized by the powertrain ECM, oil flows to passage (12) from carrier (14). Oil in passage (12) forces clutch piston (10) against clutch plates (8) and clutch discs (9). Clutch piston (10) and clutch plates (8) are connected

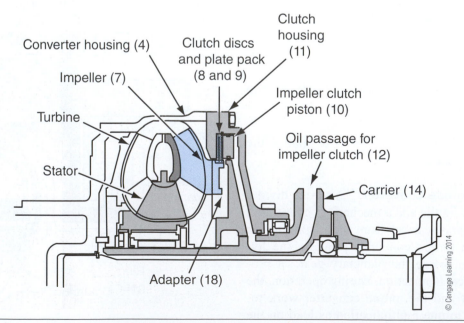

Figure 11-19 Impeller clutch torque converter.

to clutch housing (11) with splines. Clutch discs (9) are connected to adapter (18) with splines. Adapter (18) is fastened to the impeller (7) with bolts. The friction between clutch discs (9) and clutch plates (8) causes the impeller to rotate at the same speed as converter housing (4). This is the maximum torque output from the torque converter drive. As the amount of current to the solenoid increases, oil pressure to clutch piston (10) is decreased by draining some of the oil to the sump. The friction between the clutch plates and the clutch discs decreases. This causes the impeller to slip from the torque converter housing. When the impeller slips, less oil is forced to the turbine. With less force on the turbine, there is less torque output from the converter turbine shaft.

When current to the solenoid is at the maximum, there is minimum oil pressure against clutch piston (10). The clutch plates and the clutch discs now have only a small amount of friction, which causes the impeller to slip within the torque converter housing. Therefore, the impeller forces only a small amount of oil toward the turbine. There is a minimum amount of torque at the converter output shaft.

The impeller clutch modulates oil pressure at each directional shift, which allows the impeller clutch to absorb energy during all directional shifts. This reduces the amount of energy that is absorbed by the directional clutch packs when an operator shifts transmission direction while the machine is still moving in the opposite direction or under acceleration. This results in an easier shift and less wear of the transmission clutch packs.

The following conditions affect the operation of the impeller clutch:

- The position of the left service brake pedal
- Directional shifts
- Engine speed
- Selected gear
- Direction of rotation of the torque converter output shaft
- Torque converter output speed
- Position of the rim-pull selector switch in reduced position
- Position of the reduced rim-pull on/off switch

Left Service Brake Pedal

The left service brake pedal controls the amount of brake pressure used to apply the service brakes. It also controls the amount of pressure that actuates the impeller clutch. The impeller clutch is positioned between the engine and the torque converter. By using the left brake pedal, the operator can divert engine power to the implement hydraulic circuit without putting the transmission in the neutral position. As the operator depresses the pedal, the impeller clutch pressure drops quickly to a working pressure. The pressure is then modulated to a reduced pressure, causing the impeller to slip within the torque converter. This sequence occurs for 10 degrees of pedal travel. Depressing the left pedal past the 10-degree point applies the service brakes.

Shift Modulation of the Impeller Clutch

When shifting between forward and reverse, the powertrain ECM will override the setting of the left brake pedal. The ECM reduces the pressure of the impeller clutch to create smoother transitions. When this shift in direction is detected, the powertrain ECM reduces the circuit pressure of the impeller clutch to an impeller clutch-hold pressure. If lockup for the transmission directional clutch occurs, the ECM will increase the circuit pressure of the impeller clutch to maximum pressure, providing maximum torque output from the torque converter. If the left brake pedal is not depressed, the impeller clutch pressure is not regulated and pressure automatically increases to maximum. The transmission speed sensors and the torque converter output speed sensor are used to determine when the transmission directional clutches have locked up and increase the impeller clutch pressure to maximum.

Function of Reduced Rim-Pull Switches

The reduced or maximum rim-pull switch (shown in **Figure 11-20**) and the reduced rim-pull selector switch provide four levels of reduced rim-pull. Reduced rim-pull is enabled only when the transmission is in first speed forward. The reduced or maximum rim-pull switch is a two-position toggle switch found on the vehicle console. The reduced rim-pull selector

Figure 11-20 Reduced or maximum rim-pull switch.

switch is located in the right-hand console above the operator. In maximum position, the rim-pull switch commands maximum rim-pull. Maximum rim-pull is provided regardless of the position of the rim-pull selection switch. The reduced rim-pull selector switch works with the reduced or maximum rim-pull switch in order to change the torque output of the torque converter. The powertrain ECM increases the current to the impeller clutch solenoid valve to reduce the impeller clutch pressure, which results in reduced rim-pull. The position of the reduced rim-pull selector switch regulates the amount of current that is sent to the impeller clutch solenoid valve. The reduced rim-pull indicator lamp is on when reduced or maximum rim-pull switch is in the reduced position and the transmission is in first speed forward. When the transmission is shifted to second speed or higher ratios, the reduced rim-pull indicator lamp is off and reduced rim-pull is disabled.

MODULATING VALVE (TORQUE CONVERTER IMPELLER CLUTCH)

The impeller clutch solenoid valve is a three-way control valve that modulates pressure proportionally. When the powertrain ECM increases the amount of current to the solenoid, the impeller clutch pressure is reduced. When the amount of current from the powertrain ECM is at zero, the impeller clutch pressure is at the maximum, locking the impeller to the torque converter housing.

PLANETARY GEARING

The term *planetary gearing* was introduced because the gears operate in a manner similar to that of our solar system. A simple planetary gear set consists of four main components. A sun gear is located in the center of the gear set (much like the sun is the center of our solar system) and planetary gears, a planetary carrier, and the ring gear rotate around it.

A planetary carrier provides the support and centralization of the planetary gears as they rotate around the sun gear. Thus, the planetary gears are allowed to rotate around the sun gear as well as rotate on their own axis within the carrier. The carrier becomes the input, the output, or stationary member of the gear set depending on the transmission configuration.

The ring gear provides the final component; it supports the complete planetary gear set and is meshed with all the **planetary pinion gears** on their outer perimeter. Planetary gears are in constant mesh with each other. Because the gearing is enclosed and self-supporting, planetary gears can handle higher torque loads than regular external gears in mesh. Gear forces

are divided equally among the sun, planetary pinions, and ring gears.

Planetary gear sets are compact and can provide many different gear ratio combinations and directional changes. Compounding planetary gear sets increases the ratio combinations. When two or more planetary gear sets are linked by any of the three planetary components to achieve a desired range of gear ratios, they are called compound planetaries. Heavy-equipment transmissions have a multitude of varying transmission configurations, so it is critical that a technician understand planetary gearing and how it works.

Principles of Planetary Gearing

There are some basic laws of planetary gearing. For a planetary gear set to produce an output there must be at least one input drive member and one stationary or held member to provide an output. Any of the three main components (sun gear, planetary carrier, or ring gear) can be the input, the output, or the held member of the gear set. If this combination cannot be provided, a neutral or non-driving condition occurs. When any combination of held member and input member is created, the third component automatically becomes the output of that planetary gear set. Because of the gearing configuration of planetary gears and depending on which component or member is held or driven, a torque or speed increase can be produced. Seven different gear ratio combinations are possible through a single planetary gear set. External gears in mesh rotate in opposite directions and internal gears in mesh rotate in the same direction (see **Figure 11-21**). Planetary gear sets implement a combination of both principles. The sun gear and planetary pinion gears are both external gears, and because they are in constant mesh with each other, they rotate in opposite directions. The ring gear is an internal gear and is in constant mesh with the planetary gears; they will rotate in the same direction.

To calculate gear ratios with planetary gears, the same formula applies as with external gears in mesh, except for calculating the number of gear teeth on the planetary gears. Because the planetary gears are supported within the planetary carrier, there can be as few as three or as many as five planetary pinion gears. This is dictated by the torque loading or application of the transmission. The greater the number of planetary pinion gears within the gear set, the greater torque the gear set can withstand. The formula for calculating gear ratios is divide the number of gear teeth on the drive member by the number of teeth on the driven gear. Because there are three different members in a

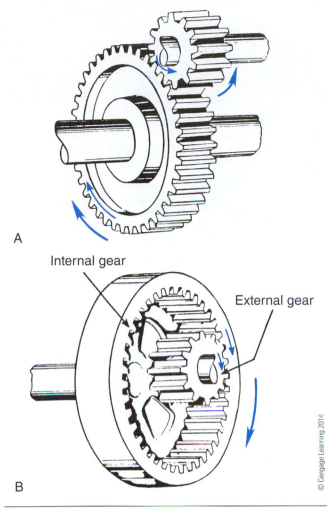

A

Internal gear

External gear

B

© Cengage Learning 2014

Figure 11-21 Comparison of external and internal gears in mesh.

planetary gear set that can either be the drive or the driven member, it is important to use the correct number of teeth and component during our calculations. Because we cannot use the number of gear teeth on the individual planetary pinion gears to calculate the gear ratio, we must add the gear teeth of the ring gear plus the gear teeth of the sun gear for the sum of the planetary carrier. For this reason the carrier will always act as the largest gear. Therefore, it will provide the greatest gear reduction as the output member and an overdrive as the input member.

Figure 11-22 shows a table of seven different combinations plus neutral that can be achieved through a single planetary gear set. Study each individual combination and calculate the gear ratio for that combination. To calculate the gear ratios we must first determine the number of gear teeth on each component (see **Figure 11-23**).

The following eight sections correspond to the table in **Figure 11-22**:

1. **Maximum Speed Reduction.** For maximum speed reduction of a simple planetary gear set, the sun gear must be the input member and the ring gear must be the stationary or held member. This automatically makes the carrier the output member. According to the chart, this will produce the maximum torque increase and cause the carrier to rotate in the same direction as the input or sun gear. Refer to **Figure 11-24**.

   To calculate the gear ratio, divide the number of gear teeth of the input (sun gear 18 teeth) into the number of gear teeth of the output (planetary carrier 61 teeth).

   61 teeth ÷ 18 teeth = Gear ratio of 3.3888"∞:1

2. **Minimum Speed Reduction.** The planetary carrier is still the output, but we reverse the roles of the sun gear and ring gear. The sun gear is now the

LAWS OF SIMPLE PLANETARY GEAR OPERATION					
Sun Gear	**Carrier**	**Ring Gear**	**Speed**	**Torque**	**Direction**
1. Input	Output	Held	Maximum reduction	Increase	Same as input
2. Held	Output	Input	Minimum reduction	Increase	Same as input
3. Output	Input	Held	Maximum increase	Reduction	Same as input
4. Held	Input	Output	Minimum increase	Reduction	Same as input
5. Input	Held	Output	Reduction	Increase	Reverse of input
6. Output	Held	Input	Increase	Reduction	Reverse of input
7. When any two members are held together, speed and direction are the same as input. Direct 1:1 drive occurs.					
8. When no member is held or locked together, output cannot occur. The result is a neutral condition.					

© Cengage Learning 2014

Figure 11-22 Laws of planetary gearing.

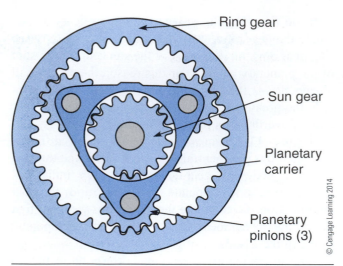

Figure 11-23 Calculating gear ratios in a planetary gear configuration.

Sun gear   18 teeth
Ring gear   43 teeth
Planetary carrier   61 teeth (Sun gear 18 teeth + Ring gear 43 teeth)

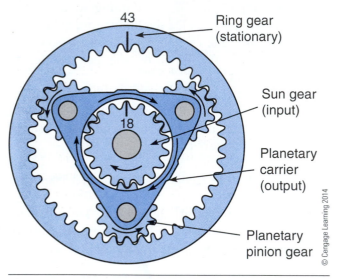

Figure 11-24 Gear ratio: maximum speed reduction/ greatest torque increase.

stationary member and the ring gear is the input. Because the carrier is still the output, the direction of rotation is the same as the input with a minimized gear reduction. This is the second lowest gear ratio with a torque increase achievable through a simple planetary gear set. Refer to **Figure 11-25**.

To calculate the gear ratio, divide the number of gear teeth of the input (ring gear 43 teeth) into the number of gear teeth of the output (planetary carrier 61 teeth).

61 teeth ÷ 43 teeth = Gear ratio of 1.4186:1

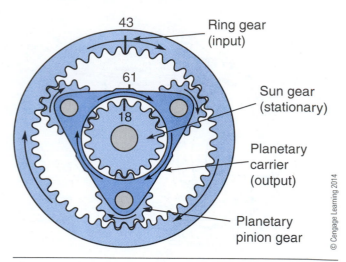

Figure 11-25 Gear ratio: minimum speed reduction/ torque increase.

3. **Maximum Speed Increase.** To achieve a maximum speed increase, the gear with the highest number of teeth must be the input. The same principle applies to planetary gearing. The planetary carrier is always the component with the greatest number of teeth on a planetary gear set, so it must be the input drive member to achieve an overdrive ratio. Depending on which is the held member, either the sun gear or the ring gear will achieve the higher overdrive ratio. **Figure 11-23** shows combination 3, with the sun gear as the output and the ring gear held. This achieves the highest speed increase with the direction of rotation the same as the input. Refer to **Figure 11-26**.

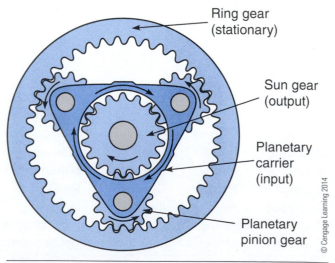

Figure 11-26 Gear ratio: maximum overdrive, speed increase/torque decrease.

To calculate the gear ratio, divide the number of gear teeth of the input (planetary carrier 61 teeth) into the number of gear teeth of the output (sun gear 18 teeth).

18 teeth ÷ 61 teeth = Overdrive gear ratio of 0.2951:1

4. **Minimum Speed Increase.** The planetary carrier remains the input and we reverse the roles of the sun and ring gears. Because the planetary carrier is still the input member and is the largest gear, an overdrive gear ratio is the outcome. With the sun gear as the held gear, the ring gear automatically becomes the output member. This results in a slightly lower overdrive condition with the direction of rotation being the same as the input. Refer to **Figure 11-27**.

To calculate the gear ratio, divide the number of gear teeth of the input (planetary carrier 61 teeth) into the number of gear teeth of the output (ring gear 43 teeth).

43 teeth ÷ 61 teeth = Overdrive gear ratio of 0.7049:1

5. **Speed Reduction Reverse Direction.** To achieve a reverse direction from the input of a simple planetary gear set, the planetary carrier must be the held member. To achieve a speed reduction in a reverse direction, the sun gear must be the input, which automatically makes the ring gear the output. Because the planetary carrier is held, the pinion gears, which are allowed to rotate on their axes, will act as idler

gears to the sun gear. As the sun gear rotates in a clockwise (CW) direction, the pinion gears rotating on their axes will rotate in the opposite direction of the sun gear (counterclockwise [CCW]) because they are external gears. The ring gear, an internal gear, will rotate in the same direction (CCW) as the planetary pinion gears and opposite to the sun gear. Reverse direction causes a speed reduction because the sun gear (input) has fewer teeth than the ring gear (output). Refer to **Figure 11-28**.

To calculate the gear ratio, divide the number of gear teeth of the input (sun gear 18 teeth) into the number of gear teeth of the output (ring gear 43 teeth).

43 teeth ÷ 18 teeth = Gear ratio of 2.3888:1

6. **Speed Increase Reverse Direction.** Remember, to achieve reverse direction, the planetary carrier must be the held member. By reversing the roles of the sun gear and ring gear we can achieve a reverse direction in an overdrive condition. The ring gear is now the input member, which automatically makes the smaller sun gear the output member. With the ring gear as the input member rotating in a CCW direction, the planetary pinion gears again act as idler gears because the planetary carrier is held and the pinion gears are forced to rotate on their axes. They will rotate in the same direction as the input ring gear but will force the output sun gear to rotate in the opposite direction (CW). Because the larger ring gear is driving the

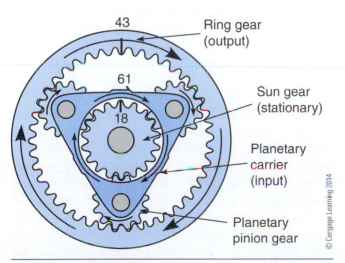

Figure 11-27 Gear ratio: minimum overdrive, speed increase/torque decrease.

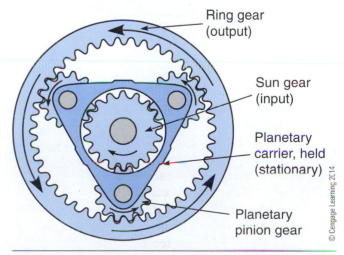

Figure 11-28 Gear ratio: reverse, speed decrease/torque increase.

smaller sun gear, an overdrive condition occurs in reverse direction. Refer to **Figure 11-29**.

To calculate the gear ratio, divide the number of gear teeth of the input (ring gear 43 teeth) into the number of gear teeth of the output (sun gear 18 teeth).

18 teeth ÷ 43 teeth = −Reverse overdrive gear ratio of 0.4186:1

7. **Direct Drive.** To achieve a direct-drive combination on a planetary gear set we must lock any two members together so that they can rotate at the same speed in the same direction. By locking two members together the third member is forced to follow along in the same direction and at the same speed. Refer to **Figure 11-30**.

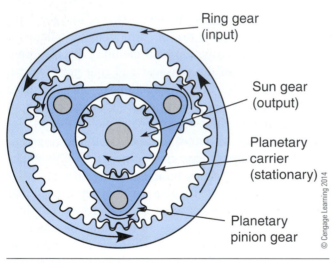

Figure 11-29 Gear ratio: reverse, speed increase/torque decrease.

Ring gear (input)
Sun gear (output)
Planetary carrier (stationary)
Planetary pinion gear

© Cengage Learning 2014

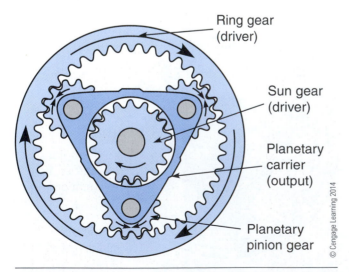

Figure 11-30 Gear ratio: direct drive. Speed and direction are the same as the input.

Ring gear (driver)
Sun gear (driver)
Planetary carrier (output)
Planetary pinion gear

© Cengage Learning 2014

In the example shown, if the sun gear and ring gear are rotated in a CW direction at the same speed, the planetary pinions will not rotate on their axes and, therefore, force the carrier to rotate in unison with the ring gear and sun gear at a ratio of 1:1. The sun gear, rotating in a CW direction, will try to force the pinion gear to rotate in a CCW direction. The ring gear, rotating in the same direction as the sun gear, will also try to force the planetary pinion gears to rotate clockwise. These two opposing forces cancel each other and prevent the pinion gears from rotating. The carrier is forced to rotate together with the sun and ring gears.

8. **Neutral Condition.** There must be at least one input member and one held member to achieve an output from a planetary gear set. If any two members are allowed to freewheel, a neutral condition results. A force (torque in this case) always takes the path of least resistance: if a planetary gear set has the sun gear as the input drive member and the carrier as the output member, the ring gear has to be the held member. If the ring gear is allowed to freewheel, the path of least resistance would be the ring gear, and it would be forced to rotate instead of the carrier. Because the carrier is the output member, the resistance on it would be greater than the resistance on the ring gear. Therefore, the path of least resistance would be the ring gear forcing it to rotate. This combination can occur even when the 1:1 ratio is provided by two inputs or by locking two members together and forcing them to rotate at the same speed. If one input is disconnected or the two drive members are unlocked, the result is the same: the torque will take the path of least resistance and a neutral condition will result. If there are two freewheeling members, neutral will always result. Refer to **Figure 11-31**.

Summary of Planetary Gearing. Here are some simple notes to remember. When you know what the planetary carrier is doing, you will be able to decipher speed, torque, or direction of a planetary gear set. Remember that the planetary carrier has no teeth and the pinion gears always act as idler gears; their number of teeth is not related to the gear ratio calculations of the planetary gear set. However, an arbitrary number needs to be assigned to the carrier in order to calculate the ratios. Simply count the number of teeth on the sun gear and the ring gear. Add these two numbers together

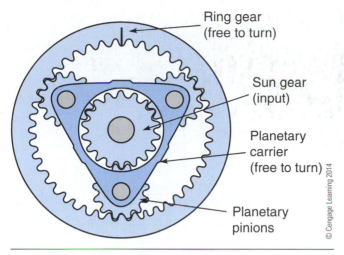

Figure 11-31 Gear ratio: neutral.

and you have the carrier gear number for calculation purposes.

- When calculating gear ratios, the carrier is always considered the largest gear in a planetary gear set.
- To calculate gear ratios, divide the number of gear teeth on the drive gear into the number of gear teeth of the driven gear. The held member plays no part in the ratio calculations.
- When the carrier is the output, a speed decrease/torque increase occurs in the same direction as the input.
- When the carrier is the input, a speed increase/torque decrease occurs in the same direction as the input.
- When the carrier is held, one of two reverse directions can occur to the output: One with a speed increase/torque decrease, or one with a speed decrease/torque increase.
- When two inputs move at the same speed or two members are locked to rotate at the same speed, a 1:1 gear ratio occurs.

CLUTCH PACKS

With planetary gear sets, there must be a means of holding certain components in a planetary gear set to produce an output. **Planetary shift transmissions** use a series of clutch packs to hold or provide an input to a particular planetary component to achieve a desired gear ratio through the planetary gear sets. In heavy-duty equipment, the clutch packs used are a multi-disc clutch comprised of friction discs and plates. The friction discs are made of ceramic/bronze compounds, elastomer compounds, or graphite compounds to withstand high heat. These clutch discs are allocated

within the transmission, and the oil provides cooling for these wet discs and plates. Some smaller equipment applications may still utilize organic/paper fibers on their friction discs. Four basic types of friction discs are shown in **Figure 11-32**.

The number of friction discs and plates used within a single clutch pack is strictly dictated by the application of the clutch pack and the torque loading applied to it. For example, first-gear clutch pack may utilize four discs and three plates and second-gear clutch pack may utilize only three discs and two plates of the same diameter. This is because first gear is susceptible to higher torque loads and requires a higher coefficient of friction than second gear.

As like any multi-disc clutch, the friction discs may have their teeth on either the outer edge of the disc or the center edge of the disc. This is dependent on the component (sun gear, ring gear, or planetary carrier) that it is either holding stationary or imparting drive torque. When the friction discs are toothed on their outer perimeter, the friction plates are toothed on their inner perimeter and vice versa. In many applications, friction plates use "dog teeth" or tangs instead of spur-type teeth on their plates. Within each clutch pack, an actuation piston engages the clutch using hydraulic oil pressure. Return springs are typically used to return the actuation piston and release the clutch pack. These components are shown in **Figure 11-33**.

Clutch packs are typically one of two types: rotating clutch packs and **stationary clutch packs**. A **rotating clutch pack**, shown in **Figure 11-34**, provides an input to a planetary component; this means that whenever the clutch pack is engaged, it will transfer engine torque to a specified planetary component and will rotate that component at a specific engine speed.

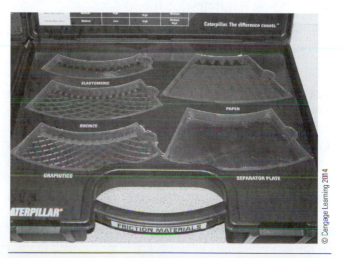

Figure 11-32 Four types of clutch disc material.

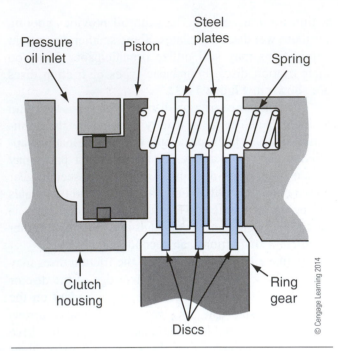

Figure 11-33 Components of a multiple-disc clutch pack.

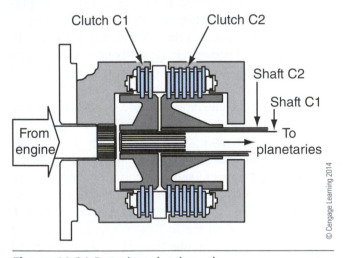

Figure 11-34 Rotating clutch packs.

When engaged, a stationary clutch pack keeps a specific planetary component from rotating (see **Figure 11-35**). Typically, stationary clutch packs are toothed to the transmission housing and lock the component to it. For example, if we want maximum torque/minimum speed ratio from a planetary gear set we know that the ring gear must be held and the sun gear must be the input. Therefore, a stationary clutch pack is used to lock the ring gear to the transmission housing, holding the ring gear stationary. A rotating clutch pack can then be used to transfer engine torque to the sun gear; the clutch pack will rotate with the sun gear when engaged. With the two clutch packs

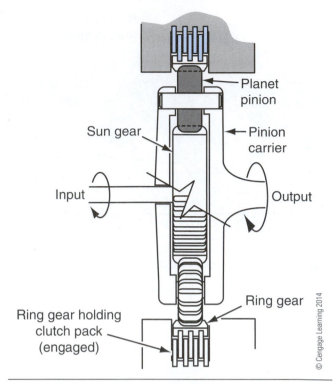

Figure 11-35 Stationary clutch pack.

engaged, the planetary carrier automatically produces the output at maximum torque/minimum speed.

COMPOUND PLANETARY TRANSMISSIONS

Many heavy-duty applications require equipment to have a multitude of forward and reverse speeds. A single planetary gear set may not be able to accomplish this; therefore, compound planetary gear sets are used. Compounding planetary gear sets allows manufacturers to design transmissions to meet their specific speed and torque ratios for any application. This is achieved by locating multiple planetary gear sets one behind the other(s). Each gear set transfers engine torque through to the transmission output shaft and each gear set is connected by any one of the three planetary components. For example, the ring gear of the first planetary may be connected to the carrier of the second planetary, and the second planetary sun gear may be connected to the sun gear of the third planetary. An illustration of multiple or compounded gear sets is shown in **Figure 11-36**.

The combinations of planetary gear ratios are limited only by the number of planetary gear sets used. In some scraper applications, an eight-speed planetary transmission is used. In the heavy-duty equipment industry the configurations of transmissions seem almost infinite. However, we will describe only two popular

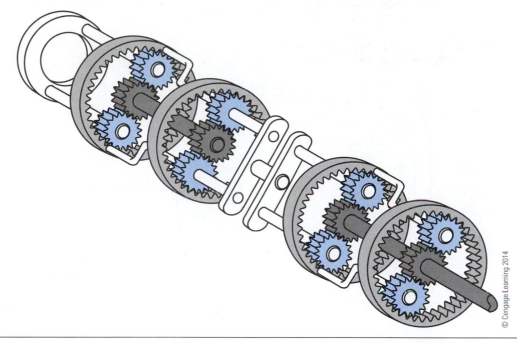

© Cengage Learning 2014

Figure 11-36 Compound planetary transmission using four interconnected planetary gear sets.

transmissions used today. It is the responsibility of the technician to become familiar with the specific transmissions. This chapter will provide some basics to understanding powershift transmissions and their principles; it should not be used as a service manual.

Planetary Powerflows

The transmission chosen to describe planetary powerflows is from a Caterpillar 988G wheel loader. Wheel loaders are very popular in the heavy-duty equipment field, so we will evaluate powerflows and the hydraulic control system used in this machine and transmission as an example. The 988G loader has a bucket capacity of approximately 8.2 to 9.2 cubic yards, depending on the application. The operating weight is approximately 110,000 pounds. It is powered by a 3456 engine rated at 475 hp. It has a four-speed forward, three-speed reverse, planetary shift transmission that can be operated in automatic or powershift mode (see **Figure 11-37**).

Six hydraulically activated clutches (one rotating clutch pack, five stationary clutch packs) are used to achieve this combination with five planetary gear sets. When energized, six electronically controlled solenoids control the oil flow to the clutch packs. A speed clutch and a directional clutch must be engaged to provide powerflow through the transmission. The speed clutch engages before the directional clutch. Number 1 clutch pack and planetary gear set provide reverse direction. Number 2 clutch and planetary gear

set provide forward direction. The number 6 clutch pack provides first speed, number 5 clutch pack provides second speed, number 4 clutch pack provides third speed, and number 3 clutch pack provides fourth speed. For speed control purposes in reverse, number 3 clutch pack cannot be engaged for fourth-speed reverse. The torque converter has an impeller clutch with the option of a lockup clutch as well.

First Speed Forward Powerflow. Refer to **Figure 11-38**. When the operator preselects first speed forward, number 6 and number 2 clutch are engaged. Number 2 clutch is the forward speed clutch, which stays engaged for all the forward speeds; it holds ring gear (6) stationary. Number 6 clutch is the first-speed clutch, which holds ring gear (17) stationary. Input shaft (23) rotates forward drive planetary sun gear (4) at torque converter speed. Because forward drive planetary ring gear is held and the sun gear is the input drive, planetary gears (25) rotate around the ring gear. Because number 2 and 3 planetary carrier (7) occupy the same housing, it is forced to rotate in the same direction as the input shaft and sun gear. The carrier becomes the output at low speed/high torque. Because number 3 sun gear is attached to the output shaft (22), it acts as the stationary member to planetary gear set 3. This will cause the number 3 ring gear (9) to rotate. The number 3 planetary gears will rotate the ring gear, which is splined to number 4 planetary carrier, causing it to rotate. The number 4 sun gear (11) is also splined to the output shaft (22), which acts as a

TRANSMISSION

6P3761 TRANSMISSION (SERIAL NO. 2SA1-UP)

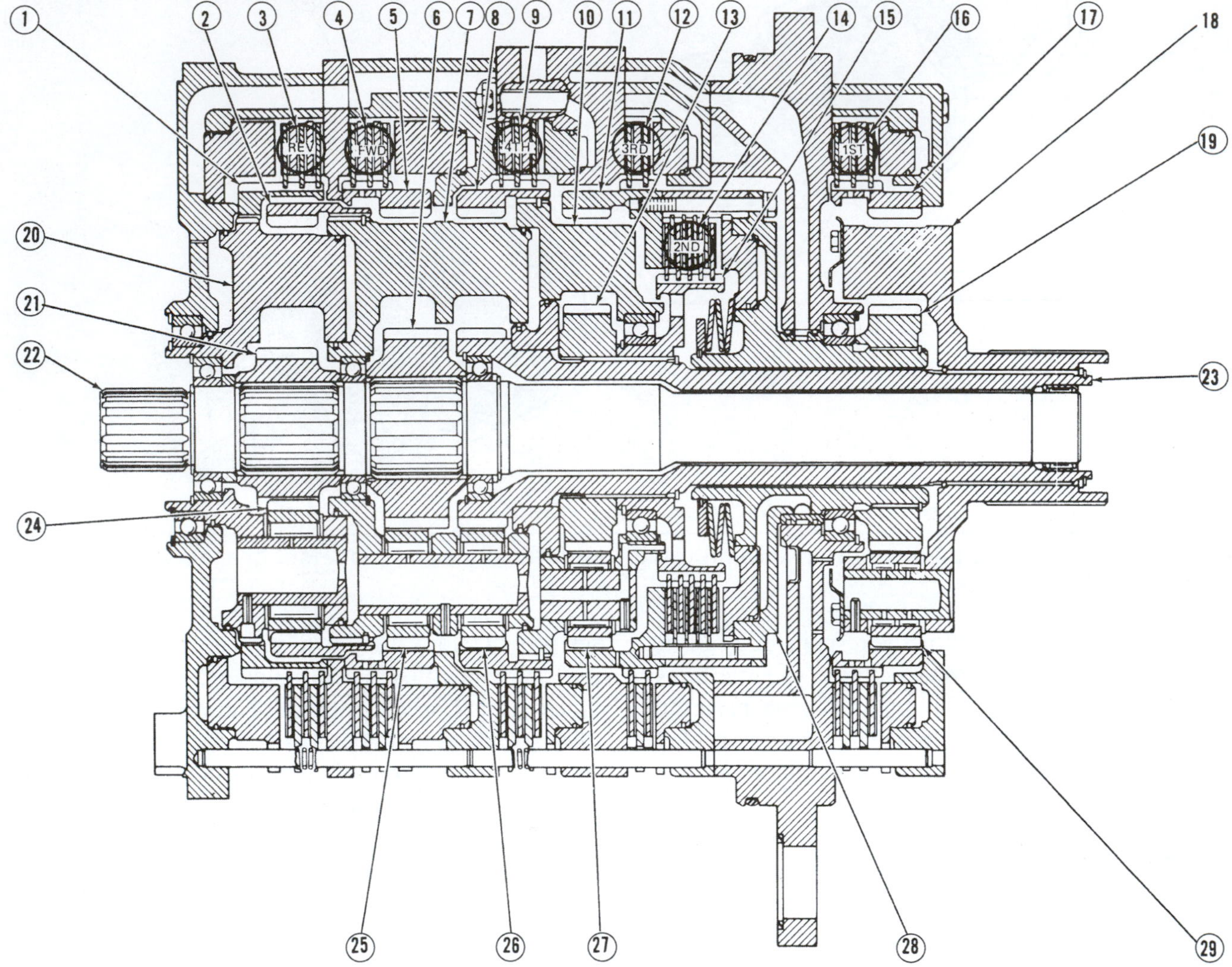

TRANSMISSION COMPONENTS

1. Ring gear for No.1 clutch.	9. No.3 clutch.
2. Coupling gear.	10. No.4 carrier.
3. No.1 clutch.	11. Ring gear for No.4 clutch.
4. No.2 clutch.	12. No.4 clutch.
5. Ring gear for No.2 clutch.	13. No.4 sun gear.
6. No.2 sun gear.	14. No.5 clutch.
7. No.2 and No.3 carrier.	15. Rotating hub.
8. Ring gear for No.3 clutch.	

16. No.6 clutch.	23. Output shaft.
17. Ring gear for No.6 clutch.	24. No.1 planetary gears.
18. No.6 carrier.	25. No.2 planetary gears.
19. No.6 sun gear.	26. No.3 planetary gears.
20. No.1 carrier.	27. No.4 planetary gears.
21. No.1 sun gear.	28. Housing assembly.
22. Input shaft.	29. No.6 planetary gears.

Figure 11-37 Four-speed forward, three-speed reverse, planetary shift transmission.

stationary member for the number 4 planetary. This will cause number 4 ring gear (13) to rotate as the output. Number 4 ring gear is fastened to the rotating clutch housing that has sun gear from the number 5 planetary gear set (19) splined to it. Because stationary clutch pack number 6 (16) is holding number 5 ring gear (17), the planetary carrier becomes the output through output shaft (22). Output shaft (22) and number 5 carrier (18) are splined together and rotate at equal speeds.

As a result, a multiple gear reduction (low speed/high torque) output is divided between number 3 and 4 planetary gear set and number 5 planetary gear set.

Second Speed Forward Powerflow. Refer to **Figure 11-39**. When the operator selects second speed forward, number 5 and number 2 clutches are engaged. Number 2 clutch is the forward speed clutch, which holds ring gear (6) stationary. Input shaft (23) rotates forward drive planetary sun gear number 4 at torque

FIRST SPEED FORWARD

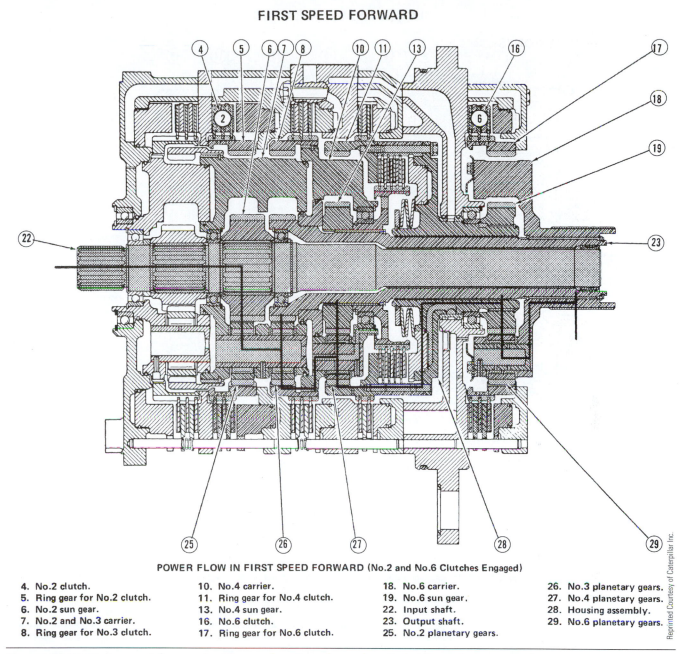

POWER FLOW IN FIRST SPEED FORWARD (No.2 and No.6 Clutches Engaged)

4. No.2 clutch.	10. No.4 carrier.	18. No.6 carrier.	26. No.3 planetary gears.
5. Ring gear for No.2 clutch.	11. Ring gear for No.4 clutch.	19. No.6 sun gear.	27. No.4 planetary gears.
6. No.2 sun gear.	13. No.4 sun gear.	22. Input shaft.	28. Housing assembly.
7. No.2 and No.3 carrier.	16. No.6 clutch.	23. Output shaft.	29. No.6 planetary gears.
8. Ring gear for No.3 clutch.	17. Ring gear for No.6 clutch.	25. No.2 planetary gears.	

Reprinted Courtesy of Caterpillar Inc.

Figure 11-38 First speed forward with number 2 and number 6 clutch packs engaged.

converter speed. Because forward drive planetary ring gear is held and the sun gear is the input drive, planetary gears (25) rotate around the ring gear. Because number 2 and 3 planetary carrier (7) occupies the same housing, it is forced to rotate in the same direction as the input shaft and sun gear. The carrier becomes the output at low speed/high torque. Because number 3 sun gear is attached to the output shaft (22), it acts as the stationary member to planetary gear set 3. This will cause the number 3 ring gear (9) to rotate. The number 3 planetary gears will rotate the ring gear, which is splined to number 4 planetary carrier. Because rotating clutch pack number 5 is engaged, it will

prevent number 4 ring gear and sun gear from rotating at different speeds. This causes the input from number 4 carrier to rotate the planetary gear set as one unit or at a ratio of 1:1. The number 4 sun gear (11) is splined to the output shaft (22).

The result is a single reduction (low speed/high torque) through planetary gear set number 3 splitting the torque path with number 4 planetary gear set.

Third Speed Forward Powerflow. Refer to **Figure 11-40**. When the operator selects third speed forward, number 4 and number 2 clutches are engaged. Number 2 clutch is the forward speed clutch, which

SECOND SPEED FORWARD

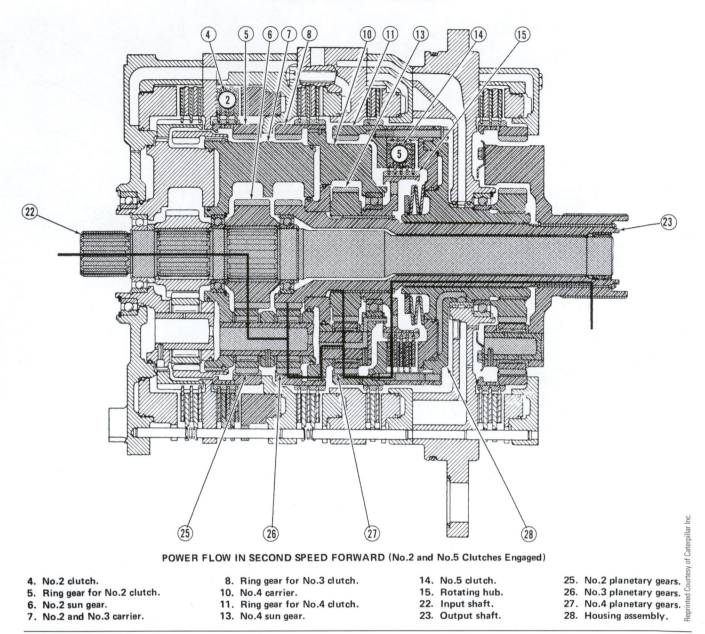

POWER FLOW IN SECOND SPEED FORWARD (No.2 and No.5 Clutches Engaged)

4. No.2 clutch.	8. Ring gear for No.3 clutch.	14. No.5 clutch.	25. No.2 planetary gears.
5. Ring gear for No.2 clutch.	10. No.4 carrier.	15. Rotating hub.	26. No.3 planetary gears.
6. No.2 sun gear.	11. Ring gear for No.4 clutch.	22. Input shaft.	27. No.4 planetary gears.
7. No.2 and No.3 carrier.	13. No.4 sun gear.	23. Output shaft.	28. Housing assembly.

Figure 11-39 Second speed forward with number 2 and number 5 clutch packs engaged.

holds ring gear number 6 stationary. Input shaft (23) rotates forward drive planetary sun gear number 4 at torque converter speed. Because forward drive planetary ring gear is held and the sun gear is the input drive, planetary gears (25) rotate around the ring gear. Because number 2 and 3 planetary carrier (7) occupies the same housing, it is forced to rotate in the same direction as the input shaft and sun gear. The carrier becomes the output at low speed/high torque. Because number 3 sun gear is attached to the output shaft (22), it acts as the stationary member to planetary gear set 3. This will cause the number 3 ring gear (9) to rotate. The number 3 planetary gears will rotate the ring gear, which is

splined to number 4 planetary carrier. Number 4 planetary carrier becomes the input to its planetary gear set. Because the number 4 clutch pack is engaged, it is holding stationary ring gear (13). Planetary number 4 sun gear becomes the output drive through output shaft (22). Note that number 4 planetary is driving in an overdrive condition compared to its input. The output shaft is still in an underdrive condition due to number 3 planetary providing the input to number 4 planetary at a maximum torque minimum speed condition multiplied by the reduced speed of the rotating sun gear. Because number 3 sun gear is attached to the output shaft and acts as a stationary member for the number 3 planetary

THIRD SPEED FORWARD

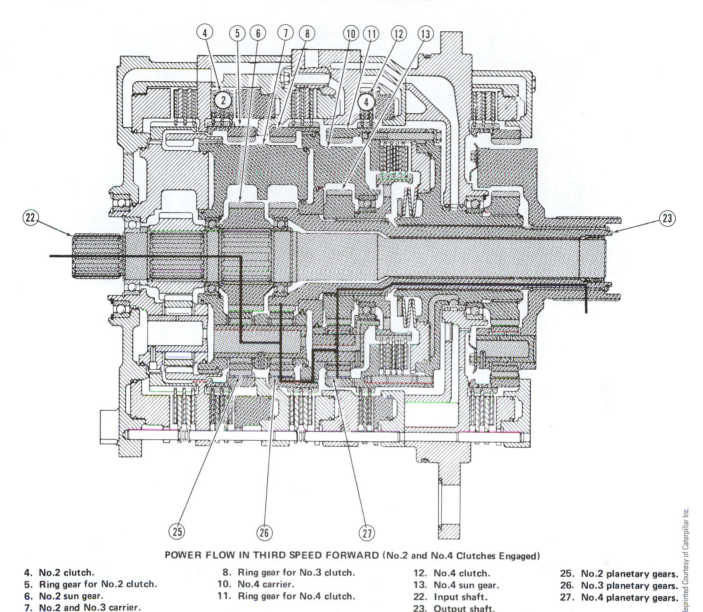

POWER FLOW IN THIRD SPEED FORWARD (No.2 and No.4 Clutches Engaged)

4. No.2 clutch.	8. Ring gear for No.3 clutch.	12. No.4 clutch.	25. No.2 planetary gears.
5. Ring gear for No.2 clutch.	10. No.4 carrier.	13. No.4 sun gear.	26. No.3 planetary gears.
6. No.2 sun gear.	11. Ring gear for No.4 clutch.	22. Input shaft.	27. No.4 planetary gears.
7. No.2 and No.3 carrier.		23. Output shaft.	

Figure 11-40 Third speed forward with number 2 and number 4 clutch packs engaged.

gear set, it reduces the output speed of the ring gear even further as the output shaft begins to rotate. This increases torque output from the transmission.

The result is a single reduction (low speed/high torque) through planetary gear set number 3, splitting the torque path with number 4 planetary gear set in an overdrive condition.

Fourth Speed Forward Powerflow. Refer to **Figure 11-41**. When the operator selects fourth speed forward, number 3 and number 2 clutches are engaged. Number 2 clutch is the forward speed clutch, which holds ring gear number 6 stationary. Input shaft (23)

rotates forward drive planetary sun gear number 4 at torque converter speed. Because the forward drive planetary ring gear is held and the sun gear is the input drive, planetary gears (25) rotate around the ring gear. Because number 2 and 3 planetary carrier (7) occupies the same housing, it is forced to rotate in the same direction as the input shaft and sun gear. The carrier becomes the output at low speed/high torque. Because number 3 ring gear is held stationary by the number 3 clutch pack, it acts as the stationary member to planetary gear set 3. This will cause the number 3 sun gear to rotate as the output member due to the input of the number 2–3 carrier. It is a direct output with no

FOURTH SPEED FORWARD

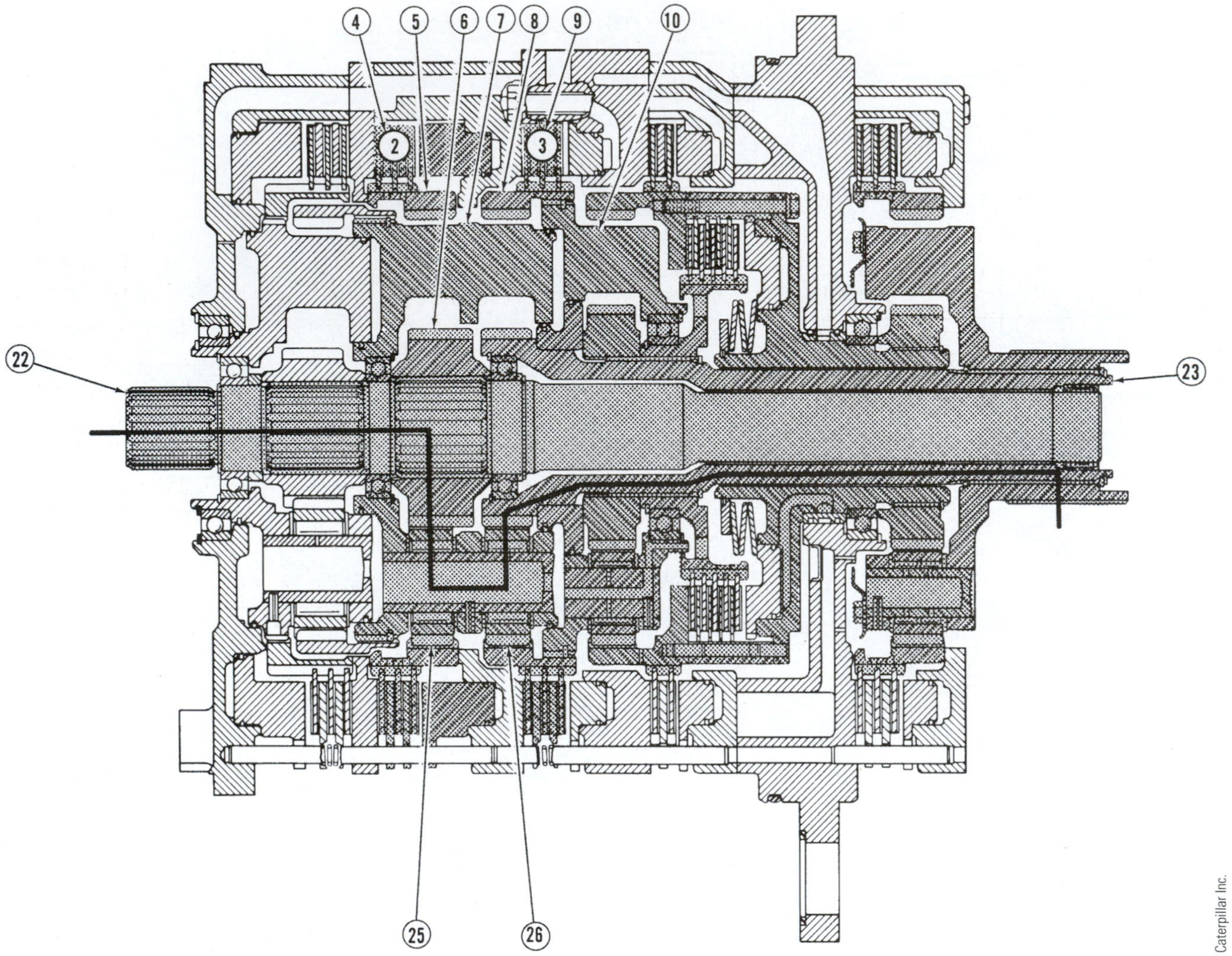

POWER FLOW IN FOURTH SPEED FORWARD (No.2 and No.3 Clutches Engaged)

4. No.2 clutch.	7. No.2 and No.3 carrier.	10. No.4 carrier.	25. No.2 planetary gears.
5. Ring gear for No.2 clutch.	8. Ring gear for No.3 clutch.	22. Input shaft.	26. No.3 planetary gears.
6. No.2 sun gear.	9. No.3 clutch.	23. Output shaft.	

Reprinted Courtesy of Caterpillar Inc.

Figure 11-41 Fourth speed forward with number 2 and number 3 clutch packs engaged.

reduction as the speed of the output shaft increases. This is the highest speed ratio at maximum torque/minimum speed through two compound planetary gear sets.

Note: Within this specific machine application, no overdrive condition occurs between the transmission input shaft and output shaft. There can be as many as four compound gear reductions for first speed.

First Speed Reverse Powerflow. Refer to **Figure 11-42**. When the operator selects first speed reverse, number 6 and number 1 clutches are engaged. Number 1 clutch is the reverse speed clutch, which holds clutch coupling for number 1 planetary carrier stationary. Number 1 clutch will stay engaged for all the reverse speeds. Number 6 clutch is the first speed clutch, which holds ring gear (17) stationary.

Input shaft (23) rotates reverse drive planetary sun gear (21) at torque converter speed. According to the basic principles of planetary gearing, "whenever the planetary carrier is the held member, a reverse direction from the input occurs." This automatically makes the ring gear (2) the output member (maximum torque/minimum speed) for the number 1 planetary in a reverse direction to the input.

FIRST SPEED REVERSE

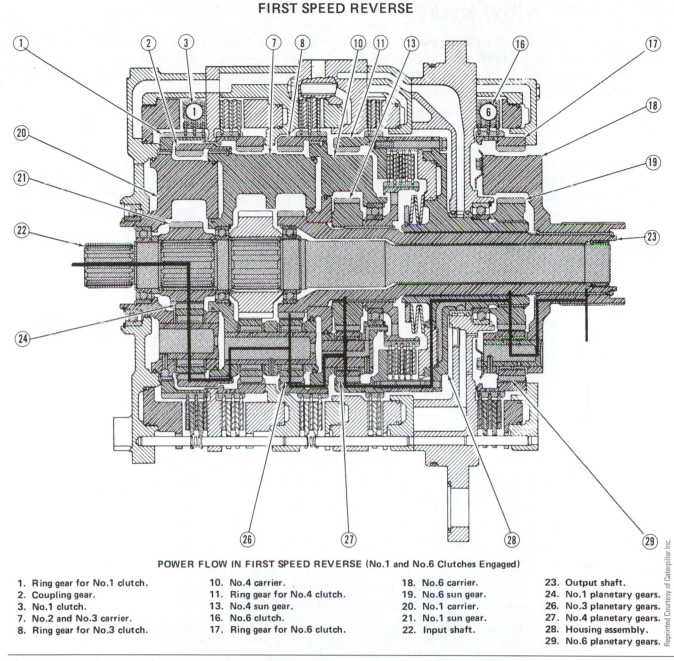

POWER FLOW IN FIRST SPEED REVERSE (No.1 and No.6 Clutches Engaged)

1. Ring gear for No.1 clutch.
2. Coupling gear.
3. No.1 clutch.
7. No.2 and No.3 carrier.
8. Ring gear for No.3 clutch.

10. No.4 carrier.
11. Ring gear for No.4 clutch.
13. No.4 sun gear.
16. No.6 clutch.
17. Ring gear for No.6 clutch.

18. No.6 carrier.
19. No.6 sun gear.
20. No.1 carrier.
21. No.1 sun gear.
22. Input shaft.

23. Output shaft.
24. No.1 planetary gears.
26. No.3 planetary gears.
27. No.4 planetary gears.
28. Housing assembly.
29. No.6 planetary gears.

Figure 11-42 First speed reverse with number 1 and number 6 clutch packs engaged.

Because ring gear number 1 is interconnected to planetary carrier 2 and 3, they will provide input to the number 2/3 planetary in a reverse direction. Because number 3 sun gear is attached to the output shaft (22), it acts as the stationary member to planetary gear set 3. This will cause the number 3 ring gear (9) to rotate. The number 3 planetary gears will rotate the ring gear, which is splined to number 4 planetary carrier, causing it to rotate. This becomes the input to number 4 planetary. The number 4 sun gear (11) is also splined the output shaft (22), which also acts as a stationary member for the number 4 planetary. This will cause

number 4 ring gear (13) to rotate as the output. Number 4 ring gear is fastened to the rotating clutch housing that has sun gear from the number 5 planetary gear set (19) splined to it. Because stationary clutch pack number 6 (16) is holding number 5 ring gear (17), the planetary carrier becomes the output through output shaft (22). Output shaft (22) and number 5 carrier (18) are splined together and rotate at equal speeds.

As a result, a multiple reduction (low speed/high torque) output is divided between number 3 and 4 planetary gear set and number 5 planetary gear set.

Second Speed Reverse Powerflow. Refer to **Figure 11-43**. When the operator selects second speed reverse, number 5 and number 1 clutches are engaged. Number 1 clutch is the reverse speed clutch, which holds clutch coupling for number 1 planetary carrier stationary. Number 5 clutch is the second speed clutch, which holds number 4 ring gear and sun gear together.

Input shaft (23) rotates reverse drive planetary sun gear (21) at torque converter speed. Because the carrier is held, this automatically makes the ring gear (2) the output member (maximum torque/minimum speed) for the number 1 planetary in a reverse direction to the input.

Because ring gear number 1 is interconnected to planetary carrier 2 and 3, they will provide input to the number 2–3 planetary in a reverse direction. Because number 3 sun gear is attached to the output shaft (22), it acts as the stationary member to planetary gear set 3. This will cause the number 3 ring gear (9) to rotate. The number 3 planetary gears will rotate the ring gear, which is splined to number 4 planetary carrier, causing it to rotate. The carrier becomes the output at low speed/high torque. Because number 3 sun gear is attached to the output shaft (22), it acts as the stationary member to planetary gear set 3. This will cause the number 3 ring gear (9) to rotate. The number 3

SECOND SPEED REVERSE

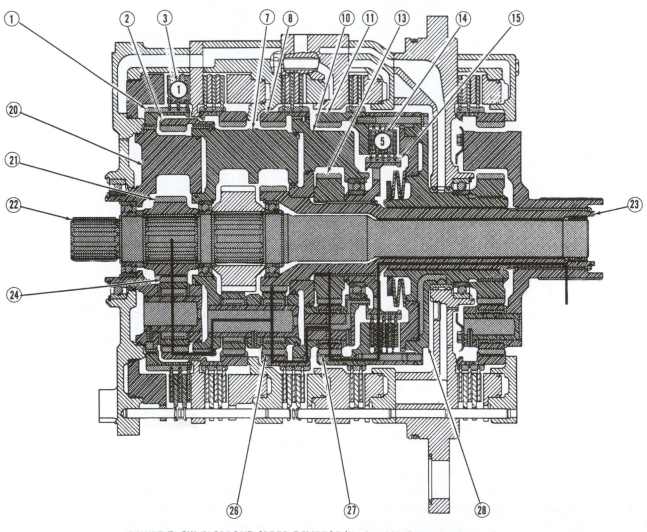

POWER FLOW IN SECOND SPEED REVERSE (No.1 and No.5 Clutches Engaged)

1. Ring gear for No.1 clutch.	10. No.4 carrier.	20. No.1 carrier.	24. No.1 planetary gears.
2. Coupling gear.	11. Ring gear for No.4 clutch.	21. No.1 sun gear.	26. No.3 planetary gears.
3. No.1 clutch.	13. No.4 sun gear.	22. Input shaft.	27. No.4 planetary gears.
7. No.2 and No.3 carrier.	14. No.5 clutch.	23. Output shaft.	28. Housing assembly.
8. Ring gear for No.3 clutch.	15. Rotating hub.		

Figure 11-43 Second speed reverse with number 1 and number 5 clutch packs engaged.

planetary gears will rotate the ring gear, which is splined to number 4 planetary carrier. Because rotating clutch pack number 5 is engaged, it will prevent number 4 ring gear and sun gear from rotating at different speeds. This causes the input from number 4 carrier to rotate the planetary gear set as one unit or at a ratio of 1:1. The number 4 sun gear (11) is splined to the output shaft (22).

The result is a single reduction (low speed/high torque) through planetary gear set number 3, splitting the torque path with number 4 planetary gear set.

Third Speed Reverse Powerflow. Refer to **Figure 11-44**. When the operator selects third speed reverse, number 4 and number 1 clutches are engaged. Number 1 clutch is the reverse speed clutch, which holds clutch coupling for number 1 planetary carrier stationary. Number 4 clutch is the third speed clutch, which holds number 4 ring gear and sun gear together.

Input shaft (23) rotates reverse drive planetary sun gear (21) at torque converter speed. Because the carrier is held, this automatically makes the ring gear (2) the output member (maximum torque/minimum speed) for the number 1 planetary in a reverse direction to the input.

Because ring gear number 1 is interconnected to planetary carrier 2 and 3, they will provide input to the number 2–3 planetary in a reverse direction. Because number 3 sun gear is attached to the output shaft (22), it acts as the stationary member to planetary gear set 3. This will cause the number 3 ring gear (9) to rotate. The number 3 planetary gears will rotate the ring gear, which is splined to number 4 planetary carrier. Number 4 planetary carrier becomes the input to its planetary gear set. Because the number 4 clutch pack is engaged, it is holding stationary ring gear (13). Planetary number 4 sun gear becomes the output drive through output shaft (22). Note that number 4 planetary is driving in an overdrive condition compared to its input. The output shaft is still in an underdrive condition due to number 3 planetary providing the input to number 4 planetary at a maximum torque

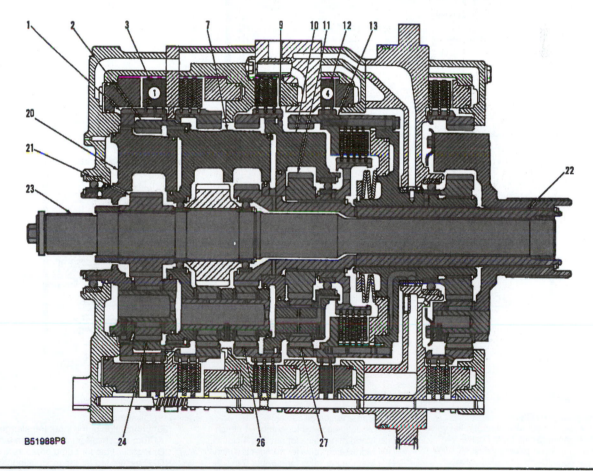

Powerflow In Third Speed Reverse (No. 1 And No. 4 Clutches Engaged)
(1) Ring gear for No. 1 clutch. (2) Coupling gear. (3) No. 1 clutch. (7) No. 2 and No. 3 carrier. (9) Ring gear for No. 3 clutch. (10) No. 4 carrier. (11) No. 4 sun gear. (12) No. 4 clutch. (13) Ring gear for No. 4 clutch. (20) No. 1 carrier. (21) No. 1 sun gear. (22) Output shaft. (23) Input shaft. (24) No. 1 planetary gears. (26) No. 3 planetary gears. (27) No. 4 planetary gears.

Figure 11-44 Third speed reverse with number 1 and number 4 clutch packs engaged.

minimum speed condition multiplied by the reduced speed of the rotating sun gear. Because number 3 sun gear is attached to the output shaft and acts as a stationary member for the number 3 planetary gear set, it reduces the output speed of the ring gear even further as the output shaft begins to rotate. This increases torque output from the transmission.

The result is a single reduction (low speed/high torque) through planetary gear set number 3, splitting the torque path with number 4 planetary gear set in an overdrive condition.

Transmission Hydraulic System

The transmission hydraulic system is responsible for providing the median for the torque converter, the pressure used to activate all the clutch packs, effective cooling and lubrication within the transmission, and to direct and control oil pressure and flow within the system.

Refer to **Figure 11-45**. The system consists of a positive-displacement external gear pump (8), a transmission filter (1), a transmission hydraulic control-relief

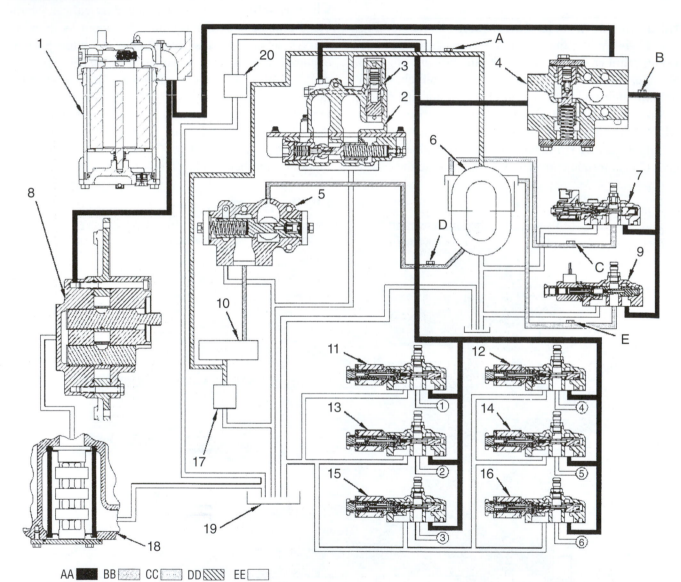

(1) Transmission oil filter
(2) Transmission hydraulic control relief valve
(3) Torque converter inlet relief valve
(4) Torque converter priority valve
(5) Torque converter outlet relief valve
(6) Torque converter
(7) Modulating valve for the lockup clutch
(8) Transmission oil pump
(9) Modulating valve for the impeller clutch

(10) Transmission oil cooler
(11) Modulating valve for number 1 clutch
(12) Modulating valve for number 4 clutch
(13) Modulating valve for number 2 clutch
(14) Modulating valve for number 5 clutch
(15) Modulating valve for number 3 clutch
(16) Modulating valve for number 6 clutch
(17) Lubrication of the transmission planetary
(18) Suction screen and magnet

(19) Oil sump
(20) Lubrication of the input transfer gears
(A) Torque converter inlet pressure tap
(B) Pressure tap for pump pressure
(C) Pressure tap for the modulating valve (lockup clutch)
(D) Torque converter outlet pressure tap
(E) Pressure tap for the modulating valve (impeller clutch)

Reprinted Courtesy of Caterpillar Inc.

Figure 11-45 Transmission hydraulic control valve body.

valve (2), a torque converter priority valve (4), a torque converter outlet relief valve (5), a lockup clutch modulating valve (7), an impeller clutch modulating valve (9), modulating valves for clutch packs (11 through 16), a transmission oil cooler (10), and a pump suction screen and magnet assembly (18). The transmission hydraulic control valve group is mounted on the top side of the transmission planetary gear sets. The transmission housings provide passageways for oil to be directed to the appropriate clutch packs and lubrication paths throughout the transmission.

Hydraulic Oil Flow in Neutral. Refer to **Figure 11-46**. When the engine is started, the transmission pump is driven by the torque converter housing at engine speed. The pump (8) picks up the oil from the transmission sump through the inlet screen and magnet assembly (18). Oil flows first through the oil filter (1) to remove any contaminants and then to the torque converter priority control valve (4). From the priority control valve, the oil is directed to the transmission hydraulic control relief valve (2) and to the clutch solenoid modulating **shift valves** (11 to 16). Clutch modulating

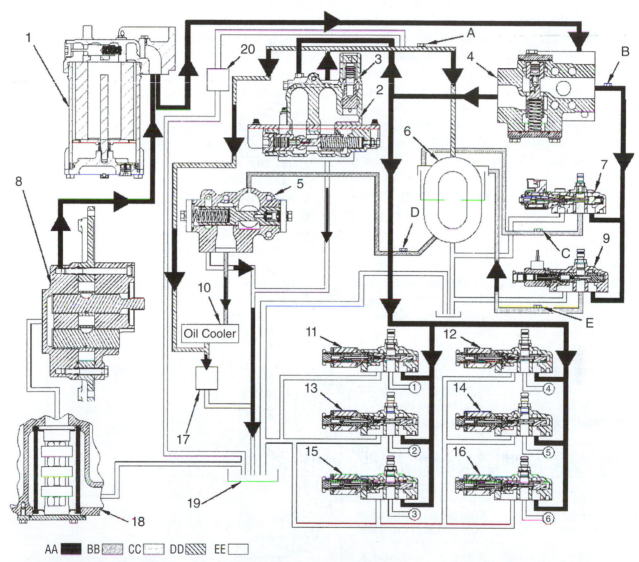

(1) Transmission oil filter
(2) Transmission hydraulic control relief valve
(3) Torque converter inlet relief valve
(4) Torque converter priority valve
(5) Torque converter outlet relief valve
(6) Torque converter
(7) Modulating valve for the lockup clutch
(8) Transmission oil pump
(9) Modulating valve for the impeller clutch

(10) Transmission oil cooler
(11) Modulating valve for number 1 clutch
(12) Modulating valve for number 4 clutch
(13) Modulating valve for number 2 clutch
(14) Modulating valve for number 5 clutch
(15) Modulating valve for number 3 clutch
(16) Modulating valve for number 6 clutch
(17) Lubrication of the transmission planetary
(18) Suction screen and magnet

(19) Oil sump
(20) Lubrication of the input transfer gears
(A) Torque converter inlet pressure tap
(B) Pressure tap for pump pressure
(C) Pressure tap for the modulating valve (lockup clutch)
(D) Torque converter outlet pressure tap
(E) Pressure tap for the modulating valve (impeller clutch)

Figure 11-46 Engine running/transmission in neutral position.

valve number 3 is energized in the neutral position and will fill the clutch pack on start-up. This helps lubricate the clutch packs by allowing the ring gears of the directional clutches to rotate within the transmission. Oil is also directed to the lockup clutch modulating valve (7) and the impeller clutch modulating valve (9). The transmission has an individual clutch solenoid modulating valve for each clutch pack. There must be at least two energized solenoids for the transmission to drive: one directional solenoid and one speed solenoid. Refer to **Table 11-1** for the combination of solenoid valves and engaged clutch packs for each forward and reverse speed.

Oil exiting the torque converter may be prone to low-pressure drops. A torque converter outlet relief valve is used on the outlet side of the torque converter; this valve is set approximately 50 psi lower than the inlet relief valve to maintain a pressure of approximately 60 psi in the torque converter and prevent possible cavitation within it. Oil from the torque converter can become extremely hot. This hot oil from the outlet relief valve is directed to the oil cooler that can be an air-to-oil cooler or an air-to-water cooler. Exiting cool oil is combined with oil from the transmission hydraulic control relief valve and is directed through the oil lubrication passages within the transmission planetaries, clutch packs, bearings, and shafts. The restrictions inside the oil galleries of the transmission keep the lubrication pressure of the warm oil at approximately 20 psi. This oil is allowed to drain into the transmission sump and is then recirculated through the pump and transmission.

Torque Converter Priority Valve

The torque converter priority valve (4) splits the oil flow between the transmission hydraulic control relief valve and the lockup clutch (7) and impeller clutch

modulating valves (9) (refer to **Figure 11-46**). The priority valve splits the flow with the priority of flow given to the transmission hydraulic control relief valve when a "low flow" condition occurs within the hydraulic system. Because torque converter oil is supplied from the transmission hydraulic control valve, the priority valve will supply oil to it first—this protects the torque converter from a low oil supply and possible damage. During this condition, the lockup clutch and impeller clutch modulating valves will cease to work, but this will not damage the torque converter.

Transmission Hydraulic Control Relief Valve

Refer to **Figure 11-47**. The transmission hydraulic control relief valve controls the maximum pressure of oil that is supplied to the clutch solenoid modulating valves. Within this valve is the torque converter inlet relief valve (3); it is set lower than the main relief valve pressure of 380 psi (2620 ± 35 kPa) and will maintain an inlet pressure of 154 ± 15 psi (1060 ± 100 kPa) within the torque converter. Although this valve is referred to as a relief valve, it acts as a pressure regulating valve due to the fluctuation of oil flow through the system. The valve tries to maintain a constant pressure regardless of flow rate and resistance to flow. Once the torque converter pressure is reached, excess oil is directed through the oil lubrication passages within the transmission planetaries, clutch packs,

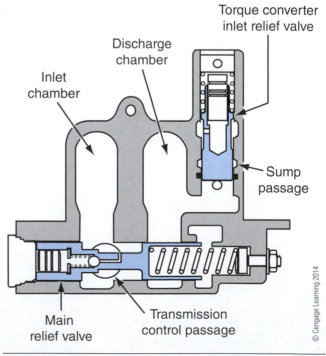

Figure 11-47 Transmission/torque converter relief valve.

TABLE 11-1: SOLENOIDS AND CLUTCHES TO SPEED AND DIRECTION		
Speed Range and Direction	**Energized Solenoids**	**Engaged Clutches**
Fourth Speed Forward	3 and 2	3 and 2
Third Speed Forward	4 and 2	4 and 2
Second Speed Forward	5 and 2	5 and 2
First Speed Forward	6 and 2	6 and 2
Neutral	3	3
First Speed Reverse	6 and 1	6 and 1
Second Speed Reverse	5 and 1	5 and 1
Third Speed Reverse	4 and 1	4 and 1

© Cengage Learning 2014

bearings, and shafts. Thus separate constant pressures for both the clutch solenoid modulating valves and the torque converter are maintained.

CLUTCH SOLENOID MODULATING VALVES

There are six clutch solenoid modulating valves (see **Figure 11-48**). Valves 3 and 5 supply oil to the directional clutches: reverse and forward, consecutively. Valves 1, 2, 4, and 6 supply oil to the speed clutches: first, second, fourth, and third, consecutively. The clutch solenoid modulating valves are the outputs of the powertrain ECM. The clutch solenoid modulating valves are proportional solenoid valves. This means

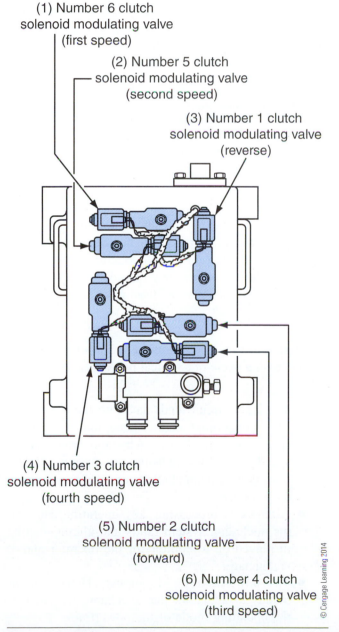

(1) Number 6 clutch solenoid modulating valve (first speed)

(2) Number 5 clutch solenoid modulating valve (second speed)

(3) Number 1 clutch solenoid modulating valve (reverse)

(4) Number 3 clutch solenoid modulating valve (fourth speed)

(5) Number 2 clutch solenoid modulating valve (forward)

(6) Number 4 clutch solenoid modulating valve (third speed)

© Cengage Learning 2014

Figure 11-48 Transmission clutch solenoid modulating valves.

that the valves will supply oil to the clutch packs directly proportional to the current supply from the ECM. If there is no current supply, the clutch pressure will be zero; if the current supply is at the maximum, the clutch pressure will be at the maximum.

The clutch solenoid modulating valves are used by the powertrain ECM to directly modulate the oil pressure that is sent to each individual clutch pack. The ECM sends a pulse width modulated (PWM) signal to vary the current flow to the solenoid valve that controls the plunger travel. The farther the plunger moves within its bore, the greater the amount of oil passage directed to the clutch pack. The powertrain ECM activates the appropriate solenoid valves based on the input from the operator. When the modulating valves are deactivated by the ECM, oil pressure within the clutch pack is drained through the valve back to the transmission sump.

TORQUE CONVERTER LOCKUP CLUTCH MODULATING VALVE

The **torque converter lockup clutch modulating valve** is a three-way proportional pressure-control valve. It is also PWM solenoid, controlled by the powertrain ECM. It functions the same as the modulated shift valves in respect to proportion of pressure to current supply. Refer back to **Figure 11-46**, component number 7. The lower the ECM current, the lower the pressure signal to the lockup clutch. When the ECM sends a current signal to the solenoid valve, oil is directed to the lockup clutch in the torque converter to lock the turbine to the torque converter housing, thus providing an output ratio of 1:1 or direct drive. The torque converter is more efficient at lockup during rotary flow. The ECM can activate the solenoid lockup valve at any speed except first speed. The following conditions must be met before the ECM will activate the lockup solenoid:

- The torque converter output speed is greater than 1,400 rpm.
- The torque converter output speed is less than 2,120 rpm.
- The left brake pedal must not be depressed.
- The lockup clutch solenoid valve has been deactivated for at least 4 seconds after a fault for a slipping clutch has been logged.

The lockup clutch will engage when these four conditions are met again in a newly selected gear. The lockup clutch is disengaged when the torque converter output speed drops below 1,250 rpm. This prevents engine lugging and damage to the friction discs and plates of the lockup clutch. Torque multiplication in a vortex flow will resume within the torque converter.

In the event of a machine accelerating down a long slope, the lockup clutch will disengage when the torque converter output speed exceeds 2,400 rpm.

Torque Converter Impeller Clutch Modulating Valve.

The torque converter impeller clutch modulating valve is a three-way proportional pressure-control valve. It is also a PWM solenoid controlled by the powertrain ECM. It functions the same as the lockup modulated valve in respect to proportion of pressure to current supply except in reverse proportion. The lower the ECM current, the higher the pressure signal to the impeller clutch. Refer back to **Figure 11-46**, component number 9. When the ECM sends a current signal to the solenoid valve, oil pressure is reduced at the impeller clutch piston. This causes the friction discs and plates to slip between the torque converter housing and the impeller. When the impeller slips, less oil is forced to the turbine. With less oil force on the turbine, there is less torque output from the torque converter, reducing rim-pull and wheel slip.

The following conditions determine what ECM signal is received by the impeller clutch modulating valve:

- The position of the left service brake pedal
- Directional shifts
- Engine speed
- Selected gear
- Direction of rotation of the torque converter output shaft and torque converter output speed
- The position of the rim-pull selector switch in reduced position
- The position of the reduced rim-pull on/off switch

TORQUE CONVERTER OUTLET RELIEF VALVE

The purpose of the torque converter outlet relief valve is to maintain a minimum pressure of approximately 60 psi in the torque converter. As the impeller blades slice through the oil, they create a low-pressure condition within the oil and cause it to vaporize; tiny vapor bubbles form and implode in a condition known as bubble collapse. As the oil is re-pressurized, the vapor bubbles can cavitate components within the torque converter. Cavitation is highly erosive but by pressurizing the torque converter, the low-pressure conditions that cause it are minimized.

POWERTRAIN ELECTRONIC CONTROL SYSTEM

The powertrain electronic control system manages the powertrain. It consists of an ECM, a monitoring circuit, and actuators.

Electronic Control Module (ECM)

The powertrain ECM manages the transmission. The transmission speed sensor and direction control switch (operator input) signal data to the powertrain ECM. The operator input sets the desired speed and direction instructions. The powertrain ECM computes outcomes based on this input information and the program instruction logged in ECM memory. After the powertrain ECM receives the input information and processes it with the instructions logged in memory, the powertrain ECM uses drivers to switch the outputs (modulating valve solenoids).

Inputs. Input devices monitor the powertrain operating conditions and signals the ECM. Typical inputs to the powertrain ECM are shift status, shaft speed, temperature, and pressure. The sensors may be analog or digital and provide real time status signals to the powertrain ECM.

Outputs. After processing, the powertrain ECM uses internal drivers to electrically switch actuators. Actuators may be simple on-off solenoids, proportioning solenoids, or piezo-actuators depending on the machine.

Data bus. A data bus is used to communicate with the other ECMs on the machine. The common data bus is a two-wire serial bus used to communicate data packets or messages from any module with an address on the bus. Optical buses are used in some more recent equipment.

Processing loop. The following is a brief example of the processing cycle of a typical powertrain processing cycle:

- The input circuit feeds data into the powertrain ECM concerning engine speed, machine ground speed, parking brake switch status, transmission speed selection, transmission direction selection, transmission shaft speed, and transmission temperatures.
- After ECM processing, key operating data status, including system faults and alarms, can be displayed on a monitor for the operator and/or technician.
- The powertrain ECM outputs electrical commands to actuators that mechanically changes the operating mode of the powertrain.

The powertrain ECM can be read and reprogrammed by means of a data link and electronic service tool (EST) loaded with OEM software. Connecting an EST to the powertrain ECM permits the technician to read and clear fault codes, and perform system programming.

A simple powertrain electronic control system has six basic functions:

- Neutral start function
- Manual shift function
- Automatic shift function
- Parking brake function
- Back-up alarm function
- Diagnostic and failure strategy functions

The 988G wheel loader, shown in **Figure 11-49**, is equipped with a joystick steering control lever. The transmission controller is built into the joystick lever. Forward/reverse direction as well as first, second, third, and fourth speeds can all be selected by pressing the upshift and downshift switches or the directional control switches. Joysticks are covered in more detail in Chapter 19.

Neutral Start Function

The start relay, an output of the powertrain ECM, supplies electrical power to the starter solenoid switch that controls the starting motor. When the operator turns the key switch to the start position, the powertrain ECM determines whether the starting conditions are met. Starting conditions include whether the transmission directional control switch is in neutral and the engine is not running. Providing the starting conditions are met, the ECM activates the start relay, which switches battery voltage to the starter solenoid, to crank the starter motor.

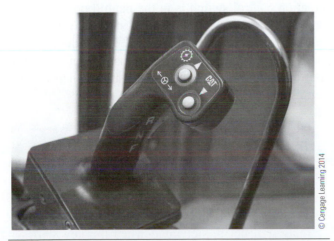

Figure 11-49 Joystick steering and transmission control.

Manual Shift Function

The manual shift function allows the operator to use the shift control inputs to select the transmission speed range. In this operational mode, it also allows the operator to change the direction of travel regardless of engine or machine speed. The torque converter impeller clutch allows smooth transitions when directional changes are made.

When an operator decides to upshift or downshift, the transmission ECM selects the next higher or lower speed range, depending on which switch is pressed. A new gear is selected on the switch transition from the depressed to released position. Holding one switch in the depressed position will not disable the other switch. The downshift switch is disabled when the transmission is in first gear. The upshift switch is disabled when the transmission is in the highest gear. Refer to **Table 11-2** for the shifting sequence.

Automatic Shift Function

The automatic shift function allows the transmission ECM to automatically shift the transmission. The operator must enable this feature by selecting AUTO mode from the Auto/Manual switch (4). There are three auto modes available: 2, 3, and 4 (see **Figure 11-50**).

The powertrain ECM processes the inputs that are signaled from the engine speed sensor, transmission speed sensors, and the torque converter output speed sensor in order to regulate transmission shifts. The ECM does not permit upshifts into a gear ratio higher than the gear that is selected with Auto/Manual switch (4). If Auto/Manual is set to the number 3 position, this signals the ECM that a gear speed higher than 3 is not desired and prevents the transmission from upshifting to fourth speed. There is no automatic downshift from second speed to first speed; this can only occur only when the downshift switch is pressed.

Momentarily pressing the downshift switch instructs the ECM to downshift the transmission by one gear. Holding down the downshift switch instructs the ECM to continue to downshift the transmission as speed decreases. The transmission remains in the downshifted gear for 3 seconds after the downshift switch is released. Then, automatic shifting is resumed.

Tech Tip: In the automatic mode of autoshift, the transmission automatically shifts from first gear to second gear. This happens when the operator shifts the direction from

TABLE 11-2: MANUAL SHIFTING SEQUENCE (988G WHEEL LOADER)

Speed Range and Direction	If REVERSE is selected, then the transmission shifts to the following speed.	If NEUTRAL is selected, then the transmission shifts to the following speed.	If FORWARD is selected, then the transmission shifts to the following speed.	If UPSHIFT is selected, then the transmission shifts to the following speed.	If DOWNSHIFT is selected, then the transmission shifts to the following speed.
4F	3R	Neutral [1](4N)	No Change	No Change	3F
3F	3R	Neutral [1](3N)	No Change	4F	2F
2F	2R	Neutral [1](2N)	No Change	3F	1F
1F	1R	Neutral [1](1N)	No Change	2F	No Change
Neutral [1](1N, 2N, 3N, 4N)	Last[2]	No Change	Last[2]	Next[3]	Prior[3]
1R	No Change	Neutral [1](1N)	1F	2R	No Change
2R	No Change	Neutral [1](2N)	2F	3R	1R
3R	No Change	Neutral [1](3N)	3F	No Change	2R

[1] The speed and the direction that is shown in **bold print** is the gear readout.

[2] A directional shift out of neutral will cause a speed shift to the last speed that was selected in either forward or reverse.

[3] Speed shifts are allowed when neutral is selected. The gear readout will change the speed that is shown when neutral is selected and a speed shift is made. However, none of the solenoids are activated until forward or reverse is selected.

© Cengage Learning 2014

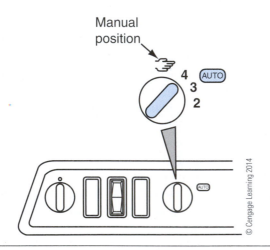

Manual position

Figure 11-50 Auto shift control switch setting.

© Cengage Learning 2014

forward to reverse. When the operator then shifts from reverse to forward, the transmission remains in second gear. This feature functions only when the machine is in first gear. A transmission quick-shift can only occur when the operator shifts the direction from forward to reverse. This feature does not affect any directional shift in any of the other gears.

Table 11-3 shows the approximate rpms at which the ECM will initiate a speed shift.

When the key start switch is turned from the off position to the on position, the powertrain ECM is activated. When the powertrain ECM is initially activated, all the modulating valves (transmission clutch) are de-energized regardless of the position of the transmission direction control switch (3). The powertrain ECM then determines if the transmission direction control switch (3) is in the neutral position. The powertrain ECM disables the transmission if the neutral position is *not* selected; the transmission direction control switch must be returned to the neutral position before a direction can be selected. If the transmission direction control switch is in the neutral position, then the powertrain ECM activates the start relay, cranking the engine. The powertrain ECM allows the selected speed to be changed if the transmission direction control switch is in the neutral position when the engine is started.

Tech Tip: Transmission direction control switch (3) must be in the neutral position when the key start switch is turned to the start position. If the transmission direction control switch is not in the neutral position, the start relay will not be activated when the key start switch is turned.

TABLE 11-3: AUTOMATIC UPSHIFT AND AUTOMATIC DOWNSHIFT			
Speed Range and Direction Shift	Transmission Speed (rpm) Standard (non-LUC)	Transmission Speed (rpm) Lockup Clutch Enabled	Transmission Speed (rpm) Lockup Clutch Disabled
1F to 2F	344	354	354
2F to 3F	613	717	631
3F to 4F	1086	1269	1116
4F to 3F	999	1180	1027
3F to 2F	564	667	580
2F to 1F	316	326	325
1R to 2R	393	459	404
2R to 3R	701	819	721
3R to 4R	1241	1450	1276
4R to 3R	1141	1349	1174
3R to 2R	645	762	663
2R to 1R	362	427	372

© Cengage Learning 2014

Parking Brake Function

The parking brake function prevents the operation of the machine with the parking brake engaged. Although most equipment has enough torque to drive through the parking brake, operating a machine with the parking brake engaged will cause accelerated wear and lead to possible damage of the parking brake components.

If the operator places the directional switch in forward or reverse with the parking brake engaged, the ECM will place the transmission into neutral by de-energizing all solenoids in the transmission hydraulic control valve. The ECM will activate the warning system. A parking brake indicator lamp flashes and an action alarm sounds to alert the operator that the parking brake is engaged. When the parking brake is disengaged, it sends a signal to the ECM indicating the parking brake is released, and the ECM energizes the appropriate solenoids based on the directional and speed switch positions.

Back-Up Alarm Function

The back-up alarm function is an output of the ECM used to alert surrounding workers and operators that the machine is reversing direction. When the operator places the directional control switch into reverse, the ECM activates the back-up alarm relay, which energizes the back-up alarm. Back-up alarms send an audible signal of between 85 and 115 decibels (dB), which should be higher than the ambient noise level. Automated (self-adjusting) and switchable models are available.

The switchable model has as many as three settings that are preset by a toggle switch at 97, 107, and 115 dB. A self-adjusting model sounds at 5 dB above the ambient noise level, to a maximum of 115 dB.

Diagnostic Operation Function

The powertrain ECM detects faults that occur in the input and output circuits within its system. A fault is detected when a signal at the contact of the ECM is outside a specific range. When this occurs, the ECM records and stores the fault data for future reference by the technician to retrieve and repair. If the fault is intermittent, it is still logged as an "intermittent fault" and can be retrieved for future reference. The ECM can provide diagnostic information to assist with troubleshooting the stored faults within the system.

A SAE J1939 data bus may incorporate J1587 protocols. The purpose of any data bus is to network electronic systems with an address on data bus backbone to enable data sharing, and significant reduction of hard wire circuits.

Electronic Service Tools (ESTs). ESTs are a requirement for communicating with machine ECMs. Most OEMs are using a Windows-based operating environment. Manufacturer-specific programs are loaded onto the PC; a data link connector (6- or 9-pin Deutsch connector) connects the machine data bus and the PC. Together they provide the interface and communication with the ECM. A communication adapter (CA) is required between the PC and the machine data bus.

Once communication is established with the data bus, the technician can select the ECM he or she wants to communicate with. Each ECM on the data bus is assigned an address known as a MID.

Data Link Connector

The data link connector (19), shown in **Figure 11-51**, is located in the right-hand corner of the console near the floor. The connector is used to connect to the data communication equipment that is used for diagnostic purposes.

Service codes are used to communicate fault paths. Service codes are displayed on the machine monitoring screen or an EST. Code paths use the following identifiers:

- Module Identifier (MID)
- Parameter Identifier (PID)
- Component Identifier (CID)
- Subsystem Identifier (SID)
- Failure Mode Identifier (FMI)

Module Identifier (MID). A MID indicates a controller with an address on the machine data bus. Each MID is assigned a three-digit code. When a fault is detected, the MID is displayed for approximately 1 second before the fault code is displayed. The following are examples of some Caterpillar SIS MIDs.

- Monitoring System: MID-030
- Powertrain ECM: MID-081
- Implement/Hydraulic ECM: MID-082
- Payload Control System: MID-074
- Engine ECM: MID-036

Component Identifier (CID). Components that are defective are tracked by the CID. Examples of these

© Cengage Learning 2014

Figure 11-51 Nine-pin data link connector.

components are the start relay and the reverse solenoid.

Parameter Identifier (PID). Means of identifying components and branch circuits within a MID.

Failure Mode Identifier (FMI). The FMI tells the type of failure that has occurred. The FMI is a two-digit code that is shown in the display area. The CID and the FMI are shown together after the MID has been displayed. A decimal point "." precedes the FMI. **Table 11-4** shows an example of FMI codes adapted from the SAE Standards, Section J1587 diagnostics.

TABLE 11-4: FAILURE MODE IDENTIFIERS (FMIs)

FMIs are important to the technician because they attempt to define the nature of a failure. Twenty are identified here but on the most recent machines, these have been extended so that imminent failures can be tagged and repaired before outright failure occurs.

FMI	Failure Description
00	Data valid but above normal operating range
01	Data valid but below normal operating range
02	Data erratic, intermittent, or incorrect
03	Voltage above normal or shorted high
04	Voltage below normal or shorted low
05	Current below normal or open circuit
06	Current above normal or grounded circuit
07	Mechanical system not responding properly
08	Abnormal frequency, pulse, or period
09	Abnormal update
10	Abnormal rate of change
11	Failure mode not identifiable
12	Bad device or component
13	Out of calibration
14	N/A
15	N/A
16	Parameter not available
17	Module not responding
18	Sensor supply fault
19	Condition not met
20	N/A

© Cengage Learning 2014

Subsystem identifier (SID). identifies circuit paths within a MID input or output circuit

MULTIPLE COUNTERSHAFT TRANSMISSIONS

As mentioned earlier many heavy-duty applications require equipment to have a multitude of forward and reverse speeds. Compound planetary gearing is one method already discussed. Another method of achieving multiple speeds is the use of a series of countershafts to compound the number of ratios required for a specific machine application. Multiple countershafts are set up in a staggered-parallel formation; different gear diameters on each shaft are in constant mesh with each other. In most instances, the external-type gears that float on the countershafts only provide a drive when a clutch pack is engaged. The external gears that are fixed to the countershaft transfer engine torque to the next countershaft. Providing all the appropriate clutch packs are engaged, torque transfer continues through each countershaft until transmitted to the transmission output shaft. Rotating clutch packs connected between the countershafts and each independent gear allow for the selection and engagement of each gear ratio by the operator. The number of gear ratios attainable is only limited by the number of countershafts used. In some grader applications, an eight-speed forward, four-speed reverse, **multiple countershaft transmission** is used. In the heavy-duty equipment industry the configurations of transmissions are many and varied. For this reason, only two popular transmissions are discussed.

MULTIPLE COUNTERSHAFT TRANSMISSION POWERFLOWS

Compound planetary transmissions were discussed earlier. The second transmission chosen is a multiple countershaft powerflow transmission from a CAT® 938G wheel loader. The 938G loader has a bucket capacity of approximately 2.8 to 4.4 cubic yards, depending on the application. The operating weight is approximately 30,000 pounds. It is powered by a 3126 engine rated at 180 hp. It has a four-speed forward, three-speed reverse, multiple countershaft transmission that can be operated in automatic or manual powershift mode (see **Figure 11-52**).

Six hydraulically activated clutches are used to transfer power from the input shaft assembly (7) to the output shaft assembly (12). The torque converter used is a standard type with no lockup or impeller clutch feature.

Six electronically controlled solenoids, when energized by the powertrain ECM, control the oil flow to the clutch packs. Unlike a planetary shift transmission that uses stationary and rotating clutch packs, a countershaft transmission uses only rotating clutch packs to provide a drive when engaged. There are six clutches: three directional clutches and three speed clutches.

The directional clutches are:

- Clutch 1 (8) (Forward low)
- Clutch 2 (9) (Forward high)
- Clutch 3 (16) (Reverse)

The speed clutches are:

- Clutch 4 (11) (Second speed)
- Clutch 5 (14) (Third speed)
- Clutch 6 (13) (First speed)

For engine torque to be transferred through the transmission to the output shaft assembly (12), one direction clutch must be engaged and one speed clutch must be engaged. A modulating valve controls each clutch. The powertrain ECM controls the engagement of the clutches by sending a variable PWM signal to the solenoids on the modulating valves.

Figure 11-53 shows a cross-sectional view of the transmission countershaft arrangement with the clutch packs located on each countershaft. The transmission has a total of five shafts: input shaft assembly (3), output shaft assembly (33), and three gear shafts. The input shaft assembly (3) is driven by the torque converter turbine shaft. Shaft assembly (9) contains Clutch 1 (4) and Clutch 2 (6), shaft assembly (19) contains Clutch 3 (23) and Clutch 4 (17), and shaft assembly (27) contains Clutch 5 (30) and Clutch 6 (28).

Neutral Position. The powertrain ECM determines which clutches need to be engaged. The proper clutch engagement is based on the inputs to the powertrain ECM. When the operator turns the start key, the powertrain ECM determines if the transmission is in the neutral position and the parking brake is engaged. If these two parameters are met, the ECM allows the start relay to engage the starter motor to crank the engine. Upon start-up, Clutch 5 (third speed) is engaged to circulate lubricant in the transmission and clutch packs prior to machine operation. Should the powertrain ECM detect machine movement (parking and service brake disengaged), it will disengage Clutch 5 and no clutch packs will be engaged.

Whenever the transmission direction and speed control lever is moved to the neutral position, or the transmission neutralizer is activated, the solenoid on the modulating valve for Clutch 5 (51) is energized

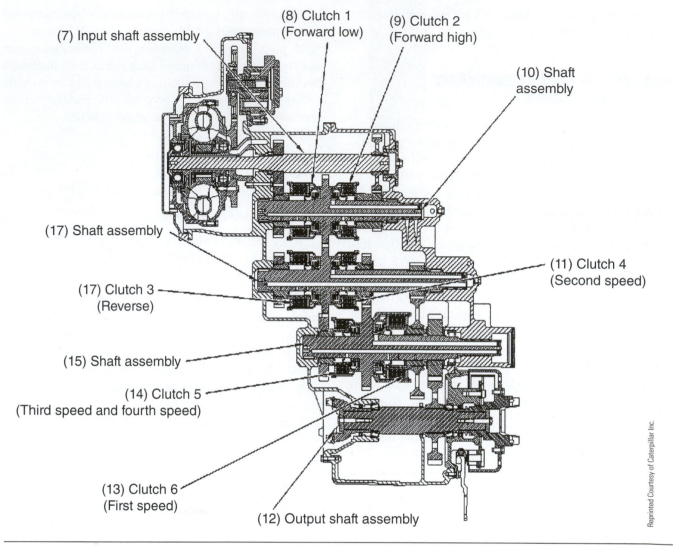

(7) Input shaft assembly

(8) Clutch 1 (Forward low)

(9) Clutch 2 (Forward high)

(10) Shaft assembly

(17) Shaft assembly

(11) Clutch 4 (Second speed)

(17) Clutch 3 (Reverse)

(15) Shaft assembly

(14) Clutch 5 (Third speed and fourth speed)

(13) Clutch 6 (First speed)

(12) Output shaft assembly

Figure 11-52 Four-speed forward, three-speed reverse, multiple countershaft transmission.

(see **Figure 11-54**). When this solenoid is energized, the modulating valve directs transmission pump oil (51) to Clutch 5 (30) in order to engage the clutch. Simultaneously, the directional and speed clutches (other than third speed Clutch 5) will disengage.

First Speed Forward Powerflow. The powertrain ECM modulates the electrical current that is sent to the selected clutch solenoids based on operator inputs to the ECM. The clutch solenoids are located on the modulating relief valves. Modulating the current to the clutch solenoid controls the clutch pressure by shifting the spool in the modulating valve.

Refer to **Figure 11-54**. When the transmission direction and speed control lever is moved to the first speed forward position, both the solenoid on the modulating valve for Clutch 6 (52) and the solenoid on

the modulating valve for Clutch 1 (54) are energized. When these solenoids are energized, the modulating valves direct transmission pump oil (41) to Clutch 1 (4) and Clutch 6 in order to engage the clutch packs. The oil in the clutches that are not engaged is returned to the sump (56).

With the transmission in the first speed forward position, Clutch 1 (4) and Clutch 6 (28) are engaged. Torque from the engine is transferred through the torque converter to the input shaft assembly (3). Because gear (1) is splined to the input shaft assembly (3), gear (1) turns gear (14) on countershaft (9). Gear (14) is splined to hub assembly (2), which causes the hub assembly to rotate. Because Clutch 1 (4) is engaged, torque is transferred from hub assembly (2) through the engaged Clutch 1 (forward low) to countershaft (9). Gear (5) is part of countershaft assembly (9)

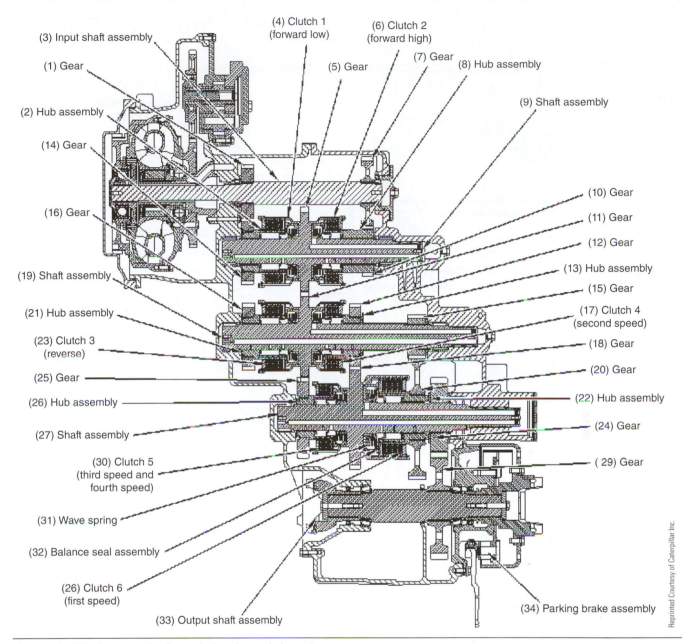

(3) Input shaft assembly
(1) Gear
(2) Hub assembly
(14) Gear
(16) Gear
(19) Shaft assembly
(21) Hub assembly
(23) Clutch 3 (reverse)
(25) Gear
(26) Hub assembly
(27) Shaft assembly
(30) Clutch 5 (third speed and fourth speed)
(31) Wave spring
(32) Balance seal assembly
(26) Clutch 6 (first speed)
(33) Output shaft assembly

(4) Clutch 1 (forward low)
(6) Clutch 2 (forward high)
(7) Gear
(5) Gear
(8) Hub assembly
(9) Shaft assembly
(10) Gear
(11) Gear
(12) Gear
(13) Hub assembly
(15) Gear
(17) Clutch 4 (second speed)
(18) Gear
(20) Gear
(22) Hub assembly
(24) Gear
(29) Gear
(34) Parking brake assembly

Reprinted Courtesy of Caterpillar Inc.

Figure 11-53 Countershaft and clutch pack locations.

and turns gear (11) on countershaft assembly (19). Gear (15) is splined to countershaft assembly (19), which turns gear (20). Gear 15 is considerably smaller than gear (20), which will provide a low speed/high torque condition. Torque is transferred from countershaft assembly (19) through gear (15) and gear (20) to hub assembly (22). First speed Clutch 6 (28) is engaged, and torque is transferred from hub assembly (22) through the engaged Clutch 6 to countershaft assembly (27). Gear (24) is splined to countershaft assembly (27), rotating gear (29), which is splined to output shaft assembly (33). Because gear (24) is smaller than gear (29) a second speed decrease/torque increase occurs through

the drive. This final torque increase is transferred from countershaft assembly (27) through gear (24) and gear (29) to output shaft assembly (33). The output shaft provides the means to connect to the front drive axle and rear drive axle.

Second Speed Forward Powerflow. Refer to **Figure 11-53**. When the transmission speed and direction control lever is placed in the second speed forward position, Clutch 4 (17) and Clutch 1 (4) are engaged. Torque from the engine is transferred through the torque converter to the input shaft assembly (3). Because gear (1) is splined to the input shaft assembly (3),

TRANSMISSION HYDRAULIC SYSTEM

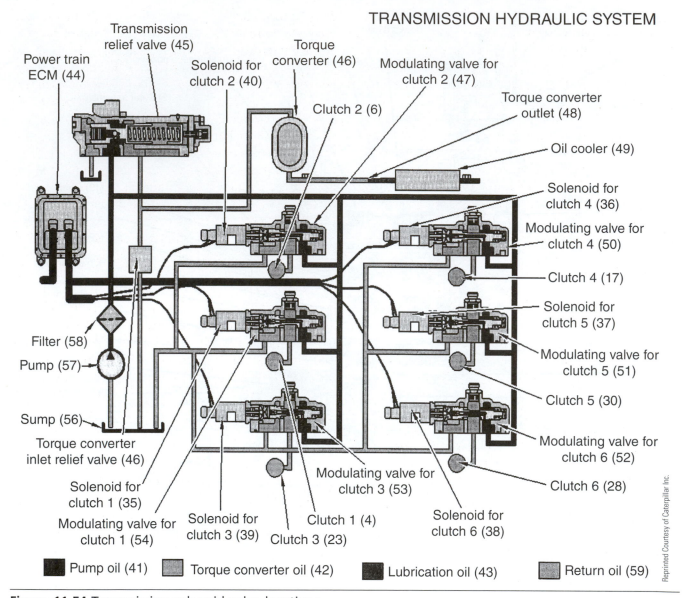

Figure 11-54 Transmission solenoid valve locations.

gear (1) turns gear (14) on countershaft (9). Gear (14) is splined to hub assembly (2), which causes the hub assembly to rotate. Because Clutch 1 (4) is engaged, torque is transferred from hub assembly (2) through the engaged Clutch 1 (forward low) to countershaft (9). Gear (5) is part of countershaft assembly (9) and turns gear (11) on countershaft assembly (19). Gear (11) is part of countershaft assembly (19), because Clutch 4 on countershaft assembly (19) (second speed) is engaged; powerflow is transferred through the clutch pack to gear (12). Gear (12) will now drive gear (16) that is in-mesh with it on countershaft (27). Gear (12) is smaller than gear (16), which will still provide a low speed/high torque condition, but not as great as for first speed because of the pitch diameter of the individual gears. Torque is transferred along countershaft assembly (27)

through gear (24) and to gear (29), which is splined to output shaft assembly (33). Because gear (24) is smaller than gear (29), a second speed decrease/torque increase occurs through the drive. This final torque increase results in a slightly greater speed increase when compared to first speed.

Third Speed Forward Powerflow. Refer to **Figure 11-53**. When the transmission speed and direction control lever is placed in the third speed forward position, Clutch 5 (30) and Clutch 1 (4) are engaged. Torque from the engine is transferred through the torque converter to input shaft assembly (3). Because gear (1) is splined to input shaft assembly (3), gear (1) turns gear (14) on countershaft (9). Because Clutch 1 (4) is engaged, torque is transferred from hub assembly (2) through the engaged clutch 1 (forward low) to countershaft (9). Gear (5) is part of countershaft assembly (9) and turns gear (11) on countershaft assembly (19). Because gear (11) is meshed with both gear (5) and gear (25), engine torque is transferred directly from gear (5) to gear (25). Gear (11) acts as an idler between the two meshed gears and provides no additional gear reduction. The gear ratio occurs between the pitch diameters of gear (5) and gear (25). Gear (25) has a larger diameter than gear (25); this results in an overdrive condition between these two gears. With Clutch 5 (30) now engaged, engine torque is transferred along countershaft assembly (27) through gear (24) and to gear (29), which is splined to output shaft assembly (33). Because gear (24) is smaller than gear (29) a speed decrease/torque increase occurs through this drive combination. The final output speed results in a speed increase over that of second gear.

Fourth Speed Forward Powerflow. Refer to **Figure 11-53**. When the transmission speed and direction control lever is placed in the fourth speed forward position, Clutch 5 (30) and Clutch 2 (6) are engaged. Torque from the engine is transferred through the torque converter to input shaft assembly (3). Because gear (7) is also splined to input shaft assembly (3), gear (7) turns gear (10) on countershaft (9). Gear (7) will drive gear (10) in an overdrive condition because of the larger diameter of gear (7). Because clutch pack (2) is engaged, engine torque is transferred to gear (5) and countershaft (9). Gear (5) is in constant mesh with gear (1), and gear (1) is in constant mesh with gear (25). Gear (1) on countershaft (19) again acts as an idler gear between gears (5) and (25) and relays engine torque directly to gear (5) and countershaft (27). Clutch pack (5) remained engaged between the 3–4 upshift and now acts as the fourth speed clutch. The speed increase occurs through Clutch 2 (6) forward high, allowing gears (7) and (10) to provide an overdrive gear ratio.

With Clutch 5 (30) engaged, gear (25) will relay engine torque to countershaft (27) and gear (24) and to gear (29) that is splined to output shaft assembly (33). Because gear (24) is smaller than gear (29) a speed decrease/torque increase occurs through this drive combination. The final output speed will result in a speed increase over that of third speed.

First Speed Reverse Powerflow. Refer to **Figure 11-53**. When the transmission speed and direction control lever is placed in the first speed reverse position, Clutch 6 (28) and Clutch 3 (23) are engaged. Torque

from the engine is transferred through the torque converter to input shaft assembly (3). Because gear (1) is also splined to input shaft assembly (3), gear (1) turns gear (16) on countershaft (19). For illustration purposes, countershaft (19) is shown below countershaft (9). In actual fact, it is located parallel to countershaft (9) and gear (16) is splined with gear (1) on input countershaft (3) as described earlier. Because reverse Clutch 3 is engaged on countershaft (19), engine torque is transferred through gears (1) and (6) and clutch pack (3) to gear (15). Gear (15) is in constant mesh with gear (20) on countershaft (27). Because first speed Clutch 6 (28) is engaged, gear (20) will now transfer engine torque to countershaft (27) at a reduced speed/high torque ratio. This is due to the gear ratio provided by driving gear (15) to driven gear (20). Through the first speed Clutch 6 (28) on countershaft (27), engine torque is now transferred to gear (24), which is splined to the countershaft. Gear (24) will provide a further gear reduction with output countershaft (33) and gear (29). Torque is split between the front and rear drive axles to provide reverse direction.

Second Speed Reverse Powerflow. Refer to **Figure 11-53**. When the transmission speed and direction control lever is placed in the second speed reverse position, Clutch 4 (17) and Clutch 3 (23) are engaged. Torque from the engine is transferred through the torque converter to input shaft assembly (3). Because gear (1) is also splined to input shaft assembly (3), gear (1) turns gear (16) on countershaft (19). Because reverse Clutch (3) and second speed Clutch 4 (17) are engaged on countershaft (19), engine torque is delivered directly to gear (12) on the countershaft. Gear (12) is in constant mesh with gear (16) and will provide a slightly higher gear ratio than first speed reverse. Gears (16) and (24) are directly splined to countershaft (27). This means that engine torque will be transferred directly through countershaft (27), to output countershaft (33). Gears (24) and (29) will provide the last gear ratio reduction to increase engine torque to the output shaft. Torque is then split between the front and rear drive axles to provide second speed reverse.

Third Speed Reverse Powerflow. Refer to **Figure 11-53**. When the transmission speed and direction control lever is placed in the third speed reverse position, Clutch 5 (30) and Clutch 3 (23) are engaged. Torque from the engine is transferred through the torque converter to input shaft assembly (3). Because gear (1) is also splined to input shaft assembly (3), gear (1) turns gear (16) on countershaft (19). Because reverse Clutch 3 (23) is engaged on countershaft (19), engine torque is transferred through gears (1) and (6) and clutch pack (3) to gear (11). Gear (11) is in constant mesh with gear (25) on countershaft (27). This will provide the highest speed ratio in reverse direction for this particular machine application. Permanently splined gear (24) on countershaft (27) provides a constant gear reduction ratio with gear (29) on output countershaft (33). Torque is then split between the front and rear drive axles to provide third speed reverse.

Table 11-5 shows all forward and reverse speed combinations with the controlling clutch pack. The powertrain ECM controls the modulating valve solenoids that direct pressurized oil to the clutch packs in order to engage the appropriate gear combinations.

Transmission Hydraulic System

The transmission hydraulic system is responsible for providing the median for the torque converter, the pressure used to activate all the clutch packs, effective cooling and lubrication within the transmission, and to direct and control oil pressure and flow within the system. **Figure 11-55** shows all the components that make up the hydraulic system for this transmission.

The tank or reservoir (15) is the sump section of the transmission. A separate reservoir is not used. Transmission pump (16) is driven by the torque converter housing and is located to the top-left of the transmission housing. Transmission filter (17) is located on the left side of the machine for easy access when servicing the machine. The transmission relief valve (20) is located with all the modulating shift valves on the top-right section of the transmission. The transmission relief valve is set for 326 ± 10 psi (2,246 ± 70 kPa). The torque converter (5) provides the input drive to the transmission and the oil cooler (7) is located to the rear of the machine directly in front of the engine radiator. It is an air-to-oil cooler and will keep the transmission and torque converter between normal operating temperatures of 168°F to 234°F (76°C to 112°C). The torque converter inlet relief valve (18) limits the oil pressure to the torque converter to 140 psi (965 kPa) maximum. This hydraulic schematic represents the clutch packs with clutch ports (9), (10), (11), (12), (13), and (14). Pump oil lines (37) depict high-pressure lines to the shift modulating valves. This oil is used to engage the clutch packs. Torque converter oil lines (38) show oil flow from the transmission relief valve to the torque converter and then to the oil cooler. Oil cooler outlet oil becomes lubrication oil (39) for the transmission bearings, shafts, and clutch packs. Return oil (40) represents oil exiting the clutch packs when the clutches are disengaged. Return oil drains back to the transmission sump where the transmission pump will recirculate it through the system.

Hydraulic Oil Flow in Neutral. Refer to **Figure 11-55**. When the engine is started, the transmission pump is driven by the torque converter housing at engine speed. The pump (16) picks up the oil from the transmission sump; oil flows first through the oil filter (17) to remove any contaminants and then to the transmission relief valve (20). The transmission relief valve is connected in parallel to the modulating shift valves and in series to the torque converter inlet relief valve. This means the oil must travel through the transmission relief valve before reaching the torque converter relief valve. Oil is then directed to the torque converter and oil cooler.

The transmission modulating valves can be considered open-center valves, which means that oil flow will travel through them when they are not activated. Because the transmission pump is a positive-displacement pump, it will continue to displace oil at its rated capacity. It must displace enough oil so that the pressure

TABLE 11-5: CONTROLLING CLUTCH PACKS FOR ALL FORWARD/REVERSE COMBINATIONS								
	First Speed Forward	Second Speed Forward	Third Speed Forward	Fourth Speed Forward	Neutral	First Speed Reverse	Second Speed Reverse	Third Speed Reverse
Clutch 1	X	X	X					
Clutch 2				X				
Clutch 3						X	X	X
Clutch 4		X					X	
Clutch 5			X	X	X			
Clutch 6	X					X		

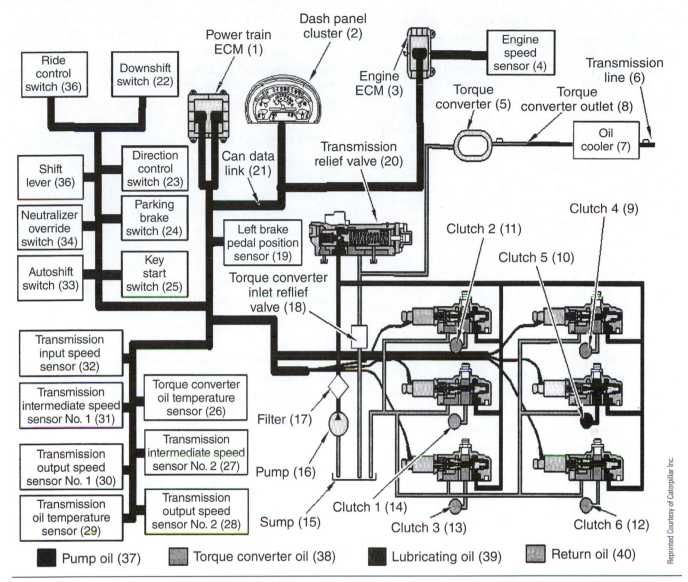

Figure 11-55 Transmission hydraulic and electronic system control.

drop across the valve spool can allow the transmission relief valve to build and maintain the pressure setting for this section of the hydraulic system. The remainder of the oil is used to supply the torque converter and lubrication circuit and, when this circuit is satisfied, the remainder of that oil is drained to the transmission sump through the transmission relief valve.

In the neutral position, the clutch modulating valve (5) is energized. This fills the clutch pack on start-up. This lubricates the clutch packs and bearings by allowing some of the clutch-pack hubs to rotate while oil pressurizes the system and saturates the components with oil.

The transmission has an individual clutch solenoid modulating valve for each clutch pack that is controlled by the powertrain ECM, based on data input from the operator and machine sensors and switches.

The ECM must energize at least two solenoids for the transmission to drive: one directional solenoid and one speed solenoid. **Table 11-6** provides the combination of solenoid valves and engaged clutch packs for each forward and reverse speed.

Clutch Solenoid Modulating Valves. This transmission is equipped with six clutch solenoid modulating valves, as shown in **Figure 11-54**. They are as follows:

- Modulating valve for Clutch 1 (54) (forward low)
- Modulating valve for Clutch 2 (47) (forward high)
- Modulating valve for Clutch 3 (53) (reverse)
- Modulating valve for Clutch 4 (50) (second speed)
- Modulating valve for Clutch 5 (51) (third speed)
- Modulating valve for Clutch 6 (52) (first speed)

TABLE 11-6: ENERGIZED SOLENOIDS AND CLUTCHES TO SPEED AND DIRECTION

Speed Range and Direction	Energized Solenoids	Engaged Clutches
Fourth Speed Forward	5 and 2	5 and 2
Third Speed Forward	5 and 1	5 and 1
Second Speed Forward	4 and 1	4 and 1
First Speed Forward	6 and 1	6 and 1
Neutral	5	5
First Speed Reverse	6 and 3	6 and 3
Second Speed Reverse	4 and 3	4 and 3
Third Speed Reverse	5 and 3	5 and 3

© Cengage Learning 2014

The clutch solenoid modulating valves are outputs of the powertrain (ECM). These valves are proportional solenoid valves. This means that the valves will supply oil to the clutch packs directly proportional to the PWM actuation signal from the ECM. The clutch solenoid modulating valves are used by the powertrain ECM to modulate the oil pressure delivered to each individual clutch pack. The ECM sends a PWM signal to manage the percentage of on-off current to the solenoid valve that controls the plunger travel. The farther the plunger moves within its bore, the greater the

amount of oil passage directed to the clutch pack. The powertrain ECM activates the appropriate solenoid valves based on the input from the operator. When the modulating valves are deactivated by the ECM, oil pressure within the clutch pack is drained through the valve back to the transmission sump.

CAUTION *The solenoid coils are not designed to be operated at a 24 direct current voltage (dcv). The powertrain ECM sends a 24-volt PWM signal at a specific duty cycle, which provides an average voltage of about 12 volts to the solenoid coils. Testing of solenoid coils should not be done with a 24-dcv supply because it can damage the solenoid. If the solenoid coils must be energized by a source other than the powertrain ECM, 12 dcv should be used to energize the solenoid coils.*

Figure 11-56 shows the components that make up the solenoid modulating valves in the de-energized position. Pump oil (16) flows into the valve body from pump passage (6). Oil then flows into passage (7), through orifice (2), and into chamber (13). Since there is no signal that is sent to solenoid (12), pin (11) cannot hold ball (1) against orifice (10). The oil flows through orifice (10) past ball (1) to the tank passage (9)

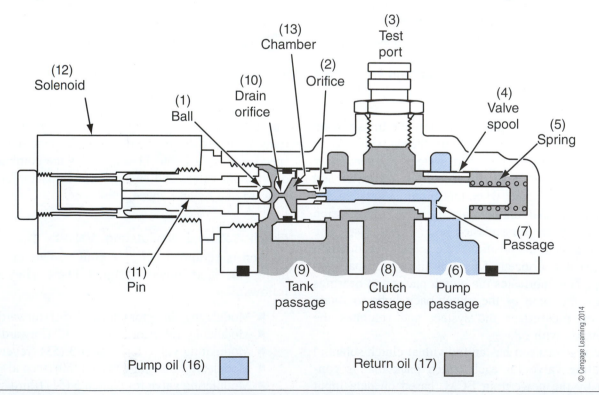

Pump oil (16)   Return oil (17)

Figure 11-56 Modulating valve with solenoid de-energized.

at a pressure drop close to zero. Oil in tank passage (9) becomes return oil (17). Spring (5) holds shift valve spool (4) to the left. When valve spool (4) is shifted to the left by the spring force, the oil in clutch passage (8) is allowed to flow through tank passage (9). When the oil in clutch passage (8) is vented to the tank, the clutch cannot be engaged. The oil in pump passage (6) is blocked from entering clutch passage (8). Pump oil in passage (7) drains through orifice (10) and to tank passage (9) together with oil from clutch pack passage (8). Even though this oil may enter the clutch passage, the clutch will not engage due to the low pressure passage to the tank. The clutch return springs will force the piston back to drain the oil.

Figure 11-57 shows the components that make up the solenoid modulating valves in the initial energized position. When the powertrain ECM requires a clutch to be engaged, a signal is sent to the solenoid (12). The signal is proportional to the desired clutch pressure. When the signal is sent to the solenoid, pin (11) moves to the right and forces ball (1) toward orifice (2). When the check ball moves toward the orifice, the flow of return oil (18) into tank passage (9) is restricted. This causes oil pressure in chamber (13) to increase and move valve spool (4) to the right against the force of spring (5). This oil becomes the pilot control pressure

for the valve spool. When valve spool (4) moves to the right, pump oil (16) in pump passage (6) enters clutch passage (8). The valve spool blocks the clutch passage from draining to tank passage (9). This causes the clutch pressure to increase due to pump oil flow directed into the clutch passage (8). The oil in the clutch passage also flows through passage (14) into chamber (15). This helps meter and control the oil flow through the valve.

Initially, the powertrain ECM sends a high signal to the solenoid in order to quickly fill the clutch with oil. Then, a reduced signal is sent to the solenoid in order to allow the clutch to engage smoothly. This helps to prevent shock loading of the driveline and to achieve smooth transitions between speed selections.

Once the clutch engages, the powertrain ECM begins to increase the voltage signal to the solenoid to completely engage the clutch pack (see **Figure 11-58**). As the signal to the solenoid increases, pin (11) moves to the right and pushes check ball (1) toward orifice (2), completely restricting the flow of oil back to tank passage (9). The oil pressure increase in chamber (13) will push the valve spool to the right. This causes the clutch pressure to increase again. With the clutch pack fully engaged, oil pressure will rise to the transmission relief valve setting. Oil directed

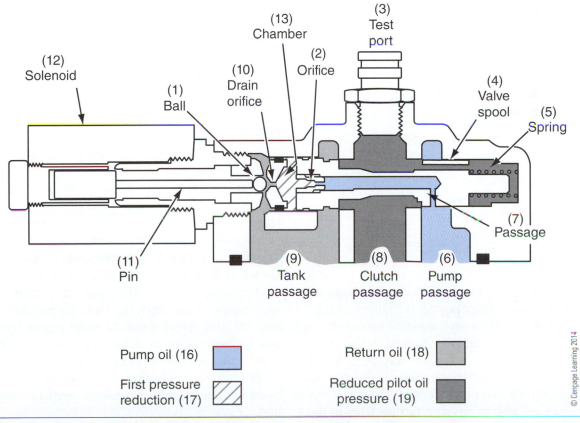

Figure 11-57 Solenoid modulating valve in the initial energized position.

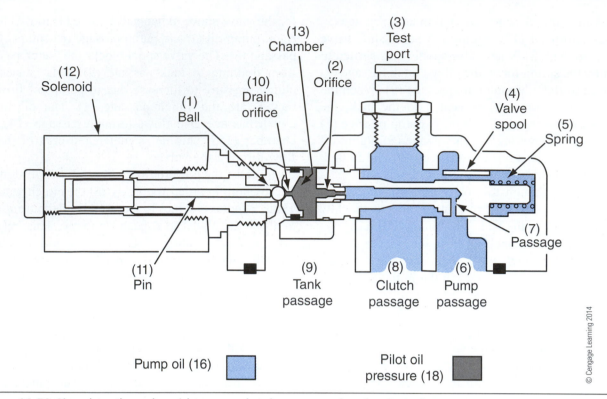

Figure 11-58 Signal to the solenoid to completely engage the clutch pack.

through passage (14) together with spring force (5) will meter oil flow through the spool valve to maintain maximum clutch pressure and allow for any leakage within the clutch circuit.

Multiple Countershaft Control Electronics

The powertrain electronic control system has six basic functions:

- Neutral start function
- Manual shift function
- Automatic shift function
- Parking brake function
- Back-up alarm function
- Diagnostic operation

These functions are identical to the functions described earlier under "Planetary Shift Transmissions," used by the CAT 988G wheel loader. Please refer to this section for more details. Although the transmissions within these two machines may be different, the ECMs will have similar operating parameters set up within their own electronic control systems. Diagnostic operations will be covered further later in this chapter.

MULTIPLEXING

Multiplexing is the term used to describe data exchange between two or more modules on a machine. A serial data bus is used to broadcast and receive messages known as data packets between multiple MIDs with addresses on the bus. A data link is used to allow a technician to communicate with the MIDs networked to a data bus.

Whenever there is more than one ECM on a vehicle, networking the ECMs is crucial to optimize machine management. An engine communicating with a transmission will function more efficiently when one knows what the other is doing. To accomplish this, the ECMs must use a common communication language. The data buses used on heavy mobile equipment are known as SAE J1587/1708 and J1939, both of which are based on the Controller Area Network (CAN) 2.0 (generation 2) architecture. Most current diesel-powered off-road vehicles use the more recent J1939 data bus for powertrain management, but may use additional buses for non-powertrain operation.

In machine applications with multiple ECMs, such as engine, transmission, hydraulics, payload control, and braking system controllers, it is important that they all communicate with each other through a CAN network.

A CAN network is a message-oriented network system that defines content rather than receiver stations and addresses. There is more information on multiplexing in Chapter 19.

POWERSHIFT TRANSMISSION MAINTENANCE AND REPAIR

Transmissions used in heavy-duty applications have system specific maintenance procedures. It is important to use and follow OEM service literature when servicing machine equipment. The following maintenance procedures are a guideline to those used in the industry today; they are by no means meant to replace OEM service literature. The focus will continue to be on the two transmissions (988G planetary shift and 938G countershaft) used as examples in this chapter.

Powershift Transmission Lubrication

Proper lubrication is one of the keys to a good preventative maintenance program. Maintaining the correct transmission oil level and ensuring that the oil is of OEM-specified quality are both critical to ensuring that a transmission achieves its expected service life. Following OEM oil change intervals and taking routine oil analysis samples are also key to a good maintenance program. Because the application of heavy-duty equipment is such that it does not accumulate mileage, servicing is based on the number of working hours the machine accumulates or the time period between service intervals. Hour meters log time based from the moment the machine is started until the engine is shut down. To keep an accurate hour meter reading, an engine oil pressure signal or even an alternator pulse activates some hour meters. Direct power from the key switch is avoided to eliminate registering hours when the engine is not in operation with the key switch on.

Here are some examples of service charts:

Maintenance Interval Schedule. All safety information, warnings, and instructions must be read and understood before you perform any operation or any maintenance procedure. Before each consecutive interval is performed, all of the maintenance requirements from the previous interval must also be performed.

The transmission oil level must be checked daily prior to operating the equipment. A dipstick (see **Figure 11-59**) or a sight glass is used in many types of

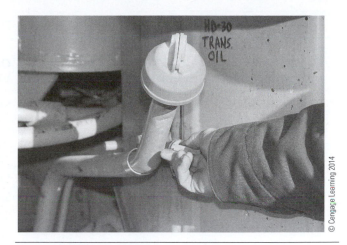

Figure 11-59 Transmission fill tube and dipstick level check.

powershift transmissions that make daily inspection quick and easy. There is no crawling under a machine and no plug to remove to check oil levels as there is on manual shift transmissions. The proper oil viscosity to be used may vary between machine applications and manufacturers.

Selecting the Proper Viscosity Grade. Ambient air temperature is the temperature of the air in the immediate vicinity of the machine. This may differ due to the machine application and from the generic ambient temperature for a geographic region. In order to select the proper oil viscosity, review the regional ambient temperature and the potential ambient temperature for a given machine application. Use the higher temperature of the two in order to select the oil viscosity and use the highest oil viscosity that is allowed for the ambient temperature for that machine. In arctic applications, the preferred method is the use of heaters for machine compartments and a high viscosity oil.

The proper oil viscosity grade is determined by the minimum (lowest) outside ambient temperature. This is the temperature when the machine is started and while the machine is operated. In order to determine the proper oil viscosity grade, refer to the "Min" column in **Table 11-7**. This information reflects the coldest ambient temperature condition for starting and operating a cold machine. Refer to the "Max" column in **Table 11-7** to select the oil viscosity grade for operating the machine at the highest temperature that is anticipated.

Machines that are operated continuously, in 24/7 operations (24 hours a day/7 days a week), should use oils that have higher oil viscosity in the final drives and in the differentials. Use of these oils will maintain the highest possible oil film thickness. Because the

TABLE 11-7: LUBRICANT VISCOSITIES FOR AMBIENT TEMPERATURES

Compartment or System	Oil Type and Classification	Oil Viscosities	°C Min	°C Max	°F Min	°F Max
Engine Crankcase	Cat DEO Multigrade	SAE 0W20	−40	+10	−40	+50
	Cat DEO SYN	SAE 0W30	−40	+30	−40	+86
	Cat Arctic DEO SYN	SAE 0W40	−40	+40	−40	+104
	Cat ECF-1	SAE 5W40	−30	+30	−22	+86
	API CG-4 Multigrade	SAE 5W40	−30	+50	−22	+122
		SAE 10W30	−18	+40	0	+104
		SAE 10W40	−18	+50	0	+122
		SAE 15W40	−9.5	+50	−15	+122
Powershift Transmissions	Cat TDTO	SAE 0W20	−40	+10	−40	+50
	CAT TDTO-TMS	SAE 0W30	−40	+20	−40	+68
	Commercial TO-4	SAE 5W30	−30	+20	−22	+68
		SAE 10W	−20	+10	−4	+50
		SAE 30	0	+35	+32	+95
		SAE 50	+10	+50	+50	+122
		TDTO-TMS	−10	+43	+14	+110
Hydraulic Systems	Cat HYDO	SAW 0W20	−40	+40	−40	+104
	Cat DEO	SAE 0W30	−40	+40	−40	+104
	Cat MTO	SAE 5W30	−30	+40	−22	+104
	Cat TDTO	SAE 5W40	−30	+40	−22	+104
	Cat TDTO-TMS	SAE 10W	−20	+40	−4	+104
	Cat DEO SYN	SAE 30	−10	+50	+50	+122
	Cat ECF-1	SAE 10W30	−20	+40	−4	+104
	Cat BIO HYDO (HEES)	SAE 15W40	−15	+50	+5	+122
	API CG-4	Cat MTO	−25	+40	−13	+104
	API CF	Cat BIO HYDO (HEES)	−40	+43	−40	+110
	Commercial TO-4					
	Commercial BF-1	TDTO-TMS	−20	+50	−4	+122
Drive Axles	Cat TDTO	SAE 0W20	−40	0	−40	+32
	Cat TDTO-TMS	SAE 0W30	−40	+10	−40	+50
	Commercial TO-4	SAE 5W30	−35	+10	−31	+50
		SAE 10W	−25	+15	−13	+59
		SAE 30	−20	+43	−4	+110
		SAE 50	+10	+50	+50	+122
		TDTO-TMS	−30	+43	+22	+110

© Cengage Learning 2014

machines are operated continuously, there is no opportunity for the oil to cool to a dangerous level.

WARNING *DO NOT use only the "Oil Viscosities" column when determining the recommended oil for a machine compartment. The "Oil Type and Classification" column MUST also be referred to.*

Tech Tip: SAE 0W oil and SAE 5W oils are not recommended for heavily loaded machines or machines that are operated continuously. Some machine models and/or machine compartments do not allow the use of all available viscosity grades. Therefore, proper selection of oil must be made using OEM recommendations. Some machine compartments allow the use of more than one oil type or viscosity grade; it is recommended that you do not mix these oil types. Also, different brands of oils may use different oil additive packages; it is recommended that different oil brands not be mixed due to chemical incompatibilities.

> **WARNING** *Care must be taken to ensure that fluids are contained during routine inspection, maintenance, testing, adjusting, and repair of equipment. Be prepared to collect the fluid with suitable containers before opening any compartment or disassembling any component containing fluids that might spill. Dispose of all fluids according to Federal and local regulations.*

Transmission Oil Service. When a transmission oil service is due on a machine, it is important to follow these steps:

1. Operate the machine for a few minutes in order to warm the transmission oil.
2. Park the machine on level ground. Lower all implements to the ground and apply slight downward pressure. Engage the parking brake and stop the engine.
3. Remove the transmission oil drain plug and allow the transmission oil to drain into a suitable container. (In some applications, it may be necessary to remove a skidpan or access plate.) Inspect the drain plug for any metal or foreign material that might indicate a possible transmission failure. Clean and reinstall the drain plug (see **Figure 11-60**).

4. Replace the transmission oil filter.
5. Use a strap-type wrench to remove the filter element. Dispose of the used filter element properly.

Note: In cases where a canister-type filter element is used, remove bolts from the filter housing cover and lift-off cover. Some filter housing covers are spring loaded and should be removed carefully. In some canister-type filters, the canister must be rotated or unscrewed to be removed (see **Figure 11-61**). Remove and dispose of the filter element. If the filter housing is equipped with a drain valve, this valve should be opened to drain the residual oil contained in the housing. If the housing is not equipped with a drain valve, it is recommended that the oil be extracted using a pump.

---

6. Clean the filter-mounting base and make sure that the entire seal is completely removed.
7. Apply a light coat of clean transmission oil to the seal of the new filter element.
8. Install the new filter and hand tighten until the seal contacts the base. Note the position of the index marks on the filter in relation to a fixed point on the filter base.

Tech Tip: The rotation index marks on the transmission oil filter are spaced 90 degrees or ¼ turn away from each other. When you tighten the transmission oil filter, use the rotation index marks as a guide.

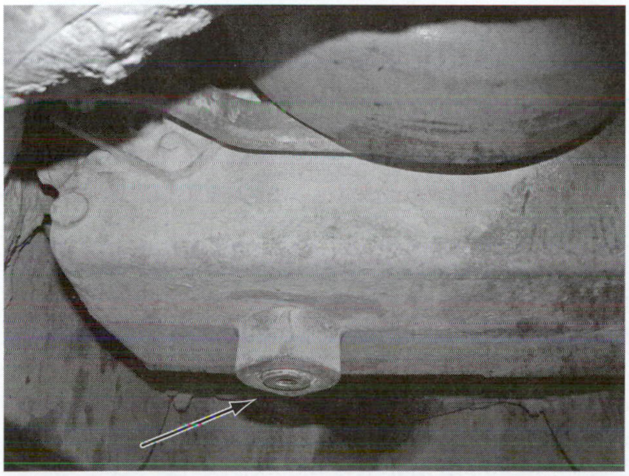

Figure 11-60 Transmission drain plug located at the base of the transmission housing.

Figure 11-61 Canister filter: (4) disposable filter element, (2) spin-on canister.

9. Tighten the filter according to the instructions printed on the filter. As a rule, tighten the filter one-quarter turn after the filter seal has contacted the filter base. You may need to use a strap wrench, or another suitable tool, in order to properly torque the filter. Make sure that the installation tool does not damage the filter.

10. Remove the magnetic strainer cover from the transmission housing. Remove the magnets, the tube, and the screen from the housing and inspect the strainer and magnets for any metal or foreign material that might indicate a possible transmission failure.

11. Wash the tube and the screen in a clean, non-flammable solvent. Use a lint-free cloth, a stiff bristle brush, or pressurized air to clean the magnets.

12. Install the magnets and the tube into the screen. Reinstall the magnetic screen and strainer back into the transmission.

13. Clean the magnetic strainer cover and inspect the seal for damage or flatness. Replace the seal as necessary. Install the magnetic strainer cover and tighten the bolts to proper specifications.

14. Remove the filler cap and fill the transmission with transmission oil.

15. Install the filler cap.

Note: Refer to the appropriate OEM service literature for the correct type and amount of oil.

Start and run the engine at low idle with the service brake applied. Slowly operate the transmission control lever from forward to reverse and through each indicated speed in order to circulate the oil within the transmission. Move the transmission control lever to the neutral position, apply the parking brake, and inspect the transmission and transmission filter for leaks with the engine running. Check the transmission oil level and add oil as necessary to obtain proper oil level (see **Figure 11-62**).

Figure 11-62 Two-sided dipstick with low-idle check on one side and engine safe start on reverse side.

Note: Certain machines require that the transmission oil level be checked with the engine running; others require that the level be checked with the engine stopped. Refer to the OEM service literature for proper procedures to check oil levels.

In many instances a transmission oil sample is required at every service interval. If an oil sample must be taken, it is recommended that the sample be taken as the oil is draining from the drain plug and not from the filter element. Obtaining an oil sample from the filter will give false readings because the filter contains most of the contaminants. If an oil sample syringe is used, an oil sample should be obtained with the engine running and the oil warm.

POWERSHIFT TRANSMISSIONS DIAGNOSTICS

This following section describes some systematic methods of troubleshooting powershift transmissions by using problem and potential cause lists. We will also address troubleshooting and the use of electronic diagnostic equipment. One of the most common complaints a technician receives is the "No Power" complaint. This complaint is one of the hardest problems to solve because it does not lead the technician to a specific component or area of the machine. Therefore, it is critical that the technician involve the operator when trying to isolate this complaint.

Stall Testing

One method of isolating a low power complaint is to perform a stall test (see **Photo Sequence 3** on pages 55–56). The torque converter stall test is designed to test the engine, the torque converter, the drive train, and the brake system as a unit. A stall test may identify engine, converter, and transmission problems. A stall test creates a situation that places the turbine in a stalled or stopped condition with the impeller at maximum engine throttle position. The torque converter will be at maximum **vortex oil flow**. To achieve this condition, the machine must be kept from moving while the transmission is in a drive mode. The vehicle service brakes must be used to prevent the vehicle from moving and in many cases, it is recommended that wheel chocks be used. Make sure that the service brakes are fully operational and that the transmission oil is at the normal operating

PHOTO SEQUENCE 3
Transmission/Torque Stall Test Procedure

A torque stall test may be performed anytime the transmission/torque converter circuit is suspected of not meeting performance standards. The test must be performed in a suitable location so as to not compromise the safety of anyone in the immediate area. Ensure that there is adequate room, both front and rear, to allow for any unintended equipment movement. The equipment must be tested on level ground. This procedure should only be performed by qualified personnel. The equipment must be brought up to operating temperature with the brake system fully charged, and all fluids in the transmission/torque circuit must be up to the recommended levels and temperatures. A low oil level may produce faulty data and could cause internal damage to the transmission/torque converter.

The following is a general procedure used to check the stall speed on off-road equipment that uses a powershift transmission. It is not intended to replace the manufacturer's recommended procedures and is only an example of the steps required to perform this type of procedure. Always refer to the OEM's service literature for the actual step-by-step instructions, as they may vary on different types of equipment.

> **CAUTION** *Never use your hands to check for leaks in a high-pressure hydraulic circuit. Use a small piece of cardboard; a pin-size leak may penetrate the skin. If oil is accidentally injected into the skin, immediate medical attention is required. Oil in a hydraulic circuit can become extremely hot during operation. This can lead to severe burns if accidentally splashed onto yourself or other personnel around the work area.*

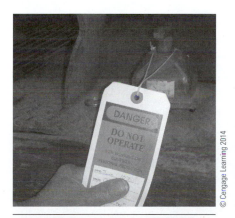

P3-1 Depending on the employer, some form of tagging or lockout procedure must be used to protect the technicians while work is being performed on the equipment. There are two different procedures used to protect the technician.

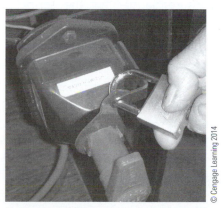

P3-2 Some employers require you to actually lock out the main power disconnect on the equipment. This lock can only be installed and removed by the technician who is performing the work, and it must be removed on completion of the work.

P3-3 Place the appropriate wheel chocks under the tires.

© Cengage Learning 2014

PHOTO SEQUENCE 3 **(Continued)**

© Cengage Learning 2014

© Cengage Learning 2014

© Cengage Learning 2014

P3-4 Install the lock bar if the equipment has articulating steering.

P3-5 Before performing a torque stall test, the brakes must be able to pass the manufacturer's recommended brake test. Apply the park brake. Ensure that the brake system is fully charged before continuing. Next, depress the brake pedal fully and check the brake pressure gauges for the correct pressure. Refer to the equipment specifications in the shop manual for the correct pressures.

P3-6 The equipment must be placed in the appropriate gear to perform a brake test. Refer to the equipment shop manual for the correct gear selection and rpm range. Not all brake tests on equipment are checked the same way.

Once the brakes have passed the brake test, place the equipment in the correct gear. Refer to the manufacturer's specifications on transmission/torque oil temperature limits before beginning this test. Repeat the stall test, slowly increasing the engine rpm until maximum rpm is reached and record the values. Do not allow the converter to stall for more than 30 seconds at a time without a cooldown period.

temperature of 168°F to 234°F (76°C to 112°C) when tests are performed.

It is critical to use specific stall test data from the OEM when performing this test. A specific gear selection must be used for a specified time duration. Torque converter overheating and damage can result by not following proper test procedures. This test is best performed outside the shop in an isolated area due to the extreme noise and possible danger involved. Personnel should stay clear of the machine during this test. The transmission auto/manual switch must be in the manual position in order to prevent any unexpected shifts during the stall test.

WARNING *To ensure that the transmission oil does not overheat, do not keep the torque converter in a full stall condition for more than 30 seconds. After the torque converter stall test, reduce the engine speed to low idle. In order to cool the oil, move the transmission control to neutral position and run the engine between 1,200 rpm and 1,500 rpm.*

The following test outlines the procedure as it is performed on the Cat 938G® loader, equipped with a hydraulic driven fan, four-speed countershaft transmission with a non-lockup torque converter.

1. Turn the engine start switch to the off position.
2. Set the hydraulic fan to the maximum rpm. In order to run the hydraulic fan system at the maximum rpm, the connection to the electro-hydraulic control valve for the hydraulic fan system must be disconnected.

Note: The electro-hydraulic control valve is located on the right side of the engine and below the hydraulic fan pump. Turn the engine start switch to the off position.

3. Disconnect the electrical connection from the connection on the electro-hydraulic control valve.
4. Engage the parking brake and shift the transmission to neutral.
5. Start the engine. Warm the transmission oil to normal operating temperature of approximately 168°F to 234°F (76°C to 112°C).
6. Fully depress the right-side service brake pedal. Hold the right service brake pedal in the depressed position for the duration of the test. Ensure that the parking brake is applied.

7. Shift the transmission direction and speed control lever to the fourth speed forward position. Shift the transmission direction control switch and shift the upshift gear speed switch to the fourth speed forward position. Fully depress the accelerator pedal. Allow the engine rpm to stabilize. Then, observe the tachometer and record the result.
8. The correct stall speed for the 938G Series II wheel loader is 2,085 ± 65 rpm.

Stall speeds that are lower than the specified rpm indicate incorrect engine performance. Stall speeds that are higher than the specified rpm indicate incorrect transmission performance or incorrect performance of the torque converter. If you must repeat the test, keep the transmission in neutral and wait for 2 minutes between tests to allow the transmission oil to return to its normal operating temperature.

Problems and Possible Causes. The following six problems and possible causes are provided as examples, but many problems can create diagnostic codes or event codes in the EMS. These codes can help find the root cause of the problem. A technician should be comfortable in his or her ability to retrieve the logged and active codes. Six problems and possible causes are provided as examples.

Problem 1: Low stall speed.

Possible Causes:
1. Engine performance is not correct.
2. The oil is cold.

Problem 2: High stall speed in both directions.

Possible Causes:
1. The oil level is low.
2. There is air in the oil.
3. Clutches are slipping.
4. There is torque converter failure.

Problem 3: High stall speed in one gear or in one direction.

Possible Causes:
1. There is a leak in the clutch circuit.
2. There is a clutch failure in the gear that has a high stall speed or in the direction that has a high stall speed.

Problem 4: There is high torque converter pressure.

Possible Causes:

1. There is a restriction inside the torque converter.
2. The oil passage is restricted or the oil cooler is restricted.

Problem 5: There is low torque converter pressure.

Possible Causes:

1. The converter inlet relief valve is stuck.
2. There is low pump flow or pressure.

Problem 6: The torque converter gets too hot.

Possible Causes:

1. There is too much load, which causes the torque converter to slip.
2. The transmission gear is incorrect for the load on the machine. Shift to a lower gear.
3. The oil level is incorrect in the transmission.
4. The water level is low in the engine radiator.
5. There are restrictions in the oil cooler or lines.
6. There is not enough oil to the converter because the converter inlet relief valve is bypassing too much flow.
7. There is not enough oil to the converter because of low pump flow or pressure.

Transmission Problems

Many of the following problems can log diagnostic codes or event codes in the ECM. These codes can help find the root cause of the problem. Technicians should retrieve logged active and historic codes.

Four sample problems and possible causes are included here as examples of situations you will encounter. You will be able to diagnose most problems by using resource materials provided by your instructor, and OEM service literature specific to the transmission you are working on.

Problems and Possible Causes

Problem 1: The transmission does not operate in any gear.

Possible Causes:

1. The fuse for the transmission shifter is open or blown.
2. There is a problem in the electrical circuit. Check ET for active diagnostic codes.
3. There is low system voltage.
4. The transmission shift control unit is faulty.

5. There is low oil pressure.
6. The torque converter has failed.
7. There is mechanical failure in the transmission.
8. The clutch discs and plates are worn too much.
9. The parking brake is engaged.
10. The machine security system is active.
11. The transmission direction and speed control lever and the transmission direction control switch are both in forward or reverse.

Problem 2: The transmission slips in all gears.

Possible Causes:

1. There is low oil pressure.
2. The torque converter has failed.
3. There is mechanical failure in the transmission.
4. The clutch discs and plates are worn too much.

Problem 3: The transmission gets hot.

Possible Causes:

1. There is a low coolant level in the engine radiator.
2. There is too much slippage by the torque converter that is caused by too much load.
3. The transmission gear is incorrect for the load on the machine. Shift to a lower gear.
4. The temperature gauge is faulty.
5. The oil level is incorrect.
6. The oil cooler or lines are restricted.
7. The clutch slips too much. This could be caused by low oil pressure or by damaged clutches.
8. Low oil flow is caused by pump wear or leakage in the hydraulic system.
9. The oil is mixed with air.
10. The torque converter inlet relief valve is stuck open and causes low oil flow through the torque converter.
11. The clutch or clutches are not fully releasing. This can be caused by warped plates or discs. This can also be caused by a broken return spring or a weak return spring.

Problem 4: There is noise in the transmission that is not normal.

Possible Causes:

1. There are damaged gears in the transmission.
2. There are worn teeth on the clutch plates or on the clutch discs.
3. There are slipping clutch plates and there is noise from the disc.
4. There are other parts that are worn or damaged.

Electronic Diagnostics

Diagnostic software programs are used to access the data bus. The service technician can use diagnostic software to perform troubleshooting and maintenance procedures on the machine. Some of the options that are commonly available are listed below:

- View diagnostic codes
- View active and logged event codes
- View the status of parameters
- Clear active and logged diagnostic codes
- Perform calibration of machine systems
- Program the ECM (Flash)
- Print reports

Using Caterpillar ET to Determine Diagnostic Codes. Connect the Caterpillar Electronic Technician to the machine. Turn the key switch to the run position. Start the Cat ET. The Cat ET will initiate communications with the Electronic Control Modules (ECMs) on the machine. After communication has been established, the Cat ET will list the ECMs. Choose the desired ECM, in this case it will be the powertrain ECM. After the diagnostic codes have been determined by the Cat ET, see the test procedure for the corresponding diagnostic code.

Troubleshooting procedures may cause new diagnostic codes to be logged. Therefore, before any procedures are performed, make a list of all the active diagnostic codes in order to determine the system problems. Clear the diagnostic codes that were caused by the procedure when each procedure is complete.

Note: Before performing a procedure, always check that a circuit breaker is not tripped. Repair the cause of any tripped circuit breaker.

A screen, as shown in **Figure 11-63**, is provided in Cat ET for active diagnostic codes. Click the "Diagnostics" button and from the pull-down window choose "active codes." The screen will display the diagnostic codes that are currently active. Active diagnostic information will include a component identifier (CID), a failure mode identifier (FMI), and a text description of the problem.

Logged Diagnostic Codes

A screen is also provided for these codes from the pull-down window. The screen displays the diagnostic

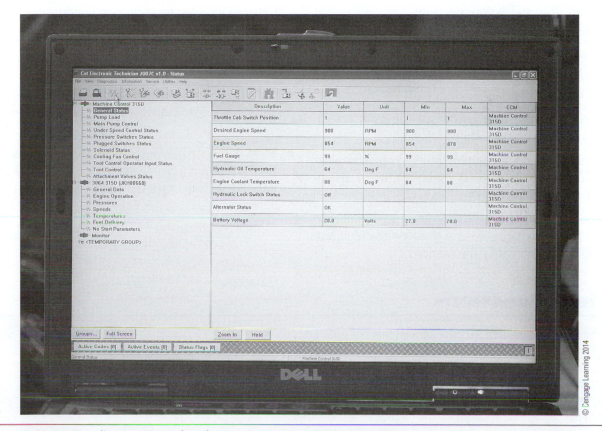

Figure 11-63 Active diagnostic codes shown on computer menu screen.

codes that are logged or intermittent. These are faults that occur from time to time or are not currently active. The logged diagnostic data includes a CID, an FMI, and a text description of the problem. Also, the logged diagnostic data includes the number of occurrences of the problem and two time stamps. The first time stamp displays the first occurrence of the problem, and the second time stamp displays the most recent occurrence of the problem.

Diagnostics are logged in non-volatile memory. On power up, the ECM will clear any diagnostic codes that have not been detected or active within the last 150 hours of machine operation.

ONLINE TASKS

1. Access a Service Information System (SIS) specific to a machine with either a planetary or a multiple countershaft powershift transmission in your shop. Locate the maintenance schedule for the transmission and the type of oil used. The SIS may require a subscription and/or password. These must be acquired by your shop or training center.

2. Access an SIS specific to a machine with either a planetary or a multiple countershaft powershift transmission. Locate the "Probable Cause" flowchart. This will give you a systematic approach to troubleshooting specific powershift transmission problems.

Shop Tasks

1. Locate a powershift transmission assembly in your shop and identify the transmission by type, planetary or countershaft, number of forward and reverse speeds, and number of countershafts, if used.

2. Locate a piece of equipment equipped with a powershift transmission. Perform an oil level check as outlined in this chapter. Refer to the OEM procedure in the appropriate manual.

3. Locate a machine equipped with a powershift transmission in your shop. With the supervision of a trained technician, perform a stall test as outlined in this chapter or by following OEM procedure. Record your findings and compare them to the OEM specifications.

Summary

- Two types of oil flow occur in a torque converter: rotary oil flow and vortex oil flow.

- A torque converter is a type of fluid coupling used to transfer engine torque to the transmission.

- A lockup torque converter can provide a direct drive to the transmission.

- A simple planetary gear set consists of a sun gear, a planetary carrier, and a ring gear.

- Depending on which component of a planetary gear set is driven, is held, or is input, torque or speed increases will occur.

- Compound planetary transmissions use combinations of planetary gear sets to produce the required gear ratios for machine applications.

- At each service interval, an oil sample should be taken to monitor transmission condition.

- A multiple countershaft transmission utilizes countershafts in constant mesh to achieve multiple gear ratios.

- The Electronic Control Module (ECM) is a computer that manages the operation of the transmission.

- Most current machines use multiplexing networks. SAE J1939-compatible hardware and software allow all ECMs with an address on the bus to network.

Review Questions

1. The three main components in a planetary gear set are

A. sun gear, ring gear, and planetary gears

B. sun gear, pinion gear, and planetary gears

C. sun gear, ring gear, and pinion gear

D. pinion gear, ring gear, and planetary gears

2. In planetary transmissions, the ring gear is applied or released by
   A. a brake band
   B. a cone-type clutch
   C. a multi-disc clutch pack
   D. None of the above. The ring gear is never held.

3. How many gear ratios can be achieved through a planetary gear set?
   A. five
   B. six
   C. seven
   D. nine

4. A stator equipped with an overrunning clutch is able to
   A. freewheel once rotary flow is achieved in the torque converter
   B. freewheel when vortex flow is achieved in the torque converter
   C. lock up once rotary flow is achieved
   D. None of the above. Stators are not equipped with overrunning clutches.

5. A lockup torque converter locks which two components?
   A. the turbine and stator
   B. the impeller and stator
   C. the impeller housing and turbine
   D. the impeller housing and output shaft

6. Vortex oil flow is:
   A. the flow of oil around the circumference of the torque converter
   B. the flow of oil across the torque converter
   C. the flow of oil due to centrifugal force
   D. the flow of oil through the impeller

7. The purpose of the stator is to:
   A. help pump the oil to the turbine
   B. redirect oil to the turbine
   C. redirect oil to the impeller
   D. assist rotary flow

8. Fluid couplings are most efficient at:
   A. high speed, no stall
   B. high speed, full stall
   C. low speed, full stall
   D. all of the above

9. A torque converter stall speed test may provide an indication of:
   A. torque converter condition
   B. engine performance condition
   C. transmission performance condition
   D. all of the above

10. If a torque converter stall speed test is slightly lower than normal, the probable cause could be:
    A. a transmission clutch pack may be slipping
    B. engine performance may be lower than normal
    C. overrunning clutch in stator is seized
    D. overrunning clutch in stator is freewheeling

11. When a torque converter stall speed test is much higher than specifications, the probable cause could be:
    A. a transmission clutch pack may be slipping
    B. engine performance may be lower than normal
    C. engine performance may be greater than normal
    D. overrunning clutch in stator is seized

12. Torque converter stall speed tests should be performed with

    A. transmission oil cold and slightly lower than normal to allow for thermal expansion

    B. transmission oil at its hottest to simulate working conditions

    C. transmission oil at operating temperature and level full

    D. the converter on a test bench

13. The purpose of the solenoid shift valves is to:

    A. relay oil pressure to designated clutches at a preset pressure

    B. develop high enough pressure to engage designated servos or clutches

    C. sense pressure regulator valve signals to determine shift points

    D. shift the selector valve to the desired position

14. Which statement best describes the electro-hydraulic valve body's operation?

    A. The valve body supplies the ECM with all the vital information needed for properly controlling shifts within the transmission.

    B. The valve body pressurizes fluid and provides cooling for hot oil.

    C. The valve allows oil flow through the valve body to the torque converter prior to the clutch pack.

    D. The valve body provides oil flow for lubrication and transmission operation through a series of solenoids controlled by the ECM.

15. Multiple countershaft transmissions differ from planetary transmissions because

    A. countershaft transmissions do not use clutch packs

    B. they are not a powershift transmission

    C. they use countershafts for multiple speed ranges

    D. they divide the input torque to the transmission

16. Multiple countershaft transmissions are limited to

    A. the number of speeds they provide

    B. the amount of torque they can withstand

    C. use with a mechanical clutch

    D. none of the above; they are as variable as planetary transmissions.

17. In a torque converter that uses a freewheeling stator, if the overrunning clutch fails and freewheels in both directions, what problem will this cause?

    A. overheating

    B. low torque output

    C. foaming oil

    D. transmission lockup and the torque converter stall

18. Which component in a torque converter always turns at engine speed?

    A. impeller

    B. turbine

    C. planetaries

    D. stator

CHAPTER

12

Hydrostatic Drive Systems and Hydraulic Retarder Systems

Learning Objectives

After reading this chapter, you should be able to:

- Describe the fundamentals of hydrostatic drive systems.
- Outline the components and construction features of hydrostatic drive systems.
- Describe the principles of operation of hydrostatic drive systems.
- Perform some recommended maintenance procedures on hydrostatic drive systems.
- Run diagnostic routines for hydrostatic drive machines.
- Outline the operation of hydraulic retarder systems.
- Describe the components and construction features of basic hydraulic retarder systems.

Key Terms

case drain	friction clutch	rotor
charge check valve	high pressure	servo control
charge pump	high-pressure relief valve	shuttle valve
charge relief valve	hydrodynamic drive	swashplate
cooling circuit	input retarder	turbine shaft
displacement control valve	low pressure	variable displacement
fixed displacement	metering	variable-displacement motor
fixed-displacement motor	output retarder	variable-displacement pump
fixed-displacement pump	output shaft	

INTRODUCTION

Hydrostatic drive systems, as shown in **Figure 12-1**, are fluid drive systems that are capable of using hydraulic fluid under pressure to convert diesel engine power to fluid power, and then convert it back to mechanical power at the wheels or tracks. This is accomplished by using a positive displacement pump and motor to transfer power through rotary motion to the drive wheels. Although there are some similarities with conventional fluid drives, such as a torque converter, the major difference is in how the power is transmitted.

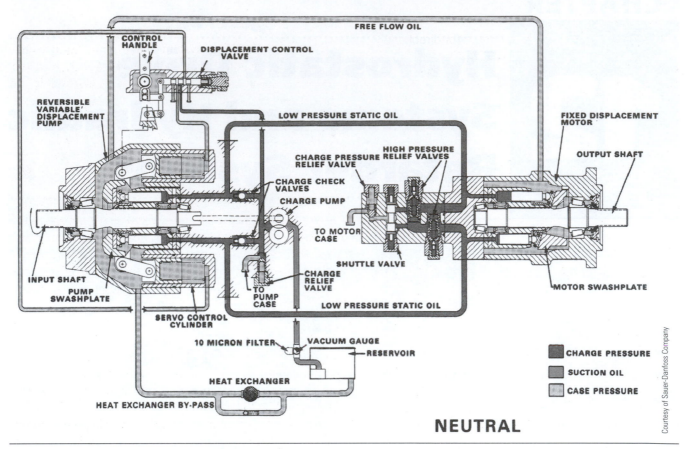

Figure 12-1 Typical layout of a hydrostatic drive system.

The difference between a hydrostatic drive system and a **hydrodynamic drive** system is the velocity of the oil flow and the pressures that are used to transmit power. In a hydrostatic drive system, oil is subjected to **high pressures** at a relatively slow speed. In a hydrodynamic drive system, oil flows at high speed and relatively **low pressure** to provide power to the wheels. In a hydrostatic drive system used on heavy equipment, the hydraulic pump, which is turned by a diesel engine, supplies the movement of oil, while the motor converts the hydraulic energy back to mechanical energy to provide drive. One of the main advantages of using this type of drive is its ability to provide infinite speed and torque range in both directions of travel.

The same rules apply to hydrostatic drive systems as to traditional mechanical drive systems but the means of delivering drive torque to the wheels or tracks differs. As with mechanical drives, in hydrostatic drivetrains the relationship between torque and speed is balanced:

- Torque is reduced as motor or drive speed increases.
- Speed is reduced as torque increases.

An advantage of a hydrostatic drive system is its ability to deliver full engine torque at the drive motor(s) when the hydraulic circuit is at or near maximum pressure. Hydrostatic drive offers considerable flexibility in component placement on a chassis. Pumps and motors may be connected or located in widely separated areas from each other. Hydrostatic drive can replace a conventional mechanical powertrain on a mobile machine, with an engine driving a hydraulic pump and motor(s) that are hydraulically looped into a circuit, as shown in **Figure 12-2**.

Hydrostatic drive provides equipment design engineers considerable flexibility when it comes to locating the drive components in equipment. Two basic design configurations are used:

- integral systems
- split systems

Integral Systems

Integral systems use a pump and motor that are mounted as a single assembly behind a diesel engine. This configuration requires no high-pressure hoses to connect the pump and motor. With this type of design,

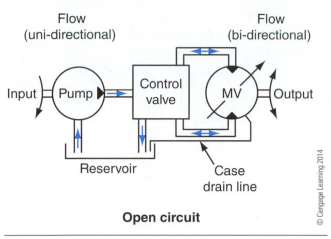

Open circuit

© Cengage Learning 2014

Figure 12-2 Schematic of an open-loop hydrostatic drive system. If the speed increases at the pump, the output power at the motor will also increase.

the output of the motor drives a conventional final drive, replacing the transmission only, as shown in **Figure 12-3**.

Split systems

Split systems may, but not necessarily, locate the hydraulic pump behind the diesel engine, and use separate motors on either side of the equipment to drive the tracks or wheels independently. Some equipment manufacturers use two pumps, one for each motor. This design requires the use of high-pressure hoses to connect the pump and motors together in a loop.

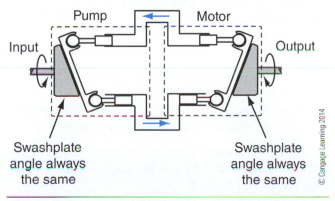

Fixed-displacement pump/ fixed-displacement motor

© Cengage Learning 2014

Figure 12-3 A fixed-displacement pump with a fixed-displacement motor. The motor has a fixed swashplate, resulting in constant torque at any given speed. The pump has a movable swashplate that can vary the speed of the pump, resulting in variable speed and constant torque.

Advantages of Hydrostatic Drive

Hydrostatic drive systems have some unique advantages over conventional drives when used to drive mobile equipment. The drive system is capable of generating a dynamic braking effect, which provides a means of slowing or stopping the vehicle, or even the ability to coast when required. In addition, hydrostatic drive can deliver 90% full torque at start-up along with providing maximum flexibility when it comes to engine placement on a machine.

The overall weight of the powertrain is significantly lower than equivalent conventional drive systems and there is increased powertrain efficiency because of reduced mass rotating parts. Finally, hydrostatic drivetrains offer infinitely variable forward and reverse speed ranges when used with variable-displacement pumps and motors.

FUNDAMENTALS

To understand how a hydrostatic drive system actually works, a basic understanding of the basic principles of hydraulics is required. A couple of important facts regarding hydraulic fluids in a confined circuit must be underlined:

- It does not compress.
- It will always adopt the shape of the component through which it flows.

There are two separate applications on mobile equipment that use hydrostatic drive technology, determined by whether the hydraulics are required to move implements or the actual machine:

- non-traction drives: used to operate implements such as mast rotators on excavators or cranes
- traction drives: used to drive the tracks or wheels

When used as motive power final drive, hydrostatic traction drive systems produce outputs that are rotary rather than reciprocating in nature.

In addition, hydrostatic drives can be categorized as either open-loop or closed-loop. Basic open-loop and closed-loop systems, as shown in **Figure 12-4**, use pumps, motors, lines, control valves, and reservoirs in their circuits. However, there is a fundamental operational difference in the way the oil is handled after it leaves the motor. In a closed-loop system, as shown in **Figure 12-5**, the oil flows from the pump to the motor and is returned directly to the inlet of the pump. In an open-loop system, the hydraulic oil flows

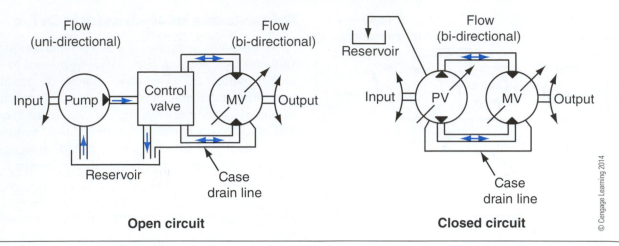

Open circuit   **Closed circuit**

© Cengage Learning 2014

Figure 12-4 Open- and closed-loop hydrostatic drive circuits.

**Variable-displacement pump/
fixed-displacement motor**

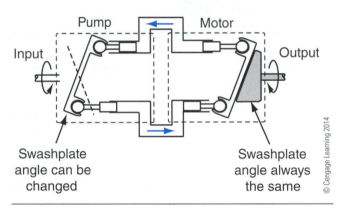

© Cengage Learning 2014

Figure 12-5 A variable-displacement pump with a fixed-displacement motor. The motor has a fixed swashplate, which results in constant torque at any given speed. The pump has a movable swashplate that can vary the speed of the pump, resulting in variable speed and constant torque.

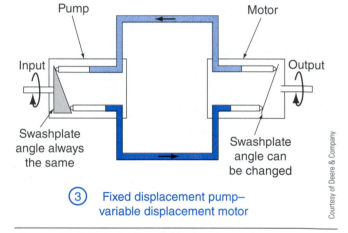

③   Fixed displacement pump–
     variable displacement motor

Courtesy of Deere & Company

Figure 12-6 How a fixed-displacement pump may be used with a variable-displacement motor. The design is more complex because the effective stroke must be controlled.

from the pump to the motor and then back to the hydraulic tank.

Depending on equipment design, conventional hydrostatic drive systems typically consist of the following drive configurations:

- A fixed- or a variable-displacement pump, with one or more fixed- or variable-displacement motors.
- Combination of a **fixed-displacement pump** with a **fixed-displacement motor**.
- Combination of a fixed-displacement pump with a variable-displacement motor, as shown in **Figure 12-6**.

- A **variable-displacement pump** to drive a fixed-displacement motor.
- Some manufacturers use a variable-displacement pump with a **variable-displacement motor**.

These components can be combined in many different configurations to suit the application in which they are to be installed.

Torque range in a hydrostatic drive system can be explained by comparing what takes place in a conventional equipment powertrain to what takes place in a hydrostatic drive loader. For example, we can compare the two different drive torque capabilities. Let's look at a hypothetical four-speed standard transmission with a 4:1 ratio in first gear, 3:1 ratio in second gear, 2:1 ratio in third gear, and a 1:1 ratio in fourth gear. Now take a look at a typical hydrostatic drive loader with a variable-displacement pump with a flow rate of 4 to 10 cubic inches per revolution and a variable-displacement motor with a maximum 10 to 40 cubic inches per revolution displacement. The ratio can be calculated as follows:

$$\frac{\text{Motor at maximum displacement}}{\text{Pump at minimum displacement}} = \frac{40}{4}$$

$$= 10:1 \text{ torque multiplication}$$

$$\frac{\text{Motor at minimum displacement}}{\text{Pump at maximum displacement}} = \frac{10}{10}$$

$$= 1:1 \text{ torque multiplication}$$

For maximum torque we divided the motor maximum output by the pump minimum output; we got a 10-to-1 ratio. For minimum torque we divided the minimum motor flow rate of 10 cubic inches per revolution by the maximum pump displacement of 10 cubic inches per revolution for a 1-to-1 ratio. When the two values are compared the result is a torque ratio of 10-to-1.

Heavy off-road equipment generally requires a variable range of torque for each type of work to be performed. A wide torque range can be obtained by using a variable-displacement pump and motor. The torque range is always based on the ratio between the pump and motor displacement. In a hydrostatic drive system that consisted of a fixed-displacement motor and pump, the torque could be varied only by changes in the system pressure.

Design Features of Hydrostatic Drive Systems

We know that pump or motor displacement can be expressed by the volume of oil that is displaced during one complete revolution. We have seen how pumps and motors with fixed or variable displacements can be combined to give the output speed and torque that are required on equipment. Although hydrostatic drive systems are considered efficient, they are subject to speed and torque losses that are inherent to their design. Some internal leakage in pump and motor circuits is built into the design and is required for lubrication and cooling.

Unintentional reductions in efficiency increase when the equipment operates at low speeds and high pressure. As the operating speed of the equipment increases, the internal leakage stays relatively the same, increasing the efficiency of the system. Some torque in a hydrostatic drive circuit is lost to friction as well as to high speed and pressure in a circuit.

Basic Hydrostatic Drive System Components

Charge Pump. The external gear pump is the most common type of pump used to charge a hydrostatic drive circuit. This pump design is ideally suited to this application because of its simplicity and low cost when compared to other pump designs. Variable displacement or high pressure is not required in the charge circuit of a hydrostatic drive, making the fixed-displacement pump a candidate well suited for this application. The purpose of a **charge pump** is to fill and make up oil to the main system as well as to provide oil flow for the lubrication and cooling of a typical hydrostatic drive circuit. In larger hydrostatic drive units it is also responsible for supplying oil to the displacement control valves.

Main Pump. The main pump generally used in a hydrostatic drive circuit is a positive displacement pump of either a variable- or a fixed-displacement design. Whether the circuit is a closed- or open-loop circuit dictates whether a fixed- or variable-displacement pump is used.

Factors like pressure and flow rating are considerations when determining what type of pump is used. If the hydrostatic drive circuit is used with a diesel engine as the power source to drive the pump, there are more options. For example, if the pump is to be driven at a constant speed, then a fixed-displacement pump with an open-center loop design is generally used. This combination allows the oil to flow back to the tank when in the neutral mode, preventing unnecessary heat build-up in the circuit.

The use of a variable-displacement pump (the most efficient pump available) in a hydrostatic drive circuit is determined by various factors; one of them is the operating speed. If the speed of the equipment will vary greatly and requires the use of hydraulic power at different operating speeds, a variable-displacement pump is used. A variable-displacement pump can decrease or increase the flow of oil in a circuit, with no change in engine speed, just by varying the displacement of the pump.

Drive Motor. The drive motor or motors selected for use in a hydrostatic drive circuit can be of either a variable- or a fixed-displacement design. Whether the circuit is a closed- or an open-loop circuit determines whether a fixed- or a variable-displacement motor is used. Factors such as torque and speed must also be considered. When a diesel engine is the power source, the options increase. A motor in a hydrostatic drive circuit is propelled by the flow of fluid delivered to it by the main pump. The displacement of the motor defines how much flow is required for a specific drive speed. To determine how much load the motor can work with, the torque and pressure rating of the motor have to be reckoned.

Filter. Contamination of the hydraulic oil must be prevented because it can lead to premature component failure. Oil supplied to the charge pump of a hydrostatic drive circuit must be pulled through a full flow filter with no bypass, and the filter must be capable of entrapping particles as small as 10 microns in size. The cleanliness of the oil used in this drive circuit is critical. It must be free from all contamination because the pumps, valves, and motors used in this circuit will be damaged by dirty oil. Flow through a filter is reduced as it becomes clogged, slowing circuit response, although it will not generally damage a system in the short term. Both the flow and entrapment ratings of filters must be considered when replacing them. Using filters with lower flow ratings than the OEM specification will result in poor system performance and/or filter collapse, with consequences of serious system damage.

Heat Exchanger. The use of a heat exchanger is a requirement if the operating temperature of the hydraulic oil will exceed 180°F (82°C) at the motor **case drain**. The heat exchanger should be located in the return circuit of the drive circuit. The size of the heat exchanger is determined by the duty cycle of the equipment. To determine what size the heat exchanger should be, calculate the capacity so that it can dissipate approximately 20% to 25% of the input heat generated by the circuit. The total system restriction should never result in a case pressure higher than 40 psi at normal operating temperature. A heat exchanger must have a bypass built into the return circuit to prevent back pressure from exceeding the maximum limit under any operating conditions that may be encountered, such as when the drive oil is cold.

Reservoir. The reservoir size needs to be calculated carefully; the minimum volume of the tank should be half of the charge pump volume measured in gallons per minute (gpm). This volume will ensure that the minimum fluid dwell time is at least 30 seconds at the maximum return flow in the circuit. The suction line outlet on the tank should be located away from the base of the tank to prevent any large particles from entering the pump inlet. A fine mesh screen (100 microns) should be located on the outlet port of the tank to prevent debris from entering the pump. The oil return line to the tank should be positioned so that the fluid has an opportunity to be properly deaerated. To facilitate a suction filter replacement, a shutoff valve should be located at the tank outlet so that the oil flow can be shut off when replacing the filter. A reservoir drain plug or drain valve should be placed in the bottom of the tank to enable oil draining should the system become contaminated.

Hydraulic Lines and Fluids. The guidelines for recommended hydraulic fluids provided by the OEM must be adhered to. Examples of service intervals will be shown for average conditions. When the equipment is operated in conditions that can be considered extreme service, the service life of the oil and filter must be monitored. Selection of hydraulic lines should also be in keeping with equipment manufacturer's recommendations for length, diameter, and pressure capabilities. Consider the bend radii of hydraulic lines that may be exposed to continual flexing during operation, keeping in mind that hoses are also rated for the number of bend cycles that they can sustain during operation.

The type of hydraulic oil used in these systems can be detrimental to the life of the seals. Modern seal materials may include synthetics. It is important that a seal made from synthetic material be properly pre-lubricated because most synthetic seal materials do not absorb oil. (Most types of conventional seal material will absorb some oil.) Some seal manufacturers recommend that you soak the seal in oil before installation. It should be noted that under no circumstances should seals be installed without pre-lubrication.

REVERSE FLOW CONTROL IN HYDROSTATIC DRIVE SYSTEMS

In order for a hydrostatic drive system to be practical in a heavy equipment application, it is necessary to be able to reverse the direction of travel without additional geartrain components being added, as shown in **Figure 12-7**. For this reason, a pump capable of reversing or stopping the flow in a circuit, without changing its direction of rotation, must be used. Reversing or stopping the flow of the pump can be

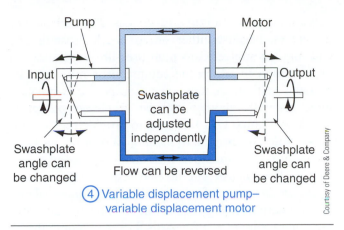

Courtesy of Deere & Company

Figure 12-7 A variable-displacement pump may be used with a variable-displacement motor. The system is much more complex as both the pump and the motor stroke must be controlled.

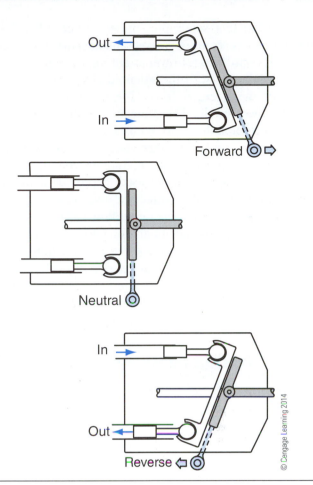

© Cengage Learning 2014

Figure 12-8 The servo piston connection to the reversible swashplate. A change in the piston position alters the swashplate angle, which varies the amount of oil delivered to the motor.

accomplished by moving the **swashplate** of a variable-displacement pump. Placing the swashplate in a vertical position, where the pistons can no longer reciprocate, stops oil from exiting the pump, depowering the hydraulic motor it feeds. This action can be compared to placing a mechanical transmission in neutral. In order to reverse the direction of travel, the swashplate angle must be reversed. This changes the direction of flow, meaning that the inlet of the pump now becomes its outlet, and vice versa. The result is that rotation of the motor is reversed.

DISPLACEMENT CONTROL OF HYDROSTATIC DRIVE SYSTEMS

When using a variable-displacement pump or motor, a displacement control system is used to control the stroke length of the pistons, as shown in **Figure 12-8**. Such a system could be as simple as a manual control, or it could be more complex and use a compensator control circuit. On mobile equipment a **servo control** can be used to change the swashplate angle; this is accomplished by adjusting the stroke length. The displacement controller that is used to vary the swashplate angle can be built into the pump housing or may be remote mounted. The control circuit uses hydraulic pressure to adjust the angle of the swashplate. As the swashplate angle is increased, the stroke is increased, increasing oil flow from the pump without an increase in the engine speed.

One variation of displacement control used by Sundstrand on their Series 90 pump is a manual displacement control that uses movement of a control lever to shift a spring-centered four-way servo valve,

which sends control pressure to either side of a dual-acting servo piston. The servo piston changes the swashplate angle, thus increasing the piston stroke and the output displacement of the pump. This change in flow provides a smooth increase in output to a variable-displacement pump.

Newer hydrostatic drive equipment uses a variation of hydraulic displacement control. This control system uses a pressure-control pilot (PCP) valve to manage the input signal pressure. The PCP valve uses a 12-volt signal to operate a spring-centered four-way servo valve that directs hydraulic oil to either side of the dual-acting servo valve. The cradle swashplate control is designed so that the angular position of the swashplate is proportional to the electric direct current (DC) input.

Some equipment manufacturers use another type of control system, which works something like a conventional automatic transmission. The motor control system uses a 12- or 24-volt electrically operated spool valve that sends pressure to either side of the **displacement control valve**, acting like a conventional forward/neutral/reverse automatic transmission control. When one of the solenoids for forward or reverse is energized, it forces the pump to go to full displacement mode. These electric control valves are factory preset and are not adjustable.

The term *stepless* is often used to describe the operation of a hydrostatic drive. Another common variation in flow control can be used to provide a *stepped* drive. Some manufacturers use two charge pumps mounted in tandem to increase oil flow capabilities. One pump is used to provide initial torque to get the machine moving; as the speed of the equipment starts to increase, the pressure in the system is allowed to drop off, starting a second pump that increases circuit flow. The result is to simulate a change in gear ratio similar to that of a shift in a conventional mechaninal transmission.

Circuit Design Features

To protect a hydrostatic drive circuit from dangerous overpressure conditions, an overpressure relief valve, as shown in **Figure 12-9**, is used in each side of the circuit, because either side of the hydrostatic circuit may be subjected to high pressure depending on the direction in which the equipment is moving. The pressure setting on these safety valves can be very high, generally about 5,000 psi. A charge pump, as shown in **Figure 12-10**, is used to provide oil flow for the **cooling circuit**, and to maintain a positive pressure on the low-pressure side of the main pump and motor circuit. The control circuit also relies on the charge pump for its supply of oil, as well as providing makeup oil for internal leakage in the circuit. The charge pump used in this circuit is a fixed-displacement pump that is generally mounted to the main pump and is often driven at the same speed.

Cooling Circuit.   As with all types of hydraulic circuits, heat is generated any time oil is circulated under pressure. This necessitates that all hydrostatic drive circuits provide a means to cool the hydraulic oil. The charge pump pulls filtered hydraulic oil from the reservoir and sends it to the main pump, as shown in **Figure 12-11**.

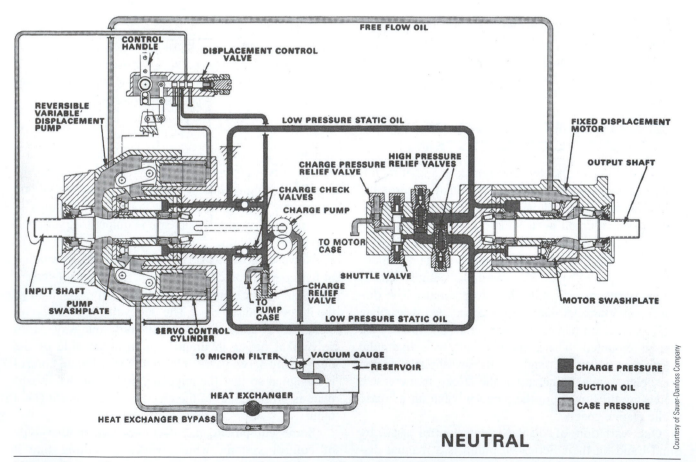

Figure 12-9 In this motor circuit, two high-pressure relief valves as well as the charge pressure relief valve are shown.

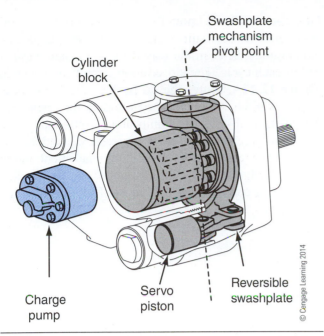

Swashplate mechanism pivot point

Cylinder block

Charge pump

Servo piston

Reversible swashplate

© Cengage Learning 2014

Figure 12-10 A variable-displacement pump that has the charge pump mounted to the back of the housing. The charge pump is driven at the same speed as the main pump.

Note: Some component manufacturers build in an internal filter element for the main pump. The oil must pass through this filter before entering the pump. Hydrostatic drive circuits cannot tolerate any contaminants in the circuit. Abrasives circulate within the circuit and can rapidly wear out the pump and motor.

When pressure in the charge pump circuit reaches its relief pressure, a check valve opens to allow the excess oil to enter the main pump housing to cool and lubricate the internal pump parts. The case drain on the main pump allows the coolant to return to the main reservoir after passing through a cooler to remove excess heat. The oil cooling circuit used in hydrostatic drive circuits has a bypass valve located in the cooler that works on pressure differential. If the pressure on the inlet side of the cooler becomes greater than that on the outlet side of the cooling circuit, the bypass valve allows the oil to bypass the

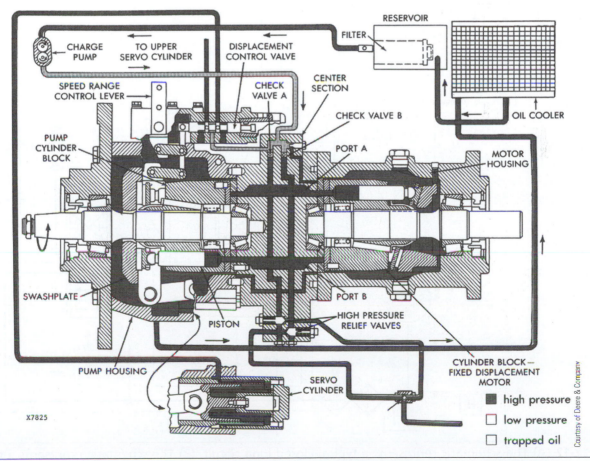

Figure 12-11 In neutral, excess charge oil entering the pump case is routed back to the reservoir.

cooler, protecting it from overpressure. When the equipment is operated in the pump neutral condition, oil from the charge pump is directed to the cooling circuit by the neutral **charge relief valve** and only flows through the pump case, not to the motor case.

Note: The terms *open loop* and *closed loop* are often used differently by manufacturers.

Closed-Loop Open-Loop Design Differences. Although these two circuit designs each use pumps and motors, they differ in the way the oil is handled at the end of each cycle. The closed-loop circuit, as shown in **Figure 12-12A**, routes the oil back to the pump inlet after it is discharged from the motor. In the open-loop design, the oil exits the pump and is returned to the hydraulic reservoir before being reintroduced to the circuit. The closed-loop variable-displacement pump circuit in **Figure 12-12** shows how the oil is handled after it is discharged from the pump. The oil is directed

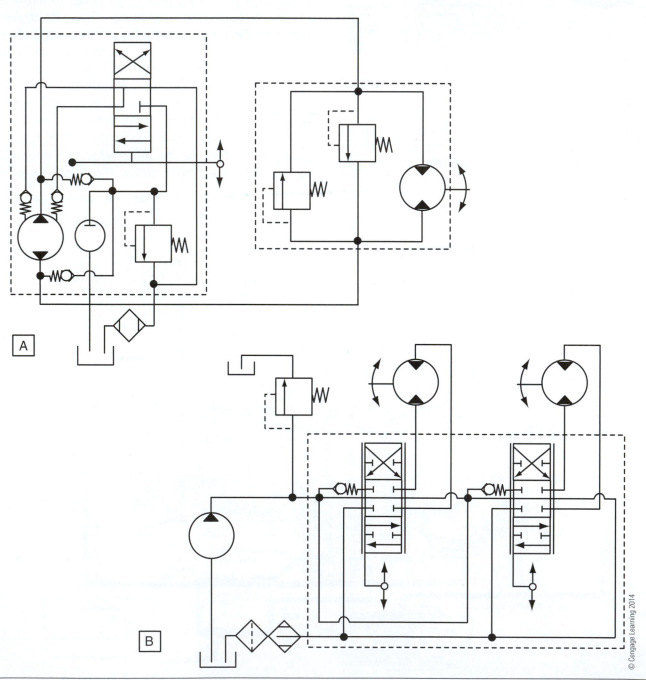

© Cengage Learning 2014

Figure 12-12 (A) The schematic represents a typical closed-loop circuit. (B) The schematic represents a typical open-loop circuit.

through the motor and then returned to the pump. In the open-loop circuit, as shown in **Figure 12-12B**, the oil leaves the pump, travels through a control circuit before being directed through the pump, and then returns to the tank. The charge pump picks up its oil directly from the hydraulic tank, unlike the closed-loop circuit where the return oil for the motor is returned to the pump. The variable-displacement closed-loop circuit design is considered the more efficient of the two circuits because the charge pump replenishes the high-pressure circuit with makeup oil and provides additional oil flow for the cooling and filtration circuit. To achieve reverse direction with this circuit design, the variable-displacement pump must be an over-center design; this allows the swashplate to move off center in both directions to provide reverse capability.

The open-loop hydrostatic drive circuit is often used where two or more different loads are driven simultaneously by either a fixed- or a variable-displacement pump and with the motor returning the hydraulic oil to the tank. Because this circuit design is often used with two or more loads, it must use a control valve to provide direction and **metering** control. Often this circuit will use a fixed-displacement pump, as it is more cost-effective to incorporate in small equipment design.

As with any circuit design, there are always disadvantages to go along with the advantages. Whenever more than one actuator is used in the circuit at the same time, load interference may become an issue in the circuit, as well as the loss of dynamic braking effect. When operating equipment that uses an open-loop circuit design, the operator must vary the engine power to provide a change in speed.

Operation of a Variable-Displacement Pump/Fixed-Displacement Motor Drive in Neutral

The variable-displacement pump/fixed-displacement motor combination offers infinite speed and direction control. The equipment operator has complete control of speed and direction by using one controller for both speed and direction. A diesel engine provides the power source for the pump. In **Figure 12-13** the engine is

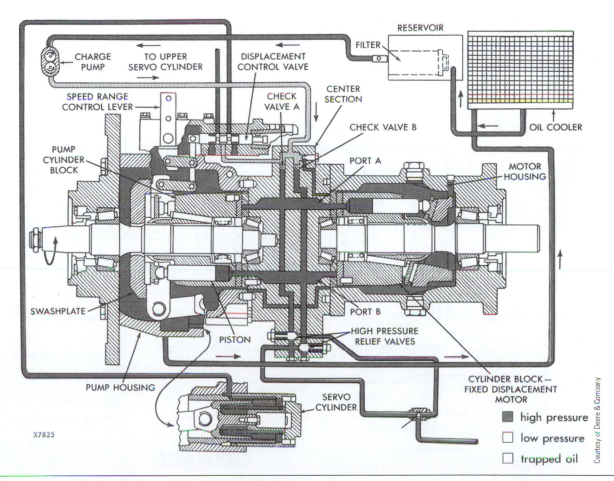

Figure 12-13 Typical hydrostatic drive circuit.

running and the control lever is in the neutral position. The displacement control valve is also in neutral position, allowing the servo control ports to send oil flow to the sump. The charge pressure exerts equal pressure on the pump pistons, forcing the swashplate into neutral.

The charge pump pulls oil from the hydraulic reservoir through a 10-micron filter. Hydraulic oil is directed to the low-pressure side of the main pump and circulates through the pump and motor housing and back through the heat exchanger before going back to the hydraulic tank. Trapped oil is held in the pump and motor block, as well as in the connecting lines, by the two **charge check valves**. Because the control lever is in the neutral position in the pump, the swashplate has de-stroked the pistons in the variable-displacement pump. With the engine running and the pump turning, no oil flows to the motor. However, an oil flow is maintained at low pressure for the control circuit and makeup oil. This compensates for any internal leakage in the pump and motor.

During operation in neutral, the **shuttle valve** in the motor remains centered and allows the charge relief valve to direct the excess oil to flow to the heat exchanger. When operating in this state, the cooling oil flows through the pump but not the motor. The heat

exchanger has a bypass, which protects the heat exchanger from high back pressure when operating with cold oil or if the heat exchanger should become plugged internally.

The **high-pressure relief valves**, located in the manifold valve assembly, protect the main circuit from high surge pressures in both main hydraulic lines during periods of quick acceleration. These high-pressure relief valves also protect the main high-pressure circuit from over-pressurization due to emergency braking or sudden changes to the load while the equipment is operational.

The shuttle valve is also located in the manifold valve assembly. Its purpose is to provide a circuit connection between either of the two main pressure lines and the charge pressure relief valve when at low pressure.

Displacement Control Valve Operation. In **Figure 12-14**, if we place the control lever in the forward drive position, the displacement control valve is moved to the forward position, allowing the control port to flow oil to the servo piston in the motor. This forces the swashplate against the reverse servo spring to a given position. After the swashplate has moved to the required angle, feedback oil is sent from the

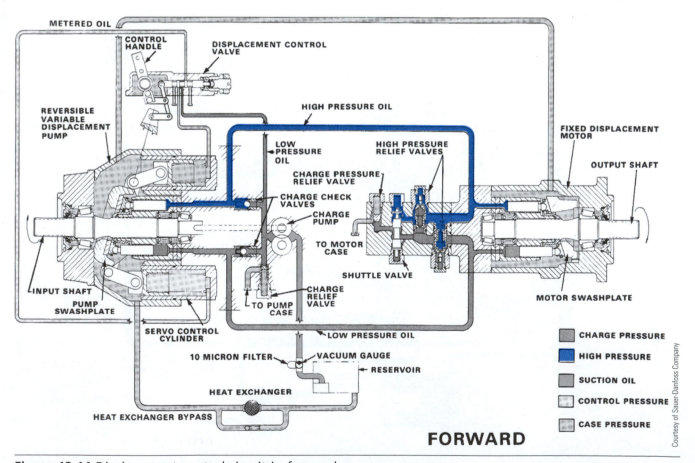

Figure 12-14 Displacement control circuit in forward.

forward servo piston in the pump back to the displacement control valve, thus shifting the spool back and leaving just enough oil flow to the forward servo cylinder to maintain the selected swashplate angle.

Operation of a Variable-Displacement Pump/Fixed-Displacement Motor Drive in Forward

Figure 12-15, a variable-displacement pump/fixed-displacement motor, shows the charge pump pulling oil from the hydraulic reservoir through a 10-micron filter. When the operator shifts the machine into the forward direction, control pressure oil is directed to the forward servo piston by the displacement control valve. This moves the swashplate off center, forcing the pump pistons to create oil flow in the circuit. The distance that the pistons reciprocate in the pump depends on the swashplate angle, which in turn determines the volume of oil displaced by the pump. High-pressure oil from the pump is sent to the fixed-displacement motor inlet, forcing the pistons in the motor to rotate in the opposite direction and giving forward direction. Note that the direction of rotation of the pump is always clockwise as viewed from

the flywheel end while motor rotation is dependent on the direction of travel selected by the operator.

During operation in forward direction, the motor sends the oil from the outlet side of the motor back to the inlet of the pump. Oil flows in a continuous loop from the pump to the motor; this is where the term *closed loop* comes from. Excess oil that bypasses the pistons in the motor for lubrication purposes can still flow back to the pump case and out through the heat exchanger to the hydraulic tank. The charge pump will make up any oil that is lost to internal leakage in the circuit. If the system pressure is exceeded while operating in forward, the high-pressure relief valve in the motor opens, allowing the oil to bypass the cylinder block assembly on the motor and return to the pump.

Operation of a Variable-Displacement Pump/Fixed-Displacement Motor Drive in Reverse

Figure 12-16 shows the charge pump pulling oil from the hydraulic reservoir through a 10-micron filter. When the operator shifts the equipment into reverse mode, control pressure oil is sent to the reverse servo piston by the displacement control valve, forcing the

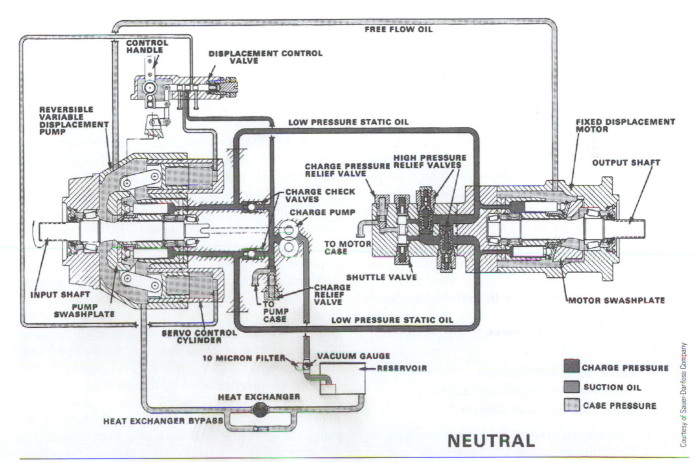

Figure 12-15 Location of 10-micron filter.

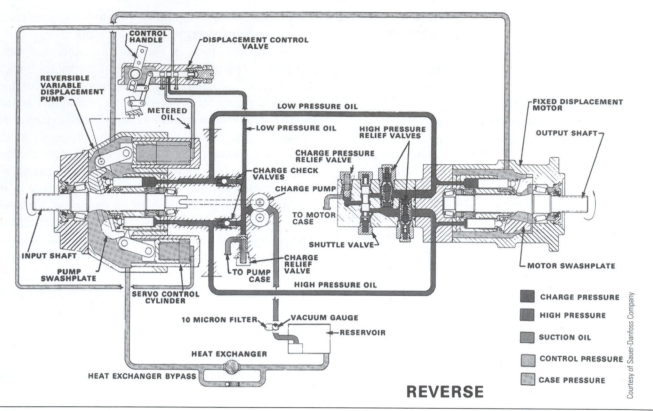

Figure 12-16 Closed-loop hydrostatic drive in reverse.

swashplate in the pump to shift off center in the opposite direction and forcing the pump pistons to create oil flow in the circuit in the opposite direction. The distance that the pistons reciprocate in the pump depends on the swashplate angle, which determines the volume of oil displaced by the pump. The high-pressure oil is sent to the fixed-displacement reverse motor inlet, forcing the motor to rotate in the reverse direction.

During operation in reverse, oil flow is reversed and the motor sends oil flow back to the opposite side of the pump. Oil flows in a continuous loop from the pump to the motor. Excess oil that bypasses the pistons in the motor for lubrication purposes can flow back to the pump case and out through the heat exchanger to the hydraulic tank. The charge pump will make up any oil that is lost to internal leakage in the circuit. If the regulated system pressure is exceeded while operating in forward, the high-pressure relief valve in the motor opens, allowing the oil to bypass the cylinder block assembly on the motor and return to the pump.

Operation of a Variable-Displacement Pump/Variable-Displacement Motor Drive

This combination of pump and motor provides the greatest flexibility in torque and horsepower combinations. The variable pump/motor combination can customize the power and torque requirements for different applications. In **Figure 12-17** the charge pump pulls oil from the hydraulic reservoir through a 10-micron filter. When the operator shifts the equipment into the forward direction, control pressure oil is directed to the forward servo piston by the pump displacement control valve and to the motor control valve. This moves the swashplate off center and forces the pump pistons to create oil flow in the circuit. The result depends on the swashplate angle in each, which determines the volume of oil displaced by the variable-displacement pump and motor. The high-pressure oil from the pump is directed to the variable-displacement motor inlet, causing the pistons in the motor to rotate in the opposite direction and giving forward direction.

Having the flexibility to vary the swashplate angle in the motor provides the ability to increase or decrease the speed and torque at the **output shaft** of the motor. Note that the direction of rotation of the pump is always clockwise as viewed from the flywheel end; the direction of the motor rotation is dependent on the direction of travel selected by the operator.

During operation in forward, the motor sends the oil from the outlet side of the motor to the inlet of the pump. Oil flows in a continuous loop from the pump to the motor and back. Excess oil that bypasses the pistons in the motor for lubrication purposes will still flow back

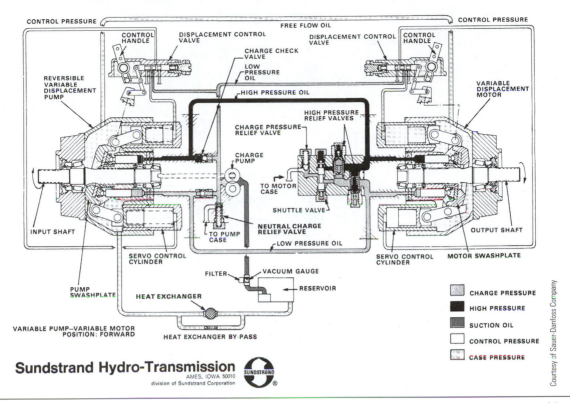

Figure 12-17 The closed-loop variable-displacement pump and variable-displacement motor combination.

to the pump case and out through the heat exchanger back to the hydraulic tank. The charge pump will make up any oil that is lost to internal leakage in the circuit. If the system pressure is exceeded while operating in forward, the high-pressure relief valve in the motor opens, allowing the oil to bypass the cylinder block assembly on the motor and return to the pump.

Operation of a Variable-Displacement Pump/Fixed-Displacement Motor Drive with Pressure Override

Figure 12-18 shows a pressure override control with a variable-displacement motor driving a fixed-displacement pump. The circuit drive pressure will normally remain below override pressure during normal operation. During this phase of operation, the pressure override valve is in the open position and does not affect the operation of the circuit. During operating conditions where the load exceeds maximum torque available, the pressure override valve takes over the normal displacement control operation. When this occurs, the pressure override control takes over the displacement control of the pump.

In **Figure 12-19**, the pressure override control valve takes control whenever the load requires a

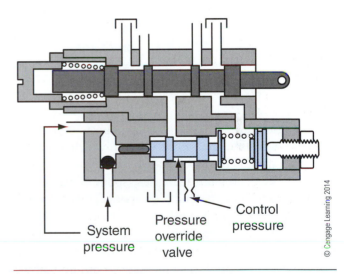

Figure 12-18 Pressure override used on a variable-displacement pump with a fixed-displacement motor.

system pressure that is higher than override pressure. This valve is a three-way valve that directs charge pressure to the displacement control valve whenever the displacement control valve is in the normally open position. In operation, this valve allows the displacement control pressure to be overridden at a given pressure. System pressure is monitored in both sides of the loop by the override control valve.

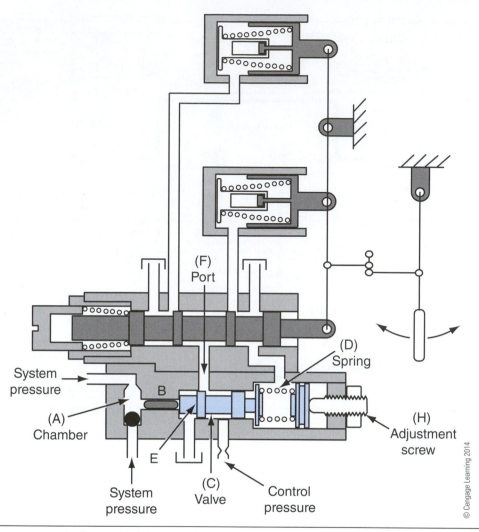

Figure 12-19 Pressure override control valve operation.

In **Figure 12-19**, the higher pressure is ported into chamber A, shifting the ball check over, which forces pin B to push the control spool against an internal spring D. As the system pressure increases, spool C pushes the spring until the control land E on spool C blocks port F. Whenever land E on control spool C blocks port F, the override pressure is in control of the displacement of the pump. The pressure in the override circuit will modulate to maintain servo pressure to ensure that the displacement will always equal a system pressure that is equivalent to override pressure. Adjustments can be made to this valve by changing the spring tension D with the adjustment screw H.

OPEN-LOOP HYDROSTATIC DRIVE SYSTEMS

An open-loop hydrostatic drive circuit, as shown in **Figure 12-20** (some manufacturers refer to it as a hydraulic drive), may be used in heavy equipment where

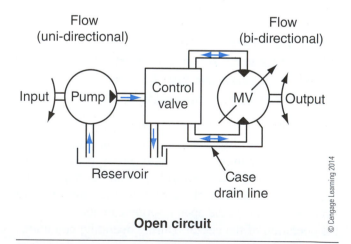

Figure 12-20 Basic open-circuit hydrostatic drive.

two or more different drive functions or loads are required. This type of drive system may be equipped with a fixed-displacement pump and may use either a fixed- or variable-displacement motor. The motor returns the hydraulic oil to the tank before it is picked up again by the pump. Often chosen for multiple-loads use, this design needs a control valve to provide direction and metering control.

Open Circuit Using a Bleed-Off Control

The open circuit design, as shown in **Figure 12-21**, may be divided into three different categories: bleed-off, meter-in, and meter-out control. Bleed-off control uses a variable-flow control valve between the pump and the motor to allow oil flow back to the tank to control the motor speed. This variation of control will not result in a loss of torque because no pressure drop occurs in this part of the circuit. Motor speed is determined by how much oil is allowed to be bled from the circuit back to tank. This design has one major advantage: The amount of torque required to drive this system is reduced due to the circuit's ability to automatically adjust the pressure to the load.

Open Circuit Using a Meter-In Control

When a meter-in control is used, a variable-flow control valve, as shown in **Figure 12-22**, must be placed between the pump and the motor to provide excellent control of the motor output speed. Due to the placement of the variable-flow control valve in this part of the circuit, we experience a pressure drop in this circuit. A corresponding loss of drive power takes place that is dependent on the pressure drop in the circuit. To compensate for this disadvantage, the circuit can be designed to use a flow-control valve that has a built-in relief valve.

Open Circuit Using a Meter-Out Control

The last variation in this type of drive control is the meter-out design, as shown in **Figure 12-23**. It uses a

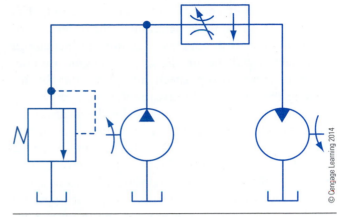

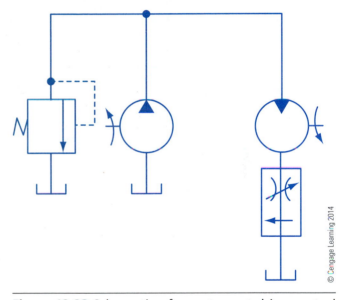

Figure 12-22 Schematic of a meter-in drive circuit.

Figure 12-23 Schematic of a meter-out drive control circuit.

flow-control valve in the return line of the circuit to achieve variable motor drive speed. As with the meter-in design, the placement of the variable-flow control valve creates a pressure drop in this circuit, thus creating a corresponding loss of drive power that is dependent on the pressure drop in the circuit.

Open Circuit with One Pump and Two Drive Motors

In this design, the pump and motors are not connected directly together; they have a directional control valve in between them. With the engine running and the control valves in the neutral position, the fluid flows from the pump through the control valves and back to the tank. When the operator activates the drive motors, the shift levers are shifted into position, allowing oil to flow through the control valves to drive

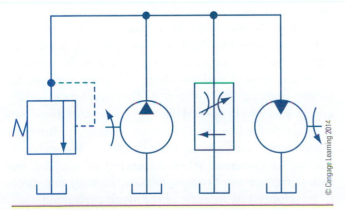

Figure 12-21 Bleed-off drive control circuit.

the motors simultaneously in the same direction. The direction of fluid flow determines the direction of motor rotation. If the operator requires an increase in equipment speed, he or she applies engine throttle, which increases pump speed and flow, thereby increasing the speed of the drive motors.

System pressure is determined by the load on the equipment. In drive mode, oil flow from the motors flows through the control valves back to the hydraulic tank. A relief valve is in the circuit to prevent overpressure in the circuit during operation. If this design is used on independent wheel or track drives, the operator can drive one motor in forward and one in reverse, allowing the unit to rotate without any forward or reverse movement. The operator can also choose to rotate one drive at a time, if required, for precise equipment control.

HYDROSTATIC DRIVE SYSTEM MAINTENANCE

Tech Tip: Before performing any maintenance to hydrostatic drive equipment, take note of any potential hazards that may expose you to the possibility of injury or equipment damage. If work needs to be performed under a piece of equipment, for example, under a boom or a part that is supported by hydraulics, it must be blocked mechanically. No one should ever attempt to work under anything that is not supported by a mechanical support. The equipment must be secured with the appropriate wheel chocks to prevent movement if the equipment were inadvertently placed in a condition where freewheeling could occur. Always be aware of the possibility of getting caught up in a pinch point when working with moving or rotating parts.

Before loosening or removing any hoses or components on a hydrostatic drive circuit, ensure that the system pressures have been relieved and that a minimum amount of oil is lost. Some system components may have heavy springs that have stored energy in them. These springs need to be properly removed with the correct tools to prevent component damage or personal injury. A complete review of the OEM service literature for each particular task must be done before work on a job can proceed. Good preventative maintenance practices are critical to

maintaining a low-cost service life on hydrostatic drive equipment.

Movement of oil from one component to another must be performed with as little internal leakage as possible. Excessive internal leakage lowers the performance capabilities of the equipment. When the components operate as designed, there should be just enough internal leakage to provide lubrication inside the pump and motor. Oil that is lost to internal leakage flows back through a filter and cooler before being reintroduced to the main circuit. The components in the hydrostatic drive system can be damaged quickly from heat created by excessive internal leakage. Built-in internal leakage in this drive circuit prevents the volumetric efficiency of a hydrostatic drive system from reaching 100%.

Consider the type of oil that we are going to use; the wrong grade of oil or oil that is contaminated with dirt quickly increases the internal leakage of the components, and causes a dramatic loss of efficiency in the equipment's operating capabilities. One of the effects of excessive internal leakage is higher-than-normal hydraulic oil operating temperature, which may lead to an accelerated rate of oil breakdown. The natural tendency of any hydraulic oil that is exposed to excessive temperatures is to lose its viscosity. To prevent metal-to-metal contact, a hydraulic oil film must be maintained with a minimum viscosity of 47 SUS at 210°F (99°C). Choose the hydraulic fluid carefully; not all hydraulic fluids meet the requirements of equipment manufacturer's specifications. Using the equipment over time with incompatible oil may cause internal wear in the components, which increases internal leakage above normal levels. Oil that has low viscosity will leak more than an oil with a higher viscosity, so using oil with the right viscosity index is vital to long system component life.

Keeping dirt out of the hydrostatic circuits and components is easier said than done. When hydraulic oil and filters are scheduled to be replaced, ensure that the external areas around the parts that are going to be disassembled are properly cleaned before removal. Whenever possible, use a high-pressure wash on equipment to thoroughly clean the area to be worked on and to ensure that no water can enter the system before opening the system. If the hydrostatic drive circuit has to be opened for any reason, it must be properly primed to prevent serious damage to internal components. This damage can start in "low oil" or "no oil" conditions. When equipment is serviced, be sure to replace any O-rings, gaskets, or seals that have to be removed as a result of the work performed.

System Maintenance

Oil replacement in a hydrostatic drive system is dependent on a number of operating parameters. In general, the following guidelines should be considered:

- If the equipment uses a sealed hydraulic tank, the recommended oil change interval can be as long as 2,000 hours of operation. This recommended interval may shorten to as little as 500 hours if the system has a vent to atmosphere (open reservoir), such as an air filter on the filler cap. When considering the oil change intervals for equipment that you are going to service, always consult the equipment manufacturer's specifications.
- The maintenance intervals that are developed by equipment owners are often a result of operating experiences and may differ somewhat from the manufacturer's recommendations. An example of this is in the mining industry, where operating conditions would be considered extremely severe compared to the same equipment operating in a logging operation. Service intervals in an underground mining environment necessitate intervals that are often only 200 hours apart.
- Ten-micron filters on hydrostatic drive inlet systems are recommended on most systems.
- The level of oil in the reservoir should be checked daily by the equipment operator.
- Fluid added to the hydraulic tank of a hydrostatic drive circuit should be filtered before it is poured into the tank, and the hydraulic lines and fittings need to be checked daily for leakage and abrasion.
- The heat exchanger should be cleaned externally at a regular interval to maintain safe oil operating temperatures.

Start-Up Procedures after Service. The procedures outlined here are intended for use as general examples of steps to follow; they should not be used for actual start-up procedures on any particular drive system. Always refer to the OEM service literature for the equipment that you are working on for the proper procedures and specifications.

Tech Tip: Before starting hydrostatic drive equipment after servicing or repairing, we need to become familiar with a few general safety precautions. In many cases, when equipment needs to be serviced or repaired, it is not always necessary to remove the parts for repair.

Access to the area being serviced should be high-pressure washed before work begins and after it is completed to minimize the chance of contamination and to prevent potential slipping conditions. Washing also minimizes fire hazards from oil on or near hot surfaces.

- After performing certain types of repairs or service work, it may be necessary to have the vehicle raised off the ground on approved safety stands before start-up to prevent possible unexpected equipment movement, which may pose a safety hazard. To determine whether this procedure must be followed, always consult the OEM service literature.
- When working around areas of the equipment that may pose a fire hazard, only use cleaning agents that are not flammable.
- Use approved wheel chocks on both sides of the drive wheels to prevent accidental movement of the equipment. After performing some types of repairs or service work, starting the equipment for the first time may cause an unexpected loss of drive, which can cause a loss of dynamic braking and lead to an unexpected freewheeling condition on some types of equipment.
- When starting up the equipment for the first time after a repair or service procedure, extreme caution must be used when dealing with high-pressure hydraulic circuits. Never use your hands to feel for a leak in a hose or component. High-pressure fluid can puncture the skin and get into your blood stream, which may lead to blood poisoning. Always seek medical attention if you suspect this has occurred to you or to another technician.
- If you have serviced a charge pump, main pump, or hydraulic motor, the component must be filled with clean, filtered hydraulic oil before it is operated. Check the manufacturer's service literature for information as to where and how much fluid is required.
- Ensure that the reservoir is filled to the proper level with the correct type of hydraulic oil.

Note: Depending on the design of the system, if gravity does not fill the charge pump line, then the line must be filled by hand.

- After performing service work to a main pump or motor, it is advisable to disconnect the linkage to the pump control on initial start-up to allow the pump to find neutral. Be sure equipment is blocked properly to avoid unexpected movement.

- Before initial start-up on a hydrostatic drive unit, disable the engine from starting. Use the engine cranking system to prime the hydraulic circuit system. Crank the engine over until a charge pressure gauge mounted on the charge pump circuit registers a pressure build-up of at least 20 to 30 psi. This ensures that oil will get to the main components immediately and prevent initial start-up damage from occurring.

- After completing the priming steps, start the engine and let it run for a few minutes at low idle rpm. This will ensure that the system gets filled correctly. Observe the charge pressure to ensure it is stabilized to a normal pressure. Refer to the manufacturer's specifications to obtain the correct values. If charge pump pressure values are not within specifications, shut down the engine to troubleshoot the problem.

- If the charge pump pressure reading is normal, the engine should be shut down and the pump control linkage reconnected. At this time check the oil level in the main hydraulic tank and top up, if necessary.

- Now run the system through some basic tests. Start the engine and increase the rpm to approximately 1,500 to 1,800 rpm. Observe the charge pump pressure at this point to ensure it remains within specifications.

- After ensuring that the charge pump pressure is normal, shift the unit back and forth for a few minutes in forward and reverse directions, and observe the charge pump pressure. If the charge pump pressure falls below normal operating pressure while performing this test, shut down the unit and refer to the manufacturer's troubleshooting guide to resolve the problem.

- Although there are some differences in the pressures in hydrostatic drive system designs, the basic pressure test points are obtained in approximately the same locations. Following is a list of test points that should be recorded. Refer to the manufacturer's specifications for test points and values for the equipment you are

testing. The values given are typical values for the Sundstrand hydrostatic drive and are not necessarily the values for the equipment you are working on.

1. Charge pump pressure at idle and 1,800 rpm.
2. Pressure test point A: 210 to 240 psi above pump case pressure in neutral at 1,000 rpm.
3. Pressure test point B: 300 to 385 psi above pump case pressure in neutral at 1,500 to 1,800 rpm.
4. Pressure test point C: 230 to 250 psi above motor case pressure pump; must be in the stroked position at 1,500 to 1,800 rpm.

Tech Tip: Always refer to the manufacturer's service literature for the correct specifications for the equipment on which you are working.

Figure 12-24 shows typical pressure connection points used to test system pressures.

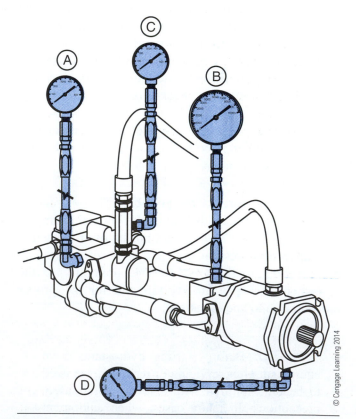

Figure 12-24 Diagram showing typical connection points to test system pressures.

GENERAL TROUBLESHOOTING (CLOSED-LOOP) PROCEDURE

The procedures outlined are intended for use as general examples of steps to follow, and should not be used for actual troubleshooting pressures and procedures on any particular hydrostatic drive system. Always refer to the OEM service literature for the equipment that you are working on for the proper pressures, procedures, and specifications.

Before you look at the troubleshooting charts, you need to familiarize yourself with some basic troubleshooting facts and procedures. Hydrostatic drive systems must maintain defined pressures and flows in various parts of the circuit at different times to operate efficiently. Changes in these performance values can have an adverse effect on system operation. Certain pressures and flows or restriction in the circuit need to be recorded and monitored. To perform pressure and vacuum reading on a hydrostatic drive circuit, you will need four different gauges with pressure snubbers, as shown in **Figure 12-25.**

To properly identify the correct pressure levels in a hydrostatic closed-loop circuit, you must be able to correctly measure and interpret the values. Be sure that the reading taken is accurate; the pump must be running at maximum rpm. We will discuss the four basic pressures and restrictions that all hydrostatic system operation depends on.

Starting with the charge pump, we may need to measure and record the inlet vacuum by using a vacuum gauge that records values in inches of mercury (Hg). The vacuum gauge used should have a range of 0 to 30 inches of mercury (Hg). Typical vacuum readings measured on the inlet side of the charge pump should be approximately 10 inches of mercury (Hg) at normal operating temperatures. If it is not possible to bring the equipment up to operating temperature, performing this test with cold oil will result in a reading that is slightly higher than the one at operating temperature.

The charge pressure in a typical hydrostatic drive circuit needs to be measured with an accurate gauge with a range from 0 to 400 psi. Typical values for this test are given at two different operating conditions. To explain the relationship of pressure readings, we will use an example showing the values for a Sundstrand closed-loop hydrostatic drive circuit. The minimum charge pressure is given as a minimum value that should not be less than a given pressure above the case pressure at any time. A reading of 170 psi charge pressure with a case pressure of 40 psi is within the acceptable range of 130 psi above case pressure. The second value we need to examine is the charge pressure when the motor output shaft is turning. This value on this system should be 160 psi above the motor case pressure. Another reading to be checked is the charge pressure when the system is in neutral. The minimum reading here should be 190 psi above pump case pressure. The next reading to check is the system pressure, sometimes called the high pressure. This is measured with a high-pressure gauge with a range from 0 to 10,000 psi. The gauge is installed in the manifold on the motor housing. This reading should never exceed the maximum value listed in the manufacturer's specifications under any normal operating conditions and temperatures.

The final pressure value is the case pressure. This pressure is measured with a gauge with a range from 0 to 400 psi, and the reading should not exceed 40 psi under normal operating conditions. The pressure values used in the preceding discussion were used to show the relationship of each reading to each other reading and may vary somewhat with other manufacturer's systems. Regardless of the values, the four measured values will be identified by the same names on most closed-loop hydrostatic drive circuits, and the relationship of each to each other will be compared in any troubleshooting process. As in all troubleshooting, the process must start at the beginning, in this case, the hydraulic tank.

Testing Charge Pressure

The charge pressure in a hydrostatic drive system is responsible for many jobs within the drive system. It has to replenish oil that is lost through the case drain, it supplies oil for controlling the servos, and it provides enough oil flow for cooling. It must also maintain a specific pressure within the low-pressure side of the hydrostatic system. Testing the charge pressure is a common practice in diagnosing hydrostatic drive system problems. **Photo Sequence 4** outlines a general procedure used in testing a charge pressure. It is not intended to replace specific manufacturer's procedures or guidelines. Note that equipment using hydrostatic drives is infinitely variable and that this procedure may not be applicable to all applications.

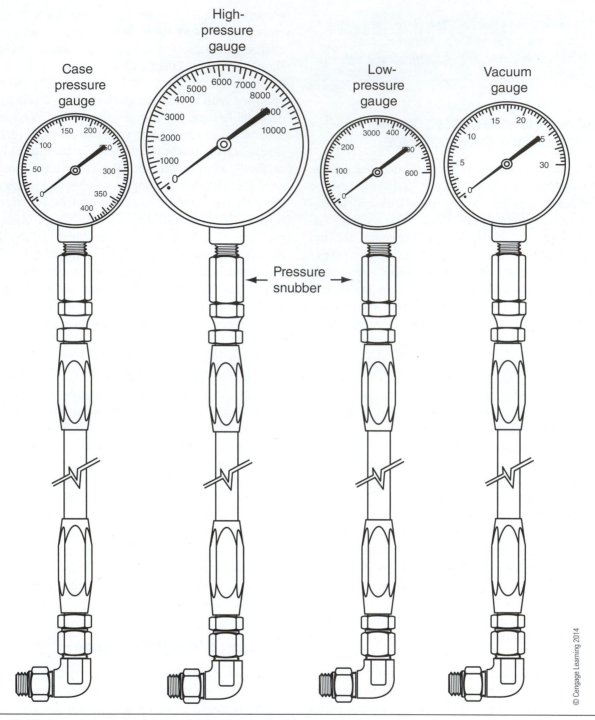

Figure 12-25 Gauges required to test a hydrostatic drive system.

HYDROSTATIC DRIVE (CLOSED-LOOP) TROUBLESHOOTING PROCEDURE

Note: The following troubleshooting guidelines are intended for use as a learning guide and are not intended to be a substitute for the manufacturer's specific troubleshooting guide for a particular application. Always consult the OEM service literature for each equipment-specific troubleshooting guide before attempting to diagnose a system fault.

PHOTO SEQUENCE 4
Testing Charge Pressure in a Hydrostatic Drive

P4-1 This procedure is implemented when the machine has a low performance complaint. The example we are using is on a Caterpillar 248B Skid Steer Loader.

This procedure outlines the order in which to perform a charge pressure test on the hydrostatic drive system for this machine. Before any work is started on a machine, proper safety procedure and lockout systems must be implemented. Before any testing, the machine must be at operating temperature for accurate readings, and oil levels must be at their proper levels.

P4-2 First, raise the boom high enough to properly place the safety arm to lock the boom in the raised position. Remove the holding pin and slowly lower the safety arm into place. **Note:** Do not let the safety arm fall into place; it may damage the cylinder rod.

P4-3 Because the hydrostatic drive system is located below the operator's cab, it must be raised to access the pumps and hydraulic components. To raise the cab, first remove the cab bolts located at the front corners of the cab.

(Continued)

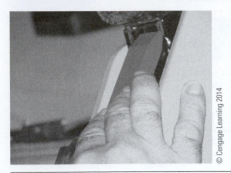

P4-4 and P4-5 The cab on machine is equipped with gas-filled cylinders and will raise with very little effort. With machines not equipped with gasfilled cylinders, an overhead crane is necessary to raise the cab. Once the cab is in the raised position, make sure the safety latch is in place to prevent any sudden drop of the cab.

P4-6 Using the OEM service literature, find the specification for the specific charge pressure setting (450 ± 30 psi/3,100 ± 200 kPa @ low idle) (480 ±30 psi/3,300 ± 200 kPa @ high idle). Locate the pressure tap port for testing the charge pressure.

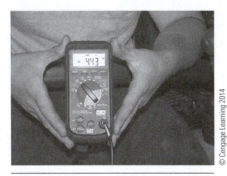

P4-7 Insert a proper pressure gauge or transducer that will fit the range of the charge pressure (0 to 1,000 psi). Failure to do so will damage the gauges.

P4-8 Lower the cab and place the pressure gauge in a position that can be read but cannot be damaged from the running engine. Start the engine and take a reading at both low and high idle. Compare your readings with the specification in the service literature. Repeat the test to confirm your readings.

Hydrostatic Drive System Will Not Operate in Either Direction

Cause	Symptoms	Remedy
Hydraulic fluid level low Faulty control linkage Drive coupling disconnected	Below normal charge pressure, fluctuating charge pressure, or both	Find and repair leaks. Refill hydraulic tank. Check all the linkage for binding. If required, adjust the linkage to pump control arm. Do not move the control arm to accommodate the linkage length. Inspect the coupling that drives the pump and check the motor drive coupling for slippage.
Inlet line from the charge pump or the filter is plugged or has collapsed.	Below-normal charge pressure Vacuum test shows high restriction at the charge pump.	Clean or replace the filter or replace the inlet line.
Charge pump relief valve faulty or stuck open	Below-normal charge pressure If the pressure is low when the pump is in neutral, charge pump is faulty. If the pressure is low when the system is stroked, the problem is in the manifold charge relief valve.	Replace the charge valve assembly.
Charge pump drive is severed.	No charge pressure in neutral or when stroked Low charge pressure Charge pressure fluctuates rapidly when maximum system pressure is reached. Charge pressure drops to zero or near zero. Maximum system pressure is below normal high-pressure relief setting.	The charge pump should be replaced. Check for excessive internal leakage at the pump or motor to determine which one is defective.
Pump or motor has internal wear.	Hydraulic tank and filter contaminated with brass Pump or motor noisy in operation	Remove the charge relief valve spring in the motor manifold and install a solid pin to replace the spring. This blocks the relief valve in the closed position. Remove the line from the motor case drain connection at the pump case port and install a threaded plug to seal the hole. Place a flow meter with the capacity to handle charge pump flow. Attach to the motor case drain line and route the output to the tank. Run the equipment at full rpm in neutral and engage the pump control to achieve the highest system pressure you can. Observe the flow in gpm coming out of the motor case drain.

(continued)

Cause	Symptoms	Remedy
		If the flow coming out of the motor is less than 50% of the charge pump flow, the motor has to be replaced. Since the pump and motor operate using the same hydraulic oil, the pump should be replaced also. If the motor is not damaged, the pump should still be replaced.
Control valve linkage is disconnected internally.	Charge pressure in neutral will be normal. When the control handle is moved, the pump does not stroke.	Control linkage at the directional control arm must be disconnected. Move the control arm.
Stuck or faulty check valve in the charge circuit	Charge pressure is low or erratic.	Check for air in the system. Check for damaged charge pressure control valve.
Control orifice plugged at the control housing to pump mount.	Charge pressure will be normal in neutral, but the pump will not stroke.	Remove blockage from the control orifice behind the control housing.

Hydrostatic Drive System Operates in One Direction Only

Cause	Symptoms	Remedy
Control linkage is faulty.	Loss of drive in one direction	Control linkage is out of adjustment or binding.
High-pressure relief valve sticking or stuck open	System pressure is lower than normal in only one direction.	Repair or replace the damaged relief valve.
System has one check valve that is faulty.	System pressure is low or zero in only one direction. The charge pressure may be higher than specification.	Repair or replace the two check valves located in the pump under the charge pump. Check for damaged check valve seats.
Faulty directional control valve Stuck spool or sticking in one direction	Pump will not center itself to neutral when linkage is disconnected.	Control valve should be replaced.
Motor manifold shuttle valve spool is jammed.	Lower than normal system pressure in only one direction	Motor manifold assembly must be replaced.

Hydrostatic Drive System Does Not Return to Neutral

Cause	Symptoms	Remedy
Control linkage sticky or faulty	System does not return to neutral.	Check neutral operation with control linkage disconnected. If it appears normal, the control linkage may be binding. Servo cylinders may need adjustment. Refer to manufacturer's technical documentation for adjustment instructions. Some pumps may need to be factory calibrated.
Control valve out of adjustment		The displacement control valve should be replaced or readjusted.

System Operating Temperature Runs Above 180°F

Cause	Symptoms	Remedy
Low fluid level in the hydraulic system	Low charge pressure reading	Refill hydraulic system with the correct grade of oil.
Hydraulic oil cooler plugged		Blow out or power wash cooler air passages.
Hydraulic oil by passing the cooler	Check temperature drop of the oil across the hydraulic oil cooler.	Repair or replace cooler bypass valve.
Plugged filter or suction line	Charge pump vacuum above normal limits; system exhibits low charge pressure.	Replace hydraulic filter or charge pump inlet line.
System exhibits excessive internal leakage.	System pressure will be lower than normal in one or both directions.	Verify the operation of the high-pressure relief valve and replace defective valve.
	Charge pressure may be erratic or drop to near zero when the maximum system pressure is reached.	Pump or motor, or both, may need replacement.

Hydraulic System Noisy in Operation

Cause	Symptoms	Remedy
Air in the system	Hydraulic oil in the tank is foaming.	Fill hydraulic tank to normal level.
Air entering the suction side of the charge pump circuit	Charge pressure is low or erratic.	Check for leaks on the suction side of the pump, filter, and hose connections.

Sluggish Acceleration and Deceleration

Cause	Symptoms	Remedy
Air entering the system	Poor equipment operating performance	Check the orifice behind the control valve to pump housing for blockage.
Control orifice is blocked with foreign material.		Remove the charge pump and remove plug from the charge pump gauge port.
Internal wear or damage to pump and motor		Blow air through it to dislodge any foreign material.
		Remove the charge relief valve spring in the motor manifold and install a solid pin to replace the spring.
		This blocks the relief valve in the closed position.
		Remove the line from the motor case drain connection

(continued)

Cause	Symptoms	Remedy
		at the pump case port and install a threaded plug to seal the hole. Place a flow meter with the capacity to handle charge pump flow to the motor case drain line and route the output to the tank. Run the equipment at full rpm in neutral and engage the pump control to achieve the highest system pressure possible. Observe the flow in flow gpm coming out of the motor case drain. If the flow coming out of the motor is less than 50% of the charge pump flow, the motor has to be replaced. Because the pump and motor operate using the same hydraulic oil, the pump should be replaced. If the motor shows no sign of damage, then only the pump should be replaced.

HYDRAULIC RETARDER SYSTEMS

Large capacity heavy-duty equipment operating on long, steep grades such as pits and quarries often can reach speeds as high as 50 mph. Equipment manufacturers design brake systems with average operating conditions in mind. Often these large haulers can benefit from the use of an additional auxiliary brake system (hydraulic retarder), which some equipment manufacturers include on their equipment. As an example, a haulage truck operating in an open pit or mining environment may be required to operate on grades as steep as 20%; this will quickly overheat a conventional brake system and lead to premature brake lining wear. The design intent of a hydraulic retarder in this application is to keep the equipment speed in check when operating on steep grades by using the conventional brakes only to bring the vehicle to a full stop. As the name suggests, a retarder uses a fluid to create resistance in an enclosure that consists of two stationary parts forming a housing with a rotating member in the middle (a **rotor**). The hydraulic retarder can best be defined as a device that can convert the kinetic energy of a moving vehicle into heat energy, as shown in **Figure 12-26.** The resulting heat energy is then sent to an auxiliary cooler, which dissipates it to the atmosphere. Selective use of a hydraulic retarder by an

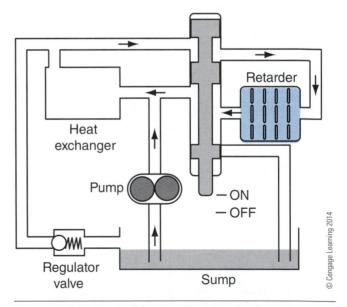

Figure 12-26 Typical hydraulic retarder layout.

© Cengage Learning 2014

operator can maximize the intervals between service brake overhauls, which in turn will reduce the operating downtime and operating costs of the equipment.

Although there are many different designs and manufacturers of hydraulic retarders, we will focus on two of the most common types used on heavy equipment. There are some key differences in retarder design and mounting location. The integral retarder design typically has two distinctly different design configurations. The retarder may be located either at the front of the transmission between the torque converter housing (input retarder), as shown in **Figure 12-27**, or mounted to the rear (output retarder) of the transmission, as shown in **Figure 12-28**.

To explain the advantage of using this type of auxiliary braking system on heavy equipment, we will use as an example of a hypothetical 80,000-pound haulage truck (36,290 kilograms) operating in an open pit environment. When the truck descends an 8% grade at approximately 30 mph, it could require approximately 500 hp of continuous braking to prevent a speed increase while descending the grade. In our example, this haulage truck's brakes are designed to absorb approximately 250 hp of kinetic energy. If we take into account the truck's rolling resistance and parasitic losses of the engine and powertrain (estimated at approximately

100 hp), the truck's braking capacity could fall short by up to 150 hp on this type of grade. If the equipment were required to make an emergency stop, the heat energy dissipation would be a lot higher, and the brake system would most likely not stand up to the job. Although this example is a hypothetical one, you can see the problems that can arise.

In these types of applications, the use of a hydraulic retarder reduces the operating braking temperatures and increases the service life of the brake system. To maintain high efficiency in hydraulic retarder design, the clearance between the housings and rotor must be minimal. Hydraulic retarder efficiency can run from 80% to as high as 115% equalling the efficiency of most diesel engine retarders. Oil temperature and pressure gauges are mounted in the operator's compartment to allow instant monitoring of these two important conditions. We will focus on two common systems in use in these off-road applications: the Allison and the Caterpillar designs. Some equipment manufacturers integrate the controls of the retarder with the powershift transmission controls while others are independently mounted. Some manufacturers use an independent pump and others use the existing transmission pump to provide fluid for retarder operation.

Input Retarders

The front-mounted hydraulic retarders or **input retarders**, as shown in **Figure 12-27**, are mounted to the front of a transmission between the torque converter and the transmission case. The housings are typically made from cast iron or aluminum alloys and have integrally cast vanes formed on the inside surface. These vanes are sometimes referred to as pockets. The design of the rotor vanes, with respect to width, length, and depth, depends on the intended application. The rotor is fastened to the **turbine shaft**, which allows the rotor to rotate between the front and rear housings.

The oil pressure in the retarder circuit is controlled by a pressure regulator valve, while the flow of oil to the retarder is controlled by a control valve. With the engine operating at high idle, the pressure in the retarder circuit depends on the application and can vary from 40 to 120 psi. When the operator activates this circuit, the oil flow enters the retarder near the center of the rotor. The rotor pumps pressurized oil to the pockets in the two stationary members (stators). This causes a high resistance in the retarder, which forces the rotor to slow down, absorb kinetic energy, and turn it into heat energy. Since the rotor is connected to the transmission, the retarding action is transmitted to the drive wheels. To remove the excess heat that is

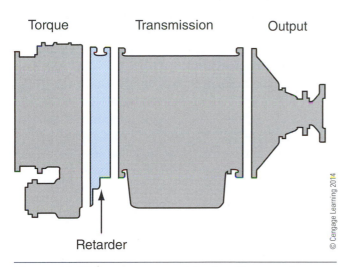

Figure 12-27 Front-mounted hydraulic retarder.

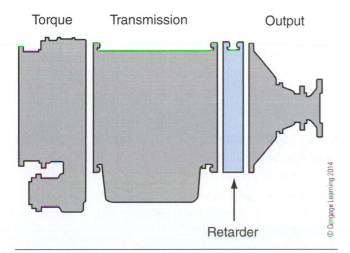

Figure 12-28 Rear-mounted hydraulic retarder.

produced when retarder operation is required, engine speed is reduced to an idle to allow the cooling system to absorb the heat from the retarder circuit. The amount of retarder braking depends on the flow control, which is determined by the operator's use of either a hand or foot control.

Output Retarders

Rear-mounted or **output retarders**, as shown in **Figure 12-28**, are mounted to the rear of a transmission between the main gear box and the output shaft. The housings are typically made from cast iron or aluminum alloys and have integrally cast vanes (pockets) formed on the inside surface. The rotor is splined to the output shaft, which allows the rotor to turn with the output shaft at the same speed. Operation of the rear-mount retarder is similar in principle to the front-mount design. One additional feature has been added to this design: a multi-plate **friction clutch**. Although the retarder operates similar to the front-mounted design, it has the ability to quickly apply braking effect through the initial application of the friction brake and while the oil charges the retarder. The main advantage of this design is that the forces generated by the retarder's action do not get transmitted through the transmission components because the retarder is located behind the transmission. Therefore, only the output shaft receives the braking force. Because the rear-mounted retarder turns at output shaft speeds, the braking effect can be enhanced due to varying speeds that can be achieved by selecting different gear ranges.

ONLINE TASKS

1. Use an Internet search engine to research the operation of an off-road hydrostatic drive system. Identify a piece of equipment in your particular location, paying close attention to the hydrostatic drive circuit operation, and identify whether it is an open-loop or closed-loop drive circuit. Outline the reasoning behind the manufacturer's choice of drive circuit design for this particular application.
2. For more information on hydrostatic drives, check out this URL: *http://www.carldyke.com/ courses/hydrostatic-drive-troubleshooting-boot-camp?gclid=CKbZ8Mvk4bUCFcdDMgodx2AAyA*.

Shop Tasks

1. Select a piece of equipment in your shop that has a hydrostatic drive system and identify the type of components that make up the drive circuit. Determine whether it is a closed-loop or open-loop design, and outline how oil is directed to this circuit along with the charge pressures used.
2. Select a piece of mobile equipment in your shop. Identify the type and model of the equipment and list all of the major components and the corresponding part numbers of a typical hydrostatic drive circuit. Identify and use the correct parts book for the particular piece of equipment being worked on.

Summary

- The hydrostatic drive system is a fluid drive system that uses hydraulic fluid under pressure to convert diesel engine power to fluid power and then to convert it back to mechanical power at the wheels or tracks.
- The basic differences between a hydrostatic drive system and a hydrodynamic drive system are the velocity of the oil flow and the pressures that are used.
- The integral system uses a pump and motor mounted as a single assembly on the back of a diesel engine.
- The hydrostatic drive system is capable of providing a dynamic braking effect as a means of slowing and stopping the vehicle or even the ability to coast when required.

- Hydraulic fluid will not compress and it adopts the shape of any hydraulic system component it happens to flow through.
- The available torque on start-up in a typical hydrostatic drive system can be as high as 90%.
- Internal leakage in the pump and motor circuit is required for lubrication and cooling.
- The charge pump supplies the oil flow for the main pump, as well as oil flow for the lubrication and cooling of a hydrostatic drive system.
- The main pump used in a hydrostatic drive circuit is a positive displacement type pump either variable or fixed displacement.

- The use of a heat exchanger is a requirement if the operating temperature of the hydraulic oil will exceed 180°F (82°C) at the motor case drain.

- The reservoir is sized to a minimum of half of the charge pump displacement volume rating in gallons per minute (gpm).

- The displacement control valve controls the swashplate angle in the pump.

- In a closed-loop system, the oil flows from the pump to the motor and is returned to the inlet of the pump. In an open-loop system, the hydraulic oil flows from the pump to the motor and back to the hydraulic tank.

- The term *stepless* is often used to describe the operation of the hydrostatic drive system.

- Hydrostatic drive circuits have low tolerance for contaminants which result in excessive wear in the pump and motor circuit.

- An open-loop hydrostatic drive circuit is a hydraulic drive that may be used in heavy equipment where two or more different drive functions or loads are necessary.

- The selection of hydraulic lines should be in keeping with equipment manufacturer's recommendations for length, diameter, and pressure capabilities.

- Before performing maintenance on hydrostatic drive equipment, note of any potential hazards that may result in injury or equipment damage.

- Hydraulic oil that is exposed to excessive temperatures loses its viscosity.

- When work is performed on equipment, use wheel chocks on both sides of the drive wheels to prevent accidental movement of the equipment.

- A hydraulic retarder is a device that converts the kinetic energy of a moving vehicle into heat energy.

- A retarder may be located either at the front of the transmission between the torque converter housing (input retarder) or at the rear mounted to the rear of the transmission (output retarder).

- To maintain high efficiency in hydraulic retarder design, the clearance between the housings and rotor has to be minimal. Hydraulic retarder efficiency can run from 80% to as high as 115% equalling the efficiency of most diesel engine retarders.

- On steep grades, the use of a hydraulic retarder reduces the operating braking temperatures and increases the service life of the brake system.

- A rear-mounted retarder turns at transmission output shaft speeds; the braking effect can be enhanced by selecting different gear ranges.

Review Questions

1. Which statement describes a hydrostatic drive system operation?

   A. high flow velocity with low pressure   C. high flow velocity with high pressure

   B. low flow velocity with high pressure   D. low flow velocity with low pressure

2. What occurs to the pressure in the drive system in a hydrostatic drive when the load on the motor increases?

   A. Drive pressure in the system increases.   C. Drive pressure in the system stays the same.

   B. Drive pressure in the system decreases.   D. Charge pressure in the system increases.

3. What is the maximum potential torque available at start-up on a hydrostatic drive loader?

   A. 100%   C. 70%

   B. 90%   D. 50%

4. What occurs in a variable-displacement pump when the swashplate angle is increased?

    A. More oil is displaced per stroke.   C. Charge pressure increases.

    B. Less oil is displaced per stroke.   D. Motor speed decreases.

5. What determines the rate and direction of flow in a hydrostatic drive system?

    A. piston diameter   C. pump speed

    B. swashplate angle   D. motor speed

6. Which component supplies makeup oil in a hydrostatic drive system?

    A. crossover relief valve   C. low-pressure relief valve

    B. swashplate angle   D. charge pump

7. Which component supplies oil to the main pump in a hydrostatic drive system?

    A. charge pump   C. low-pressure relief valve

    B. swashplate   D. main relief valve

8. Which component directs the low-pressure side of the drive loop to the low-pressure charge relief valve?

    A. shuttle valve   C. low-pressure relief valve

    B. swashplate   D. charge pump

9. Which system is capable of offering independent track control and infinite variable speed control?

    A. hydrostatic drive   C. belt drive

    B. hydrodynamic drive   D. chain drive

10. Which component must always be used in a closed-loop hydrostatic drive circuit?

    A. charge pump   C. fixed-displacement pump

    B. variable-displacement pump   D. fixed-displacement motor

11. What controls the direction and the amount the swashplate moves?

    A. servo pistons   C. low-pressure relief valve

    B. charge pump   D. shuttle valve

12. What function does the main relief valve serve in a hydrostatic drive circuit?

    A. protect against excessive pressure   C. protect the charge pump from excessive
       spikes   pressure

    B. protect the motor from overspeed   D. maintain maximum torque output at high speeds

13. What regulates the system when a hydrostatic drive system is in neutral pressure?

    A. charge relief valve   C. low-pressure relief valve

    B. shuttle valve   D. charge pump

14. What is the main design feature difference between a closed-loop system and an open-loop system?

    A. The return oil from the motor goes directly back to the pump on the closed-loop system.

    B. The return oil from the pump goes back to the motor on the closed-loop system.

    C. The return oil from the pump goes back to the tank on the closed-loop system.

    D. The return oil from the pump goes back to the charge pump on the closed loop system.

15. What occurs in a variable-displacement pump when the pistons and barrel rotate with the valve plate at an angle?

    A. The pump is in neutral.

    B. The charge pressure will decrease.

    C. Pistons will stroke.

    D. The motor will stop turning.

16. What occurs when the swashplate is moved to the maximum angle for maximum torque?

    A. high-speed operation

    B. neutral operation

    C. low-speed operation

    D. reverse operation

17. Which component determines the rate and direction of flow inside a variable-displacement pump?

    A. swashplate

    B. pistons

    C. control valve

    D. charge pump

18. Which component controls the servo piston movement in a variable-displacement pump?

    A. displacement control valve

    B. charge pump

    C. retraction plate

    D. shuttle valve

19. What occurs when the swashplate is moved to the minimum angle for minimum torque?

    A. high-speed operation

    B. neutral operation

    C. low-speed operation

    D. reverse operation

20. What occurs in a hydrostatic drive system when the load on the motor decreases?

    A. The pressure in the system increases.

    B. The pressure in the system decreases.

    C. The pressure in the system remains constant.

    D. The speed of the system decreases.

21. Which type of hydraulic retarder is located at the front of a power shift transmission between the converter and the main gear box?

    A. input retarder

    B. friction clutch retarder

    C. output retarder

    D. range retarder

22. Which component is found only in the rear-mounted retarder?

    A. rotor

    B. stator

    C. friction clutch

    D. pressure-regulating valve

23. What determines the degree of braking in a hydraulic retarder?

    A. oil flow and pressure   C. rotor speed

    B. stator speed   D. turbine speed

24. Where does the oil enter the hydraulic retarder during retarder operation?

    A. At the outer edge of the stator   C. At both sides of the stator.
       housing.
                                          D. It is always charged with oil.
    B. Near the center of the rotor.

25. What component is the rotor mounted to in a rear-mounted hydraulic retarder?

    A. turbine shaft   C. stator

    B. output shaft   D. countershaft

CHAPTER

13 Driveline Systems

Learning Objectives

After reading this chapter, you should be able to:

- Describe the fundamentals of drivelines.
- Outline the components and construction features of drivelines.
- Describe the operation principles of drivelines.
- Perform the recommended maintenance procedures for drivelines.
- Outline the diagnostic procedures to troubleshoot drivelines.

Key Terms

brinelling	inclinometer	slip joint
broken-back driveshaft	low-block	slip yoke
compound driveline angle	needle bearings	spalling
cross	non-parallel driveshaft	trunnion
delta-wing	parallel driveshaft	universal joint
driveshaft	phasing	weld yoke
end yoke	pitting	Welch plug
galling	runout	wing-type U-joints
high-block	slip assembly	working angle

INTRODUCTION

Driveline technology has evolved considerably in the last 100 years. The concept of the modern **universal joint** was first discovered a few centuries ago, but it was not applied until early in the twentieth century when a fellow by the name of Clarence Spicer pioneered the development of the modern driveshaft U-joint. Anyone working in the heavy-equipment trade these days has come across the company name "Spicer," made famous by their pioneering development and manufacturing of modern driveline and universal joints. The modern driveline and U-joint arrangement shown in **Figure 13-1** has evolved to what could be considered an engineering marvel.

Heavy equipment found at construction sites, open-pit mines, and quarries these days is capable of hauling hundreds of tons at a time. The power developed

335

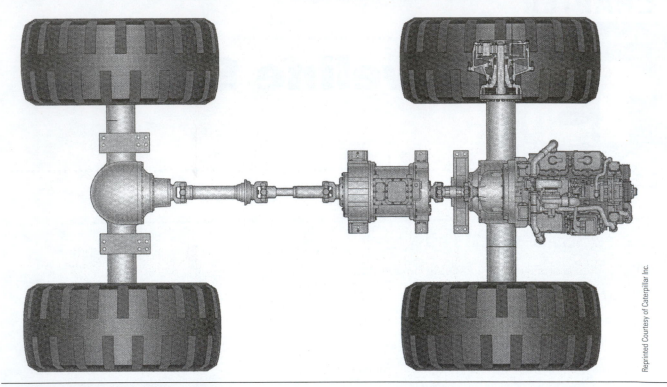

Reprinted Courtesy of Caterpillar Inc.

Figure 13-1 Modern driveline and U-joint arrangement.

by the diesel engine must pass through a series of **driveshafts** and powertrain components on its way to the drive wheels. The torque generated by the use of large gear reductions found in heavy equipment powertrain components (often as high as 100,000 lb-ft) is generated at the wheel ends. This high torque must be handled by a series of drivelines on its way to the final drives. On heavy equipment, an overall gear reduction of 100 to 1 from the engine to the drive wheels is common.

In this example, we study the gear reduction of a typical loader used in a quarry operation. Let's assume the engine is capable of producing 1,200 lb-ft of torque at the flywheel, and the typical torque converter used in this application has a 2-to-1 internal gear reduction between the turbine shaft and the output shaft; this would send 2,400 lb-ft of torque to the transmission. A typical four-speed heavy-duty powershift transmission will often use a first gear with a 6-to-1 reduction. If you do the math, you will see that the overall torque produced at the output shaft of the transmission is on the order of 14,400 lb-ft. This torque now flows through a differential where it is increased by a 4-to-1 reduction to 57,600 lb-ft. The final reduction at the wheels is through a 4-to-1 planetary gear set, producing an output torque of 230,400 lb-ft. This example shows the torque potential that can be transmitted through the drivelines between powertrain components on heavy equipment.

The transmission of power through a drivetrain as shown in **Figure 13-2** is not necessarily in a straight line; the position of components often requires drivelines to transmit power at odd angles. Modern suspension systems pose another problem; final drives are often mounted on flexible suspension systems that cause the drive axles to move when the equipment is driven over uneven terrain. A **slip joint** allows the driveshaft to shorten and lengthen as the drive axle moves over a roadway. A final hurdle was overcome by the addition of universal joints on both ends; this allowed the driveshaft to adapt to speed changes, produced by driveline angle changes between the powertrain components while it is in operation.

FUNDAMENTALS

Driveline designs can be divided into two categories: **parallel driveshaft** installations and **broken-back driveshaft** installations (non-parallel). Either of these designs may be used on a typical piece of heavy equipment. In a loader that has the torque converter mounted to the back of the engine and the transmission separately mounted rigidly to the frame of the loader, it is common to find a rigidly mounted driveshaft connecting the two components with a universal joint at each end. This design is suitable for use where both input and output sources are rigidly mounted.

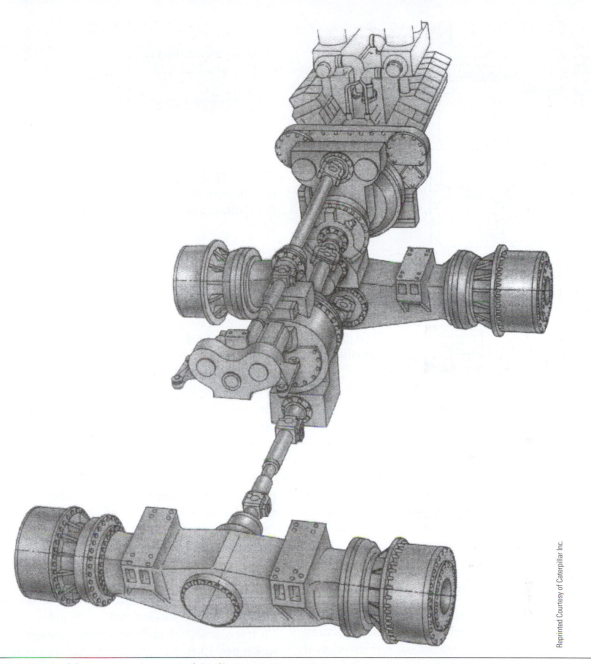

Figure 13-2 Typical heavy equipment driveline arrangement.

In an application where one of the components is able to move, this design would not be suitable and a slip joint would be required to allow for movement. An example of this is a typical transmission-to-drive axle connection with a suspension-mounted drive axle. In **Figure 13-3**, we see a typical off-road application of an axle that is mounted rigidly to the frame and does not have any significant movement other than frame and component flexing. The rigid design commonly found on heavy equipment generally has a long trouble-free service life. It is reliable, low cost, and easily maintained—because there are no suspension alignment concerns as the equipment wears.

In applications where the equipment has a live suspension, such as a large haulage truck, the use of a parallel driveshaft arrangement is common. As the haulage truck moves over uneven terrain, the driveshaft is able to change length by the use of a splined slip joint. A parallel design is better suited to this application because the operating angle tends to remain constant as the axles move through their intended design travel. The yokes must remain parallel to each other while in operation, and the **working angles** must remain equal to minimize vibration and wear of the U-joints.

In certain applications, it is better to use a non-parallel design installation of a driveshaft. The yoke

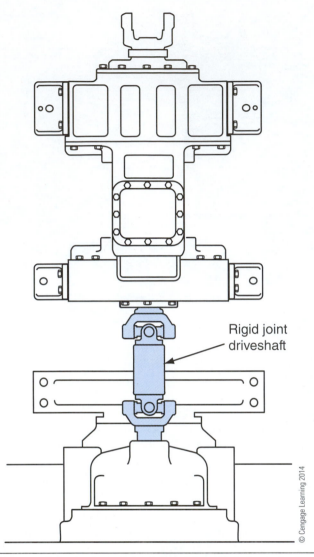

Figure **13-3** Typical rigid joint driveshaft installation found on heavy equipment.

angles are not parallel to each other in this installation, but the driveline working angles must remain equal in order to maintain vibration-free operation. Due to the design restrictions on some equipment, it becomes necessary not only to have a vertical plane offset, but also to have a horizontal plane offset, as shown in **Figure 13-4**. Due to the inherently slow-speed operation of most heavy equipment, working angles on driveshafts are often much higher than on high-speed vehicles. Vibration does not usually become a problem on heavy equipment; the equipment often operates on rough terrain and the operator would probably not notice vibration until it was extreme.

As working angles increase, the life of the driveshaft U-joints decreases. Driveline speed plays an important role in determining the maximum allowable working angles that can be used. The use of large gear ratios on heavy equipment does not allow the vehicle

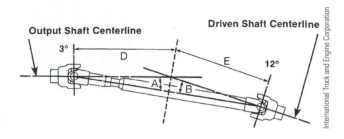

Figure **13-4** Typical non-parallel driveline design installation. Note the unequal yoke angles between the output shaft yoke and the driven shaft yoke.

too much opportunity for high road speeds. Typical road speeds range from 1 to 15 mph (2 to 25 kph). A universal joint has a maximum speed at which it can operate; this speed is always proportional to the working angle of the installation. Heavy equipment generally does not operate at speeds above 30 mph (50 kph), which limits the maximum speed at which the drivelines turn. This allows equipment manufacturers to use driveline angles of 8 to 12 degrees on large, slow-moving equipment. Typical driveline speeds on final drives can be up to 100 to 800 rpm. **Figure 13-5** shows a general example of speed versus the maximum working angles that are often used.

Universal joints connect the drivelines to the powertrain components and carry the full torque that the equipment is capable of generating, while at the same time allowing the driveshaft to operate at different angles. Because of this, the U-joint **cross** or spider, as some manufacturers call it, is constructed of a high-quality forged casting. The four ends, or trunnions, where the **needle bearings** are located, have a case-hardened surface that is ground to a fine finish.

Interconnected internal grease passages, as shown in **Figure 13-6**, are drilled into the trunnions to allow the needle bearing to be greased. Although a few manufacturers are starting to use a sealed U-joint that requires no greasing during its service life, the majority of universal joint manufacturers still provide a

Driveshaft RPM	Maximum Working Angle
4000	4 degrees of 15'
3000	5 degrees of 50'
2000	8 degrees of 40'
1500	11 degrees of 30'
1000	12 degrees of 30'

Figure **13-5** Typical driveline working angle and speed relationship found on heavy equipment.

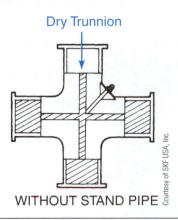

Dry Trunnion

WITHOUT STAND PIPE

Courtesy of SKF USA, Inc.

Figure 13-6 Cross-drilled trunnion used to distribute grease to the bearing end.

means of greasing the **trunnion** bearings. Grease fittings are located close to the **center** of the cross, allowing for periodic lubrication of the U-joint bearings.

Many different cap designs are available for heavy-duty universal joints; we will discuss the three most common cap designs. The **low-block** design is a standard type of cap and is suitable for use on smaller equipment that does not regularly achieve high torque levels in service. A good example of this application is a typical 2- to 4-yard loader. The **delta-wing** design is generally used on 4- to 8-yard loaders between the torque converter and the transmission where torque values are minimal. The **high-block** cap (high wing), as shown in **Figure 13-7**, is often found on large equipment between the transmission and the final drive. Its large block design can carry high torque values that are generated by equipment used for severe-duty applications. A good example of this is an 8- to 12-yard loader used in quarries and open-pit mines.

As discussed earlier, we know that driveshaft speeds are not very high on a typical piece of heavy equipment; this does not mean that we will ignore working angle geometry. Due to the high torque loads on U-joints used on heavy equipment, working angles must be held to within 1 degree of each other to maintain equal U-joint speed at each end of the driveshaft. This prevents uneven speed fluctuations, as

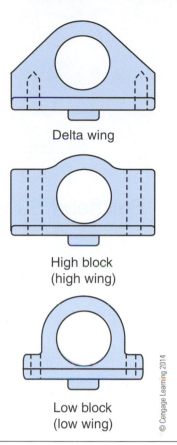

Delta wing

High block
(high wing)

Low block
(low wing)

© Cengage Learning 2014

Figure 13-7 Three common U-joint designs.

shown in **Figure 13-8**, which can lead to premature failure of the U-joint needle bearings.

Heavy equipment designs often use some form of driveshaft support when driveshafts pass through equipment bulkheads as shown in **Figure 13-9**. The distance between the engine, transmission, and drive axle on heavy equipment can be substantial, and may require the use of driveshaft supports. The use of a bulkhead support bearing (a pillow block bearing) supports the driveshaft and decreases its natural tendency to flex while transferring torque. Flange-type bearings are generally used in this type of installation and are generally flange mounted to the frame. Usually, a lube fitting is provided for lubrication, but on some larger units where accessibility to the bearing is

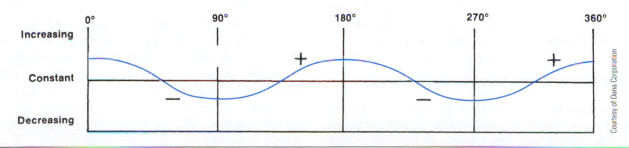

Figure 13-8 Speed fluctuation of a typical driveshaft during one revolution.

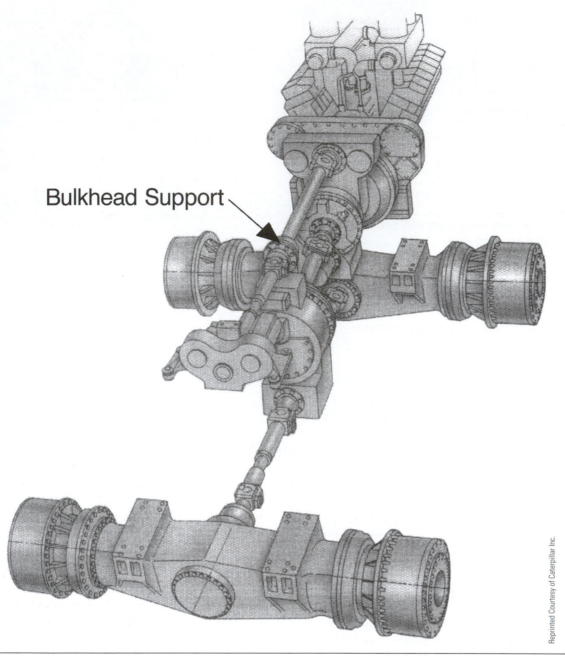

Bulkhead Support

Figure 13-9 Typical driveline bulkhead support-bearing assembly found on heavy equipment.

limited due to the equipment design, a remote greasing point is provided. On smaller equipment, this support bearing may be sealed, thus making lubrication unnecessary.

CONSTRUCTION FEATURES

Figure 13-10 shows an exploded view of a typical haulage truck drivetrain that consists of three separate driveline assemblies. The coupling shaft assembly connects the transmission to the standard slip assembly

through a center bearing to the front axle. The short coupling **slip assembly** connects the two axles. We will look at each of these assemblies and learn how each contributes to the transfer of torque to the drive wheels. As we learned earlier in this chapter, the torque carried through the drivelines on off-road equipment is significantly higher in most cases than what you would normally have with on-road equipment. Because of this, off-road drivelines are generally designed to handle the expected slow speeds and high torque found on off-road equipment.

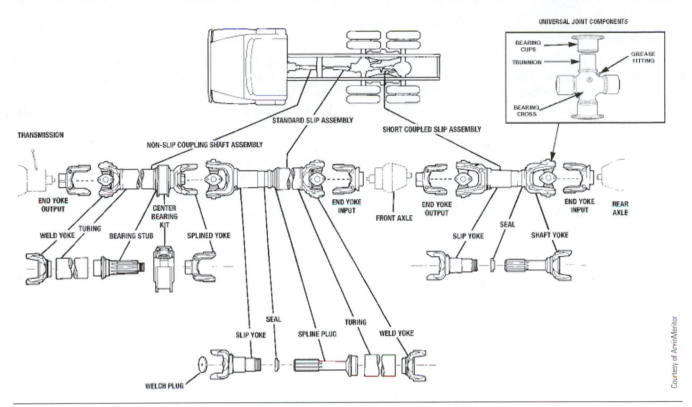

Figure 13-10 Exploded view of typical haulage truck driveline configuration.

Tubing

The most common tubing used in the construction of heavy-duty drivelines is cold-rolled electric-welded; this is generally considered standard factory build on most off-road equipment. The tubing is manufactured by rolling the steel into a tube shape and welding it together. The fabrication process does not always produce a perfectly true running shaft, but it is generally suitable for low-speed applications like those found between the transmission and the drive axles on most equipment. The wall thickness and driveline diameter determine the torque-carrying capabilities of the driveshaft. A higher quality of construction used to fabricate driveshafts takes the conventional rolled-tube, electric-welded shaft one step further. The tubing is drawn over a mandrel to ensure that it is straight and runs true. This extra step produces a much straighter tube and also adds to the strength of the tubes by stress relieving the shaft. The third type of tubing used for driveshaft construction is known as seamless tubing, which is heavy-wall tubing and is generally used in heavy-duty applications such as large haulage trucks found in open-pit mining and quarry operations. Because of the thickness of the tubing, it is suited for slow-speed applications. When selecting driveshaft tubing, consider the factors of speed, length, operating angle, and torque-carrying capabilities.

Driveshafts

The driveshafts are balanced by the manufacturer but given the slow speeds at which drivelines rotate on off-road equipment, vibration is seldom a concern. When driveshaft replacement is required, be sure to observe the equipment manufacturer's recommendations and use only the recommended shaft for the application. Using a replacement driveshaft with a different diameter size may shorten the life expectancy of the driveline and U-joints. Oversizing is not necessarily better; a larger rotating mass may overload the component bearings. Forged yokes are welded to each end of the driveshaft to accept the U-joints, which are required to compensate for the offset that exists between the engine, transmission, and drive axles. The yokes at each end of the driveshaft are connected through the U-joint to a yoke on the output or input shaft of the transmission and drive axles. In most cases, the driveline connection will be a two-piece connection between the components to allow for differences in component mounting variations on equipment and to compensate for driveshaft length changes due to suspension movement during operation.

Slip Joints

The splines on slip joints are made from high-quality hardened steel; the male and female ends have matching splines to provide a tight-fitting connection between the two ends of the driveshaft, as shown in **Figure 13-11**. The machined slip joints are welded to the ends of each shaft. The male end of the slip joint has external splines, which are an exact match to the internal splines found on the female end of the driveshaft. When connected together, the splines of each fit snugly yet allow the overall length of the driveshaft to shorten or lengthen during operation. The sliding fit must be held to tight tolerances because the full torque developed must be transmitted through this connection. Some manufacturers coat the splines with an anti-friction substance, such as nylon or high-phosphate grease. When mounted together, the splines are sealed with a threaded dust cap that holds a seal to keep dirt out of the splines on the female end and a blind plug (**Welch plug**) or cap at the other end. These require periodic lubrication with high-quality grease through a grease fitting located on the female end of the shaft.

Universal Joints

Universal joints (U-joints) are necessary on each end of a driveshaft to connect powertrain components that are not mounted on the same plane. U-joints are designed to accommodate the rotational speed changes that occur as a driveshaft rotates: the greater the working angle of the U-joint, the greater the differential between the inside angle (faster) and the outside angle (slower) of the joint assembly. **Figure 13-12** shows the parts breakdown of a typical universal joint.

The trunnions have bearing caps with internal needle bearings mounted on the ends that allow the caps to rotate. A rubber seal, as shown in **Figure 13-13**, is located at the end of each cap to keep dirt and water out of the needle bearing during operation. Needle bearings are ideally suited for use in this application because they require very little radial space and are capable of carrying high radial loads, thus giving them a long service life if properly maintained. The bearing caps are mounted to corresponding forged yokes at each end of the driveshaft. Although they come pre-lubricated, needle bearings must be lubed with high-quality grease after initial installation and again at regular service intervals.

Some manufacturers locate a standpipe (check valve) at each end of the bearing cross as shown in **Figure 13-14**. This check valve prevents the grease

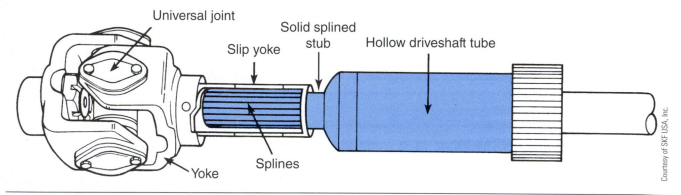

Figure 13-11 Slip joint connection on a typical two-piece driveshaft.

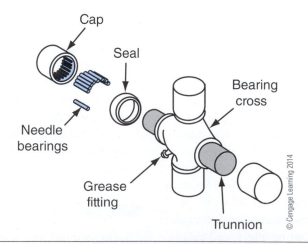

Figure 13-12 Typical universal joint construction showing the major parts.

Figure 13-13 Typical heavy-duty universal joint showing one bearing cap removed.

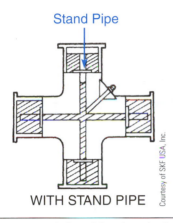

Figure 13-14 A standpipe (check valve) prevents the hot grease from flowing down away from the vertical trunnion.

from running out of the trunnion when the vehicle is parked or the trunnion is stopped in the vertical position. Without this check valve action, the bearing would quickly become damaged from lack of lubrication.

Wing Design. U-joints come in many different design configurations. **Figure 13-15** shows three styles that are common on some smaller off-road equipment, such as small two-yard front-end loaders. The bearing caps come in different design configurations, which depend on the service application of the U-joint. Generally, heavy equipment U-joints come in different wing configurations. Although these are not the only type of bearing caps used on off-road equipment, they are by far the most common.

- Low-block and delta-wing bearing caps are used on drivelines in light-duty applications (for example, on a 2-yard loader).
- High-block bearing caps are used in severe duty applications such as transmission to axle installations (for example, on a 20-ton loader).

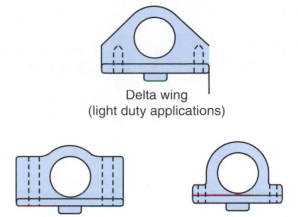

Delta wing
(light duty applications)

High block (high wing)
(severe duty applications)

Low block (low wing)
(light duty applications)

Figure 13-15 Three bearing cap designs used on heavy-duty U-joints.

All wing-style heavy-duty bearing caps have a keyway that fits into a corresponding keyway in the yoke, as shown in **Figure 13-16**, allowing for an extremely high-torque carrying connection. The wing joints come with special high-grade cap screws and should be torqued to the manufacturer's specifications. Some manufacturers recommend that Loctite® be applied to the bolt threads before they are torqued to specification. Always follow the manufacturer's recommended installation procedures when installing U-joints on a driveshaft.

OPERATING PRINCIPLES

Off-road equipment has two driveshaft configurations that are used in the majority of equipment. The parallel type is the more popular of the two. When the components are mounted solidly to the frame of the equipment, it is easy to maintain parallel angles of the flanges and yokes. The working angles should be equal because the angles are directly dependent on the speed of the driveshaft.

If the correct working angles for a given speed cannot be met with the component installation parallel, each component can be shimmed to achieve the required working angle. For example if the transmission centerline is offset 3 degrees down from horizontal, and the drive axle is offset 3 degrees up from the horizontal, shimming can be used to reduce the working angles, if necessary.

The non-parallel design is used when the zero vertical angles of the component installation cannot be met. **Figure 13-17** shows that the transmission is offset

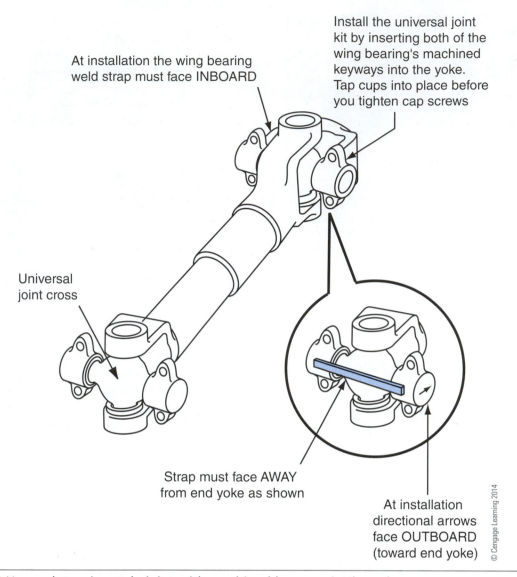

Install the universal joint kit by inserting both of the wing bearing's machined keyways into the yoke. Tap cups into place before you tighten cap screws

At installation the wing bearing weld strap must face INBOARD

Universal joint cross

Strap must face AWAY from end yoke as shown

At installation directional arrows face OUTBOARD (toward end yoke)

© Cengage Learning 2014

Figure 13-16 Heavy-duty wing-style joint with machined keyways in the yoke.

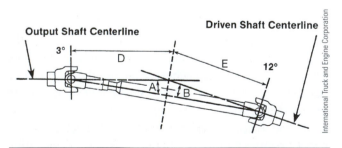

Output Shaft Centerline

Driven Shaft Centerline

3°

D

E

12°

A

B

International Truck and Engine Corporation

Figure 13-17 Non-parallel driveline installation shows the transmission angle is 3 degrees off vertical and the driven shaft is 12 degrees off vertical.

3 degrees from the vertical, and the driven shaft is 12 degrees off vertical. Angles A and B need to be equal to avoid speed differences in each of the U-joints on the same shaft. If the working angles are held to within 1 degree of each other, the driveshaft will slow down at the front U-joint and speed up at the rear U-joint by an equal amount, cancelling out any speed fluctuations. The longevity of universal joints is affected by the angles and speeds at which they operate; exceeding the working angle and recommended speed of the driveshaft will almost certainly lead to problems. Off-road equipment operates at slow speeds and working angles are often higher than those found on high-speed on-road equipment. Placement of power-train components in equipment often requires the use of higher working angles. Fortunately, we have speed (or the lack thereof) working in our favor.

Compound Driveline Installations

Off-road equipment component placement can result in drivelines having **compound driveline angles**. **Figure 13-18** shows a typical driveline installation with

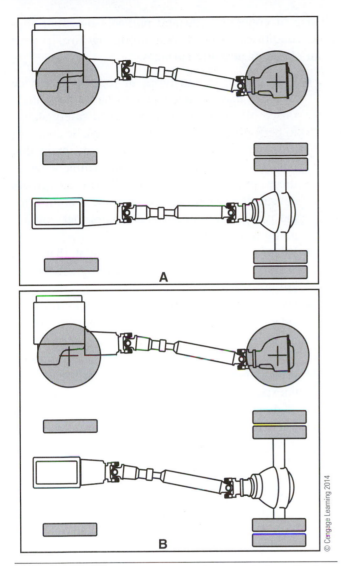

© Cengage Learning 2014

Figure 13-18 Single plane driveline angle installation.

the driveline installed with a vertical plane offset. This is by far the most common type of installation and results in a long, trouble-free driveshaft U-joint life. Often, manufacturers will offset the installation in the horizontal plane but maintain a 0-degree vertical plane.

On some equipment it is not possible to install the components so that either a vertical or a horizontal offset is achieved. Design limits on some equipment require that the components are installed in such a way that a compound driveline installation is unavoidable.

Compound driveline installations cannot be engineered out of the equipment design, and compromises will always exist. Vibration on off-road equipment may not be noticeable to the operator, but the effect of vibration on the needle bearings used in the U-joints will take its toll on U-joint life. All U-joints must operate at an angle of at least a ½ degree. This sets up a speed change that causes the shaft to slow down and speed up during each revolution.

These changes take place at each end of the driveshaft. Speed fluctuations in the driveline cause torsional vibrations to be set up in the driveshaft as well as cause problems in the transmission itself and possibly in the drive axles as well. If the working angles are set up correctly within 1 degree of each other, these speed changes cancel out each other, minimizing this effect.

Driveline Speed

If it were possible to watch a driveshaft U-joint move through an angle from an end view, we could see that the U-joint at the driven end moves through an ellipse. The U-joint moves through each quadrant in a fixed amount of time; in other words, the driveshaft increases and decreases in speed are predictable for each revolution. When we examine a conventional two U-joint driveshaft, we can see that if the second U-joint had an equal working angle it would decelerate at the same time that the first U-joint was accelerating and at about the same rate. The acceleration and deceleration rate is directly proportional to the working angle of the U-joints. Since the working angles can change because of suspension movement, the velocity of the driveshaft constantly fluctuates. If the driveshaft was installed one tooth out, resulting in an out-of-phase condition, the resulting forces would be similar to a person snapping a long rope; the whipping action would produce a snapping action at the opposite end of the rope. If this can be reproduced at both ends simultaneously, the forces will cancel each other out. This is how driveline **phasing** works; the forces generated at each end of the driveshaft cancel each other out if the driveshaft is correctly assembled.

Phasing

When removing a two-piece driveshaft, look for the phasing marks, as shown in **Figure 13-19**. If none are present, scribe a mark on the two connecting halves before removing the driveshaft. When the time comes to install the driveshaft, this will ensure that it is done correctly. When determining driveshaft installation parameters, one other factor that must be considered is the driveshaft's length and diameter (as well as the rpm). Most manufacturers restrict driveshaft length to a maximum of 70 inches. The greater the weight and length of the shaft, the greater the radial forces generated. As the speed of the driveshaft goes up, balance becomes more important. The overall length of the driveshaft is measured across the centerline of the U-joints. If proper maintenance has been adhered to, U-joint operating life, as rated by most manufacturers, is calculated for a

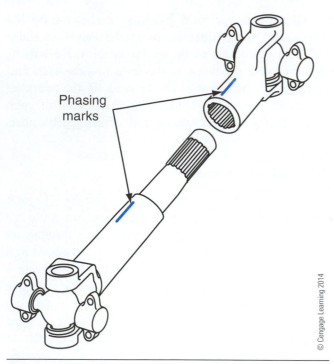

Phasing
marks

© Cengage Learning 2014

Figure 13-19 Where to place the phasing marks on a driveline.

continuous operating load at 3,000 rpm for 5,000 hours with a 3-degree working angle. If you were to double the working angle, the life expectancy would be reduced by 50%. If you reduce the load by 50%, you double the life. Because the drivelines on off-road equipment run at various loads and speeds, it is impossible to predict the life of a U-joint.

MAINTENANCE PROCEDURES

This section contains maintenance procedures for drivelines in general. Some procedures require special tools to service drivelines safely. Failure to use the special tools and follow the correct procedures as outlined by the tool manufacturer may result in personal injury or equipment damage. The procedures outlined here are intended for use as general examples of the steps to follow and should not be used for actual maintenance procedures on any particular drive system. Always refer to the manufacturer's service literature for the equipment that you are working on for proper procedures and specifications. Before you start any work on equipment, whether it is servicing or repair, you need to become familiar with a few general safety precautions. In many cases, when equipment needs to be serviced or repaired, it is not always necessary to remove the parts. Access to the area being serviced should be high-pressure washed before work begins and after it is completed; this minimizes the

chance of contamination and prevents potential slipping conditions as well as minimizing fire hazards from oil on or near hot surfaces.

- To perform certain types of repairs or service work, it may be necessary to have the vehicle raised off the ground on approved safety stands to perform tests and repairs. Always be aware of possible unexpected equipment movement that may pose a safety hazard. To determine whether these procedures must be followed, always consult the manufacturer's service literature for the type of equipment and repairs that you are going to perform.
- When cleaning areas on equipment that may pose a fire hazard, use only cleaning agents that are not flammable.
- Use appropriately sized wheel chocks on both sides of the drive wheels to prevent accidental movement of the equipment. After performing some types of repairs or service work, starting the equipment for the first time may cause an unexpected loss of drive, which can cause a loss of dynamic braking and lead to an unexpected freewheeling condition on some types of equipment. Always consult the manufacturer's service literature for the proper procedures.
- When starting up the equipment for the first time after a repair or service procedure to a driveline, use extreme caution when dealing with rotating shafts. Keep your hands clear of possible rotating shafts on initial start-up.

Driveline Inspection

Driveshaft inspection needs to be performed on a regular basis because driveline failures on equipment often lead to a loss of drive, exposing the operator to potential injury, and/or damaging the equipment. When equipment comes in for service or repair, you can limit driveline performance issues by performing a few inspection procedures each time the driveshafts are lubed.

- Driveline vibrations and U-joint failures are often caused by loose **end yokes**, slip spline play, bent driveshaft tubing, and excessive radial play.
- Check the drivelines visually for twisting, dents, missing balance weights, and build-up of dirt on the shaft, which can lead to balance problems.

Regular inspection of the driveshaft during routine lube intervals will pay off in fewer unexpected breakdowns. The most common cause of U-joint and slip joint problems can be traced to lack of proper lubrication. The importance of proper lubrication must be emphasized, as many service personnel do not follow the recommended procedures. Enough grease must be injected to ensure that the old grease from the U-joint or splines is flushed out because it may contain grit that acts as an abrasive when mixed with the grease.

> *Tech Tip:* Before you attempt to grease driveshaft fittings, clean the zerk nipple carefully. This prevents any dirt from being forced into the bearing during re-lubing.

Lubrication

The importance of using the correct type of lubrication on drivelines cannot be overemphasized on off-road equipment. Lithium soap-based extreme pressure (EP) grease that meets the National Lubricating Grease Institute (NLGI) classification grade 1 or 2 specifications should be selected. Typically, grade 3 and 4 greases are not recommended for use in cold weather. If the bearings are sealed, they are pre-lubed with synthetic grease and do not require any additional greasing. When replacing a U-joint, note that the grease in the U-joint is there only to protect it during shipping and storage. Lubing the U-joint after installation is a service requirement and should never be put off. Lubrication schedules can vary greatly depending on operating conditions. Off-road equipment generally requires that the severe-duty cycle recommendations on lube charts be followed. Off-road classifications are generally considered to apply to equipment that operates on unpaved roadways for 10% or more of the time.

U-Joint and Slip Joint Lubrication

Proper lubrication of the driveshaft U-joints and slip joint is extremely important to the life expectancy of the driveshaft (see **Table 13-1**). Pump grease through the fitting slowly until the new grease

TABLE 13-1: LUBRICATION SCHEDULE

Type of Service	Miles	Time
City driving	5,000 to 8,000	3 months
On highway	10,000 to 15,000	1 month
Extended haul	50,000	3 months
Severe usage	2,000 to 3,000	1 month

© Cengage Learning 2014

flows from each of the four caps. This ensures that any dirt that has gotten into the caps has been flushed out and replaced with fresh grease. If grease does not flow from all four bearing caps, try moving the driveshaft U-joint from side to side while applying grease.

If this does not resolve the problem, grease passages may be plugged and you may have to remove the U-joint to check the condition of the bearing cap and needle bearings. Under no circumstances should this condition be ignored because it will lead to a driveshaft failure. The cost of a U-joint is a small price to pay compared to the cost of a driveshaft. If driveshaft separation occurs while the equipment is moving, extensive damage may occur to not only the driveshaft, but also to any surrounding components, such as transmissions and drive axles. Whenever U-joints are lubed, it makes sense to lube the slip joint also. The same technique used on the U-joints applies to the slip joint.

Flush fresh grease through the grease fitting until the grease comes out the relief hole in the end cap. Block the relief hole and continue to flush the grease through the slip joint until the grease appears at the **slip yoke** seal area. It may become necessary to remove the end cap to completely flush the slip joint.

Using the appropriate grease for the application is important. For example, in extremely cold weather the use of a cold-weather-rated grease will ensure that the grease does not freeze in the splines and cause unnecessary wear or damage. Pillow block bearings or hanger bearings usually come sealed. If no grease fitting is present, refer to the manufacturer's recommendations for the proper service procedure. If it becomes necessary to replace the hanger bearing, be sure to note the shim pack thickness used under the hanger bearing or pillow block. The shim pack ensures the correct location of the bearing assembly to maintain correct driveshaft angles.

U-Joint Replacement

Replacement of universal joints is considered a routine procedure in a service shop and does not require any special tools other than a U-joint puller and press, which should be part of a service shop's tool inventory. In some cases, replacement of the U-joint may require you to support the driveshaft to prevent it from dropping unexpectedly. Before performing this work refer to the manufacturer's service literature to ensure that the work is performed correctly.

CAUTION *To prevent the possibility of an eye injury, always wear safety glasses when performing any work on the equipment. When it becomes necessary to use a low blow hammer to dislodge a bearing cap or U-joint from a yoke, serious personal injury or damage can occur if safe procedures are not followed.*

U-Joint Installation

Hardened hammers have a place in a technician's toolbox, but they should not be used to remove a U-joint from a yoke. Damage will often result from the use of this tool. When a U-joint replacement is performed, serious damage can occur to the driveshaft assembly if excessive force or blows are used. When replacing U-joints with the driveshaft mounted in a vise, be sure not to exert excessive clamping force on the drive tube; this will distort or bend the tube, causing premature driveline or U-joint failure.

There are a number of U-joint mounting designs that are used on equipment, as shown in **Figure 13-20**; only the most common types used will be discussed in detail in this section.

The high **wing-type U-joint**, as shown in **Figure 13-21**, is the most common design used on off-road driveshafts; the installation procedures will be discussed here in depth. Proper blocking of the equipment is essential to performing the work safely. If you intend to remove the driveshaft, it is necessary to gather the correct blocking or slings to support the shaft once it is disconnected from the yoke.

- Support the driveshaft with the appropriate blocking or sling, loosen the four cap screws

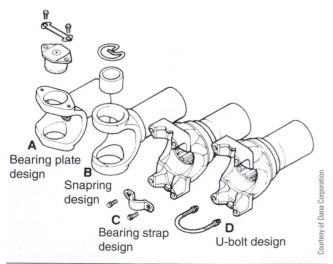

Figure 13-20 Four types of mounting joints used on heavy equipment.

A Bearing plate design
B Snapring design
C Bearing strap design
D U-bolt design

Courtesy of Dana Corporation

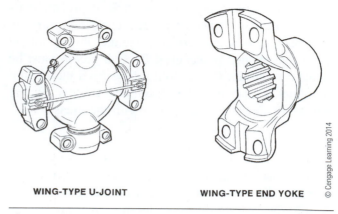

WING-TYPE U-JOINT   WING-TYPE END YOKE

© Cengage Learning 2014

Figure 13-21 Typical high wing-type U-joint used on off-road equipment.

and remove them from the yoke end of the driveline.

- While supporting the driveline yoke, pry the U-joint from the component end yoke. If a hammer blow is required to loosen the driveline from the end yoke, use only a brass, copper, or soft blow hammer.

- Do not use heat to loosen a U-joint cap from the yoke; it will ruin the U-joint. If a slide hammer is required to remove the U-joint from the driveshaft, bear in mind that damage to the driveshaft will occur if excessive force is used.

- High wing-type U-joints are generally permanently assembled with steel straps welded by the manufacturer during assembly. Unless you are discarding the U-joint, do not cut these straps; they protect the U-joint from misalignment when fitting to the yoke.

- Permalube U-joints have no grease fittings and do not require greasing. If a grease fitting is present, then the U-joint must be greased with the appropriate grease following the lube procedure outlined by the manufacturer.

- Before installing the new U-joint, measure the distance between the yoke ears with a yoke gauge tool to ensure that the correct U-joint is being installed (see **Figure 13-22**). This fit is critical to the proper operation and life of the U-joint assembly.

- Install the U-joint back onto the driveshaft; lightly tap the bearing caps with a soft blow hammer until the keyway is bottomed out in the yoke. Install new cap screws and hand tighten them to the yoke. Do not apply lubrication or anti-seize compound to the yoke saddle of the connection. Always use the new bolts that are supplied with the U-joint.

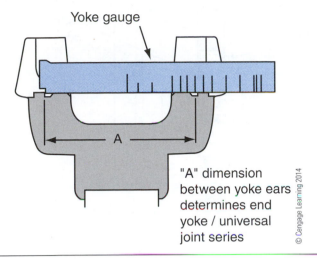

Yoke gauge

"A" dimension between yoke ears determines end yoke / universal joint series

© Cengage Learning 2014

Figure 13-22 Measuring the distance between the yoke ears to determine correct U-joint dimension.

Note: Some manufacturers apply a lock patch to the bolt threads, making it necessary to wrench the bolts in. Do not remove this coating from the threads. Use a torque wrench to alternately tighten the bolts to the proper specifications.

The torque chart shown in **Table 13-2** is provided as an example: refer to the U-joint manufacturer's specifications for the proper bolt torque. After installation, be sure to lubricate the U-joints until the grease flows from all four trunnion bearings. If the grease will not flow from all four trunnions, shake the assembly up and down and side to side while trying to apply grease. If this does not work, loosen the bearing cap-screws and apply grease until all four trunnion bearings

purge grease. Re-torque the capscrews to the correct torque specifications to complete the installation.

Bearing Cap U-Joint Removal

CAUTION *To prevent any possibility of eye injury, wear safety glasses when performing this type of work.* **Figure 13-23** *shows an example of removing a U-joint from a driveshaft yoke with a specialized U-joint two-jaw puller.*

- Expose the bolt heads by bending back the lock tabs before removing the bolts that hold the plate to the cap.
- Install a suitable U-joint two-jaw puller, as shown in **Figure 13-23** to remove the cap plates.
- Push the yoke to one side to facilitate the removal of the bearing cap from the yoke bore.
- Slide the yoke to the opposite end and remove the opposite bearing cap from the yoke bore.

Snap Ring Design U-Joint Replacement

As outlined in the previous procedure, a technician should follow the safety precautions listed below.

- Remove the snap rings from the bearing caps on all four U-joints. It may be necessary to lightly tap on each of the bearing caps to relieve the tension on the snap rings.

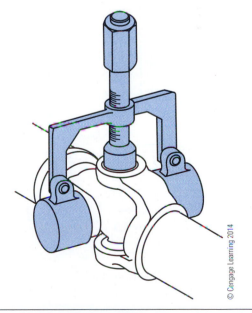

© Cengage Learning 2014

Figure 13-23 Universal joint puller used to remove plate-type U-joint cap plates.

TABLE 13-2: TORQUE CHART			
U-Joint	**Name**	**Description**	**Torque**
Wing style	Capscrew	½ – 20 × 2½	115–135 lb-ft
			155–183 nm
	Capscrew	½ – 20 × 1½	115–135 lb-ft
			155–183 nm
		⅜-24	40–50 lb-ft
			54–68 nm
		⁷⁄₁₆-20	63–83 lb-ft
			85–112 nm
	Jam nut	½-H.D.	63–83 lb-ft
			85–112 nm

© Cengage Learning 2014

- Use the correct removal tool for round bushing caps; using a different cap installation tool will damage the bearing caps. The use of the following three approved removal and installation procedures is highly recommended to prevent damage to the cups.

Removing Bearing Caps with a Bridge and Bearing-Cup Receiver Tool

- Place the U-joint into the bearing-cup bushing receiver being careful to align the bearing cups correctly in the receiver.
- Use the press to force the bottom cup out of the yoke, as shown in **Figure 13-24**.
- Turn the shaft end for end and repeat the last procedure until the cup is free of the yoke.
- The U-joint is now free to be removed from the yoke.

Removing a U-Joint with a Universal Joint Press

- Carefully place the universal joint press on the U-joint, as shown in **Figure 13-25**.
- With an appropriate wrench, turn the screw clockwise until the bearing cap becomes free, then turn counterclockwise until the bearing cap can be removed.

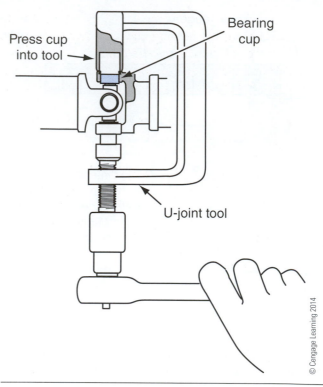

Figure 13-25 Removing bearing caps using a universal joint press.

- Turn the U-joint end for end and repeat the procedure to remove the other two bearing caps and remove the U-joint from the yoke.
- Using a wrench, turn the adjustment screw clockwise until the bushing loosens, and then turn it counterclockwise until the bushing can be removed.
- Turn the U-joint end for end and repeat the procedure for the other two bushings and remove the U-joint from the yoke.

Installing Replacement Universal Joints

- Carefully remove two bearing cups and thread the joint through the yoke.
- Replace one bearing cup on the trunnion and position the cross to accept the second bearing cup from the outside of the yoke.
- At this point, you can use either the U-joint press or the yoke bearing-cup installation tool to press the first bushing into the yoke just past the snap ring groove. This will allow you to install the first snap ring in position.
- After the second cup is pushed into the appropriate location, install the second snap ring. Be sure the snap ring is fully seated into position.

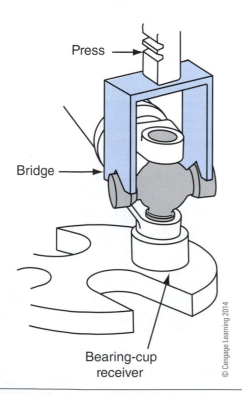

Figure 13-24 Removing bearing caps using a press, bridge, and bearing-cup receiver.

- Use a soft blow hammer to strike the yoke ears to ensure the bearing cups are centered and the universal joint is not binding.
- Hand tighten the cap screws on the bearing plate or strap design and torque up to manufacturer's specifications.
- Although the U-joint is factory pre-lubed for shipping purposes, this is inadequate for actual operating conditions. You must be sure to lube it until grease appears at all four bearing caps.

Driveshaft Installation

Once the U-joints have been installed, the driveline should be assembled correctly before installation on the equipment. Examine the driveshaft tubing for any damage that may have occurred while it was disassembled for servicing. Check for things like missing balance weights, evidence of twisting, or dirt build-up, which may cause operating problems. The following is a general list of potential problems to look for when installing a driveline on equipment:

- Dents or irregularities on the tube surface of the driveline need to be examined and a determination on its continued use should be made.
- The splines should not bind when moved in and out in the driveline's normal travel range.
- Verify that the slip joint seal is in good condition and will not become a source of contamination to the slip splines. (If in doubt, replace the seal.)
- The U-joint crosses must not be binding or have tight spots in their normal travel range.
- Check the yoke flanges carefully for any signs of damage, such as burrs, which will interfere with a proper fit. Often, parts are painted for aesthetic reasons and to prevent rusting. The paint should be removed from the mating surfaces of the yoke; it prevents proper seating of the parts during assembly.
- If there is evidence of driveline twisting on the tube, alignment of the yokes needs to be verified.

Driveline Balance

As mentioned earlier in this chapter, driveline vibration is not as apparent on off-road equipment as it is on high-speed road equipment. Therefore, it is important for a technician to spot any problems during servicing that would lead to early driveline component failure. Although driveline speeds are relatively low due to low road speeds, imbalance can still cause problems for components such as U-joint trunnion

bearings, driveshaft hanger bearings, and bulkhead bearings. A badly unbalanced driveline can cause bending movements to occur, which may lead to early driveline failure. If imbalance is suspected on a particular driveline, it may become necessary to have the driveshaft dynamically balanced at a shop that specializes in this type of procedure. Vibration on off-road equipment with an active suspension makes it next to impossible to make balance corrections to a driveline while on the equipment. If vibration damage shows up on off-road drivelines, the working angles should be verified on affected drivelines.

Measuring and Recording Driveline Angles

The following procedure may be used for off-road equipment.

Tools Required

- A good quality electronic **inclinometer** or spirit-level protractor is needed to measure driveline angles.
- A tape measure is needed to measure ride height or component height.
- A photocopy of a driveline worksheet is very helpful.

Procedure

- If the equipment has pressurized tires, be sure to inflate them to the recommended values before beginning this procedure. The equipment should be checked on level ground front to rear as well as from side to side.
- It is necessary to place the output yokes on driveshafts that are to be checked in the vertical position.
- Use a magnetic-base protractor or an inclinometer to measure the driveline angles. The use of a high-quality electronic inclinometer is recommended for this type of work. Accuracy is very important.
- Before you measure any angles, determine the components' angles. For example, look at the components and determine whether the front of the component is higher than the rear, or the rear is higher than the front.
- Determine the driveline phasing by checking the examples in **Figure 13-26** and record on a sheet of paper.
- The first measurement taken and recorded on a sheet of paper should be the transmission output

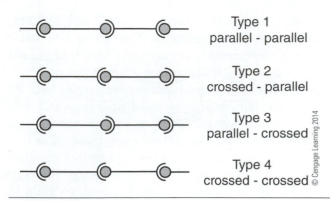

Figure 13-26 Example of a driveline phasing configuration.

yoke angle, as shown in **Figure 13-27**. If the yoke is inaccessible, use a flat section of the transmission case, such as a countershaft cover. Record the reading on a sheet of paper.

■ Measure the driveline angle by placing the inclinometer on a flat section of the driveline tubing to measure the first two driveshaft readings, as shown in **Figure 13-28**. Be sure that your inclinometer is not resting on any dirt or on a weld joint. Record the data on a sheet of paper.

■ Place the inclinometer on a flat portion of the drive axle or on the output yoke if possible (this will depend on the type of U-joint being used), and record the angle on a sheet of paper.

■ Place the inclinometer on the interaxle driveline, as shown in **Figure 13-29**, and record the angle on a sheet of paper. If there is no room to place the inclinometer on the driveline tube, then place it vertically and subtract 90 degrees from your reading.

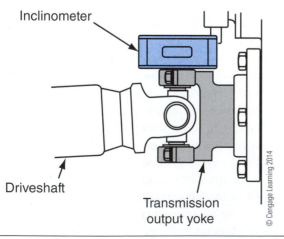

Figure 13-27 Measuring the transmission output yoke angle.

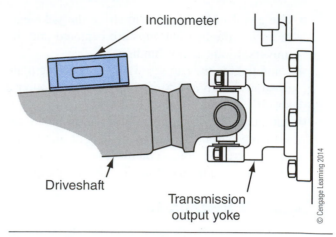

Figure 13-28 Measuring the driveline angle with an inclinometer.

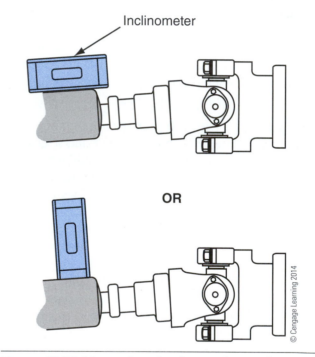

Figure 13-29 Measuring the forward rear drive axle angle on the output yoke.

■ Place the inclinometer on the input yoke of the rear axle. (This will depend on the type of U-joint being used.)

Note: A spacer may be required, or you may have to place the inclinometer on a flat portion of the axle tube to obtain a reading.

Making Corrections to the U-Joint Operating Angles

The methods that are required to make changes to the driveline working angles on off-road equipment will

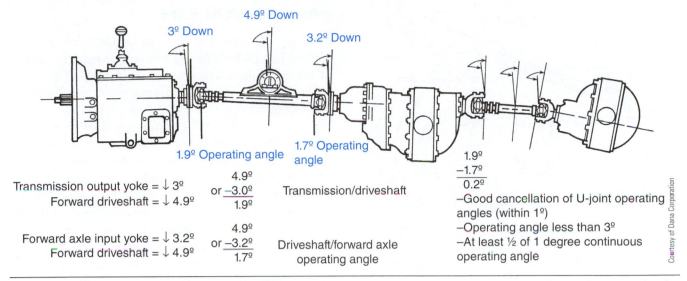

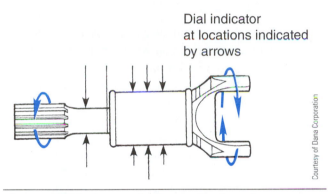

4.9° Down

3° Down

3.2° Down

1.9° Operating angle

1.7° Operating angle

1.9°
−1.7°
0.2°

Transmission output yoke = ↓ 3°	4.9°	Transmission/driveshaft
Forward driveshaft = ↓ 4.9°	or −3.0°	
	1.9°	

Forward axle input yoke = ↓ 3.2°	4.9°	Driveshaft/forward axle
Forward driveshaft = ↓ 4.9°	or −3.2°	operating angle
	1.7°	

−Good cancellation of U-joint operating angles (within 1°)
−Operating angle less than 3°
−At least ½ of 1 degree continuous operating angle

Courtesy of Dana Corporation

Figure 13-30 Example of a driveline system showing completed measuring procedure along with the calculations necessary to evaluate the working angles.

depend totally on the type of powertrain used. If the equipment has an active suspension system such as Air Ride®, a slightly different procedure must be followed. Before measuring the working angles on equipment, check the manufacturer's recommendations for performing this type of driveline evaluation. **Figure 13-30** shows the location of angles that need to be measured.

Equipment that has leaf-spring suspension uses axle shims to change the working angle of the components. These are generally installed on the axle saddle between the leaf springs and the axle. This adjustment tilts the axle to change the working angles on the driveline.

On some off-road equipment, the components are solidly mounted to the frame; in this case the component is shimmed between the mounting points to change the working angles. Equipment that has components that are solidly mounted to the frame, such as front-end loaders, generally has very few problems with changes in working angles. The same can't be said for equipment that uses some form of live suspension.

The U-joint operating angles can be changed by things such as worn bushings in the spring hangers or incorrect air bag height dimensions. Changes to the chassis from overloading or wear and tear on the engine and transmission mounts can account for undesirable working angle changes.

Driveline **runout** has a negative effect on vibration and driveline component wear, as shown in **Figure 13-31**. Checking the driveline runout is a relatively easy procedure and can be done in the field with only a few simple tools. The one stipulation that must be adhered to is that the driveline must be clean

Dial indicator at locations indicated by arrows

Courtesy of Dana Corporation

Figure 13-31 Arrows indicate the locations to perform measurements required to access the driveline runout condition of the driveline.

in the area where the runout readings are to be performed.

If runout is detected while measuring driveline uniformity, a maximum tolerance of 0.010" (0.254 mm) is generally allowable, but as always, refer to the manufacturer's service literature for specific applications.

When trying to pinpoint a runout problem on a driveline, one factor that could play a major role in maintaining ideal working angle is yoke runout. A dial indicator is mounted to the transmission housing close to the yoke positioned on a machined surface of the yoke near the yoke shoulder, as shown in **Figure 13-32**. Different designs of U-joints will necessitate a creative approach to obtaining this measurement. Take a reading on one end of the yoke; then rotate the yoke 180 degrees and take a second reading. A good rule is not to exceed 0.005" (0.127 mm) runout. Always check the manufacturer's service literature for specifications.

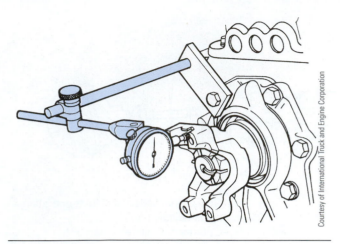

Figure 13-32 When performing these runout checks, be sure that the pointer is on a machined surface.

Vertical Alignment Check of a Yoke

Checking the vertical alignment is performed by using a protractor held against the machined face of the yoke, as shown in **Figure 13-33**. Ensure that the equipment is on level ground and that the component and its yoke are properly torqued. A loose mount will give a false reading, leading to a bad diagnosis. To perform this check, the driveshaft must be disconnected. Access to the yoke face is required, as well as the ability to rotate the yoke to a vertical position to obtain the reading. A protractor with a magnetic base and attached to the yoke should give a reading of 90 degrees minus the component offset angle. The reading should be within ½ degree of angle of the component to which the yoke is attached. Should the reading be higher than specified, the yoke must be replaced.

FAILURE ANALYSIS

Many driveline failures can be identified and explained just by performing a detailed visual inspection of the failed parts. However, certain types of driveline failures require specialized equipment and technical knowledge to determine the cause. Often the failures can be attributed to poor equipment operation, which can be the result of operators not being aware of the damage that can occur from certain operating practices. The second issue we need to be aware of is the use of driveline components that do not meet the application requirements for a specific use. Driveline components are manufactured to meet a large variety of applications; they need to be selected on this basis. A third issue to be discussed is the prevalence of poor maintenance practices, which have sometimes become accepted standards in some shops.

We will examine the most common types of failures and attempt to explain what occurred to cause the failure. Driveline failures occur in many different forms, but in the majority of cases, U-joints are to blame. Whenever driveline service is performed, a technician should spend some time inspecting the U-joints and yokes for any signs of problems.

A technician should look for the following conditions when examining a U-joint for reuse:

- **Galling**: This occurs when friction between two surfaces causes metal to be displaced.
- **Spalling**: Fatigue causes flakes of metal to break off, as shown in **Figure 13-34**. This condition is associated with splines as well as trunnion races.
- **Brinelling**: Surface fatigue causes grooves to wear into a surface, as shown in **Figure 13-35**.

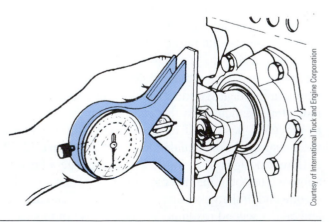

Figure 13-33 Magnetic-based protractor used to check vertical alignment of the yoke.

Figure 13-34 Example of spalling on a U-joint trunnion bearing surface.

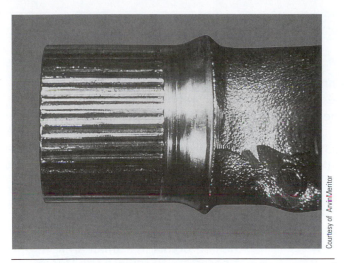

Courtesy of ArvinMeritor

Figure 13-35 Example of brinelling on the surface of a U-joint trunnion bearing surface.

Courtesy of ArvinMeritor

Figure 13-36 Hairline cracks in metal finally lead to complete fracture of the U-joint cross.

It is often associated with improper U-joint installation. Watch for false brinelling, which looks like a polished surface rather than actual grooves worn into a surface.

- **Pitting**: This is caused by corrosion that forms little pits in the metal leading to surface wear and parts failure.
- **Cracks**: These are an indication of metal fatigue, which starts out as tiny stress cracks. Eventually this weakens the metal until it finally breaks, as shown in **Figure 13-36**.

Tables 13-3 through **13-9** show various conditions, their possible cause, and suggested corrective actions used to address driveline failures.

ONLINE TASKS

1. Use a suitable Internet search engine to research the configuration of an off-road driveshaft system. Identify a piece of equipment in your particular location, paying close attention to the driveline arrangement, and identify whether it is a parallel or a non-parallel design. Outline the reasoning behind the manufacturer's choice of driveline design for that particular application.
2. Check out this URL: *http://www.meritor.com/ customer/northamerica/training/default.aspx* for more information on driveline configuration and design.

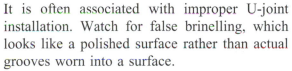

TABLE 13-3: PREMATURE WEAR PROBLEMS		
Condition	**Cause**	**Corrective Action**
Premature U-joint wear	Cross hole misaligned on end yoke	Check the alignment of the cross holes on the end yoke with an alignment bar. Replace if misaligned.
	Excessive angularity	Measure U-joint operating angles. Reduce the angles if excessive.
	Lubrication frequency Inadequate or incorrect lube for the application	Check lubrication and application requirement to ensure correct lube is being used.
	Damaged or worn seals	Replace U-joint.
Repeated U-joint wear	Excessive heavy operating loads	Use a higher capacity U-joint to carry heavy load.
	Operating at excessive angles and speeds	Install a high-capacity heavy-duty driveline and U-joint.
	Worn or damaged seals	Replace universal joints.
	Inadequate lubrication intervals or incorrect grade of lubricant	Lubricate according to manufacturer's recommendations.

(continued)

TABLE 13-3: (CONTINUED)

Condition	Cause	Corrective Action
End galling of trunnion and bearing assembly	Driveline operating at excessive angle	Measure driveline angles and reduce if they are excessive.
	Torque load too high for driveline U-joint size	Install a higher capacity driveline U-joint.
	Lubrication intervals are inadequate or wrong grade of lubricant being used	Lubricate according to manufacturer's recommendations.
Needle rollers brinelled into bearing cup and cross	Extremely heavy running loads	Install a higher capacity driveline U-joint.
	Insufficient operating angles	Increase the operating angles to at least 2 degrees.
	Normal appearing bearing wear	If brinelling occurs in a small area, it is not necessary to replace the U-joint.
	Lubrication intervals are inadequate or wrong grade of lubricant being used	Lubricate according to manufacturer's recommendations.
Spalling of needle bearing into trunnion surfaces	Lubrication intervals are inadequate or wrong grade of lubricant being used	Lubricate according to manufacturer's recommendations.
	Contamination from dirt	Check U-joints for wear; replace if necessary.
		If brinelling occurs in a small area, it is not necessary to replace the U-joint.
	Normal bearing wear	Worn components should be replaced.
Bearing or cross assembly broken	Excessive torque loads for the size of driveline and U-joints	Replace with a higher capacity driveline U-joint.

© Cengage Learning 2014

TABLE 13-4: YOKE FRACTURE PROBLEMS

Condition	Cause	Corrective Action
Yoke broken or cracked	Yoke lug interference at full jounce or rebound	Use high angle yokes. Check the design application.
	Excessive torque loads for the size of driveline and U-joints	Replace with a higher capacity driveline U-joint.
	Fatigue from bending caused by high secondary couple loads	Reduce the universal joint running angles.

© Cengage Learning 2014

TABLE 13-5: UNIVERSAL JOINT CENTER PROBLEMS

Condition	Cause	Corrective Action
Trunnion or U-joint cross fracture	Excessive high loading	Overloading or abuse of vehicle
Fractured bushing	Excessive loading of U-joint	Check for excessive driveline torque in low gear. Use high-capacity U-joints.
	Excessive operating angles	Check for high working angles. Reduce if necessary.
	Worn or damaged parts	Replace parts if worn excessively.

© Cengage Learning 2014

TABLE 13-6: WING-STYLE U-JOINT PROBLEMS

Condition	Cause	Corrective Action
Loose mounting bolts	Dirt or paint on the mounting pads	Check to see if the mounting pads are fretted or loose drive tang.
Broken mounting bolts	Overtorqued or undertorqued mounting bolts	Yoke mounting pads must be free of dirt and bushing must be fully seated before torque is applied.
	Fretting on the mounting pads or around bolt holes with fretting on the drive tang	
	Fretting on the mounting pad or bolt hole	Loose bolt
	Excessive driveline angularity	Check for excessive driveline operating angle and reduce the angle if necessary.

TABLE 13-7: ROUND BUSHING PROBLEMS

Condition	Cause	Corrective Action
Difficult to remove or replace bushings	Yoke bushing hole distorted Corrosion caused by fretting with rust build-up	When removing the U-joints be careful not to distort the cup holes when using a hammer on the center cross.
New center cross will not install in the yoke	Yoke ears are distorted from abuse and are binding the parts.	Yoke must be replaced.

TABLE 13-8: SLIP YOKE SPLINE WEAR PROBLEMS

Condition	Cause	Corrective Action
Seized splines	Not using correct lubrication on splines	Lubricate according to manufacturer's recommendations.
	Worn or damaged splines	Splines need to be replaced.
	Contamination	Lubricate the splines according to manufacturer's recommendations.
Galling	Splines are worn or damaged	Mating splines need to be replaced.
	Contamination	Lubricate the splines according to manufacturer's recommendations. Check the condition of the spline seal.
Outside diameter is worn excessively at the extremes.	Wrong lubrication	Should use a larger diameter tube
	Extremely loose outside diameter fit	Spline components need to be replaced.
Spline shafts or tube is broken in torsion.	Tube size inadequate	Replace with a larger capacity driveline and U-joints.

© Cengage Learning 2014

TABLE 13-9: SHAFT AND TUBE PROBLEMS

Condition	Cause	Corrective Action
Support bearing shaft wear	Driveline is too long for the operating speeds.	Replace with a two-piece driveline with a shaft support bearing.
	Incorrect lube of bearings	Center bearing should be replaced.
Support shaft rubber insulator wear	Bending fatigue due to excessive coupling loads	Reduce the U-joint operating angles.
	Too heavy a torque load for the U-joint and driveline size	Replace with higher capacity driveline and U-joint
	Support shaft bearing misaligned	Mounting bracket needs to be realigned to frame cross member.
Tube circle weld is fractured.	Balance weight located in the apex of the weld yoke	Tube needs to be replaced and rebalanced.
	Balance weight is too close to the circle weld.	Tube needs to be replaced and rebalanced.
	Circle weld is defective.	Tube needs to be replaced and rebalanced.
Shaft is broken from bending.	Driveline is too long for the operating speeds.	Replace with a two-piece driveline with a shaft support bearing.
	Bending fatigue due to excessive coupling loads	Reduce the U-joint operating angle.

© Cengage Learning 2014

Shop Tasks

1. Select a piece of equipment in your shop that has a driveline assembly and identify the type of components that make up the driveline arrangement. Identify whether or not the driveline has excessive runout, and what type of U-joint is used on this arrangement.

2. Select a piece of mobile equipment in your shop that uses a driveline, identify the type and model of the equipment, and list all the major components and the corresponding part numbers of the driveline. Identify and use the correct parts book for this task.

Summary

- Typical driveline assembly components consist of yokes, U-joints, slip splines, slip assembly/ies, hanger bearings, and propeller shafts.

- Driveline designs can be split into two categories: parallel driveshaft installations and broken-back driveshaft (or non-parallel) installations.

- Non-parallel design is used when the zero vertical angles of the component installation cannot be met.

- Off-road equipment component placement can result in drivelines having compound driveline angles.

- Driveline speed plays an important role in determining the maximum number of allowable working angles that will be used.

- Ensure that working angles are within 1 degree of each other to maintain equal U-joint speed at each end of the driveshaft.

- The most common tubing used in the construction of heavy-duty drivelines is cold-rolled, electric-welded tubing.

- The splines on the slip joints are made from high-quality hardened steel.

- A maximum tolerance of 0.010" (0.254 mm) is generally allowable when measuring driveline uniformity.

- A driveshaft should not exceed 70 inches in overall length.

- The universal joint trunnions have bearing caps with internal needle bearings mounted on the ends that allow the caps to rotate.

- Some manufacturers locate a standpipe (check valve) at each end of the bearing cross.

- Wing joints come with special high-grade cap screws and should be torqued to manufacturer's specifications.

- If you double the working angle, the life expectancy would be reduced by 50%. If you reduce the load by 50%, you double the life of the U-joint.

- Although a U-joint is factory pre-lubed for shipping purposes, this grease is inadequate for actual operating conditions.

- Use a soft metal hammer (copper or brass) to prevent damage to the hardened parts.

- Magnetic-base protractor or an inclinometer are used to measure the driveline angles.

Review Questions

1. Why must a driveshaft maintain flexibility while continuously transferring torque between powertrain components?

   A. to allow the driveline to lengthen during cornering

   B. to allow for continuous length changes in reaction to road changes

   C. to allow for contraction during load changes

   D. to allow for speed fluctuations

2. Which component consists of a hardened, splined stub, welded to the end of a driveshaft tube?

   A. universal joint

   B. slip yoke

   C. hanger bearing

   D. cardan joint

3. Which component contains a forged journal cross and trunnions?

   A. slip yoke

   B. center support

   C. universal joint

   D. hanger bearing

4. What describes the purpose of a slip yoke on a driveshaft?

   A. slips to dampen clutch engagement and road shock during deceleration

   B. allows the driveshaft to change length in response to the road profile

   C. permits the transmission to slide back for service

   D. all of the above

5. Which factors apply to determining driveline diameter and tube wall thickness?

   A. operating torque load

   B. operating environment

   C. equipment size

   D. equipment speed

6. Which design feature prevents the reverse flow of hot grease in a universal joint?

   A. zerk fittings

   B. trunnions

   C. bearing cap seal

   D. standpipe

7. What term is used to identify the universal joint cross in technical documentation?

   A. spider

   B. trunnion

   C. yoke

   D. spline

8. Which angles must be equal on a one-piece parallel joint driveshaft installation with two U-joints?

   A. All working angles on all driveshafts must be equal.

   B. Angles of both mating yokes must be equal.

   C. Angles of the transmission output shaft and the axle pinion shaft must be equal.

   D. Working angles of its two U-joints must be equal.

9. Which statement indicates the correct installation specification for a set of U-joints on a driveshaft?

   A. maximum operating angle of 10 degrees

   B. minimum operating angle of 0 degree

   C. operating angles of each U-joint should be within 3 degree

   D. operating angles of each U-joint should be within 1 degree

10. Which operating application would be correct when determining more frequent driveline lubrication intervals?

   A. off-road operation

   B. pick-up and delivery

   C. linehaul trucking

   D. short-haul trucking

11. Which failure would likely occur from an incorrectly installed U-joint?

   A. damage from galling

   B. damage from spalling

   C. damage from pitting

   D. damage from brinelling

12. Which statement would not be considered a cause of driveline vibration?

   A. driveshaft out-of-balance

   B. equal operating angles

   C. improper phasing

   D. unequal U-joint working angles

13. Which is the correct procedure when lubricating U-joints?

   A. Grease until you see grease come out of a trunnion seal.

   B. Pump grease until it exits from all four trunnion seals.

   C. Apply only two shots of grease per trunnion.

   D. Hand pack each bearing cap when first installing a U-joint.

14. What must have occurred if a driveshaft is out of phase?

   A. U-joint has been assembled incorrectly.

   B. Transmission yoke has been damaged.

   C. Differential carrier yoke has been damaged.

   D. Driveshaft has been separated at the slip splines.

15. What tool must be used to remove a universal joint from a driveline yoke?

   A. two-jaw U-joint puller

   B. 12-pound sledge hammer

   C. 25-pound slide hammer on yoke

   D. hydraulic jack on yoke hub

16. When inspecting a U-joint failure you see metal flakes separating from the bearing surfaces on the cross. What has occurred to cause this?

   A. brinelling

   B. spalling

   C. galling

   D. false brinelling

14 Heavy-Duty Drive Axles

Learning Objectives

After reading this chapter, you should be able to:

- Describe the types of heavy-duty drive axles used on off-road equipment.
- Identify the construction features of heavy-duty drive axle assemblies.
- Describe the operation of drive axles used on off-road equipment.
- Identify the different differential designs used on heavy-duty drive axles.
- Identify the components of a heavy-duty off-road differential.
- Describe the operation of a heavy-duty off-road differential.
- Describe the operation of a "Torque Hub Planetary."
- Identify the maintenance and adjustment procedures on heavy-duty drive axles.
- Describe the common failures that occur on heavy-duty drive axles and analyze the causes.

Key Terms

amboid gear	double reduction axle	planetary gear set
banjo housing	full-floating axle	semi-floating axle
bevel gear	hypoid gear set	single reduction axle
carrier assembly	interaxle differential	spiral bevel gear set
crown gear	limited slip differential	tandem drive
differential	live axle	torque hub planetary drive
differential carrier	no-spin differential	
differential lock	pinion gear	

INTRODUCTION

Heavy-duty drive axles are the final phase in a powertrain and are used to deliver torque to the drive wheels. Drive axles can be categorized into three distinct designs. The first design propels the vehicle and supports the weight of the load. The second design carries the load and steers the vehicle. The third design only supports the load; this type is found on the rear section of articulating haulage trucks. Drive axles are usually mounted directly to the equipment frame and must be capable of supporting the full weight of the equipment, as well as the torque and shock loads that are generated during operation.

Off-road equipment axles fall into two categories, the more common being the axle assembly that only transmits torque to the wheels and has no steering capability. The second category provides steering capability.

Articulating steering is used on equipment with an axle that has no steering capability. Articulating wheel loaders have a **differential** in each of the two axle assemblies. Larger articulating haulage trucks can have two rear axles, as shown in **Figure 14-1** with a single axle on the front section. All drive axles (sometimes called **live axles**), whether steering or straight drive, are required to change the direction of the flow of power and reduce the speed while increasing the torque to the drive wheels. The **differential carrier** is required to accommodate the change in wheel speeds when cornering, allowing the outer wheel to travel farther than the inner wheel when a vehicle corners. Axles extend out from the differential carrier to the wheels that drive the wheel hubs. On large equipment with outboard planetary drives, the axles do not carry any weight and are called **full-floating axles**. Axles that are used only to steer the equipment or carry the weight of the load are known as dead axles. Most configurations of off-road equipment use all-wheel-drive due to the terrain they operate on; they rarely use a dead axle.

The second category of drive axle assembly used on off-road equipment has steering capabilities and is used on equipment where an articulating frame is not required. Because of the high torque requirement and slow travel speeds, axle assemblies used on off-road equipment come with a conventional differential **carrier assembly** and often use a second speed

reduction, through the use of a planetary drive, that can be located inboard or outboard on the axle assembly. On large axle assemblies, the differential carrier is considered the primary reduction with the axles driving a secondary reduction using a planetary drive. The primary and secondary drives usually are lubricated with the same type of oil, but the drives often have their own oil level checks.

FUNDAMENTALS

Although the drive axle's primary role is to carry the vehicle load, it must also be able to propel the equipment. The role of the differential is crucial to the operation of the drive axle. It must split the torque coming from the transmission and send each wheel the appropriate amount of power to propel the equipment and at the same time allow the wheels to individually rotate at different speeds when the equipment needs to turn. Each drive axle, or half shafts as they are sometimes called, sends power independently to the wheels. The power flows from the driveshaft to the pinion shaft to the ring gear (or **crown gear**), which changes the flow of power 90 degrees and splits it, sending equal amounts of power to each drive wheel. The design of the crown and **pinion gear** set allows the flow of power to be redirected at a 90-degree angle. The term *differential* is derived from the term *to differentiate*—to show a difference. The actual speed needs of each wheel can be split by the differential without interrupting the power sent to the wheels.

Tech Tip: Manufacturers do not use standard terminology for the components of a drive axle assembly. For example, some manufacturers describe the ring gear as a crown gear, or call a drive axle a half shaft. The same thing occurs with the term *driveshaft*. Some manufacturers use this term to identify the shaft that transmits power from the differential to the wheel ends. Others refer to the driveshaft as the shaft that transmits power to the differential carrier from the transmission. We attempt to use consistent terms in this book, but you should get used to the lack of standardization as it is unlikely to change in the near future. Most technicians will encounter equipment literature from many different manufacturers, and will become accustomed to each manufacturer's technical terms.

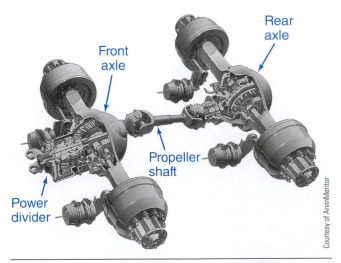

Courtesy of ArvinMeritor

Figure 14-1 Typical tandem drive used on large articulating haulage trucks.

CONSTRUCTION FEATURES

The carrier assembly consists of seven major sub-components as shown in **Figure 14-2**:

- Ring gear
- Input shaft
- Pinion gear
- Differential spider
- Differential side gear
- Differential pinion gear
- Drive axles

To understand how subcomponents work together, we need to examine the components that make up the carrier assembly. **Figure 14-2** shows the input yoke mounted to the splined pinion shaft and secured in place with a large nut that must be torqued to specification.

Note: The torque on this nut is critical to the pinion shaft's tapered roller bearing preload, and must be performed properly. The flow of power is transmitted to the crown gear at right angles from the pinion gear, which is in constant mesh with it.

The crown gear is bolted to the differential case halves, as shown in **Figure 14-3**. The breakdown of the parts in the differential assembly is shown in the exploded view. The four legs of the spider are bolted into corresponding holes in the differential case halves. Four pinion gears are mounted to the spider legs,

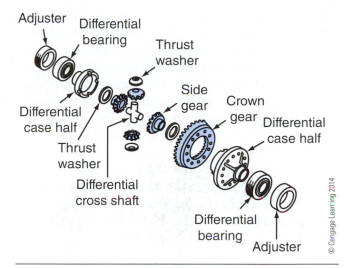

Figure 14-3 Typical breakdown of components in a differential carrier.

which are in mesh with the teeth of the two side gears. The differential case is held in place with tapered roller bearings on each end that are adjusted with a preload when the running pattern is set up. The flow of power is sent to the carrier assembly from the crown gear where the flow of power is divided and sent to the axles. The side gears have internal splines that mesh with the internal splines of the axles. The carrier assembly is flange-mounted to the **banjo housing**. The axles are inserted into the ends of the banjo housing and mesh internally with the side gears.

OPERATING PRINCIPLES

In the past, solid axles were used extensively and worked well when the equipment moved in a straight line with little or no turning. Tire wear became a problem if the equipment had to be turned frequently. To help prevent excessive tire scuffing, a method was devised to enable the wheels to travel at different speeds when cornering and still allow torque to be transferred to the wheels.

To understand how this differential action works, we examine what takes place during operation. When the equipment is driven forward in a straight line, torque is transmitted through a driveshaft to the pinion shaft of the differential carrier. The torque is then sent through the ring gear, which changes the direction of power flow by 90 degrees, allowing the axles to be driven at right angles to the input power.

When torque rotates the crown gear, which is bolted to the differential case, it causes the spider to rotate. Differential pinions are mounted to each of the four spider shafts. The pinions, which are in mesh with the

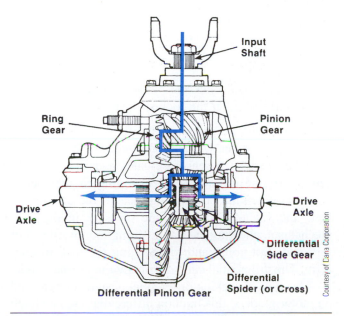

Figure 14-2 Typical differential identifying the main components and showing the flow of power to the drive axles.

side gears, rotate thereby allowing the axles to transmit torque to the drive wheels. The differential pinion gears are at right angles to the side gears; when the differential gears rotate with the spider gears, the side gears move along with them. The side gears are free to rotate because they are not splined to the differential housing. They can rotate independent of the differential housing. If the axles are both going the same speed, then the differential pinion gears will not rotate on their axes, as shown in **Figure 14-4**. The torque applied to the side gears will be equal as long as the equipment moves in a straight line. There will be no motion inside the differential housing; the differential gears, ring gear, and drive axles all rotate at the same speed.

Differential Operation

If a U-turn is performed with a loader, you would see the differential action that takes place inside a conventional differential. If the inside wheel slows down 10%, the outside wheel will speed up 10%.

As an example, a loader with a 10-foot width makes a 180-degree turn. The inside wheel is traveling on a 10-foot (3.05 m) radius and the outside wheel is traveling on a 20-foot (6.1 m) radius. If you perform the calculations, you will see that the inside wheel actually travels a distance of 31.5 feet (9.6 m), while the outside wheel travels over twice that distance, 63 feet (19.2 m). This shows why a differential is necessary on off-road equipment.

Without differential action, the tires would take a real beating—not to mention the strain on the drive axle components. The inside wheel has more resistance when making the turn; this causes unequal torque to be applied to the side gears in the differential. This change in resistance forces one axle to slow down thus forcing the differential pinions to walk around the side gear. This movement, in turn, causes the opposite side gear to speed up an equal amount.

There is a downside to this design. If one wheel loses traction, the differential will operate the same as if in a turn. The same amount of torque is sent to both wheels, but it can only be equal to the amount that is required to turn the wheel with the least amount of resistance.

Locking Differential Operation

Both the solid axle and conventional differential have design limitations that do not work well individually on equipment. Therefore, a design had to be developed that combined the benefits of a solid axle with the turning capabilities of a conventional differential. The first locking differentials were designed to be locked and unlocked by the equipment operator. The operator would lock the differential when going in a straight line or when wheel slip occurred, and then unlock it when cornering.

The operator could choose to engage the locking differential when mud, snow, ice, or other slippery surface conditions present themselves. The lockout could be engaged mechanically by using air or hydraulic pressure. Depending on the equipment manufacturer, when the **differential lock** is activated, a solenoid sends air or hydraulic oil to the shift mechanism, which acts on the shift fork to engage the splines on the differential case. This locks the differential case, gearing, and axle shafts to provide equal torque to both wheels.

Figure 14-5 shows a single reduction differential carrier with lockout. The carrier consists of the following parts:

- Actuator assembly
- Shift fork
- Sensor switch
- Outer splines—axle shaft to shift collar
- Shift collar and differential case splines
- Axle shaft
- Mechanical interface between fork and detent
- Shift collar in the locked position

Figure 14-5 also shows the differential action with the lock disengaged (A) and engaged (B). When in the engaged position, a sensor actuates a dash light indicating that the differential is operating in the locked position. When the operator disengages the differential lock, the solenoid exhausts the pressure in the shift mechanism, allowing the return spring to move the shift fork and shift collar and disengaging the lock mechanism. The differential can now operate in normal mode.

This type of locking differential design is often used on motor graders; this traction differential uses a clutch between the left side gear and the differential housing. Depending on the manufacturer of the differential, either oil or air pressure is used to activate the clutches.

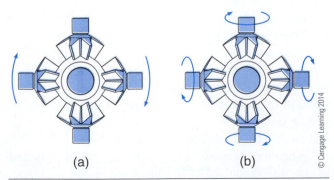

(a)   (b)

© Cengage Learning 2014

Figure 14-4 (a) Differential action in straight line motion. (b) Differential action during cornering.

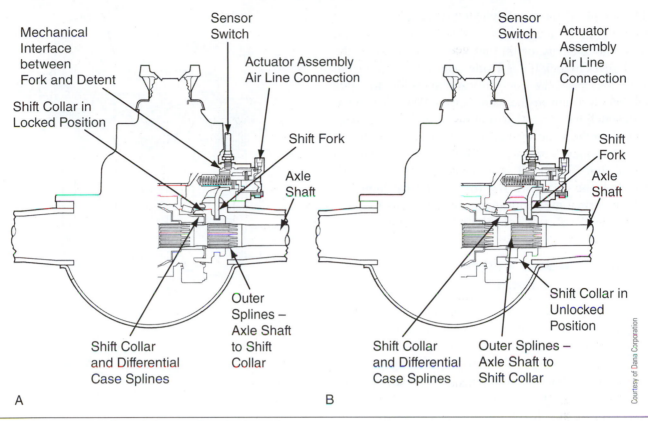

Figure 14-5 A parts breakdown of a single reduction differential carrier showing (A) lock in the disengage position, and (B) lock in the engaged position.

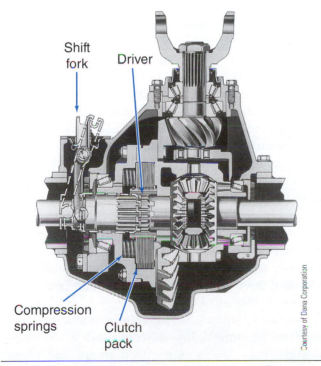

Figure 14-6 Heavy-duty controlled traction differential.

Figure 14-6 shows how a typical traction control works. When engaged, this design locks the left side gear to the differential case with the use of a clutch pack that, in this case, consists of 21 discs. The 10 friction discs are splined to the driver, while the 11 steel discs have tangs on the outside and engage the differential case while spring pressure keeps the discs clamped together. The shift fork is controlled by an actuator that when engaged forces the internal splines and the splines on the end of the side gear to lock. This action locks the side gear to the differential housing and causes both axles to turn at the same speed.

With this design, the wheels can slip based on a predetermined load, such as when the equipment must negotiate a turn. This slipping is based on the clutch disc's coefficient of friction. The increased torque on the outside wheel forces the plates in the clutch pack to slip and allows the outside side gear to increase its speed, forcing the differential to operate in normal mode. This allows the outside axle to turn faster than the inner axle. This design is available with either an operator-controlled shift controller or with a permanent traction control. The latter design can only operate in the conventional differential mode when the clutch discs slip; it will always be in the lockup mode.

Hydraulic Lock No-Spin Operation

Another variation of this design uses hydraulic pressure to activate the clutch discs. The operator can lock out the differential whenever the equipment is operating in a straight line. When engaged the oil flows into a cavity behind a piston which forces the clutch discs to lock the left side gear to the differential case. This causes the side gears and the spider to rotate together as a unit at the same speed. The differential pinions will lock the other side gear, forcing both axles to rotate at the same speed. This operating mode forces the differential to operate as if it were a solid axle. To release, the lockup oil is exhausted from the piston cavity by the operator. The differential can now operate in normal mode, allowing the side gears to rotate independently of each other as when cornering.

Mechanical Jaw Clutch No-Spin Operation

This simple mechanical differential lock is widely used on smaller equipment such as backhoes. A jaw clutch locks the differential, providing equal torque to both wheels. This type of lock is usually engaged with a foot control: the operator steps on a pedal, operated through a mechanical linkage, and constant torque is applied to a lock lever. The lever forces a fork to engage a coupling on the side gear into an adapter on the differential housing. This action pushes together the two halves of the jaw clutch, locking the side gear to the differential housing and providing equal torque distribution to each wheel. Once engagement has taken place, the lock pedal can be released. While the wheels are locked together, the side force generated by torque transfer keeps the jaw clutch engaged. When the torque becomes equal at each wheel, the reduced side force automatically disengages the jaw clutch. This simple design provides long, trouble-free life on smaller equipment.

Limited Slip Differential

This design of differential lock, as shown in **Figure 14-7**, is commonly used on smaller wheel loaders and backhoe loaders. It is very compact and lends itself to installation in tight spaces such as those found on small equipment. This design takes advantage of gear tooth separating forces to apply the clutches. The **limited slip differential** applies torque equally to both wheels during normal operation. **Figure 14-7** shows the two separate clutches, one on each side of the differential housing, connecting the side gear to the differential case. A small amount of side clearance is left in each clutch during assembly to ensure that each clutch engages correctly during operation. Under normal operating conditions with good traction at both

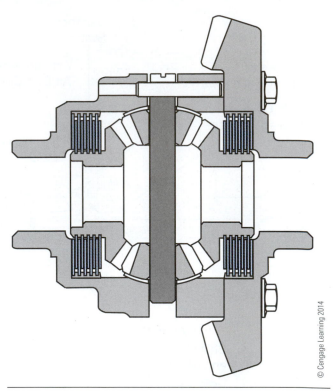

© Cengage Learning 2014

Figure 14-7 Typical limited slip differential.

wheels, equal torque will be distributed to each wheel during straight-line operation. In a conventional differential, if one wheel loses traction and speeds up, the other wheel slows down in proportion to the speed increase on the free wheel. This affects forward motion with speed and torque losses. If a limited slip differential is used in this same situation, the clutches make it more difficult for this to occur because as more input torque is applied to the differential, greater internal friction develops in the side clutch. The locking effect takes place due to high internal friction generated by the increased torque being applied. If a speed difference at the axles occurs, such as during cornering, the separating forces that occur in the differential during this process compress the clutch packs and allow the torque in the faster turning wheel to be redirected to the wheel that has greater traction. During turning, the wheels generate enough torque to overcome the clutch packs, allowing the wheels to change speed just as in a conventional differential.

Unlike a conventional differential spider, the spider used in the limited slip differential is a three-piece design. This design is used on small loaders and backhoes. The longer shaft holds two pinion gears, while each of the short shafts holds one pinion gear. The ends of each shaft fit a corresponding bore in the connector, as shown in **Figure 14-8**. Dowels are used to fasten the shafts to the rotating housing. Internal slack is present in the arrangement, which allows the

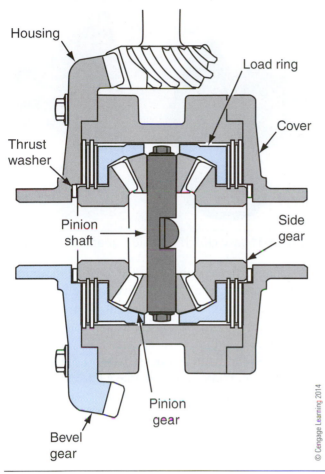

Housing

Load ring

Cover

Thrust
washer

Pinion
shaft

Side
gear

Pinion
gear

Bevel
gear

© Cengage Learning 2014

Figure 14-8 Limited slip differential that uses a load ring between the connector housing and the differential housing.

gear tooth separating forces to be transferred to the clutch discs. A side gear sits on top of each connector and meshes with the differential pinion gears. The clutch packs sit against the side gear contact face. The discs are splined to the side gears, while the plates have tangs that connect them to the housing.

Whenever a speed difference is present in the differential, gear tooth separating forces act on the side gears by axially squeezing the clutch packs together. The amount of torque is totally dependent on the input torque applied to the differential. When a clutch pack is engaged on the faster wheel end, the total input torque will increase, allowing the torque to be split when the clutches are engaged. The clearance in the clutch packs is adjusted by the use of shims when they are assembled.

On larger equipment a slightly different design is often used. Load rings are used to provide additional separating forces when a speed difference is present between the wheel ends. This design uses a two-piece spider that is notched in the middle to allow the two shafts to intersect with each other. Both ends of the

spider shafts are squared off and mate with wedges in two pressure rings inside the differential housing.

This design uses a pressure ring to hold the side gear in place. The pressure ring has lugs on the outside diameter that fit into corresponding slots in the differential housing. The design is such that the pressure ring is allowed to move axially in the differential housing whenever the clutch packs are engaged. The spider shafts are driven by the pressure rings and have a natural tendency to ride up in the wedge during operation.

During axle speed differences, the separating forces of the gear teeth will force the pressure rings to rotate the spider shafts. The square end of the spider shafts in the wedge-shaped cutout try to resist the rotational forces that are present, increasing the axial forces on the pressure rings. This force engages the clutch packs, resulting in torque transfer. The spiders are not connected to the differential housing directly; they are held in place by the pressure ring.

Figure 14-9 shows the discs and plates being installed. The plates have protruding notches on the outside diameter that match notches in the rotating housing, and the discs have internal teeth that mesh with the side gears. Axial forces that are generated by unequal axle speed force the clutch packs to engage. When fully assembled, the side gears sit on both sides of the spider shafts and pinion gears. They will mesh together at all times. The notches in the pressure rings hold the spider shafts in place. The clutch plates sit against the pressure rings, rotating with the differential housing, while the discs rotate with the side gear.

In **Figure 14-10**, the equipment is operating in a straight line with equal traction on each wheel. The torque is divided equally between the two wheels. The differential pinion gears rotate as a complete unit and

Reprinted Courtesy of Caterpillar Inc.

Figure 14-9 Note that the steel plates are held on the outside by the rotating housing and the discs are held by internal teeth to the side gears.

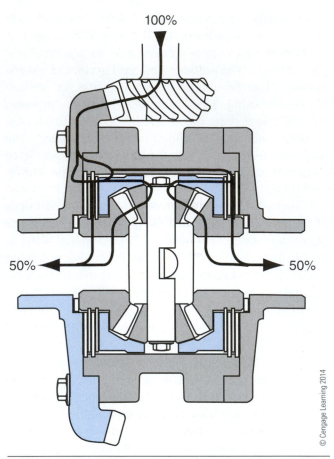

Figure 14-10 Limited slip differential operating with equal axle speed. Torque is divided equally between both wheels.

do not rotate on their own axis. If wheel slip occurs, separation forces will engage the clutch packs, forcing the faster wheel to transfer torque to the slower wheel. This torque transfer will be proportional to the input torque.

No-Spin Differential Operation

The **no-spin differential** uses jaw clutches that are splined to the side gears. The jaw clutches stay engaged with the side gears as long as the equipment is operated in a straight line. When the equipment turns, the outside wheel will speed up disengaging the outside jaw clutch, allowing the outside wheel to freewheel at a faster speed than the inside wheel. All the torque is sent to the inside wheel for traction. During straight-line operation, the no spin splits the power equally between the two wheels and operates much like a solid axle.

Figure 14-11 shows how the spider fits into the jaw clutch, which also has internal teeth that are splined to the side gear. The axles are splined internally and mesh with the side gear. During cornering, the jaw clutches

Figure 14-11 Jaw clutch and side gear.

move axially, disengaging from the teeth on the spider. While this takes place, all the drive torque goes to the inner wheel to propel the equipment. The springs are used to keep the spider and clutch engaged during straight-line operation.

Although the name *spider* is used to describe this subcomponent, it does not look like a traditional spider in a conventional differential. This unit still uses four cylindrical projections spaced at right angles to each other that fit corresponding spaces in the differential housing, which forces the spider assembly to rotate with the differential housing. A center cam ring with square teeth meshes with the jaw clutch, which is held in place by a snap ring in the center of the spider assembly. One tooth on the spider is longer than the rest and is called the spider key. The spider key mates with a corresponding notch in the cam ring. When the equipment changes direction, the notch feels the force from the spider key. When the equipment is cornering, the jaw clutch disengages the faster turning outside wheel, forcing the holdout ring to keep the jaw clutch disengaged until the outside wheel speed synchronizes with the spider. When this occurs, the jaw clutch is allowed to engage once again with the spider teeth.

When the jaw clutch and spider are engaged, the holdout ring is driven by the center cam. The holdout ring sits in a groove between the two layers of teeth on the jaw clutch. The only connection between the jaw clutch and the holdout ring is friction. Friction forces the holdout ring out when the jaw clutch disengages, forcing the teeth of the holdout ring to disengage from the center cam. Friction forces the holdout ring around until it contacts the spider key. The spider key holds the holdout ring so that the teeth of the ring stand on the teeth of the center cam and turn with the spider. The top of the holdout ring slips into the groove of the jaw clutch. When the jaw clutch is disengaged and turning counterclockwise, the end of the holdout ring contacts the spider key and keeps the holdout ring in the raised position. A slight clockwise movement will allow the teeth of the holdout ring to mesh with the center cam, and to be driven by the center cam.

Figure 14-12 shows the spider with the center cam ring in place and each of the jaw clutches on either side with the holdout rings in place. When the equipment is moving in a straight line with equal traction on both wheels, the outer teeth of the spider mesh with the outer teeth of the jaw clutches. The center cam teeth mesh with the holdout ring as well as with the inner teeth on the jaw clutches. The teeth of the spider drive the jaw clutch when the outer teeth are in mesh, sending equal torque to both wheels. When the equipment corners, one of the jaw clutches will begin to rotate faster due to the larger turning radius of the outer wheel. This forces the jaw clutch teeth to move up in the slot, initiating clutch disengagement.

Whenever the outer teeth of the jaw clutch move up in the slot, the inner teeth will disengage from the center cam. The faster speed of the jaw clutch forces the inner teeth of the jaw clutch to ride up the teeth of the center cam. When this happens the holdout ring comes out of its groove on the center cam, and the jaw clutch moves back on the splines of the side gear. As long as the jaw clutch is trying to rotate faster than the spider, the two components will remain disengaged. When the wheel speed becomes synchronized, the ground resistance exerts a negative force on the free wheel, causing the jaw clutch to slow down. Friction between the jaw clutch and holdout ring moves the holdout ring back into mesh with the center cam teeth. When this occurs, the spring forces the jaw clutch back into engagement with the spider.

A snap ring holds the cam ring in place and allows the cam ring to turn on the snap ring until it contacts the spider key and notch. The notch is required because the

equipment can be driven in both directions. When the spider is rotated in the direction of the arrow, shown in **Figure 14-13**, the spider's outer teeth push against the left edge of the bottom jaw clutch slot, driving the jaw clutch in the same direction as the spider. The teeth of the center cam and inner jaw clutch teeth are in mesh and force the jaw clutch to move to the end of the slot. Although the center cam can move, a snap ring holds it in place in the spider. The notch in the cam ring limits the amount of movement when it contacts the spider key. In **Figure 14-13**, we see that the spider key is in contact with the right side of the cam ring. The shape of the inner teeth forces disengagement to take place if the jaw clutch is forced to rotate at a higher speed than the spider.

Off-road equipment manufacturers often use a no-spin or limited slip differential to replace a standard differential. Equipment manufacturers do not recommend using two no-spin differentials in both front and rear axle assemblies. In good road conditions, the limited slip differential works well, but on slippery road conditions the maneuverability of the equipment is compromised because the clutch packs do not slip easily. This limits differential action, which increases the turning radius of the equipment. This design also tends to increase operating costs because of drag associated with these types of differentials.

A no-spin differential provides better traction than a limited slip differential, and has an increased turning radius because the inside wheel does not slow down as in a standard differential. Only the outside wheel speeds up in a no-spin differential. If two no spins were used on a loader, the turning radius would be even worse than with one. Although there are drawbacks to no-spin differentials, if good traction is a must, then they are required in some applications. If the use of a no-spin or limited slip differential is a requirement, then tire size must be matched on the same axle. Never operate a

Reprinted Courtesy of Caterpillar Inc.

Reprinted Courtesy of Caterpillar Inc.

Figure 14-12 Spider/cam ring assembly. Note the holdout rings on the jaw clutches.

Figure 14-13 Left jaw clutch in the disengaged position.

no-spin or limited slip differential with two different tire diameters on the same axle; damage will result from the constant clutch action. One tire will turn faster if diameter difference is excessive, forcing jaw clutch disengagement or clutch slippage as is the case with the limited slip differential.

Planetary Differentials

This design is often found on articulating, off-road equipment where you often need to separate the drive shafts on a tractor from the trailer. The use of standard differentials between two axle shafts is necessary to allow for speed differences in wheels due to turning or uneven terrain. Planetary differentials operate much like standard differentials. This design is used to minimize driveline stress as the driveshafts are allowed to turn at different speeds. Like the standard differential, the planetary differential can vary the torque split between each wheel end. In an articulating truck, equal torque can be sent to each axle or be split between the two axles, depending on design. If an articulating haulage truck has three axles, the torque can be split by the planetary differential so that 40% of the torque goes to the tractor and 60% goes to the trailer. Each axle of the trailer receives 30% of the total torque. Planetary differentials can be locked to provide maximum torque when road conditions are poor.

The planetary differential input torque is provided by a gear in the transmission transfer drive that meshes with the carrier of the planetary differential. The two sets of planetary gear shafts are held by the carrier and drive the planet gears. There is a clutch in the carrier that can be used to lock the planetary differential, if required. When operated in a straight line, the planet gears do not rotate on their own axes; they rotate as a unit, transferring all the torque to the wheels, as shown in **Figure 14-14**.

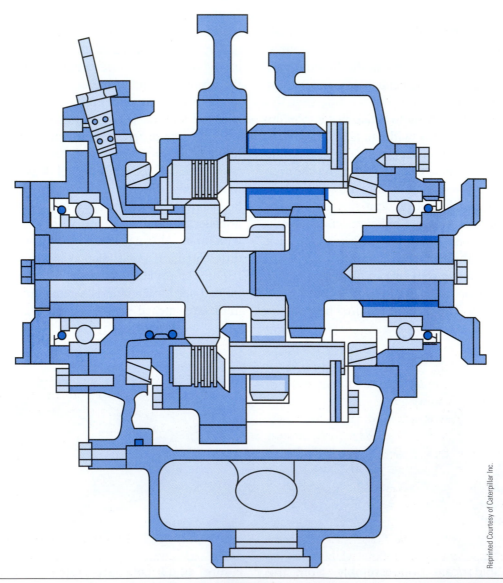

Reprinted Courtesy of Caterpillar Inc.

Figure 14-14 Cutaway view shows the lockup clutch location in relation to the planetary differential assembly.

If a change in resistance is felt at one of the wheels, the planet gears on the axle that is turning slower begin to rotate on their own axes. This causes the teeth of the planet gears, which are meshed with the other set, to speed up the other axle, allowing the torque to be split between the two axles proportionally. If the differential is placed in the locked position, a clutch pack locks the carrier to one axle shaft, which provides equal torque to each drive wheel.

DIFFERENTIAL GEAR DESIGNS

A thorough explanation of gear types can be found in Chapter 10, "Introduction to Powertrains." Many different designs of gears are used in off-road differentials; in this chapter we outline the most popular types. The following is a list of the most common types:

- Spiral bevel gears
- Spur bevel gears
- Hypoid gears
- Amboid gears

The crown and pinion gear sets on off-road equipment often use spiral bevel, hypoid, or **amboid gear** designs. The manufacturer's equipment design and component layout and application will determine the type of gears used. **Bevel gears** are used extensively in the differential where they transfer power at right angles to the drive axles. The teeth can be either straight cut or spiral, depending on the application. Remember, the longer the tooth contact area, the greater the load carrying capacity will be; this rule applies to any gear design.

Spiral bevel gears are set up so that the pinion sits on the centerline axis of the ring gear as shown in **Figure 14-15**. This gear design provides good, continuous tooth contact with one tooth in contact at all times during operation. It also generates less noise than the straight cut spiral design. Hypoid gears look similar to spiral bevel gears, but they are easily identified by the offset that is present between the pinion gear and ring gear. In this configuration the pinion gear sits below the centerline of the ring gear, which requires a longer tooth design. This is an advantage in the contact area because there are two teeth in contact in this design compared to one-tooth contact in the **spiral bevel gear sets**. Hypoid gears are considered to be a stronger gear set. Given an equal diameter size of ring gear, the **hypoid gear set** will use a bigger pinion due to its greater torque carrying capacity. Hypoid gear sets drive forward on the convex side of the ring gear tooth, which provides greater strength and quieter operation when compared to spiral bevel gears.

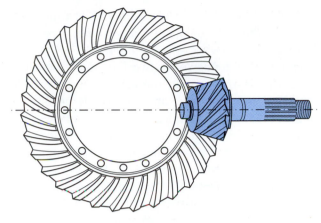

Spiral bevel gear set

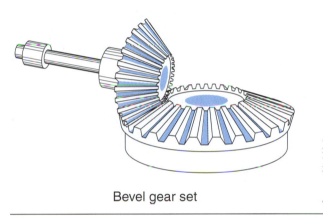

Bevel gear set

Courtesy of ArvinMeritor

Figure 14-15 Examples of straight cut and spiral bevel gear sets.

In a hypoid gear set, the gears slide across each other, wiping away any lubrication that is present between the contact teeth; due to this operating characteristic, the use of EP-rated lubrication is a must. Most equipment manufacturers recommend the use of synthetics with this design because of the superior shear strength of synthetic lubricants.

Amboid gears are similar in design to hypoid gears; the difference is the location of the pinion in relation to the ring gear, which is located above the centerline of the ring gear. This requires that the driveshaft be mounted higher than on the hypoid design, and the drive side of the gear is on the concave side of the ring gear tooth. This arrangement is commonly used on the rear axle of a **tandem drive** installation.

POWER DIVIDERS

Many manufacturers use a tandem drive arrangement on larger off-highway haulage trucks. This allows the equipment to haul heavier loads when compared to single-axle designs. This design requires the use of extra gearing to split the drive torque between the two axles. The additional gearing must also allow the axle to operate at different speeds if required.

A power divider performs two functions. It must be able to divide the driving torque between two axle assemblies, and it must be able to provide differential action between the two axles when necessary.

An interaxle is located in the forward rear axle and uses most of the same components that are found in a conventional differential. **Figure 14-16** shows a cross section of the differential. Note the location of the interaxle differential, between the input shaft and the output shaft. Helical gears are often used to send torque to the front differential pinion. A lockout is used on the forward drive axle to lock the interaxle differential thereby providing maximum traction when road conditions are poor. The lockout mechanism is engaged by the operator, who implements a sliding clutch that locks together the helical and differential side gears to give a positive drive to both axles. The lockout should be applied only during slow speed operation and never engaged during wheel slip. This feature should be used only when road conditions are poor; continuous use of this feature will lead to excessive tire wear.

Operation

The flow of power comes from the transmission and enters the input shaft of the forward axle assembly, where it flows to the forward and rear axles. The interaxle differential, which is mounted on the input shaft of the forward differential, operates the same as a conventional differential providing

differential action between the forward and rear axle assembly. A conventional differential is located in each of the two axle assemblies, which provide differential action in each of the drive axles on both axle assemblies. **Figure 14-17** shows the flow of power directed through the input shaft that is splined to the interaxle differential spider. From here power travels through the interaxle differential to the forward and rear axles. The power flows into the forward axle and is transferred from the spider through the floating helical side gear on the input shaft to the drive pinion and on through the ring gear to the forward rear axle.

The rear axle has the power coming in through the spider to the output shaft side gear, and moving on to the propeller shaft and to the drive pinion and ring gear of the rear axle when the unit is operating in the normal mode (unlocked). If the interaxle is not in operation, the helical side gear and interaxle differential are turning together as a unit. During this phase of operation torque is transferred equally to the forward and rear differentials. The forward axle pinion is driven from the input shaft by the floating helical side gear. The rear axle is driven from the output shaft side gear, through the output shaft side gear, to the output shaft, and to the propeller shaft.

When lockout operation is required, the operator engages the lockout mechanism, which allows a sliding clutch to lock the differential side gear to the helical gear. This locks the interaxle to produce a positive drive to each axle. Whenever there is a wheel speed difference between the forward axle and rear axle, the interaxle differential splits the torque and

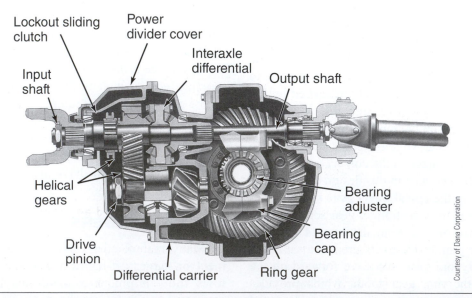

Figure 14-16 Sectional view of an interaxle differential showing how the torque is split between the forward and rear differential.

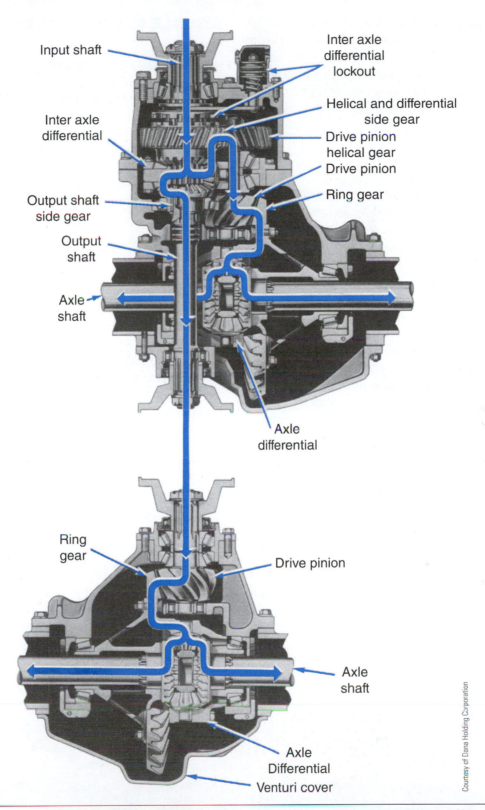

Figure 14-17 Sectional view showing the flow of power through the forward and rear differential.

speed between the two axle assemblies. The same thing can occur whenever there is a speed difference between wheels on the same axle assembly. Each differential is capable of splitting the torque and speed between each wheel on each axle assembly.

The lockout mechanism can be air or hydraulically actuated, depending on the application and manufacturer. The mechanism consists of a shift fork and pushrod assembly and a sliding lockout clutch. The actuating force extends the pushrod and shift fork mechanism, which forces the clutch against a helical side gear. The clutch is splined to the input shaft, which is splined to the differential. When shifted, the clutch locks the helical side gear to the differential to stop any differential action from occurring. If the operator locks the interaxle differential, power is divided equally between the two axle assemblies.

LUBRICATION

A large percentage of interaxles are lubricated by the splash method: Oil is distributed by the gears turning and splashing the oil internally, and directional baffles are used to direct oil to critical areas of the differential assembly. Getting oil to the interaxle differential is essential; the splash method, however, does not ensure that oil gets into this area. Some manufacturers provide a lubrication pump that ensures positive oil distribution to all areas of the interaxle differential.

External gear pumps are often chosen for this purpose. **Figure 14-18** illustrates a cutaway of a typical drive axle carrier showing the mounting location of the lube pump. The oil pump picks up oil in the sump and sends it through a filter. From there it is pumped through an external oil line in some cases, or it can be routed internally to the top of the carrier assembly where it is distributed to spray nozzles that spray oil on critical areas such as the differential gears and the front pinion bearing assembly. Additional lubrication is provided by rotation of the ring gear, which splashes oil around the inside to lubricate the drive gears and all internal bearings. If the vehicle is equipped with a positive lubrication system, oil filter replacement is required on a regular basis. Always refer to the manufacturer's documentation to determine the service interval between oil filter replacements.

DOUBLE REDUCTION AXLES

Off-road equipment is often required to haul large, heavy loads. An additional gear reduction may be required to get the job done. A **double reduction axle** uses two gear sets to provide additional torque, as shown in **Figure 14-19**. The design uses a conventional pinion ring gear to get the first gear reduction. The second gear reduction is provided by a helical spur gear and pinion set. The pinion and ring gear drive work the same as a conventional differential; however, the differential case is not bolted to the ring gear, instead it is mounted to the spur gear. The spur pinion is keyed to a shaft that holds the ring gear to provide drive. The spur pinion and the spur gear are in constant mesh. The torque flows into the input shaft and goes through the ring gear to provide the first reduction.

Power now flows to the spur pinion, which is driven directly by the shaft that the ring gear is mounted to. The second reduction is achieved through the spur pinion driving the spur gear mounted to the differential carrier, and power flows through a conventional differential to the drive axles.

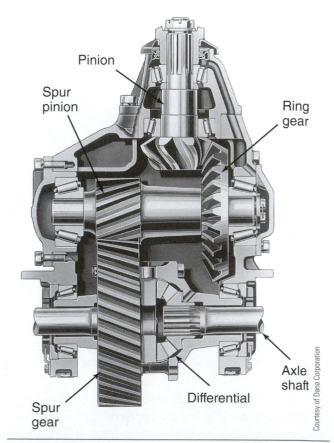

Figure 14-19 Cutaway of a double reduction drive axle assembly.

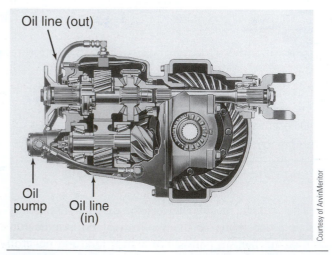

Figure 14-18 Cutaway of a power divider. Note the location of the lube pump.

Some manufacturers use a **planetary gear set** in place of helical spur gears to achieve double reduction. Input torque is sent to the drive pinion and flows through the ring gear for the first reduction. The ring gear has internal teeth that hold four planetary idler pinions. These pinions revolve around a stationary sun gear to provide the second gear reduction. Torque is transferred through the planetary gears to the differential and out to the drive axles. Power flows through the drive pinion, ring gear, pinion idlers, and out through the differential to the axle shafts.

TWO-SPEED DRIVE AXLES

Two-speed axle assemblies provide the advantages of two axles in one unit, as shown in **Figure 14-20**. This design is capable of operating as a **single reduction axle**, and can be shifted to a low-speed operating mode by the use of an external shift mechanism, which combines the use of planetary gears to provide a second reduction. Although it is similar in construction to the double reduction axle, this axle can be operated in two different gear ratios independent of each other, while the double reduction can only operate through both gear reductions at the same time. Thus, this type of drive axle is extremely flexible since the unit can operate in high-speed single reduction, and can be shifted into a low-speed reduction when required. These can be used on both axles of a tandem drive. When operating in high range, the flow of power only goes through the ring gears and pinions on both axles. When operated in low range, the flow of power goes through the planetary gears to produce the second reduction.

Equipment with two-speed axles can take advantage of the extra speed range and use it to get more effective gearing. An example of this is the use of low range in each of the two axle ratios: first gear in low range, then first gear again in high range, and then continue up the gear ranges using two stages for each range.

If disassembly becomes necessary when servicing equipment, always refer to the manufacturer's service literature to identify the equipment's axle design. For example, look at two different axle designs manufactured by the same company. Although both are classified the same, they have significant differences that must be noted or mistakes can be made when servicing. Two different types of differential carrier designs are used by a manufacturer: a ribbed carrier and a conical carrier assembly. Lubrication for each is quite different and should be noted as such. With the conical differential carrier, oil is picked up by the gear support case and is sent to the pickup trough. The oil is channelled to the carrier passage to the right differential bearing, then passes through the differential and planetary units to the left-hand differential bearing, and then flows back to the reservoir. With the ribbed differential carrier, the oil is picked up by an oil collector drum and sent to the oil distributor; from there it is sent in two different directions. Half flows to the drive pinion bearings, while the other half flows to the right-hand differential bearing, and then through the differential to the planetary gears and on to the left-hand differential bearing before returning back to the case.

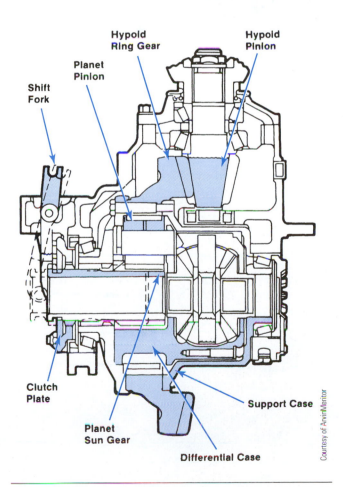

Figure 14-20 A two-speed planetary drive axle. Note the shifting mechanism used to engage the second speed reduction through the internal planetary drive gears.

Courtesy of ArvinMeritor

DRIVE AXLE CONFIGURATIONS

Off-road equipment drive axles, in most cases, come with an extra planetary drive reduction that is generally not found on highway vehicles. Planetary drives are compact in design and can be made as small as necessary to fit into almost any space. They are inherently capable of carrying large amounts of torque. The design of this drive allows the torque to be split evenly throughout all the gears. The planetary reduction drive may be located either inboard or outboard on the drive axle. Inboard planetary drives sit right next to the differential and receive torque from a short axle shaft that serves as a sun gear or input for the planetary drive. Outboard planetary drives are located in the wheel end of the axle assembly, as shown in **Figure 14-21**, and are driven by a much longer axle from the differential, which sends torque to the input or sun gear of the drive. Each design has its advantages and disadvantages.

The inboard planetary drive, as shown in **Figure 14-22**, has room at the wheel ends for a much larger range of wheel widths. The disadvantage is that the axle needs to be built stronger because it carries the full weight of the load on the wheel ends and absorbs end thrust. The axle has two tapered roller bearings at the wheel end to support the load and is driven by the planet carrier for a final torque increase.

The outboard planetary drive, on the other hand, has the planetary reduction in the wheel ends, which

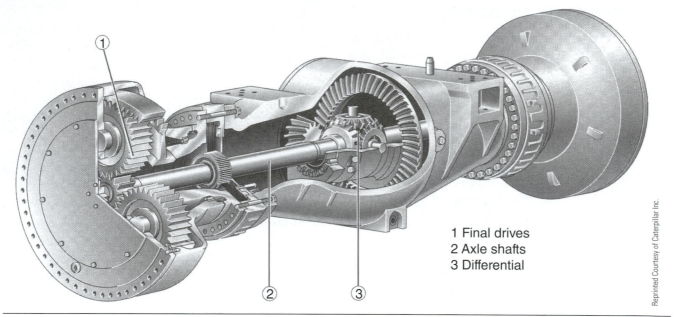

1 Final drives
2 Axle shafts
3 Differential

Reprinted Courtesy of Caterpillar Inc.

Figure 14-21 Typical heavy-duty drive axle with outboard planetary reduction.

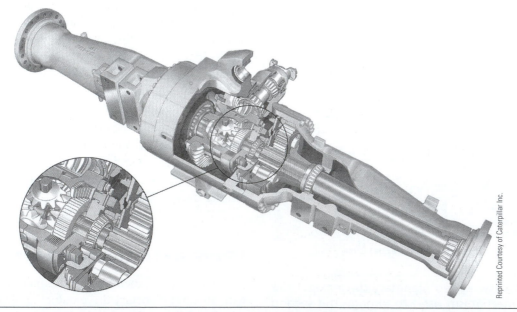

Reprinted Courtesy of Caterpillar Inc.

Figure 14-22 Typical heavy-duty drive axle with inboard planetary reduction. Note the axle carries the full weight of the load supported by two large tapered roller bearings on the axle.

gives the planetary drive its own rigid housing. The axle is driven by the differential sending torque through the axle to the other end, which acts as the sun gear or input on the planetary drive. An advantage of this design is that the axles do not support the weight of the equipment and can be easily removed if the equipment needs to be towed.

Planetary Drive Operation

Planetary drives receive torque from the differential through an axle shaft, which serves as a sun gear (input gear). This shaft has splines at both ends; one end fits in the differential, and the other end (sun gear) sends torque to the planetary drive gears. The sun gear meshes with the three planet gears, which are mounted together in a planet carrier. The ring gear is held or locked in place; the planet carrier acts as the output, resulting in a gear reduction. Torque from the sun gear forces the planet pinions to rotate around the inside of the ring gear, which is held stationary. The planet pinion carrier rotates in the same direction as the sun gear, sending engine torque to the wheels at a slower speed. This increases torque.

Regardless of the design, both types of planetary drives operate the same: the sun gear is the input, the ring gear is held, and the planet carrier becomes the output. The one main difference between the two designs is that the axle in the outboard design is a full-floating axle and does not carry any weight, while the inboard design axle carries the full weight of the load through large tapered roller bearings mounted at the wheel end of the axle housing.

Torque Hub Final Drives

In some off-road applications where slow speed and high torque are necessary, the use of an additional gear reduction can be found by incorporating a **torque hub planetary drive** as shown in **Figure 14-23**. This final drive can be found in a single, double, and triple planetary drive configuration. The torque reduction can range from 15:1 to as high as 132:1. This allows for a flexible speed and torque range, which can be custom tailored to applications such as construction equipment, conveyors, augers, mixers, and winches. These torque reduction final drives can be found on wheel end applications as well as shaft or spindle drives. An example of each is shown in **Figure 14-24**.

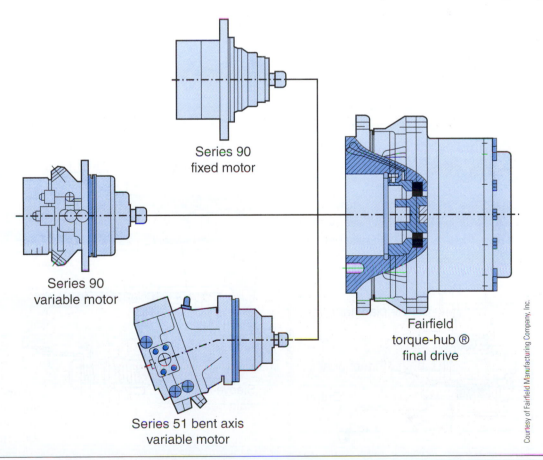

Series 90
fixed motor

Series 90
variable motor

Series 51 bent axis
variable motor

Fairfield
torque-hub ®
final drive

Courtesy of Fairfield Manufacturing Company, Inc.

Figure 14-23 Illustration of a Fairfield torque hub planetary drive showing an assortment of drive options available.

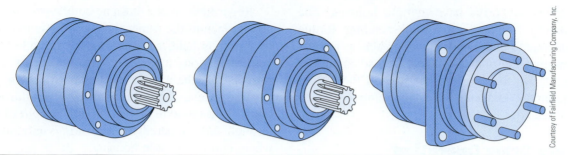

Figure 14-24 Illustration of Fairfield torque hub planetary drive showing three different drive options for wheel end applications.

Operation

A simple planetary gear set as shown in **Figure 14-25** has one ring gear, a number of planet gears mounted on a planet carrier, and a sun gear. This configuration of gears can achieve a maximum of 7.5:1 torque reduction. If you stacked two planetary gear sets together in one final drive, you could achieve a greater reduction. The configuration shown in **Figure 14-26** is used to achieve a greater reduction in a differential gear set without increasing the differential size in a significant way. In the example of a hybrid differential planetary drive, the sun gear, planet clusters, and two internal ring gears show how greater torque reduction is developed. The input gear drives the planet clusters, which turn on a fixed ring gear, which sends out output torque to the second ring gear. Each planet cluster has two different sized planet gears on the same shaft. This arrangement achieves an extremely high gear ratio in a small space compared to a simple planetary gear set.

Full-Floating Axle

Full-floating axles are used on large, slow-moving off-road equipment with a drive wheel on each end of the axle housing. **Figure 14-27** shows two large

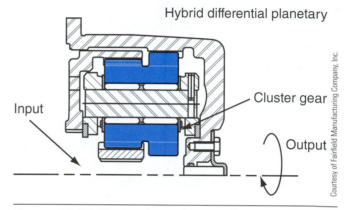

Figure 14-26 Illustration of a Fairfield torque hub planetary drive showing how reduction works.

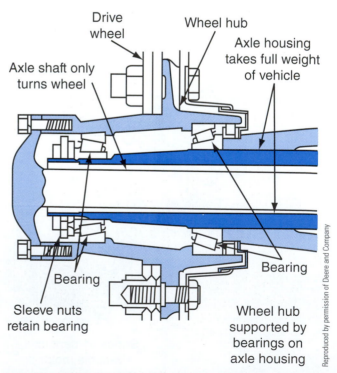

Figure 14-27 Typical heavy-duty full-floating drive axle.

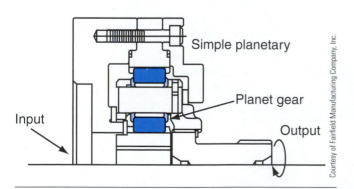

Figure 14-25 Illustration of a simple planetary drive showing how the gear reduction works.

tapered roller bearings that are mounted to the wheel hub and carry the full weight of the equipment and load. On larger equipment, the bearings are mounted on a spindle that bolts to the axle housing. The axle itself only transmits the torque of the engine through it and does not carry any weight; hence, the term *floating* is used to describe its function. On smaller equipment, the axle is connected to the drive wheels through a bolted flange that can be removed to gain access to the axle without removing the wheels.

On larger equipment with outboard planetary drives, the axle is floating and held in place with a thrust plate that can be removed to gain access to the planetary drive gears for servicing. The axles are easily removed if and when the equipment needs to be towed in a non-run situation. Caution should be used when removing the axles to allow the wheels to freewheel. Using either wheel chocks or a tow hook, the equipment should be secured to another piece of towing equipment to prevent accidental movement.

Semi-Floating Axle

The term *semi-floating* is derived from the way the axle is mounted in the assembly, as shown in **Figure 14-28**. The axle at the differential is splined to the side gear and floats at this end. At the wheel end of the **semi-floating axle**, the shaft carries one or two bearings, depending on the axle load carrying capacity. At the wheel end of the axle, the load is carried by the axle, which also transmits the engine torque. This design forces the axle to handle all of the torque and strain caused by uneven road conditions. This design comes with two types of bearing configurations: tapered roller bearings or heavy-duty ball bearings. If tapered roller bearings are used, a means of adjustment is provided for preloading the bearings, either with

shims or with a threaded nut that is torqued to a manufacturer's recommended torque specification. One of the main disadvantages of this design is that, if the axle were to break, the wheel could fall off. If that happened to the full-floating design, the result would be only a loss of drive to the wheel end. These are often found on light-duty applications where weight and load is often less than 5,000 pounds per axle.

Drive/Steer Axles

Drive/steer axles (see **Figure 14-29**) are used on off-road equipment that does not use an articulating steering arrangement. Some manufacturers use this type of drive/steer configuration because of limitations in size, weight, length, and width of equipment design. The drive axle section of this drive/steer axle is similar to a conventional drive axle. The difference is in the wheel end design.

Kingpins are used to mount the wheel end to the axle housing, allowing the wheel end to turn from side to side. The steering knuckle that supports the kingpin at the top and bottom has bushings that have recessed grooves to retain grease. Each steering knuckle is connected to the other by way of a tie-rod that is threaded at each end to allow wheel adjustment. The ends of the cross tubes have a left-hand thread at one end and a right-hand thread at the other end, which provides a means to adjust the toe in or out during alignments. Engine power is sent through the differential and flows through the Cardan axle shaft universal joint to the outer axle shaft to drive the wheels. The outer ends of the shaft are splined to a drive flange, which is bolted to the wheel hub. Depending on the manufacturer, a single- or double-Cardan joint allows the equipment to transfer torque and steer at the same time, as shown in **Figure 14-30**.

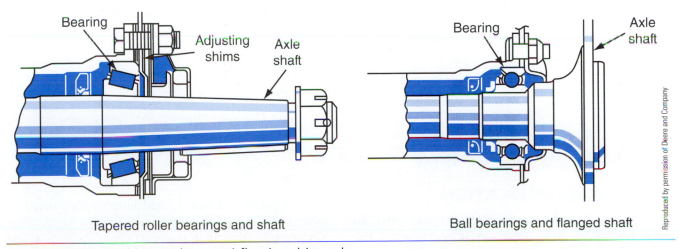

Tapered roller bearings and shaft   Ball bearings and flanged shaft

Figure 14-28 Typical heavy-duty semi-floating drive axle.

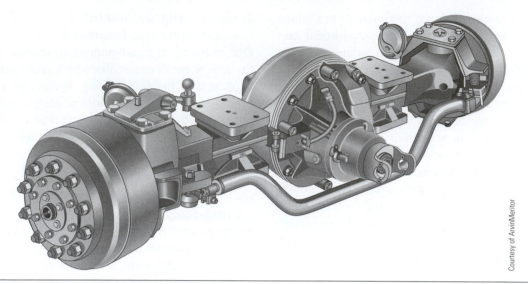

Figure 14-29 Typical heavy-duty drive/steer axle.

Courtesy of ArvinMeritor

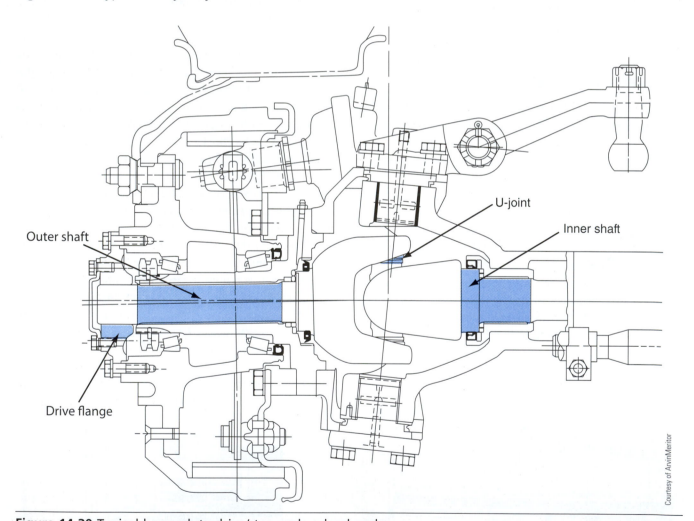

U-joint

Inner shaft

Outer shaft

Drive flange

Figure 14-30 Typical heavy-duty drive/steer axle wheel end.

Courtesy of ArvinMeritor

DRIVE AXLE LUBRICATION

Technicians who service and repair off-road equipment must understand the importance of following the manufacturer's maintenance procedures. A good number of axle-related failures are caused by using a lubricant that is not ideal for the application. This results in a shorter component life than could otherwise be achieved. Many failures could be prevented by following sound maintenance practices concerning lubrication selection and use. Axle failures can and have produced many undesirable consequences in the past. When lubrication levels are not properly maintained in axles, the service life of the bearings and gears will be adversely affected, often leading to catastrophic failures. Regardless of how well the equipment is designed and operated, if it is not properly maintained, premature axle component wear will occur and will most definitely lead to early failure. Although some failures are caused by improper installation of components, technicians need to make themselves aware of correct methods and ensure that the proper replacement parts and tools are used to assemble the components.

Lubrication-related damage may take place even if the correct lubricant is used. Often the intervals between servicing are inadequate for the application, resulting in component damage. Lubricants that are used in axle assemblies have the following three general functions:

- Providing an adequate lubrication film quickly and reducing friction between sliding and rolling surfaces
- Removing excess heat from high-friction areas and maintaining correct operating temperatures
- Removing dirt and wear particles from the bearings and other high-friction areas

If and when lubrication damage occurs to axle components, three basic problem areas are generally responsible:

- Contaminated lubrication
- Low lubrication levels
- Incorrect lubricant for the application or depleted additives

Lubrication practices discussed in this chapter are general examples of what should be done and when it should be performed. In actual practice, maintenance should be performed based on the OEM's recommended servicing literature. Many manufacturers allow the use of synthetic lubricants in axles; this extends service intervals significantly. The initial cost of synthetic lubricants is much higher upfront, but in the long run can be significantly less costly. Drain intervals can often be increased by two to three times when compared to conventional crude-based lubricants. Synthetic lubricants have many qualities that conventional oils do not have, such as a much higher boiling point when compared to conventional oil (600°F [316°C] versus 350°F [177°C]). When deciding whether to use synthetic oil, consider the weather conditions to which the equipment will be exposed. Synthetic oil has much better fluidity than conventional oil in extremely cold weather; this could mean the difference between having lubrication instantly or operating for a few minutes before the oil can flow properly through a typical heavy-duty drive steer axle wheel end assembly as shown in **Figure 14-31**.

Using the Proper Lubricant

Many differentials have been ruined by a phenomenon called *channeling*, which occurs in cold weather: Thickened oil is parted by the rotating ring gear and is too thick to flow back. This results in inadequate lubrication until the oil heats up. Often, by then the gears and bearings have sustained damage from lack of lubrication. A good solution to this type of problem is to use a high-quality synthetic lube—one that meets API-GL-5 specifications. Use of synthetic lube also provides the benefit of extended drain intervals, which reduces the cost of labor and ensures better protection. All lubricants used in axle assemblies must meet the American Petroleum Institute (API) and the Society of Automotive Engineers (SAE) Gear Lubrication (GL) standards. Today the use of GL-1, GL-2, GL-3, GL-4, and GL-6 is no longer approved for modern axle assemblies by many manufacturers and should be discontinued. The recommended lubricant to use in drive axles today is API-GL-5, which is rated as an extreme pressure lubricant and is required for use in axle assemblies that use hypoid gears. Using the correct viscosity will be operating-temperature dependent, and should be adhered to. **Table 14-1** shows the proper grade of lubricant for a given operating temperature.

Most OEMs approve the use of synthetic lubricants providing they meet API-GL-5 classification standards. Due to the wide variety of operating extremes that off-road equipment is subjected to, it is difficult to determine ideal drain intervals for a lot of equipment. Off-road equipment service intervals are usually not measured in miles but are measured in hours of operation.

Contaminated Lubrication

Some equipment shops have a policy of mixing different brands of lubricants; this is not recommended. Not all lubricants are compatible with each other. Every effort should be made to prevent the mixing of different lubricants in axle assemblies. Some additives used by certain manufacturers are not compatible with the additives of other manufacturers. Never mix different brands of gear lube. Some synthetic oils, when mixed with petroleum-based oil,

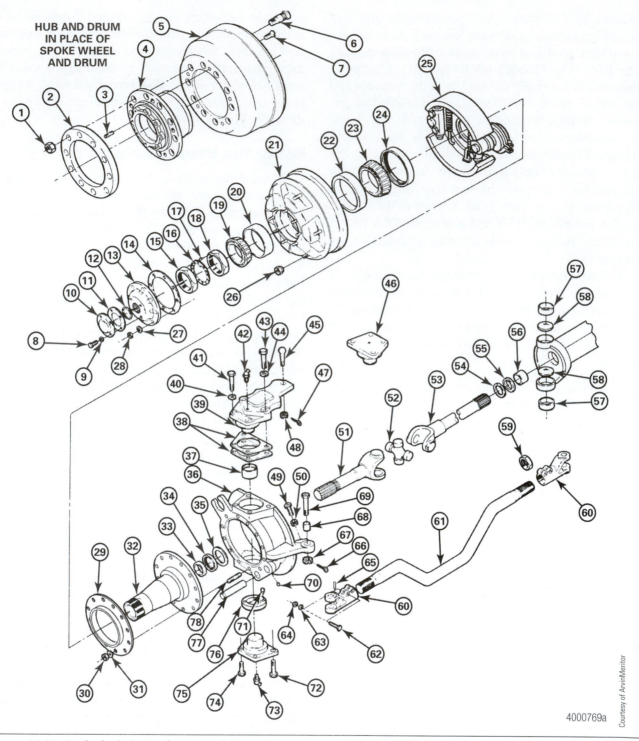

HUB AND DRUM
IN PLACE OF
SPOKE WHEEL
AND DRUM

Courtesy of ArvinMeritor

4000769a

Figure 14-31 Exploded view of a typical heavy-duty drive/steer axle.

will thicken. This can lead to foaming, which can result in early component failure. Water, dirt, and wear particles can cause extensive damage in a short period of time. Water can enter the axle assembly through a faulty shaft seal or through the carrier to a banjo housing joint. Wear particles are generally a result of normal wear and can be minimized by the use of magnetic drain plugs. The magnets collect metal particles that settle to the bottom of the axle housing during shutdown periods. Some manufacturers use an oil pump in the drive axle to distribute oil and a filter in the circuit to filter out dirt particles.

Results of Lubrication Failures

When a lubricant does not perform up to the equipment's application standards, the life of the internal components will be shortened. Following is a

TABLE 14-1: PROPER GRADE OF LUBRICANT	
Ambient Temperature Range	**Correct Grade**
−40°F(4.4°C) to −15°F(−9.4°C)	75W
−15°F to 100°F (37.7°C)	80W–90W
−15°F and above	80W–140W
10°F (−12.2°C) and above	85W–140W

list of causes and effects that may occur due to lubrication failures:

- Lubrication that is contaminated with water can cause scoring, etching, or pitting on the running surfaces of the gears and bearings.
- Under-filling the axle assembly may lead to metal-to-metal contact, which can lead to high friction that can cause overheating and result in oil breakdown. This will usually lead to severe scoring and galling that can actually melt metal surfaces. The internal parts will often be blackened from the heat.
- The use of an improper lubricant, which includes the practice of mixing different oils, may lead to surface etching or lubricant breakdown, leading to metal-to-metal contact.
- Using a viscosity that is not suitable for the application temperatures may result in a lubrication film breakdown, leading to severe scoring and galling. The internal gears will often be blackened from the heat.

Tech Tip: When filling the axle housing and planetary wheel ends, allow enough time for the oil to flow through the components and find its natural level. When the filling is completed, allow a few minutes to pass, then recheck the level. The lubricant should be warm when drained; this ensures that contaminants are suspended in the oil and get flushed out.

Low Lubricant Levels

Lubrication levels on off-road equipment are generally checked every 250 hours of operation. Oil levels in drive axles must be checked properly; the oil level must be even with the bottom of the oil filler plug hole. A good rule is to replace the oil every 2,500 hours of operation or after 1 year, whichever comes first.

Remove the plug in the axle banjo housing; the oil should be even with the bottom of the oil level plug hole, as shown in **Figure 14-32**. Note that if the pinion angle

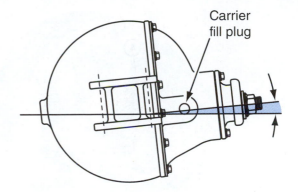

Pinion angle less than 7°
fill to carrier fill plug hole

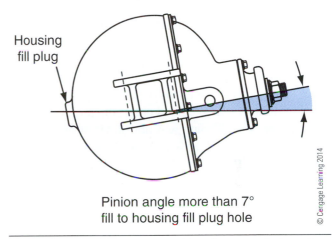

Pinion angle more than 7°
fill to housing fill plug hole

Figure 14-32 Checking the oil level in an axle housing.

is more than 7 degrees, a different oil level is used. Refer to **Figure 14-32** for details. To properly check the oil levels in a differential drive axle and the planetary wheel ends, the axle first should be run and then allowed to stand for approximately 5 minutes on level ground before checking the oil level. This allows time for the oil to drain back down to the sump. Some axles have seals in the axles separating the oil in the wheel ends from the planetary drive. In this case, the oil level is checked independently in each wheel end and the differential banjo housing. The wheel end seal prevents the oil in the wheel ends from entering the differential banjo housing. When checking the oil level in a top-mount or inverted pinion drive axle, a slightly different procedure must be followed. The oil hole on these units is located in the axle banjo housing. Note that some axles have another smaller hole located just below the oil level hole; this hole is for a temperature sensor and should not be used to fill or check the oil level.

Checking the Planetary Wheel Ends

To check the oil level of the planetary wheel ends, the wheel hub oil level indicator plug (see **Figure 14-33**)

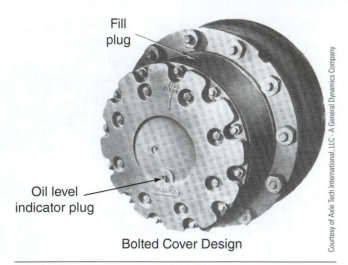

Bolted Cover Design

Courtesy of Axle Tech International, LLC - A General Dynamics Company

Figure 14-33 Checking the oil level in the planetary wheel ends.

must be in the down position. Note that you may need to move the vehicle to line up the oil level plug in the correct location. Remove the plug and check the oil level; if it is low, remove the fill plug and add lube oil as required. New planetary wheel ends should be drained of initial fill after 75 to 100 hours of operation; this prevents any break-in material from causing excessive wear. Oil should be drained when warm to ensure that any wear particles will drain with the oil and will not have time to settle to the bottom. Some axles do not have an oil check plug at the wheel ends; in this case, the wheel ends of the axle must be raised one at a time to allow oil to flow to the opposite wheel end. After completing this procedure, the oil level in the banjo housing may have to be topped up.

Lubrication Change Intervals

Lubricants used in axle assemblies are necessary to provide lubrication to the internal components and, equally important, they distribute additives throughout the axle and wash away small wear particles that would otherwise cause accelerated wear to the gears and bearings. The oil also carries away excess heat from high-friction areas. All these conditions, along with constant exposure to heat, lower the performance properties of the oil. Oil analysis should be used to determine optimum oil change intervals. Guidelines that are given in the manufacturer's documentation are based on average operating conditions. The average temperature of lubricants in axles is between 160°F and 220°F (71°C to 104°C) in summer months. Operating equipment with temperatures above 200°F (104°C) causes the oil to oxidize at a higher rate, depleting the additives that give the oil its increased load-carrying capacity. This reason alone makes more frequent oil change intervals desirable.

DRIVE AXLE INSPECTION

Extensive dirt build-up on the outside of the axle housing can contribute to high lubricating oil temperatures and should be cleaned off regularly. Power washing the axle assembly when dirt is noticeable makes good sense. If the axle is to be repaired, it is mandatory to power wash the area that will be dismantled for servicing. During the inspection procedure, certain off-road equipment axles may require a technician to pressurize the axle center housing and wheel ends to check for leaks before disassembly. An axle housing that has seals that isolate the wheel ends (some wheel ends with internal wet disc brakes) need to be air-pressure-tested at each wheel end separately in addition to testing the banjo housing. This air-pressure testing procedure requires the use of a 30-psi (206 kPa) air-pressure gauge, an air shutoff valve, and an air regulator. Regulated air pressure (12 psi [82 kPa]) is applied to the planetary wheel end fill hole. The pressure should be maintained for 15 seconds. If the pressure does not hold, the wheel hub oil seal is leaking and must be replaced. Pressure check the banjo housing the same way but use the air vent hole in the banjo housing to perform this test. If 12-psi (82 kPa) pressure does not hold, find the source of the leak and make the necessary repairs.

AXLE REPAIR PROCEDURES

The procedures outlined below are intended to be used as a guide and are not intended to replace the manufacturer's technical documentation or repair literature. Extreme care should be exercised when working on axle assemblies because the subassemblies are, in some cases, very heavy and may require the use of a lifting device. Before beginning work on an axle assembly, the technician should review the manufacturer's recommended repair procedures carefully to avoid any possibility of damaging the subcomponents through improper disassembly and assembly and to prevent any possibility of injury. Before beginning any repairs, ensure that the vehicle is properly blocked up and the tires and wheels have been removed and stored in a safe location to facilitate the safe removal of the subassemblies. Drain all lubricants from the subassembly that is to be removed.

Planetary Wheel End Removal

1. Once the oil has been drained, match mark the planet cover, spider, and hub, as shown in **Figure 14-34**. Remove the bolts and planet cover.
2. Separate and remove the planetary spider assembly from the wheel hub. The use of puller screws may be necessary to complete this task.

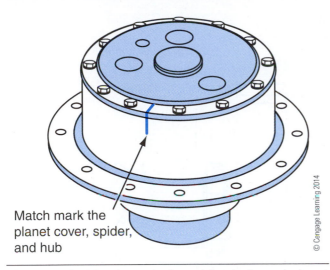

Figure 14-34 Match mark the spider, hub, and planet cover before removing.

The adhesive that seals the spider to the hub may make it necessary to use a pry bar to loosen the spider assembly.

3. Depending on the model of the axle assembly, removing a set screw (if present) may be required before removing the planet pins.

4. In some cases, a press may be required to remove the pins, as shown in **Figure 14-35**; in other cases, they may be drifted out with a block of hardwood or brass drift.

CAUTION *Do not strike the planet pins with a metal hammer; they are hardened steel and may splinter and cause personal injury.*

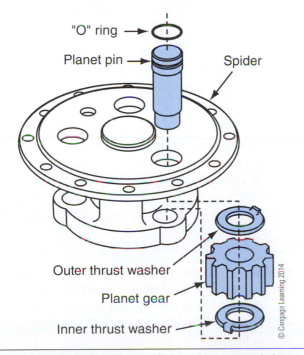

Figure 14-35 Breakdown of the planetary pinion.

5. Remove the planet pins, thrust washers, and needle bearing if present.
6. Remove the ring axle shaft and sun gear.
7. Remove the wheel bearing adjusting nut lock.

Note: Depending on the model, some units may use a double nut with lock and may be locked to the hub in a different manner.

8. Before removing the wheel bearing adjusting nut, use a sling to support the wheel hub, and then remove the adjusting nut.
9. Remove the ring gear hub from the wheel hub. Be careful not to pinch your fingers in the process.
10. On some models, the ring gear to ring gear hub may be bolted together with cap screws and lock plates on the back of the assembly.
11. Remove the outer wheel bearing from the hub.
12. Remove the inner seal and inner bearing from the hub.

Planetary Wheel Ends Installation

1. Install new bearing cups with a suitable driver if the inner wheel bearings are to be replaced.
2. Install a new oil seal in the rear of the hub and lubricate the lip.
3. Depending on the axle model, a suitable lifting device may be required to reinstall the hub assembly on the axle. Damage to the inner lip seal may occur if alignment is not maintained during the assembly process.
4. Reinstall the ring gear assembly (see **Figure 14-36**) into the hub and install the adjustment nut as shown in **Figure 14-37** to prevent the hub from slipping off the spindle.

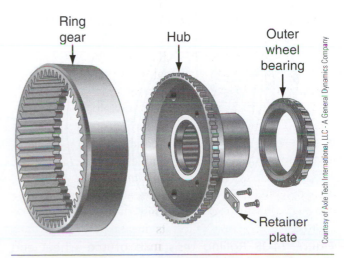

Figure 14-36 The ring gear assembly.

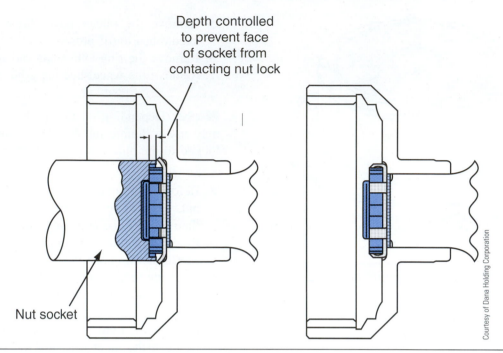

Depth controlled
to prevent face
of socket from
contacting nut lock

Nut socket

Courtesy of Dana Holding Corporation

Figure 14-37 Illustration shows the position the wrench must be in to avoid shearing the inner tang on the lock plate.

Wheel Bearing Adjustment Procedures

Because a large number of adjustment methods are used on different types of wheel ends, it is important to understand that the following adjustment procedures are examples of typical adjustment procedures and should not be used to actually adjust all wheel bearing preloads. Always refer to the manufacturer's recommended procedures for a particular type of preload wheel end adjustment. **Figure 14-38** shows how to check the rolling resistance of the wheel end assembly.

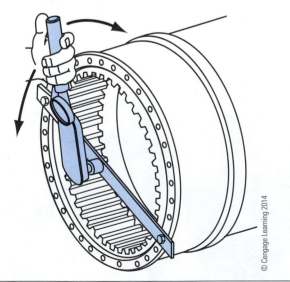

© Cengage Learning 2014

Figure 14-38 Rolling resistance of the wheel end measurement.

Two basic bearing adjustment procedures that are used by most manufacturers will be outlined in this section—the nut and lock plate design and the retainer plate and shim pack design. Before any wheel bearing assembly can be adjusted, the tapered bearing and cups must be seated properly in the hub assembly. To bottom out the bearing cups in the hub assembly, the adjusting nut must be torqued up to approximately 500 lb-ft (677 Nm) repeatedly until the torque does not advance the adjusting nut any further.

Nut and Lock Plate Adjustment. The nut and lock plate design has many variations. Always refer to the manufacturer's installation procedure for the particular type that is being worked on.

1. Loosen the nut a quarter turn until a slight bearing endplay is felt and the hub rotates freely.
2. Measure the no-load rolling resistance of the wheel hub assembly. Refer to **Table 14-2** for the correct values. Now refer to **Figure 14-38**.

Note: The rolling resistance is the force being shown on the dial torque wrench while the wheel is turning and is not to be confused with the start-up torque of the wheel, which is always higher.

TABLE 14-2: TORQUE CHART FOR WHEEL BEARING ADJUSTMENT		
Condition	Rolling Torque Range	Jam Nut Torque
New tapered roller Bearing adjustment	7 to 12 lb-ft (10–16 Nm) torque greater than no-load rolling torque reading	500 lb-ft (677 Nm)
Adjusting used bearings	3 to 5 lb-ft (4–6 Nm) torque greater than no-load rolling torque reading	500 lb-ft (677 Nm)

© Cengage Learning 2014

TABLE 14-3: RECOMMENDED TORQUE VALUE	
Cap Screw Size	Recommended Torque Value
½–13 Grade 8	90–100 lb-ft (123–135 Nm)
¾–10 Grade 8	300–330 lb-ft (407–447 Nm)
1¼–10 Grade 8	1475–1625 lb-ft (1999.9–2711.8 Nm)
1¼–7 Grade 9	1850–2000 lb-ft (2508.4–2711.8 Nm)

© Cengage Learning 2014

TABLE 14-4: EXAMPLE OF CALCULATIONS TO DETERMINE SHIM PACK THICKNESS	
Plate Thickness	0.994″ (25.25 mm)
Depth measured at three places, 120 degrees apart, add together and divide by 3 to get an average reading.	1.127″ (28.62 mm)
	1.125″ (28.57 mm)
	1.126″ (28.60 mm)
	3.378″/3 = 1.126″ (28.60 mm)
Subtract plate thickness.	1.126″ (28.60 mm)
	0.994″ (25.25 mm)
	0.132″ (3.35 mm)
Add 0.005″ (0.2 mm).	0.005″ (0.12 mm)
	0.137″ (3.47 mm)
Select shim pack using micrometer.	**Total Shim Required** 0.137″ (3.47 mm)

© Cengage Learning 2014

3. Torque the adjusting nut to the recommended value above the rolling resistance value. Refer to **Table 14-2** for the correct value.

4. Once the correct preload is reached, apply Never-Seez™ or a similar product to the threads before installing the jam nut against the adjusting nut lock. Be sure that the socket does not contact the outer tangs of the lock when torque is applied to the jam nut as this could shear the inner lock tang. Torque the jam nut to 500 lb-ft (677 Nm).

5. Recheck the rolling torque value; be sure that it is not above the recommended torque value shown in **Table 14-2**.

6. If the torque value has not been achieved, repeat the procedure until satisfactory values have been reached.

7. After the correct rolling resistance values have been achieved, bend the lock tangs over the jam nut to prevent loosening.

Plate and Shim Pack Adjustment. The following steps are to be used as examples only. They should not be used as actual procedures. Always refer to the manufacturer's specifications for the recommended adjustment procedures and specifications for a particular type of wheel end bearing.

1. Before beginning, measure the retaining plate thickness and record the value as shown in **Figure 14-39**.

2. Install the retainer plate without the shims.

3. Tighten the cap screws to the recommended torque values as outlined in **Table 14-3**.

4. Strike the retainer with a soft-blow hammer to ensure bearings are seated properly, then loosen the cap screws two turns, and rotate the wheel end four or five times.

5. Measure the rolling resistance using the correct dial indicator torque wrench and adapter.

6. Use a depth micrometer to measure the distance from the face of the retainer plate to the end of the spindle.

7. Subtract the retaining plate thickness (step 1) from the average reading and add 0.005″ to the shim pack thickness to get the correct amount of shim.

8. Measure the shim pack thickness and record the value used for the first attempt.

9. Install the shim pack under the retainer plate and torque up the retainer bolts to recommended value in **Table 14-3**, following the manufacturer's recommended procedures. See **Table 14-4** for an example calculation.

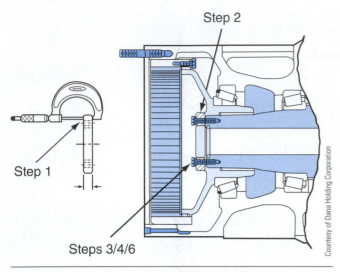

Step 2

Step 1

Steps 3/4/6

Courtesy of Dana Holding Corporation

Figure 14-39 The steps required to calculate correct shim pack installation.

10. Before rechecking the rolling resistance, roll the wheel end over four or five times.
11. The rolling resistance with a new bearing should be 30 to 50 lb-ft (41–67 Nm) above the no-load rolling resistance.
12. Find the rolling resistance for the used bearings (a used bearing is one that has been in use for 300 hours).
13. If the rolling resistance is high, add a 0.001" shim and recheck. If the reading is low, remove a 0.001" shim and recheck.

Differential Removal

Before attempting to lift the equipment, be sure that wheel chocks are placed on the opposite set of wheels to prevent any accidental movement. The following steps to remove a differential from a drive axle assembly are an example of this procedure. Any attempt to remove a differential should be preceded by reading the manufacturer's recommended procedures for each specific axle.

1. Use a jack that is rated for the load you are lifting to raise the equipment, and block it with approved stands or blocking.
2. Drain the lubricant from the axle assembly differential housing and disconnect the driveline universal joint from the carrier.
3. Drain the wheel ends and remove the planetary covers and axle shafts from each end.
4. Before removing the differential mounting bolts, place the differential jack under the differential assembly for support.
5. Remove all the cap screws except the two top ones, and take up the weight with the differential jack.

6. Back off the last two fasteners to the end without actually removing them completely. Loosen the differential carrier with a pry bar.
7. Carefully remove the two fasteners and gently slide the differential jack out with the differential.
8. Remove the differential housing gasket, if used, and discard it.
9. After the differential is overhauled or replaced, reinstall the differential following the above procedure in reverse.

Differential Overhaul

The differential carrier should be secured to an overhaul stand before attempting to disassemble it. Inspect the crown and pinion for physical damage before attempting to remove it. If there are broken teeth or indication of excessive wear measuring the backlash is not necessary. If the gear set is determined to be usable, then the backlash and runout must be measured and recorded. The crown and pinion must be reinstalled with the same backlash to maintain the correct tooth-contact pattern. If the gear set is damaged and must be replaced, this last step is not required.

1. If the differential carrier has a thrust screw adjustment, it must be backed off before disassembly, as shown in **Figure 14-40**.
2. Rotate the differential carrier so that the crown gear is at the top to facilitate easy removal.
3. Match mark each bearing cap to its corresponding carrier leg as shown in **Figure 14-41**. Use a center punch; paint marks will wash off when the parts are cleaned.

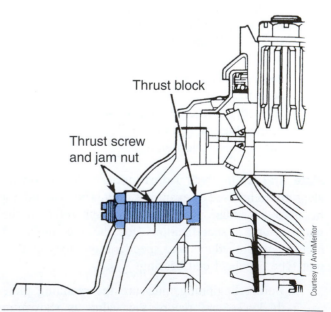

Thrust block

Thrust screw and jam nut

Courtesy of ArvinMeritor

Figure 14-40 Location of the thrust screw assembly.

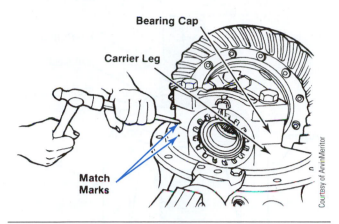

Figure 14-41 How to correctly match mark the caps before removal.

4. Remove the cotter pins or lock plates that hold the side bearing adjustment rings in place.
5. Loosen and remove the two adjusting rings from the carrier. If the gear set will be reused, keep the bearing caps and adjusters together for reuse in their original location.
6. Once the bearing caps are removed, lift the differential crown and pinion out of the housing. (Don't forget to remove the thrust block at this time if used.)

Differential/Crown Gear Disassembly

1. If match marks are not present on the carrier case, use a center punch to identify correct orientation. The carrier halves must be installed in the same location when reassembled.
2. Separate the case halves and remove the spider, four pinion gears, two side gears, and the thrust washers, as shown in **Figure 14-42**.
3. If you need to replace the differential side bearings, a puller is required to remove them without damaging the housing.
4. If the crown gear needs to be replaced, unbolt it from the differential casing. In some cases, it may need to be pressed out of the flange case.

Removing the Pinion Drive. The pinion drive on heavy-duty differentials is mounted to a pinion carrier, which houses two tapered roller bearings and the drive pinion and yoke.

1. First remove the pinion nut with an impact. Sometimes a pinion nut will have a cotter pin locking it to the shaft; remove it first.
2. In some cases, such as on some larger differentials, the pinion nut is torqued as high as

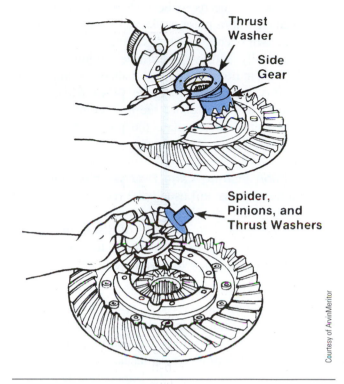

Figure 14-42 Removal of the spider and pinion gears.

1,000 lb-ft and must first be loosened with a large ¾-inch drive power bar.
3. Refer to **Figure 14-43** to view the procedure for this type of pinion nut removal.
4. Once the pinion yoke is removed, remove the cover and seal assembly along with the gasket from the bearing cage.

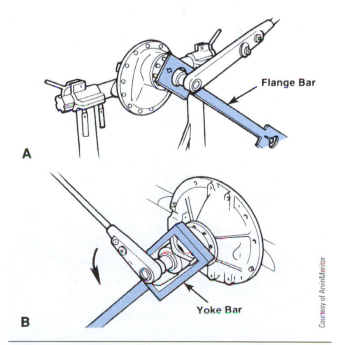

Figure 14-43 Removing the input (A) flange or (B) yoke.

5. If the seal needs to be replaced, it must be installed with a press or sleeve driver or seal damage will occur.

6. Carefully remove the drive pinion assembly from the housing; do not damage or loosen the shims, as shown in **Figure 14-44**. Measure the shim pack and record the measurement. It will be needed to calculate the pinion depth when reassembling.

Drive Pinion Bearing Cage Disassembly. On large differentials a press will usually be required to remove the pinion from the bearing carrier, as shown in **Figure 14-45**. As with any type of disassembly, first review the specific repair literature for the type of differential that is being disassembled.

1. Use a suitable press to remove the pinion from the bearing carrier, as shown in **Figure 14-45**. Be careful not to drop the pinion to the ground once it has cleared the bearing; this could damage the pinion gear faces.

2. If you are replacing the pinion bearings, use a press to remove the inner bearing from the pinion shaft, as shown in **Figure 14-46**.

3. If the spigot bearing is to be replaced, remove the snap ring and use a suitable bearing puller to remove the bearing.

Differential Carrier Assembly

The steps outlined below are basically the reverse steps that are required to disassemble a differential carrier. This procedure is intended to be an example showing the general steps required to assemble a differential. The procedures that are used may vary extensively

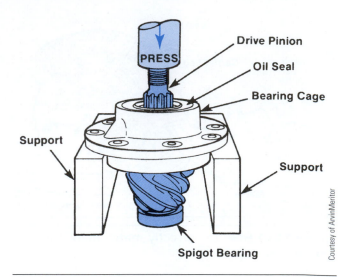

Figure 14-45 Pressing the drive pinion from the bearing cage.

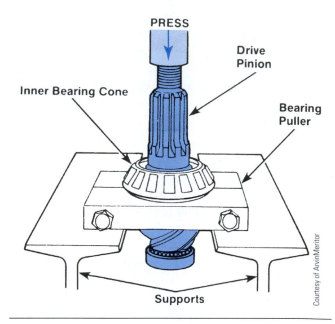

Figure 14-46 Pinion bearing removal.

from differential to differential and should be attempted only after you have consulted the manufacturer's repair literature for a specific product.

1. If the pinion bearings have been replaced, the new bearing cups must be pressed into place in the bearing cage on a press. Installing the cups with a metal hammer is not recommended; the cups are made from hardened, high-quality steel and may splinter from impact blows.

2. Press the inner bearing cone onto the drive pinion using a press. Be sure to apply pressure only on the inner race of the cone, not on the roller cage, or damage to the bearing will occur.

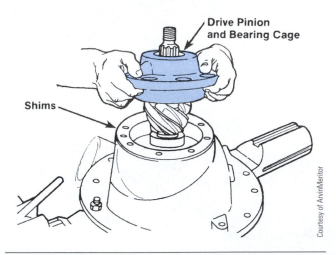

Figure 14-44 Removing the drive pinion and bearing cage.

3. Use a press to install the spigot bearing to the end of the drive pinion and install the snap ring in place, if applicable. Be sure that the bearing is installed correctly. Some bearings have a radius on one shoulder of the inner race. This radius faces toward the drive pinion teeth. The other side may have no radius and will prevent the bearing from seating properly.

4. Lubricate all the bearings and cups with axle lube before installing the drive pinion in the cage.

5. Install the bearing spacers and shims onto the pinion and press the outer bearing into place using a press. Refer to **Figure 14-47** for correct orientation.

6. Set the correct preload to the drive pinion assembly. To adjust the bearing preload, refer to the section under testing and adjustment later in this chapter.

7. To adjust the shim pack thickness between the pinion cage and the differential housing, refer to the section under testing and adjustment later in this chapter.

Installing the Drive Pinion Assembly. After completing the preload adjustments to the drive pinion assembly, install the assembly into the differential housing as indicated below:

1. If a new crown and pinion is to be installed, follow the calculation procedure outlined later in this chapter to determine the shim pack thickness. If the same crown and pinion is reinstalled, then install the shim pack that was removed during disassembly.

2. Be sure to line up the oil holes in the shim pack with the oil holes in the housing using the guide studs as shown in **Figure 14-48**.

3. To install the drive pinion assembly in the housing, it may be necessary to use a soft-blow

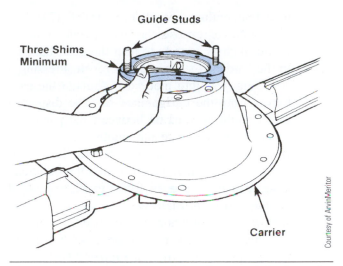

Figure 14-48 Installation of the shim pack before the drive pinion assembly is installed.

hammer to force the parts together. Install the seal cover and gasket to the assembly. Refer to **Figure 14-49** for the correct orientation.

4. Install and torque all the cap screws to the correct torque values recommended by the manufacturer's specifications.

5. Install the input yoke and torque up to the manufacturer's recommendations; use the yoke or flange bar during the torque procedure.

Installing the Crown Gear. The steps outlined here are shown as examples only. Crown gears, or ring gears as they are sometimes called, may have different installation procedures that are dependent on brand name. Always refer to the manufacturer's technical documentation for the actual procedures and specifications to perform this type of work.

1. To facilitate ease of installation, the ring gear should be heated for a few minutes, using water

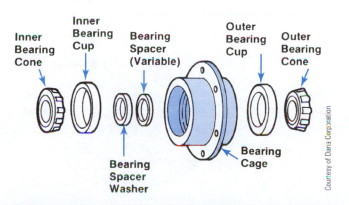

Figure 14-47 Correct orientation of parts in a pinion cage assembly.

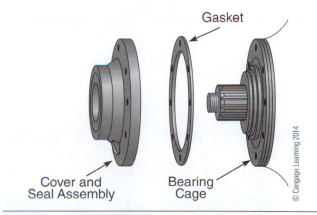

Figure 14-49 Correct orientation of the seal cover and gasket.

to a temperature of 160°F to 180°F (71°C to 82°C). Never expose the gear to any type of flame, such as a torch, to heat the gear; damage to the hardened surface could result. Pressing a cold ring gear into place can damage the case halves from the interference fit. The displaced metal could be trapped between the parts and cause excessive ring gear runout.

2. Install and torque the fasteners to the manufacturer's recommendation so that the bolt head is against the ring gear. Refer to **Figure 14-50** for the correct orientation.

3. If new side bearings are to be used, install the bearings to both sides of the case halves by using a press, as shown in **Figure 14-51**.

4. Lubricate both sides of the case halves along with all the internal parts before assembly.

5. Place the case half with the crown gear facing up on a suitable bench to begin assembly.

6. Install one thrust washer and one side gear into the case, as shown in **Figure 14-52**.

7. Install the spider, differential pinions, thrust washers, and side gear. Refer to **Figure 14-53** for correct orientation.

8. Install the other case half into position, being careful of alignment. Tighten the fasteners in a cross pattern to the manufacturer's torque specifications.

9. Check to see that the differential spiders rotate freely when assembled.

Crown Gear Assembly Installation. The following steps are general examples of installation procedures. The actual procedures for this type of installation require that you follow the manufacturer's service

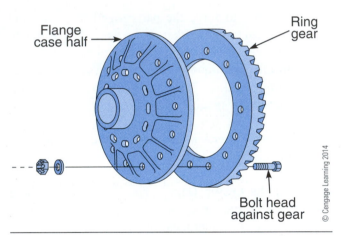

Figure 14-50 Correct orientation of the crown gear.

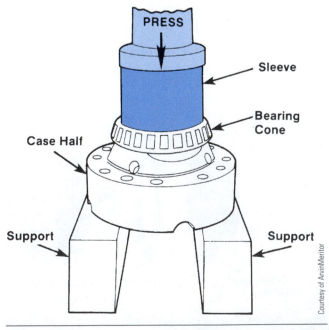

Figure 14-51 Bearing cone installation in case half.

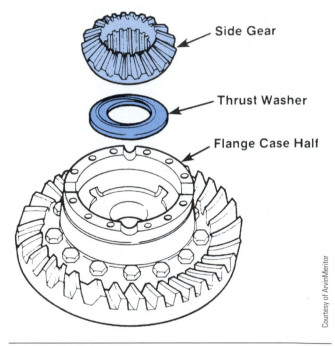

Figure 14-52 Correct orientation of the thrust washer and side gear. Lubricate all internal parts with axle lube before installing into place.

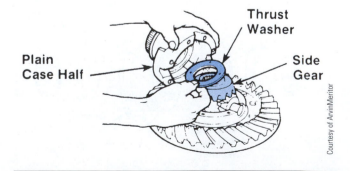

Figure 14-53 Correct orientation of the spider, differential pinions, thrust washer, and side gear.

literature on installing the crown gear assembly into the differential housing.

1. Be sure that the cups and bearing bores of the differential carrier are clean and dry. Apply lube to the inside surface of the cups, and be sure to lube the roller bearings in the process.
2. Install the bearing cups onto the side bearing before installing the crown assembly into the housing. Lower the assembly carefully into the housing, being careful that the cups stay in place on the side bearings.
3. Install the bearing cups following the orientation marks that were made during disassembly, and torque up to specifications.
4. To perform all the necessary adjustments in the right sequence, refer to the testing and adjusting section later in this chapter.
5. Perform these steps in the following order: Adjust the preload on the differential bearing. Check the runout of the ring gear. Adjust the backlash of the ring gear. Check and adjust the tooth-contact pattern. Adjust the thrust screw.

Installing the Differential Carrier into the Axle Housing

After completing the differential carrier assembly, it is now ready to install into the axle housing. The steps outlined here are only examples. Always consult the manufacturer's technical documentation before attempting to install the carrier into the axle housing; not all installations are exactly the same. The installation procedure is the reverse of the disassembly procedure and is outlined below:

1. Thoroughly clean all mating surfaces with clean rags and solvent before attempting to assemble the components.
2. Inspect the mating surfaces for burrs or any other abnormality that may interfere with the fitment. Use a file to dress down any burrs that may be present.
3. Apply a thread locking compound, such as Loctite® to the studs and torque up to specification.
4. In some cases, the manufacturer does not require a gasket to be installed between the differential carrier and the mounting flange of the banjo housing. Apply a small bead of silicone to the banjo housing surface to form a seal.
5. Install the differential carrier carefully, making sure the mating surfaces are in contact all the way around. Use only one fastener at each corner (at this time) to ensure that the assembly is drawn in evenly.

6. Inspect the mating area carefully to ensure a good fit before installing the rest of the fasteners. Torque all fasteners to manufacturer's specification.
7. Reinstall the axle, sun gear, and planetary carrier into each wheel end.

Note: Clean the mating surfaces of the wheel hub and planetary carrier and apply a bead of silicone to the wheel hub contact area before assembling.

---

8. Be sure to reinstall the planetary carrier to the correct orientation, using the alignment marks that were made during disassembly. Clean the mating surfaces of the wheel hub and planetary carrier and apply a bead of silicone to the wheel hub contact area before assembling.

Axle Assembly Adjustments and Checks

Due to the large number of differential designs, this section covers the adjustments made to a conventional single reduction differential assembly. The adjustment procedures that are shown in this section are to be used as examples; any adjustments a technician would make on an actual repair will require the use of the manufacturer's shop literature for the differential that is being overhauled.

Pinion Bearing Preload Adjustment. Newer differential carriers use a press-fit outer bearing on the drive pinion. Although not as common, some older units use a slip-fit outer bearing that slips over the drive pinion. This section outlines both adjustment procedures.

Press-Fit Method of Adjustment

1. Assemble all the parts that are required to perform this procedure. Refer to **Figure 14-54** to view the correct order of assembly. Be sure to center the spacers between the bearing cones.
2. If a new gear set or new bearings are used, choose a nominal spacer based on the manufacturer's specifications. If the original parts are used, use the original spacers that were removed when it was disassembled.
3. Place the cage assembly in a suitable press and apply and hold a load onto the bearings. As pressure is applied, rotate the cage back and forth; this ensures seating the bearings.

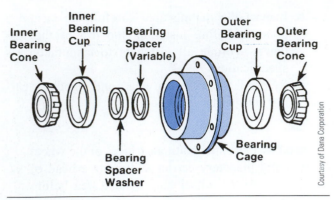

Figure 14-54 Correct orientation of parts in a pinion cage assembly before adjusting the preload.

4. Hold pressure on the bearing assembly. Use a string that is wound around the bearing cage and attached to a spring scale to measure preload on the bearings.

5. The scale should be read while the cage is actually turning; this will avoid getting the start-up torque reading, which would be higher.

6. When using this method of preload calculation, the radius (half the diameter of the cage) must be used to perform these calculations.

7. Depending on the specification, it may become necessary to adjust the pinion preload, in which case you either add or remove shims or spacers to achieve the desired preload.

8. When using this method it is important to select a spacer that is 0.001" larger than specified to compensate for bearing expansion when it is pressed onto the pinion shaft. This method will result in preload that will be correct *most* of the time.

The following is an example of formulas used to calculate preload in Standard Measurement:

Pull (lb) × radius (inches) = preload (lb-in.)

To convert to Newton meters use:
Preload (lb-in.) × 0.113 ; (conversion value) = preload (Nm)

The following is an example of formulas used to calculate preload in Metric Measurement:

Pull (kg) × radius (cm) = preload (kg-cm)

To convert to Newton meters use:
Preload (kg-cm) × 0.098 (conversion value) = preload (Nm)

In this example we will use actual values and use the formulas to calculate preload on a drive pinion. The pull on the scale is 10 pounds and the pinion cage diameter is 10 inches.

The following is an example in Standard Measurement:

10 lb × 5 in. = 50 lb-in. (preload)
To convert to Newton meters use:
50 lb-in. × 0.113 = 5.65 Nm

The following is an example in Metric Measurement:

4.6 kg × 2.7 cm = 58.42 kg-cm (preload)
To convert to Newton meters use:
58.42 kg-cm × 0.098 = 5.73 Nm (preload)

Yoke Method of Preload Adjustment. This method uses the pinion bearing assembly, fully assembled. The prior adjustment procedure did not use the drive pinion in the equation.

- Completely assemble the drive pinion cage including the input flange, as shown in **Figure 14-55**, by using a press.
- Assemble the yoke, flange washer, and nut and torque to specification. Note the position of the yoke on the splines of the drive pinion; it must rest against the outer bearing before it is torqued.
- Two methods can be used to perform this. If you are working on a small differential assembly (12-inch ring gear or smaller) and have a large vise available, you can place the lip of the pinion carrier in a vise equipped with soft jaws to keep it from turning while you torque the pinion nut to specification. If you are working on a larger differential assembly, you will temporarily install it in the banjo housing and secure it with two hand-tightened bolts to hold it in place during the torquing procedure. Do not install the shims at this time since it is only a temporary installation.

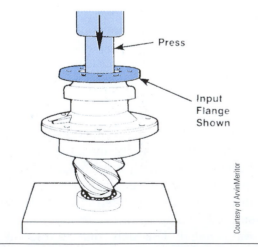

Figure 14-55 Procedure for installing the input flange with a press.

- Lock the yoke with a flange bar and torque up the pinion nut to specification.
- Place a dial indicator torque wrench with the appropriate socket on the pinion nut and rotate it slowly while checking the reading on the dial.
- If adjustment of the preload is necessary to change the spacer thickness, disassemble the unit and repeat the entire procedure.

Checking/Adjusting Pinion Cage Shim Pack

If the crown and pinion is being reused, start with the shim pack that was removed when the unit was disassembled. If a new gear set is used, then use this procedure to determine the shim pack thickness.

1. Measure the old shim pack with a micrometer and record that number along with the number on the new drive pinion. Refer to **Figure 14-56**.
2. If the number on the old pinion was a plus number, subtract it from the shim pack thickness. If the number was a minus, add it to the shim pack thickness.
3. Check the number on the new pinion and record it for reference later.
4. If the new pinion number is a plus, add the number to the shim pack thickness that was calculated previously. If the number on the new pinion is a minus, subtract it from the previously calculated shim pack.

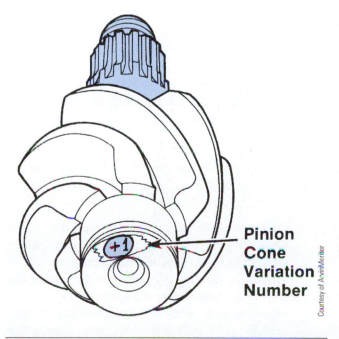

Figure 14-56 Location of the pinion cone variation number.

5. The answer to the previous two steps now yields the new shim pack thickness to be installed with the new gear set.

See **Table 14-5** and **Table 14-6** for examples of the calculations.

Differential Bearing Preload Adjustment

Although there is more than one method for making this adjustment, only the most common one will be

TABLE 14-5: CALCULATIONS OF SHIM PACK THICKNESS

	Inches	mm
Old Shim Pack Thickness	0.030	0.76
Old PC Number PC 2	−0.002	−0.05
Standard Shim Pack Thickness	0.028	0.71
New PC Number PC + 5	+.005	+.13
New Shim Pack Thickness	0.033	0.84
	Inches	**mm**
Old Shim Pack Thickness	0.030	0.76
Old PC Number PC − 2	+0.002	+0.05
Standard Shim Pack Thickness	0.032	0.81
New PC Number PC + 5	+0.005	+0.13
New Shim Pack Thickness	0.037	0.94

© Cengage Learning 2014

TABLE 14-6: CALCULATIONS OF SHIM PACK THICKNESS

	Inches	mm
Old Shim Pack Thickness	0.030	0.76
Old PC Number PC + 2	−0.02	−0.05
Standard Shim Pack Thickness	0.028	0.71
New PC Number PC − 5	−0.005	−0.13
New Shim Pack Thickness	0.023	0.58
	Inches	**mm**
Old Shim Pack Thickness	0.030	0.76
Old PC Number PC − 2	+0.002	+0.05
Standard Shim Pack Thickness	0.032	0.81
New PC Number PC + 5	−0.05	−0.13
New Shim Pack Thickness	0.027	0.68

© Cengage Learning 2014

shown in this example. As with all adjustments, this example is to be used only as a general guide; always consult the manufacturer's service literature to get the exact procedures and specifications for each individual design application.

1. It is important to start with a small amount of endplay in the carrier assembly. Place a dial indicator onto a flat surface, such as the mounting flange of the carrier, and place the plunger of the dial indicator so that it sits 90 degrees to the back of the ring gear, as shown in **Figure 14-57**.
2. To begin this procedure, loosen the adjusting ring opposite the dial indicator until a small amount of endplay is registered on the indicator. Use an appropriate T-bar wrench for this adjustment.
3. To adjust the ring gear to the left or right, use a pry bar, as shown in **Figure 14-58** as you read the

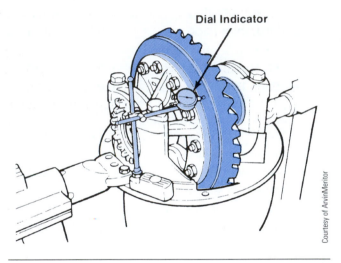

Figure 14-57 Location of the dial indicator used to measure carrier endplay.

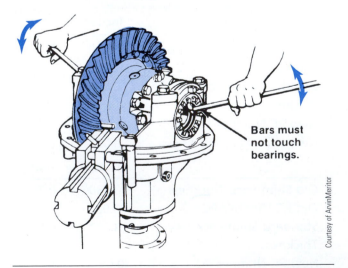

Figure 14-58 Adjustment procedure used to achieve preload to the differential assembly bearings.

dial indicator. Use the differential case as a pinch point when performing this adjustment. Tighten the same adjusting ring until no endplay shows on the dial indicator, moving the ring gear to the right or left as required to achieve no endplay. Keep repeating until zero endplay is reached.

4. To complete this adjustment procedure, tighten each adjusting ring one notch farther in from zero-lash to preload the bearings.

Ring Gear Runout Check

To check the runout on the ring gear assembly, the following steps must be followed carefully or an inaccurate reading will result. Follow the steps below in sequence to measure the ring gear runout.

1. Place a magnetic dial indicator on the flange mount of the differential housing and position the pointer 90 degrees to the back face of the ring gear with a slight preload on the pointer.
2. Adjust the dial indicator to zero and slowly rotate the ring gear. Record the total runout during one complete revolution of the ring gear. A nominal value will typically be around 0.008". Always consult the manufacturer's service literature for the specific differential that is being adjusted.
3. If excessive runout is recorded, then the differential must be disassembled and checked carefully for any burrs that might have been missed during assembly. The root cause of excessive runout must be corrected as this will lead to a short crown and pinion life.

Ring Gear Backlash Adjustment.   If the old ring gear set is being reused, adjust the backlash to the values that were measured before disassembly. If a new ring gear set is used, adjust the backlash according to the manufacturer's specifications.

....

Note: Backlash readings should be taken in at least three different places on the ring gear to get an average reading.

1. Place a magnetic dial indicator on the flange mount of the differential housing and position the pointer 90 degrees to the face of a ring gear tooth with at least one revolution of preload on the pointer and set it to zero, as shown in **Figure 14-59**.
2. Move the ring gear back and forth slightly and read the dial indicator movement; repeat this in three places spread out evenly around the ring gear. If the reading is not within

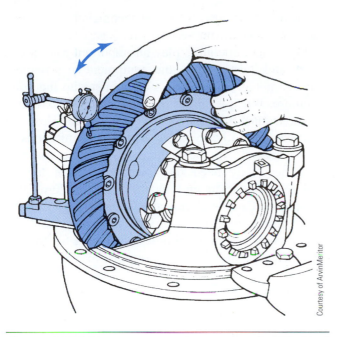

Figure 14-59 How to correctly measure backlash of the ring gear. Note: Check in at least three places to get an average reading.

specifications, adjust the backlash as outlined in the following step.

3. Depending on which way the backlash must go, move the adjusting ring one turn at a time by loosening one bearing and tightening the opposite one. This will maintain bearing preload. Backlash will increase when the ring gear is moved away from the pinion and decrease when moved toward the pinion. Refer to **Figure 14-60**.

Ring Gear Tooth-Contact Adjustment. As with any adjustments on a differential, this one is critical to the

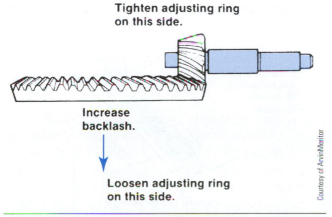

Tighten adjusting ring on this side.

Increase backlash.

Loosen adjusting ring on this side.

Figure 14-60 What happens when the ring gear is moved away from the pinion. Note: Backlash increases.

life of the gear set and, if not done correctly, can lead to a catastrophic failure. The tooth-contact pattern consists of a contact patch that runs lengthwise along the tooth of the ring gear as well as the profile of the tooth.

1. Paint at least 8 to 12 teeth on the ring gear before rolling the gear teeth together to check the pattern. A good pattern, as shown in **Figure 14-61**, will have the contact mark centered on the ring gear teeth lengthwise. This test is in an unloaded condition and does not represent what takes place under load. The contact pattern in this case will be about one-half to two-thirds of the tooth.

2. When checking the pattern on a used gear set, the contact will not usually display square as is the case with a new gear set. The contact pattern will have a pocket at the toe of the gear that forms a tail at the root of the tooth. This pattern would look better if a load could be placed on the gear set. The more wear there is on the gear set, the more pronounced the tail will be.

3. Before you disassemble the gear set, perform a tooth contact check and sketch the pattern for future reference. A good pattern will not extend to the toe, and it will be centered evenly along the tooth face between the top land and the root. At no time should the pattern run off the tooth. If required, adjust the contact pattern by moving the ring gear and pinion gear either away from each other or toward each other. Pinion position in relation to ring gear position is determined by the thickness of the shim pack between the pinion cage and the differential housing. This adjustment controls the tooth depth.

4. These adjustments are interrelated because one will affect the other; they must be considered together even though they are two distinct adjustments.

Whenever these adjustments must be made, adjust the pinion first, and then adjust the backlash. Continue this process until a satisfactory

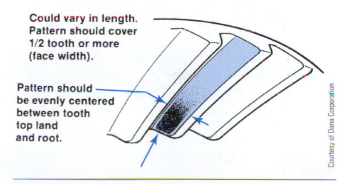

Could vary in length. Pattern should cover 1/2 tooth or more (face width).

Pattern should be evenly centered between tooth top land and root.

Figure 14-61 Contact pattern on a new gear set.

pattern is achieved. Here is where experience comes in: An experienced technician will set a good pattern in a few minutes, whereas an inexperienced technician will often take longer before figuring it out. **Figure 14-62** shows an example of a contact pattern that needs to be corrected.

Thrust Screw Adjustment

If the differential has a thrust screw, it must be adjusted properly. Follow the steps listed below to complete this procedure correctly. Before beginning this adjustment, the carrier must be rotated in the stand until the ring gear is facing down toward the floor.

1. Place the thrust block on the back side of the ring gear and rotate the ring gear until the thrust block and the thrust screw hole are lined up. Install the thrust screw until it contacts the thrust block, and then back out approximately one-half turn.
2. Install the lock nut onto the thrust screw and secure it, making sure that the screw does not turn during this procedure. Torque the jam nut to specifications.

FAILURE ANALYSIS

Drive axles are subject to failure from a wide variety of conditions; some are caused by inappropriate equipment operation and others are caused by poor maintenance practice. A technician must understand the limitations of drive axle components and their lubricants in order to assess failure conditions and come up with reasonable conclusions. Component damage can lead to expensive repairs and downtime. If a technician replaces the failed components without identifying and correcting the root cause, more problems associated with the failure will occur. It is a

considerable challenge for the technician to achieve maximum component service life.

Equipment operators play an important role in this effort, and must operate the equipment in a manner that will not stress the axle's components to the point of failure. When a failure occurs on a drive axle, two questions must be answered: What occurred? How can a similar failure be avoided? There are limitations to failure analysis in the field and, in some cases, a component may need to be analyzed in a lab. A qualified technician can identify most failures that occur and recommend procedures that will prevent the same failure from occurring again. This section will look at some common failures and show what to look for to successfully identify corrective measures.

Failure Types

Axle assemblies fail due to one or more of the following conditions:

- Normal wear
- Improper lubrication
- Fatigue
- Shock loading
- Improper operation

Normal Wear. Drive axles are designed to operate for a reasonable amount of time in the field with nothing more than regular maintenance. All components eventually fail, but there are preventative measures that can be performed by the operator and service technician that will minimize premature failures. During the break-in process, a small amount of wear takes place. In some cases, this wear has a positive effect on component life. Gears that mesh together benefit from this type of wear. Components sometimes come from the manufacturer with a defect in the surface finish. These conditions can be recognized by a qualified technician as normal wear, thus eliminating unnecessary component replacement.

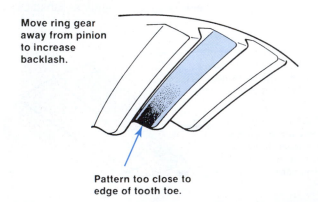

Move ring gear away from pinion to increase backlash.

Pattern too close to edge of tooth toe.

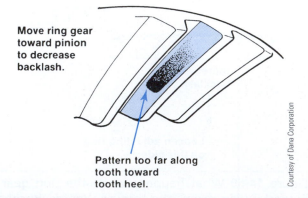

Move ring gear toward pinion to decrease backlash.

Pattern too far along tooth toward tooth heel.

Courtesy of Dana Corporation

Figure 14-62 Contact pattern too far toward the heel and too close to the toe.

Some parts should always be replaced when the axle components are overhauled.

Improper Lubrication. Using the incorrect type of lubricant can have a negative effect on the axle component life. Contamination from things like moisture or dirt can cause scoring and pitting of mating surfaces. Dirt in the lubrication acts like sandpaper, wearing out mating components and shortening their service life. **Figure 14-63** shows the damage that can occur when the lubricant additive package becomes depleted. The following list addresses more conditions that are likely to cause damage:

- Overheating damage results from high oil operating temperatures, which can be caused by several operating conditions. Operating the equipment with low oil in the axle assemblies will lead to higher operating temperatures.
- Overfilling the axle assembly with oil may also lead to higher than normal operating temperatures.
- Excessive increase in engine power or torque rating and restricted air flow over oil coolers used on some equipment will lead to high oil temperatures.
- Using the wrong grade or viscosity of lubrication may lead to overheating of internal components.
- Depleted additives in the lubricant can cause a wear pattern that is referred to as *crow's feet*. These look like scoring lines or ridges on the gear teeth.

- Mixing different types of lubricants can lead to premature failure.

Fatigue (Torsional Bending). This condition is a result of torsional loading and bending of a shaft or component that leads to progressive destruction of the component. The contoured lines often seen on the face of a broken component represent typical characteristics that are called steps. These occur before total breakdown of the part. Refer to **Figure 14-64** for an example of this condition. This condition usually can be attributed to overloading of the axle assembly.

Repeated overloading of the axle shafts can lead to a star-shaped torsional fracture in the splined area of an axle shaft. Refer to **Figure 14-65** to see this symmetrical star-shaped break.

Shock Loading. Shock loading happens to components that are stressed beyond the strength of the material, which sometimes causes instantaneous fracture, as shown in **Figure 14-66**. This occurrence may not always happen instantly; it sometimes initiates a crack, which eventually leads to breakage. Shock loads can subject axle shafts to high torsional or twisting forces that can lead to torsional shear. When initial shock impact is not great enough to shear the tooth or shaft, a crack will develop and, in time, lead to failure. One of the common causes associated with this type of failure is an operator who changes direction of a loader without bringing the equipment to a complete stop. Losing traction on one side of an axle drive, and then

Figure 14-63 How depleted lubricant additive damage causes ridging on the gear teeth.

Figure 14-64 How torsional fatigue can appear after a failure.

Figure 14-65 A star-shaped torsional fatigue fracture.

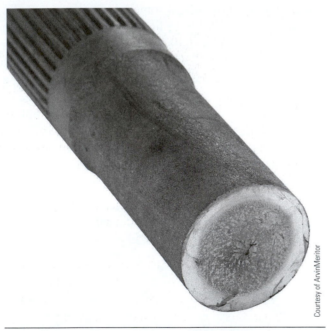

Figure 14-66 How a high-impact shock load can lead to a clean break of an axle shaft.

suddenly regaining traction often results in this type of failure. In most cases, this type of failure does not produce immediate catastrophic failures; the failure can take place days or weeks later.

Spinout and Overload. Overload can have a devastating effect on many components in a drive axle assembly, which can result in axle housing damage. Operating equipment that exceeds the Gross Axle Weight Rating (GAWR) can often cause extensive

component damage. Identifying these causes presents a teachable moment in which the equipment operators are shown the results of this type of abuse in order to prevent further occurrences.

Spinout is another common condition that occurs frequently on equipment such as haulage trucks, as shown in **Figure 14-67**. On a single rear axle haulage truck differential spinout occurs when the equipment loses traction on one side of the drive, which causes one wheel to spin excessively. In the case of a tandem drive, the spinout occurs in the interaxle differential; one axle spins while the other remains stationary. This condition may lead to differential spider and pinion damage or breakage.

ONLINE TASKS

1. Use a suitable Internet search engine to research the configuration of an off-road drive/steer axle system. Identify a piece of equipment in your particular location paying close attention to the drive/steer axle arrangement, and identify whether or not it uses an additional planetary gear reduction in the wheel end. Outline the reasoning behind the manufacturer's choice of this particular design for that particular application.

2. Check out this URL: *http://www.meritor.com/customer/northamerica/training/default.aspx* for more information on drive/steer axle configuration and design.

Figure 14-67 Damage on a differential spider.

Shop Tasks

1. Select a piece of equipment in your shop that has a drive/steer axle assembly and identify the type of components that make up the wheel end arrangement. Identify whether or not the drive axle has excessive wear in the wheel ends and what type of U-joint is used on this arrangement.

2. Select a piece of off-road equipment in your shop that uses a drive/steer axle and identify the type and model of the equipment, listing all the major components and the corresponding part numbers of the drive/steer wheel end assembly. Identify and use the correct parts book for this task.

Summary

- The drive axle's primary roles are to support the vehicle load and provide torque to propel the equipment.

- Axles can be categorized into three distinct designs. The first is capable of propelling the vehicle and carry the weight of the load. The second is designed to carry the load and steer the vehicle. The third is designed to only carry the load; this type is found on the rear section of an articulating haulage truck.

- A jaw clutch is used to lock the differential, providing equal torque to both wheels.

- The limited slip differential has been designed to provide equal torque application to both wheels during normal operation.

- The no-spin differential uses jaw clutches that are splined to the side gears. The jaw clutches stay engaged with the side gears as long as the equipment is operated in a straight line.

- The planetary differential input torque is provided by a gear in the transmission transfer drive that meshes with the carrier of the planetary differential. The two sets of planetary gear shafts are held by the carrier and drive the planet gears.

- A large percentage of interaxles are lubricated by the splash method; oil is distributed by the gears turning and splashing the oil internally.

- The function of a power divider can be categorized in two ways. It must be able to divide the driving torque between two axle assemblies, and it must be able to provide differential action between the two axles when required.

- Equipment with two-speed axles can take advantage of the extra speed range and use it to get more effective gearing. An example of this is the use of low range in each of the two axle ratios: First gear in low range, then first gear again in high range, and continue up through the gear ranges.

- Planetary axle drives receive torque from the differential through an axle shaft, which serves as a sun gear (input gear). The sun gear meshes with the three planet gears, which are mounted together in a planet carrier (output). The ring gear is held or locked in place.

- In many off-road applications where slow speed and high torque are necessary, the use of an additional gear reduction can be found by incorporating a torque hub planetary drive.

- In a hybrid differential planetary drive, the sun gear, planet clusters, and two internal ring gears provide a greater torque reduction. The input gear drives the planet clusters, which turn on a fixed ring gear, which sends output torque out to the second ring gear.

- Full-floating axles use two large, tapered roller bearings mounted to the wheel hub and carry the full weight of the equipment and load.

- Semi-floating axles are splined to the side gear and float at this end. At the wheel end the axle shaft carries one or two bearings, which carry the full weight of the load.

- *Channelling* occurs in sub-zero temperatures when the cold thickened oil is sheared by the rotating ring gear and is too thick to flow back, resulting in inadequate lubrication until the oil heats up.

- If the pinion bearings are replaced, the new bearing cups must be pressed into place in the bearing cage on a press.

- If the crown and pinion is being reused, start with the shim pack that was removed when the unit was disassembled.

- If excessive ring gear runout is recorded, the root cause must be corrected because this can reduce crown and pinion life.

- When checking the pattern on a used ring gear and pinion set, the contact will not usually display square as on a new gear set.

Review Questions

1. What is the final drive section of an articulating haulage truck drive axle usually known as?

   A. rear transaxle   C. differential axle

   B. dead axle   D. articulating axle

2. On a single reduction differential, what is the drive axle ring gear bolted or riveted to?

   A. axle shafts   C. differential case flange

   B. bearing cage   D. pinion gear

3. What takes place in a conventional differential during cornering?

   A. Outside wheel rotates faster than the inside wheel.

   B. Outside wheel rotates slower than the inside wheel.

   C. Inside wheel is accelerated to the speed of the outside wheel.

   D. Outside wheel is braked to the speed of the inside wheel.

4. Which drive configuration uses a drive axle configuration that uses two gear sets to provide additional gear reduction and has greater torque at the drive wheels?

   A. single reduction axle   C. tandem drive axles

   B. power splitter axle   D. double reduction axle

5. Which one of the following statements is accurate?

   A. All two-speed axles use double reduction.

   B. Double reduction axles are always two-speed axles.

   C. All tandem drive axles are double reduction.

   D. Single reduction axles are seldom used in haulage trucks.

6. Which statement describes an amboid gear arrangement?

   A. Axis of the drive pinion is at the centerline of the ring gear.

   B. Axis of the drive pinion is below the centerline of the ring gear.

   C. Axis of the drive pinion does not identify the amboid gear design.

   D. Axis of the drive pinion is above the centerline of the ring gear.

7. Which components must rotate at the same speed when the differential lock on a single reduction axle is engaged?

   A. differential case and both axles

   B. both axles and the drive pinion

   C. drive pinion and crown gear

   D. drive pinion and both side gears

8. Which axle hub arrangement carries the weight of the load on the wheel end of the axle?

   A. semi-floating   C. tandem axles

   B. full-floating   D. trailing axles

9. Which axle type would use a kingpin?

   A. tandem drive axles

   B. steering axles

   C. trailer lift axles

   D. articulating axles

10. What components change the direction of power flow in a drive axle carrier?

    A. differential gearing

    B. crown and pinion

    C. planetary reduction gearing

    D. axle shafts and side gears

11. Which term is used to identify a differential carrier housing?

    A. dead axle

    B. pusher axle

    C. banjo housing

    D. trailing housing

12. Where does the gear reduction actually take place on a single reduction differential carrier?

    A. differential gearing

    B. planetary carrier

    C. propeller shaft and pinion

    D. pinion and crown gear set

13. Which gear design uses a spiral cut pinion and ring gears mesh so that the pinion intersects the ring gear just below its center axis?

    A. amboid

    B. hypoid

    C. spiral bevel

    D. spur gear

14. What component does the differential pinion drive in an interaxle differential?

    A. spider

    B. ring gear

    C. pinions

    D. propeller shaft

15. What connects the steering knuckles on a drive steer axle?

    A. steering differential arm

    B. Pitman arm

    C. tie rod

    D. drag link

15 Final Drives

Learning Objectives

After reading this chapter, you should be able to:

- Describe the types of final drive systems used on off-road equipment.
- Identify the purpose of final drives used on off-road equipment.
- Identify the final drive system designs used on off-road equipment.
- Outline the construction features of final drive systems.
- Describe the operation of final drive systems on off-road equipment.
- Identify the maintenance and adjustment procedures on final drive systems.
- Describe the common failures that occur on final drive systems and analyze the causes.

Key Terms

backlash	mechanical seals	side bar
beach marks	pinion drive	single reduction drive
bull-type final drive	pinion shaft	sprocket shaft gear
chain drive	planet gear	spur gear
double reduction drive	planetary drive	sun gear
elongation	preload	viscosity
final drive	ring gear	
herringbone gears	roller link	

INTRODUCTION

The term *final drive*, as the name suggests, is the final stage in a series of reductions that begin at the engine flywheel, flow through the torque converter and transmission, and end at the **final drive** (wheels or tracks). **Figure 15-1** shows a typical final drive assembly used on a motor grader. The term can actually be applied to four types of drive systems: **chain drives**,

pinion drives, **planetary drives**, and straight axle drives. Straight axle drives and planetary drives on wheel ends are covered in detail in another section of this book. We may refer to them when discussing final drive designs because some manufacturers include straight axles and planetary drives in their designs.

Final drives receive torque from the transmission, reducing its speed and increasing the torque through the use of various gear and chain drive arrangements.

Figure 15-1 Final drive on a Caterpillar 12H motor grader.

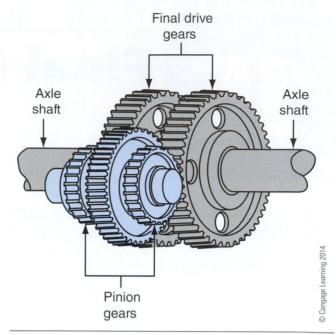

Figure 15-2 Typical pinion drive used in a differential case drive.

The main purpose of a final drive is to increase torque and decrease speed; however, some manufacturers take advantage of this arrangement to drop the power flow lower to the ground, such as in a bulldozer or grader. The advantages of this include reducing the drive speed and stresses, and increasing the available torque at the drive wheels to move large loads.

Transmissions could be designed to accommodate the speed reductions and torque increases that are required on off-road equipment, but that would require using heavier components, such as shafts and gearing, which would make the transmissions too bulky and heavy. By using a final drive to increase torque, the other powertrain components can be lighter because they do not have to carry huge torque loads. Final drives are designed to support the weight of the machine and have to absorb the torque and shock loads at the wheels or tracks. **Figure 15-2** shows a typical example of a pinion drive used in many final drive applications. In the case of a bulldozer, the final drive assembly includes steering clutches that vary the final drive speed on each track for steering control.

FUNDAMENTALS

Final drive designs used on newer off-road equipment are classified as single reduction or double reduction. The most common method of achieving reductions is transferring power through gears. Chain drives are used extensively on off-road equipment and are suitable for this application. Because speeds are relatively low, this type of drive is a suitable alternative.

In many cases, straight cut gears transmit power in final drives. A few exceptions are the use of spiral bevel or straight bevel gears, worm gears, or **herringbone gears**. The purpose of a final drive is to change the

direction of powerflow and to provide a means of either lowering or raising the output to match equipment mobility requirements. **Figure 15-3** shows an example of a pinion drive that drops the output lower to the ground to improve ground clearance. Speed reductions through final drives can vary from low, 3:1, up to as high as 12:1. Most final drives use various gear designs to transmit torque, but they are not limited to this method.

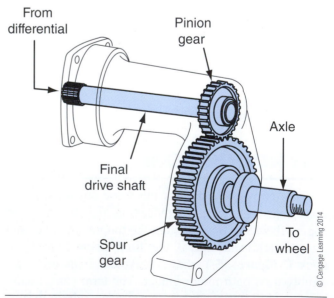

Figure 15-3 Typical pinion drive used in a final drive of a bulldozer.

Often manufacturers use chain drives to change speed and torque on equipment, as is the case on many road graders. Chain drives provide an economical way to transmit torque at low speeds over long distances (e.g., tandem drives found in road graders).

Worm gears are another popular method of transmitting torque in final drives. Worm gear drives permit a large speed reduction accompanied by a very large increase in torque, within a compact space compared to gear trains.

CONSTRUCTION FEATURES
Pinion Drives

Two basic designs of pinion drives are used on off-road equipment. The first has the drive assembly located in the differential housing (refer to **Figure 15-2**) while the second type has the final reduction incorporated into the outer end of the final drive (see **Figure 15-4**).

The pinion drive is possibly the most common of all the final drives used on off-road equipment. Most manufacturers use pinion drives on the majority of small track equipment, such as bulldozers. This drive utilizes a single gear reduction and is often called a **bull-type final drive**. The **spur gear** and pinion drive design is commonly found on smaller equipment; it has the advantage of containing all the gears in the differential transmission case, making it more compact and cost-effective to service and maintain because only one lubrication system is necessary. This design is quite popular on small tractors as the power is transmitted in a straight line to the wheel ends, minimizing cost by requiring fewer parts. The axle shafts generally have two tapered roller bearings mounted to

the shaft to carry the weight of the machine and absorb end thrust; they connect to the final drive gears by splines on the axle shaft ends. The bearings are generally located at either end of the axle shaft facing each other. Bearing **preload** can be altered either by shimming or by an adjusting nut with a locking mechanism to maintain the proper preload setting.

Bull-Type Final Drive

This drive design is common on track equipment where ground clearance may be a concern. Locating the pinion drive at the ends of the axle assembly permits the gears to be separate in their own housing, necessitating a dedicated reservoir. If a failure occurs, the damage is contained in the outer housing and does not destroy the remaining gear train components, as in an integral design. This design has the advantage of dropping the power flow closer to the ground, and is often referred as drop-axle housing. The weight of the equipment is supported by the entire final drive assembly and can supply an increase in torque through the use of only one gear set—a small pinion driving a large bull gear. **Figure 15-5** shows an example of a **single reduction drive** commonly used on bulldozers.

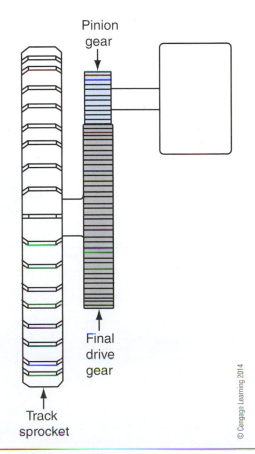

Pinion gear

Final drive gear

Track sprocket

© Cengage Learning 2014

Figure 15-5 Single reduction final drive on a dozer.

Reprinted Courtesy of Caterpillar Inc.

Figure 15-4 Typical pinion drive used in a final drive of a bulldozer. Note the use of the straight spur gear design in this application.

On smaller dozers, the housings that contain this gear set are usually manufactured from a one-piece casting and are bolted to the steering clutch housing. The pinion gear is splined to the **pinion shaft** and held in place with roller bearings; the larger final drive gear is usually splined to the track drive sprocket, which is supported by large tapered roller bearings. This design uses the time-proven splash method for lubrication; the housing acts as a sump for the lubricating oil, which is splashed around by the rotating final drive gear. Newer equipment uses **mechanical seals** to contain the oil and keep dirt out. This type of seal is better at keeping the dirt out and outlasts conventional lip-type seals in this application.

Double Reduction Final Drives

On larger track equipment **double reduction drives**, as shown in **Figure 15-6**, are commonly used in the final drive in an elevated position. This concept is utilized to reduce shock loading and implement loads that result in gear and bearing misalignment. It also prevents the final drive from coming into contact with work surfaces or any other abrasive material that could accelerate wear.

A double reduction drive has two separate gear reductions in line with each other. They are generally located in a separate housing at the track end of the drive.

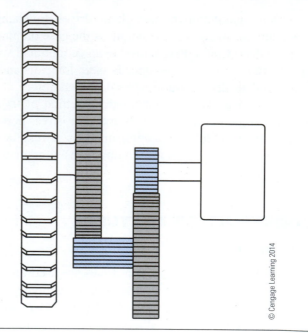

Figure 15-6 Double reduction final drive used on large dozers.

Figure 15-7 shows a typical example of a double reduction final drive used on dozers. The housings are often made in two parts with the inner one being part of the steering housing assembly. The outer housing is bolted to the steering housing, which encloses the final drive components.

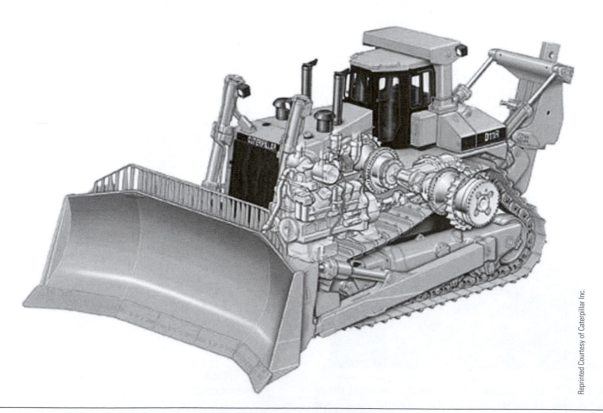

Figure 15-7 Position of planetary final drive used in an elevated design.

The bearings used to support the drive pinion are tapered roller bearings capable of supporting the high loads that are generated at this end of the drive. As with most tapered roller bearings, shims are used to adjust the bearing preload. Most manufacturers also incorporate the brakes and steering clutches into this section of the drive.

The intermediate pinion and gear set are splined together on the same shaft and are usually supported by a large tapered roller bearing at each end (see **Figure 15-8**) with the preload adjustment set by shimming the bearings. The **sprocket shaft gear** is splined to the sprocket shaft, which also holds the drive sprocket. The sprocket is mounted to the sprocket hub, which is held in place with either a large nut or a series of bolts. In most cases, newer double reduction drives are pressure lubricated; some older units continue to use the splash method of lubrication.

Planetary Final Drives

Planetary final drive designs also come in single reduction and double reduction configurations and function much the same way as previously noted. The main difference is the way in which they transmit. **Figure 15-9** shows an example of a typical planetary final drive used on some types of dozers.

Most modern equipment uses these types of final drives for many good reasons, the most important of which is size. Planetary drives are inherently compact, especially when compared to conventional pinion drives. This reduces the overall size of the final drive and not only is it more compact, but it can transmit more torque than a conventional pinion drive of similar size because it uses multiple gears. This design equalizes torque and minimizes loads on support bearings and shafts, which translates into longer bearing and component life.

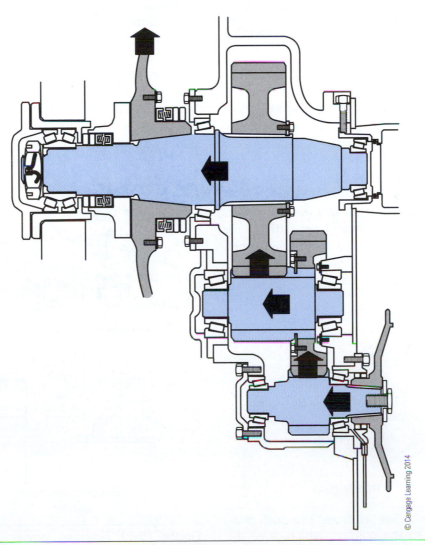

© Cengage Learning 2014

Figure 15-8 Flow of power through a typical final drive on large dozers.

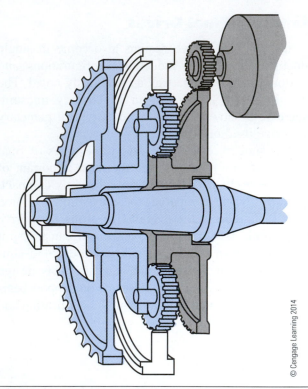

Figure 15-9 Cutaway of a flow of power through a typical planetary final drive.

Figure 15-10 shows simple planetary gears. The torque flows from the axle to the **sun gear**, which transfers the torque to the **planet gears** that are mounted together on a carrier. Depending on the size of the final drive, there can be up to four planets mounted on a carrier rotating within a stationary **ring gear**. The ring gear is held stationary by either bolts or held on splines. The principle adheres to one of four planetary gear laws: Whenever the planet carrier is the output and the ring gear is held, a forward reduction drive results. *Note:* The more planets there are, the higher the capacity of the gear set.

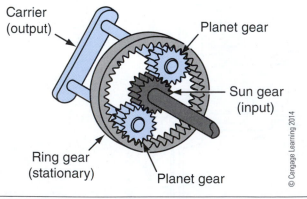

Figure 15-10 Power flows in from the sun gear, transferring torque to the planets, which drive the track sprocket.

Manufacturers produce variations of double reduction planetary drives to suit many applications. Some manufacturers use a combination of a pinion drive with a planetary drive at the drive sprocket end to drive the track. **Figure 15-11** shows one variation of this popular drive. Two sets of planetary gears are used in this application with the inboard sun gear receiving torque from the axle. This forces the inboard carrier, which is splined to the outboard sun gear, to walk around the ring gear, causing the outboard sun gear to turn at the same speed as the inboard planetary carrier. The second reduction is achieved by the outboard planetary carrier walking around the ring gear, which transmits torque to the track. In this way, high torque increase is produced from a light, compact drive. Tapered roller bearings are used to support the weight of the assembly as well as the load.

CHAIN DRIVES

Chain drives are used in many heavy equipment applications, such as road graders, where torque is transferred from the final drive sprocket to the wheel sprockets. Using chain drive in place of gears is an inexpensive and efficient way to transmit torque over a specific distance. Chain drives can transmit torque for long periods of time with little maintenance, so long as dirt is kept out of the equation. The roller chain links mesh with the teeth on the sprockets to provide a

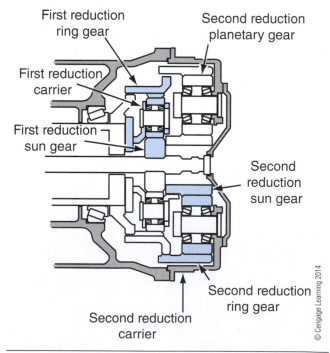

Figure 15-11 Flow of power through a double reduction planetary final drive.

positive drive between the front and rear wheels of the tandem drive, the same as would be provided by a gear drive. **Figure 15-12** shows a typical final drive arrangement used on a grader.

To distribute the wear evenly and prevent the **roller link** from contacting the same sprocket teeth each revolution, roller chains have an even number of links and the sprockets they run on have an odd number of teeth. Some applications incorporate an idler sprocket on the slack side of the chain to take up excessive chain slack, with the added disadvantage of having to make periodic adjustments to the chain tension. A typical grader tandem drive has a drive sprocket that drives two wheel sprockets, one at either end of the tandem drive, with a heavy-duty roller chain. To get long chain life with this application, alignment is critical and must be checked during any repair work.

Figure 15-13 shows torque coming from the axle shaft into the sun gear of the planetary final drive reduction. The two drive sprockets are mounted to the planet carrier and transmit power to the two drive chains. A driven sprocket at each end of the tandem housing is mounted to a wheel spindle. Power is transmitted by the two chains to each of the wheels on either side of the grader's final drive. The wheel spindle housing contains the components that make up the brake system. Lubricating oil in the tandem housing is also used to cool the brakes. The pivot housing bolted to the tandem housing allows the tandem drive to oscillate. This connection allows the wheels to follow the road contour, providing traction to all four wheels continuously.

Roller chains are made up of a series of pin links and roller links connected to each other with **side bars**; rollers and bushings are fitted to the pin links and are

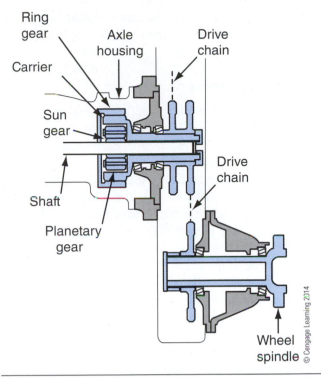

Figure 15-13 Flow of power through a tandem motor grader chain drive.

also held together with side bars. Roller chains come preassembled from the manufacturer with provision for a master link to connect them together. **Figure 15-14** shows the components that make up typical roller chain construction. (Each pin link interconnects to a roller link, with the pin links consisting of two pins with two side bars, one for each side. The roller links consist of two side bars, two bushings, and two rollers.) These components are interconnected to form a roller chain. A master link is used to connect them; the only

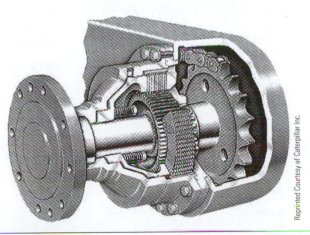

Figure 15-12 Flow of power through a grader tandem final drive.

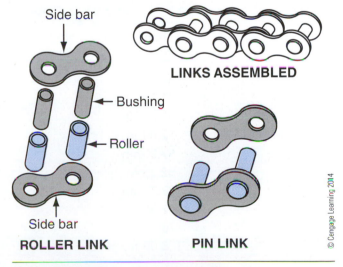

Figure 15-14 Parts of a roller chain.

difference between a pin link and a master link is that the latter can be easily removed to facilitate chain replacement.

Chain Adjustment

Chain tension can vary greatly on final drives, due to equipment application. Always consult the manufacturer's service literature for each application to determine the amount of tension or slack that is required. Some manufacturers use an idler sprocket to regulate chain tension; others use a guide shoe. A good rule to use when measuring chain tension is to have about one quarter inch of slack for every foot between shaft centers when the chain tension is removed from the drive side of the chain. The key to preventing chain and sprocket wear is to ensure that sprocket alignment is not compromised and that lubricating oil is removed and filtered or replaced as recommended by the equipment manufacturer. Chain lubrication is generally provided by the drive chains themselves, which run through the oil in the housing and distribute oil along the drive chain's path.

Chain Wear

To assess drive chain wear, the chain should be checked when mounted in the respective housing. The drive is designed to tolerate normal wear that takes place over a period of time; up to a maximum of 3% **elongation** is considered normal. There are exceptions to this rule. Always consult the manufacturer's service literature before determining the allowable elongation specifications for the equipment that you are checking. Chain replacement is required when the chain reaches elongation specifications. Extensive damage to the

tandem drive may result if the chain breaks during operation.

Some manufacturers recommend checking chain length under tension as an indicator of drive chain wear. Follow the manufacturer's recommended chain removal procedure to remove the drive chains. Once the chain is removed, clean it using the manufacturer's approved cleaning procedure, and inspect the chain for cracks, broken rollers and pins, and deformation. If any of these conditions are found, the chain must be replaced. After determining the drive chain length is higher than specifications permit, you need to determine whether the sprockets are worn.

CAUTION *Always replace both chains in tandem housings at the same time. Splicing in a new section is not recommended; this will cause improper engagement with the sprockets, leading to sprocket damage.*

To determine chain elongation dimensions, hang the chain from a hoist and attach a scale to measure the chain tension as shown in **Table 15-1**. Attach the free end of the scale to a heavy object, such as the grader. Refer to **Table 15-1** for the correct tension to apply to the scale. To achieve accurate dimensions the chain must be placed under tension to displace the oil that may be trapped between the rollers and pins of the chain. Once the correct tension is applied, measure the distance across the number of chain links recommended in **Table 15-1** to determine elongation limits. If the chain passes this wear test, it can be reused. **Table 15-1** shows an example of measurements for Caterpillar grader final drives recommended chain

	TABLE 15-1: CHAIN TENSION MEASUREMENTS				
Model	Chain Pitch	No. of Chain Links	New Measurement	Maximum Wear	Chain Tension
120H	44.5 mm	14	622.3 mm	641.1 mm	173 kg
135H	1.75 in.		24.50 in.	25.24 in.	381 lb
12H,140H,	50.8 mm	12	609.6 mm	628.7 mm	277 kg
143H,160H,163H	2.00 in.		24.00 in.	24.72 in.	610 lb
14H	57.2 mm	11	628.7 mm	647.7 mm	286 kg
	2.25 in.		24.75 in.	25.50 in.	630 lb
16H	63.5 mm	10	635 mm	654.1 mm	354 kg
	2.50 in.		25.00 in.	25.75 in.	780 lb
24H*	76.2 mm	10	762 mm	773.4 mm	782 kg
	3.00 in.		30 in.	30.45 in.	1724 lb

© Cengage Learning 2014

*Manufacturer recommends 1.5% maximum chain elongation.

tension and measurement specifications. Always refer to the manufacturer's specification for the equipment that is being worked on.

In some cases, chain and sprocket wear cannot be determined visually without some disassembly of the tandem final drive. Consult the manufacturer's service literature to determine the proper course of action. Sometimes it is possible to check for chain wear by looking at the chain wrapped around the largest sprocket in the drive that is accessible. The chain links should not ride up the tooth, as shown in **Figure 15-15**, but should sit at or near the bottom of the sprocket. If the sprocket teeth are worn, the chain should be replaced.

If chain elongation is not a problem, it may be possible to reuse the chain if it passes the visual inspection when wrapped around a sprocket. Removing a link may prolong the life of the chain drive. In some cases, the sprockets can be reversed on some equipment, extending drive life. New links should not be used on used chains; the new links will have a shorter pitch. This causes shock loading of the chain during operation and should be avoided. In most cases, it is advisable to change the sprockets when installing a

new chain. Any sprocket tooth wear, which may look good to the eye, will shorten the life of the new chain considerably. A determination must be made based on the manufacturer's recommended procedure for each specific application.

Table 15-2 outlines common problems and their likely causes; it is a helpful guide when evaluating chain wear and operating conditions.

MAINTENANCE AND LUBRICATION

Modern final drive assemblies can operate for extended periods of time with very little maintenance. Technicians are often required to make a determination as to how much work can be done economically in

TOOTH WEAR

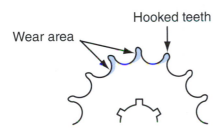

STRETCHED CHAIN

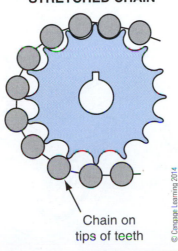

Chain on tips of teeth

© Cengage Learning 2014

Figure 15-15 Chain link wear and tooth wear.

TABLE 15-2: DETERMINING CHAIN WEAR AND OPERATING CONDITIONS	
Problem	Likely Cause
Excessive wear on the side bars and link plates	Misalignment
Pins, bushings, or rollers broken	Sudden shock loads
	Insufficient lubrication
	Mismatched chain and sprocket
	Too high a chain operating speed
Chain whip	Incorrect slack adjustment
	Seized chain joints
	Chain wear uneven
Chain climbing the sprockets	Worn-out chain
	Excessive chain slack
	Sprockets worn out
	Mismatched chain and sprocket
Broken sprocket teeth	Excessive shock loading
	Chain climbing the sprockets
Chain drive lubrication overheats	Too high operating speed
	Insufficient lubrication
	Oil level too high in housing
Chain sticks to sprockets	Mismatched chain and sprocket
	Incorrect lubricant
	Excessive chain slack

© Cengage Learning 2014

the field on final drives and where that work will take place. A thorough knowledge of operating principles related to final drives is required to be able to make that assessment. A technician must be able to determine whether to rebuild a final drive assembly or replace it. The importance of maintaining and servicing the final drive systems cannot be overstated, as the cost of repairing and replacing the components is time-consuming and expensive. Equipment technicians require a good understanding of modern lubrication requirements for final drive systems because a large percentage of problems stem from improper servicing procedures.

By their very nature, final drives support the weight of the equipment as well as transmit torque to the tracks or wheels. Oil leaks from final drives invariably lead to low oil levels, which in turn lead to inadequate lubrication conditions and ultimately to early failure. The longevity of a final drive is dependent on using the correct lubricant to aid in reducing friction between moving parts; there must be a means of removing excess heat as well as suspending dirt and wear particles in the lubricant between drain intervals. Effective lubrication is dependent on using the correct lubricant and changing it at the appropriate intervals. The recommendations discussed in this chapter are general and provide a guideline only. Always refer to the manufacturer's service literature for specific procedures and information as to the correct oil for the equipment you are servicing.

The practice of mixing different brands of oil is discouraged; it often results in compatibility issues. Experts in the equipment service field caution that mixing lubricants may cause accelerated breakdown of the additives used in the oil, which may lead to increased wear. In some cases viscosity changes and oil foaming occur from mixing synthetics and petroleum-based lubricants. Most manufacturers today approve the use of synthetic oils in final drives. Synthetic lubricants are more effective in both extreme cold weather and high temperature conditions. The ability of the lubricant to adhere to components is better with synthetics than with conventional oils. This is often referred to as throw-off resistance in manufacturer's lubrication literature. Lube change intervals may sometimes be extended by using synthetic oil, but always refer to the manufacturer's recommendations when determining oil change intervals.

Maintaining correct oil levels in a final drive is critical. Oil foaming or operating with too low an oil level can be prevented. Refer to the manufacturer's service literature for the proper procedures and techniques used to determine the correct oil level.

Some manufacturers use magnetic-type drain plugs to capture metal wear particles. Excessive metal particles found on drain plugs can be an indication of imminent final drive failure. These symptoms should be investigated before a costly breakdown occurs. Following manufacturer's recommendations, such as draining the lubricant when it is warm, will eliminate most particles and dirt because they will be suspended in the oil. Once the oil cools, contaminants settle to the bottom of the housing, so some will remain.

Check lubrication requirements before servicing equipment; conventional gear oils may not be compatible with friction materials that are often used in brakes or steering clutches, which may be located internally in final drives. Some gear oils are not compatible with the seal materials used in final drives and may cause brake performance problems. Not all multigrade oils will meet the requirements of manufacturers, because some multigrade oils use viscosity index improvers with high-molecular-weight polymers, which can lead to poor viscosity performance. Final drive oil should be able to handle high temperature and load conditions and be able to develop a sufficient coefficient of friction to meet the demands of brakes and steering clutches used in final drives. Choosing the correct oil for this application is critical to brake and steering clutch performance. The lubricant should have good rust and corrosion protection as well as a low rate of oxidation. It must be highly resistant to foaming and protect against copper corrosion.

Tandem Drive Breather Cleaning/Replacement

Although the tandem drives are sealed from the environment, they must have a means of equalizing the pressure that is built up internally during operation. A breather is located on each drive (see **Figure 15-16**) to

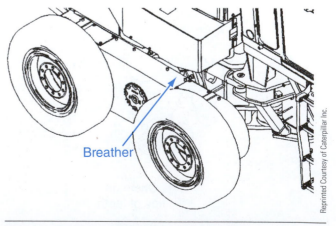

Breather

Figure 15-16 Location of the breather.

prevent pressure build-up during operation. Periodic cleaning or replacement of this breather is required and should not be neglected; seal damage will result from excessive pressure build-up in the drive if it becomes plugged.

Tandem Drive Oil Change

Before performing any service work to the tandem drive, make sure that it is cleaned thoroughly around the fill plug and drain plug.

1. Place a suitable container under the drain plug and remove the drain plug and oil check plug to drain the oil. Refer to **Figure 15-17** for the location of plugs.
2. It may be necessary to flush the drive with diesel fuel to get rid of sludge build-up. To inspect the inside of the drive housing, remove the inspection cover on the top of the drive.
3. Clean and reinstall the drain plug when the oil drain is completed.
4. Refill the tandem drive with an approved lubricant until oil flows from the check plug hole in the outer cover, as shown in **Figure 15-18**.
5. Install the inspection cover, operate the equipment, and check the drive housing for any leaks.
6. Recheck the oil level refill until it is even with the bottom of the oil check hole, as shown in **Figure 15-18**.

FINAL DRIVE BEARING ADJUSTMENT

Whenever a final drive bearing adjustment procedure has to be performed, the equipment must be properly blocked up to ensure that the work can proceed safely. Refer to the manufacturer's recommended blocking procedure. The following general steps

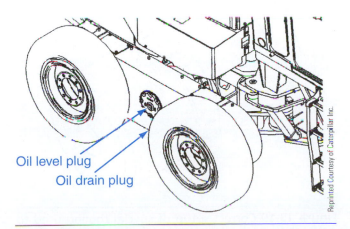

Figure 15-17 Location of oil level and drain plug.

Oil level plug
Oil drain plug

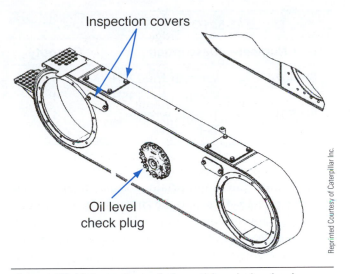

Inspection covers

Oil level check plug

Figure 15-18 Location of the oil level check plug.

provide a guide to the actual steps to be performed before checking and adjusting the final drive. Consult the manufacturer's procedures for more specific instructions on your off-road equipment. Be sure to dispose of used final drive oil following locally accepted regulations and procedures. Never reuse the old fluid; dirt may have been accidentally introduced into the oil. Rapid wear results when used oil is reused.

1. Be sure to park the grader on a flat, level surface if possible, and ensure that the transmission control lever is in the "park brake on" position.
2. Place the wheel lean pin into the front axle. This prevents the front of the machine from unexpectedly shifting sideways.
3. If the grader has an articulating frame, make sure to install the articulating lock pin. This will prevent the machine from unexpectedly moving sideways due to the articulating frame.
4. Place wheel chocks at the front and rear tires.
5. Be sure to choose a jack that is rated for lifting the intended weight.
6. Use only approved stands that are rated to support the intended load.
7. Before attempting to raise the back of the grader (tandem drive end), repeat steps 3 through 6.
8. Drain the oil level in the tandem housing to below the base of the outer cover.
9. Remove the outer cover and store the cover and bolts in a secure location.
10. Remove the lock that holds the adjusting nut in place and store in a secure location.
11. Refer to the manufacturer's service literature to select the correct spanner wrench for the final drive you are attempting to adjust. Refer to **Table 15-3** for the example of a tool selection chart.

TABLE 15-3: TOOL SELECTION CHART

Part Number	Description	Quantity
6V-4074	Wrench for 120H	1
	Wrench for 135H	1
9U-5015	Torque wrench ¾-in. drive	1

© Cengage Learning 2014

12. Using the appropriate torque wrench (see **Table 15-4**), rotate the tires while adjusting to the appropriate torque values also shown in **Table 15-4**. If the final drive has been disassembled, strike the hub with a large brass punch and hammer to ensure correct seating of the sprockets.

13. Recheck the nut torque *again* and repeat the last two steps, if required, until the torque remains constant.

14. Back off the adjusting nut one lock position.

15. Once again, retighten the adjusting nut to the final torque as specified in **Table 15-4**.

....

Note: It is important to differentiate between the rolling torque specifications in the disassembly and the assembly instructions; they are not the same.

....

16. Once torqued to the final specifications, check to see that the lock will fit against the adjusting nut, then install and torque into place. If it does not fit properly, tighten further until it fits properly. *Note:* Do not back off the adjusting nut to install the lock nut; the resulting torque will be incorrect.

17. Reinstall the outer cover to the tandem housing and refill with new lubricating oil to the correct level.

TABLE 15-4: EXAMPLE OF A BEARING PRELOAD FOR FINAL DRIVE CHART

Final Drive Group	Part No. for Nut	Adjustment Torque	Final Torque
194-6773	2G-6396	170 Nm	102 Nm
194-6774		(125 lb-ft)	(75 lb-ft)
8W-4368	8D-3540	204 Nm	136 Nm
202-7593		(150 lb-ft)	(100 lb-ft)

© Cengage Learning 2014

Differential and Bevel Gear Check

At some point in a technician's career, he or she will be asked to assess the condition of a final drive differential without removing it for inspection. The following is a general procedure that will provide a guide to determining excessive **backlash** in the differential bevel gear in a grader final drive. This example is based on a typical procedure used by Caterpillar. Before proceeding with this test, be sure to follow the manufacturer's blocking procedure in the service literature for the specific equipment being worked on. Once the equipment is properly blocked, the general steps outlined below will lead to determining backlash in the final drive.

1. Remove the driveshaft and store in a secure place. Be careful not to drop any of the U-joint bearing caps.

2. Use the appropriate ¾-in. drive power bar and socket to remove the yoke from the differential housing.

3. Attach the specialized tooling to the pinion shaft (3) as shown in **Figure 15-19**. Refer to **Table 15-5** for the tool number.

4. Securely attach the magnetic base of the dial indicator to the carrier housing.

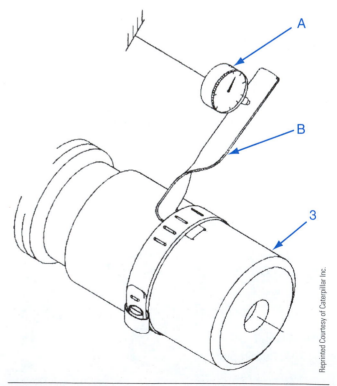

Reprinted Courtesy of Caterpillar Inc.

Figure 15-19 Correct setup of special tooling used to measure backlash in a final drive differential assembly.

TABLE 15-5: EXAMPLE OF REQUIRED TOOLING

Reference Number	Part Number	Description	Quantity
A	8T-5096	Dial Indicator Gauge Assembly	1
B	127-8072	Backlash Gauge	1

© Cengage Learning 2014

TABLE 15-6: EXAMPLE OF SPECIFICATIONS USED FOR ADJUSTMENT

Model	Assembled Backlash
12H	0.025 mm + 0.10 − 0.07 mm (0.010 + 0.004 − 0.003 in.)
120H	0.025 mm + 0.10 − 0.07 mm (0.010 + 0.004 − 0.003 in.)
135H	0.025 mm + 0.10 − 0.07 mm (0.010 + 0.004 − 0.003 in.)
140H	0.025 mm + 0.10 − 0.07 mm (0.010 + 0.004 − 0.003 in.)
143H	0.025 mm + 0.10 − 0.07 mm (0.010 + 0.004 − 0.003 in.)
160H	0.025 mm + 0.10 − 0.07 mm (0.010 + 0.004 − 0.003 in.)
163H	0.025 mm + 0.10 − 0.07 mm (0.010 + 0.004 − 0.003 in.)

© Cengage Learning 2014

5. Perform three evenly spaced readings approximately 120 degrees apart. Refer to **Figure 15-20**.

Note: Each model of grader has a slightly different procedure that must be followed to obtain accurate readings. For some models use the position "A." If the reading at "2A" is used, you have to divide the results by two. To take readings on some models, position "B" must be used. This example illustrates the importance of getting and using the correct procedures to perform this task.

6. If readings are to be taken at position "2B" you must divide the readings by two to calculate the correct value.

7. In this example refer to **Table 15-6** for the correct specifications for each application. If the backlash readings are not within specification, further disassembly will be required to determine the course of action.

Differential Lock Check on a Caterpillar 120H Road Grader

The differential used in this example, shown in **Figure 15-21**, has a lock that is used with a conventional differential to force both tandem drives to be locked together to prevent differential action from occurring during low-traction conditions. The differential locking mechanism is switched from the dash by a solenoid-activated hydraulic valve that delivers pressure oil from the transmission to the differential lock valve located on the differential carrier. This system uses a series of clutch discs that, when activated, lock the left side gear to the bevel gear, preventing the pinions from turning on the spider shaft, and causing the spider assembly to rotate as a unit. This action forces the two output axles to turn at the same speed, which provides equal power to both tandem drives.

FAILURE ANALYSIS

Final drives are subject to more extreme torque loading and shock abuse potential than any other component in the powertrain system in off-road equipment. When a final drive failure occurs, it often results in costly parts replacement and long periods of downtime. Failure analysis can be a complex procedure and may require the use of sophisticated lab equipment to determine the root cause. Fortunately, most failures can be accurately assessed in the field. Determining the root cause of a failure is the primary objective of this section of the book. Often, you can determine the root cause of the failure by careful visual inspection of the failed component.

Why and what led up to the failure may be difficult to determine. A common error in failure analysis occurs when a technician mistakenly assumes that the first damaged component found is responsible for breakdown. In many cases, the damaged component is

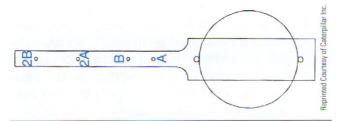

Figure 15-20 Example of special tooling used to measure backlash on the drive pinion of a final drive differential.

Reprinted Courtesy of Caterpillar Inc.

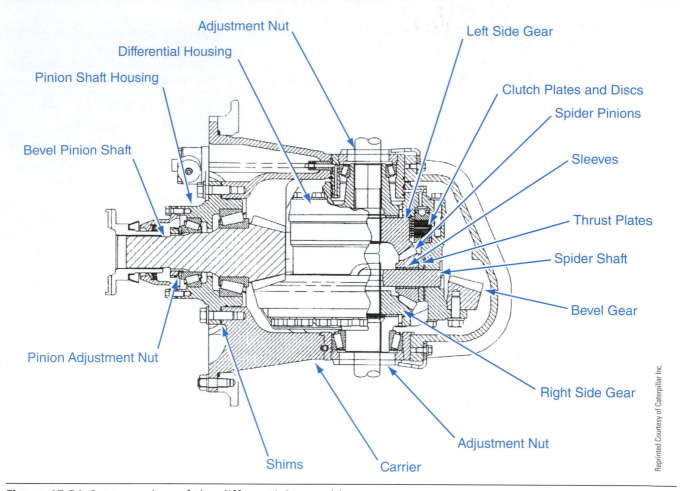

Figure 15-21 Cutaway view of the differential assembly.

a result, rather than a cause, of the failure. An investigative approach helps to determine the root cause of a failure. The progressive steps that follow can be used as a guide to developing good diagnostic and analytical skills for failure analysis.

1. Record all the information collected during the assessment stage.
2. Talk to the operator who was using the equipment at the time of the failure.
3. Assess whether the failure was instantaneous or occurred gradually.
4. If there are multiple failures, determine which occurred first and attempt to establish a sequence of events. For example, did a lubricating pump fail first, which led to a major component failure, or was it the other way around?
5. Was the equipment operating above its designed torque capabilities?
6. Determine whether there were any recent repairs or parts replaced on the equipment. If so, talk to the technician who performed the work.

7. Determine how the equipment was brought to the repair facility. Was it towed or loaded on a trailer and trucked in?
8. Establish the number of equipment hours or length of service since the previous maintenance, and what the equipment was used for.
9. Review the equipment's service records to determine whether the equipment has been properly serviced with the recommended lubricants.
10. Examine the equipment's overall condition. Look for things like oil leaks, operating abuse, and tire or track damage. Is the equipment's exterior clean or is it caked with dirt?
11. Do not disturb the failed components; leave them in their original positions. Do not clean the parts before inspection. If they are left outside, make sure they are not exposed to the environment; cover them with a tarp or other suitable cover.
12. Inspect the obvious damage first, keeping in mind that there may be more than one damaged component related to this failure.

13. If possible, establish the failure type. Was it lubrication-related failure or was it caused by a shock load? Was it operating-related, such as a failure caused by spinout (common in final drive failures)? Attempt to establish whether equipment abuse led up to the failure.

14. In the case of final drives, if it is possible, check and record backlash, endplay, and runout. It may even be possible to check tooth contact.

Keep in mind that this information is intended to be used as a guide and not to be absolute. Quite a few types of failures can occur; some of them are related to the design, and others to the operational application of the equipment. Final drive components usually fail as a result of the following:

- Normal wear
- Lubrication failures
- Shock loading
- Fatigue failure

Normal Wear

Although final drive components are designed and built to last for the life of the equipment, they eventually wear out during normal use. Initially, the components will have a slightly higher level of wear due to "break-in." When new components mesh for the first time, a certain amount of break-in wear takes place; however, this adds to the overall life of the equipment. Gears meshing with each other improves their contact patterns during the break-in period. A technician should recognize what normal wear looks like, and not get it confused with other types of abnormal wear. Some manufacturers recommend that the lubricant be replaced after an initial break-in period, because this removes break-in wear particles. Some companies use an oil analysis program to establish benchmarks for determining acceptable wear in final drive systems.

Lubrication Failures

Finding the root causes of lubrication failures is one of the most difficult diagnoses. Premature wear can occur from using the wrong type or insufficient lubricant. This can be confused for wear resulting from equipment abuse. The following list outlines problems associated with lubrication failures:

- Inadequate lubrication is a common problem on off-road equipment; oil leaks from gaskets and seals in final drives may lead to low oil levels and should always be repaired as soon as possible. When lubrication changes are performed on equipment, ensure that the oil level is correct. Many final drives rely on the splash method to distribute the lubricant around the inside of the housing. If the oil level is low, an inadequate amount of oil is circulated, resulting in possible metal-to-metal contact, overheating, and ultimately, oil breakdown. **Figure 15-22** shows burned lubricant on a differential case. Eventually, the mating components and bearings become scored, resulting in component failure. When this takes place, the internal parts become black and discolored and the oil thickens and smells burnt.

- Using a lubricant with a viscosity that is not suited for the operating temperature range to which the equipment is subjected will result in shorter component life. Refer to the manufacturer's recommended viscosity rating to determine the correct oil viscosity for the operating environment. Using oil that is not designed for cold weather operation may result in insufficient lubrication due to the thickening of the lubricant, resulting in metal-to-metal contact. Mixing different types of oils also has the potential for undesirable viscosity changes. When the correct oil viscosity is chosen, oil gets circulated immediately to the bearings and gears, lessening the effect of a dry start-up. **Figure 15-23** shows damage caused by using non-EP-rated lubricant.

- Lubrication contamination is probably the most common problem that occurs in heavy equipment. Dirt, water, and even break-in debris can lead to component failure. This often starts as scoring or etching and progresses to pitting of the mating surfaces. If dirt or any other abrasive material is in the oil, it acts as an abrasive and

4004566a

Figure 15-22 Burned lubricant on a differential case.

Figure 15-23 Effects of using oil that is not EP-rated. Extensive wear has taken place on the teeth resulting in sharp edges.

1   ROUGH CRYSTALLINE AREA
2   SMEARED AREA

Figure 15-24 Broken teeth that resulted from shock loading.

accelerates the wear of mating parts. Moisture in the sump, in combination with conventional oil in a final drive, promotes acid formation assisted by the small amounts of sulphur found in conventional oil. This results in corrosion and acid etching of mating surfaces, which eventually leads to component failure.

Shock Loading

Shock loading can often be traced to improper equipment operation. A sudden load applied to a final drive component instantly or continuously over a longer period of time results in cracks or immediate breakage, depending on the severity and frequency of the shock load. When shock loading becomes excessive, gear teeth and shafts are overloaded past the point that the material can handle, resulting in cracking or immediate breakage, as shown in **Figure 15-24**. These types of impacts subject components to excessive twisting forces that can eventually lead to fractures. Shock loading indicators on failed components can show up as small curved rings, known as **beach marks**, on the surface of the component. When this shock loading continues, it can initiate progressively larger beach marks that eventually result in complete breakage. Shock loading can be distinguished by a rough crystalline surface texture evident on the broken parts.

Fatigue Failure

Another common condition that can be mistaken for shock loading (see **Figure 15-25**) is fatigue failure, as shown in **Figure 15-26**. This condition occurs

Figure 15-25 Pinion gear set from a final drive damaged by fatigue due to overloading.

when a component is subjected to torsional or bending forces resulting from overloading of final drives. Several types of fatigue will be discussed briefly in this section. *Surface fatigue* can be defined as a condition that occurs between the surfaces of mating parts that are subjected to high torque loads. It starts with pieces of metal flaking off and progressively

Figure 15-26 Results of overload fatigue that has led to cracks and spalling.

leads to pitting and finally to complete breakage. *Bending fatigue*, if severe enough, will also cause mating gears to distort the running patterns, which can move the contact pattern to the ends of teeth and lead to concentrated loading. This type of fatigue will eventually lead to shaft breakage as shown in **Figure 15-27**. Bending fatigue in a shaft will generally cause a spiral-type fracture. *Overloading* of axle shafts is a common condition that occurs in final drives. A crack that develops from extremely high loads on axles may progress to the center core of the shaft, eventually breaking it.

Surface pitting is another condition that is caused by overloading on final drive bearings and gears. Surface pitting starts as spalling on the metal surface of mating parts, such as tooth surfaces. The teeth of gears are case hardened to a predetermined depth by the manufacturer. The underlying metal is not as hard as the surface of the teeth; this is necessary to allow the core of the gear to absorb shock loads. Continually applying heavy loads to the gear teeth surfaces causes the hardened surfaces on the teeth to break down, leading to pitting. The mating surfaces of the teeth flake off to expose a layer of softer metal, which leads to metal breakdown and results in pitting.

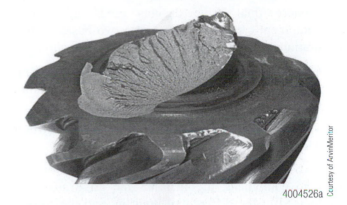

Figure 15-27 Pinion gear broken by shock loading. Note the crystalline appearance and the angle of the break.

ONLINE TASKS

1. Use a suitable Internet search engine to research the configuration of an off-road chain drive system. Identify a piece of equipment in your particular location, paying close attention to the chain drive arrangement, and identify whether or not it uses an additional planetary gear reduction. Outline the reasoning behind the manufacturer's choice of this design for that application.

2. Check out this URL: *http://science.howstuffworks. com/skid-steer2.htm* for more information on chain drive configuration and design.

Shop Tasks

1. Select a piece of equipment in your shop that has a chain drive axle assembly and identify the components that make up the chain drive arrangement. Identify whether the chain has excessive wear, and what type of chain drive link design is used on the equipment.

2. Select a piece of off-road machinery in your shop that uses a chain drive axle and identify the type and model of the equipment, listing the major components and the part numbers of the chain drive assembly. Identify and use the correct parts book for this task.

Summary

- Final drives provide the last stage of torque increase and speed reduction before power is transmitted to the ground through the tires or tracks.

- Final drives accept power from the transmission, reducing its speed and increasing the torque through the use of various gear and chain drive arrangements.

- Final drives support the weight of the equipment and absorb the torque and shock loads at the wheels or tracks.

- Modern final drive designs used on off-road equipment can be classified into two basic configurations: single reduction and double reduction.

- Chain drives are used extensively on slow-moving off-road equipment. Speeds are quite low, making chain drives a realistic alternative.

- Worm gear drives allow for a large speed reduction and a very large increase in torque.

- Two basic designs of pinion drives are used on off-road equipment; the first has the drive assembly located in the differential housing, and the second has the final reduction incorporated into the outer end of the final drive.

- On larger track equipment, double reduction drives are commonly used in the final drive.

- The bearings used to support the drive pinion are tapered roller bearings capable of supporting the high loads that are generated at this end of the drive.

- Planetary final drive designs come in single reduction and double reduction configurations and basically work in much the same way.

- Chain drives can transmit torque for long periods of time without much maintenance as long as dirt is kept out of the equation.

- To distribute the wear evenly and prevent the roller links from contacting the same sprocket teeth all the time, roller chains have an even number of links and the sprockets they run on have an odd number of teeth.

- On grader final drives, the two drive sprockets are often mounted to the planet carrier and transmit power to two drive chains (one at each end of the tandem housing).

- A chain drive is designed to tolerate some normal wear over a period of time. Three percent elongation of the chain is considered normal.

- New chain links should not be used on old chains because the new link will have a shorter pitch, which will cause shock loading of the chain during operation.

- Conventional gear oils may not be compatible with friction materials that are used in brakes or steering clutches located internally in most final drives.

- When assessing overhaul and service intervals, progressive companies use a reputable oil analysis program to establish benchmarks.

- Shock loading can often be traced to improper equipment operation where a sudden load is applied to a final drive.

- Fatigue failure occurs when the components in question are subjected to torsional or bending forces caused by overloading of final drives.

- Lubrication contamination is probably the most common problem that occurs in heavy equipment.

Review Questions

1. Where does the final gear reduction occur in a single reduction final drive on a bulldozer?
   A. differential gearing
   B. drop axle
   C. propeller shaft
   D. pinion drive

2. What would indicate that the lube level in a dozer final drive housing was acceptable?
   A. Oil level is exactly even with the bottom of the filler hole.
   B. Lube level pours out the filler hole when the plug is removed.
   C. Lube level is 1 inch below the filler hole.
   D. Lube level is ½ inch below the filler hole.

3. When is the correct time to drain the final drive oil during a regularly scheduled lube change?
   A. when it has cooled to ambient temperature
   B. at operating temperature
   C. after cooling for 1 hour
   D. after cooling overnight

4. Which type of final drive failure would likely produce a relatively smooth, flat fracture pattern?

   A. shock load

   B. fatigue

   C. lubrication failure

   D. constant torque overload

5. What indicates a fatigue-type failure in final drive gear components?

   A. pitted and spalled tooth surfaces

   B. cone-shaped fracture

   C. flat fracture pattern

   D. brinelling

6. Which type of final drive failure can result in a component becoming blackened and discolored?

   A. fatigue loading

   B. lubrication

   C. shock-load

   D. torsional loading

7. In a final drive on a dozer, if a sun gear in the planetary final drive is driven and the ring gear is stationary, which of the following statements would be correct?

   A. The planet pinions will rotate with the ring gear.

   B. The planet pinion carrier will rotate in the same direction as the sun gear.

   C. The speed of the planet pinion carrier will be increased.

   D. The planet pinion carrier will rotate in the opposite direction as the sun gear.

8. Which statement describes the term *backlash*?

   A. the space between the bearings on the pinion

   B. the clearance between the crown teeth and the pinion teeth

   C. the preload on the carrier bearings

   D. the preload on the backlash bearing carrier

9. Which type of chain is used in grader final drives?

   A. plain

   B. roller

   C. gear

   D. silent

10. What must be done to determine grader drive chain elongation dimensions?

    A. Lay the chain flat on a bench, and measure its length.

    B. Hang the chain from a hoist, attach a scale, and place the chain under tension.

    C. Hang the chain from a hoist, attach and measure its overall length.

    D. Wrap the chain around a sprocket to determine overall wear.

11. When replacing a grader's drive chain, which of the following is an acceptable practice?

    A. Replace only the rollers that are worn.

    B. Replace only stiff rollers.

    C. Never add new roller links to a used chain.

    D. Replace the complete chain.

12. When checking the backlash on a grader final drive differential how many readings are required?

    A. two

    B. three

    C. one

    D. four

16 Steering Systems

Learning Objectives

After reading this chapter, you should be able to:

- Describe the fundamentals of steering systems for off-road equipment.
- Identify the steering components used on off-road steering systems.
- Define the terms *toe, caster, camber,* and *kingpin inclination*.
- Outline the principles of operation of front-wheel and all-wheel steer systems.
- Outline the principles of operation of all-differential steer systems.
- Describe the principles of operation of clutch steer/brake systems on track equipment.
- Outline the principles of operation of articulating steer systems.

Key Terms

articulating steering	differential steer	tie-rod assembly
camber	drive steer	toe-in
caster	kingpin	toe-out
coordinated steer	Pitman arm	
crab steer	steering clutch	

INTRODUCTION

Using off-road steering systems, equipment operators can convert steering wheel or joystick movement into precise angular control of equipment with little physical effort. Without some form of power assistance, steering heavy equipment would be next to impossible; the physical effort required to articulate or control wheel angles would be challenging. At one time, mechanical linkages, in conjunction with worm gears, were used to steer mobile heavy equipment. On modern machines, such as the dozer shown in **Figure 16-1**, electric solenoids and hydraulics make the operator's role in steering equipment effortless.

Steering systems on off-road equipment come in a variety of designs; most on-road equipment requires only the front wheels of the vehicle to be steered. This chapter covers the most common forms of steering systems found on off-road equipment, including conventional front drive axle steering systems, **articulating steering** systems, differential steering systems, and **steering clutch**/brake systems commonly found on crawler tractors. This chapter focuses on common types of steering found on heavy equipment; conventional

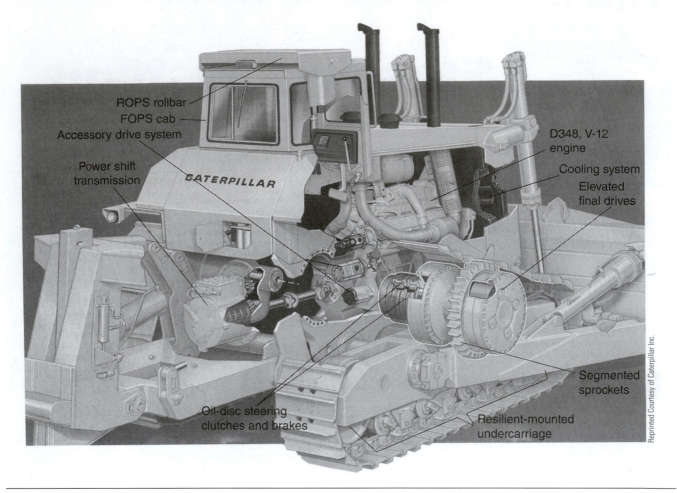

Figure 16-1 Crawler tractor showing the steering clutch and brake assembly.

on-road steer axles are covered in detail in *Heavy Duty Truck Systems*, 5th edition by Sean Bennett. No single steering system will work well on all of the different types of off-road equipment; however, most heavy-duty steer systems are operated with some form of hydraulic assist.

Although front-wheel **drivesteer** is common, off-road equipment such as backhoes often use the front wheels only for steering, as shown in **Figure 16-2**, while the rear wheels propel the equipment. Front-wheel steer used on this type of equipment incorporates the same basic components used by on-road equipment. Equipment designed for some applications requires the use of all-wheel-drive, in which case the front axle is equipped with a differential as well as steering capability. Other types of equipment, such as front-end loaders and off-road haulage trucks, steer by controlling the angular relationship of a two-piece frame. The two sections are connected by a center hinge pin with hydraulic cylinders attached to both the front and rear sections of the equipment that controls articulation.

Track-based equipment takes a different approach to steering. Manufacturers use a steering clutch/brake combination on each track. These combinations can operate independently of each other to provide directional control. Steering is accomplished by stopping or slowing one track with a brake while maintaining forward or reverse motion. Differential steering is used on equipment that requires more precise control of the drive tracks. This system uses a series of planetary gears to provide torque input to the drive tracks, and a third input from a steering motor that can increase the speed of each track. When the speed of one track is increased, the other track slows down proportionally. The term **differential steer** comes from the action that takes place when the equipment is steered; it uses the same principle of operation as a conventional differential—when one side speeds up, the other slows down in equal proportion.

Crab steer is found on equipment such as large forklifts that require precise maneuverability in tight quarters such as in lumberyards or steel plants. The operator uses this option when it is necessary to shift

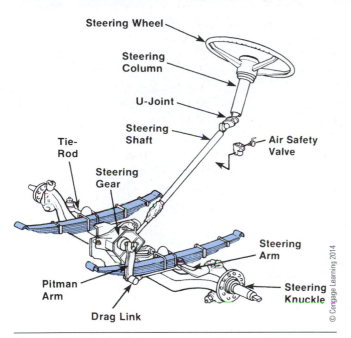

Figure 16-2 Typical front steer axle assembly.

the load left or right and still remain at right angles to the drop-off location. All four wheels turn in the same direction at the same time, allowing the equipment to move to the left or right while maintaining load alignment. Normal operation of the forklift is performed with the front wheels doing the steering, but in cases where side shifting with forward motion is required, the operator can switch to crab steering by simply selecting it with a lever in the cab.

Coordinated steer is often used on equipment such as heavy-duty forklifts or purpose-built lifters, such as the Caterpillar TH83 and Cat TH63 Telehandler. Coordinated steer is when the front and rear wheels articulate in unison to achieve a tighter turning radius. For example: if an operator wanted to turn left, the front wheels steer to the left and the rear wheels steer to the right, tightening the turning radius. Crab and coordinated steer are often found on purpose-built equipment such as the Caterpillar TH83 giving it the ability to navigate in precision material handling situations such as lumberyards and construction sites.

FUNDAMENTALS

Steering off-road equipment is accomplished in a number of different ways; all of them depend on input from an operator who controls the steering process with a steering wheel or mechanism. Steering can be accomplished with a pilot-operated steering control consisting of a joystick with an electro-hydraulic control subsystem to actuate steer cylinders. Safe and predictable steering response is dependent on a number

of factors that will be explained in detail later in this chapter. Each system will be dealt with separately because each is unique in its construction and operation, although they use the same basic principles.

The track or wheel geometry and the frame alignment contribute to safe operation and to long component life. A heavy-equipment technician must be familiar with the basic principles of operation on the various systems found on off-road equipment, and should rely exclusively on the manufacturer's service literature for procedures and specifications. The equipment and tools used to perform this type of work must be checked at regular intervals to ensure accuracy. Work on steering systems must be done accurately; loss of steering during the operation of the equipment may result in a serious injury or equipment damage. In conventional front-wheel steering assemblies, alignment of the steering components ensures maximum component life. Alignment also has a significant effect on handling and tire life. In pilot-operated hydraulic steering, whether clutch-based or differential steer, a sound knowledge of the basic operation of each system is essential.

FRONT-WHEEL STEER SYSTEM

Off-road steering systems that use front-wheel steer generally use a steering wheel or a "stick" steering control valve and steering linkage. Although mechanical steering was used on older equipment, newer systems use hydraulics to assist the operator in steering control. Whether the equipment comes with a steering wheel or stick steering, some form of hydraulic assist is used on off-road equipment.

Steering Control Arm

The steering lever is attached to the steering knuckle on the driver's side of the equipment at one end and connects to the drag link, which is used to turn the front wheels. Different manufacturers use different terms for this lever, such as *steering lever*, *control arm*, and *steering arm*. As with all steering components, the steering arm is made from alloyed, drop-forged tempered steel, and should never be heated for any reason. Heating any steering components can draw out the temper and significantly reduce the strength.

Steering Knuckle

The steering knuckle is the movable portion of the front axle that pivots on a knuckle pin or kingpin, which provides the steering action as shown in **Figure 16-3**.

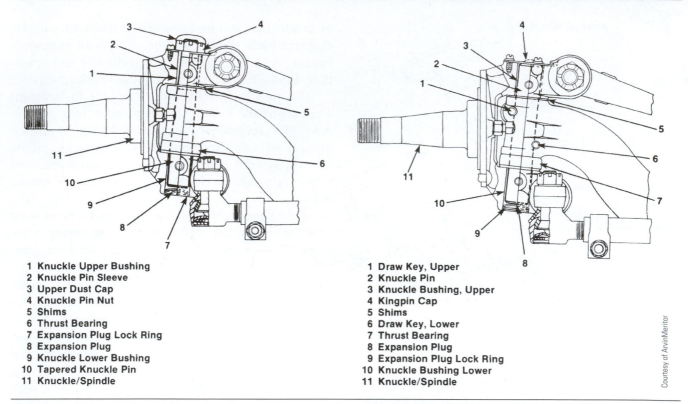

1 Knuckle Upper Bushing
2 Knuckle Pin Sleeve
3 Upper Dust Cap
4 Knuckle Pin Nut
5 Shims
6 Thrust Bearing
7 Expansion Plug Lock Ring
8 Expansion Plug
9 Knuckle Lower Bushing
10 Tapered Knuckle Pin
11 Knuckle/Spindle

1 Draw Key, Upper
2 Knuckle Pin
3 Knuckle Bushing, Upper
4 Kingpin Cap
5 Shims
6 Draw Key, Lower
7 Thrust Bearing
8 Expansion Plug
9 Expansion Plug Lock Ring
10 Knuckle Bushing Lower
11 Knuckle/Spindle

Courtesy of ArvinMeritor

Figure 16-3 Two styles of kingpins that are used on front-wheel steer equipment.

Kingpins can be either straight or tapered. Tapered pins are sealed and do not generally require lubrication; straight kingpins have grease fittings fitted to the knuckle caps at each end. Refer to the manufacturer's service literature for the correct type of grease and service intervals when maintenance is to be performed on the equipment.

New equipment often comes with sealed steering knuckles and does not require any additional lubrication during its service life. Most steering system knuckles require periodic lubrication and will have grease fittings on the knuckles. As with any part of the steering system, take the weight off the wheel ends when lubricating; this helps to distribute grease through the joint. The wheel spindle is mounted to the steering knuckle, which holds the wheel assembly and brake spider. The steering control arm is connected to the top of the left steering knuckle. The Ackerman arm, or cross-steering lever, is connected to the bottom of the steering knuckles of both wheel ends by means of tie-rod ends located on a cross tube, as shown in **Figure 16-4**.

Ackerman Arm

The Ackerman arm connects the right- and left-hand steering knuckles to the tie-rod ball studs located on either end of the cross tube. One end of the arm is bolted through the steering knuckle, while the other end is bolted to the tie-rod and held with a castellated nut and a cotter pin to lock it in place. As with any steering system component, never attempt to straighten or heat the Ackerman arm if it is bent. If it is damaged, it must be replaced.

Cross-Steering Tube Assembly

The cross tube assembly holds the tie-rods at each end. The tie-rod shanks are threaded into the cross tube and connected to an Ackerman arm on each steering knuckle. One end is left-hand thread and the other is right-hand thread. The threaded tie-rod end lengthens or shortens the cross tube, providing a means of adjusting the **toe-in** and **toe-out** settings.

Tie-Rod Assembly

Tie-rods are required to connect the steering knuckles and steering linkage; this allows the two wheel ends to work in unison to steer the equipment. The tie-rods are ball sockets that allow the steering linkage to move freely in all directions, preventing binding. Each tie-rod end is threaded at its shank, which provides a means of connecting them to the cross tube. The cross tube has matching threads at each end to accept the tie-rods; a clamp locks them in place to set the steering toe dimension. Each tie-rod has oppositely cut threads that are used to lengthen or

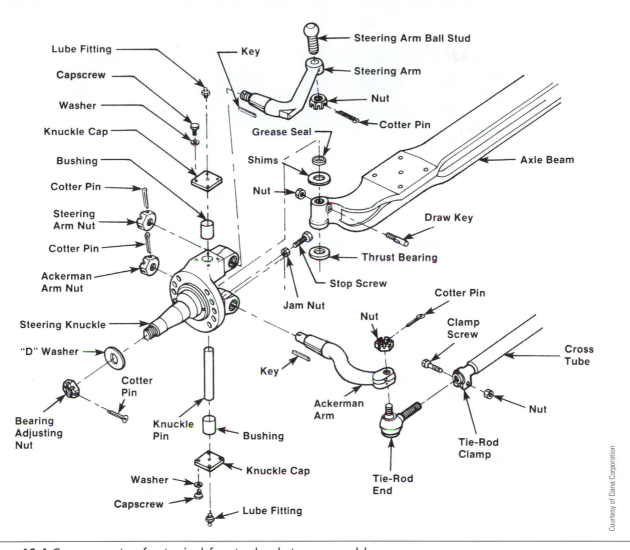

Figure 16-4 Components of a typical front-wheel steer assembly.

shorten the cross tube to set the toe dimension. After adjusting the toe setting, tie-rod clamps are used to lock the tie-rod ends into the cross tube shaft.

Drag Link

The drag link is the mechanical connection between the **Pitman arm** and the steering lever. Depending on the manufacturer, the drag link can be of one-piece or two-piece construction and is made from drop-forged steel for strength and durability. The connection between the drag link, the Pitman arm, and the steering arm is made through ball joints at each end, which absorb most of the axle motion from the steering gear.

Pitman Arm

The Pitman arm or steering gear lever is used on steering systems that use conventional steering gear boxes. The Pitman arm changes the rotary motion of

the steering gear to linear motion. The length of the Pitman arm will affect the steering response of the system; the longer the arm, the faster the steering response will be. Some manufacturers scribe a timing mark on the arm, which must be lined up with a corresponding mark on the steering gear sector shaft to center the steering properly.

Steering Gear

Although mechanical steering gear systems are a rare find on equipment nowadays, they will be explained briefly. We focus on steering systems that use hydraulic pressure to steer the equipment through an elaborate steering control valve that is still operated by a conventional steering wheel, such as found on backhoes.

Two general categories of mechanical steering boxes are the recirculating ball-type and the worm sector shaft-type, as shown in **Figure 16-5** and **Figure 16-6**.

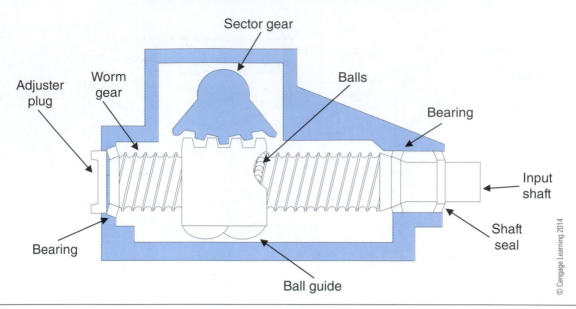

Figure 16-5 Typical mechanical recirculating ball-type steering gear.

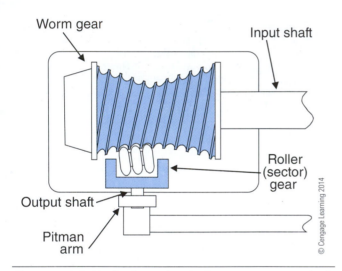

Figure 16-6 Typical mechanical worm sector shaft steering system.

Both systems multiply steering input torque from the operator and change the direction from rotary to linear motion. The worm and sector shaft steering uses a worm gear that rotates with the steering column. The sector shaft is located at 90 degrees to the worm shaft and converts the rotary motion of the worm gear into linear movement of the Pitman arm.

The recirculating ball-type steering gear uses a worm gear with a ball nut that has internal grooves that match those threads of the worm gear. The ball nut has exterior teeth that mesh with the sector gear teeth. The balls used in the grooves of the ball nut allow the mechanism to move freely with minimal friction. The ball nut moves up and down on the worm gear, rotating

the sector shaft to output steering control by means of the Pitman arm, that connects to the steering arm and steering knuckle.

Hydraulic Steering Valves

Although not all manufacturers use the type of steering valve shown in **Figure 16-7**, they use a similar principle for steering the front wheels of a backhoe. One type of John Deere system will be explained in detail to show how steering control is achieved with this design. The steering valve utilizes a rotary spool and sleeve assembly that is housed in a valve body, which has a gerotor mounted at the bottom of the valve body. When the steering valve is stationary, it defaults to a neutral position. In the neutral position, oil flow through the valve is blocked by the spool and sleeve. When the operator turns the wheel to the left, the spool rotates to allow oil to flow in the valve.

Oil flows through the sleeve into spool passages and in the housing to one side of the gerotor, forcing the gerotor gear to rotate. This forces the oil out of the gerotor into the left outlet passages in the housing and through the sleeve and spool; it then flows out the outlet port marked "L" for left. From there the oil is routed to the rod end of the steering cylinder. The oil from the cylinder is forced back into the steering valve and out to the tank. When the steering wheel stops rotating, the gerotor gear continues to turn the valve sleeve until the oil flow is blocked, returning the valve to the neutral position. When the operator turns the wheel to the right, the same movement takes place in

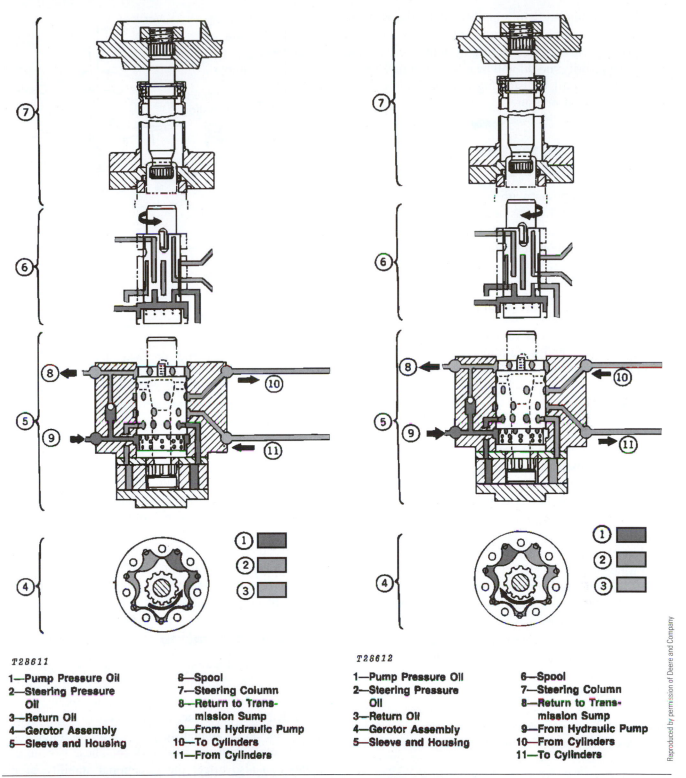

T28611

1—Pump Pressure Oil
2—Steering Pressure Oil
3—Return Oil
4—Gerotor Assembly
5—Sleeve and Housing
6—Spool
7—Steering Column
8—Return to Transmission Sump
9—From Hydraulic Pump
10—To Cylinders
11—From Cylinders

T28612

1—Pump Pressure Oil
2—Steering Pressure Oil
3—Return Oil
4—Gerotor Assembly
5—Sleeve and Housing
6—Spool
7—Steering Column
8—Return to Transmission Sump
9—From Hydraulic Pump
10—From Cylinders
11—To Cylinders

Figure 16-7 Typical hydraulic steering circuit on a John Deere 310 backhoe.

the valve, but in the opposite direction. Oil is routed to the right outlet passages and exits the port marked "R" to actuate the steering cylinder.

In the event of engine power or pressure loss, steering can still be achieved manually, because this is a power-assist system. If no oil pressure is available, a cross pin comes in contact with the spool to provide a mechanical means of rotating the gerotor gear. This provides oil flow to the steering cylinder for emergency steering.

The hydraulic system supplies oil to the steering system by means of a pressure-control valve that acts as a priority valve. The steering system and backhoe circuit get priority; no oil will be allowed to flow to the loader circuit until the steering and backhoe circuits have 1,300 psi of pressure.

STEERING GEOMETRY TERMINOLOGY

Off-road equipment with front-wheel steer uses the same concepts and terms that are used to describe conventional on-road steering systems. Several common steering system components are discussed for clarification along with descriptions of alignment terms and troubleshooting techniques. The steering system problems encountered by off-road equipment are different from those faced by on-road equipment, so alignment procedures vary and are dependent on the road conditions and speed that the equipment will operate on.

Camber

Camber is the amount the front wheels tilt outward or inward at the top. The wheels have positive camber if they are inclined outward at the top. If the wheels are inclined inward at the top, they have negative or reverse camber. Camber setting can vary considerably on off-road equipment and is determined by the terrain and equipment operating speed. Camber is defined by the kingpin inclination angle. Incorrect camber can have a negative effect on tire life. Excessive positive camber causes tires to wear more on the outside of the tire, and negative camber causes more wear on the inside of the tire. Unequal camber angles (from side to side) can have a negative impact on steering, causing the equipment to lead to the left or right. Do not bend the axles to correct camber; this practice can lead to axle weakening and breakage. Camber angle (see **Figure 16-8**) is dependent on the manufacturer's axle design and does not normally become a problem unless the axles have been bent in service.

Caster

Caster is the rearward or forward tilt of the knuckle pin (kingpin), either toward the front or toward the rear of the equipment. It is expressed in degrees and as positive, negative, or zero caster. Heavy-equipment steer axles are often set at a positive caster, which produces improved tracking and tends to make the steering axle return to a straight steer position more

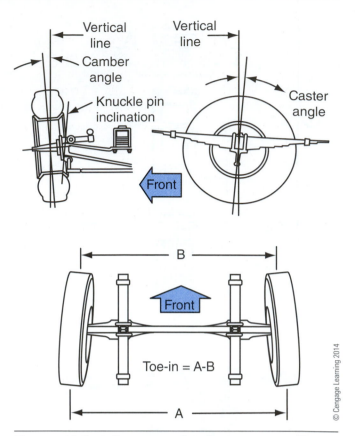

Figure 16-8 Adjustment for front steer off-road equipment.

easily after a turn is made. Positive caster (refer to **Figure 16-8**) is the rearward tilt of the kingpin as viewed from the side of the equipment.

Caster setting has little or no effect on the tire life of off-road equipment. Adjustment is made by placing shim wedges between the front springs and axle spring seats. Always use the correct shim width; the width should be equal to the spring width.

Kingpin Inclination

Kingpin inclination is defined as the number of degrees the knuckle pins (kingpins) are inclined inward at the top when viewed from the front of the equipment. Kingpin and caster angles (refer to **Figure 16-8**) are set up on equipment to produce an optimum footprint contact with the road surface. These two adjustments can also affect the directional stability of the equipment. Adjustment should not have to be changed after the initial setup unless the front axle has been bent, in which case the damaged components should be replaced.

Toe-In/Toe-Out

Toe-in is the relationship of distance between the front of the tires (when measured between the centerlines) to the same measurement at the rear of the tires (refer to **Figure 16-8**). This difference in distance is required to compensate for the natural tendency of the front wheels to toe-out during normal operation due to the effect of load and speed. The concept of toe-in adjustment is based on the premise that ideal toe-in should be zero when the equipment is moving in a straight line during normal operating conditions.

CAUTION

- *Do not attempt to straighten any bent steering axle component; always replace the damaged component.*
- *Do not weld or heat steering components because either practice can alter the structural integrity of the tempered metals.*

DRIVE STEER AXLE

Off-road equipment must be able to travel on terrain that is less than ideal, requiring all-wheel-drive in many instances. Front-drive steer axles must be able not only to support the weight of the equipment, but also to drive the front wheels and house the front brakes. The spindle is mounted on the knuckle, which is mounted to the axle housing through the knuckle caps. The knuckle caps swivel on bushings mounted in bores on the axle housing. Steering control is accomplished through the upper knuckle cap, which is connected to the steering arm. Both ends of the wheels are connected by a cross tube, which has ball ends (tie-rods) on each end to provide a swivel connection between them.

The swivel joints (tie-rod ends) are protected from dirt and road debris with boots or dust seals made from rubber or synthetic materials that require periodic greasing and adjustments. There are numerous configurations for steering linkage for off-road equipment drive steer axles and all are dependent on equipment application.

Linkages on most off-road equipment may be arranged differently from this chapter's example, but all use the same general components that are usually located behind the front-drive steer axle. The cross tube and tie-rod ends form a linkage to connect the two wheel ends and provide a means of maintaining alignment between the wheel ends. The steering arm is mounted to the wheel end, links the assembly to the power steering system. Tie-rod ends do wear out, and clearance in the ball-and-socket assembly and when it becomes excessive, it produces poor steering control and increased tire wear. In many cases, the operator may not be able to detect wear while operating the equipment.

STEERING SYSTEM MAINTENANCE

Technicians must inspect the steering system each time equipment comes in for service. When the equipment comes in for service, the tie-rod ends should be inspected for wear or possible damage from road debris. **Table 16-1** shows examples of the various categories of operation for off-road equipment and should not be used as a service guideline. Refer to the equipment manufacturer's service recommendations to determine actual intervals for servicing equipment.

TABLE 16-1: TIE-ROD SERVICE INTERVALS				
Designation	**Application**	**Inspection Interval**	**Lube Interval**	**Operating Conditions**
Light service on-road	Linehaul only	Each oil change	100,000 miles (161,000 km)	More than 50,000 miles per year, 95% on-road
Medium duty	Heavy haul	20,000 miles (32,200 km)	40,000 miles (64,400 km)	Less than 50,000 miles a year
Harsh duty	Logging, construction	10,000 miles (16,100 km)	10,000 miles (16,100 km)	Low mileage operation Mostly off-road operation
Severe duty	Mining, landfill operations	5,000 miles (100 hours)	5,000 miles (100 hours)	Heavy duty service
Very severe duty	Mining, logging, construction	48 hours	48 hours	Severe duty 80%–100% off-road

Checking Drive Steer Axle Linkage Condition

Tech Tip: Before attempting to lift and block equipment, make sure it is parked on a level surface. Set the park brake and place wheel chocks to prevent unwanted equipment movement. Mechanically support the vehicle using safety stands. Whenever the steering system is to be checked for wear, the equipment must be raised and blocked in such a way that the steering components are unloaded. Never perform any type of work on equipment that is supported by a hydraulic device alone.

Performing inspections of the steer axle components is part of a technician's role in servicing equipment. Checks should include wear of linkage components, as well as checking clearances between the steering knuckles and axle centers, as shown in **Figure 16-9**. Inspection techniques used to determine the wear in steering knuckles require that the equipment be jacked and blocked up properly. Shake the wheels vertically to check for excessive clearance between the knuckle pins and bushings or bearings. Inspect the front axle looking for indications of bending or twisting, as well as cracks that may have developed. Refer to the manufacturer's service literature for clearance specifications for knuckle joints. Check the condition of the front axle fasteners for indications of movement, loose, or broken axle mounting bolt. The wheels should turn freely from stop to stop. Ensure that the steering axle stops are adjusted correctly. Refer to the equipment service literature for the correct procedures and specifications.

Do not grease the **tie-rod assembly** before an inspection is performed; the grease may disguise wear in the joint. The wheels must be in the straight-ahead position. If possible, turn the wheels through a full left-to-right turn and then back to center; this ensures that the linkage is properly unloaded.

1. Carefully check the tie-rod boots for any damage or cracks. Inspect the boot seals for integrity and tears. If damage is discovered, replace the complete tie-rod end.
2. Check that the cotter pin is properly installed through the tie-rod end. If the cotter pin is missing, retighten the tie-rod end nut to the correct torque, following the manufacturer's specifications. Install a new pin. Never back off the nut to install the cotter pin.
3. Check the cross tube carefully for any sign of impact damage. Look for cracks. If in doubt, replace the cross tube with an original equipment tube of the same size and diameter. Never try to straighten the tube after it has been bent. Never use heat on the tube; it may affect the strength of the metal. If the tie-rod clamps are damaged, replace them.
4. Check for excessive play in the tie-rod joint by grasping the cross tube with both hands and pulling and pushing on the cross tube.
5. Check the tie-rod thread depth in the cross tube slot. The threaded tie rod end stub must be turned in beyond the depth of the cross tube slot. Note that any time a tie rod end is rotated in the cross tube threads, the toe-in and toe-out setting must be reset. Refer to specific manufacturer's recommended procedure to perform this adjustment.

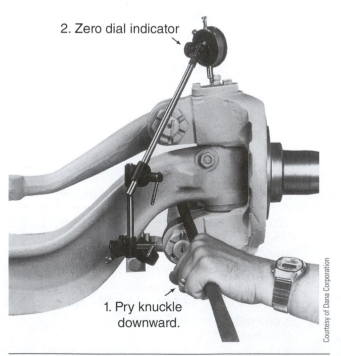

2. Zero dial indicator

1. Pry knuckle downward.

Courtesy of Dana Corporation

Figure 16-9 Checking steering knuckle for excessive vertical clearance.

CAUTION *Before proceeding with any type of service or repair work to equipment that has a steer axle assembly, make sure that the equipment park brake has been applied and the rear wheels are blocked with wheel chocks. Use an appropriate jack to lift the equipment so that safety stands can be placed in an appropriate* *location to support the equipment with the wheels raised off the ground. Under no circumstances should anyone attempt to perform work on a piece of equipment that is supported with a hydraulic jack.*

Steering Knuckle Inspection

Steering knuckles can be a high-maintenance item. The life of this connection depends on proper service and adjustment procedures. Steering knuckles should be lubricated according to the manufacturer's recommended time intervals, which can vary widely. The lubrication intervals and the type of grease used depend on the service application of the equipment. Steering knuckle wear accounts for the majority of steering problems found on equipment with this type of steering system. **Figure 16-10** shows the procedure used to check lower kingpin bushing wear on the knuckle. The following procedure can be used as a guide to determine the amount of wear in the steering knuckle joints. Always refer to the manufacturer's recommended inspection procedures and specifications.

Steering Knuckle Clearance Checks. Because of the extreme road conditions that off-road equipment must endure, adequate amounts of lubrication are very important to prevent wear in the steering components that rotate or turn frequently, such as steering knuckles.

1. With the dial indicator base mounted securely on the axle, as shown in **Figure 16-9**, set the dial indicator vertically on the end cap with the correct amount of preload.
2. Using a suitable pry bar, force the assembly downward and adjust the dial to zero.
3. Use the pry bar to force the assembly up and note the dial indicator reading or lower the front axle to check for excessive clearance. Refer to the manufacturer's recommended clearance in the shop service literature to determine clearance limits.
4. To determine kingpin bushing clearances, position the dial indicator on the upper part of the knuckle as shown in **Figure 16-9**.
5. Vigorously push and pull on the wheel assembly and observe the dial indicator reading.
6. Again, refer to the manufacturer's recommended clearance in the service literature to determine clearance limits.
7. Check the lower kingpin bushing by following the same procedure used to check the top kingpin bushing clearance.

Lubrication. Steering linkage joints often have grease fittings, although newer equipment often has sealed joints that are factory pre-lubed and never require grease. When performing maintenance work on equipment, you may find a grease fitting that refuses to take grease; replace the fitting immediately. Refer to the manufacturer's recommended lubrication chart to determine the greasing intervals on equipment.

DIFFERENTIAL STEER SYSTEMS

Differential steer systems are commonly used on a wide variety of equipment. Although there are differences in designs from manufacturer to manufacturer, the basic operating principles are similar. The following example uses a Caterpillar design to explain the construction and operating principles, as shown in **Figure 16-11**.

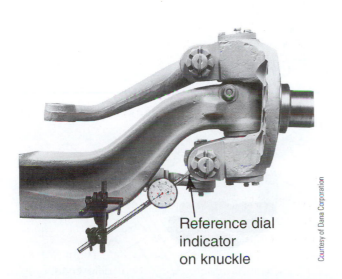

Courtesy of Dana Corporation

Reference dial indicator on knuckle

Figure 16-10 Checking lower kingpin bushing clearance using a dial indicator.

Reprinted Courtesy of Caterpillar Inc.

Figure 16-11 CAT typical differential steer assembly.

Precise steering control, while maintaining equal power distribution to the drive wheels or tracks, is often a requirement on off-road equipment. Differential steer can deliver equal amounts of power to both axles while the equipment is driven in a straight line. When the operator needs to correct steering direction, a steering motor drives one set of planetary gears that speeds up one axle and slows down the other by an equal amount, hence the term *differential steer*. The system operates on the same principle as a conventional differential, except that it uses planetary gears to achieve the differential action. Steering control is achieved without any change in power input speed, as shown in **Figure 16-12**.

Fundamentals

The Caterpillar planetary differential steer system uses a drive planetary that sends torque to a planetary on each end of the assembly. The left planetary set is called the steering planetary; the set on the right is called the equalizing planetary. Power to the assembly can be sourced from either the transmission or a steering motor, which can drive the steering planetary. When the operator activates the steering controls, the hydraulic steering motor receives power from the equipment's closed-loop hydraulic system.

In **Figure 16-12** we see all three sun gears connected together by a common center shaft, which allows them to turn at the same speed. The equalizing planetary ring gear is bolted to the right brake housing and never rotates in this configuration. The carrier of the steering planetary drives the left axle shaft, while the drive for the right axle comes from the carrier in the equalizing planetary.

With the equipment operating in the straight forward direction, as shown in **Figure 16-13**, the flow of power comes into the steering assembly through a pinion bevel gear set. The planetary drive carrier is connected directly to the bevel gear. Power flow is split between the drive planetary ring gear and the sun gear. The drive ring gear reduces the speed and increases the torque that is transmitted to the left outer axle. Power flows to the right side of the drive through a drive planetary sun gear/center shaft to the equalizing sun gear, which then drives a set of planets, resulting in reduced speed and increased torque to the right side drive of the equipment. Equipment engineers design the gear-tooth count to produce equal speeds and torque at both left and right sides of the drive when the equipment is operated in a straight forward direction. While operating in this mode with no other steering inputs, the steering planetary ring gear does not rotate.

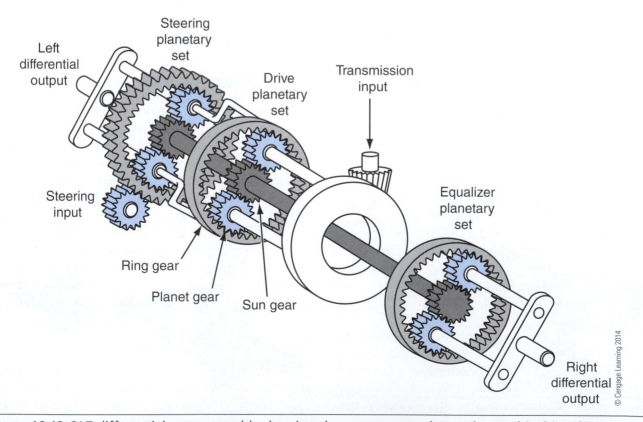

Figure 16-12 CAT differential steer assembly showing the components that make up this drive design.

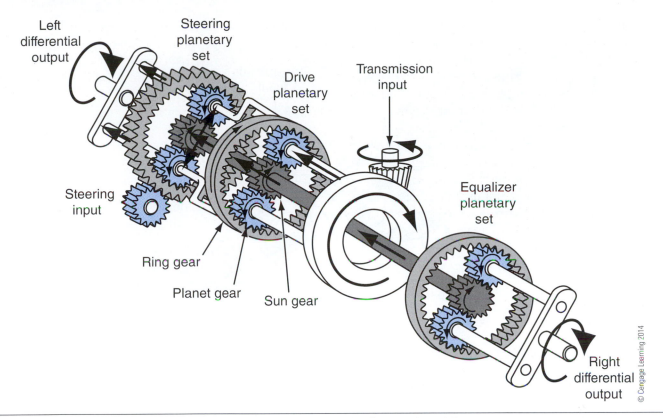

Figure 16-13 CAT differential steer assembly in the straight forward position.

In order for the equipment to make a left turn, the operator needs to speed up one track and slow down the other by an equal amount. This results in a track speed differential between the left and right side. This side-to-side speed control turns the equipment with extreme precision. Steering motor output is controlled by a steering control lever, which determines the direction and speed of the steer motor. During a left turn, the steering motor rotates the steering planetary ring gear in the opposite direction of the planetary carrier, as shown in **Figure 16-14**. This operational change causes the ring gear to slow down the speed of the left side drive. At the same time, it speeds up the planets that drive the right side sun gear, which drives the equalizing planets on the right side faster thereby producing a left turn. The majority of power to propel the equipment comes from the transmission during a turn. Any input from the steer motor results in a change in speed and direction of the steering planetary ring gear and causes speed differences between tracks.

When the operator needs to turn right, he or she activates the steering controls that cause the steering motor to turn the steering planetary ring gear in the same direction as the carrier. The ring gear input speeds up the left carrier, causing the left axle to turn faster. Because the ring gear and carrier turn together,

the planets are forced to slow down. Because the planets now drive the sun gear more slowly, the right axle slows down and forces the equipment to make a right turn.

This steering maneuver allows the operator to control the counter-rotation procedure with extreme precision. Although this function is not generally used in the actual working operation of the equipment, it does give the operator increased maneuverability in areas where space is limited. In order to perform this maneuver, the equipment must be placed in neutral because this procedure is controlled entirely by the steering motor input. Ground conditions must be similar under both tracks for this to work properly. The steering motor turns the steering planetary ring gear in the opposite direction, forcing the steering planets to turn in the same direction as the ring gear. Power flows to the sun gear causing it to rotate in the opposite direction of the steering planetary ring gear, which in turn drives the right side track in the opposite direction.

Operation

To understand the operation of differential steering, you must understand how the three planetary gear sets interconnect with each other, as shown in **Figure 16-15**, and how direction and speed are manipulated with

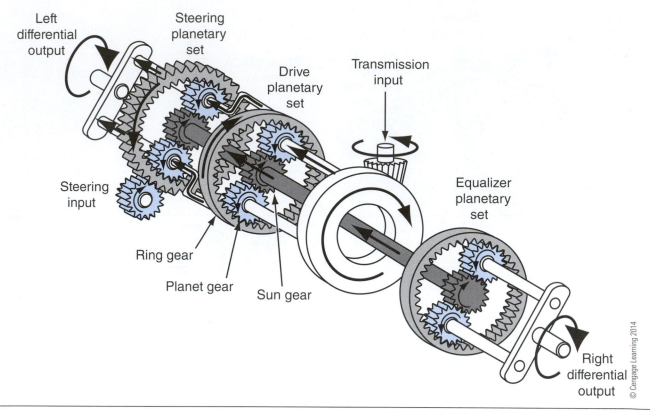

Figure 16-14 CAT differential steer assembly in the left-turn mode of operation.

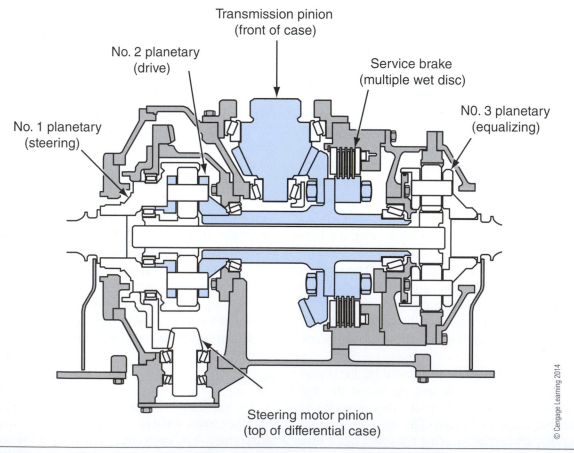

Figure 16-15 CAT differential steer assembly showing the parts that make up this drive system.

this design. The main flow of power comes from the transmission, which drives the transmission pinion. Power flows through the crown gear and drive planets, which are connected directly to the crown gear. A hydraulic steering motor, which drives the steering planetary ring gear, is driven by a variable-displacement pump. The drive oil is part of a closed-loop hydraulic system used to power the circuit. This circuit requires a counterbalance valve to prevent the motor from rotation when traveling in a straight line. It also provides makeup oil should the system overspeed and regulates the pressure from excessive pressure spikes.

When the operator activates the steering control lever, the variable-displacement pump swashplate angle is altered, forcing the drive motor to rotate the drive planetary ring gear. To achieve straight forward drive, the oil in the steering circuit is blocked from flowing in either direction, which effectively locks the motor in place and prevents the drive planetary ring gear from rotating. The power flows from the transmission to the pinion drive. The planetary drive carrier is connected directly to the bevel gear. Powerflow is split between the drive planetary ring gear and the sun gear. The drive ring gear reduces the speed and increases the torque, which is transmitted to the left outer axle. At the same time power flows to the right side drive through the drive planetary sun gear/center shaft to the equalizing sun gear (which in turn drives a set of planets to produce speed reduction and torque increase equal to the left side drive).

When the operator moves the steering controller to initiate a right turn (see **Figure 16-16**) the counterbalance valve is forced into the right steer position. This allows the flow of oil to enter a passage on the right side of the counterbalance valve through a metering orifice and to fill the right chamber. Once this chamber is filled, the oil flow forces the check valve off its seat, thus allowing oil to flow around the right crossover relief to the steering motor and causing it to rotate. This action forces the oil in the return side to return to the left side of the counterbalance valve, where it must overcome a predetermined spring tension before it forces the stem to shift to the left. This shift opens a cross-drilled port, and allows oil flow to return through the steering control valve. In order to protect the circuit from excessively high pressure, the crossover relief valves sense excessive pressure on either side of the steering circuit. During normal high-pressure operation, oil is present in the return side crossover relief valve chamber; if excessive pressure develops during normal operation, the left crossover relief valve will open to allow the excess oil to flow to the return side.

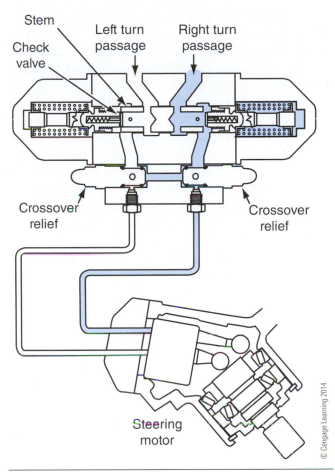

Figure 16-16 CAT counterbalance valve used in the differential steer circuit showing oil flow during a right turn.

To initiate a left turn (refer to **Figure 16-16**), the counterbalance valve is forced into the left steer position. This allows the flow of oil to enter a passage on the left side of the counterbalance valve through a metering orifice and to fill the left chamber of the counterbalance valve. After this chamber is filled, the oil flow forces a check valve off its seat, thus allowing oil to flow around the left crossover relief to the steering motor, which forces it to rotate. This forces the oil in the return side to return to the right side of the counterbalance valve, where it must overcome a predetermined spring tension forcing the stem to shift to the right. The shift exposes a cross-drilled port and allows oil flow to return through the steering control valve. In order to protect the circuit from excessively high pressure, the crossover relief valves can sense excessive pressure on either side of the steering circuit. During normal high-pressure operation, oil is present in the left side crossover relief valve chamber; if excessive pressure develops during normal operation, the right crossover relief valve opens to allow the excess oil to flow to the return side.

Overspeed can occur when equipment is operating on steep slopes. Steering control can be lost if overspeed occurs during a turn on a steep slope. The counterbalance valve is designed to prevent this condition from occurring. Operating pressure will decrease rapidly in an overspeed condition. The pressure in the right side of the steer motor drops. This causes the pressure to drop in the right side of the counterbalance valve. When this pressure drops below a preset amount, the stem moves to the right to stop the flow of oil back to the return side. When this occurs, high backpressure builds in the steer motor, which limits overspeed. To protect the circuit from excessive backpressure, the left crossover relief valve forces any excess return oil to flow back to the supply side to prevent any chance of cavitation. Some operating conditions may result in extreme overspeed; in these cases, the steering control valve has a makeup valve that allows makeup oil to enter the supply side of the

circuit to maintain a positive oil supply for continued steering control.

STEERING CLUTCH AND BRAKE STEER SYSTEMS

Manufacturers of track-type equipment use independent control mechanisms to steer and brake each track separately, as shown in **Figure 16-17**. There are many design variations for these systems. This section features the Caterpillar system as an example to explain the basic operating principles of this design.

A brake and a steer clutch are mounted together on each side of the final drive assembly. The steering brake and clutch assemblies are designed as modular units and are bolted together. Depending on the manufacturer and model, the equipment's brake assembly is either a band or multi-clutch design. The operating

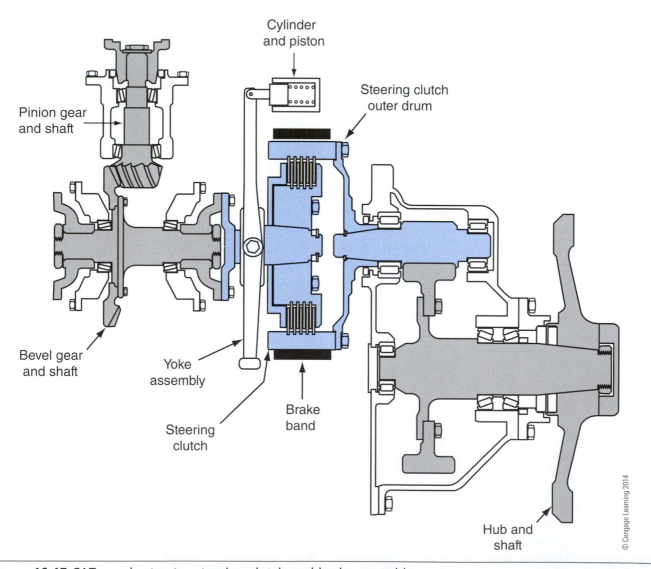

Figure 16-17 CAT crawler tractor steering clutch and brake assembly.

principle is the same in both cases; the braking action is just accomplished by using different hardware designs. In operation, the steering clutch disengages gradually to slow down the track on one side.

When the equipment has to make a hard turn, the steering clutch is disengaged and the brake is applied, forcing the track to stop turning or slow down depending on the amount of brake application. Power flows in through the pinion and bevel gear set to drive a coupling that is connected to the yoke assembly. The clutch and brake can be controlled by either a pedal or lever. On some track-type equipment, two steering pedals or two steering levers may be used to control spring-applied, hydraulically released clutches. The clutch outer drum also incorporates the brake drum. To execute a slight turn, the operator partially disengages the clutch on one track to slow it, causing the equipment to turn slightly because of the unequal track speeds.

The pedal-steer clutch control system (see **Figure 16-18**) uses the top valve to control the left side of the equipment and is shown in the hold position. Pressure flows into both control valves; no pressure is directed to the clutch from the left control valve in this position. The oil is diverted to the right side control valve, where it flows past the spool to the right side steer clutch, disengaging it. The right control valve

spool is depressed, allowing oil to flow to the steer clutch. This design allows only one clutch to be disengaged at a time. The flow of oil to the left steer control valve is sent to the right side control valve where it flows through the right control valve spool to the steering clutch. These control valves are designed so that each also functions as a pressure reducing valve. This allows the operator to get feedback from the pedals; when the pressure increases, the combination of the spring pressure and oil pressure tries to move the spool to the right against an inner spring.

The steering clutch control valve operation has no crossover supply passages in this design; this prevents the operator from disengaging both clutches at the same time. The right side lever is activated, sending oil from the pump to the right steer clutch and disengaging it. Clutch operation is proportional: the more the pedal is depressed, the more the clutch slips. The left side steering control valve blocks oil from flowing to the left steer clutch, maintaining drive speed.

If the equipment uses a steer clutch/brake band combination, when the clutch pedal is depressed, oil is sent to a steering clutch cylinder. This forces the yoke and pressure plate assembly to release some of the clutch spring pressure and allows the clutch discs to slip slightly. If full pressure is allowed into the steering clutch cylinder, the clutch disengages the track from the powerflow completely. When the operator releases the clutch completely, a mechanical brake band is applied to the outer drum to stop or slow down the track.

On newer equipment many manufacturers use a multi-disc brake, shown in **Figure 16-19**. This replaces the drum-style brake used on older equipment. Instead of using levers, modern equipment usually features fingertip steering controls that use electro-hydraulics to control the flow of oil to the steer clutch brake assembly. The assembly is lubricated and oil cooled with oil that is circulated through an oil cooler with a pump. The steering clutches are applied by hydraulic pressure, and the brake is applied by spring pressure and released with hydraulic pressure. When the operator activates the controls for a hard turn, the steering clutch is fully disengaged and the brake is applied.

The systems are designed to apply the brakes automatically when hydraulic pressure is lost. Power flows from the transmission through a bevel gear to the inner axle shaft to the input hub on each side. The input hub and the output hub are connected through the steering clutches. The brake housing is attached to the output hub through the brake clutch and does not rotate. A slight activation of the steering clutch causes the clutches to slip, resulting in a change in track speed. Whenever the brake is applied, it forces the output hub

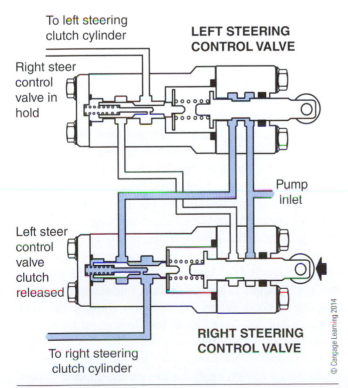

Figure 16-18 CAT crawler tractor clutch pedal steer control valves.

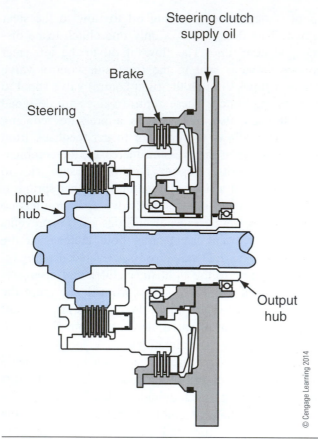

Steering clutch supply oil

Brake

Steering

Input hub

Output hub

© Cengage Learning 2014

Figure 16-19 CAT crawler tractor multi-disc brake assembly.

to stop rotating; it is connected mechanically to the stationary brake housing when the brakes are applied. When the service brakes are activated by the foot pedal, the oil flow to the brakes is gradually reduced and will vary with the amount of pedal travel the operator applies. When the steering control is pulled past detent, brake pressure decreases by 285 psi immediately; this action allows the brakes to be gradually applied from this point on. Full brake engagement occurs at maximum brake pedal travel, at which point the pressure is reduced to approximately 28 psi. This allows Bellville springs to apply the brakes.

Steering/Brake Hydraulic Circuit Operation

To ensure safe operation of the equipment under all possible operating conditions, the steering and brake control circuit gets priority over the equipment hydraulic circuit. The circuit incorporates a priority control valve to ensure that the steering and brake circuit gets oil first, as shown in **Figure 16-20**. If hydraulic pressure is lost for any reason, the steering clutches will disengage and the brakes will apply because they are

spring-applied. Four pressure-reducing valves are located in the steering brake valve control. Because the steer circuit is hydraulically actuated and the brake circuit is spring-applied, the oil requirements are different for each circuit. The steering brake control valves have four separate circuits: a separate circuit for each steering clutch and brake, and a parking brake circuit for each brake. Whenever the operator applies the park brake, oil flow to the brakes is cut off. If the operator performs a hard turn, the oil to the appropriate steering and brake circuit is drained. A small steering correction, such as when making a large radius turn, results in a gradual draining of a left or right steering circuit.

Steering and Brake Valve Operation

Figure 16-21 shows straight-line operation flow through the steer brake valve. To achieve straight-line operation, the brake/steering levers and brake pedals do not require any input from the operator. Oil flows in from the priority flow control valve and enters into the cavities for the steering clutch reducing valves. From there the oil flows through a check valve to the cavities around the brake reducing valves. Whenever the steering clutches or brakes are activated, oil will flow to the clutch cavities. During this process the oil also flows from the outlet cavity and opens a ball check valve and fills a reaction chamber. After maximum clutch pressure is reached, a spool in the reaction chamber moves to the left against spring pressure into the metered position. This restricts the flow of oil to the clutch and is responsible for controlling the clutch pressure. When the clutch pressure in the reaction cavity drops, a spring shifts the spool to the right; during this phase more oil can flow into the clutch until maximum system pressure is achieved. The operation of the brake circuit is quite different from the clutch, as the brakes require oil pressure to release them.

When the operator applies the brakes, cam lobes force the plungers located in the brake spool cavities to compress the springs against the brake spools. Once plunger-spring force is sufficient to overcome the opposing forces, the spools move to block a port that supplies pump oil through the park brake check valve from flowing to the brakes. At this point the oil from the spring-applied brakes drains, automatically applying the brakes. When the operator applies the park brake, a poppet valve blocks the flow of oil by closing a check valve and preventing oil from flowing to the brake spools. This causes the oil to drain from the brake spool cavity, applying the spring-applied brakes.

This valve design prevents the brakes from applying before the clutch is fully disengaged. When the operator

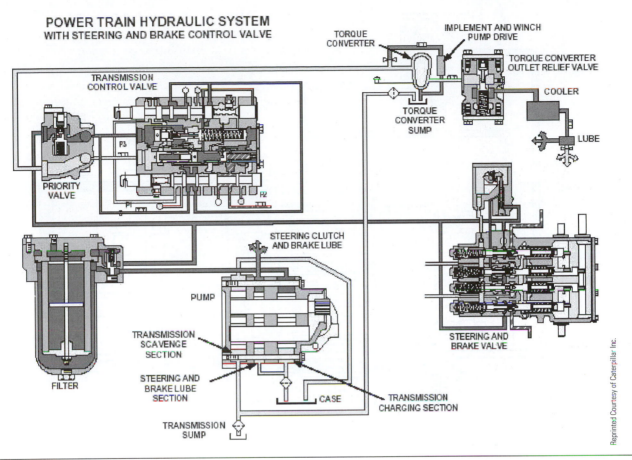

Figure 16-20 CAT crawler tractor powertrain hydraulic system.

activates the steering control, a lever forces the steering shaft to rotate, causing a cam lobe to shift the steering clutch spool. This action blocks the flow of oil to the clutch and at the same time allows the oil from the clutch to drain. If the operator continues to move the lever past its halfway point, a cam lobe on the shaft contacts the brake plunger and starts to apply the brakes. In this application, no brake application can occur until the steering clutch is in the fully released position.

ALL-WHEEL STEER

Off-road equipment often requires more maneuverability than front-wheel steer can provide. This higher level of maneuverability can be provided by all-wheel steer systems. In this section, the basics of all-wheel steer will be covered using some examples of material handlers used in the construction industry. The three steering modes used in all-wheel steer are:

- two-wheel steer
- circle steer
- crab steer

All-wheel steer systems incorporate front and rear axles with similar designs; both use conventional steering components such as tie-rods and steering cylinders. The tie-rods are connected to each wheel by means of a steering case on each side. When the operator turns the steering wheel, hydraulic oil is delivered to the steering cylinder to move the tie-rods, changing the steering direction of the wheels. The steering cases are mounted on pivot pins on the axle housing, enabling them to turn from side to side. The components used in this steering system include a variable-displacement steering pump, a selector valve, front and rear steer cylinders, a steering metering pump, and a load-sensing valve bank.

Fundamentals

In normal operating situations, two-wheel steer allows the operator to use the equipment much like any conventional material handling machinery. This mode is used for most operating conditions, and it should always be used when the unit is operated on a road. Only the front wheels will steer in this mode. The rear

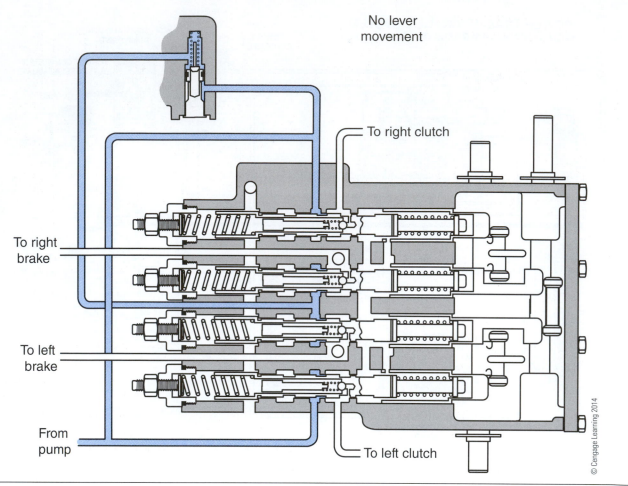

No lever
movement

To right clutch

To right
brake

To left
brake

From
pump

To left clutch

© Cengage Learning 2014

Figure 16-21 CAT crawler tractor steering brake valve.

wheels must be in the straight-ahead position for safe operation in this mode. Some equipment manufacturers install a self-aligning feature that makes it easier and quicker for an operator to switch back and forth between steering modes. Equipment sometimes has to be operated in areas with limited space. In these cases, the operator can select a circle-steer option in order to turn in a shorter radius. Both sets of wheels steer in this mode, with the rear wheels turning in the direction opposite to the front wheels. Switching back and forth between any of the steering modes requires that the rear wheels first be aligned. On equipment without the self-aligning feature, the operator must align the rear wheels manually before a switch can be made. Equipment that comes with the self-aligning feature uses a mode selector switch along with a master steering selector switch to select the operating mode.

Crab steer provides the ability to move sideways while at the same time moving forward or backward, similar to a crab, hence the name *crab steer*. The rear wheels are steered in the same direction as the front wheels causing the equipment to move either left or

right while advancing, a real advantage when working in tight quarters. These features will be explained in more detail later in this section.

STEERING OPERATION

The hydraulic pump that is used to provide oil flow in this circuit is a variable-displacement axial piston pump. A load-sensing system in the valve bank supplies oil to the metering pump, which supplies oil to steer the equipment. The rear steer axle is controlled by a three-position solenoid-operated steer mode selector valve, as shown in **Figure 16-22**. On models that are equipped with the self-aligning rear steer, additional hardware is required to self-align the rear axle.

A proximity switch is located on the right side of the rear axle, with a corresponding target attached to the right side of the steering yoke. When switching to conventional front-wheel steer, a position sensor is used to indicate straight alignment, which locks the rear wheels in the required position. This sensor is de-energized when the operator selects crab or circle steer. When operating in front-wheel steer mode, oil is

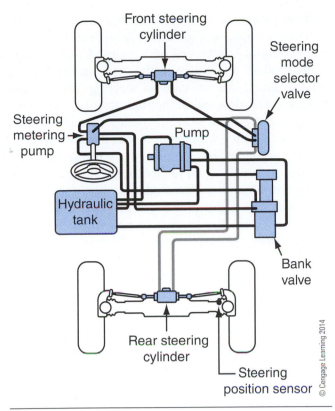

Figure 16-22 CAT telehandler steering system component layout.

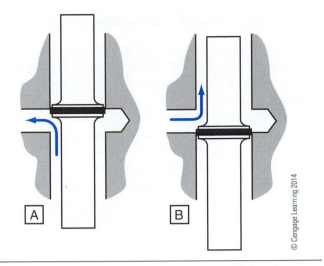

Figure 16-23 Flow compensator spool in the two modes of operation.

blocked from circulating to the rear wheels by the master steering selector switch, which de-energizes the mode selector valve. Springs center the spool to block the flow of oil to the cylinder locking it in place. The variable-displacement axial piston pump incorporates a pressure and flow compensator valve to meet the demands of the circuit. This valve controls the swashplate angle to regulate the pressure that acts on the actuation piston.

As the actuation pressure decreases, the bias spring increases the stroke, sending more oil to the steer circuit. Signal pressure from an implement and supply pressure determine how much oil is sent to the actuator piston in the pump. In this circuit, the pressure difference between the supply pressure and the signal pressure is referred to as the *margin pressure*. The spring acting on the flow compensator spool determines the actual margin pressure. Supply pressure forces the flow compensator spool to push against a spring and signal pressure on the other side of the spool.

Position "A" in **Figure 16-23** shows oil flow when the pump pressure is greater than the combination of spring and signal pressure. In this position the oil flow decreases the stroke in the pump, lowering the pressure in the circuit and forcing the flow compensator spool

into position "B," which sends oil flow from the actuator piston back to drain. An orifice in the signal passage regulates the speed of the actuator piston by constantly bleeding oil back to drain. The pressure compensator spool and spring will limit maximum system pressure when necessary, for example, if the signal relief valve becomes jammed or was improperly adjusted.

During periods of idle with no steering input, the system goes into standby mode and the only oil flow would be to compensate for internal leakage, as shown in **Figure 16-24**. No-load sensing pressure is present at the spring side of the flow compensator spool (pressure de-strokes the pump, and only a minimal amount of oil is necessary due to internal circuit leakage) and an adequate amount of supply pressure is maintained.

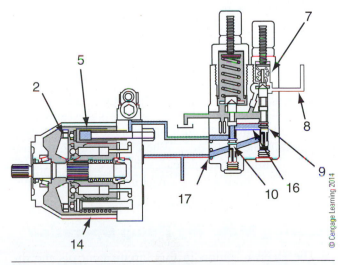

Figure 16-24 Steering circuit in the standby mode.

Pump Up-Stroking/De-Stroking

When the operator activates a circuit, the pump immediately gets stroked; signal pressure (along with spring pressure from the flow compensator valve) becomes higher than the pump standby pressure, blocking the flow of oil to the actuator piston and at the same time allowing the oil in the actuator circuit to return to the tank through the constant bleed orifice. The bias spring forces the pump to increase its stroke and send more oil to the circuit, increasing the pressure until the flow compensator spool moves into the metering position. An increase in pump stroke can be caused by a number of conditions; a change in engine speed, steering requirements, and implement activation. Although the signal pressure does not change, a drop in supply pressure can also cause the pressure to the actuator piston to drop, causing a stroke increase in the pump.

When oil flow requirements drop, signal pressure, along with spring pressure in the flow compensator spool circuit above the spool, drops to below the pump pressure. This causes the flow compensator spool to move up, forcing the oil pressure to increase to the actuator piston and causing a decrease in stroke length. This action decreases the pump output as well as system pressure. Once the flow requirements are met, the flow compensator spool moves back into the metering position. If no further input from the operator is detected, the system reverts to low-pressure standby mode.

Activities such as increased engine speed, decreased steering demand, or implement activity can activate the de-stroke mode of operation. Note that the signal pressure does not necessarily have to decrease to de-stroke the pump; an increase in supply pressure will force the flow compensator spool to shift up, causing a pressure increase to the actuator piston and overcoming bias spring pressure, which would de-stroke the pump.

Certain operating conditions often result in a high-pressure stall, for example, when a cylinder reaches the end of its stroke. To protect the circuit, the signal relief opens to limit maximum signal pressure. Supply pressure at the pump inlet forces the pressure and flow compensator spools to shift up to the metered position to maintain the flow rate. When the circuit goes into stall, the pump stroke increases the flow. The pressure compensator spool acts as a backup to the signal relief valve, limiting maximum system pressure.

Steering Metering Pump Operation

The steering pump in this circuit controls the direction and speed of the turn with two separate sections within the pump; the control section and the metering section, as shown in **Figure 16-25**. The direction control is achieved with a closed center rotary control valve, and the metering is controlled by a gerotor pump. An increase in steering wheel input increases the flow of oil to the steering cylinders. When steering input stops, centering springs position the spool in such a way that it blocks the flow of oil to the steering cylinders. Turning the steering wheel to the right forces the spool to turn, overcoming centering spring tension; travel is limited by a drive pin coming in contact with a slot in the spool. This action allows oil to flow from the control section to the metering section through a check valve in the inlet port of the metering section.

Continued turning of the wheel causes the drive pin to turn a driveshaft, forcing the rotor to turn inside the stator. This action causes the oil flow to be metered back to the control section of the pump and out the right turn port to the corresponding steering cylinder. Return oil flows back through the left turn port into the metering pump and then back to the tank. In the event of a pressure spike in the steering circuit, a check valve protects the system from overpressure. If the equipment loses power or incurs a pump failure, the steering metering pump can still be manually operated. Two check valves in the steering valve allow the equipment

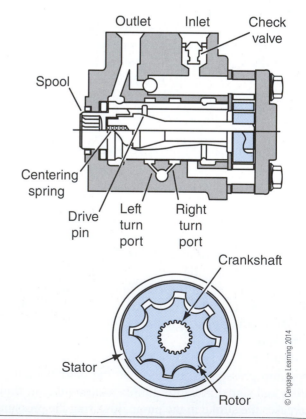

Figure 16-25 Steering metering pump cross section.

to be steered manually. A makeup valve opens between the metering pump and the steering cylinder, and a check valve (refer to **Figure 16-25**) prevents the oil from entering the steering pump.

Two-Wheel Steer Operation

When the operator executes a right turn in two-wheel steer mode, oil flows from the metering pump to the mode selector valve. From there it flows to the right side of the steering cylinder located on the front axle. The pressurized oil enters the right side of the steer cylinder and forces the tie-rods to move to the left, initiating a right turn. The oil from the return side of the steer cylinder is displaced back to the tank. In the event that the operator bottoms out the steering on either side, pressure will rise until the signal relief valve, located in the end of the valve bank, opens to prevent any buildup of excess pressure in the circuit and returning excess oil back to the tank. To perform a left turn, the oil flow is reversed. In the event of an unexpected loss of power while driving, the circuit has provisions for emergency steering capability. The operator can still use the steering wheel because the metering section of the metering pump acts as a hydraulic pump and provides enough flow to steer the equipment until it comes to a complete stop.

Crab Steer Operation

When the operator wants to steer the equipment slightly to the left while still moving ahead, he or she has to select the appropriate position on the master selector switch to energize a solenoid on the mode selector valve. This action offsets the selector valve spool to the crab steer position. The operator can now rotate the steering wheel counterclockwise. Oil will flow to the metering pump and into the left side of the front steer cylinder, and oil from the right side of the front steer cylinder will flow to the mode selector valve and on to the right side of the rear steering cylinder. Excess oil from the left side of the rear steer cylinder flows back to the tank through the mode selector valve and metering pump. This forces the two axles to steer in the same direction (left) at the same time. When the operator wants to move the equipment slightly to the right while moving ahead, the reverse occurs.

Coordinated (Circle) Steer Operation

When equipment is required to initiate a tight right turn, the appropriate position on the master selector switch is selected to energize a solenoid on the mode

selector valve. This places the selector valve spool in the circle-steer position. Turning the wheel clockwise sends pressurized oil through the mode selector valve to the right side of the rear steering cylinder. Oil from the left side of the rear steering wheel is forced to flow out through the mode selector valve to the right side of the front steering cylinder. At the same time, oil from the left side of the front steering cylinder is directed back to the tank through the metering pump. This action causes the front wheels of the equipment to steer to the right, which forces the rear cylinder rod to the left and causes the wheels to turn left. To perform a left turn while operating in the circle-steer mode, the operator turns the steering wheel to the left, forcing oil to flow from the metering pump into the left side of the front steering cylinder. Oil from the opposite end of the front steering cylinder flows through the mode selector valve to the left side of the rear steering cylinder and forces the rear wheels to steer to the right.

ARTICULATING STEERING

Articulating steering systems, as shown in **Figure 16-26**, are the most common form of steering found on off-road equipment. Construction, mining, and forestry industry equipment uses articulating steering in many off-road applications. Slow speeds and the terrain that the equipment operates on make this system ideal. Since no axle wheel steer system is

Reprinted Courtesy of Caterpillar Inc.

Figure 16-26 Typical articulating steering joint (CAT 535B skidder).

necessary, the system design is simple and often has very few complex components. A change in steering direction is made by pivoting the unit at the swivel joint that connects the two halves together. Large, specialized bearings are used to absorb the weight and load at this joint. The steering system uses two hydraulic cylinders that are connected to both halves of the equipment, as shown in **Figure 16-26**. When the operator activates the steering to change the direction of travel, either with a steering wheel or with a joystick, hydraulic oil is sent to one end of one steering cylinder and the opposite end of the other steering cylinder, forcing the equipment to articulate at the joint. Depending on the size and type of equipment, off-road vehicles can come equipped with one or two axles on the load bearing end.

Fundamentals

Although not all hydraulic steering systems use the same components, they do share common components including a hydraulic pump(s), steering cylinder(s), and a steering control valve. Our example uses the steering system from a Caterpillar 992 loader, which is typical of an articulating steering system for off-road equipment. A tandem pump, or in some cases two separate pumps, provides oil flow for this steering circuit design that consists of two individual circuits:

- pilot circuit
- steering circuit

A relief valve protects the system from overpressure. The steering control valve directs oil to the steering cylinders when the operator turns the steering wheel or joystick. Oil flows to one side of one steer cylinder and to the opposite side of the other steering cylinder.

The steering control valve controls the oil flow to the steering cylinders and allows the oil to flow to the steering system filter and oil cooler when no steering input is present. The oil cooler is protected from excess pressure by a cooler flow-control valve, which limits the pressure by sending excess oil back to the tank. When the operator initiates a turn, oil flows from the steering control valve to the steering cylinders, causing the equipment to turn. During steering wheel or joystick input, oil is allowed to flow to the steering cylinders; however, whenever steering input is stopped, oil flow stops.

The pilot circuit in this example consists of a hand metering unit (HMU) and two neutralizer valves. When the equipment is running, oil flow is always

available at the HMU. When the steering system is activated, the pilot system is responsible for movement of the spool in the steering control valve. The two sections of the HMU are a metering section and a directional control section. The metering section is actually a small gear pump that supplies a metered oil flow to the steering control valve. The more steering input there is, the quicker the oil flows to the steering control valve, which causes the valve spool to move farther and allows more oil flow from the steering pump to the steering cylinders to initiate faster steering response. A neutralizer valve is used in the pilot circuit to stop the oil flow when the equipment reaches maximum steer position in either direction. In the event of an accidental engine power loss, the HMU cannot provide emergency steering. In order to have steering capability, the pilot circuit must be able to send oil flow to the steering valve to control the valve spool position.

The smaller of the two pumps used in this circuit is also responsible for charging the accumulator through a charge valve located in the circuit. Once the accumulator is charged, the excess oil is sent to the steering control valve where it combines with the oil flow of the main pump. The return oil from the steering circuit flows through a return filter and cooler before returning to the tank. A bypass valve, located at the filter, allows the oil to bypass the cooler and go directly to the tank in the event of a high restriction at the cooler. The HMU control section operates as a closed-center valve, which stops the oil flow through the valve when there is no steering input.

Steering input forces the internal gear to turn inside the outer gear, sending a controlled flow of pilot oil through the valve body to the neutralizer valve. While there is steering input, the oil flow will continue to be metered. When steering input stops, springs force the sleeve back to the neutral position, closing the passage between the metering section and the control section of the HMU. In **Figure 16-27**, the flow of oil is blocked by the valve spool keeping the valve in the neutral position. A bleed slot in the spool allows a small amount of oil to flow to the return. This facilitates a faster response time from the spool, allowing pump pressure to unload more quickly when it returns to neutral position. In the event of an overpressure condition in the neutral position, the pilot valve will open, limiting maximum pressure to the relief valve setting. In order for the spool to return to the neutral position when steering input is ended, the oil that acts on the end of the spool must be forced to flow to the opposite end of the spool through the use of a spring on one end of the spool.

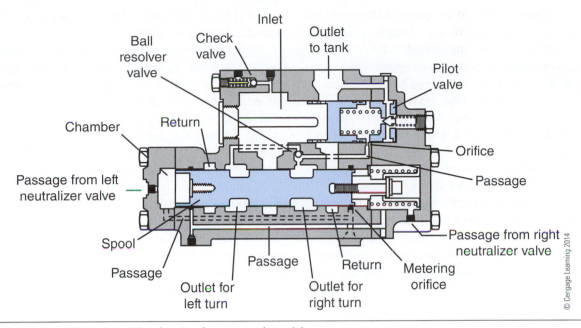

Figure 16-27 Steering control valve in the neutral position.

Operation

When steering input is initiated to make a right turn, oil flows from the pilot circuit through the neutralizer valve to the right side (spring side) of the valve spool in the steering control valve. A pressure drop occurs across the metering orifices, forcing the spool to the left. The amount of spool movement is dependent on the metered oil flow from the HMU. Oil can now flow from the pump to the head end of the left steering cylinder and the rod end of the right steering cylinder to initiate a right turn. The speed of the turn depends on the continuous input from the HMU.

In the event of an overpressure condition, the pilot valve circuit will bleed off any excess pressure above the relief valve setting to protect the system from overpressure. Oil will force the relief valve off its seat, allowing oil to flow through an internal metered passage and lowering the pressure in the spring chamber. The pressure on the inlet side of the flow control valve

moves it to the right so that oil can be dumped through holes in the flow control valve.

To prevent the equipment from bottoming out on the steering stops, a neutralizer valve is used in the pilot circuit to stop the oil flow to the steering valve, as shown in **Figure 16-28**. When the equipment is at the end of a right turn, a striker located on the frame comes in contact with the stem of the neutralizer valve and shifts it to the right to allow pilot oil to flow from the inlet to the outlet. This stops the flow of oil to the steering valve spool and returns it to the neutral position.

In order for the equipment to steer to the left, the oil must flow from the opposite end of the steering valve spool through the right neutralizer valve and through a check ball into the steering valve spool passage. The minute the striker moves back into the normal operating position, the check ball passage closes, and normal operation of the steering circuit resumes.

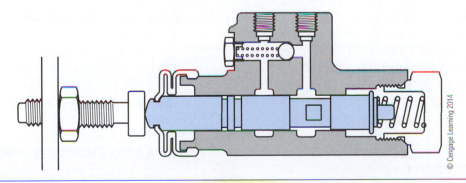

Figure 16-28 Neutralizer valve in the closed position.

Initial adjustment is set up by the manufacturer, but may need to be readjusted whenever a neutralizer valve or striker needs to be replaced.

In many operating conditions, a supplemental steering system is a necessity for obvious safety reasons, as shown in **Figure 16-29**. The HMU cannot provide emergency steering if engine power is lost. The supplemental steering system has a few additional components that are necessary to provide emergency steering capabilities. Addition of a supplemental steering pump, a diverter valve, and a check valve is required.

The steering pump must be mounted to the wheel side of the equipment so that it will turn when the equipment is in motion. This supplementary pump is generally driven by the wheel end driven gears in the transmission. When the engine is operating, the main steering pump will supply oil through the check valve to the steering control valve. The small section of the steering pump sends oil flow to the accumulator charge valve first. After the charge valve circuit is charged, the oil is routed to the diverter valve to the steering control valve, combining with the primary oil flow supplying the steering valve. When the equipment is in motion and the engine rpm is above a predetermined level, usually around 1,300 rpm, the

diverter valve sends the oil flow from the supplemental pump back to the tank.

If the rpm drops below a predetermined speed, the oil flows from the supplemental steering pump and the smaller primary steering pump are combined in the diverter valve and sent to the steering control valve. At this point, the oil flow combines with the primary pump output. As soon as the rpm exceeds a predetermined value, the diverter valve sends the supplemental oil pump output back to the tank.

Under normal operating conditions the oil flow for the steering circuit comes from the two sections of the primary steering pump. If a pump failure occurs or engine power is lost, the diverter valve sends the supplemental oil flow back into the steering control valve circuit to provide emergency steering until the equipment comes to a stop.

The diverter valve operation above a predefined rpm (usually around 1,300 rpm) is simply to send the oil flow from the supplemental steering pump back to the tank. In **Figure 16-30**, the diverter spool is shifted to the right and oil flows into the diverter spool cavity at the top, around the spool, and back to the tank. During this operating condition, only the oil flow from the small section of the primary pump goes to the

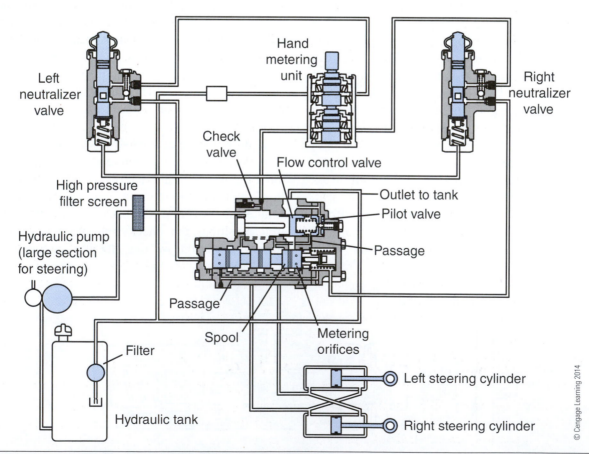

Figure 16-29 Supplemental steering system components.

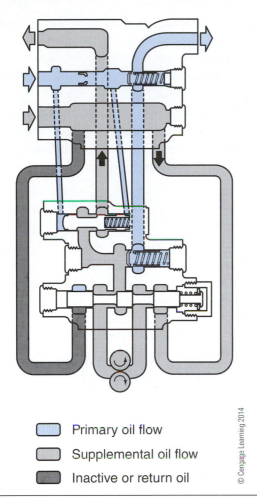

Primary oil flow

Supplemental oil flow

Inactive or return oil

© Cengage Learning 2014

Figure 16-30 Diverter valve in low idle position.

steering control valve to combine with the oil flow from the larger primary pump. When the rpm of the equipment drops below a predefined level (usually around 1,300 rpm), the diverter spool moves to the left, stopping the flow of oil back to tank. Pressure in the diverter valve passage increases until the check valve in the inlet cavity of the inlet passage opens, allowing oil from the supplemental pump to combine with the oil from the small section of the primary pump. This oil now flows to the steering control valve where it combines with the flow of the larger section of the primary pump.

While the equipment is in motion, the supplemental pump supplies oil to the steering control valve. In the event of reverse movement, the reversing spool in the bottom of the diverter valve (refer to **Figure 16-30**) shifts to the left to allow oil to flow from the supplemental pump, past a check valve, and into the diverter spool cavity.

If the equipment is moving forward, the oil is sent to the left of the reversing spool, shifting it to the right. The oil flow through the diverter valve passage increases until the check valve in the inlet cavity of the inlet passage opens, allowing oil from the supplemental pump to combine with the oil from the small section of the primary pump. This oil now flows to the steering control valve where it combines with the flow of the larger section of the primary pump.

ONLINE TASKS

1. Use an Internet search engine to research the configuration of a typical off-road steering system. Identify a piece of equipment in your particular location, paying close attention to the steering system design and arrangement, and identify whether it has its own separate hydraulic system or uses an existing pump and hoist circuit to supply oil for the steering. Outline the reasoning behind the manufacturer's choice of this design for a particular application.

2. Check out this URL: *http://www.meritor.com/ customer/northamerica/training/default.aspx* for more information on steering system configuration and design.

Shop Tasks

1. Select a piece of equipment in your shop that has a steer axle assembly and identify the type of components that make up the steer axle wheel end assembly. Check the kingpins for excessive wear and identify the kingpin bushing design.

2. Select a piece of off-road equipment in your shop that uses a conventional steer axle and identify the type and model of the equipment, listing all the major components and the corresponding part numbers of the steer axle assembly. Identify and use the correct parts book for this task.

Summary

- Off-road steering systems enable equipment operators to convert steering wheel or joystick movement, either through a series of electro-hydraulic or gear systems, into precise angular control of equipment.

- Many types of off-road equipment, such as front-end loaders and off-road haulage trucks, steer by controlling the angular relationship of a two-piece frame.

- All steering components are generally manufactured from alloyed, drop-forged, tempered steels and should never be heated.

- On front-wheel steer systems, the hydraulic system provides oil to the steering circuit by means of a pressure control valve that acts as a priority valve.

- Camber is defined by the outward or inward tilt at the top of the wheel.

- Caster is defined as the rearward or forward tilt of the knuckle pin (kingpin) either toward the front or toward the rear of the equipment.

- Kingpin inclination can be defined as the number of degrees the knuckle pins (kingpins) are inclined inward at the top when viewed from the front of the equipment.

- Front drive steer axles must be able to support the weight of the equipment and must also provide drive capability to the front wheels as well as house the front brakes.

- Differential steering systems provide the ability to deliver equal amounts of power to both axles while the equipment is driven in a straight line.

- The planetary differential steer system contains a drive planetary. The one on the left is referred to as the steering planetary; the one on the right is called the equalizing planetary.

- A planetary differential steer system uses a hydraulic motor to drive the steering planetary ring gear with oil flow coming from a variable-displacement pump.

- Steering clutch and brake systems on track-type equipment use independent control mechanisms to steer and brake each track separately.

- Equipment that uses all-wheel steer generally incorporates three different steering modes; two-wheel steer, circle-steer, and crab-steer modes.

- The rear steer axle on all-wheel steer systems is controlled by a three-position solenoid-operated steer mode selector valve.

- The variable-displacement pump used on all-wheel steer systems incorporates a pressure and flow compensator valve to supply oil flow for the circuit.

- All-wheel steer equipment has the ability to use the circle steer option, which enables the equipment to turn in a shorter radius.

- On all-wheel steer equipment when oil flow requirements drop, signal pressure (along with spring pressure in the flow compensator spool circuit above the spool) drops to below the pump pressure.

- When the steering circuit goes into stall, the pump stroke decreases the flow. Note that the pressure compensator spool acts as a backup to the signal relief valve by limiting maximum system pressure.

- The steering pump in an all-wheel steer circuit controls the direction and speed of the turn with two separate sections within the pump.

- In the crab steer mode, if the steering wheel is turned counterclockwise, oil flows to the metering pump and into the left side of the front steer cylinder, while oil from the right side of the front steer cylinder flows to the mode selector valve and on to the right side of the rear steering cylinder.

- In the coordinated (circle) steer mode, turning the wheel clockwise sends pressurized oil through the mode selector valve to the right side of the rear steering cylinder. Oil from the left side of the rear steering wheel is forced to flow out through the mode selector valve to the right side of the front steering cylinder.

- The articulating steering system contains two individual control circuits: the pilot circuit and the steering circuit.

- The pilot circuit consists of a hand metering unit (HMU) and two neutralizer valves.

- In the event of an overpressure condition, the pilot valve circuit bleeds off excess pressure above the relief valve setting to protect the system.

- To prevent the equipment from bottoming out on the steering stops, a neutralizer valve is used in the pilot circuit to stop the oil flow to the steering valve when the equipment is at the end of a turn.

- The supplemental steering pump must be mounted to the wheel side of the equipment so that it turns whenever the equipment is in motion.

- Diverter valve operation above a specified rpm (usually around 1,300 rpm) sends the oil flow from the supplemental steering pump back to the tank.

- When the rpm of the equipment drops below a specified value, the diverter spool moves to the left, stopping the supplemental flow of oil back to the tank.

Review Questions

1. What component does a steering knuckle pivot on a two-wheel steer backhoe?
   - A. Pitman arm
   - B. Ackerman arm
   - C. kingpin
   - D. steering control arm

2. Which component is responsible for controlling steering toe-out on turns on a front steer backhoe?
   - A. Pitman arm
   - B. drag link
   - C. Ackerman arm
   - D. steering control arm

3. What occurs when you increase the length of a tie-rod on a front-wheel steer backhoe steering system?
   - A. Toe-out increases.
   - B. Toe-in increases.
   - C. Toe-in decreases.
   - D. Toe remains neutral.

4. When the front tires of a backhoe tilt slightly outward away from the frame at the top, which of the following conditions would be correct?
   - A. positive camber
   - B. positive caster
   - C. negative camber
   - D. negative caster

5. How is steering knuckle vertical clearance adjusted on a front steer backhoe?
   - A. adding or subtracting shims
   - B. using the correct tapered wedge shims
   - C. selecting the correct steering knuckle
   - D. adjusting a stop bolt and jam nut

6. After making a tie-rod end adjustment, what should be observed when tightening the fastening clamps?
   - A. Ensure that the tie-rod end threads extend into the tie-rod beyond the split slot.
   - B. Ensure the equipment weight is off the wheels.
   - C. Ensure that there is no space between the clamp and the end of the cross tube.
   - D. Straighten the cross tube before adjusting the tie-rods with a torch.

7. Which component is driven by the transmission input shaft on equipment with differential steering?
   - A. sun gear
   - B. ring gear/planetary carrier
   - C. steering motor input gear
   - D. equalizing planetary gear set

8. When a right turn is performed on equipment with differential steering, which component initiates the turning process?
   - A. hydraulic steering motor
   - B. equalizing planetary gear set
   - C. steering planetary gear set
   - D. sun gear

9. Which component is driven by the hydraulic steering motor on equipment with differential steer?
   - A. transmission pinion gear
   - B. equalizing planetary gear set
   - C. steering planetary ring gear
   - D. sun gear

10. How many sun gears are used on equipment with differential steering?

    A. one   C. three

    B. two   D. four

11. Which gear is always held and never rotates in the differential steer system?

    A. drive planetary ring gear   C. steering planetary ring gear

    B. equalizing planetary ring gear   D. right planetary drive gear

12. Which gear receives equal torque with the drive planetary ring gear, when moving in a straight forward direction?

    A. right differential output gear   C. steering planetary ring gear

    B. equalizing planetary ring gear   D. sun gear

13. Which component turns the planetary ring gear in the opposite direction on a track loader during counter-rotation?

    A. hydraulic steering motor   C. steering planetary ring gear

    B. equalizing planetary ring gear   D. sun gear

14. Which component receives oil from the counter-balance valve after a right turn is initiated on a differential steer dozer?

    A. steer motor   C. right clutch

    B. left brake   D. right brake

15. How many steering and brake clutches are used on a track loader with a clutch steer/brake system?

    A. one of each on each track   C. three of each on each track

    B. two of each on each track   D. four of each on each track

16. When a track loader with a steer clutch/brake system makes a hard right turn, which component is disengaged first?

    A. left steering clutch   C. left brake

    B. right steering clutch   D. right brake

17. What occurs when the operator depresses a clutch pedal on a track loader with a steer clutch/brake system?

    A. Oil is sent to the brake piston.   C. Oil is sent to the brake control valve.

                           D. Oil is sent to the steering clutch cylinder.

    B. Oil is sent to the steering clutch control valve.

18. How many pressure reducing valves are used on track loaders that utilize clutches for the steering and braking of the tracks?

    A. one   C. three

    B. two   D. four

19. Which component receives oil from the priority flow control valve on a track loader with a steer clutch/brake system while traveling in a straight line?

    A. steering clutch pressure reducing valves

    B. the steering clutch control valve

    C. the brake control valve

    D. the steering clutch cylinder

20. When the park brake is applied on a track loader with a steer clutch/brake system, which valve blocks the flow of oil to the brake spools?

    A. pressure reducing valves

    B. steering clutch control valve

    C. poppet valve

    D. steering clutch cylinder

21. Which component is responsible for self aligning the rear axle on equipment with all-wheel steer?

    A. position sensor

    B. pressure sensor

    C. equalizing sensor

    D. pressure reducing sensor

22. What occurs in the all-wheel steering circuit when no load sensing pressure is present at the spring side of the flow compensator spool?

    A. crab steer mode

    B. standby mode

    C. coordinated steer mode

    D. two-wheel steer

23. On equipment with all-wheel steer, under what operating condition will the pressure compensating spool act as a relief valve?

    A. high-pressure stall condition

    B. standby mode

    C. high speed mode

    D. low speed mode

24. Which component receives oil flow from the metering pump on equipment when all-wheel steer operating in crab steer mode steers to the left?

    A. right side front-steer cylinder

    B. left side front-steer cylinder

    C. right side rear-steer cylinder

    D. left side rear-steer cylinder

25. Which component receives oil flow from the mode selector valve on equipment with all-wheel steer operating in coordinated steer mode while steering to the right?

    A. right side front-steer cylinder

    B. left side front-steer cylinder

    C. right side rear-steer cylinder

    D. left side rear-steer cylinder

26. When a Caterpillar 992B loader with a hydraulic steering system performs a right turn, where does the oil flow after it leaves the neutralizer valve?

    A. steering control valve

    B. left side front-steer cylinder

    C. right side rear-steer cylinder

    D. left side rear-steer cylinder

27. When a Caterpillar 992B loader reaches its safe left or right steering limits, what occurs?

    A. Neutralizer valve stops oil flow to the steering cylinders.

    B. Neutralizer valve stops oil flow to the steering valve.

    C. Pilot valve will relieve excess pressure.

    D. Flow control valve will block flow to the steering cylinders.

28. What drives the supplemental steering pump on a Caterpillar 992B loader equipped with a supplemental steering circuit?

A. engine driven

B. transmission output gear driven

C. transmission input gear driven

D. torque converter driven

29. What happens to the oil from the diverter valve at high engine rpm on a 992B Caterpillar loader?

A. flows to the steering cylinder valve

B. flows to the steering control valve

C. flows to the neutralizer valve

D. returns to the tank

30. Which circuits are normally part of the articulating steering system on a wheel loader?

A. Pilot and steering circuit

B. Pilot and Brake circuit

C. Brake and hoist circuit

D. Steering and Hoist circuit

17 Tires, Tracks, and Rim Systems

Learning Objectives

After reading this chapter, you should be able to:

- Describe the fundamentals of track and undercarriage systems.
- Describe the components that make up track and undercarriage systems.
- Describe the principles of operation of track and undercarriage systems.
- Describe the maintenance and repair procedures associated with track and undercarriage systems.
- Describe the troubleshooting procedures for track and undercarriage systems.
- Describe the fundamentals of tire and rim systems used on off-road equipment.
- Describe the maintenance and repair procedures associated with tires and rims.

Key Terms

bead expander

Belleville washer

biodegradable

broaching

calcium chloride

carrier roller

dry joint

elevated sprocket

equalizer bar

extra-long (XL) (front) track system

extra-long (XR) (rear) track system

flotation

galled

grouser

hydraulic track adjuster

I.D.

idler

link pitch

lip seal

liquid ballast

low ground pressure (LGP) system

master link

master pin

multi-piece rim

oscillating undercarriage

pitted

ply rating

restraining device

rim lip

sealed tracks

spalling

split rim

sprocket

tire cord

tire deflection

track guard

track link

track pin

track pitch

track roller

track shoes

wear step

wheel assembly

INTRODUCTION

Off-road track equipment operates at slow speeds on rough terrain that is not generally considered a roadway when compared to on-road applications. Off-road heavy equipment operating in a quarry, mine, or forestry application uses specialized tracks and undercarriages to meet the extreme duty operating cycles of the environment. Forestry equipment often requires a specialized track and undercarriage unique to their operating environment. Due to the multitude of off-road equipment tracks, only some of the most popular types of track systems and undercarriages will be covered in this chapter. Always refer to the manufacturer's service literature for information on servicing and repairing tracks and undercarriages.

Tracks and undercarriages on off-road equipment are extremely heavy and require specialized equipment to handle them safely. Tracks account for a high percentage of maintenance and repair costs when compared to the rest of the equipment. Because the tracks are in constant contact with the roadway, they must be built to absorb shock loads and to operate submerged in muddy or loose soil conditions. As a result of the extreme operating conditions to which the tracks are exposed, wear is inevitable. The technician must evaluate the wear and perform the required maintenance on a regular basis, making the necessary adjustments to minimize wear. Good operating and maintenance techniques can extend the life of the track.

FUNDAMENTALS

Track and undercarriage systems can be categorized into four distinct designs with a number of variations in each category, as shown in **Figure 17-1**. These are the standard track system, **low ground pressure (LGP) system, extra long (XL) (front)**

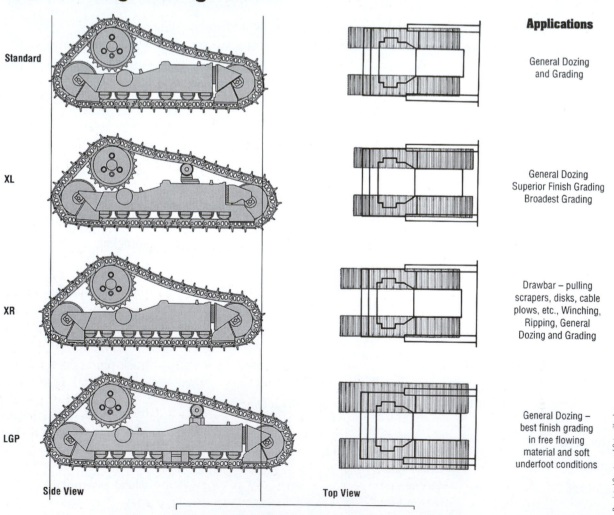

Figure 17-1 Typical undercarriage and track configurations.

track system, and extra long (XR) (rear) track system. The standard track system is designed to perform well on firm ground and can be equipped with different shoe widths. The LGP system is designed with a wider track, which increases ground contact for better flotation and stability on loose surfaces. Shoe width on the LGP track is generally a lot more than the standard track. The XL track system has an extra long front track design, which provides better balance with extremely good flotation characteristics. It is better suited for finish grading on free flowing material. The XR track system has an extra long rear section, which is more suitable for pulling scrapers, as well as for winching or ripping operations. In agricultural applications, a belt track is often used to reduce the zone of compaction. This design increases ground contact significantly over conventional LGP track systems.

To understand the construction of a track, think of it as a large endless chain with shoes that are held in place with bolts. The chain section of the track is made up of track links, pins, and bushings. A considerable number of these interconnect with each other to form the track. In addition to the chain system, other components that form the complete track and undercarriage include the drive sprocket, front idler, track rollers, tension mechanism, roller guards, and the frame.

The final drive on a crawler tractor drives a complete track and undercarriage system on each side of the equipment. Each link of the track is held together with a press-fit pin and bushing to maintain proper alignment with other link sections during operation. This combination of parts is called a link assembly. Pressed together, each section acts like a hinge, which allows the chain flexibility when rotating on the undercarriage. The track links also provide a means for attaching the track shoes that are bolted to the links. The parts of each link assembly are induction hardened to provide good wear characteristics. Pins and bushings are machined to provide a smooth bearing between them, which also increases durability. A master link or master pin completes the track assembly.

The master link is slightly different from the rest of the links. The bushing used in the master link assembly is slightly shorter than the rest of the bushings, and the master pin has a smaller diameter to facilitate easy installation and removal. Some manufacturers' master pins can be identified by a mark or drill hole on the end of the outside face of the pin (see Figure 17-2). Other manufacturers use a two-piece master link that bolts together (see Figure 17-3); this type of master link uses

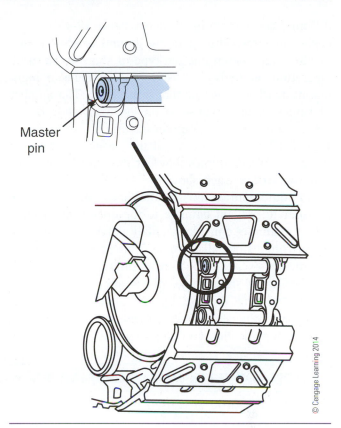

Figure 17-2 Master pin identification mark.

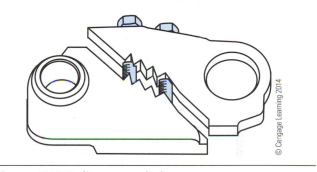

Figure 17-3 Split master link.

the same pins and bushings as the rest of the track. As a result, splitting the track is easier because no special tools are required to disassemble the track. Before attempting any maintenance or repairs to a track and undercarriage system, always refer to the manufacturer's service literature.

Track systems also come in what is commonly called a sealed system, where the pin and bushing are lubricated on assembly and are sealed to avoid dirt entering between them. This popular design eliminates internal wear.

Track shoes are bolted to the link assembly to support the equipment and improve traction and flotation during operation. They are generally constructed from a hardened metal plate and come in many different configurations. Refer to **Table 17-1** for more design features. The operating ground conditions and machine type determine the type of shoes to be used. Operation on highly abrasive soil demands a high-quality hardened surface that resist wear. The number of ridges (called **grousers**) on the track shoe designates the terrain the shoe will operate on, as shown in **Figure 17-4**. The hardness of the surface material on the grousers determines whether they are standard or extreme-service track shoes.

In some applications, rubber shoes are necessary in order to prevent damage to the operating surface. As an example, consider a small excavator that must work from a sidewalk, which should not be damaged, in order to excavate the surrounding area. The width of the track shoes also plays an important role when selecting the appropriate shoe. Working on snow, as in the case of ski hill groomers, requires an extremely wide shoe to maintain maximum flotation.

Shoe width should provide the appropriate flotation for the terrain.

TRACK AND UNDERCARRIAGE COMPONENTS

Track Links, Pins, and Bushings

Each track section is made up of two track links, a hardened pin, and a bushing. These individual track sections are interconnected to the link assembly (see **Figure 17-5** and **Table 17-2**). The two track links used in each section have provisions for attaching a track shoe and also provide a rail for the track rollers to maintain accurate track alignment. **Sealed tracks** have a solid pin. Sealed and lubricated tracks have a hollow pin, which provides a path for lubricating the pin and bushing of the next track section. When installing a center-drilled pin, the cross drill hole must be installed

TABLE 17-1: TRACK SHOE DESIGN FEATURES

No. 1	The grouser is used to penetrate the ground for traction.
No. 2	The plate surface provides a means for flotation. The wider the plate the more flotation capabilities are available.
No. 3	The design of the leading edge is curved downward. This provides the trailing edge, which is curved upward, interference-free operation. The curve also adds strength to the shoe.
No. 4	These circular cuts are called "link reliefs." They ensure that the leading edge of the shoe will not interfere with the links when bending occurs around the sprocket or idlers during operation.

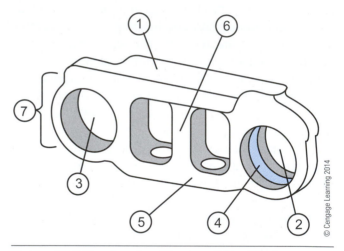

Figure 17-5 Track link with parts identified.

TABLE 17-2: TRACK LINK COMPONENTS

No. 1	The track roller runs on this section of the track link.
No. 2	Pin bore.
No. 3	Bushing bore.
No. 4	Counterbore required to hold the sealing rings on sealed or lubricated tracks.
No. 5	The shoe strap has bolt holes to accept mounting bolts for track shoes.
No. 6	Struts are used to help support the link rail.
No. 7	The bushing strap is located at the end of the track link behind the bushing bore.

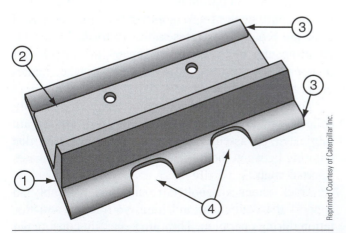

Figure 17-4 Typical track shoe.

Reprinted Courtesy of Caterpillar Inc.

© Cengage Learning 2014

toward the rail of the link, which keeps the pin in compression to resist the possibility of crushing. The pins and bushings are press-fit into the links.

Note: Be sure to follow the correct orientation; there are both right-hand and left-hand links. Clearance between the pins and bushings is minimal, providing only enough clearance for lubrication. In the case of a sealed track, the seal fits over the pin against the track link before installing another link.

Drive Sprockets

The sprocket, as shown in **Figure 17-6**, is not mounted to the track frame but is attached to the final drive. Its job is to transfer drive torque to the track. The teeth on the outside of the sprocket act like gear teeth. They engage the track links and propel the equipment on the continuous track loop. Older equipment often has a one-piece sprocket assembly; the use of individual bolts on segments is a recent innovation. On older models, the old sprocket had to be cut off and a new one welded in its place. Many equipment owners replaced the welded version with a weld on adapter ring; a new sprocket could then be bolted on, saving a considerable amount of time. Operating on different types of terrain requires that the segments be quickly interchangeable.

On modern equipment, changing sprocket segments is quick and easy; position the track so that the segment that requires changing is on the inside (away from the track links), unbolt the old segment, and bolt in and torque the new one. Rotate the track a few feet and repeat the procedure until all the segments are replaced.

Tech Tip: When equipment must operate in mud or snow, a special type of segment is available with a slotted tooth design to prevent foreign material from building up and over-tightening the track. To maintain long track life, track tension must be maintained properly during operation.

By design, the sprocket has an odd number of teeth to provide a change in tooth-to-sprocket contact every revolution of the sprocket. This is commonly called *hunting tooth*. This allows for even distribution of wear among the sprocket teeth.

Track Rollers and Carrier Rollers

Modern track design utilizes two different types of track rollers to maintain track alignment. The bottom rollers are track rollers; they support the weight of the equipment and ensure that the weight of the equipment is distributed evenly over the bottom of the track. The track rollers are spaced closely together, and generally a large number of them are mounted to the bottom side of the track frame.

The two types of track roller designs are single flange and double flange. Single flange rollers are used closest to the sprockets. Double flange rollers maximize track stability and alignment. The surface of the track rollers is hardened to the same Rockwell scale as the track links. Lubrication and cooling are provided by oil sealed in the housing. Because of the twin flanges, double flange track rollers are used to maximize track alignment.

There are several different design configurations of track rollers. Always refer to the specific service literature of the equipment you are going to work on for detailed instructions. **Figure 17-7** shows the parts breakdown of a typical Caterpillar track roller. Refer to the **Table 17-3** for a detailed description of each part.

Carrier rollers are located above the frame rail. Their purpose is to support the weight of the top section of track as it rolls between the idler and the sprocket. These rollers generally have a single flange at the center of the roller that aids in controlling track sag and whipping during operation. A secondary function of the carrier rollers is to maintain alignment on the upper portion of the track between the idler and sprocket.

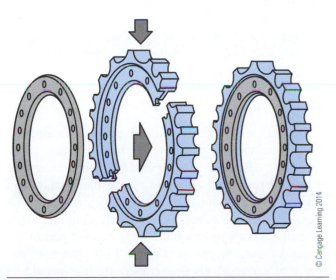

Figure 17-6 Components that make up the drive sprocket.

© Cengage Learning 2014

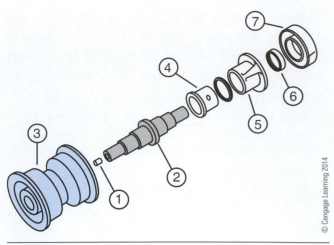

Figure 17-7 Exploded view of the components used in track rollers.

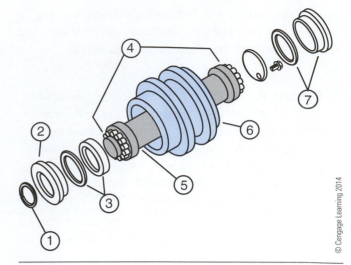

Figure 17-8 Exploded view of the components used in Cat carrier roller.

TABLE 17-3: TRACK ROLLER COMPONENTS

No. 1	A plug is used in the end of the support shaft to seal in the lubricating oil.
No. 2	The support shaft is used to support the roller shell, which can rotate freely on the shaft.
No. 3	The shell has a hardened outer surface that rolls on the track links.
No. 4	A bronze bushing is used between the shaft and shell to provide an excellent wear surface.
No. 5	This bushing is made from cast steel and is used to support the bearing to the end collar.
No. 6	Rigid seals are used to keep the dirt out and lubrication in.
No. 7	The end collar holds the roller seals in place and has provision for mounting the rollers to the track frame.

TABLE 17-4: PARTS OF A CARRIER ROLLER

No. 1	Retaining ring clips
No. 2	End collar
No. 3	Seals
No. 4	Two roller bearings and cups
No. 5	Support shaft
No. 6	Carrier shell
No. 7	Plug and O-ring

These rollers also have a hardened surface that is equal in hardness to the track link surfaces; this can have a slightly negative effect on the life of the track links. Some equipment manufacturers cantilever mount the carrier rollers to minimize material build-up.

Figure 17-8 shows an exploded view of a typical carrier roller. **Table 17-4** identifies the parts.

The retaining ring on a carrier roller fastens the end collar to the roller shell and is held in position on the support shaft by a groove. The bearings are held in place in the carrier shell by the end collar and seals, which also prevent oil from leaking out of and dirt getting into the bearings. The two roller bearing

assemblies allow the carrier roller to roll freely on the shaft. The shaft has mounting points to secure the carrier roller to the track frame. A plug and O-ring seal the lubrication opening in the shaft. The retainer plate, which mounts to the shaft, holds the tapered roller bearings inside the carrier housing. A roller cover provides a seal on one end and is fastened with bolts to the end of the carrier shell.

Idlers

Depending on the track design, the equipment could have one or two idlers per track section. If the equipment has an elevated sprocket system, there will be two idlers: a front idler and a back idler. Equipment that has a conventional oval track requires only one idler. The purpose of an idler is to help guide the track through the track rollers and to help support part of the weight of the equipment. **Figure 17-9** shows an example of a typical idler assembly. Besides providing alignment, the idler maintains the correct tension and slack on the track. The tension on the track is adjusted

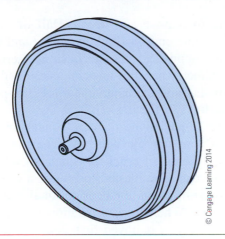

Figure 17-9 Typical Cat idler assembly.

by moving the idler back and forth on the track frame with the use of hydraulic pressure.

Equipment with oval tracks can have a two-position idler, with one position that is used when drawbar work is required. This position minimizes the amount of track that is in contact with the ground, resulting in less track wear. When equipment must operate with heavy implements attached to the front, the use of the lower mount for the idler places more track on the roadway; this makes the equipment more stable but accelerates track wear. Too much track tension contributes to accelerated wear due to increased loading on the track components.

Figure 17-10 shows a breakdown of a typical idler assembly. **Table 17-5** identifies the components that make up the assembly.

The bearing used between the idler shaft and shell in **Figure 17-10** is a plain bimetallic bearing with excellent load and wear characteristics. A retainer ring holds the end collar to the idler shell. The end collar holds the seals and the other parts inside the assembly. The outer circumference of the idler has a raised center section on which the track rides. The surface hardness of the idler, as with all track components, is the same as the track links.

The tensioning mechanism connects to the idler to maintain the correct track tension. In older equipment, the tensioning mechanism used a recoil spring with a mechanically adjusted rod to adjust track tension. More recently manufactured equipment uses hydraulic pressure to balance spring tension.

Seals

Tracks operate in extremely abrasive conditions and would quickly wear out the pins and bushings if they were not protected with seals. Seals keep lubrication in and dirt out, providing a much longer service life. Not only do the seals keep dirt out, but they also carry a certain amount of side load, preventing unnecessary wear of the track link's counterbore. Conventional track seals use **Belleville washers** that can maintain constant sealing pressure as they wear, which helps keep dirt out.

Some manufacturers use rigid seals, which consist of two individual parts like those used on lubricated tracks. These are made up of a **lip seal** and stiff wear-resistant

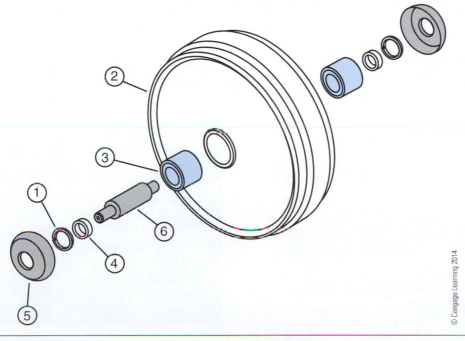

Figure 17-10 Expanded view of the components used in the idler.

TABLE 17-5: COMPONENTS OF AN IDLER ASSEMBLY	
No. 1	Retaining ring
No. 2	Idler shell
No. 3	Idler shaft
No. 4	Seals
No. 5	End collar
No. 6	Idler bearing

© Cengage Learning 2014

material on the inside called the "can." The seal shape integrity is maintained by the extremely durable urethane material from which it is constructed. The seal is held against the bushing by the load ring. These seals can withstand operating temperatures as high as 170°F (70°C). The seal is protected from thrust loads by a steel thrust ring, which limits seal compression.

The initial cost of this type of seal system is considerably higher than the conventional Belleville washer design, but the result is a much longer pin and bushing life. Over the long haul, these types of seal systems yield lower operating costs. Wear is generally not an issue with this type of seal system, and operation is quieter due to the lubricated pins and bushings. Frictional losses are reduced with a lubricated track as compared with a conventional track, resulting in a lower parasitic load which reduces operating fuel costs.

Track Guards

To help prevent the tracks from getting packed with dirt, roller guards are often used to keep rocks and foreign debris from getting between the track rollers and track links. The track guard also helps guide the track. Good practice requires that the area around the track guard be cleaned regularly, especially after operating in mud. Operating with packed debris in the track mechanism results in an excessively tight track, which greatly accelerates wear. Track tension should be checked periodically when operating in mud, sand, or snow.

The frame is the backbone of the track assembly; it holds the front idler in place and the track rollers and guards are bolted to the underside of the frame. The recoil assembly and the track carrier roller(s) are all mounted to the frame. Generally the frame causes very few problems when compared to the rest of the track components.

Split Master Links/Master Pin and Bushing

The split master link design is used by many manufacturers because it requires no specialized equipment to remove and install. Half of the link is installed on one end of a link assembly and the other half on the opposite end. The links are serrated and bolted together when the track is assembled. A split master link has approximately the same life span as the remaining links in the track.

Another popular type of track connection requires the use of a master pin and link. The master pin is longer than a normal track pin and has a machined step on one end. The master bushing is a little shorter than the other bushings and does not fit all the way into the link bore. When track separation is required, the master pin must be removed first, and then the master bushing is forced between the master pin track sections.

Equalizer Bars

The track frames on each side of the equipment are connected by an equalizer bar near the front of the machine, as shown in **Figure 17-11**, and are mounted to the sprocket that is part of the final drive at the rear of the machine. All the weight of the equipment is carried by the track rollers and frame. With this type of mounting, each side can move independently on uneven terrain to maintain contact with the ground. Diagonal braces maintain track alignment when on uneven terrain. One end of each brace is welded to the track frame while the other end is mounted to the sprocket shaft and swivels on a bearing. These bearings must be lubricated at regular service intervals.

Often older track loading equipment does not have a pivoting equalizer bar, so independent track movement is not possible. This equipment often has a rigid frame and is not able to operate on uneven terrain effectively. **Figure 17-12** shows the undercarriage of a typical track loader with a rigid frame. This type of equipment is best suited to operation on flat, hard surfaces such as those found in quarries and pits.

Reprinted Courtesy of Caterpillar Inc.

Figure 17-11 Equalizer bar in the undercarriage.

Figure 17-12 Track loader with a rigid track system.

Figure 17-14 Elevated sprocket track dozer.

Some manufacturers use an **oscillating under-carriage**, as shown in **Figure 17-13**, as an alternative to independent track movement. This has the added benefit of being able to pivot from front to rear, which can increase stability when the equipment is operating on uneven surfaces. The track frames have an equalizer bar mounted between them to allow each track frame to pivot independently on a pivot shaft. With this design, an idler swing link is used to permit the idler to move horizontally to absorb shock loads it may encounter during operation. This ensures that correct track tension is maintained. Pins are used to prevent the equalizer bar from excessive oscillating travel. Rubber pads dampen vibration between the equalizer bar and the main frame.

Elevated Sprocket

Improvements in track equipment led to the introduction of the elevated track design, which offers some advantages over a conventional track sprocket arrangement. The major benefit of this design is that it isolates the final drive from excessive shock loads. The position of the final drive also creates a better balance arrangement that also provides better traction, as shown in **Figure 17-14**. This design requires two idlers and a separate roller frame for the front and rear.

Every design has a disadvantage, and this one is no exception. Compared to a conventional undercarriage design, which has a single travel-loaded flex point, the **elevated sprocket** undercarriage has three flex points subject to travel-loading, which reduces its service life.

The two rear track frames are connected by a pivot shaft; this allows the shock loads to be absorbed by the main frame. The front roller frames fit inside the rear roller frames and are connected to each other with an equalizer bar that is pinned in the center with limited side to side movement. A recoil spring provides "recoil" should the track over-tighten from mud packing or any type of impact the track may be subjected to. Some manufacturers build tractors with a suspended undercarriage, which can improve ground contact because it allows the track to oscillate in the roller frames. This type of specialized equipment comes with longer, wider tracks that increase stability.

Specialized equipment, such as pipe layers, uses an elevated sprocket track but has a unique setup in that the roller frame does not oscillate. This unique design allows the shock loads to be absorbed by the undercarriage. The roller frames connect to the main frame two-point mounting system to provide the good stability required in pipe laying applications. A pivot shaft is still used to connect the right and left rear roller frames, allowing ground shock to be transmitted directly to the main frame. The equalizer bar is replaced with a hard bar, which connects the two front roller frames together. The front roller frames slide in and out the same way they do on a conventional elevated sprocket track, and track tension is maintained with an adjustable, grease filled hydraulic ram. A counterweight is mounted on the right side of the equipment to offset the boom bolted on the left side of the equipment.

Figure 17-13 Track loader with an oscillating under-carriage.

Track Shoes

Track shoes are used to provide a gripping function on many different surfaces, as shown in **Figure 17-15**. Track shoes come in many varieties. They are rated by a vertical designation to indicate the shoe's ability to withstand abrasion and a horizontal designation to indicate impact resistance. Impact resistance is determined by the hardness of the material. A high impact can be defined as hard concrete or rock, with continuous exposure to 6-inch bumps. The opposite end of the spectrum is low impact, which can be defined as completely penetrable material such as landscaping soil, with minimum potential for high bumps. The width of the track shoe can be tailored to suit the application. The shoe's ability to provide flotation, a large contact area, bending resistance, penetration, and self-cleaning are all points that must be considered.

The standard single grouser shoe is probably the most popular shoe in use. It is well suited for dozer and drawbar applications and provides excellent life when used in landscaping loam, gravel, or clay applications. This shoe is best for operation in low- and moderate-impact and abrasion conditions. These shoes have good penetration with excellent resistance to bending and wear and are available in a variety of widths. In applications that require working on high-impact material, an extreme-service grouser is available. This shoe has a much harder surface for high-impact conditions. When the links outlast the shoes, you probably need extreme-service grousers.

For operating conditions that require constant turns in tight places, such as on excavations, a double grouser shoe works best (see **Figure 17-16**). The use of two short grousers instead of one tall grouser causes less ground disturbance and makes the shoes extremely rigid, which also makes them highly resistant to bending. This shoe is commonly found on track-type

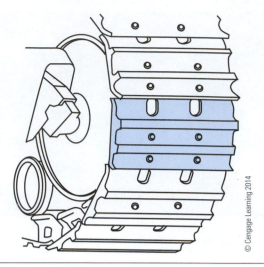

Figure 17-16 Typical double grouser shoe.

loaders and often used on hydraulic excavators where less ground penetration is desirable.

When equipment must be operated in loose, muddy ground, the use of a self-cleaning, low ground pressure shoe, as shown in **Figure 17-17**, is desirable. These shoes should never be used in high-impact or abrasive conditions. As the track rotates around the sprocket or idler, they separate slightly to allow the dirt to drop out. This design has very high flotation capability on loose, muddy ground.

In operating conditions where packing of the track becomes a problem, the use of an open-center grouser, as shown in **Figure 17-18**, is recommended. This design allows the dirt to be pushed out by the sprocket as the track rotates around the sprocket. This lessens the possibility of the track becoming too tight by forcing out dirt that tends to pack around the inside of the track.

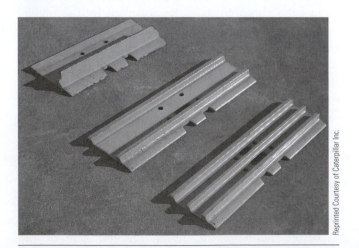

Figure 17-15 Typical track shoe designs.

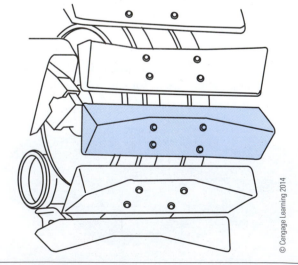

Figure 17-17 Typical self-cleaning low ground pressure shoe.

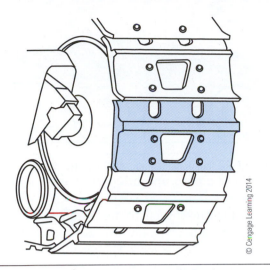

Figure 17-18 Trapezoidal center hole shoe.

Operation in snow also requires the use of an open-center grouser shoe because snow will quickly pack inside the track, causing the track to become too tight. Some manufacturers supply an all-purpose grouser shoe with three small grousers. This type of shoe is very versatile and can be used on hard smooth surfaces as well as in soft muddy conditions.

Some manufacturers make a grouser shoe that has the ability to cut and chop the surface of the ground. This design (shown in **Figure 17-19**) is intended to be used where debris can get lodged in the shoes, such as in demolition applications and at landfill sites. The shoe looks similar to a single open-center grouser but has additional diagonal side grousers that help chop the material it is operating on. Other specialty shoes also come in a variety of materials including rubber, which is often used in applications on smooth finished surfaces that would otherwise be damaged by conventional steel shoes. The downside to rubber shoes is their short life span when compared to steel shoes.

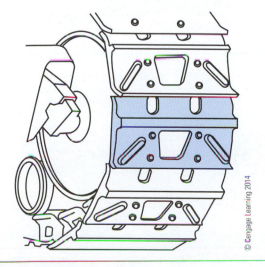

Figure 17-19 Chopper shoe.

UNDERCARRIAGE OPERATION AND TERMINOLOGY

Well-designed track components work and wear together and should wear out at about the same time. If components such as links wear out before the shoes, the type of link design chosen for that application should be examined. Track terminology used to describe the operation and wear characteristics of track systems will be used in this section. Some common terminology is similar between manufacturers. **Link pitch** is a term used to define the distance between the bushing bore center and the pin bore center. This dimension does not generally change during the life of the track.

Track Pitch

Track pitch change is an indication of wear. *Track pitch* is the distance from the center of one pin to the center of an adjacent pin in an adjoining track link section. When the track pins and bushings wear, the length of the track is increased, causing it to become loose. When the track tension can no longer be adjusted, the pins and bushings need to be turned or replaced depending on the track design. *Sprocket pitch* is the distance between the centerline of one bushing sitting in a sprocket tooth to the center of the adjacent bushing sitting two teeth over in the sprocket.

On a new track, the pitch of the track and sprocket are the same dimension. During the life of the track, the track pitch increases due to wear, but the link and sprocket pitch will not change. As the track pitch increases, the teeth of the sprocket do not mesh properly with the track, causing excessive wear to take place, as shown in **Figure 17-20**. If the bushings are allowed to wear excessively, all the related track components will wear rapidly and eventually lead to sprocket jumping, causing excessive sprocket wear. When the bushings wear externally, the worn section covers about one-third of the surface of the bushing. For the most part, using a lubricated and sealed track design eliminates

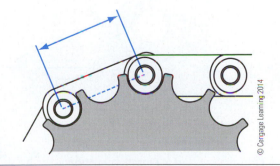

Figure 17-20 Sprocket pitch dimension measured center to center between two adjacent bushings.

this problem. Once the sprocket has excessive wear, rotating the pins and bushings will cause continued wear at a higher rate than normal: the only solution is to replace the sprocket.

Track wear occurs when the parts that are in contact with each other during heavy loading slide against each other. Wear on the track components is accelerated when you factor in dirt between the components, horsepower, and the weight of the equipment. Operation of the equipment in reverse at a high rate of speed can cause accelerated wear in the pins and bushings. This also applies when the equipment is operated with a tight track due to dirt packing between the sprocket teeth and bushings. External wear on the bushings along with internal wear of both pins and bushings indicates that the pins and bushings need to be turned.

Look for wear lines on the drive side of the sprocket teeth; this indicates the need to rotate the pins and bushings. In order to rotate the pins and bushings successfully, the wear must not extend completely around the pins. On a conventional track with a master pin and bushing, faster wear occurs first at the master pin since the pin is shorter and the bushing does not always extend into the counterbore of the master link. Replacing the master pin on a regular basis helps minimize the wear on the rest of the pins and bushings and minimizes sprocket tooth wear. On tracks that use the split master link, excessive wear is not a problem because the pins are identical throughout the track.

On a new track during forward operation two types of loads are imposed on the parts; these occur as the two track sections hinge, as shown in **Figure 17-21**. When the bushing enters or leaves the sprocket teeth in forward, the bushing rotates slightly as it leaves the sprocket. The opposite occurs in reverse, as shown in **Figure 17-22**; when the bushing enters the sprocket, it rotates slightly. This movement between the bushing and sprocket is similar to that of gear teeth when they mesh.

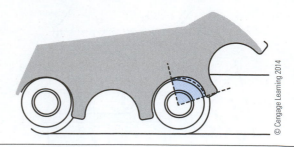

Figure 17-22 How wear occurs in the first track pin and bushing during operation in reverse.

When the bushing enters the sprocket in forward, the first bushing to come into contact with the sprocket takes about 85% of the load. The remainder of the load is absorbed by the next three bushings. When there is a mismatch between the sprocket and track pitch during operation, the bushing will slide and turn as it seats in the sprocket, causing wear on the forward drive side. An internally lubed track has an advantage over a conventional track; it prevents wear from taking place in this situation. In reverse, the first bushing comes into contact with the sprocket teeth about 30 to 60 degrees from the center of the forward drive side. This forces the bushing to rotate through the tooth root area and over to the reverse drive side of the tooth. The bushing wears externally and causes excessive wear to the sprocket tooth.

Some manufacturers use a design that allows the bushings to rotate; this reduces the wear significantly. The relative motion of the sprocket, pin, and bushing is shown in **Figure 17-23**. The lines demonstrate how much movement occurs as the sprocket engages the

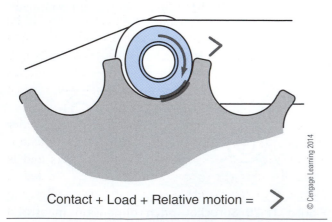

Contact + Load + Relative motion = >

Figure 17-21 How wear occurs in a track.

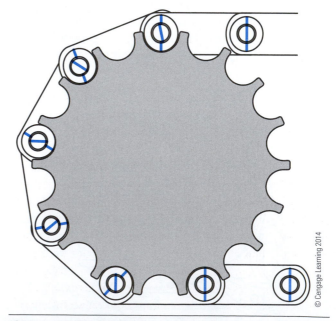

Figure 17-23 Relative motion between the pins, bushings, and sprocket as it rotates.

pins and bushings during operation. When the equipment moves in the forward direction, the pin enters the sprocket first; the first bushing is then picked up by the forward drive side of the sprocket loading it with approximately 80% of the load. Because the pin and bushing are lubricated, negligible wear results.

In reverse, if the track is in good condition and has been properly adjusted, the bushing end of the track enters the sprocket first. The forward drive side of the track picks up the first bushing and rotates it as it moves to the reverse drive side of the sprocket; this can cause more wear than travel in the forward direction. If the track has been improperly adjusted and is too tight, the load will be increased significantly. During forward operation, with good track tension and no significant packing occurring, no load is present at the top of the sprocket. With a tight track, loading will occur, causing excessive wear. In reverse, there is normally some wear while operating with a properly adjusted track. If the track tension is tighter than recommended, the wear increases significantly.

Unfortunately, some degree of packing occurs on track equipment during operation on certain types of terrain. Excessive packing can increase the track tension significantly and causes excessive wear. Tracks should always be adjusted after operating in an area to establish a normal amount of packing in the tracks. When packing gets between the track and sprocket, the bushings sit on top of the packing and cause the tension on the track to increase significantly. This increases the sprocket pitch and results in a mismatch between the track and sprocket pitch because contact between the track and sprocket occurs in the wrong area.

In the forward direction the bushing is picked up by the reverse side and slides over to the forward side. This prevents the sprocket from driving the bushing until it moves to the forward drive side. This sliding action dramatically increases the wear on the sprocket and external sides of the bushings. When operating in reverse with a packed track, the sprocket engages the bushing closer to the reverse drive side of the sprocket, causing excessive wear from the sliding action.

Inevitably, track wear will occur over time, causing the track pitch to increase significantly. In forward, the bushing will contact the drive side of the tooth higher up on the tip, initiating a sliding action between the two parts and causing excess wear. In reverse, the bushing contacts the sprocket tooth higher on the forward drive side. As the sprocket rotates, the bushings rotate through the bottom of the sprocket tooth and over to the reverse drive side of the tooth, again increasing the wear significantly.

When the track pitch is increased due to wear, packing can have a positive effect on the overall track pitch. Packing the track lengthens the sprocket pitch, and when the track wear increases the track pitch and is equal to the sprocket pitch, good contact is maintained. The down side to this scenario is that abrasive wear occurs rapidly from the packed material in the track. To minimize wear, packing the track must be kept to a minimum; too much packing tightens the track so much that significant bushing wear will occur.

When operating in conditions that are less than ideal, track packing can be minimized if the shoe design allows material to be forced out through holes in the shoe. It is probably a good idea not to use roller guards in conditions that promote packing. The undercarriage must be cleaned more often when operating in conditions that cause excessive track packing. Adjusting the track correctly for the operating conditions is beneficial to track life; loose track *will* cause excessive wear to roller flanges and sprocket teeth. If track tension is left slack, the track roller frame can be damaged. Selecting the correct track shoe cannot be over emphasized because the wrong shoe can promote excessive wear. For instance, wider shoes are excellent in muddy conditions because they provide greater flotation. But selecting a shoe wider than is necessary is not desirable because it is more susceptible to bending and breakage. Use of a wide shoe in a high-impact operating conditions results in excessive track wear and undercarriage damage.

Roller Wear

The rollers on track equipment do not all wear at the same rate or equally; they should be rotated, if possible, to prolong their service life. Rollers in certain positions will wear more on one flange than another. The flanges on the lead rollers of the track should be kept in relatively good condition because they guide the track onto the sprocket. The track rollers have a surface hardness that is about the same as the track links, and generally wear at about the same rate. Unless broken from impact damage, the rollers should be replaced at the same time as the track links, pins, and bushings. Carrier rollers can last significantly longer than the track rollers because they only act as guides and support the track weight, while track rollers must support the weight of the equipment.

A typical track system can have a number of roller arrangements depending on the track type. For example an XL track can have seven track rollers and is generally arranged so that the front roller (located at the front of the dozer) is a single flange roller, the second roller is a double flange roller, the third roller is a single flange roller, and so on, ending with the last roller as a single flange roller. The number of track rollers depends on the track design and can vary from manufacturer to manufacturer. The tightness of a track can affect the life of the carrier rollers and has a negative effect on idler life.

MAINTENANCE AND ADJUSTMENTS

It may be impossible to stop track wear altogether on a track equipped machine, but wear can be slowed to a great extent by good maintenance practice and operation. This section helps you better understand the wear and adjustments that take place on a track and undercarriage system. Always follow the manufacturer's repair literature to carry out maintenance and repair procedures on track systems as they may vary from this example. Operating track equipped machines in abrasive conditions leads to a short life due to components rubbing together in an abrasive environment. With older track systems, the pins and bushings are a constant source of maintenance, but in modern sealed and lubricated systems, pin and bushing wear has been virtually eliminated. Track systems and undercarriages wear whenever they are operated; the severity of the wear depends on the environment. Understanding what causes tracks to wear allows a technician to maintain and adjust the track system to optimum specifications.

Track shoes that are wider than necessary will affect the life of track links, rollers, and the idler flange. Track adjustment greatly affects track life. Not only is track tension easily corrected, it is also easily adjusted on modern track systems. Track tension adjustment must be performed on the terrain that the equipment will work on. Certain conditions, like mud, will pack the track more quickly than other conditions, such as crushed rock. In some instances, a certain amount of track packing is necessary before a track adjustment is performed. Always refer to the manufacturer's recommended procedures for track adjustment. Operation with too tight a track, as shown in **Figure 17-24**, leads to excessive bushing and sprocket wear, which is caused by increased loads on components. Increased loads from a tight track increase wear on the idlers and carrier rollers because of the excessive tension placed on them. High-speed operation should be avoided as much as possible; the increase in track revolutions accelerates the wear on track components tremendously. If equipment will be operated in conditions that will lead to excessive track packing, the roller guards should be removed to aid in expelling the material, which will help minimize packing.

Reprinted Courtesy of Caterpillar Inc.

Figure 17-24 Track adjusted too tight.

Good Maintenance Practice

The following list is a quick reference for inspection and maintenance that reduces damage from wear.

- Pick the correct shoe for the application; use the narrowest shoe you can, keeping in mind the importance of maintaining flotation.
- For track adjustment, be sure to perform the adjustments in the conditions under which the equipment will operate. Always follow the manufacturer's recommended track adjustment procedures.
- Get in the habit of inspecting the equipment for loose bolts and components. Check for leaks and excessive wear.
- Whenever track components need to be re-torqued, follow the manufacturer's recommended torque procedures.
- The use of roller guards should be avoided in conditions such as mud or snow, which can lead to excessive track packing.
- When inspecting and servicing equipment, be sure to inquire about the operating environment. The type of work that the equipment is to perform dictates the wear severity and location. For example, when the equipment is performing drawbar work or ripping, the weight is shifted to the back of the equipment, increasing the wear on the rear rollers and sprockets.

Generally, track wear can be measured by using either of the procedures shown in **Figure 17-25** and **Figure 17-26**. Specialized measuring tools are used by some manufacturers. Measuring tools include tape measures, depth gauges, and vernier callipers (**Figure 17-25**). Readings can be compared to the manufacturer's wear limit specifications to determine

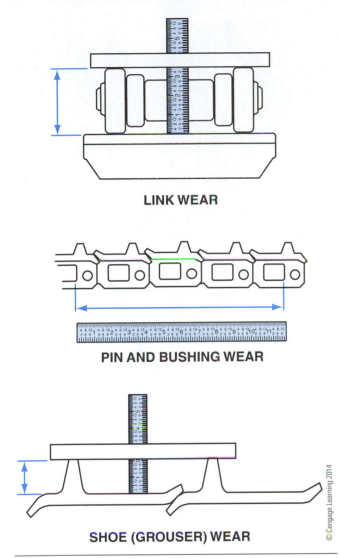

LINK WEAR

PIN AND BUSHING WEAR

SHOE (GROUSER) WEAR

© Cengage Learning 2014

Figure 17-25 Track measuring technique using measuring tools.

if replacement is necessary. Some manufacturers supply wear gauges to measure track component wear. The problem with some types of tools is that they do not give you an actual measurement; rather, they provide a visual indication that is subjective.

Tech Tip: Before performing any type of wear evaluation or taking measurements of the tracks, make sure the equipment is properly blocked and tagged with the engine off. Never perform any measuring of track components while the equipment is in operation. Some measurements require that the track be tightened before performing any measurements.

Before trying to determine the amount of pin and bushing wear on a track, the track must be tightened by

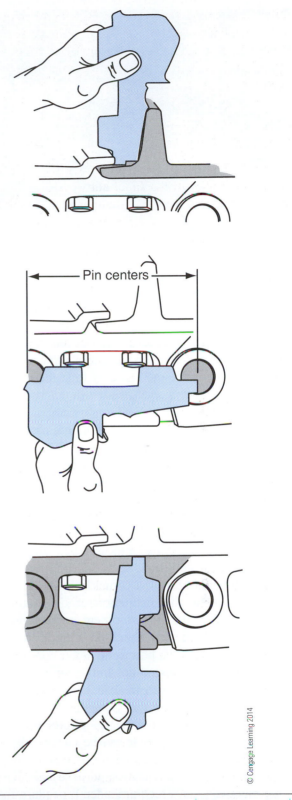

Pin centers

© Cengage Learning 2014

Figure 17-26 How gauges are used to assess track wear.

placing an old track pin, as shown in **Figure 17-27**, between the sprocket teeth and the closest link. Back up the equipment until the slack is taken up. The pin should be at approximately the 12 o'clock position.

Figure 17-27 How to tension the track to check pin and bushing wear.

Figure 17-25 shows how to measure across several links to determine wear. When using this measurement technique, do not include the master pin in any of the measurements. If you have measured across four links, then you must divide the value by four and compare it to the manufacturer's specification to determine actual wear.

A wear gauge is another common tool used by some manufacturers to determine the pin and bushing wear. Before any wear measurements can be accurately made, the track must first be cleaned. **Figure 17-26** shows how to use a wear gauge to determine pin and bushing wear.

Shoe wear is determined by measuring the height of the grouser (refer to **Figure 17-25**). The shoe bolts should be checked for tightness when checking for wear. The most common cause of shoe looseness is that they were not torqued correctly in the first place. Always refer to the manufacturer's recommended torque procedure for the equipment in question.

When installing new shoes on a track, it is good practice to re-check the bolt torque after approximately 50 to 100 hours of operation. Initial torque specifications are generally given in lb-ft or Nm, and require that an additional torque turn be applied to the bolts to ensure the bolt's maximum clamping force is achieved.

Roller and bushing wear can be checked by using several different methods. The procedure used in **Figure 17-28** is for clarification purposes; always refer to the manufacturer's recommended procedure for this check. To check the rollers for wear by using the gauge method (see **Figure 17-29**), be sure to use the correct gauge for the equipment in question. Using an incorrect gauge may lead to expensive repair work that may not be necessary.

Track rollers will wear on the outer portion of the flanges and need to be checked accurately by using an

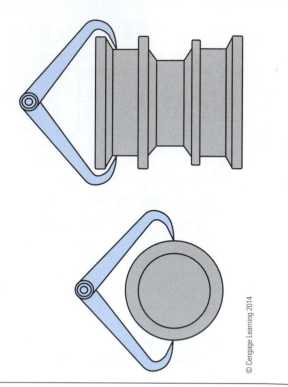

Figure 17-28 Measurement procedure for roller and bushing wear.

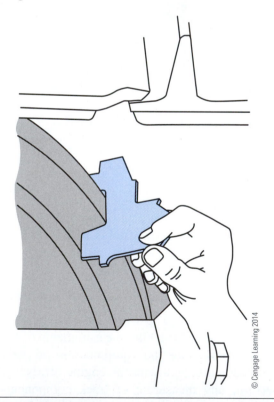

Figure 17-29 Measurement procedure to determine front idler flange wear using a wear gauge.

accepted method to determine maximum wear. The most common method of checking roller wear requires two measurements (indirect method) to be taken to determine wear. **Figure 17-30** shows how to perform

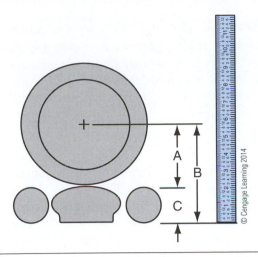

Figure 17-30 Measurement procedure to determine roller wear using the indirect methods with a ruler.

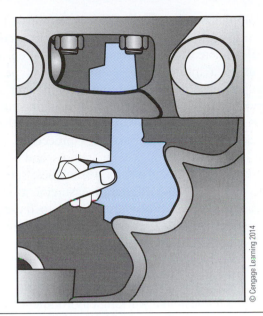

Figure 17-32 Measurement procedure to determine sprocket wear using a manufacturer-supplied wear gauge.

this procedure accurately. After taking the readings, subtract measurement C from measurement B to determine the roller radius. If the equipment has roller guards in place, the radius of the rollers must be measured with a ruler.

Front idlers tend to wear on the side and outer edges of the flanges, so proper idler wear measurements are needed to determine the wear on the flanges. This measurement can be made in one of two ways: the use of a specialized gauge, as shown in **Figure 17-29**, or with a straight edge and ruler, as shown in **Figure 17-31**. Always refer to the manufacturer's specifications to determine the allowable limits.

Determining sprocket wear with a gauge is recommended by many manufacturers. The round section of the manufacturer-supplied gauge goes into the root of the sprocket, as shown in **Figure 17-32**. Use the manufacturer's specifications for the specific sprocket to determine the wear.

CAUTION *Under no circumstance should anyone attempt to repair a track without having been given the proper instructions and special training that is required to perform this type of work safely. The track and its components are often very heavy and bulky to handle by hand. The only way to avoid potential accidents when working on track equipment is to familiarize yourself with the manufacturer's recommended procedures and to follow basic safety rules and precautions. A technician's first responsibility is to ensure that the work can be performed safely and that no one is exposed to any unnecessary hazards. No manufacturer can anticipate all potential hazards and repair procedures. Therefore, it becomes the responsibility of the technician to make sure that the procedures and techniques used to perform any repairs or maintenance to the equipment that are not covered in the service literature will not cause personal injury or damage to the equipment.*

Track Adjustment

Correct track adjustment is necessary to maintain long track life with low operating costs. Nothing shortens the life of a track faster than incorrect adjustments. If the track is adjusted tight, excessive loads are placed on the mating components, accelerating

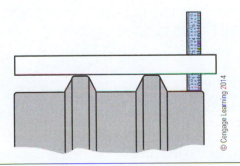

Figure 17-31 Measurement procedure to determine front idler flange wear using a straight edge and ruler.

track component wear. If the track is too loose, it will cause track whip at higher speeds, leading to excessive wear of undercarriage components. To keep these problems to a minimum, a technician must perform these adjustments according to the manufacturer's recommendations. This explanation of track adjustment is shown as an example and may vary based on the age and make of the equipment. This procedure is for Caterpillar D3 to D9 track-type tractors and loaders. Not all equipment uses the same techniques; some excavators and loaders use different measurements and, in some cases, different procedures. Refer to the manufacturer's recommended procedures for the equipment in question.

1. To ensure that the adjustments have been performed correctly, the equipment must be located in the actual working environment. Do not clean the track to make the adjustments; in some cases the track will pack slightly and any attempt to clean it may lead to an adjustment that will become too tight when the equipment is back in operation. **Figure 17-33** indicates that A is the track adjustment mechanism location and B is the location of the front idler bearing assembly.

2. Before beginning any adjustment procedure, make sure you are wearing approved safety glasses and gloves, if required. Clean any debris from the top of the adjustment mechanism plate before opening. After wiping the fitting, use a grease gun to adjust the track until the track idler is at its farthest forward position. Ensure that the relief valve remains closed.

CAUTION *At any point in the adjustment procedure, never attempt to remove any packing material between any of the track components. Once this step is completed the track between the idler and the carrier roller will be tight and look almost straight.*

3. Clean the track frame and place a mark behind the rear edge of the idler bearing. The distance will depend on whether the track has one carrier roller per side or more than one. Tracks with one carrier roller per side will use ⅜" (10 mm) as the correct dimension for this step; tracks with two or more carrier rollers will use ½" (13 mm).

4. Before performing this step, make sure you are wearing safety glasses. Once step 3 is completed on both tracks, open the relief valve slowly.

5. Locate a drawbar or track pin and insert it in the sprocket tooth near a link on both tracks, as shown in **Figure 17-34**.

6. Ensure that the relief valve is in the open position on each track; start the equipment, and move the equipment in the reverse direction until the front idler backs up one-half inch or more. This should place the pin between the track link and sprocket in the 12 o'clock position, as shown in **Figure 17-35**. At this point, move the machine forward slightly and remove the track pin from the sprocket.

7. Ensure that the hydraulic relief valve is closed on each track adjustment mechanism. Adjust the **hydraulic track adjuster** until the mark on each roller frame is visible. The track will sag

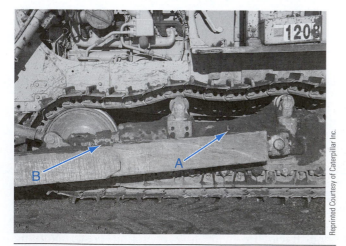

Figure 17-33 Cat dozer in the correct adjustment condition. Track tension should be adjusted in the work environment.

Figure 17-34 Technician inserting a track pin in a sprocket tooth.

Figure 17-35 Pin in the proper location (12 o'clock).

between the idler and the carrier roller. This sag will provide the correct amount of track tension for the application in which the equipment is operating. Secure the track adjustment protection plates and return the equipment to service.

Track Removal

This general overview of the track removal procedure is to be used as an example and is not intended to replace the manufacturer's service literature and supplemental bulletins. In many cases, it may be necessary to remove the track from the equipment to perform major repair work to the track. Two methods are used to connect tracks together: the master link and the master pin. A general summary of the track removal process on a master link track will be given so that the procedures involved in this type of work are well understood. Never attempt to perform any track repairs to equipment unless you have been properly trained and have the necessary equipment and manufacturer's recommended procedures on hand to perform the work safely.

The split master link uses a procedure that is unique to that design; the technician must follow the manufacturer's recommended procedures.

Tech Tip: High pressure may be present in the adjusting cylinder in the form of high grease pressure. Never risk personal injury by visually inspecting to see if the grease is released. Visually inspect the rod to see if it has moved into the front support of the track roller frame. When removing the track from the equipment, do not stand anywhere near the front of the

equipment; this will minimize any chance of injury if something went wrong. The track is heavy and can weigh upward of 1,500 pounds.

1. The equipment must be placed in a suitable position so that when the track is removed, it is on flat ground.

Note: The track must be loosened by opening the relief valve located under a protection cover on the roller frame. If the track will not slacken on its own, place a piece of round steel bar stock on the inside of the links between the bushing and the sprocket and drive the equipment in reverse slowly until the track loosens.

2. Place the track in position so that the master link is on the centerline of the rear idler, as shown in **Figure 17-36**. Place suitable blocking under the grouser bar, as shown just under the master link, and remove the four bolts (6) that hold the master shoe in place. Separate the track by removing the bolts that hold the master link together.

3. This step requires that one end of a chain (7) be fastened to the end of the top section of track, as shown in **Figure 17-37**, and the other end be fastened to a wheel loader. Start the unit and rotate the final drive in the forward direction. Allow the track to move forward so that it is pulling the wheel loader in a forward direction. Ensure that there is enough room in the front of the equipment to completely remove the track.

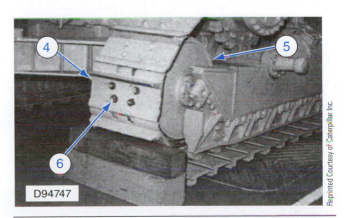

Figure 17-36 Correct location for the track blocking.

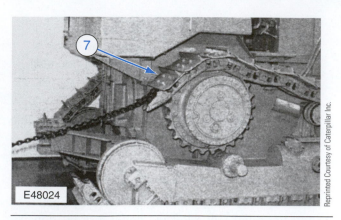

Figure 17-37 Correct procedure to remove the track.

4. In the event that it becomes necessary to raise the equipment to remove the track, use the tools listed in **Table 17-6**. Use tool A (hydraulic jack) from the table to raise the side of the equipment. Place tool B (stand) into position under the equipment. Due to the track weight, appropriate lifting devices are necessary. Use extreme caution when removing the track.

Track Installation

This general overview of the track installation procedure is for example purposes and is not intended to replace the manufacturer's service literature. The same specialty tools required to separate the track are used to reconnect it.

1. Position the track and align it under the roller frame so that the links will engage the sprocket and roller flanges correctly. Once the track is in position, carefully lower the unit onto the track. Visually inspect that the track is properly aligned with the track roller flanges.
2. Connect the same chain that was used to remove the track to the end of the split track on one end and connect the other end of the chain to a

loader or other suitable equipment that is capable of pulling the chain over the front idler. Have the loader pull the chain over the front idler and carrier rollers until it engages the sprocket teeth.
3. Exercise caution whenever the equipment is operated to ensure that personnel working in and around the equipment are not in danger of being caught up in any of the moving parts. At this point, the equipment must be started and the final drive must be rotated in reverse to pull the track together so that the master link can be reconnected.
4. Remove the pull chain and carefully align the two halves of the master link (labeled 5 and 6) so that the bolts can be installed and torqued correctly, as shown in **Figure 17-38**. Follow the manufacturer's recommended installation procedure to install and torque the master link bolts correctly; then reinstall the shoe on the master link and tighten the bolts to specification.
5. Close the relief and fill valves, as shown in **Figure 17-39**, on the track adjuster located on the track frame and perform an initial track adjustment. Follow the track adjustment procedure as outlined in the manufacturer's service literature. Replace the protective cover and O-ring on the track frame and torque to specifications.

Note: If the track assembly has been cleaned due to the repair work performed during removal, the track adjustment procedure must be repeated once the equipment operates at a worksite long enough to be sure that the track has a minimal amount of packing. This ensures that the track is properly adjusted.

TABLE 17-6: TOOLING FOR TRACK SEPARATION

Examples of Required Tooling for Track Separation	A	B
8S-7610 Base	2	
8S-7650 Cylinder	2	
9U-6600 Hydraulic jack	2	
8S-7640 Stand		2
8S-7611 Tube		2
8S-7615 Pin		2

© Cengage Learning 2014

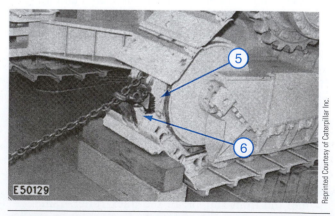

Figure 17-38 Correct procedure used to align the master link.

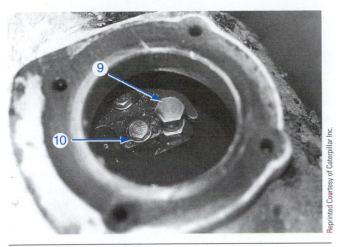

Figure 17-39 Location of the relief (9) and fill valves (10).

Track Disassembly

In some cases, it is necessary to remove and replace a track link component without removing the track from the equipment. The following procedure outlines these techniques in general and is only used to clarify the process. When performing this type of repair, always consult the manufacturer's service literature for the particular type of equipment that is being repaired. To perform this work, it is necessary to work with hydraulic pressures that can cause personal injury if proper procedures are not followed. Never inspect hydraulic lines for leaks with your hands; a small pin hole could easily cause serious injury.

1. To replace a track component, position the blocking (labeled as 1) between the front idler and the carrier roller, as shown in **Figure 17-40**. Loosen the track by relieving the hydraulic pressure at the track adjustment relief valve located in the track frame. High grease pressure may be present in the adjusting cylinder.

Never risk an injury by visually inspecting to see if the grease is released.

2. Use a suitable hoist to pull the weight of the track section to be worked on, as shown in **Figure 17-40**, and use the appropriate blocking (see **Figure 17-41**) to support the weight of the track. Under no circumstance should anyone work under a suspended load; the track is heavy and, if it were to fall, it could cause severe injury. Remove the carrier roller fastening bolts (3) from the carrier track roller frame (2), and then remove the bolts and bracket (4) that secure it to the track frame.

3. Place the appropriate blocks to support the track when it is disassembled. Place two additional blocks under the section of the track that is to be worked on. Remove the three track shoes located over the pins (4 and 5) that are to be removed. Because of the weight of the hydraulic tool required to remove the pins, a hoist attached to the tooling (A) is necessary to hold it in the correct alignment, as shown in **Figure 17-42**.

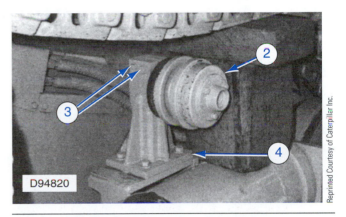

Figure 17-41 Procedure for removing the carrier roller.

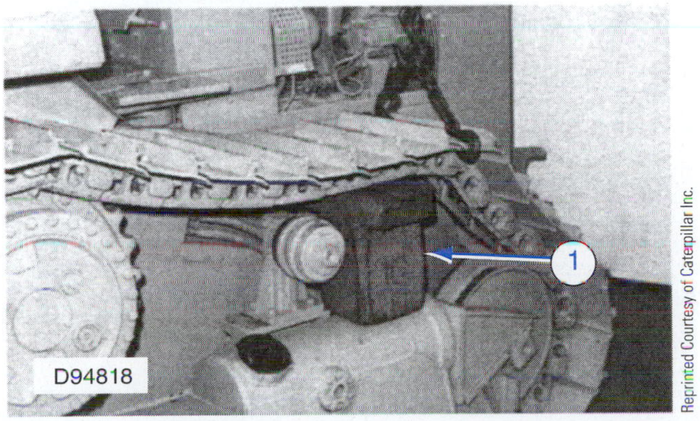

Figure 17-42 Procedure used to support the track before disassembly.

Figure 17-40 Procedure for blocking the track to remove the carrier roller.

4. Use the specialized hydraulic jack, as shown in **Figure 17-43**, to spread the link assembly so it can be removed from the track. Be careful not to damage the seal faces with the jack when forcing them apart. Use the tool to push the pin from the track and install tool B pilot pin (refer to **Table 17-7**) to keep the track together until the second pin is removed. Place the hydraulic removal tool against the second pin and remove it; install the second pilot pin. Remove the seals and rings and clean up the link assembly. Use the bushing adapter with the ram adapter to remove the bushings.

Drive Sprocket Removal and Installation

The following is an outline of the procedure used to remove and replace a drive sprocket. It is only to be used as an example. Always refer to the manufacturer's service literature when replacing the segments on track equipment. Manufacturers of track equipment generally use one of two different sprocket designs, the older being a one-piece unit that is bolted on to the final drive, and the newer being a segment-type or multi-piece design.

The newer (segment-type) is the easiest to replace because all that is required is to place the segment that needs to be replaced so that it does not contact any of the track links and unbolt it from the final drive. Before installing a new sprocket segment, be sure to wash and remove any paint or debris on the parts that may interfere with good contact between the mating parts. Locate the segment into position on the final drive, as shown in **Figure 17-44**, and install the lubricated bolts so that the head of each bolt is against the final drive. Torque the nuts to specification.

On equipment that uses a one-piece segment, the track must be disassembled before the sprocket can be replaced. Use the service literature guidelines to split the track at the sprocket before attempting to replace the sprocket.

Track Roller Removal and Installation

The following is an outline of the procedure used to remove and replace a track roller. It is only intended as an example. Before any track roller can be replaced, the track must be loosened by following the manufacturer's recommended procedure as outlined in the section on track disassembly. **Table 17-8** provides a list of tools that are required in this example; these tools may differ from the tools actually required for a specific type of track system. Use the electric hydraulic jack to lift one side of the equipment off the ground and block by using mechanical stands.

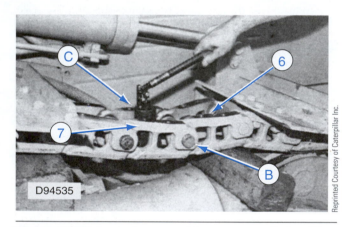

Reprinted Courtesy of Caterpillar Inc.

Figure 17-43 Procedure used to remove a link with a special hydraulic jack.

TABLE 17-7: TOOLING FOR TRACK REPLACEMENT

Disassembly Tools	A	B	C	D
9U-6600 Hand hydraulic pump	1			1
5P-2380 Track tool	1			1
8S-9978 Ram adapter	1			
5P-2188 Adapter	1			
5F-9447 Pilot pin		2		
Felco Hydraulic jack (20-ton)			1	
8S-9975 Ram adapter				1
7S-8469 Bushing adapter				1
2S-8235 Bushing pusher				1

© Cengage Learning 2014

Reprinted Courtesy of Caterpillar Inc.

Figure 17-44 Procedure used to remove a link with a special hydraulic jack.

TABLE 17-8: TOOLS FOR TRACK ROLLER REMOVAL

Required Tools for Track Roller Removal	A	B	C
8S-7610 Base	2		
8S-7650 Cylinder	2		
3S-6224 Electric hydraulic pump or jack	2		
8S-7640 Stand		2	
8S-7611 Tube		2	
8S-7615 Pin		2	
FT-1309 Lifting fork			1

© Cengage Learning 2014

Place the lifting fork under the track roller to be removed and secure with a hoist, as shown in **Figure 17-45**.

Next, remove the bolts (3) that secure the caps (1) on the roller to the track frame. Remove the track roller (2) from the track assembly.

....

Note: The weight of the roller varies (50 to 120 pounds), but they are not light. Be careful when removing them.

....

Reinstalling the track roller requires that you follow the procedure outlined in reverse order.

Idler Removal and Installation

The following procedure is a summary of the steps used to remove and replace an idler assembly. This procedure is only intended to be an example.

Before an idler can be replaced, the track must be separated by following the manufacturer's recommended procedure as outlined in the technical

documentation on track separation. The idler on dozers is generally too heavy to handle by hand; a suitable hoist should be used to secure the idler, as shown in **Figure 17-46**, to safely hold it in place before attempting to remove the mounting hardware. Once the idler is secure, remove the bolts and caps. Carefully remove the shim pack (4) and identify it for correct replacement. Move the retaining blocks (5) toward the roller frame. Use the hoist to lift the idler from its mounting position.

To perform an overhaul to the idler, refer to the manufacturer's service literature.

To reinstall the idler, place the blocks in position on the dowels on the front roller frame. Use a hoist to lower the idler into position. Reinstall the correct shim pack between the front roller frame and the blocks, as shown in **Figure 17-46**. Carefully align the holes (6) in the idler shaft with the pin in the caps. Reinstall the mounting caps and bolts and torque up to specifications.

Track Component Inspection

Once track components have been removed from the track, it is important to examine the condition of the parts to determine if they are reusable. If the equipment has been operating in an abrasive environment, the links can become worn out, resulting in stretching and cracking. This section outlines the inspection procedures used to determine the usability of the components in question. It cannot be overstated how important it is to evaluate the usability of the components because the cost of replacing all the components would be excessive. Technicians should always adhere to the manufacturer's guidelines when it comes to reusability of parts. Replace only the parts that are worn or damaged beyond use. Before any of the track parts can be properly inspected, they must be cleaned.

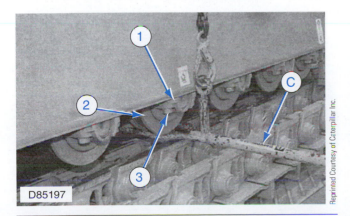

Reprinted Courtesy of Caterpillar Inc.

Figure 17-45 How to position the lifting fork before attempting to remove and install a track roller.

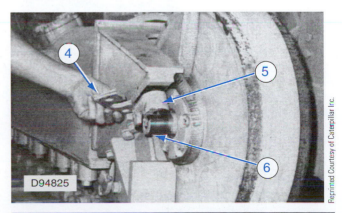

Reprinted Courtesy of Caterpillar Inc.

Figure 17-46 Shim removal process. Identify the shim packs for installation.

When new links or other track components are used for the repair, always be sure to remove the paint around the bores of a link or other components that have mating surfaces, such as track shoes and fasteners. A layer of paint prevents a proper fit and interferes with good shoe retention. An angle grinder with a wire brush or an abrasive wheel will clean up any paint in these areas quickly. Caution should be exercised whenever a high-speed grinder is used. To prevent the possibility of injury, always wear a face shield and be sure no other personnel are in the immediate area when performing this operation.

Track Link Assessment

Links can develop cracks over time; these cracks can show up near the bushing strap, in which case the links are not to be reused. There are numerous spray compounds that can be used to detect cracks in the track components. However, if none are available in the field, spray paint can be used to detect any cracks that are not easily visible. One area that needs to be examined carefully is the link strut area. Links that have developed cracks near the strut, as shown in **Figure 17-47**, should not be reused. Failure to detect cracks in this area could lead to a catastrophic track failure during operation, resulting in extensive damage to the track and undercarriage.

The toenail sections, as well as the pin boss area of the link, are other areas that are prone to developing cracks. **Figure 17-48** shows the area in question that needs to be carefully examined for cracks. Over time the bushing bores tend to become oval shaped; this results in excessive pin-to-bushing clearances and affects the sealing capabilities of the fit. When examining the condition of the pin, if you notice that the

Figure 17-48 Crack in the pin boss.

surface is rusted all the way across the bore, it can safely be assumed that the press fit is no longer present and the link should be discarded. Because the bushing bore is not required to seal, slight scoring is acceptable so long as a press fit is attainable.

Often the link rail surfaces (see **Figure 17-49**) will have pieces break off the edges; this is known as **spalling** and does not necessarily mean that the link needs to be discarded. If the link has less than one-third of the rail surface broken off, it can be reused safely; if the spalling is significantly deep and located in the bushing area, the link should be discarded.

Link counterbore damage usually results from a seal failure and, depending on the extent of damage, the link can be reused. Often a lightly **pitted** or corroded link counterbore can be cleaned up enough to reuse. An exception to this is when the equipment has been allowed to operate with a **dry joint**. The counterbore sidewall quickly becomes damaged from contact with the bushing if allowed to run dry. The track seal load ring is supported by the counterbore sidewall;

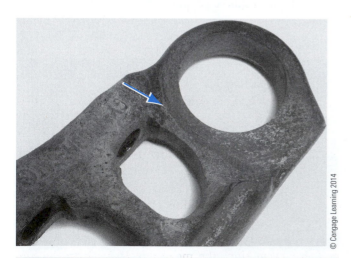

Figure 17-47 Area on a link that is susceptible to cracking.

Figure 17-49 Link with minor rail spalling. This link can be reused.

if excessive wear is present in this area, the load ring will fail in a short time. Often links with minor damage to the counterbore sidewalls can be reused as long as the counterbore height is greater than the load ring height when installed.

A link counterbore with thrust ring wear can be reused as long as the depths of the wear do not exceed a specified amount. Refer to the manufacturer's specifications for this information as it may vary. Often track roller flanges will cause metal on the sides of the links to be rolled over the track pin, as shown in **Figure 17-50**. This must be removed with a grinder before attempting to remove the pins. Occasionally, the roller flanges can cause metal to roll over the surface of the link where the track press tool sits for assembly. This metal must be removed before assembly is attempted.

Track Pin Assessment

Sometimes during disassembly, small imperfections in the pin ends can cause minor scoring in the link pin bores. As long as the scoring does not extend the full width of the bushing bore, the link can be reused. If it extends the full width of the bore, a path for lubricant to leak out is created and the link should not be reused. In **Figure 17-51** the link bore is scored from one end to the other and would provide a leak path for the lubrication so it should not be reused.

Track pin damage comes in a variety of forms and some knowledge is required to assess the usability of the used pins. Pin damage, as shown in **Figure 17-52**, can generally be categorized into three types: spalling, galling, and step wear. Because the pins are heat treated, spalling takes the form of flaking or breaking off of metal in small pieces. This condition can weaken

Figure 17-51 Link bore with major scoring that extends the full width of the bore. It should not be reused.

Figure 17-52 Track pin with minor galling damage. This pin may be reused.

the pin significantly, and most manufacturers recommend that the pins not be reused if they show signs of significant spalling.

Whenever making determinations on the reusability of track pins or other track components, always refer to the manufacturer's technical documentation for the exact parameters. Parameters differ significantly on some types of equipment; larger equipment often has less tolerance for wear in areas like track pins. Metal transfer on pins is caused by the slight movement of the pin in the bore during operation. If you encounter pins with this type of damage, it is best not to attempt to reuse them since the metal that is fused to the pin will damage the link bore and cause scoring or **broaching** as the pin is pressed into the link bore.

Figure 17-50 Link with rolled metal up over the pin boss. It may be reused but must be ground off.

Pins that develop some form of **wear step** damage generally come from the wet joint design. Minor wear under 0.005" (0.127 mm) can be tolerated for reuse in most cases, but checking the specific manufacturer's recommendations is always recommended to determine wear tolerances. In the case of pin end damage, it is generally permitted to grind a 10-degree chamfer on the end of the pin, as shown in **Figure 17-53**, to remove any burrs caused by contact with the roller flanges. If this rough edge is not removed properly, most likely it will lead to broaching or scoring to the link bore.

Some equipment manufacturers use positive pins retention (PPR). These pins have a small groove on the end that accepts a retainer ring. With this pin design you cannot chamfer the ends because it weakens the retention ring groove. These pins can have some damage to the ends, but no more than 25% of the groove should be damaged or broken.

Bushings and Counterbore Assessment

Cracks that develop in the bushing area of the link (see **Figure 17-54**) need to be identified. A crack in this area allows the lubricant to leak out of the joint, resulting in a dry joint before very long. It is important to perform a thorough inspection to ensure that no links with cracks in this area are reused.

On sealed tracks it is important to ensure that the area where the track seal fits in the link counterbore is free from any excessive wear grooves. These could affect the integrity of the seal fit. Bushing groove gauges are available from the manufacturers to measure acceptable wear on the ends. A specialized gauge, as shown in **Figure 17-55**, is used on the ends to check whether or not the seal groove is deep enough to be visible with the gauge in place. If it is, then the bushing should be discarded.

© Cengage Learning 2014

Figure 17-53 Track pin with a 10-degree chamfer ground on the end. This pin may be reused.

© Cengage Learning 2014

Figure 17-54 Track link with a small crack in the bushing area. This link should not be reused.

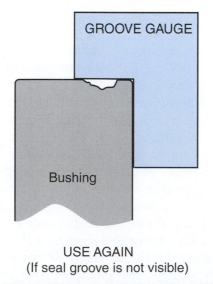

GROOVE GAUGE

Bushing

USE AGAIN
(If seal groove is not visible)

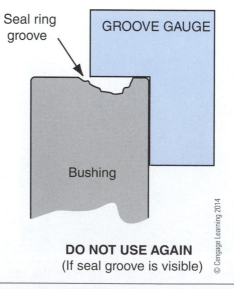

Seal ring groove

GROOVE GAUGE

Bushing

DO NOT USE AGAIN
(If seal groove is visible)

© Cengage Learning 2014

Figure 17-55 How to determine the wear limits on bushing grooves using a manufacturer's groove gauge.

If a pin shows signs of galling, then the bushings that held that pin will most likely have the same problem. A lack of lubrication on mating surfaces causes galling, which generally occurs at the same time on both mating surfaces. On certain types of equipment, slightly **galled** bushings can be reused but reference the manufacturer's usability guidelines. On some equipment if galling does not occur for more than one-half inch laterally or for more than 90 degrees of circumference, it may be reused.

If the bushings develop a slight movement in the link bore, metal transfer will occur. Metal will transfer from the link bushing **I.D.** (inside diameter) because the pin is made from a harder material than the bushing. Pins that develop metal transfer on them should not be reused; the pin-to-pin bore must seal in the oil on wet systems. This does not matter on the bushing-to-bushing bore fit as this connection does not have to seal oil in. Some metal transfer is considered acceptable on bushings for reuse by some manufacturers. Check with the specific manufacturer's replacement recommendations for the particular type of equipment being worked on.

Thrust rings are used on sealed track systems to provide the correct amount of seal compression to the track pin assembly; they must not be reused if any cracks or chips are present, as shown in **Figure 17-56**. A slight amount of wear can be tolerated on either side of the thrust rings. When assessing the reusability of the rings, if you can still see the lubrication slots on both sides of the thrust rings, then they are generally reusable. If galling is present on the thrust faces that goes all the way around on either side, the rings should not be reused.

Some manufacturers use a special track system in which the pin assembly comes preassembled. The pin, bushings, and seals are supplied as a complete unit, and should not be disassembled for inspection; it should be replaced as a complete unit. In most cases, if there is minor broaching on the link bore, it is generally reusable. The track link needs to be inspected by following the procedure used for conventional sealed tracks.

Rotating Pins and Bushing

On certain track systems it may be desirable to rotate the track pins and bushings, as shown in **Figure 17-57**, once a nominal track pitch dimension has been exceeded. Because the track pins and bushings wear in a predictable location, they can be rotated 180 degrees to

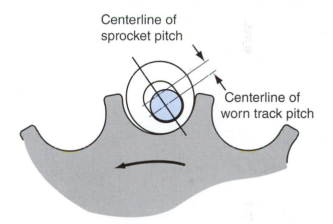

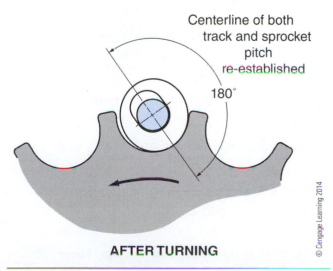

Figure 17-56 Thrust ring with a crack through it. Thrust rings with cracks should never be reused.

Figure 17-57 How track link pitch can be restored by rotating the pins and bushing 180 degrees.

bring track pitch back to acceptable dimensions. This procedure requires the use of a specialized track press, which can be portable or can require that the track be taken to a shop that has the equipment. Note that this procedure may not always be economical; the cost of labor to perform this work may exceed the cost of replacing the pins and bushing with new ones.

GENERAL GUIDE TO TROUBLESHOOTING TRACK PROBLEMS

Table 17-9 provides a quick reference guide to troubleshooting track problems.

TIRES AND RIMS

This section describes the most common types of tires and rims found on off-road equipment. Maintenance and replacement procedures can vary greatly between different types of tire and rim combinations. The explanations and examples provided in this section cover general practices and do not replace the specific manufacturers' and site-specific procedures that must be followed when servicing tires and rims.

Off-road tire and rim requirements on equipment vary greatly and are dependent on factors such as ground surface conditions. Steep hills, along with loads and speed, influence the type of tires chosen.

TABLE 17-9: TROUBLESHOOTING TRACK PROBLEMS	
Problem	**Possible Cause**
Accelerated wear on one side of the final drive sprocket teeth	Track misalignment Loose track Sprocket misaligned due to improper installation
Accelerated wear on one side of the idler(s) or rollers	Track misalignment Idler misaligned due to improper installation Track adjusted improperly or too loose
Accelerated wear to track pins, bushings, and track links	Track adjustment too tight Misaligned track Excessively worn sprocket teeth Excessive high-speed operation
Excessive wear to the final drive bearings	Track adjusted too tight Misaligned track
Operator complains of loss of drawbar power	Improper track adjustment Track adjusted too tight
Operator complains that the dozer drifts either to the right or to the left while being driven in a straight direction	Improper track adjustment One track is tighter than the other Track misalignment
Frequent packing of the track during operation	Improper track adjustment Track adjusted too loose
The track is noisy during operation	Improper track adjustment Track adjusted too loose
The track whips excessively during operation	Improper track adjustment Track adjusted too loose Idler seized in the retracted position
The track is thrown off during operation	Improper track adjustment Track adjusted too loose

The cost of off-road tire replacement can add a significant expense to the work performed if the appropriate tires are not selected and properly maintained.

Selecting the appropriate tires and rims for equipment and work performed will result in longer tire life, minimizing down time as well as reducing tire costs. The **ply rating** and tread type also play a role in safe, efficient operation of the equipment. Maintaining correct operating procedures observing speed and load limitations results in maximum effectiveness with minimal tire damage as well as preventing unscheduled down time.

Although safe handling of tires and rims will be discussed in detail in this section, it is worth noting that before beginning any type of work on or around a **wheel assembly**, ensure that you are trained properly on safe handling procedures of tires and rims. Do not place yourself or anyone else in a position that could result in injury or equipment damage. Government regulations and manufacturer's procedures and policies are in place to ensure no one performs work on tires and rims without the proper training.

Note: A typical off-road tire when inflated to 100 psi can generate the equivalent of 40,000 pounds of pressure on the **rim lip** of the wheel. When safety procedures are not followed when installing tires on **multi-piece rims**, the consequences can be damage to equipment, injury, and deaths. A tire inflated to 100 pounds of air pressure can generate enough force to throw an object a distance of 680 feet. Statistics show that when tires catastrophically fail on off-road equipment, the results may be fatal. Never stand in the trajectory of a suspected defective tire and rim.

CAUTION When a wheel assembly must be removed to perform repairs or service to equipment, visually inspect the rim assembly for cracks, bent or unseated lock rings, or other suspicious defects using the manufacturer's recommended procedure. If any defects are found or there is difficulty in removing the wheel assembly from the hub, deflate the tire before proceeding. Large tires like those found on off-road equipment can be extremely hazardous if technicians do not follow the manufacturer's instructions and procedures.

Rims

An obvious hazard with one-piece rims (**Figure 17-58**) is the sudden release of pressurized air, which can injure anyone in close proximity to the wheel assembly. Off-road rims come in a variety of designs, but can be generally classified in two categories, one-piece or multi-piece. As a rule, smaller off-road equipment is often specified with one-piece rims, whereas larger equipment more often uses multi-piece rims. There are a number of multi-piece rim designs and the use of approved procedures is required. Multi-piece rims can be identified visually on the outside by a number of

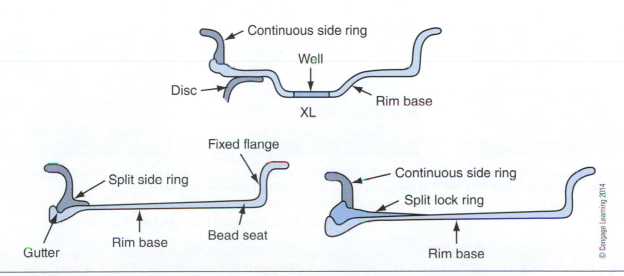

Figure 17-58 Illustration shows three types of off-road rim designs.

separate rings interlocked together and consist of up to five separate interlocked parts. Multi-piece rims must remain properly seated in order to remain sealed.

Rust, corrosion, or cracks in the metal must be evaluated to determine if the rims are in usable condition or should be replaced. Broken or bent flanges must be inspected and replaced if it is determined they are no longer safe to use. The mounting bolt holes must not be elongated, worn, or damaged. Burrs must be removed around bolt holes, as they would prevent proper contact, which could compromise wheel clamping force. Typically **split rims** do not abut when mounted; there must be a small gap in the ring as shown in **Figure 17-59** and **Figure 17-60**.

When an inspection is to be performed on tires and rims during scheduled maintenance or repair, and a tightly compressed multi-piece rim is observed (**Figure 17-61**), the tire should be deflated before

removal from the equipment and the rim must be inspected by a qualified person to determine the cause.

Physical damage to the rim as illustrated in **Figure 17-62** must be addressed immediately, as this condition could lead to a tire failure resulting in injury or death, not to mention the damage to the tire and rim, which could cost thousands of dollars to repair.

> **Note:** Left unchecked, this condition could lead to a catastrophic failure of the wheel.

Tires

Generally the type of equipment and the operating conditions that the equipment will be subjected to will determine the tire selection process. Operating conditions

Correct

Incorrect

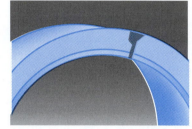

Incorrect

The components in a correctly assembled two-piece rim fit snugly and are locked together.

An incorrectly assembled or damaged two-piece rim could have a large gap inside ring. Components are not firmly locked in place.

An incorrectly assembled or damaged two-piece rim could have components not firmly locked in place.

© Cengage Learning 2014

Figure 17-59 Illustration shows correctly assembled two-piece rim along with two that are incorrectly assembled.

Correct

Incorrect

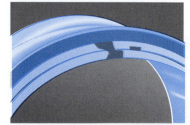

Incorrect

A three-piece rim that is correctly assembled has firmly fitted components that are locked together.

An incorrectly assembled or damaged three-piece rim could have a large gap in the lock ring. The components are not firmly locked in place.

An incorrectly assembled or damaged three-piece rim could have components not firmly locked in place.

© Cengage Learning 2014

Figure 17-60 Illustration shows correctly assembled three-piece rim along with two that are incorrectly assembled.

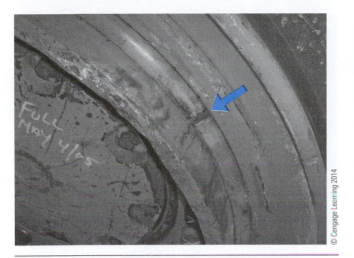

Figure 17-61 Illustration shows the gap in a split ring.

Figure 17-63 Illustration shows tire inspection procedure.

Figure 17-62 Illustration shows physical damage to the rim.

for loading and grading on rocky terrain are generally the same with short hauling distances.

Choosing the right tires will result in an increased tire life, less down time, and reduced operating costs. Tire size is only one factor that must be considered; ply rating and tire type are also equally important. Although off-road tires are designed for extreme-duty operation, they require specific care and maintenance to provide long life and achieve maximum effectiveness (see **Figure 17-63**).

Incorrect tire selection can increase tire costs and damage along with causing serious accidents. Tire speed can cause tire damage by increasing the internal temperatures resulting in heat separation failures. Tire chipping and bead damage can result from frequent abrupt braking. Tire cuts and burst damage can result from impact with sharp ground conditions and excessive tire slippage during loading operation. Rapid tire wear can be accelerated by sharp cornering and

frequent braking. To achieve maximum tire life, ensure that the weight limit for the equipment is not exceeded. Overloading can cause cord damage to the tire due to excessive **tire deflection**, which also leads to uneven wear and an increase in cut and impact damage as a result of increased cord tension.

Correct tire inflation pressure is determined by equipment type, tire size, ply rating, operating speed, and tire load. Under or over inflation will shorten the tire life dramatically. Correct inflation pressures ensure proper tire deflection as well as maximizing traction, flotation, and load carrying capacity. Over inflation increases ground contact pressures at the center of the tire, causing rapid center wear as well as increased cord tension, which leads to an increase in cuts and impact damage.

Tire Ballast

In many operating conditions it is necessary to add ballast to the tires of loaders, dozers, and graders. This additional weight can improve tire traction and equipment stability. Two common types of ballast can be used; liquid or dry ballast. The recommended volume of **liquid ballast** recommended by some manufacturers is approximately 75 percent of the tire's volume. It is not recommended by most manufacturers to fill a tire to 100 percent of its volume because it might cause tire damage due to pressure changes during operation. Tires may be filled with water to achieve extra ballast. An anti-corrosion agent should be used in tubeless tires to prevent rust damage to the inside of the rims. A solution of **calcium chloride** with an anti-corrosion agent is used in some cases; this generally results in more effective ballast. Tire weight can increase by 20 to 30 percent using this type of ballast. Some tire ballast products on the market claim to be **biodegradable** and may, in addition, eliminate rim corrosion. Such formulations of tire ballast may also be suitable for cold weather conditions, which would be a problem when using water as ballast.

Tire Inspection and Handling

Technicians must be able to properly demount tires and identify mismatched tires (**Figure 17-64**) or damaged rim components. Individuals working with off-road tires need to learn how to mount tires to rims using the correct tools and safety equipment, as well as following assembly inspection and the correct inflation procedures.

Before anyone works on large off-road tires, familiarization with the proper use of **restraining devices** and mechanical aids which are often used to handle heavy tire assemblies must be reviewed. Procedures for inflating tires while on equipment must be learned and followed.

....

Note: The following guidelines are to be used as examples of what steps are performed to remove and replace wheel rim assemblies from off-road equipment. Specific manufacturers' procedures and site-specific procedures should always be used when servicing rims and tires.

....

Guidelines for Mounting Single-Piece Rims

- It is recommended that the tire be completely deflated by removing the valve core (**Figure 17-65**).

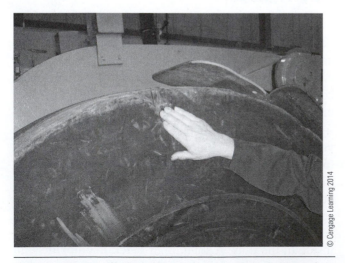

Figure 17-64 Illustration shows tire inspection procedure.

Figure 17-65 Illustration shows valve core removal procedure.

- Mounting and demounting must be performed from the narrow ledge side of the wheel if applicable.
- The tire must be matched to the rim with a compatible mating bead diameter and width.
- A rubber lubricant that is non-flammable must be used to lubricate the bead.

....

Note: In some cases the manufacturer recommends installation of tires on the rims with no lubrication.

....

- Certain size (smaller) tires must be partially inflated (if using a tire changing machine) and then mounted in a restraining device before fully inflating.
- If a **bead expander** is used, it must be removed before installing the valve core as the rim becomes airtight.
- Be sure no one is standing in the trajectory of the tire being inflated.
- Heat should never be used on one-piece rims. Damaged wheels must not be reworked or welded.

Procedures for Multi-Piece Rims

- The tire should be completely deflated by removing the valve core before a rim wheel is removed from the axle when the tire has been driven while under inflated or when obvious damage is suspected or observed.

- A rubber lubricant should be used on the bead and rim mating surfaces when the components are assembled and inflated unless the manufacturer specifically recommends against this.
- It may be possible to inflate the tire while it is on the equipment if the air pressure did not fall below 80 percent of its recommended value, providing remote-control inflation equipment is used and no one is in the tire trajectory path during the filling operation.

Note: Site-specific procedures may prohibit this type of operation.

Note: On many large machine applications, the wheel assembly is never removed from the vehicle. Specialized tire service vehicles are equipped with hydraulic cranes and grapples that remove and install the tire with the rim still bolted to the machine.

- If the tire is to be filled/inflated while off the equipment, a restraining device must be used. Inspection of the tire after inflation must be done while the restraining device is in place.
- No bent, damaged, or broken multi-piece rim components should be reused. No heating, welding, or brazing should be performed on rim components. Damaged components should always be replaced.

ONLINE TASKS

1. Use a suitable Internet search engine to research the configuration of a typical off-road track loader undercarriage. Select a piece of equipment in your particular location, paying close attention to the track design and arrangement, and identify whether it has an additional gear reduction at the drive sprocket. Outline the reasoning behind the manufacturer's choice of this design for a particular application.
2. Check out this URL: *http://www.tpub.com/content/ engine/14081/css/14081_229.htm* for more information on drive track configuration and design.

Shop Tasks

1. Select a piece of equipment in your shop that has a drive track assembly and identify the components. Inspect the track pins, note the design, and evaluate the condition of the track assembly.
2. Select a piece of off-road equipment in your shop that uses a drive track and identify the type and model of the equipment. List the major components and the corresponding part numbers of the track assembly using the correct parts book.

Summary

- Heavy equipment operating in a quarry, mine, or forestry application uses specialized tracks and undercarriages and tires to meet the extreme-duty operating cycles associated with this type of environment.
- Track and undercarriage systems can be categorized into four distinct designs: standard track system, low ground pressure (LGP) system, extra-long (XL) (front) track system, and extra-long (XR) (rear) track system.
- The chain section of the track is made up of track links, pins, and bushings.
- The drive sprocket, front idler, track rollers, tension mechanism, roller guards, and frame make up a track and undercarriage system.
- Each individual track section is made up of two track links: a hardened pin, and bushing.

- Sealed tracks have a solid pin. Sealed and lubricated tracks have a hollow pin to provide a path for lubricating the pin and bushing of the next track section.
- Track shoes are bolted to the link assembly to provide a means of supporting the equipment.
- When dozers must be operated in loose, muddy ground conditions, the use of a self-cleaning, low ground pressure shoe is desirable.
- Tracks use two different types of track rollers to maintain track alignment.
- Carrier rollers are located above the frame rail and support the weight of the top section of track.
- Rollers have a hardened surface that is equal in hardness to the track link surfaces;

- The idler guides the track through the track rollers and supports part of the weight of the equipment.
- Depending on the track design, a dozer could have one or two idlers per track section.
- Tracks operate in extremely abrasive conditions and would quickly wear out any pins and bushings if they were not protected with seals.
- A tensioning mechanism connects to the idler to maintain the correct track tension.
- Improvement in track design has led to the adoption of the elevated track design, which helps isolate the final drive from excessive shock loads.
- When the bushing enters the sprocket in forward, the first bushing to come in contact with the sprocket takes about 85% of the load.
- An internally lubed track prevents wear between the pins and bushings during operation.
- When packing gets between the track and sprocket, it causes the tension on the track to increase significantly.
- When the track pitch is increased due to wear, packing can have a positive effect on the overall track pitch.
- Adjusting the track correctly for the operating conditions lengthens track life.
- The rollers on track equipment do not all wear at the same rate equally.
- Never use a shoe that is wider than necessary; the added width causes accelerated wear to the track and undercarriage.

- Track wear can be checked using two different procedures; the use of manufacturer specialized wear gauges and general measuring devices such as tape measures, depth gauges, and vernier callipers.
- To ensure that track adjustments have been performed correctly, the equipment should be located in its actual working environment.
- Correct track adjustment is necessary to maintain long track life with low operating costs.
- A typical off-road tire when inflated to 100 psi can generate the equivalent of 40,000 pounds of pressure on the rim lip of the wheel.
- Off-road rims come in a variety of designs, but can be classified in two categories, one-piece or multi-piece.
- A tight ring gap on an multi-piece rim indicates that the tire should be deflated before removal from the equipment.
- Correct tire inflation pressure is determined by equipment type, tire size, ply rating, operating speed, and tire load.
- A technician must be familiar with the proper restraining devices and mechanical aids used in tire removal.
- The recommended volume of liquid ballast recommended by some manufacturers is approximately 75 percent of the tire's internal volume.
- It may be possible to inflate the tire while it is on the equipment if the air pressure did not fall below 80 percent of its recommended value.

Review Questions

1. On a track loader, which of the following would likely cause excessive wear on one side of the track idlers and rollers?

   A. track misalignment
   B. track too loose
   C. track too tight
   D. excessive pin and bushing wear

2. On a track loader, which one of the following would likely cause the track to jump the sprocket?

   A. track too tight
   B. track too loose
   C. track misalignment
   D. excessive track link wear

3. On a track loader, which one of the following would likely cause excessive wear to the final drive bearing?

   A. track misalignment
   B. track too loose
   C. track too tight
   D. excessive pin and bushing wear

4. Which of the following defines the term *hunting tooth* design when referring to track equipment?

   A. Every other sprocket tooth engages a different track bushing during operation.

   B. Every sprocket tooth engages the same track bushing during operation.

   C. Every other sprocket tooth engages the same track bushing during operation.

   D. Sprockets have an even number of teeth to match an even number of links.

5. When adjusting track sag, which of the following dimensions would be considered normal?

   A. 3.5 inches

   B. 1.5 inches

   C. 0.25 inches

   D. 4.5 inches

6. When pins and bushings wear on tracks, how many degrees do you rotate the pins and bushings to restore track pitch?

   A. 90 degrees

   B. 120 degrees

   C. 180 degrees

   D. 360 degrees

7. What takes place when the track pins and bushings wear?

   A. Track pitch increases.

   B. Track pitch decreases.

   C. Track tension increases.

   D. Track pitch remains constant.

8. Which of the following is an accepted method of measuring pin and bushing wear?

   A. Use a tape measure to measure the distance across several links from one pin edge to the same edge on another pin; then divide that number by the number of links.

   B. Use a tape measure to measure the distance across two links from one pin edge to the same edge on the other pin; then divide that number by the number of links.

   C. Check the wear pattern on the sprocket teeth to determine excessive pin and bushing wear.

   D. Measure the overall track length to determine pin and bushing wear.

9. When pins and bushing wear out on a track, what happens to the wear line on the mating sprocket?

   A. Wear line will be higher.

   B. Wear line will be lower.

   C. Wear line will be the same.

   D. Track tension can compensate for pin wear.

10. On a track loader, what is the purpose of a split master link?

    A. It is required to compensate for track wear.

    B. It is a two-piece link used to separate the track.

    C. It is used to identify the location of the master pin.

    D. It can be adjusted to compensate for track wear.

11. How much force can a typical off-road tire generate when inflated to 100 psi on the rim lip of the wheel?

    A. 10,000 lb

    B. 20,000 lb

    C. 30,000 lb

    D. 40,000 lb

12. What does a tight ring gap on an multi-piece rim indicate?

A. The tire should be deflated before removal.

B. This is considered normal wear on the rim face.

C. The lock ring not properly installed.

D. The tire has excessive air pressure.

13. How much liquid ballast by volume is recommended by some manufacturers in an off-road tire?

A. 40 percent

B. 60 percent

C. 75 percent

D. 90 percent

14. What is the minimum air pressure an off-road tire must have to be filled safely on equipment?

A. 40 percent of its recommended value.

B. 50 percent of its recommended value.

C. 60 percent of its recommended value.

D. 80 percent of its recommended value.

CHAPTER

18 Suspension Systems

Learning Objectives

After reading this chapter, you should be able to:

- Describe the fundamentals of off-road suspension systems.
- Identify the components that make up off-road suspension systems.
- Describe the principles of operation for off-road suspension systems.
- Outline the maintenance and repair procedures associated with off-road suspension systems.
- Identify the basic troubleshooting procedures for off-road suspension systems.

Key Terms

cross tube	load cushion	solid rubber suspension
cylinder pressure	nitrogen-charged accumulators	spring packs
drive bushing	oil/pneumatic cylinder	suspension cylinder
equalizing beam suspension	oscillating hitch	torque rods
jounce	rebound	U-bolt
leaf-spring system	saddle assembly	unsprung weight

INTRODUCTION

The main job of an off-road suspension system (see **Figure 18-1**) is to support the total weight of the equipment. If no suspension system were used, the shock loads encountered during operation would be transmitted directly to the equipment's frame. These jolts would soon cause metal fatigue, resulting in cracks in the frame structure. Equally important is the comfort of the operator, who in many cases must remain at the controls for hours on end. Modern suspension systems play a major role in providing stability to the equipment during operation on rough terrain, as well as absorbing shock loads and providing a comfortable ride for the operator. A suspension system that performs well when both loaded and unloaded generally has to compromise some performance at both ends of the spectrum. A well-designed suspension system will perform well with or without a load.

Off-road suspension systems come in many designs. Some are as simple as those found on many types of slow-moving LHDs (load haul and dump equipment), which consist of drive axles that are mounted directly to the frame with only the tires to absorb ground impact. Conventional **leaf-spring systems** are still quite common on many haulage truck

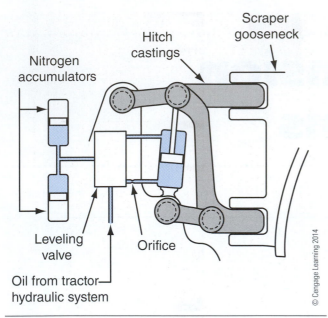

Figure 18-1 Caterpillar cushion hitch suspension system used on a 637E scraper.

applications in pits and quarries. Complex systems may incorporate nitrogen-filled shocks that are monitored and regulated by the operator, or an onboard ECM (Electronic Control Module) that monitors and controls the operation of the suspension system.

Terminology associated with heavy-duty suspension system design and performance uses terms such as, sprung and unsprung weight, jounce, rebound, and oscillation. **Unsprung weight** is the total weight of the axle plus all the components that are mounted to it, including the brake hardware and suspension components. Sprung weight is defined as the load that the suspension must support during normal operation. The term **jounce** is used to identify a spring when it is in its most compressed state. Some manufacturers refer to the rubber blocks used to prevent axle-to-frame contact as "jounce blocks." The term **rebound** describes what happens to the potential energy that is stored in a compressed spring; it kicks back. On many types of equipment, jounce and rebound are controlled with dampening devices such as shock absorbers and springs.

Suspension work on off-road equipment often involves the use of heavy lifting devices because the suspension system components are usually heavy. When working with **suspension cylinders**, be mindful that they may be under pressure and can move suddenly without warning. Never place yourself in a position that could cause you to become pinned between a tire and the frame by the sudden movement of a suspension component.

When working on suspension systems that use gas/suspension cylinders, ensure that the nitrogen in the cylinders has been discharged before checking the oil level. Never remove any plugs or valves from a suspension cylinder if the rod is not fully retracted. Do not work under a suspended load without ensuring it is properly blocked. When a jack or hoist is to be used, ensure that it has the rated capacity to lift the load. When working around heavy equipment, always wear the required personal protective equipment based on job conditions.

> **CAUTION** *The equipment that you work on will often require you to climb up to a location where the work will actually be performed. Always use the three-point contact rule when mounting or dismounting the equipment: Maintain contact with two hands and one foot, or two feet and one hand. Never attempt to mount or dismount moving equipment. Do not carry tools with you when mounting or dismounting the equipment. Use a hand line or other safe means to pull the tools and components up to where you are working.*

Use only approved lifting devices when hoisting components, and follow the correct procedure. **Figure 18-2** shows what happens to the load capacity of a chain or sling when it is used at different angles.

- **Figure 18-2A** shows the load rating capacity of the chain in a vertical position. Load capacity is 100% of its workload rating.
- **Figure 18-2B** shows the load rating capacity of the chain at a 60-degree angle from vertical. Load capacity is reduced to 86% of its workload rating.
- **Figure 18-2C** shows the load rating capacity of the chain at a 45-degree angle from vertical. Load capacity is reduced to 70% of its workload rating.
- **Figure 18-2D** shows the load rating capacity of the chain at a 30-degree angle from vertical. Load capacity is reduced to 50% of its workload rating.

Exercise caution when working with hydraulic suspension systems under high pressure. A technician is often required to locate hydraulic leaks on these systems. When checking for hydraulic leaks, never run your hands over the lines and components. High-pressure, pin-hole leaks can penetrate the skin leading to serious injury. If oil penetrates the skin, immediate medical attention is mandatory. In addition, remember that oil operating temperatures on hydraulic suspension

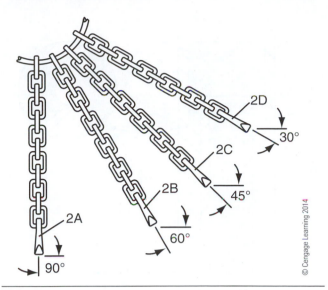

Figure 18-2 Lifting devices used at different angles will affect the load lifting capabilities. Refer to load lifting charts to determine the correct slings and angles.

Figure 18-3 How the suspension cylinder is incorporated into the front suspension in a typical haulage truck.

systems can exceed 212°F (100°C) which can cause severe burns.

Appropriate protective equipment must be used when handling hot oil or components. Hydraulic tanks are often pressurized to prevent dirt contamination and require special procedures to check the oil level safely. On equipment that uses high-pressure nitrogen in the suspension systems, spraying a soap and water solution is the only safe way to locate a leak. Compressed nitrogen has the potential to cause serious injuries.

FUNDAMENTALS

Suspension systems on off-road equipment are designed to absorb road and load impacts by attempting to isolate them from the frame. Off-road haulage trucks often use independent suspension cylinders that are capable of independent rebound rates, as shown in **Figure 18-3**. This allows the load stresses to be absorbed by the suspension instead of the frame. Off-road equipment operating on rough terrain benefits from independent suspension systems that permit axles to absorb lateral, torsional, and bending forces.

Suspension systems in off-road equipment have distinct categories: Independent oil/pneumatic suspension systems are used on haulage equipment, and solid axle suspension systems are used on LHD equipment. These systems will be discussed in depth in this chapter. Designs differ from manufacturer to manufacturer, but share basic operating principles.

Oil/Pneumatic Suspension Systems

Equipment that is used for haulage operation in pits and quarries often uses independent oil/pneumatic suspension systems. These systems on newer haulage trucks are complex and require a good understanding of both hydraulics and electrical/electronics fundamentals to service them. This family of large trucks use a suspension cylinder at each front wheel end (instead of a conventional kingpin) as part of the steering linkage. The **oil/pneumatic cylinders** also act as shock absorbers. The front suspension cylinder barrels are attached to the frame of the equipment at the top and connect to the steering knuckle at the bottom, as shown in **Figure 18-4**. When the equipment is in motion, the front wheels are permitted to move up and down as the terrain changes.

The front suspension cylinders contain nitrogen gas in chamber 2 (see **Figure 18-5**), which reacts to wheel movement during operation. As the wheel moves up, the gas charge in chamber 2 is compressed, forcing oil to flow to chamber 6 through orifice 4 and ball check 5. When the front suspension cylinder moves down during operation (see **Figure 18-6**), the compressed nitrogen gas forces the rod to move down. The oil flows from cavity 6 through orifice 4 to chamber 2. At the same time, the increase in pressure forces check ball 5 to close off the port, forcing all the oil to flow through orifice 4 on its way to chamber 2. Orifice 4 is

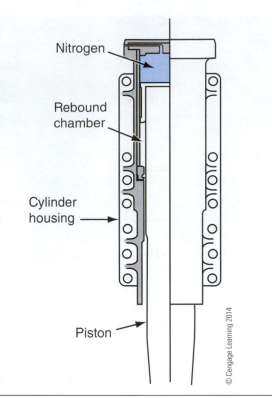

Nitrogen

Rebound chamber

Cylinder housing

Piston

© Cengage Learning 2014

Figure 18-4 Cutaway of a suspension cylinder in a typical haulage truck.

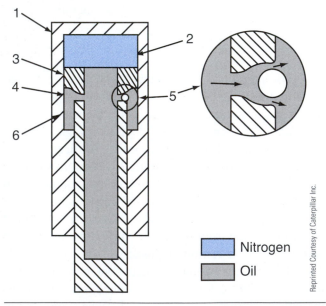

Nitrogen

Oil

Reprinted Courtesy of Caterpillar Inc.

Figure 18-5 Illustration shows a cutaway of a front suspension cylinder with the rod moving up. 1. Cylinder, 2. Nitrogen chamber, 3. Cylinder rod, 4. Orifices (2), 5. Ball check, 6. Cylinder.

gradually closed off as the rod extends downward to prevent the rod from bottoming out in the cylinder.

The rear suspension is similar to the front in that it also uses oil/pneumatic suspension cylinders. The cylinder barrels are connected to the equipment frame

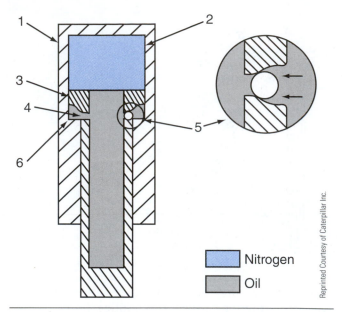

Nitrogen

Oil

Reprinted Courtesy of Caterpillar Inc.

Figure 18-6 Cutaway of a front suspension cylinder with the rod moving down: 1. Cylinder, 2. Nitrogen chamber, 3. Cylinder rod, 4. Orifices (2), 5. Ball check, 6. Cylinder.

at the top with the stem end of the cylinder mounted to a ball joint at the front of the rear axle housing. This allows the suspension cylinder to pivot on the differential housing during operation. In other words, the set-up is opposite from the arrangement used on the front suspension. The rear suspension cylinders support the entire weight of the rear of the haulage truck, including the load.

During operation, the rod is able to react to terrain changes and move up and down in the housing. The compressed nitrogen in cavity 2 provides a cushion during operation, while the ability to absorb shock loads is controlled by orifice 5. Check ball 6 controls the flow rate of the oil between chambers 2 and 4. When one of the wheel ends of the rear axle moves up during operation, the cylinder housing is forced to move up on the rod (see **Figure 18-7**). As a result of this motion, the nitrogen pressure in cavity 2 increases rapidly to provide a cushioning action. This increase in nitrogen pressure also forces the oil in cavity 2 to flow to cavity 4 by way of orifice 6 and ball check 5.

When the rear suspension cylinder housing changes direction and moves down, as shown in **Figure 18-8**, the combined weight of the vehicle and load along with the nitrogen in cavity 2 forces the rod to move out. The oil in cavity 4 is forced to flow to chamber 2 through orifice 6 as check valve 5 closes due to reverse oil flow. Because of the restrictive flow of oil from cavity 4 to cavity 2, a shock absorber action is

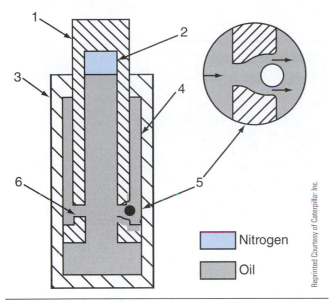

Figure 18-7 Cutaway of a rear suspension cylinder with the cylinder moving up: 1. Rod, 2. Nitrogen chamber, 3. Cylinder, 4. Cavity (2), 5. Ball check, 6. Orifices (2).

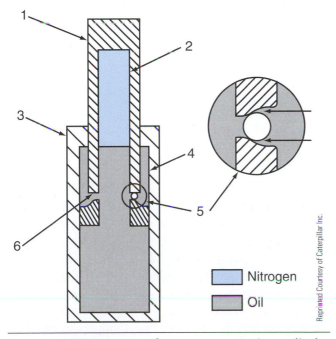

Figure 18-8 Cutaway of a rear suspension cylinder with the cylinder moving down: 1. Rod, 2. Nitrogen chamber, 3. Cylinder, 4. Cavity (2), 5. Ball check, 6. Orifices (2).

produced in the cylinder. Oil flow from chamber 2 to cavity 4 is much faster during down movement of the rod, and slows down considerably in the opposite direction because ball check 5 closes (producing a much slower flow rate).

ARTICULATING TRUCK SUSPENSION SYSTEM

Large articulating haulage trucks, as shown in **Figure 18-9**, often integrate the suspension system circuit into the vehicle's hydraulic system which, in our example, provides oil to the hoist, steering, brake, and fan drive circuits. A common hydraulic tank provides oil for all hydraulic circuits on the equipment. On older models, accumulators were used to supply stored hydraulic pressure for suspension pressure. The system was a closed circuit and isolated from the main hydraulic system by means of control valves. Newer equipment often uses a similar circuit, shown in **Figure 18-10**, that is able to utilize excess oil from a brake or hydraulic circuit to control ride height with the use of directional control valves.

Refer to **Figure 18-11** for the following suspension circuit identifiers. The suspension frame is connected to the front axle on both sides by mounting bolts. The suspension frame is mounted to each side of the **oscillating hitch** by a pin on the upper side at both ends of the front axle. At the bottom it is attached to the front frame by the lower suspension cylinders' mounting points along with the rod. Pins, located at the lower end of the suspension cylinders, connect the cylinders to the suspension frame. Through the use of suspension frame pivots connected to the oscillating hitch, independent vertical motion can be achieved. The dampening action of the suspension cylinders is achieved using **nitrogen-charged accumulators**.

Suspension System Hydraulic Circuit Operation

In order to raise the suspension system to the desired operational height, two control valves must be activated, as shown in **Figure 18-12**, with the engine running so that oil from the main hydraulic circuit can flow to the suspension cylinders. Once the correct ride height is reached, the operator closes control valves 9, which isolates the suspension cylinders so that they can operate independently. To lower the ride height, the operator opens control valve 10, also shown in **Figure 18-12**, which drain off oil from the suspension cylinders to lower the ride height. The nitrogen charge in the accumulators acts like a shock absorber by cushioning the ride during operation. As the load on the suspension cylinders changes with vertical suspension movement, the pressure in each independent circuit varies up or down, depending on which way the suspension cylinder is moving. The gas in the suspension cylinders increases when compressed during

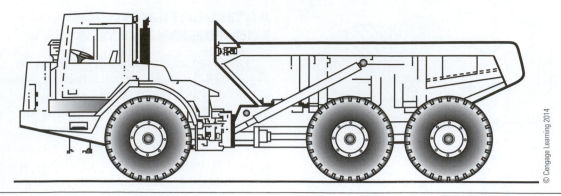

Figure 18-9 Articulating off-road haulage truck.

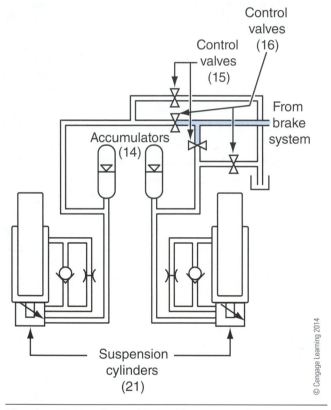

Figure 18-10 Illustration of a typical suspension circuit on an articulating truck.

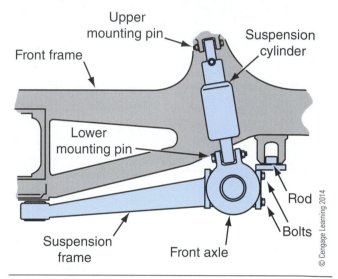

Figure 18-11 Typical suspension setup on an articulating haulage truck.

piston-up travel and decreases during piston-down travel. If a leak develops in the nitrogen side of the accumulator, a change in ride height on the side where the nitrogen leak has occurred will result.

Independent Gas-Charged Suspension Cylinder Maintenance

Many haulage trucks come equipped with independent gas suspension systems. Each suspension cylinder works independently so if one cylinder develops a leak, it does not affect the operation of the other three suspension cylinders. Gas-charged suspension systems minimize road shock transfer to the truck frame. Each wheel is connected to the truck frame by a gas/hydraulic suspension cylinder. The suspension cylinders act as kingpins for the steering system on the front axle. The suspension cylinders also provide the spring and shock absorber action for the rear suspension system and dampen suspension oscillations.

Key advantages to independent gas/hydraulic suspension systems are their simplicity of design and their ability to absorb changing load and road conditions. Gas/hydraulic suspension systems require little maintenance or adjustment when compared to the older conventional leaf-spring systems that they are rapidly replacing. Gas/hydraulic suspension systems along with operator seat suspensions provide operators with a smooth ride and minimal driver fatigue.

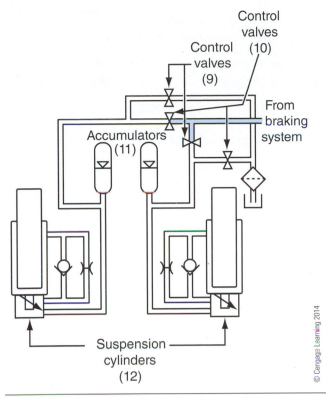

Control
valves
(10)

Control
valves
(9)

From
braking
system

Accumulators
(11)

Suspension
cylinders
(12)

© Cengage Learning 2014

Figure 18-12 Typical suspension hydraulic circuit used on a CAT 725 articulating truck.

CAUTION *Although nitrogen is not flammable, it can cause serious injuries due to the high pressures that are used so observe manufacturer's service literature when working on gas cylinders. Do not check the oil level in a suspension cylinder until all the nitrogen has been bled off. Never remove any valves or plugs from the suspension cylinder unless the rod is in the fully retracted position.*

Before making adjustments to a suspension system, the technician must consider that the pressures in each suspension cylinder can be affected by changes in ride height due to leaks at opposite ends of the equipment. For example, when the left rear suspension cylinder ride height is low, it is likely that the right front suspension cylinder is riding high because of the shift in load weight. A close visual inspection of each cylinder is necessary to determine the source of the problem. Look for oil leaks at the connection points on the suspension cylinders, valves, and the system hoses and pipes. If the vehicle is equipped with an oil distribution manifold and corresponding pressure sensors, check for leaks around them. Common sources of leaks are the O-rings used on the distribution manifold. Nitrogen

may have leaked out of the suspension cylinders, which will affect ride height. Locating external leaks is often difficult and requires the use of a soap and water solution. Check for a loose valve body which can leak.

When checking suspension cylinder operation, look for excessive cylinder stem travel. If the chromed area of the stem is clean, this may indicate excessive travel. Discuss the ride quality with the operator to determine if there has been any noticeable change. With the equipment unloaded, check the strut height to determine whether the struts are fully extended; this usually indicates a suspension problem. To correctly diagnose a suspension system problem, the technician must remember that suspension cylinders operate independently of each other. Suspension cylinders should be serviced and charged in pairs and cannot be serviced individually.

Conditions for servicing suspension cylinders may vary somewhat from machine to machine, but generally will follow the following guidelines. Service is required when:

- A suspension cylinder shows signs of external leakage.
- The weight of the truck changes substantially (more than 5,000 pounds).
- The struts are fully extended when the truck is unloaded.
- The operator complains of a rough ride.
- **Cylinder pressure** varies by more than 50 psi (345 kPa) from cylinder to cylinder.
- The ride height varies more than 0.500" (1.27 cm) from the specified ride height of the equipment.

These dimensions are temperature sensitive so the air temperature difference between a shop or building where the adjustments are performed must be no greater than 20°F. **Table 18-1** lists examples of dimensions that would typically be used on haulage trucks. These dimensions are equipment specific; consult the service literature for the particular type of equipment that is to be adjusted.

In some cases, a manufacturer will recommend that a gauge block be fabricated to check the measurements during the height adjustment procedures. An example of a typical drawing showing the fabrication dimensions is shown in **Figure 18-13**. Measurements depend on the manufacturing brand of the components. Refer to the equipment manufacturer's service literature for specific instructions on the tools required to perform this adjustment procedure.

Table 18-2 is an example of a manufacturer's list of tools required to charge suspension system components on a typical haulage truck, specific to the example discussed in this chapter.

TABLE 18-1: TYPICAL DIMENSIONS FOR HAULAGE TRUCKS

Gauge Block Dimensions (Front Suspension Cylinders)

FT Gauge Block	Tractor or Truck	Gauge Block Length A	Gauge Block Length B
1680	771C, 775B	8.50 in. (216.0 mm)	6.69 in. (170.0 mm)

Rear Suspension Cylinder Nitrogen Charge Dimensions (Distance Below the Index Line)

Quarry Truck	First Charge Line	Second Charge Line
771C	5.0 in. (128 mm)	5.5 in. (140 mm)
775B	4.9 in. (124 mm)	5.4 in. (137 mm)

© Cengage Learning 2014

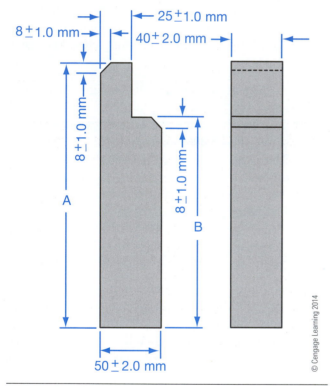

Figure 18-13 CAT suspension cylinder height adjustment tool dimensions (refer to Table 18-1 for values).

© Cengage Learning 2014

Adjusting Rear Gas-Charged Suspension Cylinders

CAUTION *When releasing the pressure in the rear suspension cylinders, the ride height will decrease quickly so ensure that you will not be trapped between the tire, fender, or frame.*

Because both rear cylinders will be charged together, remove both charge valve caps. Install a special

TABLE 18-2: EXAMPLE OF TOOLS REQUIRED FOR NITROGEN CHARGING EQUIPMENT

Part No.	Consists of	Item Description	Number Needed
FT1680		Gauge	2
7S5437		Nitrogen charge group	1
	1S8941	Hose assembly	1
	7S5106	Chuck	1
	7S8713	Gauge	1
	8S4600	Nipple	2
	8S1506	Coupling assembly	2
	1J3914	Hose assembly	1
	7S8712	Gauge	2
	7S8714	Gauge	1
	2D7325	Tee	1
	1S8937	Valve	1
	2S5244	Nipple	1
	8S4599	Coupling assembly	1
	8S1505	Regulator assembly	1
	5P3041	Nipple assembly	1
5P8610		Auxiliary fitting group	
	1S8941	Hose assembly	1
	7S5106	Chuck	1
	8S7169	Coupling	2
	2J9803	Hose assembly	2
	1S8937	Valve	2
	5P8998	Pipe nipple	2
	2D7325	Tee	1
	3D8884	90° elbow	1

Note: Auxiliary Fitting Group 5P8610 is used with 7S5437 when charging two cylinders at the same time. The 7S8712 gauge is used to check the cylinder pressure before charging.

© Cengage Learning 2014

chuck (7S5106, listed in **Table 18-2**) on each of the charge valves as shown in **Figure 18-14**. Release all the nitrogen and oil from the suspension cylinder by turning the check valve clockwise. Once the cylinders have bottomed out, allow pressure in the suspension cylinders to equalize by leaving the check valves open for approximately five minutes. Once the pressure has equalized, close the check valves by turning counter-clockwise. Before attaching the oil fill unit to begin the fill cycle, make sure the oil lines are full of oil by cycling the oil refill unit with the ends of the lines submerged in oil.

Before the charging procedure can begin, a rule must be placed on the cylinder, as shown in **Figure 18-15**, so that the cylinder extension can be measured accurately. A suitable magnet, such as a base for a dial indicator, should be used to hold the rule in place. The rule should be placed parallel to the centerline of the suspension cylinder, as shown in **Figure 18-15**. A mark must be placed on the rule (line 3) that corresponds to the top edge of the suspension cylinder head. Another line (5) must be placed 1 inch below line 3 on the rule. Once this is done, mark a corresponding line on the cylinder (line 4) that matches line 5. Using **Table 18-3** to calculate the dimensions, mark two additional lines (lines 6) below line 5 (**Figure 18-15**). Remember, if the temperature difference is greater than 20°F, then a correction must be made. For every 10°F, add 0.100" (2.54 mm) to the dimension from the example shown in **Table 18-3**.

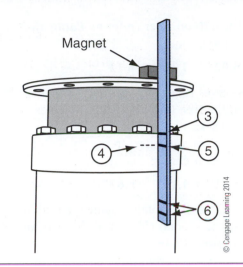

Figure 18-15 CAT rear suspension cylinder showing the location of the rule and marks.

TABLE 18-3: EXAMPLE OF NITROGEN CHARGE DIMENSIONS		
Rear Suspension Cylinders (Distance Below the Index Line)		
Quarry Truck	**First Charge Line**	**Second Charge Line**
771C	5.02 in. (127.5 mm)	5.52 in. (140.2 mm)
775B	4.87 in. (123.7 mm)	5.37 in. (136.4 mm)

To calculate the correct charge dimension when the total temperature difference is greater than 20°F, divide the distance to be added (in this case 0.100") by the temperature increment indicated in the text (10°F) and multiply the result by the Total Temperature Difference (as indicated in the following example).

Note: In the first calculation, we will use Standard values. Estimate temperatures to the nearest 10°F.

Example for 771C Quarry Truck Rear Suspension

- Shop ambient temperature: 70°F
- Outdoor ambient temperature: −40°F

Total Temperature difference between 70°F and −40°F = 110°F

Applying the formula indicated previously,

$$110°F \times (0.10"/10°F) = 1.10"$$

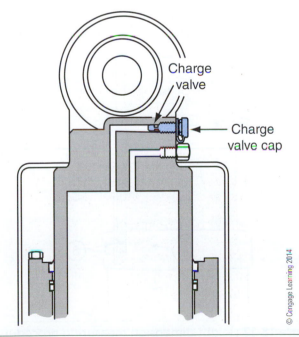

Figure 18-14 CAT rear suspension cylinder showing the location of the charge valve.

Total length on ruler (refer to **Table 18-3**) for first nitrogen charge line:

$$5.02" + 1.10" = 6.12"$$

(This gives the total distance required between line 3 and the first (upper) of the two lines 6 indicated in **Figure 18-15**.)

Total length on ruler (refer to **Table 18-3**) for second nitrogen charge line:

$$5.52" + 1.10" = 6.62"$$

(This gives the total distance required between line 3 and the second (lower) of the two lines 6 indicated in **Figure 18-15**.)

Mark the results from these calculations on the rule located on the suspension cylinder before beginning the charging process.

Rear Suspension Cylinder Oil Charge Procedure

Once the calculations have been made and the data transferred to the rule on the suspension cylinders, the oil refill unit must be connected to the suspension cylinders as shown in **Figure 18-16**. Before starting the charging procedure, make sure the gate valves are open on the lines, and then slowly open the check valves by turning them counterclockwise. Adjust the air pressure connected to the oil refill unit to 125 psi (860 kPa). Observe the movement of the cylinders; if one cylinder begins to travel at a faster rate than the other, close the gate valve to the faster cylinder so that it does not extend beyond line 5 on the rule. Continue filling the other cylinder until it reaches line 5 in **Figure 18-15**. When both cylinders (line 5) are even with the index (line 4) the oil level is correct. Close the check valves on the cylinders and close the air and gate valves before disconnecting the oil lines.

Figure 18-16 CAT oil refill unit connected to the rear suspension cylinders.

Rear Suspension Cylinder Nitrogen Charge Procedure

CAUTION *Nitrogen is the only gas that should be used to charge suspension cylinders. Gases such as oxygen must never be used because oxygen in the presence of flammable materials can be explosive. Nitrogen and oxygen cylinders can look similar so do not get them mixed up. The fitting used to connect the nitrogen bottles to the suspension cylinders should never be used on other types of compressed gas bottles such as oxygen.*

The following procedures are intended for explanation purposes only. Always refer to the specific service literature for the equipment to be serviced.

Install the quick disconnect couplings used to charge the suspension cylinders with oil to the nitrogen charge lines of the nitrogen charge kit. Recheck the line 4-to-line 5 alignment on each cylinder to make sure it is still lined up correctly before beginning the nitrogen charge procedure. For this next step, the rear suspension cylinders must be charged with nitrogen until line 4 is lined up with line 6 (see **Figure 18-17**). Before connecting the charge lines to the cylinders, make sure that the charge valves are fully closed; this prevents the possibility of oil from the suspension cylinders, which is under pressure, getting into the nitrogen charge lines.

Once the previous step has been completed, connect the charge lines to the rear suspension cylinders, as shown in **Figure 18-18**. Adjust the nitrogen regulator located on the charge bottle to 350 psi (2400 kPa). Slowly open the gate valves, and turn the chucks clockwise on

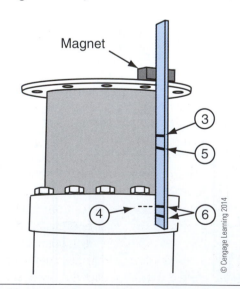

Figure 18-17 Height dimension measuring procedure used to charge rear suspension cylinders on a CAT quarry truck.

Figure 18-18 CAT nitrogen charge unit connected to the rear suspension cylinders.

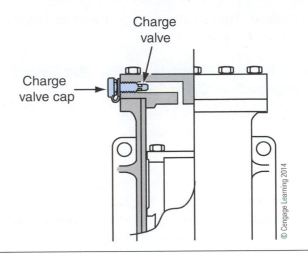

Figure 18-19 Location of the charge ports on the front suspension cylinders on a CAT quarry truck.

both suspension cylinders to begin the charging process. Continue to charge the suspension cylinders until line 4 aligns with line 6, as shown in **Figure 18-17**.

If the cylinders do not reach the correct height at the same time, close the gate valve on the suspension cylinder that has reached the correct height dimensions first and allow the other cylinder to charge until the correct height is reached. Once charging dimensions have been reached, close both gate valves before closing the check valves at the suspension cylinders. Remove the nitrogen charge hoses from the suspension cylinders and install the protective caps. Torque the caps to the recommended specification outlined in the service literature and remove the steel rules and magnets.

Front Suspension Cylinder Charge Procedures

Although the charge procedures are similar on the front and rear suspension cylinders, there are some differences. The location of the charge ports and the measuring techniques used on the front suspension cylinder are not the same as those used to adjust the height of the rear cylinders. To bleed off the oil in the front suspension cylinders, remove the charge valve caps, as shown in **Figure 18-19**, and install the chucks from the oil charge unit to the charge valves of both suspension cylinders.

Install a suitable drain line to each connection so that the oil is drained into an appropriate container. While standing in a safe location, slowly open the check valves on the charge valves by turning in a clockwise direction, allowing the oil and nitrogen to bleed off until the cylinder reaches the bottom of its travel. Note that

pressurized oil and escaping nitrogen gas can pose a safety hazard to anyone who comes in contact with either one. Give the pressure in the cylinders a chance to equalize by allowing the valves to remain open for approximately five minutes before closing them. Once this step is complete, place the charge line ends in a suitable container and cycle the oil refill pump until the lines are full of charge oil. Reattach the lines to the charge valves on the front suspension cylinders.

The rule and its corresponding height adjustment lines are not the same for the front suspension cylinders and rear suspension cylinders. **Figure 18-20** shows the location of the rule with the appropriate scribed lines in place. However, the magnet used for the rear suspension cylinder measurements can be reused for the front cylinders. Mark up line 3 on the rule to correspond with the top edge of the suspension cylinder's head. Place a second mark (line 4) on the cylinder head edge. A third line (5) must be placed 1 inch (25.4 mm) in this case below line 3 on the rule.

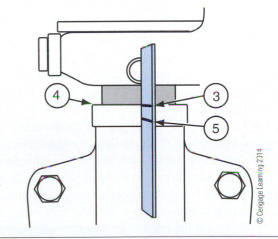

Figure 18-20 Height dimension measuring procedure used to charge the front suspension cylinders.

Once these steps are complete, you are ready to begin the oil charge procedure steps.

Oil Charge Procedure for the Front Suspension Cylinders

The following procedures are intended to be used for demonstration purposes only. Always refer to the OEM service literature for the equipment that you will be servicing.

The procedure is similar to that used to charge the rear suspension cylinders. The same specialized tools are required to perform this next step. Attach the check valves and lines from the oil charge unit to the front suspension cylinder charge valves. Adjust the air pressure to 125 psi (860 kPa) and make sure the gate valves are in the open position.

Open the check valves slowly by turning in a clockwise direction. Continue injecting oil until line 5 reaches line 4. If one cylinder reaches the line first, close the gate valve to that cylinder and allow the remaining cylinder to fill until it reaches the correct height. Close the gate valves and regulated air pressure before bleeding the air pressure to zero. If required, reuse of the quick disconnects is allowed to charge the suspension cylinders with nitrogen if they are not included with the nitrogen charge kit.

Nitrogen Charge Procedure for the Front Suspension Cylinders

The use of gauge blocks to set the correct ride height on the front suspension cylinders in this example is shown to clarify the steps required to perform this procedure. **Figure 18-21** shows an example of gauge blocks that are used to set the ride height. This requirement may vary depending on the manufacturer and model of equipment. Suspension dimensions will be affected by temperature differences between the adjustment location and the operating location. For example, if the temperature difference between the shop where the adjustment is performed and the outside temperature where the equipment is operating is greater than 20°F, a different calculated dimension is required for extra shim thickness and must be calculated based on the temperature difference. An example of this calculation is provided for a typical haulage truck but the procedures and dimensions for different manufacturer's equipment or even different models by the same manufacturer can vary substantially.

Gauge blocks are used when the total difference between the shop and the outside ambient temperatures is 7°C or less. When the total temperature difference is greater than 7°C, the nitrogen charge dimensions should

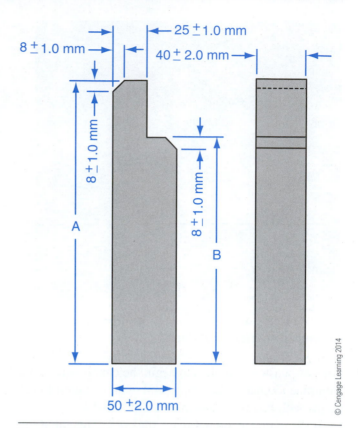

Figure 18-21 Dimensions to fabricate the gauge used to measure the front suspension cylinder height on a CAT quarry truck.

be adjusted by adding shims of a predetermined thickness (in this case, 3.3 mm). Calculate the correct charge dimension in the same way as described for rear suspensions: when the temperature difference is greater than 7°C, divide the distance to be added (in this case 3.3 mm) by the temperature increment indicated in the text (7°C) and multiply the result by the Total Temperature Difference (as indicated in the following example).

Note: When the temperature difference is greater than 7°C, the nitrogen charge dimensions should be adjusted by a correction amount by adding additional shims.

When the difference between the outside ambient temperature and temperature in the shop is greater than 7°C, then add 3.3 mm of shim thickness to each gauge block for every 7°C difference in temperature.

Example for 771C Quarry Truck Front Suspension

Note: Estimate temperatures to the nearest 7°C.

- Shop ambient temperature: 20°C
- Outdoor ambient temperature: −40°C

Total Temperature Difference between −40°C and 20°C = 60°C

Apply the formula indicated previously to determine needed shim thickness:

60°C × (3.3 mm/7°C) = 28 mm

Gauge block (refer to **Table 18-1**) plus shim thickness:

170 mm + 28 mm = 198 mm

This gives the total distance that should be used for Gauge Block length B (see **Table 18-1**) when adjusting the front suspension to the correct ride height.

Procedures. The following procedures are intended to be used for example purposes only.

Ensure that the charge valves are fully closed before connecting the nitrogen charge lines to the suspension cylinders to prevent oil from entering the nitrogen charge lines. The first step is to adjust the nitrogen regulated pressure to 600 psi (4,150 kPa). Next, open the gate valves on the high-pressure lines, allowing the nitrogen to flow to the charge valves. Slowly open the charge valve check valves by turning clockwise on both suspension cylinders to allow nitrogen to enter the cylinders. If one of the cylinders extends faster than the other and has reached slightly higher than the correct height, close the gate valve to that cylinder and continue to allow the other cylinder to reach the same height before closing the remaining gate valve. In this example, continue to inject nitrogen into the suspension cylinders until they reach the same height (6.69 in.).

Close off both gate valves first, and then close the charge valves on the suspension cylinders. Remove the lines at the quick disconnects. With the gauge blocks and shims in place, slowly open the charge valves to release some of the nitrogen until the cylinder comes to rest on the gauge block. At this point, close the charge valve.

Reconnect the nitrogen charge lines to the suspension cylinders and adjust the regulated nitrogen pressure to 350 psi (2,400 kPa). Once the previous steps have been completed, open the gate valves and charge valves and allow the nitrogen to flow into the suspension cylinders for approximately five minutes to equalize the pressure. After the pressure has been equalized, close the charge valves and gate valves and close the nitrogen supply at the tank.

Disconnect the nitrogen charge kit from the suspension cylinders, reinstall the protective valve caps, and torque to specifications. If the gauge blocks are not easy to remove, it may be necessary to start the vehicle, raise the truck body, and steer the wheels from side to side to facilitate removal.

REUSE OF GAS/HYDRAULIC SUSPENSION SYSTEM COMPONENTS

The suitability to reuse components in a suspension system can be determined by following the manufacturer's service literature. Welding and re-machining of the seal lands is often done to the head of the rear suspension cylinder housing. The rod assemblies are often sent out for re-chroming or polishing to restore the finish. The cylinders themselves can often be honed out or re-tubed. Follow the guidelines provided by the manufacturer to determine the reusability of suspension components.

When determining the reuse of suspension cylinders, take into consideration the overall length of the cylinder, the bore dimensions, the bore finish, and the internal diameter of the connecting eyes. The cylinder rods must meet specific tolerances as well. The overall length of the cylinder rod as well as the outside diameter of the rod must be within specifications. The rod eye dimensions must be within specifications; often the eye will have a replaceable bushing.

LOAD, HAUL, AND DUMP RIDE CONTROL SUSPENSION SYSTEMS

Off-road load, haul, and dump equipment, as shown in **Figure 18-22**, uses only the tires to absorb road shock. Because of the slow operating speeds of this type of equipment and the specific work that this equipment performs, conventional suspension systems are not necessary. The front axles (bucket end) on this type of equipment are bolted solidly to the frame with the only shock absorbing capability coming from the tires themselves.

Reprinted Courtesy of Caterpillar Inc.

Figure 18-22 Typical load, haul, and dump vehicle.

The rear axles are bolted to an oscillating axle or rear axle trunnion bearing support, as shown in **Figure 18-23**, which pivots on large bushings running lengthwise along the length of the equipment. This allows the equipment to maintain road contact during operation on uneven ground.

Some manufacturers offer an option for a load ride control. This allows the loader bucket to ride on a column of oil in the hoist cylinders that are connected hydraulically to an accumulator through a control valve, allowing road shock to be absorbed by the accumulator. A control switch located on the dump hoist control lever is used to activate the ride control circuit.

Figure 18-24 shows the ride control circuit in the inactive position, solenoid 7, which is a normally closed valve blocking the flow of oil from the pilot circuit to the diverter valve 8. When the lift control valve solenoid is energized, and the lift control valve is in the hold position, pilot oil flows to the diverter valve, forcing the spool inside to shift to the left. With the diverter valve in this position and the control valve in the hold position, the oil from the head end and the rod end of the lift cylinder flows through the right side of the diverter valve. The oil from the head end of the lift cylinder flows directly into accumulator 6. The rod end oil returns to the tank through the diverter valve.

When the equipment is operated on rough ground with a full bucket, the combined weight of the bucket and load act on the oil in the lift cylinder head end, which is open to the accumulator. The nitrogen pre-charged accumulator absorbs the shock loads, acting like a shock absorber. The rod end receives makeup oil since it is open through the diverter valve into the lift cylinders.

Figure 18-23 Typical loader with rear oscillating axle during operation on uneven ground.

RIDE CONTROL SUSPENSION SYSTEM TESTING AND ADJUSTING

The ride control circuit can be serviced by using the following procedure. Start the equipment and switch on the ride control option. With the joystick control lever, lower the bucket to the ground and place the lever in the float position. This step relieves the hydraulic pressure in the ride control accumulator. Once this step is completed, the equipment can be shut down and the hydraulic lines to the accumulator can be safely removed.

If the accumulator needs to be replaced, open the gas discharge valve on the accumulator one turn to release all the compressed nitrogen gas in the accumulator gas chamber.

Charging the Accumulator

The following procedures are used for example purposes only. Refer to the specific equipment service literature for the actual service procedures. Be sure that the charge valve is fully closed before connecting the nitrogen charge lines to the accumulator to prevent oil from entering the nitrogen charge lines.

The equipment must be started, warmed up, and left running to perform this procedure. To charge the accumulator with dry nitrogen, perform the following steps in order:

1. Locate the ride control accumulator and remove the guard that protects the gas charge valve and remove the cap.
2. Before you can attach the nitrogen charge adapter to the valve, the T-handle must be turned all the way out (counterclockwise) before connecting the chuck to the accumulator.
3. Connect the chuck (5) to the accumulator charge valve (see **Figure 18-25**).
4. This step requires the gas in the accumulator to be discharged to the atmosphere. Connect the quick disconnects (3) and (13) together and place the open end of the hose in such a way that the discharge will not harm anyone.
5. Open the T-handle (4) and valve (1) and raise the lift arms of the equipment. This forces the gas in the accumulator to be expelled to the atmosphere. Be sure that the open end of the hose is not pointed toward anyone.
6. Place the needle valve (1) in the open position. Turn the T-handle (4) on the gas chuck (5) clockwise all the way in.
7. Ensure that no one is positioned near the bucket end of the equipment and that adequate

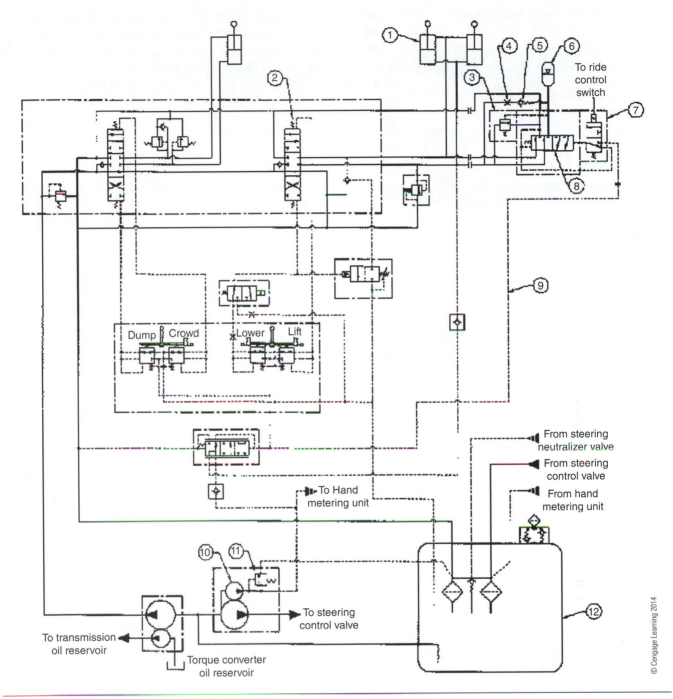

Figure 18-24 Ride control circuit used on a load, haul, and dump loader.

headroom is available to raise the bucket to its maximum height. Lift arms must be raised to their maximum height and held for five seconds. This step ensures that all the gas has been vented from the accumulator and the accumulator piston has moved all the way up.

8. Close the needle valve (1) after all the gas pressure in the accumulator has been released to the atmosphere. Place the joystick in the hold position. Lower the bucket to the ground, place

the joystick in the float position, and move the ride control switch to the on position.

9. Connect the quick disconnect couplings (10) and (11) and attach the hose to the nitrogen cylinder.

10. Adjust the nitrogen pressure (9) to 300 psi ± 5 psi (2,070 kPa ± 35 kPa). This pressure is based on an ambient temperature of 70°F (21°C). To compensate for temperature differences, consult the service literature for the equipment being serviced.

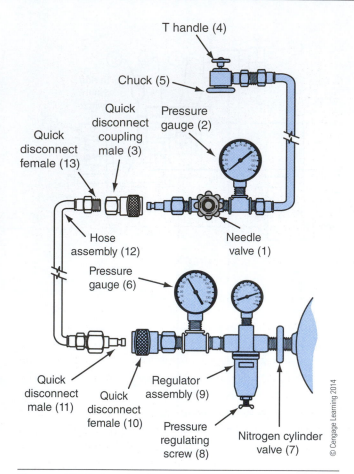

T handle (4)

Chuck (5)

Quick disconnect coupling male (3)

Pressure gauge (2)

Quick disconnect female (13)

Hose assembly (12)

Needle valve (1)

Pressure gauge (6)

Quick disconnect male (11)

Quick disconnect female (10)

Regulator assembly (9)

Pressure regulating screw (8)

Nitrogen cylinder valve (7)

© Cengage Learning 2014

Figure 18-25 Hardware required to charge the ride control accumulator.

Note: If the equipment has been in operation, the temperature of the accumulator may be higher than the ambient temperature in which case, use the accumulator temperature.

11. Open the needle valve (1) and observe the time it takes the gauge to reach charge pressure. If the gauge reads the charge pressure instantly, the gas charge was not properly relieved and all of the previous steps must be repeated. If the gauge takes 10 to 20 seconds to register the charge pressure, then the gas has been successfully removed from the accumulator.

12. Do not remove the line from the accumulator before turning off the nitrogen gas supply. This ensures that the accumulator piston is bottomed out, and the accumulator is completely filled with nitrogen.

13. Place the ride control switch in the off position, raise the lift arms, and hold the control lever in the raise position for about five seconds. This ensures that the pressure reaches and maintains the system relief valve setting, which should

register on the gas pressure gauge. If a lower pressure is observed on the gauge, then the accumulator has not been properly charged and the charge procedure must be redone.

14. If the hydraulic system relief pressure was reached, lower the bucket to the ground and place the ride control switch in the on position. Observe that the nitrogen gas pressure drops instantly to charge pressure. If the nitrogen gas pressure has not dropped to the charge pressure, repeat the charge procedure until this occurs.

15. Remove the charge hardware and install the cap and protective cover back onto the accumulator.

Installing a New Accumulator

When installing a new accumulator on the ride control system, certain steps must be observed so that all the air is removed from the nitrogen end of the accumulator. Pour approximately 2 quarts (1.9 litres) of hydraulic oil into the nitrogen end of the accumulator. The oil ensures that all the air from the nitrogen end of the accumulator is removed and lubricates the upper seal of the accumulator piston.

1. Connect the nitrogen gas chuck to the accumulator charge valve and connect the male and female quick disconnects together (3 and 13), as shown in **Figure 18-25**. Place the open end of the hose in a suitable container to collect the oil.

2. Open the needle valve (1) and screw the T-handle on the gas chuck in a clockwise direction until it is fully open. Make sure that the accumulator is in the upright position and move the piston to the nitrogen end of the accumulator.

3. The air and oil will be vented out of the accumulator as the piston moves up. When the oil flow stops, close the needle valve completely and allow any remaining oil to drain from the hose.

4. Reconnect the quick disconnect couplings (10 and 11). Connect the nitrogen charge hardware to the nitrogen bottle and adjust the pressure to 300 psi (2,070 kPa). Check the ambient temperature chart and compensate if required.

5. Charge the accumulator by opening the needle valve (1) with 300 psi (2,070 kPa). Close the gas chuck (5) and remove the nitrogen charge hardware from the accumulator. The accumulator is now ready to install.

Checking Charge Pressure

The charge pressure in the ride control accumulator may need to be checked periodically or whenever a performance problem is suspected. It is not necessary to follow all the steps outlined in the charge procedure just to check the pressure in the system. The following steps explain how to check the charge pressure quickly. Start up the equipment and place the ride control switch on the joystick in the on position. Lower the bucket to the ground and place the joystick in the float position. This ensures that all the hydraulic pressure is relieved in the ride control accumulator.

1. Shut down the equipment and remove the guard and protective cap from the accumulator.
2. Close the needle valve (1) on the nitrogen charge hardware and connect the chuck to the accumulator charge valve. Be sure to turn the T-handle completely counterclockwise before connecting the chuck to the accumulator charge valve.
3. Open the T-handle by turning clockwise and read the gauge (2) to determine whether the accumulator pressure is within specifications. If the equipment has been in operation, then the temperature charts, shown in **Figure 18-26**, must be consulted to determine the correct pressure reading.

EQUALIZING BEAM SUSPENSION SYSTEMS

Tandem drive applications that are used on articulating haulage trucks often use **equalizing beam suspension** systems, as shown in **Figure 18-27**; these

Figure 18-27 Articulating haulage truck.

© Cengage Learning 2014

systems have proven to be up to the task of operating in rough service applications. To maintain good ground contact, the equalizing beams are allowed to flex on pivots. This provides a means of balancing the loads imposed on the drive axles while still maintaining good ground contact.

There are two distinct types of equalizing beam suspension designs: the leaf-spring equalizing beam, as shown in **Figure 18-28**, and the solid rubber equalizing beam, as shown in **Figure 18-29**. The latter beam is the most common on off-road equipment. This design is ideally suited to off-road equipment as the layout actually lowers the center of gravity by placing the beam below the centerline of the axle. Torque rods are provided to absorb additional torque generated by the axle and road shock.

SOLID RUBBER SUSPENSION SYSTEMS

This design is the most popular type of equalizing beam suspension on off-road haulage trucks. A **solid rubber suspension** incorporates solid rubber **load cushions** instead of leaf springs to absorb road shock. A **saddle assembly** on each side of the suspension system holds the rubber load cushions. **Figure 18-30** shows a typical rubber cushion equalizer beam suspension system. Each of the four rubber blocks is connected to the axle assembly by drive pins. The operational loads generated by cornering and braking are all absorbed by the suspension through these pins. When the suspension is operating with no load, the outer edges of the spring load cushions are in contact with the saddle. As the load increases on the vehicle, the load cushion is compressed, further absorbing the additional loads. Alignment is maintained by the drive

CHARGING PRESSURE AND TEMPERATURE RELATIONSHIP FOR THE ACCUMULATOR	
Temperature °C (°F)	Pressure kPa (psi)*
−7 (20)	1790 (260)
−1 (30)	1835 (266)
4 (40)	1875 (272)
10 (50)	1915 (278)
16 (60)	1955 (284)
21 (70)	2070 (300)
27 (80)	2110 (306)
32 (90)	2150 (312)
38 (100)	2200 (318)
43 (110)	2250 (324)
49 (120)	2300 (330)

*Nominal allowable pressure tolerance equals ± 35 kPa (± 5 psi)

Reprinted Courtesy of Caterpillar Inc.

Figure 18-26 Temperature compensating chart used to charge an accumulator.

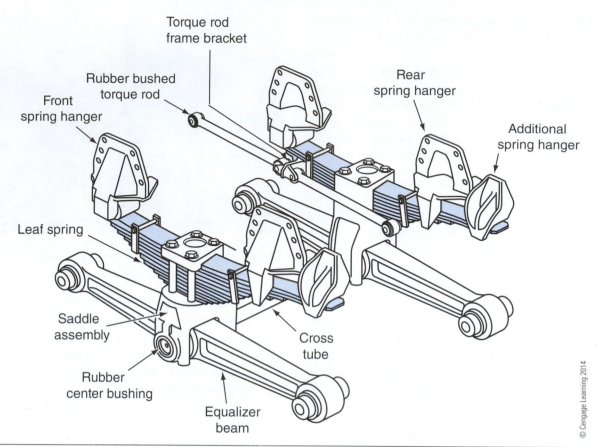

Figure 18-28 Leaf-spring equalizing suspension system.

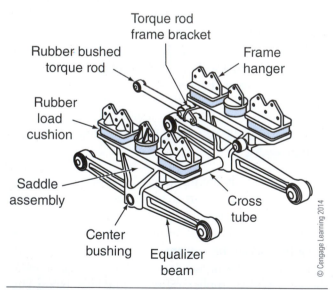

Figure 18-29 Solid rubber equalizing suspension system.

pins, which are encased in rubber bushings. The vertical **drive bushings** are used on each drive pin to allow vertical movement during loading and unloading while still maintaining suspension alignment. This design works well when loaded, but tends to give a rough ride when operating without a load.

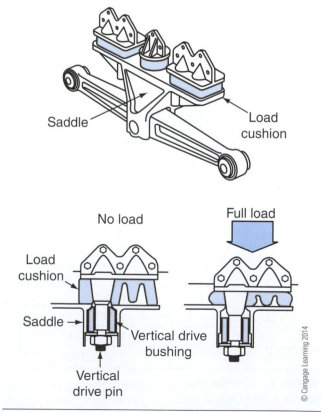

Figure 18-30 Rubber cushion-type equalizing beam suspension system.

SERVICING SUSPENSION SYSTEMS

Off-road equipment operates in severe-duty service, making regular inspections of the suspension systems essential. The equipment should be power washed before regular inspection and servicing. Power washing the suspension system allows detailed inspection of the suspension components, which includes looking for cracks or physical damage. Rubber bushings should be inspected for deterioration and cracking. If the rubber components show any signs of deterioration or cracking, they should be replaced as soon as possible. Suspension system bushings are not easily removed; specialized tools may be required to remove them. The use of a torch is not recommended because the fumes produced by the burning rubber compounds are toxic. Because a specialized hydraulic press is often required to remove the bushing and sleeve assemblies, it may be easier to replace the complete equalizer arm. Press pressures may be as high as 50 tons.

The load cushions on equalizer beam suspension systems are easily inspected because they are visible. Rubber block compounds may be damaged by cleaning solvents, so be sure to use only manufacturer-approved cleaning products. Visual inspection is all that is required to check these components.

Check to see if the unsprung height has changed significantly. Operation of equipment with worn-out load cushions is not recommended because it can damage the suspension system and frame.

U-bolts that secure the **spring packs** must be checked for tightness on a regular basis. If the bolts are corroded, cleaning them and freeing them up before checking the torque is required because corrosion will affect the re-torque values. When new fasteners are installed on suspension systems, it is recommended that the fasteners be re-torqued after a run-in period.

If a grease fitting on a spring pin does no take grease during a scheduled maintenance service, remove the weight from the spring and use a hand grease gun (capable of generating higher pressures) before replacing the fitting. Lubricating the suspension pins is critical to increasing the life span and controlling the operating costs of equipment and should always be a priority during servicing.

Figure 18-31 shows how wear to the equalizer bracket can occur when the spring ends shift. Check for loose U-bolts or problems with the U-bolt seating when these conditions are encountered. Wear can be caused by improperly mounted springs on the axle

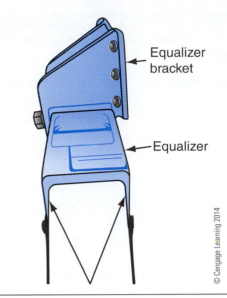

Figure 18-31 Misaligned spring rubbing against an equalizer bracket.

housing, indicating a need for a suspension alignment check.

When a broken or cracked spring is found in a spring pack, most manufacturers recommend that the complete spring pack be replaced. Never apply grease or paint to the spring leaves, because this will reduce the spring self-dampening ability. The same is true if spring leaves are rusted. Rusted spring packs should be disassembled and cleaned or replaced if cleaning is not possible.

Spring and bracket bushings are considered high-maintenance items and should be inspected for cracks and gouging on a regular basis. If the bushings are free from damage, their dimension specifications should be checked. The manufacturer's recommended dimensions and clearances should be referenced but a general rule is a maximum of 0.020" (0.5 mm) clearance between the bushing and pin. Clearance exceeding 0.020" (0.5 mm) usually requires the bushings and pins to be replaced.

TROUBLESHOOTING TIPS

The guidelines provided in this chapter are examples and are not intended to be used as a comprehensive troubleshooting guide. Follow the specific troubleshooting procedures that are outlined in the manufacturer's service literature. Take care when evaluating failures; similar but different complaints can often be attributed to the same symptoms. **Table 18-4** is a troubleshooting chart that can be used to assess the most common types of complaints that may develop on leaf-spring suspension systems.

TABLE 18-4: LEAF-SPRING SUSPENSION SYSTEM TROUBLESHOOTING GUIDE		
Symptoms	**Cause**	**Corrective Action**
Equipment leans to one side	Broken spring leaves	Replace the spring assembly
	Spring assembly weak	Replace the spring assembly
	Bent or twisted from rail	Correct the frame rail bend or twist
	Spring load capacities are not the same on both sides of the equipment	Replace the spring assemblies with the recommended assemblies
Equipment wanders	Broken spring leaves	Replace the spring assembly
	Wheels not properly aligned	Redo the wheel alignment according to manufacturer's recommendations
	Incorrect caster settings	Adjust caster to manufacturer's recommended settings
	Steering gear not centered	Center the steering
	Drive axle alignment not properly set	Align the drive axles according to manufacturer's recommendations
Equipment suspension bottoms out during operation	Equipment overloaded	Follow equipment recommended load capacity rating
	Broken spring leaves	Spring assembly should be replaced
	Weak spring assemblies	Spring assembly should be replaced
Spring breakage occurs regularly	Equipment overloaded or extremely rough service duty	Equipment load should be reduced to match operating conditions
	Insufficient center bolt torque	Re-torque center bolt to manufacturer's recommended values
	Improperly torqued U-bolt nuts	Re-torque the U-bolt nuts to manufacturer's recommended values
	Worn-out or damaged spring pin bushings allowing spring endplay	Replace the spring pins and bushings
Excessive spring noise	Improperly torqued spring center bolt or U-bolts allow excessive spring movement	Inspect components for damage and replace any damaged parts before re-torquing to manufacturer's recommended values
	Loose, bent, or broken spring shackle	Replace the damaged components as required and torque the fasteners to the manufacturer's recommended values
	Spring pin bushing worn out or damaged causing excessive spring	Replace the worn or damaged spring pins

© Cengage Learning 2014

ASSEMBLY, DISASSEMBLY, AND REPLACEMENT PROCEDURES

General Guidelines for Leaf-Spring Replacement

The guidelines shown here are for example purposes and should not replace the manufacturer's procedures.

1. Place the appropriate chocks under the wheels to prevent the equipment from moving unexpectedly.

2. Use an approved lifting device to raise the equipment to relieve the weight from the springs, and place safety stands under the frame.
3. If shocks are used, disconnect and remove them.
4. Remove the U-bolts and retainers along with the spring bumpers.
5. If the pins are lubricated, remove the grease fittings and lubricators.
6. Remove the nuts that secure the spring shackle pins.
7. Remove the bracket and shackle pins by sliding the spring off the bracket pin.

8. To install the new spring, fasten the pivot end of the spring to the frame bracket first.

9. Fasten the shackle or slipper end by aligning it with the other frame bracket.

10. Install the U-bolts to secure the springs to the axle, but do final torque until the load is on the springs.

11. Remove the frame supports to place the springs under tension, and torque the U-bolts and other hardware to specification.

12. After replacing spring components, re-check the torque values after a short break-in period.

> **CAUTION** *Caution should be exercised whenever replacing springs. Leaf-spring packs are under considerable tension and can blow apart unexpectedly if a center bolt were to break while being torqued, when the U-bolts are not in place. When spring pins are removed, ensure that the equipment is properly mechanically blocked.*

Guidelines for Equalizer Replacement

Suspension equalizer replacement requires that the front wheels of the equipment be properly chocked to prevent unexpected movement during the replacement procedure. Use an appropriate floor jack to raise the rear of the equipment so that the axles can be supported on safety stands in such a way that the weight is removed from the leaf springs. Once the equipment is properly blocked, remove the wheels from the axles and store in a safe location.

1. If the equipment being worked on has a tandem axle, the cotter pin from the outboard end of each spring retainer should be removed. Now, the retainer pins can be removed.

2. Loosen and remove the nut from the equalizer cap and tube assembly. Remove the inboard bearing washer, bolt, and outboard bearing washer as shown in **Figure 18-32**.

3. Place an appropriate bar between the tube assembly and equalizer cap and pinch the equalizer cap to remove the weight of the equalizer. Position a short piece of pipe through the inboard equalizer cap and tube assembly bolt hole. Gently tap the cap and tube assembly out of the equalizer.

4. At this point, the equalizer bushings should be cleaned so that they can be inspected for damage and wear. If the bushings show evidence of

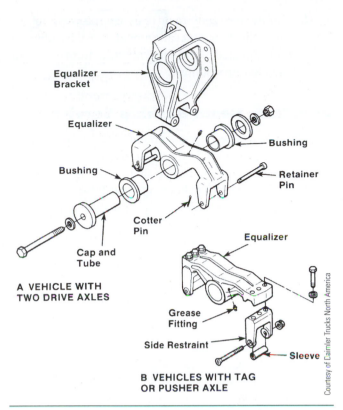

Figure 18-32 Exploded view of a typical equalizer assembly.

damage or the dimensions are excessive, they should be replaced. Refer to the service literature for the equipment for the correct specifications.

5. Before re-installing the bushing, lubricate its inside bore with chassis grease and then install the equalizer into the equalizer bracket.

6. Before installing the equalizer cap and tube assembly, apply grease to the appropriate area and insert the cap and tube assembly into the equalizer through the bracket. Force the equalizer cap and tube assembly partway into the equalizer and install the inboard wear washers between the inboard equalizer bushing and the bracket. Complete the installation by pushing the cap and tube assembly the rest of the way into the equalizer bracket.

7. Install the equalizer outer bearing washer on to the equalizer cap and tube assembly bolt *before* installing it into the cap and tube assembly.

8. Install the inboard bearing washer, locknut, and tube assembly bolt, and torque the nut to specification.

9. Reinstall the wheels and raise the equipment to remove the safety stands from under the frame and axles. Lower the equipment to the floor.

10. Before the equipment can be returned to service, the rear axle alignment must be checked and adjusted, following manufacturer's recommended procedures.

Equalizer Beam Bushing Service

When any type of replacement work on suspension systems is considered, the area must be thoroughly power washed and cleaned so that the components can be inspected for physical damage and fatigue cracks. If the rubber bushings show any signs of cracking or deterioration, they should be replaced. In most cases, specialized service tools are required to replace the equalizer beam bushings. If however, the specialized tools are not readily available, then a portable hydraulic press and the appropriate steel piping can be used to replace the equalizer beam bushings.

The load cushions or jounce blocks should be inspected for signs of distortion or damage. These blocks act as shock load insulators and are important to the operation of the suspension system. When cleaning this area, do not expose the load cushions to cleaning chemicals because they can be easily damaged by these products. When visually inspecting the load cushions, check that the blocks are still bonded to the metal insert and are not about to separate. If the unloaded height of the load cushion is less than ¼ inch (6 mm) below that of a new one, it must be replaced.

Reconditioning Equalizing Beam Suspension Systems

Clean the equipment before visually inspecting it. It may be important to ensure that the replacement parts are available to avoid unnecessary down-time. Before beginning major suspension repair work, the technician should read over the procedure in the service literature and make a note of any special tools required.

Equalizer beams are manufactured from different materials, some of which are damaged when manufacturer service procedures are not adhered to. Care should be taken to avoid damaging aluminum and cast steel equalizer beams. Nodular iron equalizer beams should be installed in so that the markings indicating "up" (or and arrow) face in the correct direction. Reinforced gate pads are built into each end and the middle section of the beams.

After inspection and the parts required to complete the repairs are in place, the repair can begin. Locate the equipment on a level, clean, and hard surface. Place wheel chocks on both axles of the equipment. Remove the driveshafts from the rear axles and mark

their location. It is good practice to reinstall parts in the same location from which they were removed.

Next, remove the four saddle caps that are used to lock the center bushing of the equalizer beams to the saddle assemblies (see **Figure 18-33**). The **torque rods** must be disconnected from the axle housing by loosening the lock nuts and driving the shaft from the bushing and axle housing brackets. A suitable hoist is required to lift the truck frame so that there is sufficient clearance for the lower part of the saddle to clear the top of the axle housing. Ensure that the axles remain properly blocked so that they do not pivot on the wheels. Use safety stands or approved blocking to support the equipment frame.

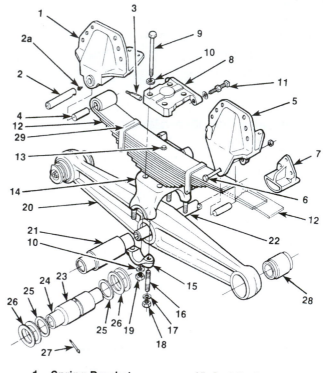

1	Spring Bracket	15	Saddle Cap
2	Spring Pin	16	Stud
2a	Zerk Fitting	17	Washer
3	Draw Key	18	Nut
4	Bushing	19	Nut
5	Spring Bracket	20	Equalizer Beam
6	Rebound Bolt with Spacer	21	Center Bushing
		22	Cross Tube
7	Spring Bracket (extended leaf only)	23	Center Bushing Bronze
8	Top Pad	24	Sleeve
9	Top Pad Bolt	25	Seals
10	Washer	26	Washer
11	Setscrew	27	Zerk Fitting
12	Leaf Spring Assembly	28	End Bushing
13	Center Bolt	29	Part Number
14	Saddle Assembly		

Courtesy of International Truck and Engine Corporation

Figure 18-33 Exploded view of an equalizing beam suspension assembly.

Leaf-Spring Equalizer Beam Disassembly

Although there are several design variations of equalizer beam mounting configurations, only one of the three most common types, the bolt-type beam end mounting, will be used as an example. This mounting is shown in **Figure 18-34**. This section explains the procedure used to disassemble the leaf-spring equalizer beam suspension.

1. Disconnect the equalizer beam from the axle housing before removing the equalizer beam end bolt.
2. Drive the bushing adapters and sleeve out of the bushing and axle housing brackets using a chisel. The notches in the adapters will help locate and keep the chisel in place during this operation. To wedge out the adapter, drive the chisel into one side first and then into the other.
3. Remove the first adapter and proceed to remove the opposite adapter with a hammer and drift.
4. Now, the equalizer beams can usually be separated from the **cross tube**. The cross tube pivots on the inner sleeve of the center bushing and slides back and forth approximately 3 inches. This results in the cross tube having a polished appearance near the center bushing, which is considered normal.
5. Once the cross tubes are separated, remove the saddle assemblies, springs, or cushions.
6. The spring and saddle assembly must be supported on a floor jack and safety stand before the spring-aligned set screws can be loosened. Next, remove the spring top saddle bolts and nuts. The top pad can now be removed safely. Lower the floor jack to remove the saddle.
7. To remove the spring assembly, place the floor jack under the spring and loosen the

locknut on the spring pin draw key, as shown in **Figure 18-33**.
8. Caution must be exercised in this step. The threads on the draw key can be easily damaged. Back off the nut just enough to protect the draw key threads before tapping the nut with a soft-blow hammer to loosen the draw key.
9. After the draw key has been removed, drive the spring pin out of the spring bracket and lower the spring assembly from the frame using the jack.

Leaf-Spring Equalizer Beam Assembly

1. Position the spring in the spring saddle in such a way that the bolt head of the spring center bolt is located in the hole in the saddle.
2. Place the spring top pad over the cup on the main spring leaf, and install the bolts and nuts that secure the saddle to the top pad. Seat the nuts on the bolts, but do not torque them yet.
3. To align the spring in the saddle, tighten the spring aligning set screws to the recommended torque; then tighten the aligning screw locknuts in place before torquing the saddle to top pad bolt nuts to the manufacturer's specifications.
4. Position a jack on the spring and saddle so that the assembly sits in the front and rear spring mounting brackets, and align the spring eye with the spring pin bore in the front bracket.
5. Install the spring pin while aligning the draw key slot in the pin with the draw key bore in the bracket, install the draw key, lock washer, and nut, and torque up to manufacturer's specifications.
6. Replace the zerk grease fitting in the spring pin and grease with the recommended lubricant.

Solid Rubber Equalizer Beam Disassembly

1. Begin by removing the four drive pin bushing retainer caps (refer to **Figure 18-34**) before removing the drive pin nuts.
2. Next, remove the load cushions and saddle from the vertical drive pins. The drive pin nuts may need to be cut to facilitate removal. Depending on the difficulty of the removal, it may be necessary to replace the drive pin bushing.
3. Remove the torque rods from the cross member frame bracket and install the new spring, cushion, and saddle assemblies.

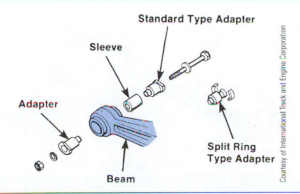

Figure 18-34 Exploded view of a three-piece adapter bolt-type beam end mounting.

Solid Rubber Equalizer Beam Assembly

1. Assemble the drive pin bushings to the saddle and install the bearing caps in place *before* installing the rubber load cushions on the saddle.
2. Lubricate the drive pins with multipurpose grease on the frame brackets, before assembling the frame brackets to the saddle. Install the drive pin nuts and washers with the flanges facing down and tighten them snugly. If the frame hangers have been removed, the nuts must be torqued to the manufacturer's specifications after they have been assembled to the chassis frame.
3. Install the preassembled saddles, load cushions, and frame hangers to the chassis frame, and install the mounting bolts, lock washers, and nuts in place. Torque the mounting bolts to the manufacturer's specifications *before* torquing the drive pin nuts.

Cross Shaft Installation

Note: This procedure applies to both the leaf-spring and rubber spring equalizer beam suspensions. The final torque values on the mounting nuts should not be applied until the wheels have been installed and lowered to the ground.

1. Before inserting the bushing adapters, align the beam end bushing with the hanger brackets. Lubricate the adapters with multipurpose grease, and place the flat surface of each adapter in the vertical position.
2. With a hammer, lightly tap the adapters into the bushing until the adapter flanges seat against the outer face of the hanger brackets. Install the bolt and locknut in place. *Do not* torque up to specifications yet. When installing rubber bushed equalizing beams to the axles and saddles, do not tighten the nuts until the torque rods are installed and the equalizer beams are level with the frame. This reduces the chance of creating torsional windup in the suspension.

Reinstalling Equalizing Beam Suspension Assemblies

1. One end of the cross tube must be positioned into one end of the center sleeve of the equalizer beam. Ensure that the tube seats correctly into the sleeve. Now, install the other equalizer beam into the cross tube and locate the beam in the spring saddle.
2. At this point, the axle assemblies can be positioned under the center of the saddle; make sure that the outer bushings are positioned in such a way that they line up with the center of the saddle legs. The frame can now be lowered carefully. Ensure that the saddles center on the beam end bushings. Install the saddle caps and nuts. *Do not* apply the final torque until the torque rods are installed and the equalizer beams are level with the frame.
3. Make sure that the cross shaft is free to float in the bushing assemblies on each side; this ensures correct alignment of the left and right equalizer beams.
4. Torque the saddle caps to ensure that the sleeve is secure to the bushing. This tightening procedure has no effect on the pivot of the cross shaft.
5. Connect the torque rods to the axle brackets and the frame brackets. During the torque sequence, tap the bracket with a hammer to seat the tapered torque rod stud into the bracket. If the torque rod ends have straddle mounts with two holes, as shown in **Figure 18-35**, a spacer is required between the bracket and cross member to get the correct axle adjustment. Once these steps are complete, torque the shaft nuts and equalizer beam end fittings to the manufacturer's specifications.
6. Reinstall the axle shafts in their original locations. Reconnect the brake lines, drive shafts, and hardware removed during the disassembly procedure.

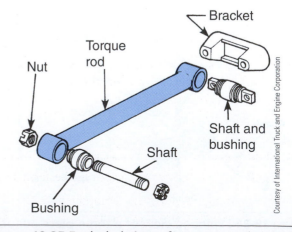

Figure 18-35 Exploded view of a torque rod assembly.

ONLINE TASKS

1. Use an Internet search engine to research the configuration of a typical off-road suspension system. Identify a piece of equipment in your particular location, paying close attention to the suspension system design and arrangement, and identify whether it has automated ride height compensation ability. Outline the reasoning behind the manufacturer's choice of design for this application.

2. Check out this URL: *http://www.meritor.com/products/suspensions/default.aspx* for more information on suspension system configuration and design.

Shop Tasks

1. Select a piece of equipment in your shop and identify the components in the suspension system assembly. Inspect the suspension components to determine whether they are excessively worn or damaged.

2. Select a piece of off-road equipment in your shop that uses a conventional spring suspension system and identify the type and model of the equipment. List the major components and note the corresponding part numbers of the major suspension system components. Identify and use the correct parts book for this task.

Summary

- An off-road suspension system's primary purpose is to support the total weight of the equipment.

- An LHD (Load Haul Dump) consists of a drive axle mounted directly to the frame with only the tires to absorb ground impact.

- Unsprung weight is the total weight of those components supported below the spring: this usually consists of the axle, brake hardware, and suspension components.

- Sprung weight is defined as the load that the suspension must support during normal operation.

- Nitrogen in suspension cylinders must be discharged before checking the oil level. Never remove any plugs or valves from a suspension cylinder if the rod is not fully retracted.

- Large haulage trucks use a gas/suspension cylinder at each front wheel end instead of a conventional kingpin as part of the steering linkage.

- A key advantage of independent gas/hydraulic suspension systems is their simplicity of design and ability to absorb changing load and road conditions.

- The pressures in each suspension cylinder can be affected by changes in ride height due to leaks at opposite ends of the equipment.

- Nitrogen gas is the only gas that should be used to charge suspension cylinders.

- Pressurized oil and escaping nitrogen gas can pose a safety hazard to anyone who comes in contact with either one.

- Be sure that the charge valves are fully closed before connecting the nitrogen charge lines to the suspension.

- Suspension dimensions can be affected by temperature differences between the adjustment location and the operating location.

- Tandem drive applications that are used on articulating haulage trucks often use equalizing beam suspension systems.

- Equalizing beam suspensions allow the machine load to be equally distributed between the axles while compensating for road irregularities.

- Caution should be exercised whenever replacing suspension springs. Multi-leaf spring packs are under massive tension and can cause serious injury if not handled properly.

- If a hydraulic leak is suspected, never run your hands over the lines to locate the leak. A high-pressure leak can penetrate the skin.

Review Questions

1. Which of the following describes the term *unsprung weight*?

   A. the total weight of the components not supported by the spring

   B. the total weight of each end of an axle

   C. the total weight of the equipment

   D. the rebound rate of the equipment

2. Which symptom would result from a broken spring pack center bolt?

   A. erratic steering

   B. U-bolt failure

   C. bouncing front end

   D. broken leaf spring

3. If you determine that a spring pack has a single cracked leaf, what is the recommended repair procedure that should be performed?

   A. Replace the spring pack.

   B. Disassemble the spring pack and replace the broken leaf.

   C. If the crack is small, weld it.

   D. Do nothing; it's only a crack.

4. While lubricating a spring pack assembly, one spring pin will not take grease. Which of the following should be done first?

   A. Change the grease fitting on the bracket spring pin.

   B. Replace the bracket spring pin.

   C. Remove the weight from the spring and attempt to grease again.

   D. Heat it up with a torch to unplug the grease passage.

5. How are the springs loaded in a spring pack under tension?

   A. U-bolts

   B. center bolt

   C. spring pins

   D. torque rod

6. What term is used to describe the most compressed condition of a suspension spring?

   A. jounce

   B. rebound

   C. deflection rate

   D. unsprung load

7. What purpose does a torque arm serve on a suspension system?

   A. controls axle torque

   B. controls rebound rates

   C. controls jounce

   D. controls the suspension height

8. Which term describes an equalizing-beam suspension?

   A. torsion bar

   B. walking beam

   C. air spring

   D. multi-leaf variable rate

9. Which of the following functions can be performed by torque rods?

   A. dampen axle windup caused by driveline torque

   B. align axles

   C. dampen axle twist under braking

   D. all of the above

10. What do large quarry trucks often use in the front suspension to replace a kingpin?

    A. suspension cylinder

    B. torque rods

    C. coil spring

    D. leaf spring

11. How are rear cylinder barrels connected to the equipment frame on a large quarry truck?

    A. torque rod

    B. kingpin

    C. cylinder stem

    D. spring pin

12. Where does the adjustable ride control on a haulage truck get its oil supply?

    A. from the brake or hydraulic
       circuit

    B. an independent hydraulic system

    C. It's a closed system, and does not need an external
       supply source.

    D. none of the above

13. What will occur if one cylinder develops a gas leak in a suspension cylinder on a haulage truck with an independent gas-charged system?

    A. It would not affect the other three
       suspension cylinders.

    B. All suspension cylinders would
       experience a pressure drop.

    C. The pressure would rise in the remaining three
       cylinders.

    D. The truck's ride height would be affected.

14. When charging a suspension cylinder with nitrogen, what procedure must be observed?

    A. Compensate for temperature if
       required.

    B. Drain the oil from the cylinder first.

    C. Support the equipment to keep it level.

    D. Take the weight off the suspension cylinder.

15. What step must be observed first when installing a new accumulator on the ride control system?

    A. Fill the accumulator with oil.

    B. Purge the accumulator of all air.

    C. Check the pressure in the accumulator.

    D. Hydraulic pressure must be present to charge properly.

19 Machine and Equipment Electronics

Learning Objectives

After reading this chapter, you should be able to:

- Understand the language of electronic controls.

- Outline the stages of a computer processing cycle.

- Define the principles of operation of thermistors, variable capacitance sensors, Hall-effect sensors, potentiometers, induction pulse generators, and machine control joysticks.

- Identify input and output devices.

- Outline the operating principles of a machine control joystick.

- Explain how haptic feedback is used in a joystick.

- Differentiate between customer and proprietary data reprogramming.

- Describe the processes used to reprogram a machine ECU with proprietary data.

Key Terms

actuator

analog joystick

central processing unit (CPU)

chopper wheel

diagnostic trouble codes (DTCs)

digital joystick

download

electronically erasable, programmable read-only memory (EEPROM)

engine/electronic control module (ECM)

engine/electronic control unit (ECU)

fault mode indicators (FMIs)

haptic feedback

joystick

message identifiers (MIDs)

microprocessor

multiplexing

optical bus

parameter

parameter identifiers (PIDs)

potentiometer

programmable read-only memory (PROM)

pulse wheel

random-access memory (RAM)

read-only memory (ROM)

SAE J1939

source address (SA)

subsystem identifiers (SIDs)

thermistor

tone wheel

upload

INTRODUCTION

Almost all current off-highway machines use computers to manage the engine, hydraulics, and a range of other chassis functions. The technician is required to troubleshoot system problems and reprogram **parameters** (values), in most cases using software loaded to personal computers (PCs), laptops, or handheld units. Onboard vehicle computers are referred to as **engine/ electronic control modules (ECMs)** (SAE recommended acronym for engine controllers) or **engine/ electronic control units (ECUs)** (SAE recommended acronym for all other chassis system controllers). Because this chapter will mainly focus on machinery management systems, ECU is the acronym that will be generally used. Because many of the input and output components are common with those used in engine management systems, we will concentrate on components that are specific to off-road machinery.

ELECTRONIC CONTROL UNIT

An electronic control unit (ECU) is a modular housing containing a **microprocessor**, data retention media, and, usually, an output or switching apparatus. The ECU is a system controller. It is usually mounted close to the components it manages, although in heavy-equipment applications, a requirement is to protect the devices from dirt. **Figure 19-1** shows ECUs mounted behind the firewall on a piece of equipment. Today's equipment uses several system controllers to manage the engine, transmissions, hydraulics, climate control, and just about everything else. All these ECUs need to communicate with each other. To facilitate this communication, each controller is connected to a communications network known as a data bus. The data bus may either be hard wire (J1939 uses a two-wire network channel) or optical.

The current heavy-duty standard data bus is known as J1939. It uses **SAE J1939** hardware and software language to allow ECUs to exchange information and work together. Each ECU is assigned an address on the chassis data bus. The data bus is the information exchange backbone of the chassis. We use the term **multiplexing** to refer to communications on a vehicle data bus.

All domestically produced, and most offshore, off-road machinery uses J1939. Although J1939 was engineered as a powertrain network, it can also be used for other chassis functions. In cases where data bus traffic is intense, J1939 can be complemented by proprietary buses, which may be either hard wire or optical.

Figure 19-1 ECUs mounted behind the firewall on a piece of equipment.

For example, the complex hybrid powertrain covered in Chapter 20 uses several **optical buses** in conjunction with J1939.

Data Processing

A machine computer manages an information-processing cycle composed of three distinct stages:

1. Data input
2. Data processing
3. Outputs

Data Input. Data is simply raw information. Most of the data to be inputted to a machine ECU comes from monitoring sensors, such as rpm signals, hydraulic pressure, and motion sensors. Input data may be in analog or digital formats.

Data Processing. A **central processing unit (CPU)** contains a control unit that executes program instructions and an arithmetic logic unit (ALU) to perform numeric calculations and logic processing such as comparing data. Random-access memory **(RAM)** is data that is electronically retained in the ECU—this data can be manipulated by the CPU because it can be accessed at high speed. RAM is often referred to as *main memory*. The ECU can also call up more permanent types of data that are retained magnetically or optically. Read-only memory **(ROM)**, programmable read-only memory **(PROM)**, and electronically erasable, programmable read-only memory (EEPROM) data can be transferred to RAM by the CPU for processing. Because RAM data is electronically retained, it is lost when its power-up circuit is opened.

Outputs. The results of ECU processing operations must be converted to action by switching units and **actuators**. In most ECUs, the switching units are integral with the ECU. ECU commands are converted to electrical signals that drive actuators. Actuators are the electrical components such as solenoids that convert ECU processing into action.

Data Retention in Vehicle ECUs

Data is retained electronically, magnetically, and optically in current agricultural and heavy-equipment machinery ECUs. Optically retained data in laser-read systems are used in some specialty machines such as those in Chapter 20. ECU memory is categorized as follows: random-access memory, read-only memory, programmable ROM, and electronically erasable PROM.

Random-Access Memory. Random-access memory (RAM) is also known as *main memory* because the CPU can manipulate data only when it is retained electronically. At start-up, RAM is electronically loaded with the system management operating instructions and all necessary running data retained in other data categories (ROM/PROM/EEPROM) so that the CPU can access it at high speed.

Volatility. RAM data is electronically retained, which means that it is volatile or temporary. It requires power-up. At power off, all RAM data is lost. Data such as temperature, oil temperature/pressure, and shaft/track speeds are also logged into RAM and, providing the values are within a threshold window defined in ROM/PROM/EEPROM, they have significance only at that specific moment; they can therefore be safely discarded when the ignition key is opened.

NV-RAM. A second category of RAM is used in some older machine systems, usually those with no

EEPROM capability. This is nonvolatile RAM (NV-RAM) in which data is retained until either the battery is disconnected or the ECU is reset. Like RAM, NV-RAM requires power-up to retain its memory so it is connected to source battery power.

Read-Only Memory. ROM data is magnetically or optically retained and designed not to be overwritten. A majority of the total data retained in the ECU is logged in ROM. In a system ECU, the master program for the system management is logged into ROM.

Programmable ROM. Programmable ROM (PROM) is magnetically retained data—usually a chip, set of chips, or card(s) socketed into the ECU motherboard. PROM can sometimes be removed and replaced. PROM's function is to qualify ROM to a specific system application. In the earliest examples of machine electronics, programming could be altered only by physically replacing the PROM chip(s).

Electronically Erasable PROM. The **electronically erasable, programmable read-only memory (EEPROM)** data category contains customer data programmable options and proprietary data that can be altered and modified using a variety of tools ranging from a generic reader programmer to a mainframe computer. It is usually magnetically retained in current systems, but this will probably change in the near future. EEPROM provides the ECU with a write-to-self capability allowing it to log fault codes, audit trails, tattletales, and failure strategies.

Proprietary Data. Proprietary data logged into EEPROM is more complex both in character and in methods of altering. An example of proprietary EEPROM data is the fuel map in an engine ECU. The procedure here normally requires accessing a centrally located mainframe computer (usually at the OEM headquarters location) via the Internet, **downloading** the appropriate files to a PC, and finally reprogramming (altering or rewriting the original data) the ECU.

INPUT CIRCUIT

ECU inputs can be divided into sensor inputs and switched inputs. Inputs to the ECU can be divided into:

- monitoring
- command

Sensors and switches are used as input circuit components. A temperature signal to the ECU is an example of a monitoring input. The throttle position sensor is an example of a command input.

Sensors

Anything that signals input data to a computer system can be described as a sensor. Sensors may be simple switches that an operator toggles open or closed to ground a reference voltage, modulate a reference voltage (V-Ref), be powered up either by V-Ref, or require power-up outside of the V-Ref circuit.

Types of Sensors

Sensors may generate their own voltage, receive a constant reference voltage (usually 5 V known as V-Ref), or be powered up at a higher voltage.

1. Thermistors. **Thermistors** precisely measure temperature. They are used to signal oil, air, fluid, and ambient temperatures. The two types of thermistor are defined by whether resistance increases or decreases when the temperature rises:
   - Negative temperature coefficient (NTC): Temperature goes up, resistance goes down.
   - Positive temperature coefficient (PTC): Temperature goes up, resistance goes up.

   The ECU receives temperature data from thermistors in the form of analog voltage values.
2. Variable capacitance (pressure) sensors. These three-wire sensors are supplied with reference voltage and usually designed to measure pressure or linear position values. The medium whose pressure is to be measured acts on a ceramic disc and moves it either closer or farther away from a steel disc, varying the capacitance of the device and thus the voltage value returned (signaled) to the ECU.
3. Potentiometers. The **potentiometer** is a three-wire (V-Ref, ground, and signal) variable resistor. Once again, these receive a reference voltage and output a signal proportional to the motion of a mechanical device. Potentiometers are voltage dividers. The moving mechanical device moves a contact wiper over a variable resistor. As the wiper is moved over the variable resistor, the resistance path is altered, with the supply voltage being divided between the signal (sent to the ECU) and ground. Potentiometers are used in some types of equipment such as joysticks and throttle position sensors (TPSs).

4. Piezoresistive pressure sensor. These are sometime referred to as Wheatstone bridge sensors because of the printed circuit used to output its digital signal. They are more accurate than variable capacitance sensors and are primarily used in engines, usually as a boost pressure sensor: for that reason, they are only briefly mentioned here.

....

Tech Tip: When a sensor logs an active fault code, try disconnecting the sensor and observing whether the fault mode indicator (FMI) changes. If there is no change, it suggests that the fault lies in the wiring circuit. If the FMI changes, it suggests a problem in the sensor.

....

Signal-Generating Sensors

Hall-Effect Sensors. Hall-effect sensors generate a digital signal as timing windows or vanes on a rotating disc pass through a magnetic field. They can be used for both linear and rotary position signaling. In a rotary Hall-effect sensor, the rotating disc is known as a **pulse wheel** or **tone wheel**. Because these terms are used to describe the rotating members of induction pulse generators, care should be taken to avoid confusion. The frequency and width of the signal provides the ECU with speed and position data. The disc incorporates a narrow window or vane for relaying position data and is powered up using V-Ref or other system voltage. The Hall-effect sensor outputs a digital signal by blocking a magnetic field with a vane from a semiconductor sensor.

Hall-effect sensors are used in most current equipment joysticks replacing potentiometers. They can also signal shaft position/speed, oil level, and transmission ratios.

Induction Pulse Generator. An induction pulse generator consists of a toothed disc known as a reluctor, pulse wheel, or tone wheel (slang but commonly used: **chopper wheel**) with evenly spaced teeth or serrations rotated through the magnetic field of a permanent stationary magnet. As the field builds and collapses, alternating current (AC) voltage pulses are generated and relayed to the ECU. Reluctor-type sensors are used variously on modern equipment in applications such as antilock brake system (ABS) wheel speed sensors, and all types of shaft speed sensors.

Switches. Switches complement the sensor circuit and can usually be classified as command inputs. Switches may be electromechanical or "smart"; that is, they use signals rather than analog voltage values to signal a change in status. Switches can be subdivided into three groups:

1. Switches grounding a reference signal (V-Ref). A good example would be that of a coolant level sensor. V-Ref from the ECU grounds through the coolant in the upper radiator tank. If the coolant level drops below the sensor level, the reference signal loses its ground, which after a preprogrammed time period, for example, 8 seconds, signals a fault alert.

2. Manual switches control electrical circuit activity. Many versions of this type of switch are used by the operator to control machine functions. A good example is the standard ignition key.

3. Smart switches use digital signals to indicate a change in status: these signals may be initiated by a ladder switch (resistor bank) or by an integral processor. The signals produced by a smart switch may be automatically generated by a change in status condition or be generated by a mechanical action such as an operator toggling a switch.

OUTPUT CIRCUIT

The switching apparatus used with each electronic management system is what truly differentiates one manufacturer's system from that of its competitors. All ECUs must effect the results of processing into action and they do this using ECU drivers to control actuators.

Actuators

A wide range of actuators are used on machine control systems. Most are fairly simple in terms of their operating principles. We take a brief look at some of them here.

Solenoids. A simple solenoid consists of a coil and an armature. The coil is usually stationary and the armature moves within it. The armature can be integral with devices such as hydraulic poppet valves, spool valves, and levers that effect mechanical movement. They are widely used in machine hydraulics and engines. A simple solenoid has two status conditions: *off* or *on*. In most cases, the armature is spring-loaded to default to the mechanical *off* position when no current is flowed through coil. When an ECU driver energizes the coil, the magnetic field that builds mechanically moves the armature. Because solenoids use electromagnetism, they respond more slowly due to the time required to build and collapse electromagnetic fields: This limits their use in applications that require fast response and frequent cycling.

Proportioning Solenoids. Proportioning solenoids are coil and armature devices that function similarly to a solenoid except that they are capable of precise linear or rotary positioning. To ensure the accuracy of the position, most proportioning solenoids have a means of signaling *actual* position while the ECU driver attempts to control current flow to the proportioning solenoid to maintain desired position. An example of a linear proportioning solenoid would be the oil control spool valves widely used in machine electronics.

Piezo Actuators. Piezo actuators have an advantage over solenoids because of their super-fast response rates. The mechanical response of a piezo actuator occurs the instant it is hit with its electrical trigger. The first generation of piezo actuators was bulky due to the size of the stack of piezo wafers required, but this is changing fast. Not only is the physical size of these actuators being reduced (to date, they tend to be a little bulkier than solenoids) but they also require lower electrical actuation pressures (voltages) and current drawn than solenoids. Piezo actuators are often used in the rumble packs integrated into joysticks for sensory feedback, covered a little later in this chapter.

Stepper Motors. A stepper motor is a brushless electric motor capable of precision positioning of a shaft (and whatever is connected to it). A normal direct current (DC) electric motor spins when current is flowed through its electromagnetic fields. Stepper motors use multiple toothed electromagnets arranged around a rotor gear integral with the motor shaft. All but one of the toothed electromagnets are slightly offset from the gear teeth on the rotor: This means that when the rotor gear teeth are perfectly aligned with one of the toothed electromagnets, they are slightly offset from the teeth on the others. When the next electromagnet in the rotation is energized, the rotor gear moves slightly to realign the magnetic field and completes a "step." Typically, four electromagnets are used spaced 90 degrees apart in the rotation. In this way the rotor can be precisely aligned to any position in rotation. The precision increases with the number of teeth in the rotor and, correspondingly, on the electromagnets.

Stepper motors are used for precision positioning of valve and door/gate controls on machines, such as those used in exhaust gas recirculation (EGR) mixers. Because a constant DC current flow creates heat (and therefore resistance variables), they are regulated by pulse width modulation (PWM) to reduce resistance-generated heat in the controller.

JOYSTICKS

The key control device in modern heavy and agricultural equipment is the **joystick**. The first generation of joysticks were analog input-only devices. That has changed. Although the primary function of a modern joystick is to drive input commands or messages, it is also an output device, enabling the management system to provide real-time, tactile feedback to the machine operator. Today's joysticks are also digital devices and although the general operating principle of a digital joystick is similar to its analog predecessor, the outputs are delivered at higher speeds and much greater precision.

Joystick Operating Principle

The machine operator interface with a joystick is the stick. The stick is mounted on a fixed ball device that permits linear and axial inputs. In addition, the handle can be fitted with switching devices in the palm hub to enhance the control and system management capacity of the device.

The simplest type of **analog joystick** is shown in **Figure 19-2**. This consists of a fixed box section within which a pair of rotatable, slotted axis shafts are mounted on pivots. Each axis shaft is diametrically opposed, as shown in **Figure 19-2**. When the lower end of the joystick, that is, its length below the fixed ball, is inserted through the axis shaft slots, moving the control (upper end) of the joystick moves the axis shafts into any position. Tilt the joystick forward from the control end and the lower end will move rearward. Tilt the control end of the joystick northeast, and the lower end will move southeast. In this way, the control end of the joystick can be moved into any position the mechanics of the axis shafts and positioning of the fixed ball allows. They are therefore configured for three-dimensional movement. Because the objective of a joystick is to signal a specific joystick position to a computer, motion sensors have to be located at the pivot points of the two axle shafts. Now we have a simple joystick.

What the joystick is doing is measuring the lower end of the stick's position on two axes. We usually call these the X-axis East-West) and the Y-axis North-South). When the control end of the joystick is moved, the X-Y coordinates pinpoint a geometrically exact location; the coordinates can then be signaled to a computer so it can plot whatever command is appropriate. In a simple joystick, a number of springs are used to center the device when the control end is released by the operator. This action will zero out the X-Y coordinates, signaling a neutral position.

While the primary function of any joystick is to signal position, we can also equip the control end of the handle with switches. For instance, while playing a computer game or flying a jet fighter plane, the joystick handle could be equipped with a fire button. When the fire button is depressed, the resulting action

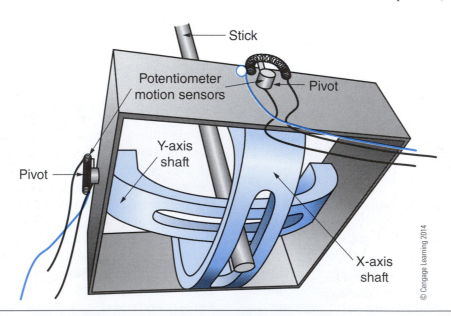

Figure 19-2 Simple analog joystick.

would be to shoot down a virtual aircraft in the computer game, or a real target from the jet fighter. Joysticks are a critical component used in today's off-road equipment.

Analog Joystick

If all we wanted to do with the joystick was to play computer games, an analog joystick would probably meet our needs. An analog joystick is one that uses a pair of potentiometers, one for each axis shaft, as motion sensors. Potentiometers are contact-type voltage dividers that consist of a variable resistor, a contact wiper, and three terminals:

- V-Ref (constant 5 volts from the computer)
- Signal voltage (depends on the wiper position on the variable resistor)
- Ground

By locating a potentiometer at the pivot points on the X and Y pivot shafts, the exact joystick coordinates can be signaled to the control computer using analog voltage values. Early mobile machinery joysticks used analog joysticks. Each potentiometer was supplied with a constant value V-Ref of 5 V-DC; tilting the joystick moves a contact wiper over a variable resistor altering the signal voltage. The signal voltage returned by each potentiometer can then be used to plot a precise position on the X-Y axes. **Figure 19-3** shows how potentiometer-type axis motion sensors are located on a typical analog joystick assembly.

Digital Joysticks

Hall-effect joysticks were introduced in the 1980s, filling a need to produce more accurate and reliable joysticks. Today, most agricultural and heavy-equipment joysticks (as well as those used on cranes, laparoscopic surgical navigation, commercial and military aircraft) use digital Hall-effect sensors. **Digital joysticks** using Hall-effect sensors can also add a rotary Z-plane as shown in **Figure 19-4**. This is useful where precise real-time responses are required of an operator.

Hall-effect sensors are described earlier in this chapter. The mechanics of a digital joystick function identically to analog versions. The difference is that the potentiometers located at each axis shaft are replaced by non-contact Hall-effect sensors. Contactless sensing provides longer operational life and greater accuracy. Strain gauges are also used in some applications; these generate force transducers so that the output signal to the computer is proportional to the physical effort applied to the joystick handle rather than tilt deflection alone. However, at this moment in time, strain gauges are more likely to be used in a machine training simulator rather than the machine itself.

Haptic Feedback. Some joysticks today are built with what is known as **haptic feedback**. A haptic joystick is an active output device that provides sensory feedback to the operator in much the same way a phone on *vibrate* mode does: the computer can deliver

Figure 19-4 Digital joystick with X and Y axes plus a rotary Z input.

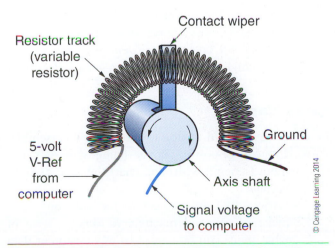

Figure 19-3 Potentiometer-type axis motion sensor.

a haptic response to the joystick handle that causes it to:

- resist tilt movement by returning an opposing force
- vibrate when implement feedback resistance is encountered

Depending on the nature of the sensory feedback required, a haptic joystick uses piezo actuators, electro-active polymers, and minute electric motors to deliver feedback to the operator. Pressure sensors within the device measure the force applied to the joystick, which is signaled back to the ECU in the form of a sensory cue that can determine the aggressiveness of the resulting response. Depending on the equipment, joysticks may be used to control hydraulic implements, steering, and powertrains.

SELF-DIAGNOSTICS AND PROGRAMMING

An advantage of SAE J1587 and J1939 compliance is the fact that current systems can at the minimum be read by generic electronic service tools (ESTs) that use RP1210 communication protocols. This is achieved by using the SAE **message identifiers (MIDs)** or **source addresses (SAs)** for modules connected to the bus, and the use of processing **diagnostic trouble codes (DTCs)** through **subsystem identifiers (SIDs)**, **parameter identifiers (PIDs)**, and **fault mode indicators (FMIs)**. In most cases, fault code read-outs can be displayed on the digital dash display. **Figure 19-5** shows the operator interface on a modern off-road machine.

Self-diagnostic capability in machine electronic systems can be grouped as onboard in the form of flash codes or digital data displays, and EST-read such as scan tools and PC-based systems. The self-diagnostic capability of most current controllers goes beyond merely identifying electronic faults. Logic maps are used to identify many common hydromechanical faults.

ECU Programming

Vehicle ECU programming can be divided into two general categories:

- customer data programming
- proprietary data programming

Customer Data Programming. Customer data programming generally refers to any information that may have to be changed on a routine basis. It is usually concerned with day-to-day running parameters that the machine owner has ownership of. Technicians commonly perform customer data programming tasks such as altering tire size data and machine travel speed limits. Customer data programming fields may be password protected.

Proprietary Data Programming. Proprietary data programming is the data over which the machine manufacturer has ownership. The technician enables proprietary data programming but does not actually perform the reprogramming. Proprietary data includes any ECU-programmed data that the OEM does not want the customer to change. If critical ECU files become damaged or corrupted, this could also require proprietary data reprogramming. Proprietary data programming normally takes place in three distinct stages: downloading, programming the ECU, and uploading verification.

ONLINE TASK

1. Check out the following OEM web sites:
2. Penny and Giles Controls *www.pennyandgiles.com*
3. DeltaTech Controls *www.deltatechcontrols.com*
4. Dearborn Group Technologies *www.dgtech.com*

Shop Tasks

1. Locate an old analog joystick and disassemble. Connect a power supply (suggest a 5 V-DC input to begin with) and chart the X-Y axes by voltage.
2. Use an EST to connect to the data bus on a piece of off-road machinery. Make a note of what fields can be read. List any customer data programming options.

© Cengage Learning 2014

Figure 19-5 Control console on a modern off-road machine.

Summary

- All current off-highway machines use computers to manage the engine, transmission, hydraulics, and many other chassis systems.

- A machine with multiple ECU-managed systems usually networks them to a chassis data bus known as J1939: a data bus networks the different chassis systems. They can be either hard wire or optical.

- A vehicle ECU information-processing cycle comprises three stages: data input, data processing, and outputs.

- RAM or main memory is electronically retained and therefore volatile.

- The master program for system management in an ECU is usually written to ROM.

- PROM data is used to qualify the ROM data to a specific chassis application.

- EEPROM is used for write-to-self capability, detailing customer and proprietary programming, holding failure strategy, logging codes, and the recording of audit trails and tattletales.

- *Multiplexing* is the term used to describe system network transactions in which two or more ECUs share data over a data bus such as J1939, and other machinery buses, which may be hard wire or optical.

- Input data is divided into *command* data and system *monitoring* data.

- Thermistors precisely measure temperature and operate on either an NTC or a PTC principle.

- Variable capacitance-type sensors are often used to measure pressure values.

- Piezo-resistive sensors using a Wheatstone bridge circuit can also be used for pressure measurement signaling.

- Hall-effect sensors generate a digital signal and are used to signal either rotating or linear position and speed data.

- Induction pulse generators are used to input shaft rotational speed data.

- Machine system ECUs are responsible for regulating reference voltage, conditioning input data, processing, and driving outputs.

- Engine management ECUs can be mounted on the engine itself or in a remote location such as under the dash.

- ECU output devices are known as actuators. Some examples of actuators are solenoids, proportioning solenoids, stepper motors, and piezo actuators.

- Joysticks may be analog or digital devices. They are primarily input devices and signal linear and rotational instructions to the ECU.

- Haptic joysticks provide sensory feedback to the operator, which increases system awareness.

- Customer data programming includes those parameters that may require changing on a day-to-day basis as well as those that may have to be reprogrammed if the vehicle application changes or critical components are changed.

- Proprietary data reprogramming is usually performed in these three stages: downloading system files, reprogramming the ECU, and uploading verification.

Review Questions

1. Which of the following data retention media is electronically retained?

   A. RAM   C. PROM

   B. ROM   D. EEPROM

2. To which of the following memory categories would customer data programming be written from an EST?

   A. RAM   C. PROM

   B. ROM   D. EEPROM

3. A Hall-effect sensor is capable of signaling data concerning:

    A. pressure   C. fluid flow

    B. temperature   D. rotational or linear position

4. Induction pulse generators are used to input data concerning:

    A. pressure   C. rotational speed

    B. temperature   D. altitude

5. Which of the following components uses a voltage divider principle and is used on some older machine joysticks?

    A. pulse generator   C. thermistor

    B. potentiometer   D. galvanic sensor

6. When the handle of a joystick is tilted southwest (SW), in which direction does the lower end of the stick move?

    A. northeast   C. southeast

    B. northwest   D. southwest

7. Technician A states that Hall-effect joysticks use a noncontact signaling principle that makes them more reliable than potentiometer types. Technician B states that a Hall-effect joystick does not require an input voltage to function. Who is correct?

    A. Technician A only   C. Both A and B

    B. Technician B only   D. Neither A nor B

8. What type of motion sensor is used in an analog joystick?

    A. yaw meter   C. bubble level

    B. potentiometer   D. Hall-effect

9. What type of motion sensor is used in a digital joystick?

    A. yaw meter   C. bubble level

    B. potentiometer   D. Hall-effect

10. How many motion sensors are required in a joystick that plots only on X and Y axes?

    A. one   C. three

    B. two   D. four

20 AC Electric Drive Systems

Learning Objectives

After reading this chapter, you should be able to:

- Explain the advantages of an AC diesel-electric drivetrain.

- Identify some of the OEMs providing AC drivetrains.

- Identify some specific machines using electric-drive (ED) technology.

- Explain how an AC trolley-drive powertrain functions.

- Identify some of the key components used in an AC drivetrain.

- Describe how rectifiers and inverters function.

- Explain how AC is generated, then rectified to DC, and then inverted back to AC.

- List the advantages of AC traction motors over equivalent DC motors.

- Outline the components housed in an inverter cabinet.

- Perform some basic level troubleshooting on off-highway ED vehicles.

- Identify critical safety practices for working on high-voltage ED vehicles.

- Understand the importance of undergoing machine-specific training before working on high-potential ED vehicles.

Key Terms

alternating current (AC)

chopper module

crowbar

direct current (DC)

electric drive (ED)

electrocution

electronic control module (ECM)

Electronic Technician (ET) software

excitation field regulator (EFR)

hybrid drive

inverter

inverter cabinet (IC)

rectifier

regenerative braking

resistance temperature detection (RTD)

rimpull

Service Information System (SIS)

trolley-drive

INTRODUCTION

This chapter covers heavy-duty, off-road, electric drivetrains. Electric drive systems are a variation of the **hybrid drive** system that has become commonplace in light-duty on- and off-road vehicles over recent years. A hybrid drive system is one that can option two drive sources (diesel-electric/diesel-hydraulic, etc.). The heavy-duty drivetrain arrangement used as the basis for this chapter consists of a diesel engine that drives a generator. The **alternating current (AC)** power produced by the generator is transmitted to an **inverter cabinet** that rectifies the AC to **direct current (DC)** similar to the way an alternator functions. The electrical power is transmitted as DC through the chassis, and then inverted back to AC by thyristor invertors to power-rugged, three-phase induction motors. These three-phase induction motors act as the drive traction motors located at each rear wheel station in trucks or track drive sprockets in dozers. The role of the traction motors is to convert AC electrical energy to mechanical drive torque.

Background

Electrical drive has been used by Komatsu, Liebherr, and Terex for a number of years, and more recently has been optioned by Caterpillar. In addition, hybrid drive powertrains have been used by Deere, Volvo, and others. In certain applications, **electric drive (ED)** offers the advantages of improved fuel economy and a smoother engine operating environment. In heavy off-road equipment, the fuel economy bonus can be up to 30% improved over equivalent mechanical drive systems. When a diesel engine is powering a generator, it can be run at consistent rpms and permits the use of a smaller engine than would be found in an equivalent mechanical or hydrostatic drivetrain. In addition, ED trucks lend themselves to trolley-assist or **trolley-drive** technology such as that used in the Komatsu rock truck shown in **Figure 20-1**. Trolley-assist is used in deep open-pit mines: it allows power to be picked up by the truck from overhead electrical lines supported by retractable trolley poles. Trolley-assist can be cost-effective in saving diesel fuel, as the overhead line powers the vehicle instead of the diesel-driven onboard generator. **Figure 20-2** shows a 2011 Caterpillar D7E and **Figure 20-3** shows the operator view from the cab and one of the two joystick control paddles.

Figure 20-1 Komatsu AC-drive rock truck that can option trolley-drive.

Figure 20-2 Caterpillar D7E bulldozer.

Key Definitions

To comprehend this chapter, you must understand basic electrical theory as it applies to motive power technology. Here are some reminders:

Alternating current (AC) is current that cyclically reverses direction: the number of times it does this per second is measured in hertz. The frequency of

Reprinted Courtesy of Caterpillar Inc.

Figure 20-3 Operator view from a Caterpillar D7E cab that features joystick paddle controls and an AccuGrade digital operator interface.

standard AC current of our grid is 60 hertz, often expressed as 60 cycles per second. AC is usually produced by rotating a coil within a magnetic field. An alternator is an AC generator.

Direct current (DC) is current that flows in one direction only: it is the standard current used in most highway vehicles. DC is stored as potential in batteries.

Diodes are two-terminal, solid-state (electronic) devices that act as electrical check valves. When used to convert AC to DC, they allow current to pass through them in forward-bias and block it in reverse-bias.

Rectifiers convert AC into DC. They are solid-state (electronic) devices that operate by grouping diodes in bridges.

Inverters convert DC into AC. In years past, electromechanical inverters were used to convert DC to AC, but today's inverters use transistors to produce the same effect. Solid-state inverters can produce a more precise AC sine wave frequency than that available from the power grid.

Applications

The AC electric drive systems described in this chapter are largely based on Caterpillar technology used in:

- D7E bulldozer with a C9.3 six-cylinder diesel engine
- 795F AC rock truck with a C175-16 diesel engine

The source power in the D7E dozer is a C9.3 common rail (CR) fueled diesel engine. **Figure 20-4** shows a left-side view of the engine through the powertrain access door on the D7E.

© Cengage Learning 2014

Figure 20-4 The Caterpillar C9.3 engine location in the D7E dozer.

Source power in the 795F is from a Caterpillar 16-cylinder C175 diesel engine. This type of rock truck is designed for open-face mining and will be the primary reference in this chapter. The AC electrical drivetrain used in the 795F allows the use of a smaller engine than might be found in an equivalent capacity truck using a conventional mechanical drivetrain, which would typically be powered by a C175-20 powerplant. **Figure 20-5** shows the layout of the major powertrain components of the Caterpillar 795F AC rock truck.

Trolley-Drive

Although trolley-drive technology requires a major upfront investment, the payoff can be enormous. Retractable electrical trolley lines currently cost around $2 million per mile. However, powering trucks hauling payloads of well over 200 tons on steep gradients out of open-face mines use a lot of diesel fuel, so the investment can make sense. Trolley lines typically use

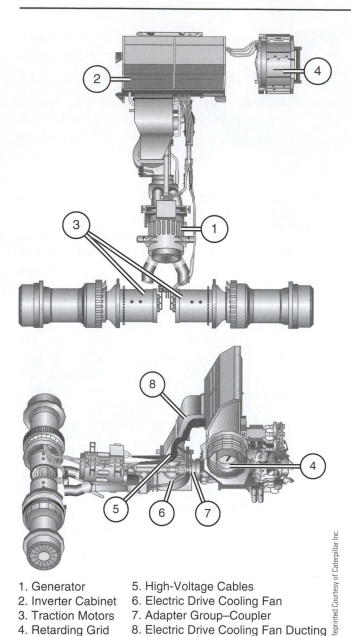

1. Generator
2. Inverter Cabinet
3. Traction Motors
4. Retarding Grid
5. High-Voltage Cables
6. Electric Drive Cooling Fan
7. Adapter Group–Coupler
8. Electric Drive Cooling Fan Ducting

Reprinted Courtesy of Caterpillar Inc.

Figure 20-5 Drivetrain components of a Caterpillar 795F AC rock truck.

voltage values of 1,600 to 1,800 V-AC so AC drivetrains lend themselves to trolley-assist where continuous heavy-haul runs over the same track are required. The truck connects to the overhead trolley line at the beginning of the haul, and disconnects when the haul is completed. Fully loaded travel speeds tend to be significantly higher when an AC drivetrain is powered directly by the trolley current.

Komatsu has always manufactured AC drivetrains capable of trolley-drive. Some examples of 200-ton plus, AC-driven, mining trucks with trolley-assist capability are:

- Komatsu 860E-1K with Cummins QSK60 power and Siemens electronics

- Terex MT 6300 AC with MTU 16V 4000 power
- Liebherr T282B with MTU 16V 4000 power
- Belaz 75302 with MTU 16V 4000 power and GE electronics
- Hitachi EH5000 AC2 with MTU 16V 4000 power
- Caterpillar 795F with Cat 175-16 power

Why Use AC-Drive Motors?

Today's AC motors can be managed by a solid-state, adjustable frequency inverter that adjusts frequency and voltage to control the rpm of an otherwise fixed-speed AC motor. Control is typically produced through pulse width modulation (PWM) of drive output, enabling voltage and frequency to be maintained in a constant ratio at any motor speed. AC drive is a better bet when in a drivetrain environment that can be explosive, wet, or otherwise hazardous. In addition, AC motors require significantly less maintenance. Modern AC motors can provide full torque at stall rpms, an application that used to require a series-wound DC motor. The advantages of AC motors include:

- Lower initial cost
- Lower maintenance cost
- Safer in moist and harsh operating environments
- Full torque at stall rpms
- Adjust to fast-changing loads
- Precise speed regulation

Electrical current is usually stored in batteries on vehicles. Batteries store potential as DC. When DC stored in batteries from regenerative braking has to be converted back into AC to power an AC motor, an inverter is required. Inverters use solid-state (transistorized) oscillators to produce a sine wave AC form that can drive traction motors.

AC-DRIVE POWERTRAIN COMPONENTS

The Caterpillar AC drivetrain uses diesel power to provide mechanical torque to an AC generator. The AC is then converted to DC using solid-state rectification that works similarly to that used on vehicle alternators. DC power is transmitted from the generator rectifier through the chassis, and then rectified back to AC by thyristor invertors to power AC induction motors. These three-phase induction motors act as the drive traction motors located at each rear wheel station in trucks or track drive sprocket in dozers. **Figure 20-6** shows the drive sprocket used on a D7E dozer.

Figure 20-6 Drive sprocket on a D7E dozer.

Generator

The drive source for the AC is a diesel engine using a torque converter as an intermediary. The torque converter is coupled to a drive shaft that connects to an adapter, and then to the generator. Because the diesel engine is driving a generator, it can be run at optimum engine and generator speeds.

The 795F AC traction generator (**Figure 20-7**) is a three-phase, two-bearing, eight-pole, synchronous generator. It is cooled by an external blower and uses a brushless excitation system. The brushless excitation system can be compared to a brushless alternator in terms of its operating principles and is said to provide lower maintenance costs and longer maintenance intervals because there are no brushes to

maintain and replace. The power output of the generator is 2,510 kW at a maximum voltage of 1,930 volts. It weighs 9,800 pounds (4,445 kg).

Generator Rotor. **Figure 20-8** shows the generator rotor. Three of the eight sets of poles and field windings have been removed from the image so that the key components can be better identified. The rotor is made up of the following:

- Shaft (callout 1)
- Spider (callout 2)
- Field poles (callout 3)
- Windings (callout 4)

The field windings use insulated copper wire that is layer wound onto the poles. The poles connect to the spider, and V-blocks (callout 5) are located between adjacent field windings. The spider and pole windings are then coated and sealed with an epoxy resin. The rotor assembly is shrink-fitted and keyed to the shaft with a pair of balance rings (callout 6) bolted to each end of the rotor.

Generator Output. The rotor field windings receive DC current from the generator's excitation system. This DC current produces a powerful magnetic field, which is cut by the stator as the rotor turns. Motion of the rotor's magnetic field as it cuts through the stator windings induces an AC voltage on the stator. This induced voltage is the generator output voltage.

Excitation System. Shown in **Figure 20-9**, the excitation system consists of an exciter stator assembly and an exciter rotor assembly.

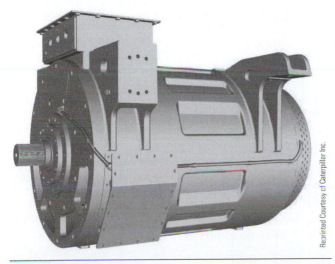

Figure 20-7 The AC traction generator used in the 795F rock truck powertrain.

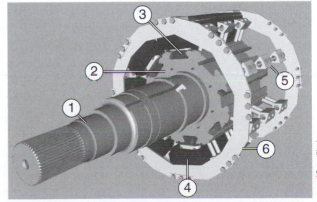

1. Shaft 3. Field Poles 5. V-blocks
2. Spider 4. Windings 6. Balance Rings

Figure 20-8 AC generator rotor assembly: three of the eight poles have been removed on the right side to enable a better view of the internal components.

Figure 20-9 (a) Exciter stator. (b) Exciter rotor.

The exciter stator produces the electromagnetic field. It encircles the exciter rotor, which forms the armature. The sole function of the excitation system is to supply DC current to the generator field windings in the generator rotor. In this way, the rotor produces the magnetic field that, when turned, induces voltage into the stator windings.

Rectifier. The rectifier rotates with the exciter rotor. It is a three-phase, full-wave bridge rectifier that converts AC from the exciter armature to DC. The DC is then conducted to the generator main rotor, which contains the field windings. The rectifier works on much the same principles as those used in an alternator. There are six diodes, three in forward (positive) bias and three in reverse (negative) bias. The field windings are arranged in a wye formation as shown in **Figure 20-10** before supplying the diode bridge. The DC waveform that results after rectification is shown in **Figure 20-11**.

Excitation Field Regulator

The **excitation field regulator (EFR)** is a module that manages the excitation field and therefore generator output. The EFR receives its input commands from the drivetrain ECM, and based on the command signal, generates the appropriate DC current. This DC current is then routed into the generator exciter coil. The location of the EFR on the 795F is shown in **Figure 20-12**.

The EFR receives a pulse width modulated (PWM) constant frequency, command signal request from the drivetrain ECM that runs between 5% and 95% on time. Based on this signal, the EFR enables module operation.

Module Operation. Module operation begins when its input supply voltage reaches 144 volts. At this point, the EFR directs current flow into the exciter: this is monitored by diagnostic feedback, which should signal normal operation mode. At the moment the EFR is enabled, the drivetrain ECM supplies the EFR with the request PWM signal that ranges between 5% and 95% on time. The EFR supplies the requested current output based on the requested PWM signal. Module operation begins at the moment POWER INPUT and ENABLE INPUT fall into an acceptable range.

The EFR operates when supplied with a power input (battery voltage) between 18 and 32 V-DC.

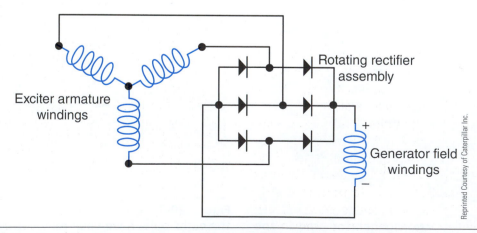

Figure 20-10 Schematic of the wye exciter winding formation and diode bridge used to rectify AC to DC.

Excitation assembly armature output

Figure 20-11 Excitation armature output before and after rectification.

Figure 20-12 EFR location under the generator assembly on the 795F truck.

The optimum power input voltage is 27.5 V-DC ± 0.5 V-DC (this equates to the charging circuit voltage of the 24-volt DC system) and although the EFR will function at 18 V-DC, it will be unable to supply full output current.

Connection Boxes. A high-voltage connection box encloses the generator output terminals. An auxiliary connection box encloses the terminal boards for the connection of the bearing temperature monitors, the stator temperature monitors, and the exciter field connections. **Resistance temperature detection (RTD)** thermometers are used to monitor bearing and stator temperature. RTDs are resistive thermal devices used to measure the change in the electrical resistance of gases that takes place with temperature change: they may also be known as platinum temperature detectors (PTDs).

Inverter Cabinet

Vehicle speed and rimpull on a mechanically driven truck are managed by the transmission. **Rimpull** is a rating of the torque available between the tire and whatever surface the vehicle is moving over; it is limited by traction. On an **electric-drive (ED)** rock truck, power from the generator is transmitted by means of high-voltage cables to an **inverter cabinet (IC)**. The IC plays a critical role in managing rimpull, direction, and speed of the truck.

Figure 20-13 and **Figure 20-14** show the front of the IC. It contains the components and controls

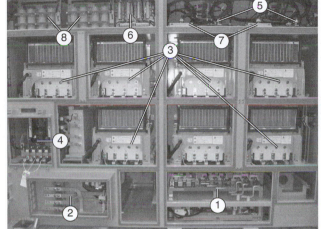

1. High-voltage connections
2. Retarding grid connections
3. Phase modules
4. Contactor assembly
5. Motor ECMs
6. Terminal board
7. Interface modules
8. Bulk capacitor assembly

Figure 20-13 A front view of the inverter cabinet on a 795F truck.

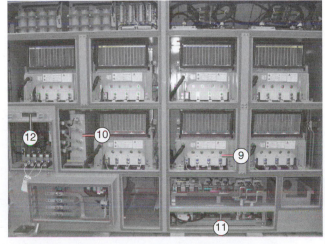

9. Chopper module
10. DC power bus capacitor
11. Crowbar assembly
12. Inverter active lamps

Figure 20-14 A front view of the inverter cabinet on a 795F truck.

necessary to rectify, control, and invert the powertrain system to both move (propel) and retard (brake/regenerate) the truck. During operation, the direction of electrical power flow depends on the mode of operation. There are two basic modes of normal operation:

- Propel mode
- Retarding (braking) mode

Propel Mode. Electrical power from the engine-driven generator is rectified into DC electrical power. DC power is then transmitted and converted (inverted) into variable voltage and frequency AC power. AC power is then routed to the AC traction motor. The actual voltage and frequency supplied to the AC traction motors is based on operator request processed through the drivetrain ECM, which generates a torque command.

Retarding Mode. When the truck is operated under retarding mode, the AC traction motors are used to generate AC power. This can happen in three ways:

- Inertial energy required to maintain travel speed
- Reducing machine speed at any time
- Braking or stopping the machine while on downhill grade

In retarding mode, the role of the traction motors is reversed and they become generators. This is known as **regenerative braking**.

IC Components. The following components are housed in the inverter cabinet (shown in **Figures 20-13 and 20-14**); some are described in more detail later in the chapter.

High-Voltage Connections. (1) contain the bus connections for the generator and traction motors. This section also contains current flow sensing devices that signal the motor ECM.

Retarding Grid Connections. (2) contain the bus connections for the retarding grid and its blower (cooling) motor. This is also known as the periscope.

Phase Modules. (3) convert the DC voltage on the DC bus to three-phase AC power for the traction motors. The gate command and feedback signals use fiber optics to message the interface and phase modules.

Conductor Assembly. (4) contains a pair of high-voltage contactors that close to direct power from the DC bus to the dynamic retarding grid circuits.

Motor ECMs. (5) the two motor ECMs use J1939 to communicate with other system ECMs and a proprietary bus to communicate with the drivetrain ECM and dynamic retarding inverter (DRI).

Terminal Board. (6) connects the current and voltage sensors to the motor ECMs.

Interface Modules. (7) connect the motor ECMs and phase modules by means of fiber optics.

Bulk Capacitor Assembly. (8) maintains a 24-volt supply to the interface modules in the event of a failure of the chassis 24-volt system. The assembly consists of a cassette bank of eight capacitors.

Chopper Module. (**Figure 20-14**) (9) regulates DC bus voltage by routing power to the dynamic retarding grid.

DC Power Bus Capacitor. (10) filters DC power from the rectifier suppressing transient voltage spikes and stores energy for proper inverter operation.

Crowbar Assembly. (11) is a solid-state device that routes power from the DC bus to the second circuit of the dynamic retarding grid during shutdown or certain fault modes.

Inverter Active Lamps. (12) are a pair of LEDs that are connected across the DC bus and illuminate until bus voltage exceeds 150 V-DC.

Traction Motors

ED trucks have no differential to propel the truck. Instead, AC power is transmitted by high-voltage cables to traction motors (see **Figure 20-5** callout #3) located inside the rear axle housing. Caterpillar 795F rock trucks use a pair of traction motors, one for each rear drive wheel.

Wheel Speed Sensors. A pair of wheel speed sensors with different functions is mounted to each traction motor. Speed sensor 1 feeds back wheel speed to the traction motor ECM on the wheel in which it is mounted. Speed sensor 2 signals wheel speed to the drivetrain ECM. Each wheel speed sensor uses a Hall-effect operating principle that outputs a constant frequency, digital output with a finite phase shift so that the direction of motor rotation can be read by the ECM receiving the signal.

Electronic Control Modules

Management of the 795F electric drivetrain system is provided by three **electronic control modules (ECMs)**. The ECMs used in the system are:

- Drivetrain ECM
- Motor 1 ECM
- Motor 2 ECM

The drivetrain ECM is located in the cab front ECM compartment. The two motor ECMs are located in the inverter cabinet (see **Figure 20-13**). The primary functions of the three control modules are described below.

Drivetrain ECM.

The drivetrain ECM is located in an ECM compartment inside the cab. The drivetrain ECM is responsible for overall control of the system. It is therefore responsible for generator output at any given moment of operation. Generator output depends on operator (driver) input and the operational mode of the truck. The power converter control modules are also managed by the drivetrain ECM.

Because the powertrain ECM is responsible for the overall management of the system, it sends torque commands to the motor control ECMs that manage the AC traction motors. The drivetrain ECM is also responsible for managing the electric drive cooling fan.

Motor ECMs.

The two motor control ECMs receive their commands from the drivetrain ECM. The motor 1 ECM controls the operation of the driver power transistors in the inverter cabinet that are used to manage the left-side traction motor (motor 1) while the motor 2 ECM does the same for the right-side traction motor (motor 2). Each motor control ECM monitors the output current and the operating temperature of the traction motors they manage. In addition, each motor control ECM has other dedicated functions:

- Motor 1 ECM controls the operation of the **chopper module**. The chopper module switches the voltage level of the DC power bus and the retarding contactors that are used for retarding (braking). It regulates DC bus voltage by sending power to the second circuit of the dynamic retarding grid. This can occur while the truck is in propel or retarding mode. It can also happen during certain fault conditions and when the drivetrain ECM is programmed for engine load testing. Command and status signals for the chopper are transmitted by fiber-optic signals to an interface module.

- Motor 2 ECM controls the operation of the **crowbar**. The crowbar is used to initiate an immediate discharge of the DC power bus during system shutdown. System self-diagnostics that identify a potential danger can also initiate a crowbar discharge of the DC power bus. The crowbar is an electronic device that directs power to the DC bus of the dynamic retarding grid during fault and shutdown conditions. Also located on this assembly are the DC bus voltage sensor, ground fault sensor, and ground relay capacitor.

The two motor ECMs communicate with other powertrain ECMs on the J1939 bus. In addition, they also communicate by means of a proprietary bus line to the drivetrain ECM and the dynamic retarding inverter (DRI). When required by the driveline ECM, the motor ECMs are responsible for the high-speed power switching patterns that produce the drive-wheel induction motor torques that move the machine. Each motor ECM controls one inverter (one-half of the phase modules) and one traction motor.

DC Power Bus Capacitor.

In addition to the capacitors in the phase and chopper modules, there is an external DC power bus capacitor. This capacitor filters the DC from the rectifier, absorbs transient voltage spikes, and provides energy storage required for proper inverter operation.

Dynamic Retarding Inverter.

The dynamic retarding inverter (DRI) controls the grid blower motor. The DRI receives DC power from the dynamic retarding grid and inverts it to AC. The three-phase AC output from the DRI controls the speed of the retarding grid blower. A DRI capacitor protects the DRI from transient voltage spikes that can occur from the contactor assembly. **Figure 20-15** shows a schematic of the retarding circuit; note the location of the chopper in the circuit.

Contactor Assembly.

The contactor assembly (see **Figure 20-16**) consists of two contactors and the contactor pilot relay. Each contactor is a single-pole, normally open (SPNO) switch. The contactors have a set of triple-face contacts that undergo a wiping action each time they open and close. This wiping action increases reliability and reduces maintenance by minimizing arcing. Under normal conditions, the voltage present on the DC bus is routed to the first circuit of resistors on the dynamic retarding grid when the contactors close.

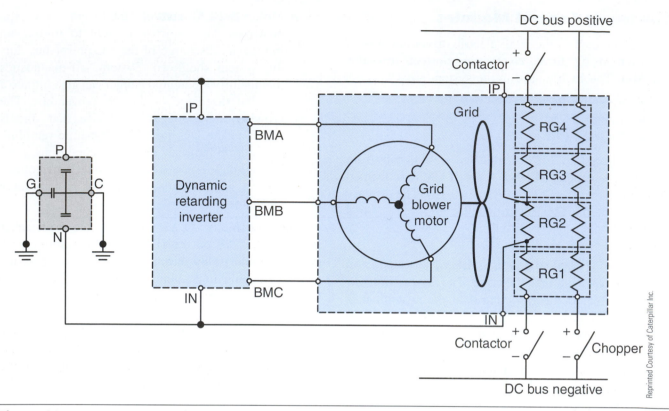

Figure 20-15 A DRI circuit schematic. Note that DC bus voltage can be increased proportionally as the retarding requirement increases.

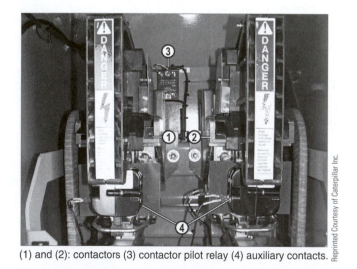

(1) and (2): contactors (3) contactor pilot relay (4) auxiliary contacts.

Figure 20-16 Contactor assembly on a 795F truck.

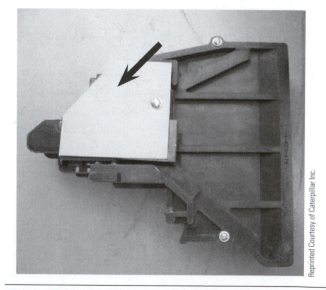

Figure 20-17 Contactor arc chute.

The contactors are controlled by the motor 1 ECM. They activate under retarding, certain fault modes, during load testing, and during shutdown. Each contactor contains an arc chute with an attached set of permanent magnets as shown in **Figure 20-17**. The arc chute contains a safety interlock so the main contacts will not operate without the arc chute in place.

Arcing occurs any time the contacts are opened under load. Arcs between the contact surfaces can reach temperatures as high as 25,000°C. As the arc is ignited between the contacts, magnets on both sides of the chute (indicated by arrow in **Figure 20-17**) draw in the arc to neutralize it. The arc chute is made of corrugated, insulating material. The corrugations of the arc chute intermesh in order to provide a serpentine path for the arc. The arc is extended, cooled, and rapidly extinguished as it is forced to travel in the arc chute. This process helps minimize replacement of the contacts

---

Tech Tip: Never attempt to clean contactor tips. When contacts are tested to be out of specification, they should be replaced.

Interface Modules

Two interface modules (IMs) connect the motor ECMs to the phase modules of the converter group by means of fiber-optic conductors. The IMs have two major subcomponents:

- Fiber-optic interface: converts electrical signals from the motor ECM to optical signals that communicate with the intelligent power module (IPM) in the phase modules and the crowbar.
- Gate drive power supply: creates the 24 V-AC required to switch the gates for each insulated-gate bipolar transistor (IGBT) located within each phase module.

Phase Modules

Phase modules are used for torque control of the AC traction motors. There are six in total and they constantly adjust the voltage and the frequency of the AC power supplied to the motors. They carry out the torque commands computed by the drivetrain ECM. The phase modules are powered up at 24 V-AC.

Intelligent Power Modules. Each phase module houses four intelligent power modules (IPMs). The IPMs contain the control circuits for interpreting optical signals from the interface module and switching gate status of the internal insulated-gate bipolar transistors (IGBTs).

The IGBTs produce the high-speed switching patterns of the DC bus voltage as dictated by the motor ECMs. It is this switching of the DC bus voltage that creates the AC waveform required to propel the traction motors at the wheels. IPMs are paired in parallel to form the AC output for each of the phases required to maintain the DC bus voltage when the IGBTs are gated. IGBTs conduct no current while their gate is in the OFF state, so each IPM contains a freewheeling diode in parallel with the IGBT. This provides transient protection rectification when the drivetrain is in retarding mode.

High-Voltage Connections

The high-voltage compartment houses the bus connections for the generator power cables, along with the power supply cables for the traction motors. The lugs that connect to the bus are double-crimped to the power cables with a hydraulic crimper. Each lug is double-bolted to the bus to reduce movement and minimize electrical problems that could result from poor connections.

Inverters

One interface module is required for each inverter in the inverter cabinet. The inverter cabinet houses two inverters (one in each side). These are known as inverter 1 and inverter 2. Interface module 1 is associated with inverter 1, and interface module 2 is associated with inverter 2.

Inverter 1 consists of six pairs of intelligent power modules (IPMs) for traction and an IPM pair for the chopper. Inverter 2 consists of six pairs of IPMs for traction and a crowbar thyristor.

Digital/Optical Signals

The interface module uses 11 digital inputs. These signals are used to switch on and off:

- the inverter IPM
- the chopper IPM
- the crowbar thyristor

During normal operation, the interface module is enabled by a driver from the appropriate motor ECM. As soon as the interface module is in enabled mode, the IGBT gate drive digital inputs ("1"/high or "2"/low) are converted to an optical output to be relayed to the appropriate IPMs for torque control. During normal operation, as long as optical feedback matches optical command, electrical feedbacks will be inverted from electrical command. The medium used for optical transmission to and from the interface module is a plastic optical fiber (POF) cable.

Command and Feedback Signal Timing

The timing of the optical outputs from the interface module and optical input signals from the IPMs is precisely monitored and compared in real time. The relationship between command and feedback signals is used to indicate fault conditions. Command and feedback signals should match under normal operating conditions. When gate drive signal does not match the feedback signal, an event code is activated and broadcast on the bus. Event codes are the same for each motor ECM.

AC POWERTRAIN DIAGNOSTICS

Because of the extreme danger represented by high-voltage drivetrains, technicians must complete the OEM-recommended training before attempting to diagnose system problems. In addition, they should have access to the OEM diagnostic software, OEM online service information system, and have all the appropriate safety resources at their disposal. A brief introduction to some electronics diagnosis is provided here as an example of the routines required.

ECM Diagnostics

Diagnostic codes can be activated by any system ECM. Codes are broadcast using J1939 (SA/MID FMI) formats and displayed on the digital dash display. Active and logged codes on the Caterpillar 795F are accessed using Caterpillar's **Electronic Technician (ET) software** on a laptop computer. A comprehensive list of the system diagnostic codes that can be launched by each of the three control modules can be viewed on Caterpillar's **Service Information System (SIS)**. **Figure 20-18** shows the inverter and ECM assemblies on a D7E dozer.

Diode (Rotating Rectifier) Continuity Test

Diodes can be accessed through the exciter access cover on the bottom of the right-hand side of the generator. The engine/rotor has to be rotated to access all of the diodes. Remove the nut and washer securing one pair of diode lead wires. When the diode has been

Figure 20-18 Inverter and ECM layout in a D7E dozer.

removed from the rotating rectifier assembly, test the diode using a digital multimeter (DMM). Refer to SIS for test specifications.

An optional continuity test may also be performed. Connect the test leads across the diode in one direction; then reverse the test leads. There should be continuity in one direction but not the other. If there is continuity in both directions (shorted diode) or no continuity in either direction (open diode), the diode is defective. Follow proper torque requirements when replacing diodes.

Tech Tip: Caterpillar states that all the diodes should be replaced if any one diode tests defective. Do not remove diodes from their bases to test, because the soft copper threads on the diode can deform.

Generator Malfunctions

Table 20-1 explains some causes and remedies for generator malfunctions. These are sourced from Caterpillar; the company underlines the fact that seemingly minor defects can cause serious system damage if not promptly repaired. The service information system (SIS) referred to is the Caterpillar SIS.

SAFE WORK PRACTICES ON ED VEHICLES

Because of the high voltages involved and number of different types of ED drivetrains, the first and most important rule is for technicians to have had manufacturer training on the specific piece of machinery they are going to work on. An AC-driven machine manufactured by Komatsu may have some commonalities with those manufactured by Caterpillar or Liebherr but to truly know the machine, OEM-specific training is mandated. Failure to observe this rule exposes a technician to **electrocution** risks. The definition of electrocution is death by electrical shock. **Attempting to work on a machine you do not understand can put your life at risk.** Furthermore, equipment manufactured offshore may use different regulatory codings than domestically manufactured machinery, so nothing should be assumed.

Personal Safety

Listed here are the general rules of personal safety when working around the high voltages generated by

TABLE 20-1: GENERATOR OUTPUT PROBLEMS		
Performance Symptom	Possible Cause	Remedy
No voltage output	Excitation lead is open, shorted, or grounded to frame.	Check lead continuity and excitation conductors for grounding. Replace lead(s) if required.
	Output short circuit.	Disconnect generator leads from converter group and then check for short or ground in generator leads. If the generator leads check OK, then check the generator field windings.
	Field coils shorted or open.	Check for short or open in the generator field. Check SIS instructions to access rotating rectifier assembly. Disconnect the generator field leads from the rotating rectifier assembly to check the generator field windings.
	No excitation to generator.	Check EFR output. Refer to SIS for correct operation.
Low generator voltage	Excitation circuit not functioning properly.	Verify correct engine rpm.
	High resistance connections (they will be warm/hot).	Check connections at rectifier. Verify correct torque.
	Weak field due to high temperatures.	Verify operation of the generator cooling circuit.
Fluctuating generator voltage when run at a constant speed	Excitation circuit not functioning properly.	Refer to SIS for regulator operation troubleshooting.
	Defective bearing resulting in uneven air gap.	Replace bearings.
High-voltage output from generator	Excitation circuit not functioning properly.	Refer to SIS for regulator operation troubleshooting.
	Overspeed.	Reduce engine rpm.
	Converter control malfunction.	Verify converter control operation.
	Blocked ventilation passages.	Clear/clean air passages.
	High ambient temperature.	Change the way the truck is operated.
Vibration	Improper mounting or misalignment of coupling.	Correct alignment.
	Defective bearing.	Replace bearings.
	Rotor out of balance (after repair).	Re-balance rotor.
	Vibration caused by other equipment.	Check other equipment.
	Rotor out of balance caused by winding failure.	Check for overspeed condition. Repair or replace rotor and re-balance.

© Cengage Learning 2014

electric drive vehicles. Some of these rules are common sense, others maybe not: they are listed here in no particular order of importance.

Eye Protection. This is a no-brainer and has become mandated in many jurisdictions. Safety glasses should be worn in any heavy-equipment service facility whether or not their use has been made compulsory by the employer, the employer's insurance company, or the local jurisdiction. When working around electricity, safety glasses are even more essential.

Jewelry and Body Piercings. Watches, jewelry, and body piercings are more often than not manufactured from metals that are good conductors of electricity. It is common sense to remove all metals from your person even when located on parts of the body that you would not normally expect to come into contact with the machine. The same rule applies to metals located in the pockets of coveralls such as wrenches, screwdrivers, and lighters.

Insulating Gloves. When working around electrical machinery, it is best to ensure that no part of your body comes directly into contact with the machine. Rubber insulating gloves can protect against high-potential voltage discharge and save lives. When working around ED vehicles, insulating gloves should always be worn, even after the vehicles have been electrically disabled.

Electric Shock Resistant (ESR) Footwear. Any mechanical technician working around high-voltage-potential chassis equipment should wear ASTM Class 75 electric shock resistant (ESR) footwear. Make sure you read and understand the section on safety footwear in Chapter 1 of this book. ESR footwear is designated by an orange omega on a white rectangle in both the United States and Canada. The standards are as follows:

- ASTM ESR footwear: must withstand the application of 14,000 volts at 60 hertz for 1 minute with zero current flow or leakage exceeding 3.0 milliamperes under dry conditions.
- CSA ESR footwear: must withstand the application of 18,000 volts at 60 hertz for 1 minute with zero current flow or leakage exceeding 1.0 milliamperes under dry conditions.

Failure to wear the appropriate footwear can mean the difference between life and death. Despite innumerable safety measures used on chassis circuits, technicians should be aware that high-potential (voltage) circuits are more prone to leakage when something goes wrong: remember that *voltage* equals *pressure*.

Moisture Awareness. High humidity, pooled water, and damp conditions all provide excellent paths for conducting electricity. Before working on ED drivetrains, pay attention to the ambient conditions on, and around, the machine you are to work on. The rubber sole on your ESR boot is not going to help much if you are standing in a puddle of water.

Key-Off. Generally, you should make sure that the ignition circuit is off before working on a machine. At times you will be prompted by diagnostic software to activate the ignition circuit; when this is part of a structured diagnostic routine, it's obviously justified, but only then. In fact, when you are working on an ED machine, it is good practice to remove the key so that the machine is unable to start.

Electrically Disable Machine. For general service work, it makes sense to electrically disable the machine. Again, it might be necessary to enable the electrical circuit(s) during a diagnostic routine, but do so only when prompted by the software. Check with the OEM service information software to determine how to safely electrically disable the machine if it is not obvious.

Fire Awareness. High-potential ED machines represent a greater fire hazard due to the danger of arcing and the increased number of batteries. Keep a fire extinguisher nearby—a multipurpose dry chemical extinguisher is the best bet.

Capacitive Discharge Precautions. Even when keyed-off and electrically disabled, ED vehicles still contain components that can unload a lethal amount of electricity into your body. Battery banks and cassettes of ultracapacitors store electrical energy and must be respected at all times. This underlines the importance of receiving OEM training on the specific machine you are working on.

Err On the Side of Caution. When you are unsure of what you are working with, make it your business to find out, using your brain. Even if you have had training in the systems you are up against, it is easy to forget, especially if you do not work with the specific machine on a regular basis. Using your brain means consulting the OEM service literature either online or in hard-copy service manuals. It may also involve seeking the advice of another technician who is more familiar with the equipment.

Summary

- AC diesel-electric drivetrains are used in off-highway applications because they provide significantly better fuel efficiency, longer engine life, and lower maintenance costs.

- Off-highway, electric drive (ED) vehicles are manufactured by Caterpillar, Komatsu, Terex, Liebherr, Hitachi, and Belaz.

- Off-road, ED technology lends itself to rock trucks used in open-face mining application due to lower emissions and better fuel economy. It is also used in some dozer applications including the new Caterpillar D7E.

- High-potential ED off-road vehicles can be easily adapted for AC trolley-drive. AC trolley-assist requires overhead trolley cables slung over a consistently used route, usually located for a continuous loaded, uphill pull of a rock truck. Trolley lines typically use voltage values of 1,600 to 1,800 V-AC so AC drivetrains run entirely off grid-supplied electricity while connected to the grid.

- When sourcing trolley power, an ED truck connects to the overhead trolley line at the beginning of the haul, and disconnects when the haul has been completed.

- Trolley-line, full-load travel speeds tend to be significantly higher than when under diesel-electric power mode.

- Rectifiers use solid-state technology to change AC to DC in exactly the same way an alternator does the job.

- Inverters convert DC into AC using transistors that produce a more precise AC sine wave frequency than that available off the power grid.

- A diesel-electric ED vehicle uses a diesel engine to drive an AC generator: the AC is rectified to DC, routed to battery storage and an inverter cabinet that alters the DC back to AC to drive traction motors.

- AC traction motors can provide maximum torque at stall rpms and generally require much less maintenance than equivalent DC drive motors.

- The inverter cabinet (IC) is the operational heart of the ED drivetrain: it contains most of the electronics required for vehicle operation.

- The IC manages drivetrain outcomes in two modes: propel mode and retarding mode.

- Troubleshooting off-highway ED vehicles requires OEM training, the OEM diagnostic software, and access to the OEM service information system.

- Personal and machine safety practices must be understood by anyone working on or around high-voltage ED vehicles.

Review Questions

1. What component is used to convert AC electrical power to DC electrical power?

   A. rectifier   C. transducer

   B. inverter   D. transformer

2. What component is used to convert DC electrical power to AC electrical power?

   A. rectifier   C. transducer

   B. inverter   D. transformer

3. Technician A says that a Caterpillar D7E uses a diesel-electric AC powertrain. Technician B says that the traction motors used in most current off-highway vehicles require DC power. Who is correct?

   A. Technician A only   C. Both A and B

   B. Technician B only   D. Neither A nor B

4. Identify the common voltage values used in a typical trolley-drive grid:

   A. 12 V-DC   C. 220 V-AC

   B. 24 V-DC   D. 1,600–1,800 V-AC

5. Technician A says that AC traction motors generally require higher maintenance than equivalent power DC traction motors. Technician B says that current AC motors are capable of producing maximum torque at stall rpms. Who is correct?

   A. Technician A only   C. Both A and B

   B. Technician B only   D. Neither A nor B

6. Where is the drivetrain ECM located on the Caterpillar 795F AC ED rock truck?

   A. inverter cabinet (IC)   C. left rear traction motor housing

   B. front cab ECM cabinet   D. right rear traction motor housing

7. Where are the motor ECMs located on the Caterpillar 795F AC ED rock truck?

   A. inverter cabinet (IC)   C. on each traction motor assembly

   B. front cab ECM cabinet   D. below the generator module

8. Technician A says that when an ED vehicle is running under trolley power, travel speeds tend to be lower than when the onboard diesel-electric generator is providing the power. Technician B says that the operating cost of a loaded rock truck is much lower than when operating under onboard diesel-electric generator power. Who is correct?

   A. Technician A only   C. Both A and B

   B. Technician B only   D. Neither A nor B

9. Technician A says that the term *electrocution* means death by electric shock. Technician B says that once a high-potential ED vehicle has been powered down, the risk of electrocution is eliminated. Who is correct?

   A. Technician A only   C. Both A and B

   B. Technician B only   D. Neither A nor B

10. If an AC generator excitation field circuit has failed, what is the likely outcome?

    A. excessive generator output voltage   C. no generator output

    B. low generator output voltage   D. voltage surges

11. Technician A says that AC motors are safer to operate than DC motors in a high humidity environment. Technician B says that DC motors are safer to operate than AC motors in an environment where there are explosion dangers. Who is correct?

    A. Technician A only   C. Both A and B

    B. Technician B only   D. Neither A nor B

12. What diesel engine is used to power a Caterpillar D7E dozer?

    A. Caterpillar C175-20   C. Caterpillar C15

    B. Caterpillar C175-16   D. Caterpillar C9.3

Glossary

absolute pressure a pressure scale with its zero point at absolute vacuum. Expressed as pounds per square inch absolute (psia).

accumulator (hydraulics) a device used to store potential energy. May refer to a pressure storage device in a hydraulic circuit or electrical energy storage devices.

active suspension a hydraulic suspension system used on off-road equipment and large haulage trucks. The system has the capability of making changes to the ride height.

actuator device that produces mechanical motion in response to an electrical, air, or hydraulic signal. Examples are solenoids and piezo stacks (electrical), brake chambers (pneumatic), and telescopic rams (hydraulic).

adaptive suspension a suspension designed to adjust to load and road conditions; this usually means an air suspension system. Adaptive suspensions can either be non-electronic or electronically controlled. The term *active suspension* is also used.

aeration the blending of a gas into a liquid such as the mixing of air into oil by the churning action of rotating gears.

aftercooler describes a variety of different heat exchangers used on engines, transmissions, and air brake systems. An aftercooler on an air brake system is used to condense water from moisture-saturated air leaving the compressor by cooling it.

air dryer a device, usually located downstream from the air compressor, designed to remove moisture from the air system. Air dryers use two principles to eliminate air from an enclosed brake circuit: cooling/condensation and dessicant media.

air/hydraulic master cylinder an air-actuated hydraulic master cylinder that enables a pneumatic brake control circuit to actuate a hydraulic brake circuit.

air-over-hydraulic brakes brake system that uses an air control circuit to actuate a hydraulic slave circuit.

air-over-hydraulic intensifier converts control circuit air pressure (from the brake treadle valve) into hydraulic slave circuit pressure to actuate the foundation brakes.

air spring suspension any one of a number of different types of pneumatic suspension.

alternating current (AC) current flow that alternates in direction producing a sine wave of nodes and antinodes. Usually produced by rotating a coil within a magnetic field.

Allen wrench hex-shaped wrench designed to fit into recessed hex Allen screws.

alloy mixture of metallic elements to create a specific metal. Steel is an alloy of iron and carbon.

amboid gear a gear arrangement in which the axis of the drive pinion gear is set above the radial centerline of the crown gear.

American National Standards Institute (ANSI) administrator and coordinator of U.S. private-sector, voluntary standardization. ANSI standards are developed by consensus.

analog use of physical variables to represent numbers such as voltage or length.

analog joystick a system control joystick that uses potentiometer-type motion sensors to signal X and Y axis commands.

annulus the largest gear member of a simple planetary (epicyclical) gear set. Located outside of the planetary gears.

anti-corrosion agent a solution or additive that inhibits chemical corrosion.

application pressure the hydraulic pressure that is used to apply the service or park brake on a hydraulic brake system.

articulating steering a steering design that uses hydraulic cylinders connected to the front and rear sections of a loader frame. The cylinders are connected together through a hinge pin, which articulates the front and rear sections of equipment to achieve steering control. The axles used on this equipment are non-steer axles.

axial load a pure tension or compression load acting along the long axis of a straight structural member or shaft.

axial runout the amount of wobble on a shaft or rotor when viewed from the front or back.

axle shaft a shaft that is splined to the differential at one end and transmits torque to the wheel ends.

back pressure the potential pressure that is created on the return side of a hydraulic circuit. This added resistance to flow increases the pressure required to move the load.

backlash the lost motion or looseness (play) between the faces of meshing gears or threads.

baffle a separating plate used in a hydraulic tank to separate the inlet side from the outlet side.

banjo housing the housing of an axle assembly that contains the differential, brake, and drive components.

bar metric unit of pressure measurement roughly equivalent to one unit of atmospheric pressure.

beach marks the term used to describe shock loading on gears which shows up as small curved rings on the surface of the component.

bead expander device used to mount tubeless tires: prevents inflation air from escaping by forcing the tire beads into the tapered bead seat.

belleville spring see *diaphragm/bevel spring*.

belleville washer a type of cone-shaped washer that provides a preload on the part to which it is attached. Used on older crawler tracks to provide a seal to keep dirt from getting between the pins and bushing.

Bernoulli's principle states that the sum of the kinetic and potential energy in a moving, incompressible fluid will remain the same. So, in a fluid moving in a steady state through a circuit, if the potential energy decreases, the kinetic energy must increase so that the sum of the two remains the same.

bevel gear a gear design used in differential spider gears in a conventional differential to split the flow of power proportionally between wheel ends.

biodegradable matter that can be decomposed by other living organisms.

boom circuit part of an excavator implement hydraulic circuit. The boom arm connects to and pivots on the upper cab structure. The boom circuit is actuated by double-acting hydraulic cylinders.

Boyle's Law states that for a given, confined quantity of gas, pressure is inversely related to the volume so that as one value goes up the other has to go down.

brake application pressure the hydraulic pressure that is used to apply the service brakes. This pressure is modulated by the foot brake valve and can vary from zero to the maximum allowable brake application pressure limit.

brake chambers foundation brake actuators that convert pneumatic pressure into linear force to actuate slack adjusters.

brake pads the friction faces and plate assembly that is used in disc brake calipers to stop a vehicle.

brake pressure regulated by either a charge valve or a pressure regulating valve, which is often part of the foot brake valve assembly.

brake pump a hydraulic pump that is used to charge the brake circuit with hydraulic oil. Used to apply the service brake or release the spring-applied park brake.

brake shoes the friction faces and lever assembly that is forced against the drum in a brake system to slow down or stop a vehicle.

brake valve a hydraulic valve that is foot operated. Modulates pressure applied to the wheel ends of a vehicle to slow down or stop it quickly or smoothly.

brinelling grooves worn into a cross by the needle rollers and caused by a lack of lubrication. Can also be caused by excessive loading.

broaching the breaking down of the walls of a pin bore during the installation of a pin.

broken-back driveshaft a driveshaft installation where the U-joint angles are equal, but the yokes are not parallel.

bucket circuit part of an excavator implement hydraulic circuit. The bucket connects to and pivots on the stick; the bucket circuit is actuated by a double-acting hydraulic cylinder.

bull-type final drive a term used to describe a drive that utilizes a single gear reduction. Often found in a final drive assembly of a bulldozer.

burst pressure the pressure at which a fluid power device is rated to fail or rupture when subjected to overpressure. Usually 2½ to 4 times specified working pressure.

bypass an optional or secondary route for fluid flow in a circuit.

calcium chloride a white deliquescent salt ($CaCl_2$) mixed with water to add weight (ballast) to machine wheels.

caliper a disc brake component that is used in conjunction with brake pads to apply force on a rotor to slow down or stop a vehicle.

cam an eccentric or oval shaped component.

cam ring an eccentric liner, usually stationary, within which mobile pump members rotate.

camber the tilt of a wheel assembly when viewed from the front of the vehicle from true vertical. A wheel that is leaning inward at the top from true vertical is said to have negative camber. A wheel that is leaning outward at the top from true vertical is said to have positive camber.

Canadian Standards Association (CSA) Canadian safety standards association that is the equivalent of our Underwriters Laboratories (UL).

cardan universal joint a device in which a cross connects the yokes of a driveshaft.

carrier assembly the drive axle assembly that holds the gearing that changes the flow of power at right angles to propel the wheels.

carrier rollers located above the frame rails. Their purpose is to support the weight of the top section of track as it rolls between the idler and the sprocket. These rollers generally have a single flange, which aids in controlling track sag and whipping during operation.

case drain (hydrostatic drive) a term used to identify the return oil flowing back to the hydraulic tank from the case or housing of a piston pump or motor. The oil flow is a result of internal leakage built into the piston pump required for internal lubrication.

caster the tilt of the kingpin in either the forward or backward direction when viewed from the side. Positive caster is when the kingpin leans to the rear of the centerline. Negative caster is when the kingpin leans forward of the centerline.

cavitation bubble collapse erosion in hydraulic or engine coolant circuits.

center of gravity the position at which a body of mass is at a balanced point from gravity.

centerbolt a bolt that is used to hold a multi-leaf spring pack together.

central processing unit (CPU) computer subcomponent that executes program instructions and performs arithmetic and logic computations.

centrifugal clutch a clutch that utilizes a centrifugal force to apply pressure against a friction disc in proportion to the rotational speed of the unit.

ceramic facing a clutch disk facing made from a mixture of ceramics and copper or iron.

chain drive a term used to describe the final drive design used on a grader tandem drive.

chain hoist device used for lifting or lowering a load by means of a drum or lift-wheel around which chain wraps manually, electrically, or pneumatically driven and may use chain, fiber, or wire rope connected to a hook as its lifting medium.

charge check valve a check valve that is internally located in the charge valve. Prevents the flow of oil in the charge valve from returning back to the tank.

charge pressure the pressure in an accumulator used in a hydraulic system that has fluid under pressure stored in it.

charge pump a positive displacement pump that is used to supply oil flow to a piston pump in a hydrostatic drive circuit.

charge relief valve a relief valve that is located internally in a brake charge valve. Used to ensure that maximum brake pressure is never exceeded.

charge valve an automatic regulating valve that is capable of regulating pressure in a brake circuit between a high and a low setting.

check valve a valve in a fluid circuit used to permit flow in one direction only.

cherry picker a portable shop hoist device consisting of a hydraulically raised lift arm or boom.

chock the blocking of a wheel on equipment before performing any work. A triangular metal object that looks like a wedge is used to chock the equipment from moving.

chopper module the chopper module switches the voltage level of the DC power bus and the retarding contactors that are used for retarding (braking).

chopper wheel rotating disc that cuts a magnetic field to produce rotation speed or position data to an ECM by signaling either an AC voltage value (analog) or a pulse width modulated (digital) signal.

closed-center a hydraulic circuit in which the pump is not unloaded to the reservoir when in the center or neutral position.

coil spring a coil of wire that contributes to the motive force of a spring. In extension and torsion springs, all the coils are active coils. In compression springs, only the coils that show daylight between them are active coils. A spiral-shaped spring that can be compressed or extended without permanent deformation.

collar shift has parallel shafts fitted with gears in constant mesh. If gear speeds are different, grinding will occur. Used for low gear on some older model transmissions.

come-along a cable or chain hook, hand-actuated, linear ratcheting device used for lifting or applying separation pressure.

compensator control a component that is used on a variable-displacement pump or motor to control the displacement as required to compensate for pressure changes.

compound driveline angle a term used to describe the angular relationship of a driveline on both horizontal and vertical planes.

compressibility the measure of the volume change in a fluid when subjected to pressure.

compressor any of a number of mechanically driven devices used to compress a gas. In air brake systems, compressors create the potential energy required to actuate foundation brakes.

conductor in electricity, any material that permits electrical current to flow; any element with less than four electrons in its valence shell. In hydraulics, enclosed circuits or pipes that route hydraulic media through the circuit.

cone-type clutch a cone clutch serves the same purpose as a disk or plate clutch. However, instead of mating two spinning disks, the cone clutch uses two conical surfaces to transmit friction and torque.

constant-mesh a transmission that contains helical gears. In this type of transmission, certain countershafts are constantly in mesh with the main shaft gears. The main shaft meshing gears are arranged so that they cannot move endwise. They are supported by roller bearings that allow them to rotate independently of the main shaft.

cooler any of a number of different types of heat exchanger.

cooling circuit a specialized oil circuit often used in a hydraulic brake system to direct a metered amount of oil through a cooler to remove excess heat.

coordinated steer both front and rear steer axles turn in or out in opposite directions, allowing the equipment to turn in a tighter circle.

counterbalance valve a valve that creates resistance to flow in one direction while permitting unrestricted flow in the opposite direction.

coupler device that links two components or circuits such as the fittings used in hydraulic circuits. In most cases, couplers have both a male and a female component.

coupling any one of a number of different types of mechanical connection. Used to transfer fluid, linear, or rotary force.

crab steer used on equipment that requires precise steering control in confined spaces. All four wheels turn in the same direction at the same time, allowing the equipment to move to the left or right while maintaining load alignment.

cracking pressure a condition that occurs in a hydraulic system when a pressure reaches a predefined value allowing excess flow to bypass a circuit.

cross the "body" of a universal joint.

cross tube used to connect the equalizing beams together. Reduces potentially damaging load transfers that occur during operation on haulage trucks with equalizer bean suspensions.

crowbar solid-state device used in AC hybrid drivetrains to divert power from the DC bus to the retarding grid during shutdown or certain fault modes.

crown gear (differential) the larger gear in a differential that is driven at right angles by a pinion gear. Some manufacturers refer to it as a *ring gear*.

cylinder (hydraulic) a component that is capable of converting fluid energy into mechanical motion.

cylinder pressure describes the existing nitrogen charge pressure found in suspension cylinders on off-road equipment.

dampen to buffer or insulate.

dampened discs clutch friction discs with built-in coaxial springs used to dampen driveline oscillation and shock loads. Many of today's clutches use dampened discs due to the run slow, gear fast, drivelines used with high-torque engines.

dash control valve control circuit valve used to manage parking and emergency braking in an air brake system-articular design of bearing cap commonly found on heavy-duty driveshaft U-joints.

diaphragm/bevel spring like coil spring equipped clutches, diaphragm springs operate under indirect pressure using a retainer and lever arrangement to exert pressure on the pressure plate. Also called a *Belleville spring*.

differential drive axle gearing used to transmit torque at right angles while at the same time capable of varying speed to each wheel during cornering.

differential carrier the component that contains the differential gearing in an axle assembly.

differential hydraulic cylinder a hydraulic cylinder that has opposed piston areas which are not of equal size.

differential lock a term used to describe what takes place when both axles on a drive axle are locked together mechanically.

differential side gear the side gears mesh with the pinions and have internal splines that connect to the axles.

differential spider the cross that is used in a conventional differential to hold the four pinion bevel gears.

differential steer this design uses a series of planetary gears to provide torque input to the drive tracks, with a third input from a steering motor that can increase the speed of each track independently. When the speed of one track is increased, the other track slows down proportionally.

digital joystick a control joystick that uses Hall-effect motion sensors to signal X-Y axes data and sometimes rotational movement (Z signal).

direct current (DC) an electric current flowing in one direction only.

direct drive gearing of a transmission so that one revolution of the engine produces one revolution of the transmission's output shaft. The final drive ratio of a direct drive transmission is 1:1.

directional control valve a valve that functions to direct or block flow through selected channels or circuits.

directional valve a hydraulic valve that can control the direction of oil flow in a circuit or stop it completely.

disc brake vehicle brake system in which squeeze pressure is applied to a rotor by a caliper. Both pneumatic and hydraulic disc brakes are used on off-highway vehicles.

disk-and-plate-type clutch a device used to break torque flow between an engine and transmission and to apply/release application force to other units. Utilizes a friction disc sandwiched between a flywheel and pressure plate.

displacement the amount of fluid that passes through a pump or motor during one revolution or stroke.

displacement control valve a valve that can vary the displacement of oil flowing in a hydrostatic drive circuit.

double-acting cylinder a hydraulic cylinder that is capable of converting hydraulic energy into mechanical motion in either direction.

double check valve used when a component or circuit must receive or be controlled by the higher of two possible sources of fluid pressure. The valve is usually designed to bias the higher source pressure.

double reduction axle a differential that uses two different gear sets to achieve two speeds at the axles.

double reduction drive the term used to describe a final drive that has two separate gear reductions in line to each other. Generally located in a separate housing at the track end of the drive of a bulldozer.

download data transfer from one computer system or network to another.

drift movement or wander from a preset position or value. Can be applied to hydraulic implement position or engine rpm.

drive bushings vertical bushings on each drive pin that allow vertical movement during loading and unloading while still maintaining suspension alignment.

drive gear a gear that drives another gear or causes another gear to turn.

driveline the power transmitting components that connect the transmission output yoke to the drive axle.

driven gear a gear that is being driven or caused to turn by another gear that is meshed with it.

driven member a gear that is driven by another gear.

driveshaft the shaft that connects the powertrain components together to transmit torque to the wheels.

drive steer type of drive axle commonly found on heavy equipment. Provides the capability to steer and propel the equipment at the same time.

drivetrain the group of components that generate power and deliver it to the road surface, water, or air. This includes the engine, transmission, driveshafts, differentials, and the final drive.

drum the rotating component of a drum brake system that is fixed to the wheel assembly, which rotates around a stationary set of brake shoes. Brake action is achieved when the shoes are forced out against the inside of the drums by mechanical force that is controlled through air or hydraulic pressure.

drum brakes a brake system used on equipment that uses a rotating drum attached to a wheel assembly that rotates around stationary friction shoes.

dryer reservoir module (DRM) an integrated reservoir, dryer, governor, and pressure protection pack assembly used in some vehicle air-supply circuits.

dry joint a condition associated with track systems whereby the counterbore sidewall becomes damaged quickly from contact with the bushing when allowed to run dry.

dryseal coupler threaded fitting in which the roots and crests of the opposing threads contact before the flanks contact and a seal is created by crushing the thread crests. Dryseal couplers are usually intended for one-off usage and usually do not require thread seal compound.

dual-circuit application valve the *foot valve* used with FMVSS 121 compliant service brakes and sometimes with off-highway equipment. Masters service applications from primary and secondary circuits. Also known as *treadle valve*.

dynamic friction kinetic (or dynamic) friction occurs when two objects are moving relative to each other and rub together. The coefficient of kinetic friction is typically denoted as μ_k, and is usually less than the coefficient of static friction.

electric drive (ED) term used to describe a vehicle that is driven by electric motors, either in a serial or parallel arrangement with an engine. ED in this book references AC powertrains in a serial drive arrangement with a diesel engine and may also be trolley-drive capable.

electric shock resistant (ESR) footwear designated by an orange omega, ESR footwear must be able to sustain a minimum of 14,000 volts with zero leakage under dry conditions.

electrocution death caused by electrical shock.

electro-hydraulic valve a hydraulic valve, controlled by an electrical signal, which controls the flow.

electro-kinetic friction applies to the forces restricting the movement of an object that is already in motion on a relatively smooth hard surface.

electro-magnetic clutch an electrically actuated clutch used for on-off cycles. Usually spring-loaded (mechanically) to engage and electrically energized to release, but the opposite can also be true.

electronic control module (ECM) an engine or system controller. ECMs may be stand-alone or networked into a proprietary or standard powertrain (J1939) bus.

electronic technician (ET) software Caterpillar's in-house diagnostic software designed to run in a Windows environment.

electronically erasable, programmable read-only memory (EEPROM) vehicle system controller memory category that can be overwritten or flashed (reprogrammed) with customer or proprietary data; includes the computer write-to-self capability.

elevated sprocket this sprocket is not mounted to the track frame but is attached to the final drive. The elevated position keeps the final drive out of the dirt. If the equipment has an elevated sprocket system, there are two idlers, a front idler and a back idler. Equipment with a conventional oval track requires only one idler.

elongation the wear that takes place on crawler tracks. As track components wear, the track experiences elongation.

end yoke the name of the output yoke on a powertrain component that connects to the driveline yoke.

endplay axial play in a shaft wheel or rotor. A dial indicator is used to measure the endplay.

engine control module (ECM) a system or engine controller. An integrated electronic module containing the processing and output hardware required to manage a system. SAE recommends the acronym be used exclusively for engine controllers.

engine/electronic control unit (ECU) a system or engine controller. An integrated electronic module containing the processing and output hardware required to manage a system. SAE recommends that this acronym be used for all system controllers other than the engine but seldom adhered to by OEMs.

equalizer bar or beam a term used to describe a beam that is used in a tandem equalizer beam suspension system.

equalizing beam suspension a suspension system used on off-road equipment that lowers the center of gravity of the axle load.

expanding shoe-type clutch a form of clutching device that utilizes a set of brake shoe-type friction material within a drum used to apply and release engine torque to a specific drive transmission.

excitation field regulator (EFR) module that manages the generator excitation field and therefore generator output. The EFR receives input commands from the drivetrain ECM, and based on the command signal, generates the appropriate DC current: the DC current is then routed into the generator exciter coil.

extra-long (front) (XL) track system: the XL track system has an extra-long front track design (distance between drive sprocket and front idler) that provides better balance with extremely good flotation characteristics, which is better suited for finish grading on free flowing material.

extra-long rear (XR) track system: the XR track system has an extra-long rear section (distance between drive sprocket and rear idler) that is more suitable for pulling scrapers, as well as for winching or ripping operations. In the agricultural applications, a belt track is often used to reduce the zone of compaction. This design increases ground contact significantly over conventional LGP track systems.

eye bolt lifting hardware consisting of a lifting ring and threaded bolt, forged as a single unit. Available in various sizes.

falling-object protection system (FOPS) structural devices mounted on mobile equipment designed to protect the operator in the event of falling objects from above such as when a rock truck is being loaded, or skids, when fork lift trucks are in operation.

false brinelling a condition that is a result of manufacturing. It appears as a polished surface usually associated with bearing surfaces. Often mistaken for a defect.

fatigue wear damage that occurs to metal and components after many hours of operation under heavy loads.

fatigue failures catastrophic failures that occur due to old age or excessive wear.

fault mode indicators (FMIs) define a component or system circuit failure by categorizing it into one of

30 electrical/electronic fault modes (SAE) the diagnostic severity of a failure condition.

ferrule a compression ring used to create a seal between the line and coupling nut. Ferrules used in pneumatic and hydraulic circuits can be made out of malleable plastic, brass, copper, or steels. Many ferrules deform to conform with the fitting when torqued and should not be reused.

fiber rope any type of rope made up of natural or synthetic fibers such as manila (natural) or nylon, polyester, polypropylene, and polyethylene (synthetic).

filter a component that captures contaminants from hydraulic fluid during operation.

final drive see *input shaft* and *output shaft.*

fitting a mechanical coupling device used to attach or connect lines and hoses; usually consists of both a male and a female member.

fixed-displacement motor a hydraulic drive motor that accepts an equal amount of oil per revolution from a hydraulic pump regardless of speed.

fixed-displacement pump displaces an equal amount of oil each stroke or revolution, regardless of speed.

flaring capable of being spread outward. A *flaring* tool is used to expand tubing. Also describes the aerodynamic panels used on some vehicles.

flotation the capability, of a vehicle tread or tire, to remain on top of a soft surface, such as sand, wet ground, or snow.

flow continuous movement of a *fluid* within or outside of a circuit. The term can also be used to describe the movement of electrons in an electrical circuit.

flow compensator a hydraulic component used to control pump output.

flow control valve used in hydraulic circuits to regulate the speed of actuators. Function by defining a flow area. Rated by operating pressure and capacity. This valve can be used in meter-in or meter-out applications; two general types are used: bleed-off and fixed orifice.

flow divider a valve that divides the flow from one source into two or more branches.

flow limiting compensator also referred to as "load sensing" compensator. In either case, they both perform the same function: they control the amount of oil displacement from the pump to the control valve.

flow rate the volume of hydraulic oil that passes through a component during a specific time.

fluid a liquid or gas substance tending to flow or conform to the outline of its container.

fluid coupling a hydrodynamic device used to transmit rotating mechanical power from one source to another.

foot valve a service brake application valve used to apply the brakes and located in the operator's compartment.

force push or pull effort measured in units of weight. Force represents the ability to do work but this ability does not necessarily result in any work done: for instance, you can push all you like on a bulldozer but it is unlikely to move the vehicle. The formula for force (F) is calculated by multiplying pressure (P) by the area (A) it acts on. $F = P \times A$.

forward/reverse shuttle a transmission component made up of clutch plates and discs used to drive a transmission in ether a forward or reverse direction.

friction the resistance to relative motion between two surfaces in contact with each other.

friction clutch the driven disc used in a hydraulic clutch pack of a powershift transmission.

friction disc a type of disc used in a wet disc brake system that has a friction material bonded to both surfaces, which is forced to rub against steel discs in the wheel end of the wet disc brake system by either hydraulic pressure or spring pressure.

full-floating axle an axle that does not carry any weight but transmits torque to the drive wheels.

full flow filter a filter which is capable of accepting the full flow of a hydraulic circuit through it during circuit operation.

galling transfer or displacement of metal. Usually occurs between two metal parts in frictional contact. Caused by a lack of lubrication, or overloading.

gear pitch the number of teeth per given unit of pitch diameter in a gear.

gear ratio the ratio of an input shaft speed to the speed of the output shaft.

gear set two or more gears that intermesh.

gears any of a number of different types of toothed or spiraled shafts designed to either impart or receive torque from another.

grouser the ridges on a track shoe that designate the terrain the shoe will operate on. The hardness of the surface material on the grousers determines whether they are standard or extreme-service track shoes.

haptic feedback the capacity of a joystick to provide sensory feedback to the operator in the form of vibration or resistance using electroactive polymers, piezo actuators, and miniature electric motors.

head pressure equal to the height of a liquid in a column multiplied by the density of the liquid.

heat exchanger a component that transfers heat from one medium (i.e., fluid to fluid) to another.

helical cut gear a gear with teeth cut at some angle other than at a right angle across the face of the gear, thus permitting more than one tooth to be engaged at all times and also providing a smoother and quieter operation than the spur gear.

herringbone gears a gear design used on heavy-duty final drive applications. A gear arrangement in which the drive gears are shaped like chevrons. They provide high torque capability with no axial loading.

high block a heavy-duty design of bearing cap used on off-road equipment U-joints.

high brake pressure any brake pressure above a predefined value in a brake system that is necessary to stop the equipment safely.

high-pressure relief valve a regulating valve used in a hydraulic circuit to protect the system from over pressure.

hoist any of a number of different types of lifting devices including chain falls, cherry pickers, scissor jacks, and cranes.

hook lifting hardware typically used with chains for making a connection to the load being lifted.

horizontal parallel or level with the horizon.

hybrid drive widely used term that describes a powertrain that can option more than one power source: examples are hybrid-diesel-electric, and hybrid-diesel-hydraulic.

hydraulic analyzer a hydraulic circuit diagnostic instrument consisting of a pressure gauge(s), flow meter, gate (choke) valve, and thermometer. May also be equipped with additional instruments.

hydraulic pump a component in a hydraulic system that converts mechanical motion into fluid power.

hydraulic track adjuster a hydraulic actuator that is used to adjust the track tension on a crawler track by increasing or decreasing the hydraulic pressure in the actuator. The actuator adjusts the position of the idler to change track tension.

hydraulics the science of using fluid power circuits that use liquids, usually specially formulated oils, in confined circuits to transmit power or multiply human or machine effort.

hydrodynamic drive in a hydrodynamic drive system, the hydraulic oil flows at high speed and relatively low pressure to provide power to the drive wheels.

hydrodynamics the science of forces acting on or exerted by liquids; usually, but not always, in confined circuits.

hydrokinetics the study of fluid behavior under flow.

hydrometer an instrument used to measure the specific gravity of a liquid, usually coolant or battery electrolyte.

hydropneumatics term used to describe a compound pneumatic and hydraulic circuit: an example is air-over-hydraulic brakes.

hydroscopic a term used to describe a fluid's ability to absorb moisture. A feature of a good brake fluid is its ability to absorb moisture to prevent corrosion.

hydrostatic drive uses a positive displacement hydraulic pump and motor to transfer power through rotary motion to the drive wheels. In a hydrostatic drive system, the oil is subjected to high pressures at a relatively slow speed.

hydrostatics the study of fluid behavior and pressure exerted by liquids at rest.

hygroscopic capable of absorbing moisture. A characteristic of some brake fluids. This is an advantage for hydraulic brake fluid because it inhibits coalesced water droplets that can cause corrosion, freeze, or boil.

hygroscopic breather hydraulic reservoir breather capable of eliminating moisture and airborne particulate through the vent.

hypoid gear set a gear set used in a differential that has the pinion mounted below the centerline of the crown gear.

I.D. acronym that refers to the inside diameter of a circular opening in a component.

idler the front idler on a crawler track is used to align, support and guide the track as it rotates. It also can be used to provide a means of tensioning the track.

impeller/pump an impeller is a rotating component of a pump, usually made of aluminum, which transfers energy from the engine fluid being pumped by forcing the fluid outward from the center of rotation.

inclinometer a device used to measure the angles of drivelines and flanges in a powertrain.

inflammable capable of being *inflamed* or burned.

input retarder a type of brake mechanism used on a driveline between a main transmission housing and a torque converter. Absorbs torque produced by the engine and converts it to heat energy.

input shaft (final drive) a shaft that transmits torque from a power source (transmission) to a final drive.

interaxle differential a gear design used to transmit power from the front axle to the rear axle on a tandem drive axle arrangement.

International Standards Organization (ISO) an international standards governing body that describes itself as ''non-governmental,'' but many ISO standards become legally enforced.

inverter term used to describe a device that transduces DC electrical current into an AC.

inverter cabinet (IC) in a Caterpillar ED drivetrain, the components responsible for managing on-vehicle electricity: this includes rectifying DC to AC, and inverting AC to DC.

jack a hydraulic or air-over-hydraulic vertical lifting device.

jaw clutch a simple mechanical lock used on drive axles to lock the differential, providing positive drive to both wheels.

Joint Industry Committee (JIC) organization that sets compatibility standards for hydraulic circuits.

jounce (travel) the greatest compressed position of a suspension system.

joystick a manual control stick that uses motion sensors to signal commands and in some cases delivers sensory feedback. See *analog joystick* and *digital joystick*.

kidney-loop machine a machine connected into a hydraulic circuit to ''loop'' the oil from the machine through a set of low-micron filters to remove contaminants.

kilopascal (kPa) metric unit of pressure increasingly used by OEMs in North America and universally in the rest of the world. One hundred kPa is equivalent to 14.56 psi.

kinetic energy the energy of motion.

kingpin a linkage pin that allows the steering knuckle to pivot from side to side. Usually made from high-quality steel.

kPa kilo-Pascal. Metric unit of pressure measurement. 101.3 kPa = 14.7 psi = 1 atm.

laminar flow streamlined flow through a hydraulic circuit. A hydraulic circuit should be designed to minimize friction caused by sharp angles and internal obstructions. Also, a desirable flow condition that can occur in a hydraulic system that minimizes friction.

land raised surface beside a groove such as is found on a spool valve.

laws of leverage the mechanical advantage obtained by the use of a lever or combination of levers. The laws of leverage can be derived using Newton's law of motion.

leaf-spring system a series of springs used in equipment suspension systems to provide shock absorbing capabilities. A single spring is called a leaf spring.

limited slip differential used in a differential to provide equal torque to both drive wheels during normal operation.

linear actuator device that converts hydraulic energy into linear motion such as COE cab lift cylinders or dump box rams. Also, a hydraulic cylinder that converts hydraulic energy into linear motion.

liner the stationary external pump member used in gear and vane pumps; also known as a *cam ring*.

link pitch the distance between the bushing bore center and the pin bore center. This dimension does not generally change during the life of the track.

lip seal a type of dynamic seal that is generally made from synthetic material. It uses a spring on the inside of the lip to form a tight seal on a shaft.

liquid ballast a mixture of calcium chloride and water used to fill tires for the purpose of adding extra weight to a vehicle.

live axle a drive axle that propels the equipment. Often used in conjunction with a dead axle in a tandem axle configuration.

load cushion when a suspension is operating with no load, the outer edges of the spring load cushions are in contact with the saddle. As the load increases on the vehicle, the load cushion is compressed, further absorbing the additional loads.

load proportioning valve (LPV) a valve used in equipment with hydraulic brakes to sense load at the rear axle and adjust the brake pressure accordingly.

low brake pressure in a brake system, any brake pressure below a predefined value that is necessary to stop the equipment safely.

low block a light-duty bearing cap design found on light-duty off-road equipment driveline U-joints.

low ground pressure system (LGP) track equipment that uses the low ground pressure system uses a wider track which increases ground contact for better flotation and stability on loose surfaces. Shoe width on the LGP track is generally a lot wider than on the standard track.

machinist's rule a steel ruler used to take shop linear measurements and may act as a straight edge.

magnetic-type clutch the torque is produced by the magnetic field across the air gap between the two sections of the clutch causing them to rotate as one unit.

main pressure the system pressure standard used as system pressure to manage the hydraulic circuits in an automatic transmission.

manifold a fluid conductor that allows for multiple connections.

manometer a single tube bent into a U-shape and mounted to a linear measuring scale. It may be filled to the zero on the calibration scale with either water (H_2O) or mercury (Hg), depending on the pressure range to be measured.

manual control valve a hydraulic control device that is actuated by the operator of the equipment such as a foot valve or lever which controls directional control.

margin pressure the pressure differential between "signal pressure" and hydraulic pump "outlet pressure."

master cylinder a hydraulic brake circuit that develops brake pressure to apply the brakes at the wheel end of the vehicle. In a hydraulic brake system, the dual circuit application control valve.

master link in order to assemble or disassemble some crawler tracks, a master link is required to complete the crawler track assembly.

master pin in order to assemble or disassemble some crawler tracks, a master pin is required to complete the crawler track assembly.

mechanical linkage a series of rigid links connected with joints to form a closed chain, or a series of closed chains. Each link has two or more joints, and the joints have various degrees of freedom to allow motion between the links.

mechanical seal a seal that uses two highly machined metal faces that rotate on each other to provide a seal between two rotating parts. Usually found at the wheel end of a drive axle to seal the oil in the axle housing and rotating wheel end.

message identifiers (MIDs) identifies an onboard electronic system or circuit using SAE standard numeric codes. In a multiplexed data bus, all MIDs have an address on the bus. Term is used by some OEMs as *module identifiers*. In recent systems, the term *source address (SA)* is preferred.

meter to regulate or control the amount of fluid in a circuit.

metering the precise measuring of fluid flow.

metering valve a valve used on a vehicle to regulate the brake pressure to the front and rear brakes. Improves brake balance.

microcontroller a module integrating a microprocessor and output switching: used for system control.

micrometer a precise measuring instrument typically consisting of a thimble, barrel, and anvil. Used to make ID, OD, and thickness measurements.

microprocessor a small processor: sometimes used to describe a complete computer module.

modulate to regulate.

modulated brake pressure varies by the amount of pressure applied to the foot pedal. The foot brake valve varies (modulates) the pressure to the service brake.

modulated foot valve a foot-operated brake valve that is capable of either raising or lowering the brake pressure by increasing the travel of the actuator.

modulating valve a valve that regulates or proportions. In air brake systems, it may have *inversion valve* features.

modulation a lowering or raising condition that occurs in a brake system wheel end pressure that is controlled by the brake valve.

module identifiers (MIDs) term used by some OEMs to describe any controller with an address on the chassis data bus. The term *source address (SA)* is used on current systems.

motor a rotary device that can change hydraulic energy into mechanical energy.

multi-piece rim a tire rim used on large equipment that consists of two or more pieces.

multiple countershaft any transmission that utilizes more than one countershaft to supply the gear ratios necessary for the transmission output shaft.

multiple countershaft transmission a housing containing at least two countershafts having parallel axes of rotation and being fixed against all movement except rotative with respect to the housing.

multiplexing term used to describe the networking of two or more system controllers on a data backbone to enable data exchange and reduce hard wire.

National Institute for Automotive Service Excellence (ASE) body that defines certification standards and manages certification testing for the motive power trades including truck, auto, schoolbus, auto-machinists, etc.

needle bearing the term used to describe the type of bearing used in universal joint cups that rotate against the universal joint trunnion during driveshaft rotation.

needle cup the part of a cross which fits over the trunnion and holds the needle rollers.

needle rollers (bearing) cylindrical bearings positioned inside the bore of the needle cup.

nitrogen-charged accumulator a dampening device used on suspension systems on off-road haulage trucks. Nitrogen-charged accumulators provide a dampening action on the suspension cylinders.

non-constant-mesh entails sliding the teeth of the gear pairs together, and hoping they engage at some point.

non-parallel driveshaft the driveline arrangement where the companion flanges of a transmission and drive axle are not parallel to each other, but the working angles of the U-joints are equal.

non-slip shaft assembly a driveline that is a fixed length. Consists of a weld yoke, tubing, center bearing, and an end yoke.

no-spin differential used on off-road equipment to provide a positive drive to both wheels during straight-line operation. When cornering, the outside wheel is disengaged by a jaw clutch allowing the outside wheel to speed up while the inside wheel supplies the drive.

Occupational Safety and Health Administration (OSHA) federal body that sets standards and monitors employee health and safety in the workplace.

OD acronym for the outside diameter of a circular object such as a bushing.

oil/pneumatic cylinders act as shock absorbers in the suspension of heavy equipment. The front suspension cylinder barrels are attached to the frame of the equipment at the top and are fastened to the steering knuckle at the bottom.

open-center a hydraulic circuit in which pump delivery fluid is allowed to recirculate freely to the reservoir in the "center" (neutral) position.

operator protective structures (OPS) general term used to describe FOPs and ROPs: reinforced safety apparatus installed on mobile equipment designed to protect the operator in the event of a falling object or rollover incident.

optical bus a network that uses optical signals and message packets in place of digital electronics.

orifice a restriction in a circuit which limits flow.

original equipment manufacturer (OEM) used to describe a chassis manufacturer or a manufacturer of major chassis subcomponents such as engines, transmissions, or brakes.

oscillate movement or cycles with rhythmic regularity. See *oscillation*. Can refer to back-and-forth movement.

oscillating hitch used on off-road equipment to provide independent vertical motion of suspension frame pivots.

oscillating undercarriage this design has the added benefit of being able to pivot from front to rear, which can increase stability when the equipment is operating on uneven ground. The track frames have an equalizer bar mounted in between them to allow each track frame to pivot independently on a pivot shaft.

oscillation the act of oscillating. The natural dynamics of frame rails, the vibrations of drivelines, wheel dynamics, and electrical wave pulses.

output retarder a hydrodynamic device located on the output shaft of a transmission that converts power to heat energy. Hydraulic driveline brakes mounted on the rear of a transmission that are designed to apply retarding force directly to the driveline.

output shaft (final drive) a shaft in a transmission that directs torque to a final drive input shaft.

overdrive gearing of a transmission so that one revolution of the engine produces more than one revolution of the transmission's output shaft.

oxidation the result of oxygen reacting with a substance such as a metal or oil. Rust is the result of oxidizing iron.

packing general term used to describe sealing media used in movable hydraulic components.

parallel driveshaft a driveline arrangement that has all the yokes and companion flanges parallel to each other. The working angles of the U-joints must also be equal.

parameter a value, specification, or limit.

parameter identifiers (PIDs) coded components and conditions within a MID or SA used to indicate system fault paths.

park brake valve an electric hydraulic valve that is used in a brake circuit to activate the park brakes.

particle counter monitors and reports the level of contaminants within a hydraulic system.

Pascal's law states that pressure applied to a confined liquid is transmitted undiminished in all directions and acts with equal force on all equal areas and at right angles to those areas.

personal protective equipment (PPE) clothing, eyewear, gloves, and footwear designed to protect workers from injury in the workplace. Most PPE must meet standards defined by the ASTM and approving agencies such as UL and CSA.

phasing the relationship of the end yokes on each end of a two-piece driveline to each other.

pilot any of a number of devices that receive signals and then effect an action. Can also be known as a relay.

pilot line a control or signal line to an actuator. In hydraulic systems, a pilot line often uses a lower pressure value.

pilot-operated relief valve consists of a main relief valve through which the main pressure line passes, and a smaller relief valve that acts as a trigger to control the main relief valve. Used to handle large volumes of fluid with little pressure differential.

pilot pressure a hydraulic valve used to provide lower pressure in a circuit that is used to control another circuit, such as the dump/hoist circuit on a loader.

pilot valve a secondary valve that is used to control the operation of another valve.

pinhole injection injury a serious form of injury caused by the injection of minute, often microscopic, droplets of hydraulic fluid into flesh. The injured person is often not aware of the injury until significant tissue damage has occurred.

pin link pins that join the individual links together on a roller chain.

pinion drive a term used to describe the type of drive used on some types of heavy equipment. Uses two bevel spur gears, a pinion gear, and a ring gear to transmit the flow of power at right angles through a final drive in a powertrain.

pinion gear (differential) a short stub shaft with gear teeth machined on one end that mates to a crown gear in a differential. The crown and pinion gears often come in a set.

pinion shaft a term used to describe a shaft that is used in the final drive of a bulldozer. The pinion shaft is held in place with roller bearings, while the larger final drive gear is usually splined to the track drive sprocket, which is also supported by large tapered roller bearings.

pitch diameter the diameter of the pitch circle. Operating pitch diameter is the pitch diameter at which the gears operate.

pitman arm the connection used between the steering gear and the drag link in a conventional steering system.

pitting surface irregularities on metal that are caused by corrosion or from abrasive wear.

plain bevel gear gears of conical form designed to operate on intersecting axes. Most commonly the shafts will intersect at a 90-degree angle. Straight bevel gears (as opposed to spiral bevel gears) have teeth that are straight, radially, from the gear's centerpoint.

planet gear the name given to the gears (usually 3) that rotate around the inside of a ring gear on the final drive wheel end of a heavy-duty drive axle.

planetary carrier typically, the planet gears are mounted on a movable arm or carrier, which itself may rotate relative to the sun gear. Holds one or more peripheral planet gears, of the same size, meshed with the sun gear and ring gear.

planetary drive a term used to describe a type of drive that consists of at least one ring gear, a sun gear, and two or more planet gears. Used to transmit torque to wheel end or final drive.

planetary gear set a gear set consisting of three elements: a sun gear, a planetary carrier (containing 3 or more planetary gears), and a ring gear (annulus). The three separate elements intermesh, and the hold/release status of each determines the output ratios. Also known as an *epicyclic gear set*.

planetary pinion gears small gears fitted into a framework called the planetary carrier.

planetary shift transmission any transmission configuration that utilizes planetary gearing.

platinum temperature detectors (PTDs) resistive thermal devices used to measure the change in the electrical resistance of gases that takes place with temperature change; they are also known as *resistive temperature detectors (RTDs)*.

ply rating the number of layers used in tire construction.

pneumatic brakes air brakes.

pneumatically controlled clutch a pneumatic release mechanism comprised of a treadle valve and slave cylinder used to engage and disengage a clutch through a medium of compressed air.

pneumatic suspension an air suspension.

pneumatics a branch of fluid power physics dealing with pressure and gas dynamics.

pop-off valve a term used to describe the safety pressure-relief valve used on air brake systems. Most pop-off valves are set to trip at 150 psi and are located on the supply (wet) tank in the system. See *safety pressure-relief valve*.

port plate an axial piston pump component that defines the oil inlet and outlet of the pump.

positive displacement a characteristic of a pump or motor that has its inlet positively sealed from its outlet so that no fluid can recirculate within the component: another way of saying this is: what goes into the pump per cycle is what comes out.

positive-displacement pump a fluid pump where the inlet and outlet are not connected hydraulically.

potentiometer a three-terminal variable resistor used to signal mechanical position by means of an analog voltage value. The terminals are V-Ref, ground, and output signal. Mechanical movement of a wiper over a variable resistor is used to signal position; also known as a contact-type position sensor.

pounds per square inch (psi) standard unit of pressure used in North America: the force acting over a sectional square inch. One psi is equivalent to 6.891 kPa.

power the rate of performing work.

powerflow the mechanical flow path that conducts motive power through a system or component. For instance, the powerflow through a transmission in a specific gear ratio would be the sequential routing of the torque flow path through those gears that are actually engaged.

powershift transmissions in which the shifting or selecting of a gear ratio does not require breaking torque from the engine or the synchronization of gears.

powertrain the components that make up the drive system on equipment. It includes the engine, torque converter, transmission, and final drive.

pre-charge pressure the pressure of nitrogen gas used in an accumulator before hydraulic pressure is added.

preload a load placed on a tapered roller bearing to maintain good tooth contact between gear teeth in a powertrain. Also, a manufacturer's suggested predetermined force that should be applied to a bearing assembly to maintain a specified shaft or gear alignment.

pressure the force applied over unit area. Usually expressed in pounds per square inch (psi) or kilopascals (kPa).

pressure-control valve used to regulate pressure in the system or in sections of a system.

pressure differential the difference in pressure values between two points of a circuit or component.

pressure differential valve a warning light used in a brake circuit to alert the operator that there is a difference in brake pressure from one side to the other in a dual circuit brake system.

pressure drop a change in pressure between two points in a hydraulic line caused by the energy lost to maintain flow in the circuit.

pressure limiting compensator devices that de-stroke the pump (reducing flow) when system pressure is achieved, after which pressure is maintained but with low flow.

pressure plate applies pressure to the clutch disc for the transfer of torque to the transmission. The pressure plate, when coupled with the clutch disc and flywheel, makes and breaks the flow of power from the engine to the transmission.

pressure protection valve a normally closed, pressure sensitive control valve. Some are externally adjustable. Can be used to delay the charging of a circuit or reservoir, or to isolate a circuit or reservoir in the event of a pressure drop-off.

pressure reducing valve a valve used in a circuit that lowers the pressure to a value that is necessary to operate an additional circuit at reduced pressure.

pressure relief valve a pressure limiting device that trips at a predetermined pressure to define circuit pressure or act as a safety device. A *safety pop-off valve* is one type of pressure relief valve.

pressure switch an electric hydraulic valve used in a hydraulic brake circuit to monitor the brake pressure. When the pressure becomes too low, it turns on a warning light in the operator's compartment.

preventative maintenance (PM) scheduled maintenance downtime designed to eliminate unscheduled equipment breakdowns.

preventative maintenance inspection (PMI) the inspection that accompanies a PM schedule.

prime mover the initial mechanical source of motive power. In the context of a hydraulic circuit, the means used to drive the hydraulic pump would be properly the prime mover, although the term is sometimes used to describe the pump itself.

priority valve a protective device on hydraulic (automatic transmissions) and pneumatic circuits. Ensures that the control system upstream from a valve or subcircuit has sufficient pressure to function.

programmable read-only memory (PROM) memory category in vehicle ECUs that may have to be overwritten or reprogrammed. May be altered by physically replacing chips (older systems) or by flashing.

propeller shaft driveshaft.

proportioning valve a hydraulic brake valve used on vehicles equipped with front disc and rear drum brakes. Its function is to reduce application pressure to the rear drum brakes to prevent premature lock-up under severe braking.

psi pounds per square inch. Standard measure of pressure values used in North America. One psi is equivalent to 6.9 kPa (metric measure of pressure).

psia pounds per square inch absolute. Measure of pressure values used in North America that factors in atmospheric pressure of 14.7 psi.

pulse wheel rotating disc used to signal rpm or rotational data to an ECU. Usually refers to the disc used in a rotary Hall-effect sensor rather than an analog sensor.

pump any of a number of different devices used to move or displace fluids.

quick-connect coupler see *quick-release coupler*.

quick-release coupler any of a number of different types of hydraulic or pneumatic connecters with spring-loaded seal couplings designed for fast connect/disconnect.

quick-release valve an air circuit valve that acts like a Tee fitting when air is applied to it from a source, but exhausts air at the valve when source air is removed. The objective is to speed release times. Used on various air brake and other vehicle pneumatic circuits.

rack-and-pinion gear a straight length of square or rectangular steel bar with teeth on one side, across which a pinion is driven. Technically a gear rack is a spur gear with an infinite pitch diameter.

radial extending from the center of an axis or center of a circle. An example is the spokes on a wheel.

radial load a load acting perpendicular or 90 degrees to a shaft.

radial play eccentric movement or runout in a rotating component. See *radius*.

radial runout the wobble of a turning wheel as viewed from the side of the vehicle. Used to describe one type of runout on rotors, shafts, and wheels.

radius a straight line extending from the center of a circle to its circumference. The radius of a circle is half its diameter.

ram a hydraulic (or pneumatic) linear actuator. Examples are a dump box and COE cab lift rams.

random-access memory (RAM) electronically retained "main" memory in a computer.

ratio valve an air brake valve used on some pre-ABS front axle brakes to proportion, on a graduated scale, application pressures to the front brakes.

read-only memory (ROM) data retained magnetically or optically that is designed not to be overwritten: designed to be permanent but may become corrupted.

rebound occurs when the energy that is stored in a spring kicks back. On many types of equipment, rebound is controlled with dampening devices such as shock absorbers and springs.

reciprocating motion the back-and-forth motion of pistons in a positive-displacement hydraulic piston pump.

rectification the conversion of AC to DC.

rectifier the component subsystem responsible for converting an AC sine wave into DC. Used in alternators and mobile AC generators.

reeving the act of fitting lifting ropes with sheaves.

regenerative braking term used to describe the role reversal of an electric motor when instead of producing output torque, it absorbs input torque and generates electrical energy to be stored in batteries or ultracapacitors.

regulator a device that limits system or subsystem pressure in a fluid power circuit.

relay valve in air brake systems, a valve that receives a control signal, then uses a local supply of air to send out to actuator devices. Most relay valves replicate the signal pressure valve to the out-circuit, but some are designed to amplify or modulate the signal.

release bearing unit within the clutch used to disengage the clutch; also known as throw-out bearing.

relief valve a hydraulic valve that sends excess oil back to the hydraulic tank.

reservoir any of a number of different types of storage chamber, both pressurized and non-pressurized, used to store fluids such as hydraulic oil or the compressed air in the air system. A tank or container that is used to store hydraulic oil.

residual pressure the potential energy stored in a hydraulic circuit as pressure when it is inactive.

resistance temperature detectors (RTDs) resistive thermal devices used to measure the change in the electrical resistance of gases that takes place with temperature change; they are also known as platinum temperature detectors (PTDs).

restraining device a restraint device that protects machine operators from hazards.

reverse modulation a condition that occurs in a spring-applied brake circuit that modulates the brake pressure in what would normally be the opposite of what takes place in conventional hydraulic brakes. Example: when the brake pedal is depressed in a spring-applied brake, the pressure is lowered in proportion to the travel of the brake pedal.

rigid anything that cannot flex or bend.

rigid disc steel clutch plates to which friction linings, or facings, are bonded or riveted.

rim lip the small part of the rim that is next to the tire bead.

rimpull rating of the torque available between a drive tire and whatever surface the vehicle is moving over; it is limited by traction.

ring gear spur gear such as that used on a flywheel for engine cranking, or the outer member of a planetary gear set.

roller link the link of a chain drive that holds the roller in place. Each link interconnects to another to form a chain.

rollover protection system (ROPS) structural apparatus fitted to mobile equipment designed to protect the operator in the event of an unintended rollover. ROPS must be of sufficient strength to prevent roof collapse in the event of a rollover, so they must be capable of supporting the entire weight of the vehicle.

root mean square (rms) term used to mean averaging when performing machining operations and also used to describe AC voltage measurement by giving it an equivalent DC value.

rotary actuator a device that converts fluid flow into incremental rotary motion as opposed to continuous cycle rotary motion.

rotary flow fluid movement, such as that produced by a paddle wheel in water or a propeller in air, that results in centrifugal force. See *rotary oil flow*.

rotary oil flow caused by the centrifugal force applied to the fluid as the converter rotates. Occurs as turbine speed increases within the torque converter.

rotate to turn or spin.

rotating clutch pack a multi-disc and plate-type clutch. When engaged, will rotate to provide an input within a transmission.

rotation the motion of a gear, shaft, or wheel.

rotor any of a number of different components that rotate. In air brakes, the rotor turns with the wheel and has braking force applied to it. In an alternator, the rotor provides the magnetic fields required to create current flow.

runout any kind of wobble in a rotating component. Runout may be axial or radial. Using the example of a wheel on a vehicle, axial runout is wobble as viewed from the front or rear of the vehicle, and radial runout is wobble viewed from the side of the vehicle.

saddle assembly an assembly on each side of the suspension system that holds rubber load cushions. Each of the four rubber blocks is connected to the axle assembly by drive pins.

SAE J1939 the heavy-duty standard multiplexing data backbone designed for powertrain management but also used for other vehicle systems networking.

safe working load (SWL) the manufacturer's recommended maximum weight load for a line, rope, crane, or any other lifting device or component.

safety pressure-relief valve the pressure-relief valve used on air brake systems, more commonly called a *pop-off valve*. Usually set to trip at 150 psi and located on the supply (wet) tank in the system.

scissor jack an air-over-hydraulic portable floor hoist that lifts equipment using a scissor action.

sealed tracks a pin design used on crawler tracks that uses a solid pin between the links. Sealed and lubricated tracks have a hollow pin, which provides a path for lubricating the pin and bushing of the next track section.

semi-floating axle the term semi-floating is derived from the way the axle is mounted in the assembly. The axle at the differential is splined to the side gear, and floats at this end. At the wheel end of the axle, the shaft carries one or two bearings depending on the axle load carrying capacity.

sensor energy conversion device used to signal a condition or state. Sensors are used to signal speed, pressure, temperature, etc.

sequence valve valve used to supply a second actuator only after the first actuator has been satisfied at a preset pressure. Outlet oil from a directional control valve divides oil flow between two actuators that function in series.

service bulletin see *technical service bulletin* (TSB).

service information system (SIS) Caterpillar's online service information portal: requires a subscription account to access. Almost all OEMs today have a comparable online service information system and the term is often used generically.

service literature term used to describe most kinds of electronically accessed service instructions, parts manuals, and the databases that retain them. Replaces hard copy service manuals and TSBs. Most OEMs no longer make hard (paper) copy versions of service literature available.

service manual a hard (paper) copy book that outlines service repair and troubleshooting procedures. Today most service manuals are electronic due to lower cost and ease of updating. Consequently, the term *service literature* is more often used.

servo a component actuated by another. One type of servo is a hydraulic slave piston that is actuated by a control valve.

servo brake a braking action that occurs due to the action of one shoe on the other in a drum brake circuit. Example: the primary shoe of a drum brake transmits a force to the secondary shoe and forces it to wedge into the drum providing braking action.

servo control a hydraulic valve that can control the flow of oil in proportion to the input signal.

shackle lifting hardware consisting of an anchor or D-ring and removable screw pin.

sheave used on lifting devices to change wire rope direction when multiple parts of line are needed.

shift bar housing transmission subassembly that houses shift rails, shift yokes, detent balls and springers, interlock balls, and pin and neutral shaft.

shift fork or yoke Y-shaped components located between gears on the main shaft that, when actuated, cause the gears to engage or disengage via the sliding clutches. Shift forks are located between low and reverse, first and second, and third and fourth gears in a standard main section.

shift valves hydraulic valves that direct hydraulic pressure to the clutch packs to engage a specific gear ratio. The shift valve determines when to shift from one gear to the next.

shock absorbers hydraulic devices used to dampen vehicle spring oscillations.

shuttle valve a prioritizing valve with two inlet ports and a single outlet port. The ball or spool shuttles to direct the higher of the two inlet pressure values to the outlet. Also, a valve located in a circuit that allows flow to a component from two different sources.

side bar a connector. Roller chains are made up a series of pin links and roller links connected to each other with side bars.

sight glass allows a technician to see the flow of a liquid in a line (such as fuel or refrigerant) for diagnostic purposes.

single-acting cylinder a hydraulic cylinder designed to produce thrust in one direction only: returned by gravity or spring force.

single-countershaft transmission any transmission with a single countershaft.

single-phase main the standard 110-120 V-AC household electrical supply.

single reduction axle a drive axle that uses a single gear reduction that is achieved by a crown and pinion gear set.

single reduction drive a term used to describe a final drive that reduces the gear ratio just once. An example is a pinion drive used on small bulldozer final drives.

slave cylinder a device in the hydraulic clutch system that is activated by hydraulic force, disengaging the clutch. See *wheel cylinder.*

slave valve any air or hydraulic actuator. A spool-type hydraulic valve that uses pressure from a pilot valve to move the position of its spool.

sliding gear gears that slide along the shaft to which they are attached when moved by the shift forks: They are splined to the shaft so they rotate at shaft speed. A sliding gear has protrusions (dogs) on its sides that engage to slots in freewheeling gears when required to transmit torque.

sling configuration describes the methods of securing a load to be hoisted and also the design features of different slings.

slip assembly the male/female splined sections of a driveshaft that mates to permit lengthening and shortening during operation.

slip joint the splined stub and socket joint of a driveshaft in which allows changes in driveshaft length experienced during operation.

slip yoke a part of a driveline assembly that provides a means for the length of a driveshaft to change during operation.

slop widely used slang term for endplay: see *endplay.*

slug see *slug volume.*

slug volume the fluid volume of a pump chamber or engine cylinder swept by the plunger/piston in one cycle.

snap ring a retaining split ring with radial spring pressure. Snap rings are usually fitted to grooves and are used to retain shims, springs, bearings, and other components.

socket a female cavity to which a pin or shaft is inserted. Electrical connectors are equipped with sockets to which terminal pins are inserted to make a connection.

solid rubber suspension a solid rubber cushion-type equalizing beam suspension system used on off-road haulage trucks.

solvent substance that dissolves other substances.

source address (SA) term used in J1939 to indicate any module with an address on the vehicle data bus. SAs are described numerically; for instance the engine is 01 and the transmission is 03.

spalling a condition that occurs on metal parts when chips or flakes of metal separate from fatigue as opposed to wear.

specification an assembly, engineering, or technical detail or value.

spike a sudden surge in hydraulic pressure or electrical voltage. Also, a slang term meaning the tractor steering column-mounted trailer service brake valve.

spiral bevel gear set a helical gear set that includes a crown and pinion that mesh at right angles with the pinion gear mounted on the centerline of the crown gear.

spline radial lands that either protrude from, or are milled into, a shaft to permit it engaging with female splines for the purpose of transmitting torque.

split rim a tire rim that consists of two or more parts.

spool valve commonly used hydraulic valve in which a sliding spool within a valve body is used to open and close hydraulic subcircuits. A cylindrical part that moves from side to side in a hydraulic valve to control oil flow in a circuit.

spring hangers specialized connectors. Springs are mounted to the top of each equalizer beam onto saddle mounts and are connected to the frame by spring hangers. The spring assemblies pivot on spring pins attached to the hanger brackets.

spring packs a series of springs bound together with a center bolt in an elliptical shape. They provide a dampening action to the equipment during operation.

sprocket a wheel with toothed cogs designed to engage with a chain or mesh with a similar shaped device. The sprocket is not mounted to the track frame but is attached to the final drive. Its job is to transfer drive torque to the track. The teeth on the outside of the sprocket act like gear teeth on a gear.

sprocket shaft gear the sprocket shaft gear is splined to the sprocket shaft, which also holds the drive sprocket. The sprocket is mounted to a sprocket hub of a final drive on a bulldozer.

spur bevel gear set a straight cut gear set that includes a crown and pinion that mesh at right angles with the pinion gear mounted on the centerline of the crown gear.

spur gears cylindrical gears with teeth that are straight and parallel, and operate on parallel axes.

standards Australia non-profit Australian safety standards association that is the equivalent of our Underwriters Laboratories (UL).

standby pressure the set pressure at the pump outlet when machine is at rest with the engine running.

static charge accumulated electrical charge potential not flowing in circuit.

static discharge arc that occurs when static electrical charge finds a flow path.

static friction occurs when two objects are not moving relative to each other. The coefficient of static friction is typically denoted as μ_r. The initial force to get an object moving is often dominated by static friction. S friction is, in most cases, higher than the kinetic friction.

static head measured pressure in a circuit in which there is no fluid flow; normally expressed in feet of water.

stationary clutch pack a multi-disc and plate-type clutch. When engaged it will hold stationary one of the three planetary gear sets (ring gear, sun gear, planetary gears).

stator located between the torque converter pump/impeller and turbine. It redirects the oil flow from the turbine back into the impeller in the direction of impeller rotation with minimal loss of speed or force.

steel disc a rotor used in a wet disc brake system that is manufactured from steel: braking occurs when clamped by friction discs actuated by either hydraulic or spring pressure.

steering clutch a clutch design found on track equipment: capable of slowing each track independently to achieve direction control.

stick circuit part of an excavator implement hydraulic circuit. The stick arm connects to and pivots on the boom and has the bucket connected to it. The stick circuit is actuated by double-acting hydraulic cylinders.

straight cut spur gears in which the leading edges of the teeth are aligned parallel to the axis of rotation: they can mesh correctly when fitted to parallel shafts.

subsystem identifiers (SIDs) branch circuit within a MID or SA used for purposes of identifying faults.

sump a reservoir.

sun gear the central gear of a planetary gear set, usually located on an axle of a wheel end planetary drive.

supply pressure pressure that is common to a circuit such as the pressure that comes from a charge valve to charge the accumulators of a brake circuit.

supply tank in a vehicle air brake circuit, the supply tank is the first reservoir in the circuit to receive air from the air compressor. The supply tank feeds the primary, secondary, and any auxiliary air tanks in the circuit.

surge a temporary rise in pressure or fluid level in a circuit.

suspension cylinder off-road haulage trucks use suspension cylinders to help absorb road shock during operation. Some suspension systems contain high hydraulic and nitrogen pressures.

swashplate a stationary canted (angled) plate used in an axial piston pump on which pistons in a rotating cylinder ride: as the pump cylinder rotates, the pistons reciprocate in the barrel bores. The swash angle can be varied to provide variable displacement.

swing circuit the means of hydraulically rotating the upper structure on a typical excavator circuit usually through 360 degrees.

synchronizer performs two functions. It brings components that are rotating at different speeds to one synchronized speed to ensure that the main shaft and the main shaft gear to be locked to it are rotating at the same speed, and it actually locks these components together.

system pressure a common pressure in a fluid circuit. An example is the main hydraulic dump and hoist pressure found in a hydraulic circuit. Usually refers to the range of pneumatic circuit pressures between cut-out and cut-in pressure values in an FMVSS 121 compliant air brake system. Can refer to other fluid power circuit pressures such as those in transmission hydraulics and engine lube circuits.

tandem drive a two-axle drive system used on off-road haulage trucks to provide four-wheel-drive at the rear of the equipment.

technical service bulletin (TSB) service literature either online or hard copy that updates or modifies a repair procedure.

tee fitting a fluid circuit valve usually set up with a single source but distributes to two branches of the circuit; often T-shaped.

temper the heat treating process used to harden (or in some cases soften) materials. Term is usually applied to heat treatments of middle- and high-alloy steels.

tensile strength the highest stress a material can withstand without separating, expressed in psi. In steels, tensile strength typically exceeds *yield strength* by about 10%.

thermistor temperature sensing device used to signal temperature data to an ECU.

three-phase main high potential electrical supply used for shop machinery available in 220 V-AC and 450-600 V-AC depending on region.

tie rod a rod or cross tube that connects both wheels indirectly to a steering box, which provides steering input to the wheels.

tire ballast liquid inserted into a tire to increase traction and reduce wheel slippage.

tire cord a textile reinforcement used to provide flex resistance.

tire deflection the reduction in section height of a loaded tire at a given inflation pressure.

tie-rod assembly steering component consisting of two adjustable metal rods that define wheel tracking on a steering axle: a tie rod links both wheels so that they can be turned simultaneously by a single steering control arm.

toe-in a steering condition in which a pair of tires on the same axle track inward when viewed from the front of the equipment.

toe-out a steering condition in which a pair of tires on the same axle track outward when viewed from the front of the equipment.

tone wheel rotating disc used to signal rpm or rotational data from a rotary Hall-effect sensor to an ECU.

torque the measure of force applied to a lever arm. Normally expressed in lb-ft (pound-feet) or lb-in. (pound-inch).

torque converter lockup clutch improves cruising power transmission efficiency. The application of the clutch locks the turbine to the pump, causing all power transmission to be mechanical, thus eliminating losses associated with fluid drive.

torque hub planetary drive consists of a sun (input) gear, three cluster (planets) gears, and two internal (ring) gears. Torque flows through the input gear driving the cluster gears within a held ring gear so output is transmitted to the second ring gear. Each cluster planet gear consists of two gears with different diameters on the same shaft producing higher ratios in a compact package.

torque limiting compensator prevents a hydraulic system from working at high pressure and maximum flow simultaneously.

torque multiplication a torque converter operating condition in which input torque from an engine is lower than the converter output torque.

torque rods used on suspension systems to absorb additional torque generated by the axle and road shock.

torricelli's tube Torricelli discovered the concept of atmospheric pressure by inverting a tube filled with mercury into a bowl of the liquid, then observing that the column of mercury in the tube fell until atmospheric pressure acting on the surface balanced against the vacuum created in the tube. His experiment proved that vacuum in the column would support 29.92 inches of mercury at sea level.

torsion twisting, especially when one end of a beam or body is held.

torsional failures a failure condition caused by extreme twisting forces. When this occurs in a powertrain, shaft fracture may result causing loss of drive.

toxic poisonous.

toxicity having poisonous characteristics.

track circuit the drive and steer hydraulic circuit used in bulldozers, excavators, and other track-driven equipment.

track guard used to keep rocks and foreign debris from getting in between the track rollers and track links.

track link one element of the chain section of the track. Track links, pins, and bushings interconnect with each other to form the track.

track pin connects the links and bushings together on a track to form a chain.

track pitch the distance from the center of one pin to the center of an adjacent pin in an adjoining track link section. When the track pins and bushings wear, the length of the track increases and the track becomes loose.

track roller the bottom rollers on a crawler track are called track rollers. They support the weight of the equipment and ensure that the weight of the equipment is distributed evenly over the bottom of the track.

track shoes bolted to the link assembly to provide a means of supporting the equipment and to provide traction and flotation during operation of track systems. They are generally constructed from a hardened metal plate and come in many different configurations.

transfer case an additional gearbox located between the main transmission and the rear axle. Its function is to transfer torque from a machine's transmission to the front and rear driving axles.

treadle valve a dual-circuit service brake actuation valve that meters air from the primary and secondary service reservoirs to the service lines and brake chambers. Also known as a *foot valve*.

trip in technology, this term is used to mean *to open*. Examples: a safety valve is *tripped* when its specified pressure is achieved, or a circuit breaker is *tripped* when its specified current is exceeded.

trolley-drive overhead AC cable used to provide electric drive to machines running a consistent uphill run over the same track. Trolley lines typically use voltage values of 1600 to 1800 V-AC. The AC rock truck connects to the overhead trolley line at the beginning of the haul, and disconnects when it has been completed.

trunnion the machined surfaces of the universal joint cross. The bearing cups fit into this cross.

tube hydraulic line or passage.

turbine the output (driven) member of a torque converter splined to the input shaft of the transmission.

turbine shaft output component of a torque converter that connects to the transmission. Also referred as the transmission *input shaft*.

two-way check valve a shuttle valve that is supplied with air from two sources and has a single outlet. The valve is designed to shuttle and bias air flow from the higher of the two sources to the outlet.

two-way valve a direction control valve used in a hydraulic system to divide the oil flow into two flow paths.

U-bolt a U-shaped steel bolt with threads at both ends that is used to fasten a multi-leaf spring pack to the axles of equipment.

underdrive any drivetrain arrangement in which an output speed ratio is less than the input speed.

underwriters laboratory (UL) organization that ensures product compliance to standards of safety.

universal joint (U-joint) a joint that provides a flexible coupling that allows power to be transmitted from one shaft to another.

unloading valve valve used to charge a hydraulic circuit to maintain a threshold pressure.

unsprung weight the weight of the equipment that is not supported by the suspension system.

upload the act of transferring data from one computer or medium to another.

vacuum the absence of matter and therefore the absence of pressure. The term is often used to refer to a partial vacuum, which means any pressure condition below atmospheric.

valve a mechanical device used to control flow in a fluid power circuit.

vapor a gas. One of the three states of matter, the other two being solid and liquid.

variable-displacement motor a hydraulic motor that is capable of varying its displacement per revolution.

variable-displacement pump a hydraulic pump that is capable of varying the displacement per stroke or revolution regardless of the speed.

variable-pressure reducing valve a pressure reducing valve that can be adjusted to specific pressures required in hydraulic brake systems.

vent a breather device.

vertical exactly upright or plumb.

vibration rhythmic or arrhythmic movement or oscillation. In vehicle technology, usually refers to an undesirable condition. Example: *driveline vibration*.

viscosity the shear resistance of a liquid. Often used to mean oil thickness or resistance to flow. Viscosity ratings are assigned by SAE (Society of Automotive Engineers) and ISO (International Standards Organization).

viscosity index (VI) a measure of the viscosity-to-temperature performance of a fluid.

volume geometric space within a chamber. The term is often applied to mean pump output, in which case it is represented in gpm.

vortex eddy force caused by spiraling fluid. A moving vehicle creates a *vortex* behind it as it cuts through the air. Term is used to describe both gas and liquid dynamics.

vortex oil flow circular oil flow in a torque converter that occurs as oil is forced from the impeller to the turbine, then back to the impeller during operation.

wear step track pins develop a specific form of damage called wear step, generally associated with the wet joint design.

welch plug a plate or cup used to seal the hole in the throat of a slip yoke.

weld yoke a term used to describe a heavy-duty yoke design used on a typical heavy equipment driveshaft application.

wet tank commonly used slang term for supply tank. See *supply tank*.

wheel assembly refers to the entire rotating assembly on the end of an axle including the tire and rim.

wheel cylinder a hydraulic piston-type valve that is used to exert force in a drum brake system to force shoes out against the drums of a vehicle to slow down or stop it.

wheel end pressure the hydraulic pressure found at the wheel end in a fluid brake circuit. This pressure is generally modulated by the brake valve.

winch a device used for the purpose of lifting, pulling, or lowering using either chains or rope.

wing a bearing cap design. For example, a delta wing is a common bearing cap design.

wing-type U-joint a U-joint cup design that is held in place by bolts that secure it to a yoke on a driveshaft or powertrain component.

wire rope steel braided rope consisting of many individual wires braided into strands which are wound around a center core.

work when effort or force produces an observable result (movement), work results. In a hydraulic circuit, this means moving a load. To produce work in a hydraulic circuit, there must be flow. Work is measured in units of force multiplied by distance, for example, pound-feet.

working angle the angle defined by the intersection of the centerlines of two driveshafts connected by a universal joint.

working load limit (WLL) the maximum load that can be applied to an assembly or component in straight tension, also known as *safe working load*.

worm gear worm gear sets run on non-intersecting, perpendicular shaft axes and provide high ratios of reduction. The worm is a threaded cylindrical shaft that drives the worm gear (also referred to as the *worm wheel*). Worm gears look somewhat similar to helical gears except that they have a curved throat recessed in their face to allow the worm access to the flanks of the gear teeth.

yield strength the highest stress a material can withstand without permanent deformation, expressed in psi. In steels, yield strength is typically about 10% less than *tensile strength*.

yoke the parts that are interconnected by a universal joint on a driveshaft.

zerk fitting a type of grease nipple that uses a spring-loaded ball to seal an orifice. It accepts grease from a grease gun but does not allow dirt or other foreign material to enter the component.

Index

Note: Page numbers followed by *f*, *t*, or *p* indicate figures, tables, or photosequences, respectively.